Springer Finance

Springer Finance

Springer Finance is a programme of books aimed at students, academics, and practitioners working on increasingly technical approaches to the analysis of financial markets. It aims to cover a variety of topics, not only mathematical finance but foreign exchanges, term structure, risk management, portfolio theory, equity derivatives, and financial economics.

M. Ammann, Credit Risk Valuation: Methods, Models, and Applications (2001)
E. Barucci, Financial Markets Theory: Equilibrium, Efficiency and Information (2003)
N.H. Bingham and R. Kiesel, Risk-Neutral Valuation: Pricing and Hedging of Financial Derivatives, 2nd Edition (2004)
T.R. Bielecki and M. Rutkowski, Credit Risk: Modeling, Valuation and Hedging (2001)
D. Brigo amd F. Mercurio, Interest Rate Models: Theory and Practice (2001)
R. Buff, Uncertain Volatility Models – Theory and Application (2002)
R.-A. Dana and M. Jeanblanc, Financial Markets in Continuous Time (2003)
G. Deboeck and T. Kohonen (Editors), Visual Explorations in Finance with Self-Organizing Maps (1998)
R.J. Elliott and P.E. Kopp, Mathematics of Financial Markets (1999)
H. Geman, D. Madan, S.R. Pliska and T. Vorst (Editors), Mathematical Finance – Bachelier Congress 2000 (2001)
M. Gundlach and F. Lehrbass (Editors), CreditRisk+ in the Banking Industry (2004)
Y.-K. Kwok, Mathematical Models of Financial Derivatives (1998)
M. Külpmann, Irrational Exuberance Reconsidered: The Cross Section of Stock Returns, 2nd Edition (2004)
A. Pelsser, Efficient Methods for Valuing Interest Rate Derivatives (2000)
J.-L. Prigent, Weak Convergence of Financial Markets (2003)
B. Schmid, Credit Risk Pricing Models: Theory and Practice, 2nd Edition (2004)
S.E. Shreve, Stochastic Calculus for Finance I: The Binomial Asset Pricing Model (2004)
S.E. Shreve, Stochastic Calculus for Finance II: Continuous-Time Models (2004)
M. Yor, Exponential Functionals of Brownian Motion and Related Processes (2001)
R. Zagst, Interest-Rate Management (2002)
Y.-I. Zhu and I.-L Chern, Derivative Securities and Difference Methods (2004)
A. Ziegler, Incomplete Information and Heterogeneous Beliefs in Continuous-Time Finance (2003)
A. Ziegler, A Game Theory Analysis of Options: Corporate Finance and Financial Intermediation in Continuous Time, 2nd Edition (2004)

Steven E. Shreve

Stochastic Calculus for Finance II

Continuous-Time Models

With 28 Figures

Steven E. Shreve
Department of Mathematical Sciences
Carnegie Mellon University
Pittsburgh, PA 15213
USA
shreve@cmu.edu

Mathematics Subject Classification (2000): 60-01, 60H10, 60J65, 91B28

Library of Congress Cataloging-in-Publication Data
Shreve, Steven E.
Stochastic calculus for finance / Steven E. Shreve.
p. cm. — (Springer finance series)
Includes bibliographical references and index.
Contents v. 2. Continuous-time models.

1. Finance—Mathematical models—Textbooks. 2. Stochastic analysis—Textbooks. I. Title. II. Springer finance.
HG106.S57 2003
332'.01'51922—dc22 2003063342

ISBN 978-1-4419-2311-0

Printed in the United States of America.

9 8 7 6 5 4 3 2

springeronline.com

To my students

Preface

Origin of This Text

This text has evolved from mathematics courses in the Master of Science in Computational Finance (MSCF) program at Carnegie Mellon University. The content of this book has been used successfully with students whose mathematics background consists of calculus and calculus-based probability. The text gives precise statements of results, plausibility arguments, and even some proofs, but more importantly, intuitive explanations developed and refined through classroom experience with this material are provided. Exercises conclude every chapter. Some of these extend the theory and others are drawn from practical problems in quantitative finance.

The first three chapters of Volume I have been used in a half-semester course in the MSCF program. The full Volume I has been used in a full-semester course in the Carnegie Mellon Bachelor's program in Computational Finance. Volume II was developed to support three half-semester courses in the MSCF program.

Dedication

Since its inception in 1994, the Carnegie Mellon Master's program in Computational Finance has graduated hundreds of students. These people, who have come from a variety of educational and professional backgrounds, have been a joy to teach. They have been eager to learn, asking questions that stimulated thinking, working hard to understand the material both theoretically and practically, and often requesting the inclusion of additional topics. Many came from the finance industry, and were gracious in sharing their knowledge in ways that enhanced the classroom experience for all.

This text and my own store of knowledge have benefited greatly from interactions with the MSCF students, and I continue to learn from the MSCF alumni. I take this opportunity to express gratitude to these students and former students by dedicating this work to them.

Acknowledgments

Conversations with several people, including my colleagues David Heath and Dmitry Kramkov, have influenced this text. Łukasz Kruk read much of the manuscript and provided numerous comments and corrections. Other students and faculty have pointed out errors in and suggested improvements of earlier drafts of this work. Some of these are Jonathan Anderson, Nathaniel Carter, Bogdan Doytchinov, David German, Steven Gillispie, Karel Janeček, Sean Jones, Anatoli Karolik, David Korpi, Andrzej Krause, Rael Limbitco, Petr Luksan, Sergey Myagchilov, Nicki Rasmussen, Isaac Sonin, Massimo Tassan-Solet, David Whitaker and Uwe Wystup. In some cases, users of these earlier drafts have suggested exercises or examples, and their contributions are acknowledged at appropriate points in the text. To all those who aided in the development of this text, I am most grateful.

During the creation of this text, the author was partially supported by the National Science Foundation under grants DMS-9802464, DMS-0103814, and DMS-0139911. Any opinions, findings, and conclusions or recommendations expressed in this material are those of the author and do not necessarily reflect the views of the National Science Foundation.

Pittsburgh, Pennsylvania, USA — Steven E. Shreve
April 2004

Contents

Introduction

Background

By awarding Harry Markowitz, William Sharpe, and Merton Miller the 1990 Nobel Prize in Economics, the Nobel Prize Committee brought to worldwide attention the fact that the previous forty years had seen the emergence of a new scientific discipline, the "theory of finance." This theory attempts to understand how financial markets work, how to make them more efficient, and how they should be regulated. It explains and enhances the important role these markets play in capital allocation and risk reduction to facilitate economic activity. Without losing its application to practical aspects of trading and regulation, the theory of finance has become increasingly mathematical, to the point that problems in finance are now driving research in mathematics.

Harry Markowitz's 1952 Ph.D. thesis *Portfolio Selection* laid the groundwork for the mathematical theory of finance. Markowitz developed a notion of mean return and covariances for common stocks that allowed him to quantify the concept of "diversification" in a market. He showed how to compute the mean return and variance for a given portfolio and argued that investors should hold only those portfolios whose variance is minimal among all portfolios with a given mean return. Although the language of finance now involves stochastic (Itô) calculus, management of risk in a quantifiable manner is the underlying theme of the modern theory and practice of quantitative finance.

In 1969, Robert Merton introduced stochastic calculus into the study of finance. Merton was motivated by the desire to understand how prices are set in financial markets, which is the classical economics question of "equilibrium," and in later papers he used the machinery of stochastic calculus to begin investigation of this issue.

At the same time as Merton's work and with Merton's assistance, Fischer Black and Myron Scholes were developing their celebrated option pricing formula. This work won the 1997 Nobel Prize in Economics. It provided a satisfying solution to an important practical problem, that of finding a fair price for a European call option (i.e., the right to buy one share of a given stock at a specified price and time). In the period 1979–1983, Harrison, Kreps, and Pliska used the general theory of continuous-time stochastic processes to put the Black-Scholes option-pricing formula on a solid theoretical basis, and, as a result, showed how to price numerous other "derivative" securities.

Many of the theoretical developments in finance have found immediate application in financial markets. To understand how they are applied, we digress for a moment on the role of financial institutions. A principal function of a nation's financial institutions is to act as a risk-reducing intermediary among customers engaged in production. For example, the insurance industry pools premiums of many customers and must pay off only the few who actually incur losses. But risk arises in situations for which pooled-premium insurance is unavailable. For instance, as a hedge against higher fuel costs, an airline may want to buy a security whose value will rise if oil prices rise. But who wants to sell such a security? The role of a financial institution is to design such a security, determine a "fair" price for it, and sell it to airlines. The security thus sold is usually "derivative" (i.e., its value is based on the value of other, identified securities). "Fair" in this context means that the financial institution earns just enough from selling the security to enable it to trade in other securities whose relation with oil prices is such that, if oil prices do indeed rise, the firm can pay off its increased obligation to the airlines. An "efficient" market is one in which risk-hedging securities are widely available at "fair" prices.

The Black-Scholes option pricing formula provided, for the first time, a theoretical method of fairly pricing a risk-hedging security. If an investment bank offers a derivative security at a price that is higher than "fair," it may be underbid. If it offers the security at less than the "fair" price, it runs the risk of substantial loss. This makes the bank reluctant to offer many of the derivative securities that would contribute to market efficiency. In particular, the bank only wants to offer derivative securities whose "fair" price can be determined in advance. Furthermore, if the bank sells such a security, it must then address the hedging problem: how should it manage the risk associated with its new position? The mathematical theory growing out of the Black-Scholes option pricing formula provides solutions for both the pricing and hedging problems. It thus has enabled the creation of a host of specialized derivative securities. This theory is the subject of this text.

Relationship between Volumes I and II

Volume II treats the continuous-time theory of stochastic calculus within the context of finance applications. The presentation of this theory is the raison d'être of this work. Volume II includes a self-contained treatment of the probability theory needed for stochastic calculus, including Brownian motion and its properties.

Volume I presents many of the same finance applications, but within the simpler context of the discrete-time binomial model. It prepares the reader for Volume II by treating several fundamental concepts, including martingales, Markov processes, change of measure and risk-neutral pricing in this less technical setting. However, Volume II has a self-contained treatment of these topics, and strictly speaking, it is not necessary to read Volume I before reading Volume II. It is helpful in that the difficult concepts of Volume II are first seen in a simpler context in Volume I.

In the Carnegie Mellon Master's program in Computational Finance, the course based on Volume I is a prerequisite for the courses based on Volume II. However, graduate students in computer science, finance, mathematics, physics and statistics frequently take the courses based on Volume II without first taking the course based on Volume I.

The reader who begins with Volume II may use Volume I as a reference. As several concepts are presented in Volume II, reference is made to the analogous concepts in Volume I. The reader can at that point choose to read only Volume II or to refer to Volume I for a discussion of the concept at hand in a more transparent setting.

Summary of Volume I

Volume I presents the binomial asset pricing model. Although this model is interesting in its own right, and is often the paradigm of practice, here it is used primarily as a vehicle for introducing in a simple setting the concepts needed for the continuous-time theory of Volume II.

Chapter 1, *The Binomial No-Arbitrage Pricing Model*, presents the no-arbitrage method of option pricing in a binomial model. The mathematics is simple, but the profound concept of risk-neutral pricing introduced here is not. Chapter 2, *Probability Theory on Coin Toss Space*, formalizes the results of Chapter 1, using the notions of martingales and Markov processes. This chapter culminates with the risk-neutral pricing formula for European derivative securities. The tools used to derive this formula are not really required for the derivation in the binomial model, but we need these concepts in Volume II and therefore develop them in the simpler discrete-time setting of Volume I. Chapter 3, *State Prices*, discusses the change of measure associated with risk-neutral pricing of European derivative securities, again as a warm-up exercise for change of measure in continuous-time models. An interesting application developed here is to solve the problem of optimal (in the sense of expected utility maximization) investment in a binomial model. The ideas of Chapters 1 to 3 are essential to understanding the methodology of modern quantitative finance. They are developed again in Chapters 4 and 5 of Volume II.

The remaining three chapters of Volume I treat more specialized concepts. Chapter 4, *American Derivative Securities*, considers derivative securities whose owner can choose the exercise time. This topic is revisited in a continuous-time context in Chapter 8 of Volume II. Chapter 5, *Random Walk*, explains the reflection principle for random walk. The analogous reflection principle for Brownian motion plays a prominent role in the derivation of pricing formulas for exotic options in Chapter 7 of Volume II. Finally, Chapter 6, *Interest-Rate-Dependent Assets*, considers models with random interest rates, examining the difference between forward and futures prices and introducing the concept of a forward measure. Forward and futures prices reappear at the end of Chapter 5 of Volume II. Forward measures for continuous-time models are developed in Chapter 9 of Volume II and used to create forward LIBOR models for interest rate movements in Chapter 10 of Volume II.

Summary of Volume II

Chapter 1, *General Probability Theory*, and Chapter 2, *Information and Conditioning*, of Volume II lay the measure-theoretic foundation for probability theory required for a treatment of continuous-time models. Chapter 1 presents probability spaces, Lebesgue integrals, and change of measure. Independence, conditional expectations, and properties of conditional expectations are introduced in Chapter 2. These chapters are used extensively throughout the text, but some readers, especially those with exposure to probability theory, may choose to skip this material at the outset, referring to it as needed.

Chapter 3, *Brownian Motion*, introduces Brownian motion and its properties. The most important of these for stochastic calculus is quadratic variation, presented in Section 3.4. All of this material is needed in order to proceed, except Sections 3.6 and 3.7, which are used only in Chapter 7, *Exotic Options* and Chapter 8, *Early Exercise*.

The core of Volume II is Chapter 4, *Stochastic Calculus*. Here the Itô integral is constructed and Itô's formula (called the Itô-Doeblin formula in this text) is developed. Several consequences of the Itô-Doeblin formula are worked out. One of these is the characterization of Brownian motion in terms of its quadratic variation (Lévy's theorem) and another is the Black-Scholes equation for a European call price (called the Black-Scholes-Merton equation in this text). The only material which the reader may omit is Section 4.7, *Brownian Bridge*. This topic is included because of its importance in Monte Carlo simulation, but it is not used elsewhere in the text.

Chapter 5, *Risk-Neutral Pricing*, states and proves Girsanov's Theorem, which underlies change of measure. This permits a systematic treatment of risk-neutral pricing and the Fundamental Theorems of Asset Pricing (Section 5.4). Section 5.5, *Dividend-Paying Stocks*, is not used elsewhere in the text. Section 5.6, *Forwards and Futures*, appears later in Section 9.4 and in some exercises.

Chapter 6, *Connections with Partial Differential Equations*, develops the connection between stochastic calculus and partial differential equations. This is used frequently in later chapters.

With the exceptions noted above, the material in Chapters 1–6 is fundamental for quantitative finance is essential for reading the later chapters. After Chapter 6, the reader has choices.

Chapter 7, *Exotic Options*, is not used in subsequent chapters, nor is Chapter 8, *Early Exercise.* Chapter 9, *Change of Numéraire*, plays an important role in Section 10.4, *Forward LIBOR model*, but is not otherwise used. Chapter 10, *Term Structure Models*, and Chapter 11, *Introduction to Jump Processes*, are not used elsewhere in the text.

1

General Probability Theory

1.1 Infinite Probability Spaces

An infinite probability space is used to model a situation in which a random experiment with infinitely many possible outcomes is conducted. For purposes of the following discussion, there are two such experiments to keep in mind:

(i) choose a number from the unit interval [0,1], and
(ii) toss a coin infinitely many times.

In each case, we need a sample space of possible outcomes. For (i), our sample space will be simply the unit interval $[0, 1]$. A generic element of $[0, 1]$ will be denoted by ω, rather than the more natural choice x, because these elements are the possible outcomes of a random experiment.

For case (ii), we define

$$\Omega_\infty = \text{ the set of infinite sequences of } H\text{s and } T\text{s}. \tag{1.1.1}$$

A generic element of Ω_∞ will be denoted $\omega = \omega_1\omega_2\ldots$, where ω_n indicates the result of the nth coin toss.

The samples spaces listed above are not only infinite but are *uncountably infinite* (i.e., it is not possible to list their elements in a sequence). The first problem we face with an uncountably infinite sample space is that, for most interesting experiments, the probability of any particular outcome is zero. Consequently, we cannot determine the probability of a subset A of the sample space, a so-called *event*, by summing up the probabilities of the elements in A, as we did in equation (2.1.5) of Chapter 2 of Volume I. We must instead define the probabilities of events directly. But in infinite sample spaces there are infinitely many events. Even though we may understand well what random experiment we want to model, some of the events may have such complicated descriptions that it is not obvious what their probabilities should be. It would be hopeless to try to give a formula that determines the probability for every subset of an uncountably infinite sample space. We instead give a formula for the probability of certain simple events and then appeal to the properties of probability measures to determine the probability of more complicated events. This prompts the following definitions, after which we describe the process of setting up the uniform probability measure on $[0, 1]$.

Definition 1.1.1. *Let Ω be a nonempty set, and let $\mathcal{F}$ be a collection of subsets of Ω. We say that $\mathcal{F}$ is a* σ-algebra *(called a* σ-field *by some authors) provided that:*

(i) the empty set $\emptyset$ belongs to $\mathcal{F}$,
(ii) whenever a set A belongs to $\mathcal{F}$, its complement A^c also belongs to $\mathcal{F}$, and
(iii) whenever a sequence of sets $A_1, A_2, \ldots$ belongs to $\mathcal{F}$, their union $\cup_{n=1}^{\infty} A_n$ also belongs to $\mathcal{F}$.

If we have a σ-algebra of sets, then all the operations we might want to do to the sets will give us other sets in the σ-algebra. If we have two sets A and B in a σ-algebra, then by considering the sequence $A, B, \emptyset, \emptyset, \emptyset, \ldots$, we can conclude from (i) and (iii) that $A \cup B$ must also be in the σ-algebra. The same argument shows that if $A_1, A_2, \ldots, A_N$ are finitely many sets in a σ-algebra, then their union must also be in the σ-algebra. Finally, if $A_1, A_2, \ldots$ is a sequence of sets in a σ-algebra, then because

$$\bigcap_{n=1}^{\infty} A_n = \left(\bigcup_{n=1}^{\infty} A_n^c \right)^c,$$

properties (ii) and (iii) applied to the right-hand side show that $\cap_{n=1}^{\infty} A_n$ is also in the σ-algebra. Similarly, the intersection of a finite number of sets in a σ-algebra results in a set in the σ-algebra. Of course, if $\mathcal{F}$ is a σ-algebra, then the whole space Ω must be one of the sets in $\mathcal{F}$ because $\Omega = \emptyset^c$.

Definition 1.1.2. *Let Ω be a nonempty set, and let $\mathcal{F}$ be a σ-algebra of subsets of Ω. A* probability measure *$\mathbb{P}$ is a function that, to every set $A \in \mathcal{F}$, assigns a number in $[0, 1]$, called the* probability *of A and written $\mathbb{P}(A)$. We require:*

(i) $\mathbb{P}(\Omega) = 1$, and
(ii) (countable additivity) whenever $A_1, A_2, \ldots$ is a sequence of disjoint sets in $\mathcal{F}$, then

$$\mathbb{P}\left(\bigcup_{n=1}^{\infty} A_n \right) = \sum_{n=1}^{\infty} \mathbb{P}(A_n). \qquad (1.1.2)$$

The triple $(\Omega, \mathcal{F}, \mathbb{P})$ is called a probability space.

If Ω is a finite set and $\mathcal{F}$ is the collection of all subsets of Ω, then $\mathcal{F}$ is a σ-algebra and Definition 1.1.2 boils down to Definition 2.1.1 of Chapter 2 of Volume I. In the context of infinite probability spaces, we must take care that the definition of probability measure just given is consistent with our intuition. The countable additivity condition (ii) in Definition 1.1.2 is designed to take care of this. For example, we should be sure that $\mathbb{P}(\emptyset) = 0$. That follows from taking

$$A_1 = A_2 = A_3 = \cdots = \emptyset$$

in (1.1.2), for then this equation becomes $\mathbb{P}(\emptyset) = \sum_{n=1}^{\infty} \mathbb{P}(\emptyset)$. The only number in $[0, 1]$ that $\mathbb{P}(\emptyset)$ could be is

$$\mathbb{P}(\emptyset) = 0. \tag{1.1.3}$$

We also still want (2.1.7) of Chapter 2 of Volume I to hold: if A and B are disjoint sets in $\mathcal{F}$, we want to have

$$\mathbb{P}(A \cup B) = \mathbb{P}(A) + \mathbb{P}(B). \tag{1.1.4}$$

Not only does Definition 1.1.2(ii) guarantee this, it guarantees the *finite additivity* condition that if $A_1, A_2, \ldots, A_N$ are finitely many disjoint sets in $\mathcal{F}$, then

$$\mathbb{P}\left(\bigcup_{n=1}^{N} A_n\right) = \sum_{n=1}^{N} \mathbb{P}(A_n). \tag{1.1.5}$$

To see this, apply (1.1.2) with

$$A_{N+1} = A_{N+2} = A_{N+3} = \cdots = \emptyset.$$

In the special case that $N = 2$ and $A_1 = A$, $A_2 = B$, we get (1.1.4). From part (i) of Definition 1.1.2 and (1.1.4) with $B = A^c$, we get

$$\mathbb{P}(A^c) = 1 - \mathbb{P}(A). \tag{1.1.6}$$

In summary, from Definition 1.1.2, we conclude that a probability measure must satisfy (1.1.3)–(1.1.6).

We now describe by example the process of construction of probability measures on uncountable sample spaces. We do this here for the spaces $[0, 1]$ and Ω_∞ with which we began this section.

Example 1.1.3 (Uniform (Lebesgue) measure on $[0, 1]$*).* We construct a mathematical model for choosing a number at random from the unit interval $[0, 1]$ so that the probability is distributed uniformly over the interval. We define the probability of closed intervals $[a, b]$ by the formula

$$\mathbb{P}[a, b] = b - a, \ 0 \le a \le b \le 1, \tag{1.1.7}$$

(i.e., the probability that the number chosen is between a and b is $b - a$). (This particular probability measure on $[0, 1]$ is called *Lebesgue measure* and in this text is sometimes denoted $\mathcal{L}$. The Lebesgue measure of a subset of $\mathbb{R}$ is its "length.") If $b = a$, then $[a, b]$ is the set containing only the number a, and (1.1.7) says that the probability of this set is zero (i.e., the probability is zero that the number we choose is exactly equal to a). Because single points have zero probability, the probability of an open interval (a, b) is the same as the probability of the closed interval $[a, b]$; we have

$$\mathbb{P}(a,b) = b - a, \ 0 \le a \le b \le 1. \tag{1.1.8}$$

There are many other subsets of $[0,1]$ whose probability is determined by the formula (1.1.7) and the properties of probability measures. For example, the set $\left[0,\frac{1}{3}\right] \cup \left[\frac{2}{3},1\right]$ is not an interval, but we know from (1.1.7) and (1.1.4) that its probability is $\frac{2}{3}$.

It is natural to ask if there is some way to describe the collection of all sets whose probability is determined by formula (1.1.7) and the properties of probability measures. It turns out that this collection of sets is the σ-algebra we get starting with the closed intervals and putting in everything else required in order to have a σ-algebra. Since an open interval can be written as a union of a sequence of closed intervals,

$$(a,b) = \bigcup_{n=1}^{\infty} \left[a + \frac{1}{n}, b - \frac{1}{n}\right],$$

this σ-algebra contains all open intervals. It must also contain the set $\left[0,\frac{1}{3}\right] \cup \left[\frac{2}{3},1\right]$, mentioned at the end of the preceding paragraph, and many other sets.

The σ-algebra obtained by beginning with closed intervals and adding everything else necessary in order to have a σ-algebra is called the *Borel σ-algebra* of subsets of $[0,1]$ and is denoted $\mathcal{B}[0,1]$. The sets in this σ-algebra are called *Borel sets*. These are the subsets of $[0,1]$, the so-called events, whose probability is determined once we specify the probability of the closed intervals. Every subset of $[0,1]$ we encounter in this text is a Borel set, and this can be verified if desired by writing the set in terms of unions, intersections, and complements of sequences of closed intervals.[1] □

Example 1.1.4 (Infinite, independent coin-toss space). We toss a coin infinitely many times and let Ω_∞ of (1.1.1) denote the set of possible outcomes. We assume the probability of head on each toss is $p > 0$, the probability of tail is $q = 1 - p > 0$, and the different tosses are independent, a concept we define precisely in the next chapter. We want to construct a probability measure corresponding to this random experiment.

We first define $\mathbb{P}(\emptyset) = 0$ and $\mathbb{P}(\Omega) = 1$. These $2^{(2^0)} = 2$ sets form a σ-algebra, which we call $\mathcal{F}_0$:

$$\mathcal{F}_0 = \{\emptyset, \Omega\}. \tag{1.1.9}$$

We next define $\mathbb{P}$ for the two sets

$$\begin{aligned} A_H &= \text{the set of all sequences beginning with } H &= \{\omega; \omega_1 = H\}, \\ A_T &= \text{the set of all sequences beginning with } T &= \{\omega; \omega_1 = T\}, \end{aligned}$$

[1] See Appendix A, Section A.1 for the construction of the *Cantor set*, which gives some indication of how complicated sets in $\mathcal{B}[0,1]$ can be.

by setting $\mathbb{P}(A_H) = p$, $\mathbb{P}(A_T) = q$. We have now defined $\mathbb{P}$ for $2^{(2^1)} = 4$ sets, and these four sets form a σ-algebra; since $A_H^c = A_T$ we do not need to add anything else in order to have a σ-algebra. We call this σ-algebra $\mathcal{F}_1$:

$$\mathcal{F}_1 = \{\emptyset, \Omega, A_H, A_T\}. \tag{1.1.10}$$

We next define $\mathbb{P}$ for the four sets

$$\begin{aligned} A_{HH} &= \text{ The set of all sequences beginning with } HH \\ &= \{\omega; \omega_1 = H, \omega_2 = H\}, \\ A_{HT} &= \text{ The set of all sequences beginning with } HT \\ &= \{\omega; \omega_1 = H, \omega_2 = T\}, \\ A_{TH} &= \text{ The set of all sequences beginning with } TH \\ &= \{\omega; \omega_1 = T, \omega_2 = H\}, \\ A_{TT} &= \text{ The set of all sequences beginning with } TT \\ &= \{\omega; \omega_1 = T, \omega_2 = T\} \end{aligned}$$

by setting

$$\mathbb{P}(A_{HH}) = p^2,\ \mathbb{P}(A_{HT}) = pq,\ \mathbb{P}(A_{TH}) = pq,\ \mathbb{P}(A_{TT}) = q^2. \tag{1.1.11}$$

Because of (1.1.6), this determines the probability of the complements A_{HH}^c, A_{HT}^c, A_{TH}^c, A_{TT}^c. Using (1.1.5), we see that the probabilities of the unions $A_{HH} \cup A_{TH}$, $A_{HH} \cup A_{TT}$, $A_{HT} \cup A_{TH}$, and $A_{HT} \cup A_{TT}$ are also determined. We have already defined the probabilities of the two other pairwise unions $A_{HH} \cup A_{HT} = A_H$ and $A_{TH} \cup A_{TT} = A_T$. We have already noted that the probability of the triple unions is determined since these are complements of the sets in (1.1.11), e.g.,

$$A_{HH} \cup A_{HT} \cup A_{TH} = A_{TT}^c.$$

At this point, we have determined the probability of $2^{(2^2)} = 16$ sets, and these sets form a σ-algebra, which we call $\mathcal{F}_2$:

$$\mathcal{F}_2 = \left\{ \begin{matrix} \emptyset, \Omega, A_H, A_T, A_{HH}, A_{HT}, A_{TH}, A_{TT}, A_{HH}^c, A_{HT}^c, A_{TH}^c, A_{TT}^c, \\ A_{HH} \cup A_{TH},\ A_{HH} \cup A_{TT},\ A_{HT} \cup A_{TH},\ A_{HT} \cup A_{TT} \end{matrix} \right\}. \tag{1.1.12}$$

We next define the probability of every set that can be described in terms of the outcome of the first three coin tosses. Counting the sets we already have, this will give us $2^{(2^3)} = 256$ sets, and these will form a σ-algebra, which we call $\mathcal{F}_3$.

By continuing this process, we can define the probability of every set that can be described in terms of finitely many tosses. Once the probabilities of all these sets are specified, there are other sets, not describable in terms of finitely many coin tosses, whose probabilities are determined. For example, the set containing only the single sequence $HHHH\dots$ cannot be described in terms of finitely many coin tosses, but it is a subset of A_H, A_{HH}, A_{HHH}, etc. Furthermore,

$$\mathbb{P}(A_H) = p,\ \mathbb{P}(A_{HH}) = p^2,\ \mathbb{P}(A_{HHH}) = p^3, \dots,$$

and since these probabilities converge to zero, we must have

$$\mathbb{P}(\text{Every toss results in head}) = 0.$$

Similarly, the single sequence $HTHTHT\dots$, being the intersection of the sets A_H, A_{HT}, A_{HTH}, etc. must have probability less than or equal to each of

$$\mathbb{P}(A_H) = p,\ \mathbb{P}(A_{HT}) = pq,\ \mathbb{P}(A_{HTH}) = p^2q, \dots,$$

and hence must have probability zero. The same argument shows that every individual sequence in Ω_∞ has probability zero.

We create a σ-algebra, called $\mathcal{F}_\infty$, by putting in every set that can be described in terms of finitely many coin tosses and then adding all other sets required in order to have a σ-algebra. It turns out that once we specify the probability of every set that can be described in terms of finitely many coin tosses, the probability of every set in $\mathcal{F}_\infty$ is determined. There are sets in $\mathcal{F}_\infty$ whose probability, although determined, is not easily computed. For example, consider the set A of sequences $\omega = \omega_1\omega_2\dots$ for which

$$\lim_{n\to\infty} \frac{H_n(\omega_1\dots\omega_n)}{n} = \frac{1}{2}, \tag{1.1.13}$$

where $H_n(\omega_1\dots\omega_n)$ denotes the number of Hs in the first n tosses. In other words, A is the set of sequences of heads and tails for which the long-run average number of heads is $\frac{1}{2}$. Because its description involves all the coin tosses, it was not defined directly at any stage of the process outlined above. On the other hand, it is in $\mathcal{F}_\infty$, and that means its probability is somehow determined by this process and the properties of probability measures. To see that A is in $\mathcal{F}_\infty$, we fix positive integers m and n and define the set

$$A_{n,m} = \left\{\omega; \left|\frac{H_n(\omega_1\dots\omega_n)}{n} - \frac{1}{2}\right| \le \frac{1}{m}\right\}.$$

This set is in $\mathcal{F}_n$, and once n and m are known, its probability is defined by the process outlined above. By the definition of limit, a coin-toss sequence $\omega = \omega_1\omega_2\dots$ satisfies (1.1.13) if and only if for every positive integer m there exists a positive integer N such that for all $n \ge N$ we have $\omega \in A_{n,m}$. In other words, the set of ω for which (1.1.13) holds is

$$A = \bigcap_{m=1}^{\infty} \bigcup_{N=1}^{\infty} \bigcap_{n=N}^{\infty} A_{n,m}.$$

The set A is in $\mathcal{F}_\infty$ because it is described in terms of unions and intersections of sequences of sets that are in $\mathcal{F}_\infty$. This does not immediately tell us how to compute $\mathbb{P}(A)$, but it tells us that $\mathbb{P}(A)$ is somehow determined. As it turns out, the Strong Law of Large Numbers asserts that $\mathbb{P}(A) = 1$ if $p = \frac{1}{2}$ and $\mathbb{P}(A) = 0$ if $p \neq \frac{1}{2}$.

Every subset of Ω_∞ we shall encounter will be in $\mathcal{F}_\infty$. Indeed, it is extremely difficult to produce a set not in $\mathcal{F}_\infty$, although such sets exist. □

The observation in Example 1.1.4 that every individual sequence has probability zero highlights a paradox in uncountable probability spaces. We would like to say that something that has probability zero cannot happen. In particular, we would like to say that if we toss a coin infinitely many times, it cannot happen that we get a head on every toss (we are assuming here that the probability for head on each toss is $p > 0$ and $q = 1 - p > 0$). It would be satisfying if events that have probability zero are sure not to happen and events that have probability one are sure to happen. In particular, we would like to say that we are sure to get at least one tail. However, because the sequence that is all heads is in our sample space, and is no less likely to happen than any other particular sequence (every single sequence has probability zero), mathematicians have created a terminology that equivocates. We say that we will get at least one tail *almost surely*. Whenever an event is said to be almost sure, we mean it has probability one, even though it may not include every possible outcome. The outcome or set of outcomes not included, taken all together, has probability zero.

Definition 1.1.5. *Let $(\Omega, \mathcal{F}, \mathbb{P})$ be a probability space. If a set $A \in \mathcal{F}$ satisfies $\mathbb{P}(A) = 1$, we say that the event A occurs* almost surely.

1.2 Random Variables and Distributions

Definition 1.2.1. *Let $(\Omega, \mathcal{F}, \mathbb{P})$ be a probability space. A* random variable *is a real-valued function X defined on Ω with the property that for every Borel subset B of $\mathbb{R}$, the subset of Ω given by*

$$\{X \in B\} = \{\omega \in \Omega; X(\omega) \in B\} \tag{1.2.1}$$

is in the σ-algebra $\mathcal{F}$. (We sometimes also permit a random variable to take the values $+\infty$ and $-\infty$.)

To get the Borel subsets of $\mathbb{R}$, one begins with the closed intervals $[a, b] \subset \mathbb{R}$ and adds all other sets that are necessary in order to have a σ-algebra. This means that unions of sequences of closed intervals are Borel sets. In particular, every open interval is a Borel set, because an open interval can be written as the union of a sequence of closed intervals. Furthermore, every open *set* (whether or not an interval) is a Borel set because every open set is the union of a sequence of open intervals. Every closed set is a Borel set because it is the complement of an open set. We denote the collection of Borel subsets of $\mathbb{R}$ by $\mathcal{B}(\mathbb{R})$ and call it the *Borel σ-algebra of* $\mathbb{R}$. Every subset of $\mathbb{R}$ we encounter in this text is in this σ-algebra.

A random variable X is a numerical quantity whose value is determined by the random experiment of choosing $\omega \in \Omega$. We shall be interested in the probability that X takes various values. It is often the case that the probability that X takes a particular value is zero, and hence we shall mostly talk about the probability that X takes a value in some set rather than the probability that X takes a particular value. In other words, we will want to speak of $\mathbb{P}\{X \in B\}$. Definition 1.2.1 requires that $\{X \in B\}$ be in $\mathcal{F}$ for all $B \in \mathcal{B}(\mathbb{R})$, so that we are sure the probability of this set is defined.

Example 1.2.2 (Stock prices). Recall the independent, infinite coin-toss space $(\Omega_\infty, \mathcal{F}_\infty, \mathbb{P})$ of Example 1.1.4. Let us define stock prices by the formulas

$$S_0(\omega) = 4 \text{ for all } \omega \in \Omega_\infty,$$

$$S_1(\omega) = \begin{cases} 8 \text{ if } \omega_1 = H, \\ 2 \text{ if } \omega_1 = T, \end{cases}$$

$$S_2(\omega) = \begin{cases} 16 \text{ if } \omega_1 = \omega_2 = H, \\ 4 \;\; \text{if } \omega_1 \neq \omega_2, \\ 1 \;\; \text{if } \omega_1 = \omega_2 = T, \end{cases}$$

and, in general,

$$S_{n+1}(\omega) = \begin{cases} 2S_n(\omega) \text{ if } \omega_{n+1} = H, \\ \frac{1}{2}S_n(\omega) \text{ if } \omega_{n+1} = T. \end{cases}$$

All of these are random variables. They assign a numerical value to each possible sequence of coin tosses. Furthermore, we can compute the probabilities that these random variables take various values. For example, in the notation of Example 1.1.4,

$$\mathbb{P}\{S_2 = 4\} = \mathbb{P}\big(A_{HT} \cup A_{TH}\big) = 2pq. \qquad \square$$

In the previous example, the random variables S_0, S_1, S_2, etc., have distributions. Indeed, $S_0 = 4$ with probability one, so we can regard this random variable as putting a unit of mass on the number 4. On the other hand, $\mathbb{P}\{S_2 = 16\} = p^2$, $\mathbb{P}\{S_2 = 4\} = 2pq$, and $\mathbb{P}\{S_2 = 1\} = q^2$. We can think of the distribution of this random variable as three lumps of mass, one of size p^2 located at the number 16, another of size $2pq$ located at the number 4, and a third of size q^2 located at the number 1. We need to allow for the possibility that the random variables we consider don't assign any lumps of mass but rather spread a unit of mass "continuously" over the real line. To do this, we should think of the distribution of a random variable as telling us how much mass is in a set rather than how much mass is at a point. In other words, the distribution of a random variable is itself a probability measure, but it is a measure on subsets of $\mathbb{R}$ rather than subsets of Ω.

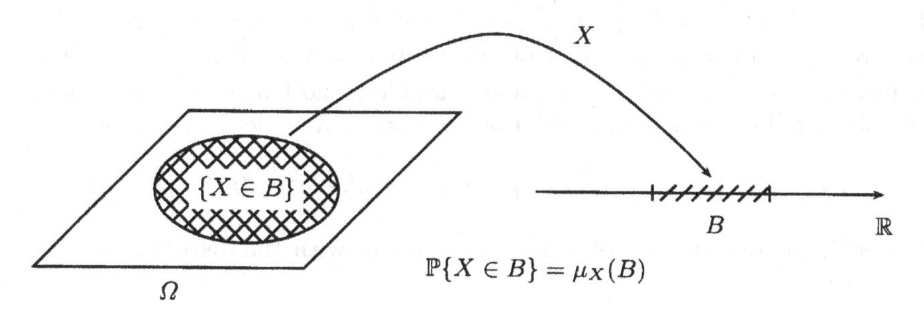

Fig. 1.2.1. Distribution measure of X.

Definition 1.2.3. *Let X be a random variable on a probability space $(\Omega, \mathcal{F}, \mathbb{P})$. The* distribution measure of X *is the probability measure μ_X that assigns to each Borel subset B of $\mathbb{R}$ the mass $\mu_X(B) = \mathbb{P}\{X \in B\}$ (see Figure 1.2.1).*

In this definition, the set B could contain a single number. For example, if $B = \{4\}$, then in Example 1.2.2 we would have $\mu_{S_2}(B) = 2pq$. If $B = [2,5]$, we still have $\mu_{S_2}(B) = 2pq$, because the only mass that S_2 puts in the interval $[2,5]$ is the lump of mass placed at the number 4. Definition 1.2.3 for the distribution measure of a random variable makes sense for discrete random variables as well as for random variables that spread a unit of mass "continuously" over the real line.

Random variables have distributions, but distributions and random variables are different concepts. Two different random variables can have the same distribution. A single random variable can have two different distributions. Consider the following example.

Example 1.2.4. Let $\mathbb{P}$ be the uniform measure on $[0,1]$ described in Example 1.1.3. Define $X(\omega) = \omega$ and $Y(\omega) = 1 - \omega$ for all $\omega \in [0,1]$. Then the distribution measure of X is uniform, i.e.,

$$\mu_X[a,b] = \mathbb{P}\{\omega; a \leq X(\omega) \leq b\} = \mathbb{P}[a,b] = b - a, \quad 0 \leq a \leq b \leq 1,$$

by the definition of $\mathbb{P}$. Although the random variable Y is different from the random variable X (if X takes the value $\frac{1}{3}$, Y takes the value $\frac{2}{3}$), Y has the same distribution as X:

$$\begin{aligned}\mu_Y[a,b] &= \mathbb{P}\{\omega; a \leq Y(\omega) \leq b\} = \mathbb{P}\{\omega; a \leq 1 - \omega \leq b\} = \mathbb{P}[1-b, 1-a] \\ &= (1-a) - (1-b) = b - a = \mu_X[a,b], \quad 0 \leq a \leq b \leq 1.\end{aligned}$$

Now suppose we define another probability measure $\widetilde{\mathbb{P}}$ on $[0,1]$ by specifying

$$\widetilde{\mathbb{P}}[a,b] = \int_a^b 2\omega \, d\omega = b^2 - a^2, \quad 0 \leq a \leq b \leq 1. \tag{1.2.2}$$

Equation (1.2.2) and the properties of probability measures determine $\widetilde{\mathbb{P}}(B)$ for every Borel subset B of $\mathbb{R}$. Note that $\widetilde{\mathbb{P}}[0,1] = 1$, so $\widetilde{\mathbb{P}}$ is in fact a probability measure. Under $\widetilde{\mathbb{P}}$, the random variable X no longer has the uniform distribution. Denoting the distribution measure of X under $\widetilde{\mathbb{P}}$ by $\widetilde{\mu}_X$, we have

$$\widetilde{\mu}_X[a,b] = \widetilde{\mathbb{P}}\{\omega; a \leq X(\omega) \leq b\} = \widetilde{\mathbb{P}}[a,b] = b^2 - a^2, \quad 0 \leq a \leq b \leq 1.$$

Under $\widetilde{\mathbb{P}}$, the distribution of Y no longer agrees with the distribution of X. We have

$$\begin{aligned}\widetilde{\mu}_Y[a,b] &= \widetilde{\mathbb{P}}\{\omega; a \leq Y(\omega) \leq b\} = \widetilde{\mathbb{P}}\{\omega; a \leq 1-\omega \leq b\} = \widetilde{\mathbb{P}}[1-b, 1-a] \\ &= (1-a)^2 - (1-b)^2, \quad 0 \leq a \leq b \leq 1. \qquad \square\end{aligned}$$

There are other ways to record the distribution of a random variable rather than specifying the distribution measure μ_X. We can describe the distribution of a random variable in terms of its *cumulative distribution function (cdf)*

$$F(x) = \mathbb{P}\{X \leq x\}, \quad x \in \mathbb{R}. \tag{1.2.3}$$

If we know the distribution measure μ_X, then we know the cdf F because $F(x) = \mu_X(-\infty, x]$. On the other hand, if we know the cdf F, then we can compute $\mu_X(x,y] = F(y) - F(x)$ for $x < y$. For $a \leq b$, we have

$$[a,b] = \bigcap_{n=1}^{\infty} \left(a - \tfrac{1}{n}, b\right],$$

and so we can compute[2]

$$\mu_X[a,b] = \lim_{n\to\infty} \mu_X\left(a - \tfrac{1}{n}, b\right] = F(b) - \lim_{n\to\infty} F\left(a - \tfrac{1}{n}\right). \tag{1.2.4}$$

Once the distribution measure $\mu_X[a,b]$ is known for every interval $[a,b] \subset \mathbb{R}$, it is determined for every Borel subset of $\mathbb{R}$. Therefore, in principle, knowing the cdf F for a random variable is the same as knowing its distribution measure μ_X.

In two special cases, the distribution of a random variable can be recorded in more detail. The first of these is when there is a *density function* $f(x)$, a nonnegative function defined for $x \in \mathbb{R}$ such that

$$\mu_X[a,b] = \mathbb{P}\{a \leq X \leq b\} = \int_a^b f(x)\,dx, \quad -\infty < a \leq b < \infty. \tag{1.2.5}$$

In particular, because the closed intervals $[-n, n]$ have union $\mathbb{R}$, we must have[3]

[2] See Appendix A, Theorem A.1.1(ii) for more detail.
[3] See Appendix A, Theorem A.1.1(i) for more detail.

$$\int_{-\infty}^{\infty} f(x)\,dx = \lim_{n\to\infty} \int_{-n}^{n} f(x)\,dx = \lim_{n\to\infty} \mathbb{P}\{-n \le X \le n\}$$
$$= \mathbb{P}\{X \in \mathbb{R}\} = \mathbb{P}(\Omega) = 1. \tag{1.2.6}$$

(For purposes of this discussion, we are not considering random variables that can take the value $\pm\infty$.)

The second special case is that of a *probability mass function*, in which case there is either a finite sequence of numbers $x_1, x_2, \ldots, x_N$ or an infinite sequence $x_1, x_2, \ldots$ such that with probability one the random variable X takes one of the values in the sequence. We then define $p_i = \mathbb{P}\{X = x_i\}$. Each p_i is nonnegative, and $\sum_i p_i = 1$. The mass assigned to a Borel set $B \subset \mathbb{R}$ by the distribution measure of X is

$$\mu_X(B) = \sum_{\{i;x_i \in B\}} p_i, \quad B \in \mathcal{B}(\mathbb{R}). \tag{1.2.7}$$

The distribution of some random variables can be described via a density, as in (1.2.5). For other random variables, the distribution must be described in terms of a probability mass function, as in (1.2.7). There are random variables whose distribution is given by a mixture of a density and a probability mass function, and there are random variables whose distribution has no lumps of mass but neither does it have a density.[4] Random variables of this last type have applications in finance but only at a level more advanced than this part of the text.

Example 1.2.5. (Another random variable uniformly distributed on [0,1].) We construct a uniformly distributed random variable taking values in $[0, 1]$ and defined on infinite coin-toss space Ω_∞. Suppose in the independent coin-toss space of Example 1.1.4 that the probability for head on each toss is $p = \frac{1}{2}$. For $n = 1, 2, \ldots$, we define

$$Y_n(\omega) = \begin{cases} 1 \text{ if } \omega_n = H, \\ 0 \text{ if } \omega_n = T. \end{cases} \tag{1.2.8}$$

We set

$$X = \sum_{n=1}^{\infty} \frac{Y_n}{2^n}.$$

If $Y_1 = 0$, which happens with probability $\frac{1}{2}$, then $0 \le X \le \frac{1}{2}$. If $Y_1 = 1$, which also happens with probability $\frac{1}{2}$, then $\frac{1}{2} \le X \le 1$. If $Y_1 = 0$ and $Y_2 = 0$, which happens with probability $\frac{1}{4}$, then $0 \le X \le \frac{1}{4}$. If $Y_1 = 0$ and $Y_2 = 1$, which also happens with probability $\frac{1}{4}$, then $\frac{1}{4} \le X \le \frac{1}{2}$. This pattern continues; indeed for any interval $[\frac{k}{2^n}, \frac{k+1}{2^n}] \subset [0, 1]$, the probability that the interval contains X is $\frac{1}{2^n}$. In terms of the distribution measure μ_X of X, we write this fact as

$$\mu_X \left[\frac{k}{2^n}, \frac{k+1}{2^n}\right] = \frac{1}{2^n} \text{ whenever } k \text{ and } n \text{ are integers and } 0 \le k \le 2^n - 1.$$

[4] See Appendix A, Section A.3.

Taking unions of intervals of this form and using the finite additivity of probability measures, we see that whenever k, m, and n are integers and $0 \leq k \leq m \leq 2^n$, we have

$$\mu_X\left[\frac{k}{2^n}, \frac{m}{2^n}\right] = \frac{m}{2^n} - \frac{k}{2^n}. \tag{1.2.9}$$

From (1.2.9), one can show that

$$\mu_X[a,b] = b - a, \quad 0 \leq a \leq b \leq 1;$$

in other words, the distribution measure of X is uniform on $[0,1]$.

Example 1.2.6 (Standard normal random variable). Let

$$\varphi(x) = \frac{1}{\sqrt{2\pi}} e^{-\frac{x^2}{2}}$$

be the *standard normal density*, and define the *cumulative normal distribution function*

$$N(x) = \int_{-\infty}^{x} \varphi(\xi)\, d\xi.$$

The function $N(x)$ is strictly increasing, mapping $\mathbb{R}$ onto $(0,1)$, and so has a strictly increasing inverse function $N^{-1}(y)$. In other words, $N(N^{-1}(y)) = y$ for all $y \in (0,1)$. Now let Y be a uniformly distributed random variable, defined on some probability space $(\Omega, \mathcal{F}, \mathbb{P})$ (two possibilities for $(\Omega, \mathcal{F}, \mathbb{P})$ and Y are presented in Examples 1.2.4 and 1.2.5), and set $X = N^{-1}(Y)$. Whenever $-\infty < a \leq b < \infty$, we have

$$\begin{aligned}
\mu_X[a,b] &= \mathbb{P}\{\omega \in \Omega; a \leq X(\omega) \leq b\} \\
&= \mathbb{P}\{\omega \in \Omega; a \leq N^{-1}(Y(\omega)) \leq b\} \\
&= \mathbb{P}\{\omega \in \Omega; N(a) \leq N(N^{-1}(Y(\omega))) \leq N(b)\} \\
&= \mathbb{P}\{\omega \in \Omega; N(a) \leq Y(\omega) \leq N(b)\} \\
&= N(b) - N(a) \\
&= \int_a^b \varphi(x)\, dx.
\end{aligned}$$

The measure μ_X on $\mathbb{R}$ given by this formula is called the *standard normal distribution.* Any random variable that has this distribution, regardless of the probability space $(\Omega, \mathcal{F}, \mathbb{P})$ on which it is defined, is called a *standard normal random variable.* The method used here for generating a standard normal random variable from a uniformly distributed random variable is called the *probability integral transform* and is widely used in Monte Carlo simulation.

Another way to construct a standard normal random variable is to take $\Omega = \mathbb{R}$, $\mathcal{F} = \mathcal{B}(\mathbb{R})$, take $\mathbb{P}$ to be the probability measure on $\mathbb{R}$ that satisfies

$$\mathbb{P}[a,b] = \int_a^b \varphi(x)\,dx \text{ whenever } -\infty < a \leq b < \infty,$$

and take $X(\omega) = \omega$ for all $\omega \in \mathbb{R}$. □

The second construction of a standard normal random variable in Example 1.2.6 is economical, and this method can be used to construct a random variable with any desired distribution. However, it is not useful when we want to have multiple random variables, each with a specified distribution and with certain dependencies among the random variables. For such cases, we construct (or at least assume there exists) a single probability space $(\Omega, \mathcal{F}, \mathbb{P})$ on which all the random variables of interest are defined. This point of view may seem overly abstract at the outset, but in the end it pays off handsomely in conceptual simplicity.

1.3 Expectations

Let X be a random variable defined on a probability space $(\Omega, \mathcal{F}, \mathbb{P})$. We would like to compute an "average value" of X, where we take the probabilities into account when doing the averaging. If Ω is finite, we simply define this average value by

$$\mathbb{E}X = \sum_{\omega \in \Omega} X(\omega)\mathbb{P}(\omega).$$

If Ω is countably infinite, its elements can be listed in a sequence $\omega_1, \omega_2, \omega_3, \dots$, and we can define $\mathbb{E}X$ as an infinite sum:

$$\mathbb{E}X = \sum_{k=1}^{\infty} X(\omega_k)\mathbb{P}(\omega_k).$$

Difficulty arises, however, if Ω is uncountably infinite. Uncountable sums cannot be defined. Instead, we must think in terms of integrals.

To see how to go about this, we first review the Riemann integral. If $f(x)$ is a continuous function defined for all x in the closed interval $[a,b]$, we define the Riemann integral $\int_a^b f(x)dx$ as follows. First partition $[a,b]$ into subintervals $[x_0,x_1], [x_1,x_2], \dots, [x_{n-1},x_n]$, where $a = x_0 < x_1 < \cdots < x_n = b$. We denote by $\Pi = \{x_0, x_1, \dots, x_n\}$ the set of partition points and by

$$\|\Pi\| = \max_{1 \leq k \leq n} (x_k - x_{k-1})$$

the length of the longest subinterval in the partition. For each subinterval $[x_{k-1}, x_k]$, we set $M_k = \max_{x_{k-1} \leq x \leq x_k} f(x)$ and $m_k = \min_{x_{k-1} \leq x \leq x_k} f(x)$. The upper Riemann sum is

$$\mathrm{RS}_\Pi^+(f) = \sum_{k=1}^{n} M_k(x_k - x_{k-1}),$$

and the lower Riemann sum (see Figure 1.3.1) is

$$\mathrm{RS}^-_\Pi(f) = \sum_{k=1}^{n} m_k(x_k - x_{k-1}).$$

As $\|\Pi\|$ converges to zero (i.e., as we put in more and more partition points, and the subintervals in the partition become shorter and shorter), the upper Riemann sum $\mathrm{RS}^+_\Pi(f)$ and the lower Riemann sum $\mathrm{RS}^-_\Pi(f)$ converge to the same limit, which we call $\int_a^b f(x)dx$. This is the *Riemann integral.*

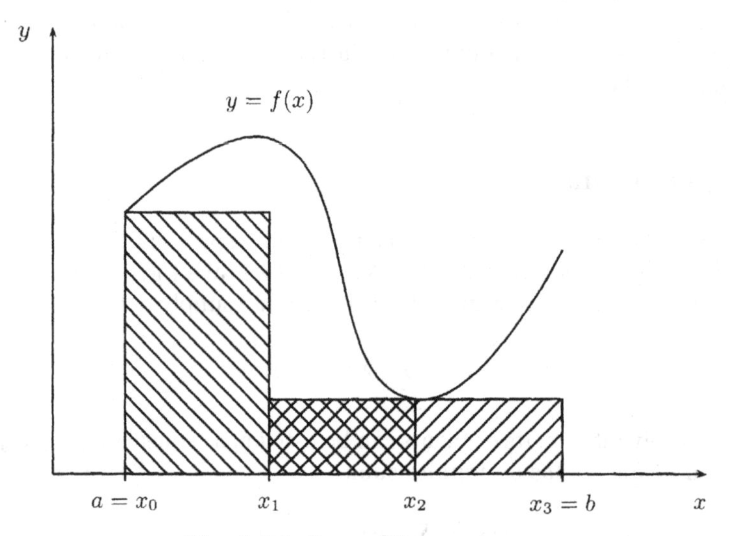

Fig. 1.3.1. Lower Riemann sum.

The problem we have with imitating this procedure to define expectation is that the random variable X, unlike the function f in the previous paragraph, is a function of $\omega \in \Omega$, and Ω is often not a subset of $\mathbb{R}$. In Figure 1.3.2 the "x-axis" is not the real numbers but some abstract space Ω. There is no natural way to partition the set Ω as we partitioned $[a,b]$ above. Therefore, we partition instead the y-axis in Figure 1.3.2. To see how this goes, assume for the moment that $0 \le X(\omega) < \infty$ for every $\omega \in \Omega$, and let $\Pi = \{y_0, y_1, y_2, \dots\}$, where $0 = y_0 < y_1 < y_2 < \dots$. For each subinterval $[y_k, y_{k+1}]$, we set

$$A_k = \{\omega \in \Omega; y_k \le X(\omega) < y_{k+1}\}.$$

We define the lower Lebesgue sum to be (see Figure 1.3.2)

$$\mathrm{LS}^-_\Pi(X) = \sum_{k=1}^{\infty} y_k\, \mathbb{P}(A_k).$$

This lower sum converges as $\|\Pi\|$, the maximal distance between the y_k partition points, approaches zero, and we define this limit to be the *Lebesgue integral* $\int_\Omega X(\omega)\,d\mathbb{P}(\omega)$, or simply $\int_\Omega X\,d\mathbb{P}$. The Lebesgue integral might be ∞, because we have not made any assumptions about how large the values of X can be.

We assumed a moment ago that $0 \leq X(\omega) < \infty$ for every $\omega \in \Omega$. If the set of ω that violates this condition has zero probability, there is no effect on the integral we just defined. If $\mathbb{P}\{\omega; X(\omega) \geq 0\} = 1$ but $\mathbb{P}\{\omega; X(\omega) = \infty\} > 0$, then we define $\int_\Omega X(\omega) d\mathbb{P}(\omega) = \infty$.

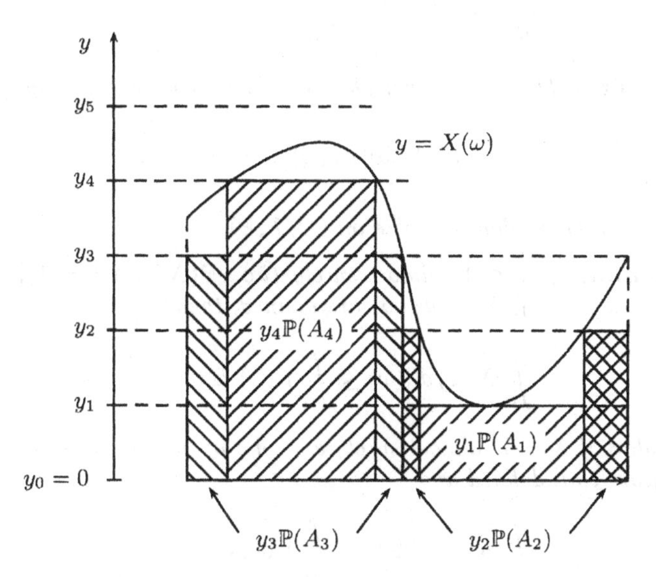

Fig. 1.3.2. Lower Lebesgue sum.

Finally, we need to consider random variables X that can take both positive and negative values. For such a random variable, we define the *positive* and *negative parts* of X by

$$X^+(\omega) = \max\{X(\omega), 0\}, \quad X^-(\omega) = \max\{-X(\omega), 0\}. \tag{1.3.1}$$

Both X^+ and X^- are nonnegative random variables, $X = X^+ - X^-$, and $|X| = X^+ + X^-$. Both $\int_\Omega X^+(\omega)\,d\mathbb{P}(\omega)$ and $\int_\Omega X^-(\omega)\,d\mathbb{P}(\omega)$ are defined by the procedure described above, and provided they are not both ∞, we can define

$$\int_\Omega X(\omega)\,d\mathbb{P}(\omega) = \int_\Omega X^+(\omega)\,d\mathbb{P}(\omega) - \int_\Omega X^-(\omega)\,d\mathbb{P}(\omega). \tag{1.3.2}$$

If $\int_\Omega X^+(\omega)\,d\mathbb{P}(\omega)$ and $\int_\Omega X^-(\omega)\,d\mathbb{P}(\omega)$ are both finite, we say that X is *integrable*, and $\int_\Omega X(\omega)\,d\mathbb{P}(\omega)$ is also finite. If $\int_\Omega X^+(\omega)\,d\mathbb{P}(\omega) = \infty$ and

$\int_\Omega X^-(\omega)\,d\mathbb{P}(\omega)$ is finite, then $\int_\Omega X(\omega)\,d\mathbb{P}(\omega) = \infty$. If $\int_\Omega X^+(\omega)\,d\mathbb{P}(\omega)$ is finite and $\int_\Omega X^-(\omega)\,d\mathbb{P}(\omega) = \infty$, then $\int_\Omega X(\omega)\,d\mathbb{P}(\omega) = -\infty$. If both $\int_\Omega X^+(\omega)\,d\mathbb{P}(\omega) = \infty$ and $\int_\Omega X^-(\omega)\,d\mathbb{P}(\omega) = \infty$, then an "$\infty - \infty$" situation arises in (1.3.2), and $\int_\Omega X(\omega)\,d\mathbb{P}(\omega)$ is not defined.

The Lebesgue integral has the following basic properties.

Theorem 1.3.1. *Let X be a random variable on a probability space $(\Omega, \mathcal{F}, \mathbb{P})$.*

(i) If X takes only finitely many values $y_0, y_1, y_2, \ldots, y_n$, then

$$\int_\Omega X(\omega)\,d\mathbb{P}(\omega) = \sum_{k=0}^{n} y_k \mathbb{P}\{X = y_k\}.$$

(ii) **(Integrability)** *The random variable X is integrable if and only if*

$$\int_\Omega |X(\omega)|\,d\mathbb{P}(\omega) < \infty.$$

Now let Y be another random variable on $(\Omega, \mathcal{F}, \mathbb{P})$.

(iii) **(Comparison)** *If $X \leq Y$ almost surely (i.e., $\mathbb{P}\{X \leq Y\} = 1$), and if $\int_\Omega X(\omega)\,d\mathbb{P}(\omega)$ and $\int_\Omega Y(\omega)\,d\mathbb{P}(\omega)$ are defined, then*

$$\int_\Omega X(\omega)\,d\mathbb{P}(\omega) \leq \int_\Omega Y(\omega)\,d\mathbb{P}(\omega).$$

In particular, if $X = Y$ almost surely and one of the integrals is defined, then they are both defined and

$$\int_\Omega X(\omega)\,d\mathbb{P}(\omega) = \int_\Omega Y(\omega)\,d\mathbb{P}(\omega).$$

(iv) **(Linearity)** *If α and β are real constants and X and Y are integrable, or if α and β are nonnegative constants and X and Y are nonnegative, then*

$$\int_\Omega (\alpha X(\omega) + \beta Y(\omega))\,d\mathbb{P}(\omega) = \alpha \int_\Omega X(\omega)\,d\mathbb{P}(\omega) + \beta \int_\Omega Y(\omega)\,d\mathbb{P}(\omega).$$

PARTIAL PROOF: For (i), we consider only the case when X is almost surely nonnegative. If zero is not among the y_ks, we may add $y_0 = 0$ to the list and then relabel the y_ks if necessary so that $0 = y_0 < y_1 < y_2 < \cdots < y_n$. Using these as our partition points, we have $A_k = \{y_k \leq X < y_{k+1}\} = \{X = y_k\}$ and the lower Lebesgue sum is

$$\mathrm{LS}_\Pi^-(X) = \sum_{k=0}^{n} y_k \mathbb{P}\{X = y_k\}.$$

If we put in more partition points, the lower Lebesgue sum does not change, and hence this is also the Lebesgue integral.

We next consider part (iii). If $X \leq Y$ almost surely, then $X^+ \leq Y^+$ and $X^- \geq Y^-$ almost surely. Because $X^+ \leq Y^+$ almost surely, for every partition Π, the lower Lebesgue sums satisfy $\mathrm{LS}_\Pi^-(X^+) \leq \mathrm{LS}_\Pi^-(Y^+)$, so

$$\int_\Omega X^+(\omega)\, d\mathbb{P}(\omega) \leq \int_\Omega Y^+(\omega)\, d\mathbb{P}(\omega). \tag{1.3.3}$$

Because $X^- \geq Y^-$ almost surely, we also have

$$\int_\Omega X^-(\omega)\, d\mathbb{P}(\omega) \geq \int_\Omega Y^-(\omega)\, d\mathbb{P}(\omega). \tag{1.3.4}$$

Subtracting (1.3.4) from (1.3.3) and recalling the definition (1.3.2), we obtain the comparison property (iii).

The linearity property (iv) requires a more detailed analysis of the construction of Lebesgue integrals. We do not provide that here.

We can use the comparison property (iii) and the linearity property (iv) to prove (ii) as follows. Because $|X| = X^+ + X^-$, we have $X^+ \leq |X|$ and $X^- \leq |X|$. If $\int_\Omega |X(\omega)|\, d\mathbb{P}(\omega) < \infty$, then the comparison property implies $\int_\Omega X^+(\omega)\, d\mathbb{P}(\omega) < \infty$ and $\int_\Omega X^-(\omega)\, d\mathbb{P}(\omega) < \infty$, and X is integrable by definition. On the other hand, if X is integrable, then $\int_\Omega X^+(\omega)\, d\mathbb{P}(\omega) < \infty$ and $\int_\Omega X^-(\omega)\, d\mathbb{P}(\omega) < \infty$. Adding these two quantities and using (iv), we see that $\int_\Omega |X(\omega)|\, d\mathbb{P}(\omega) < \infty$. □

Remark 1.3.2. We often want to integrate a random variable X over a subset A of Ω rather than over all of Ω. For this reason, we define

$$\int_A X(\omega)\, d\mathbb{P}(\omega) = \int_\Omega \mathbb{I}_A(\omega) X(\omega)\, d\mathbb{P}(\omega) \text{ for all } A \in \mathcal{F},$$

where $\mathbb{I}_A$ is the *indicator function (random variable)* given by

$$\mathbb{I}_A(\omega) = \begin{cases} 1 \text{ if } \omega \in A, \\ 0 \text{ if } \omega \notin A. \end{cases}$$

If A and B are disjoint sets in $\mathcal{F}$, then $\mathbb{I}_A + \mathbb{I}_B = \mathbb{I}_{A \cup B}$ and the linearity property (iv) of Theorem 1.3.1 implies that

$$\int_{A \cup B} X(\omega)\, d\mathbb{P}(\omega) = \int_A X(\omega)\, d\mathbb{P}(\omega) + \int_B X(\omega)\, d\mathbb{P}(\omega).$$

Definition 1.3.3. *Let X be a random variable on a probability space $(\Omega, \mathcal{F}, \mathbb{P})$. The* expectation *(or* expected value*) of X is defined to be*

$$\mathbb{E}X = \int_\Omega X(\omega)\, d\mathbb{P}(\omega).$$

This definition makes sense if X is integrable, *i.e.; if*

$$\mathbb{E}|X| = \int_{\Omega} |X(\omega)| \, d\mathbb{P}(\omega) < \infty$$

or if $X \geq 0$ almost surely. In the latter case, $\mathbb{E}X$ might be ∞.

We have thus managed to define $\mathbb{E}X$ when X is a random variable on an abstract probability space $(\Omega, \mathcal{F}, \mathbb{P})$. We restate in terms of expected values the basic properties of Theorem 1.3.1 and add an additional one.

Theorem 1.3.4. *Let X be a random variable on a probability space $(\Omega, \mathcal{F}, \mathbb{P})$.*

(i) If X takes only finitely many values $x_0, x_1, \ldots, x_n$, then

$$\mathbb{E}X = \sum_{k=0}^{n} x_k \mathbb{P}\{X = x_k\}.$$

In particular, if Ω is finite, then

$$\mathbb{E}X = \sum_{\omega \in \Omega} X(\omega)\mathbb{P}(\omega).$$

(ii) **(Integrability)** *The random variable X is integrable if and only if*

$$\mathbb{E}|X| < \infty.$$

Now let Y be another random variable on $(\Omega, \mathcal{F}, \mathbb{P})$.

(iii) **(Comparison)** *If $X \leq Y$ almost surely and X and Y are integrable or almost surely nonnegative, then*

$$\mathbb{E}X \leq \mathbb{E}Y.$$

In particular, if $X = Y$ almost surely and one of the random variables is integrable or almost surely nonnegative, then they are both integrable or almost surely nonnegative, respectively, and

$$\mathbb{E}X = \mathbb{E}Y.$$

(iv) **(Linearity)** *If α and β are real constants and X and Y are integrable or if α and β are nonnegative constants and X and Y are nonnegative, then*

$$E(\alpha X + \beta Y) = \alpha \mathbb{E}X + \beta \mathbb{E}Y.$$

(v) **(Jensen's inequality)** *If φ is a convex, real-valued function defined on $\mathbb{R}$, and if $\mathbb{E}|X| < \infty$, then*

$$\varphi(\mathbb{E}X) \leq \mathbb{E}\varphi(X).$$

PROOF: The only new claim is Jensen's inequality, and the proof of that is the same as the proof given for Theorem 2.2.5 of Chapter 2 of Volume I. □

Example 1.3.5. Consider the infinite independent coin-toss space Ω_∞ of Example 1.1.4 with the probability measure $\mathbb{P}$ that corresponds to probability $\frac{1}{2}$ for head on each toss. Let

$$Y_n(\omega) = \begin{cases} 1 \text{ if } \omega_n = H, \\ 0 \text{ if } \omega_n = T. \end{cases}$$

Even though the probability space Ω_∞ is uncountable, this random variable takes only two values, and we can compute its expectation using Theorem 1.3.4(i):

$$\mathbb{E}Y_n = 1 \cdot \mathbb{P}\{Y_n = 1\} + 0 \cdot \mathbb{P}\{Y_n = 0\} = \frac{1}{2}.$$

Example 1.3.6. Let $\Omega = [0,1]$, and let $\mathbb{P}$ be the Lebesgue measure on $[0,1]$ (see Example 1.1.3). Consider the random variable

$$X(\omega) = \begin{cases} 1 \text{ if } \omega \text{ is irrational}, \\ 0 \text{ if } \omega \text{ is rational}. \end{cases}$$

Again the random variable takes only two values, and we can compute its expectation using Theorem 1.3.4(i):

$$\mathbb{E}X = 1 \cdot \mathbb{P}\{\omega \in [0,1]; \omega \text{ is irrational}\} + 0 \cdot \mathbb{P}\{\omega \in [0,1]; \omega \text{ is rational}\}.$$

There are only countably many rational numbers in $[0,1]$ (i.e., they can all be listed in a sequence $x_1, x_2, x_3, \ldots$). Each number in the sequence has probability zero, and because of the countable additivity property (ii) of Definition 1.1.2, the whole sequence must have probability zero. Therefore, $\mathbb{P}\{\omega \in [0,1]; \omega \text{ is rational}\} = 0$. Since $\mathbb{P}[0,1] = 1$, the probability of the set of irrational numbers in $[0,1]$ must be 1. We conclude that $\mathbb{E}X = 1$.

The idea behind this example is that if we choose a number from $[0,1]$ according to the uniform distribution, then with probability one the number chosen will be irrational. Therefore, the random variable X is almost surely equal to 1, and hence its expected value equals 1. As a practical matter, of course, almost any algorithm we devise for generating a random number in $[0,1]$ will generate a rational number. The uniform distribution is often a reasonable idealization of the output of algorithms that generate random numbers in $[0,1]$, but if we push the model too far it can depart from reality.

If we had been working with Riemann rather than Lebesgue integrals, we would have gotten a different result. To make the notation more familiar, we write x rather than ω and $f(x)$ rather than $X(\omega)$, thus defining

$$f(x) = \begin{cases} 1 \text{ if } x \text{ is irrational}, \\ 0 \text{ if } x \text{ is rational}. \end{cases} \tag{1.3.5}$$

We have just seen that the Lebesgue integral of this function over the interval $[0,1]$ is 1.

To construct the Riemann integral, we choose partition points $0 = x_0 < x_1 < x_2 < \cdots < x_n = 1$. We define

$$M_k = \max_{x_{k-1} \leq x \leq x_k} f(x), \quad m_k = \min_{x_{k-1} \leq x \leq x_k} f(x).$$

But each interval $[x_{k-1}, x_k]$ contains both rational and irrational numbers, so $M_k = 1$ and $m_k = 0$. Therefore, for this partition $\Pi = \{x_0, x_1, \ldots, x_n\}$, the upper Riemann sum is 1,

$$\mathrm{RS}_{\Pi}^{+}(f) = \sum_{k=1}^{n} M_k(x_k - x_{k-1}) = \sum_{k=1}^{n}(x_k - x_{k-1}) = 1,$$

whereas the lower Riemann sum is zero,

$$\mathrm{RS}_{\Pi}^{-}(f) = \sum_{k=1}^{n} m_k(x_k - x_{k-1}) = 0.$$

This happens no matter how small we take the subintervals in the partition. Since the upper Riemann sum is always 1 and the lower Riemann sum is always 0, the upper and lower Riemann sums do not converge to the same limit and the Riemann integral is not defined. For the Riemann integral, which discretizes the x-axis rather than the y-axis, this function is too discontinuous to handle. The Lebesgue integral, however, which discretizes the y-axis, sees this as a simple function taking only two values. □

We constructed the Lebesgue integral because we wanted to integrate over abstract probability spaces $(\Omega, \mathcal{F}, \mathbb{P})$, but as Example 1.3.6 shows, after this construction we can take Ω to be a subset of the real numbers and then compare Lebesgue and Riemann integrals. This example further shows that these two integrals can give different results. Fortunately, the behavior in Example 1.3.6 is the worst that can happen. To make this statement precise, we first extend the construction of the Lebesgue integral to all of $\mathbb{R}$, rather than just $[0, 1]$.

Definition 1.3.7. *Let $\mathcal{B}(\mathbb{R})$ be the σ-algebra of Borel subsets of $\mathbb{R}$ (i.e., the smallest σ-algebra containing all the closed intervals $[a, b]$).*[5] *The* Lebesgue measure *on $\mathbb{R}$, which we denote by $\mathcal{L}$, assigns to each set $B \in \mathcal{B}(\mathbb{R})$ a number in $[0, \infty)$ or the value ∞ so that*

(i) $\mathcal{L}[a, b] = b - a$ whenever $a \leq b$, and
(ii) if $B_1, B_2, B_3, \ldots$ is a sequence of disjoint sets in $\mathcal{B}(\mathbb{R})$, then we have the countable additivity property

$$\mathcal{L}\left(\bigcup_{n=1}^{\infty} B_n\right) = \sum_{n=1}^{\infty} \mathcal{L}(B_n).$$

[5] This concept is discussed in more detail in Appendix A, Section A.2.

Definition 1.3.7 is similar to Definition 1.1.2, except that now some sets have measure greater than 1. The Lebesgue measure of every interval is its length, so that $\mathbb{R}$ and half-lines $[a, \infty)$ and $(-\infty, b]$ have infinite Lebesgue measure, single points have Lebesgue measure zero, and the Lebesgue measure of the empty set is zero. Lebesgue measure has the finite additivity property (see (1.1.5))

$$\mathcal{L}\left(\bigcup_{n=1}^{N} B_n\right) = \sum_{n=1}^{N} \mathcal{L}(B_n)$$

whenever $B_1, B_2, \ldots, B_N$ are disjoint Borel subsets of $\mathbb{R}$.

Now let $f(x)$ be a real-valued function defined on $\mathbb{R}$. For the following construction, we need to assume that for every Borel subset B of $\mathbb{R}$, the set $\{x; f(x) \in B\}$ is also a Borel subset of $\mathbb{R}$. A function f with this property is said to be *Borel measurable.* Every continuous and piecewise continuous function is Borel measurable. Indeed, it is extremely difficult to find a function that is not Borel measurable. We wish to define the Lebesgue integral $\int_{\mathbb{R}} f(x)\, d\mathcal{L}(x)$ of f over $\mathbb{R}$. To do this, we assume for the moment that $0 \leq f(x) < \infty$ for every $x \in \mathbb{R}$. We choose a partition $\Pi = \{y_0, y_1, y_2, \ldots\}$, where $0 = y_0 < y_1 < y_2 < \ldots$. For each subinterval $[y_k, y_{k+1})$, we define

$$B_k = \{x \in \mathbb{R}; y_k \leq f(x) < y_{k+1}\}.$$

Because of the assumption that f is Borel measurable, even though these sets B_k can be quite complicated, they are Borel subsets of $\mathbb{R}$ and so their Lebesgue measures are defined. We define the lower Lebesgue sum

$$\mathrm{LS}_{\Pi}^{-}(f) = \sum_{k=1}^{\infty} y_k \mathcal{L}(B_k).$$

As $\|\Pi\|$ converges to zero, these lower Lebesgue sums will converge to a limit, which we define to be $\int_{\mathbb{R}} f(x)\, d\mathcal{L}(x)$. It is possible that this integral gives the value ∞.

We assumed a moment ago that $0 \leq f(x) < \infty$ for every $x \in \mathbb{R}$. If the set of x where the condition is violated has zero Lebesgue measure, the integral of f is not affected. If $\mathcal{L}\{x \in \mathbb{R}; f(x) < 0\} = 0$ and $\mathcal{L}\{x \in \mathbb{R}; f(x) = \infty\} > 0$, we define $\int_{\mathbb{R}} f(x) d\mathcal{L}(x) = \infty$.

We next consider the possibility that $f(x)$ takes both positive and negative values. In this case, we define

$$f^+(x) = \max\{f(x), 0\}, \quad f^-(x) = \max\{-f(x), 0\}.$$

Because f^+ and f^- are nonnegative, $\int_{\mathbb{R}} f^+(x)\, d\mathcal{L}(x)$ and $\int_{\mathbb{R}} f^-(x)\, d\mathcal{L}(x)$ are defined by the procedure described above. We then define

$$\int_{\mathbb{R}} f(x)\, d\mathcal{L}(x) = \int_{\mathbb{R}} f^+(x)\, d\mathcal{L}(x) - \int_{\mathbb{R}} f^-(x)\, d\mathcal{L}(x),$$

provided this is not $\infty - \infty$. In the case where both $\int_{\mathbb{R}} f^+(x)\,d\mathcal{L}(x)$ and $\int_{\mathbb{R}} f^-(x)\,d\mathcal{L}(x)$ are infinite, $\int_{\mathbb{R}} f(x)\,d\mathcal{L}(x)$ is not defined. If $\int_{\mathbb{R}} f^+(x)\,d\mathcal{L}(x)$ and $\int_{\mathbb{R}} f^-(x)\,d\mathcal{L}(x)$ are finite, we say that f is *integrable*. This is equivalent to the condition $\int_{\mathbb{R}} |f(x)|\,d\mathcal{L}(x) < \infty$. The Lebesgue integral just constructed has the comparison and linearity properties described in Theorem 1.3.1. Moreover, if f takes only finitely many values $y_0, y_1, y_2, \ldots, y_n$, then

$$\int_{\mathbb{R}} f(x)\,d\mathcal{L}(x) = \sum_{k=0}^{n} y_k\, \mathcal{L}\{x \in \mathbb{R}; f(x) = y_k\},$$

provided the computation of the right-hand side does not require that $\infty - \infty$ be assigned a value.

Finally, sometimes we have a function $f(x)$ defined for every $x \in \mathbb{R}$ but want to compute its Lebesgue integral over only part of $\mathbb{R}$, say over some set $B \in \mathcal{B}(\mathbb{R})$. We do this by multiplying $f(x)$ by the indicator function of B:

$$\mathbb{I}_B(x) = \begin{cases} 1 \text{ if } x \in B, \\ 0 \text{ if } x \notin B. \end{cases}$$

The product $f(x)\mathbb{I}_B(x)$ agrees with $f(x)$ when $x \in B$ and is zero when $x \notin B$. We define

$$\int_B f(x)\,d\mathcal{L}(x) = \int_{\mathbb{R}} \mathbb{I}_B(x) f(x)\,d\mathcal{L}(x).$$

The following theorem, whose proof is beyond the scope of this book, relates Riemann and Lebesgue integrals on $\mathbb{R}$.

Theorem 1.3.8. (Comparison of Riemann and Lebesgue integrals). *Let f be a bounded function defined on $\mathbb{R}$, and let $a < b$ be numbers.*

(i) The Riemann integral $\int_a^b f(x)dx$ is defined (i.e., the lower and upper Riemann sums converge to the same limit) if and only if the set of points x in $[a,b]$ where $f(x)$ is not continuous has Lebesgue measure zero.
(ii) If the Riemann integral $\int_a^b f(x)dx$ is defined, then f is Borel measurable (so the Lebesgue integral $\int_{[a,b]} f(x)\,d\mathcal{L}(x)$ is also defined), and the Riemann and Lebesgue integrals agree.

A single point in $\mathbb{R}$ has Lebesgue measure zero, and so any finite set of points has Lebesgue measure zero. Theorem 1.3.8 guarantees that if we have a real-valued function f on $\mathbb{R}$ that is continuous except at finitely many points, then there will be no difference between Riemann and Lebesgue integrals of this function.

Definition 1.3.9. *If the set of numbers in $\mathbb{R}$ that fail to have some property is a set with Lebesgue measure zero, we say that the property holds* almost everywhere.

Theorem 1.3.8(i) may be restated as:

The Riemann integral $\int_a^b f(x)dx$ exists if and only if $f(x)$ is almost everywhere continuous on $[a,b]$.

Because the Riemann and Lebesgue integrals agree whenever the Riemann integral is defined, we shall use the more familiar notation $\int_a^b f(x)\,dx$ to denote the Lebesgue integral rather than $\int_{[a,b]} f(x)\,d\mathcal{L}(x)$. If the set B over which we wish to integrate is not an interval, we shall write $\int_B f(x)\,dx$. When we are developing theory, we shall understand $\int_B f(x)\,dx$ to be a Lebesgue integral; when we need to compute, we will use techniques learned in calculus for computing Riemann integrals.

1.4 Convergence of Integrals

There are several ways a sequence of random variables can converge. In this section, we consider the case of convergence almost surely, defined as follows.

Definition 1.4.1. *Let $X_1, X_2, X_3, \ldots$ be a sequence of random variables, all defined on the same probability space $(\Omega, \mathcal{F}, \mathbb{P})$. Let X be another random variable defined on this space. We say that $X_1, X_2, X_3, \ldots$* converges to X almost surely *and write*

$$\lim_{n\to\infty} X_n = X \text{ almost surely}$$

if the set of $\omega \in \Omega$ for which the sequence of numbers $X_1(\omega), X_2(\omega), X_3(\omega), \ldots$ has limit $X(\omega)$ is a set with probability one. Equivalently, the set of $\omega \in \Omega$ for which the sequence of numbers $X_1(\omega), X_2(\omega), X_3(\omega), \ldots$ does not converge to $X(\omega)$ is a set with probability zero.

Example 1.4.2 (Strong Law of Large Numbers). An intuitively appealing case of almost sure convergence is the *Strong Law of Large Numbers.* On the infinite independent coin-toss space Ω_∞, with the probability measure chosen to correspond to probability $p = \frac{1}{2}$ of head on each toss, we define

$$Y_k(\omega) = \begin{cases} 1 \text{ if } \omega_k = H, \\ 0 \text{ if } \omega_k = T, \end{cases}$$

and

$$H_n = \sum_{k=1}^n Y_k,$$

so that H_n is the number of heads obtained in the first n tosses. The Strong Law of Large Numbers is a theorem that asserts that

$$\lim_{n\to\infty} \frac{H_n}{n} = \frac{1}{2} \text{ almost surely.}$$

In other words, the ratio of the number of heads to the number of tosses approaches $\frac{1}{2}$ almost surely. The "almost surely" in this assertion acknowledges the fact that there are sequences of tosses, such as the sequence of all heads, for which the ratio does not converge to $\frac{1}{2}$. We shall ultimately see that there are in fact uncountably many such sequences. However, under our assumptions that the probability of head on each toss is $\frac{1}{2}$ and the tosses are independent, the probability of all these sequences taken together is zero. □

Definition 1.4.3. *Let $f_1, f_2, f_3, \ldots$ be a sequence of real-valued, Borel-measurable functions defined on $\mathbb{R}$. Let f be another real-valued, Borel-measurable function defined on $\mathbb{R}$. We say that $f_1, f_2, f_3, \ldots$* converges to f almost everywhere *and write*

$$\lim_{n\to\infty} f_n = f \textit{ almost everywhere}$$

if the set of $x \in \mathbb{R}$ for which the sequence of numbers $f_1(x), f_2(x), f_3(x), \ldots$ does not have limit $f(x)$ is a set with Lebesgue measure zero.

It is clear from these definitions that convergence almost surely and convergence almost everywhere are really the same concept in different notation.

Example 1.4.4. Consider a sequence of normal densities, each with mean zero and the nth having variance $\frac{1}{n}$ (see Figure 1.4.1):

$$f_n(x) = \sqrt{\tfrac{n}{2\pi}}\, e^{-\frac{nx^2}{2}}.$$

If $x \neq 0$, then $\lim_{n\to\infty} f_n(x) = 0$, but

$$\lim_{n\to\infty} f_n(0) = \lim_{n\to\infty} \sqrt{\frac{n}{2\pi}} = \infty.$$

Therefore, the sequence $f_1, f_2, f_3, \ldots$ converges everywhere to the function

$$f^*(x) = \begin{cases} 0 & \text{if } x \neq 0, \\ \infty & \text{if } x = 0, \end{cases}$$

and converges almost everywhere to the identically zero function $f(x) = 0$ for all $x \in \mathbb{R}$. The set of x where the convergence to $f(x)$ does not take place contains only the number 0, and this set has zero Lebesgue measure. □

Often when random variables converge almost surely, their expected values converge to the expected value of the limiting random variable. Likewise, when functions converge almost everywhere, it is often the case that their Lebesgue integrals converge to the Lebesgue integral of the limiting function. This is not always the case, however. In Example 1.4.4, we have a sequence of normal densities for which $\int_{-\infty}^{\infty} f_n(x)dx = 1$ for every n but the almost everywhere limit function f is identically zero. It would not help matters to use the everywhere limit function $f^*(x)$ because any two functions that differ only on a set of zero Lebesgue measure must have the same Lebesgue integral.

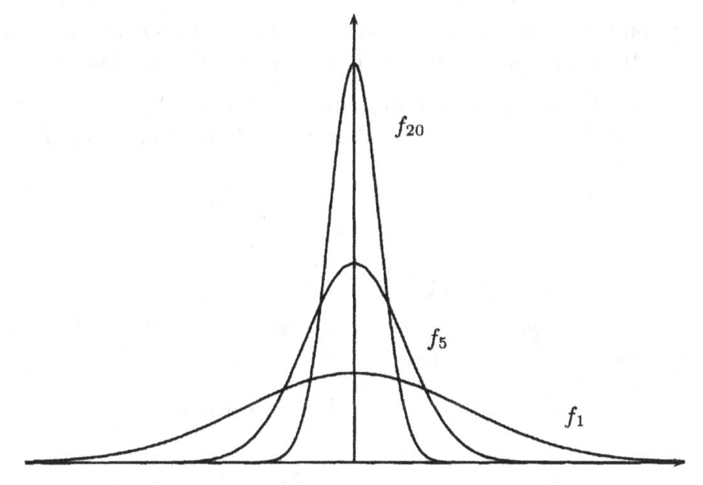

Fig. 1.4.1. Almost everywhere convergence.

Therefore, $\int_{-\infty}^{\infty} f^*(x)\,dx = \int_{-\infty}^{\infty} f(x)\,dx = 0$. It cannot be otherwise because $2f^*(x) = f^*(x)$ for every $x \in \mathbb{R}$, and so

$$2\int_{-\infty}^{\infty} f^*(x)\,dx = \int_{-\infty}^{\infty} 2f^*(x) = \int_{-\infty}^{\infty} f^*(x)\,dx.$$

This equation implies that $\int_{-\infty}^{\infty} f^*(x)\,dx = 0$. It would also not help matters to replace the functions f_n by the functions

$$g_n(x) = \begin{cases} f_n(x) & \text{if } x \neq 0, \\ 0 & \text{if } x = 0. \end{cases}$$

The sequence $g_1, g_2, \dots$ converges to 0 *everywhere*, whereas the integrals $\int_{-\infty}^{\infty} g_n(x)\,dx$ agree with the integrals $\int_{-\infty}^{\infty} f_n(x)\,dx$, and these converge to 1, not 0. The inescapable conclusion is that in this example

$$\lim_{n\to\infty}\int_{-\infty}^{\infty} f_n(x)\,dx \neq \int_{-\infty}^{\infty} \lim_{n\to\infty} f_n(x)\,dx;$$

the left-hand side is 1 and the right-hand side is 0.

Incidentally, matters are even worse with the Riemann integral, which is not defined for f^*; upper Riemann sums for f^* are infinite, and lower Riemann sums are zero.

To get the integrals of a sequence of functions to converge to the integral of the limiting function, we need to impose some condition. One condition that guarantees this is that all the functions are nonnegative and they converge to their limit from below. If we think of an integral as the area under a curve, the assumption is that as we go farther out in the sequence of functions, we keep adding area and never taking it away. If we do this, then the area under the limiting function is the limit of the areas under the functions in the sequence. The precise statement of this result is given in the following theorem.

Theorem 1.4.5 (Monotone convergence). *Let $X_1, X_2, X_3, \ldots$ be a sequence of random variables converging almost surely to another random variable X. If*

$$0 \leq X_1 \leq X_2 \leq X_3 \leq \ldots \text{ almost surely,}$$

then

$$\lim_{n \to \infty} \mathbb{E}X_n = \mathbb{E}X.$$

Let $f_1, f_2, f_3, \ldots$ be a sequence of Borel-measurable functions on $\mathbb{R}$ converging almost everywhere to a function f. If

$$0 \leq f_1 \leq f_2 \leq f_3 \leq \ldots \text{ almost everywhere,}$$

then

$$\lim_{n \to \infty} \int_{-\infty}^{\infty} f_n(x)\, dx = \int_{-\infty}^{\infty} f(x)\, dx.$$

The following corollary to the Monotone Convergence Theorem extends Theorem 1.3.4(i).

Corollary 1.4.6. *Suppose the nonnegative random variable X takes countably many values $x_0, x_1, x_2, \ldots$. Then*

$$\mathbb{E}X = \sum_{k=0}^{\infty} x_k \mathbb{P}\{X = x_k\}. \tag{1.4.1}$$

PROOF: Let $A_k = \{X = x_k\}$, so that X can be written as

$$X = \sum_{k=0}^{\infty} x_k \mathbb{I}_{A_k}.$$

Define $X_n = \sum_{k=0}^{n} x_k \mathbb{I}_{A_k}$. Then $0 \leq X_1 \leq X_2 \leq X_3 \leq \ldots$ and $\lim_{n \to \infty} X_n = X$ almost surely ("surely," actually). Theorem 1.3.4(i) implies that

$$\mathbb{E}X_n = \sum_{k=0}^{n} x_k \mathbb{P}\{X = x_k\}.$$

Taking the limit on both sides as $n \to \infty$ and using the Monotone Convergence Theorem to justify the first equality below, we obtain

$$\mathbb{E}X = \lim_{n \to \infty} \mathbb{E}X_n = \lim_{n \to \infty} \sum_{k=0}^{n} x_k \mathbb{P}\{X = x_k\} = \sum_{k=0}^{\infty} x_k \mathbb{P}\{X = x_k\}. \qquad \square$$

Remark 1.4.7. If X can take negative as well as positive values, we can apply Corollary 1.4.6 to X^+ and X^- separately and then subtract the resulting equations to again get formula (1.4.1), provided the subtraction does not create an "$\infty - \infty$" situation.

Example 1.4.8 (St. Petersburg paradox). On the infinite independent coin-toss space Ω_∞ with the probability of a head on each toss equal to $\frac{1}{2}$, define a random variable X by

$$X(\omega) = \begin{cases} 2 & \text{if } \omega_1 = H, \\ 4 & \text{if } \omega_1 = T, \omega_2 = H, \\ 8 & \text{if } \omega_1 = \omega_2 = T, \omega_3 = H, \\ \vdots & \\ 2^k & \text{if } \omega_1 = \omega_2 = \cdots = \omega_{k-1} = T, \omega_k = H. \\ \vdots & \end{cases}$$

This defines $X(\omega)$ for every sequence of coin tosses except the sequence that is all tails. For this sequence, we define $X(TTT\ldots) = \infty$. The probability that $X = \infty$ is then the probability of this sequence, which is zero. Therefore, X is finite almost surely. According to Corollary 1.4.6,

$$\begin{aligned} \mathbb{E}X &= 2 \cdot \mathbb{P}\{X = 2\} + 4 \cdot \mathbb{P}\{X = 4\} + 8 \cdot \mathbb{P}\{X = 8\} + \ldots \\ &= 2 \cdot \frac{1}{2} + 4 \cdot \frac{1}{4} + 8 \cdot \frac{1}{8} + \ldots \\ &= 1 + 1 + 1 + \cdots = \infty. \end{aligned}$$

The point is that $\mathbb{E}X$ can be infinite, even though X is finite almost surely. □

The following theorem provides another common situation in which we are assured that the limit of the integrals of a sequence of functions is the integral of the limiting function.

Theorem 1.4.9 (Dominated convergence). *Let $X_1, X_2, \ldots$ be a sequence of random variables converging almost surely to a random variable X. If there is another random variable Y such that $\mathbb{E}Y < \infty$ and $|X_n| \leq Y$ almost surely for every n, then*

$$\lim_{n\to\infty} EX_n = EX.$$

Let $f_1, f_2, \ldots$ be a sequence of Borel-measurable functions on $\mathbb{R}$ converging almost everywhere to a function f. If there is another function g such that $\int_{-\infty}^{\infty} g(x)\,dx < \infty$ and $|f_n| \leq g$ almost everywhere for every n, then

$$\lim_{n\to\infty} \int_{-\infty}^{\infty} f_n(x)\,dx = \int_{-\infty}^{\infty} f(x)\,dx.$$

1.5 Computation of Expectations

Let X be a random variable on some probability space $(\Omega, \mathcal{F}, \mathbb{P})$. We have defined the expectation of X to be the Lebesgue integral

$$\mathbb{E}X = \int_{\Omega} X(\omega)\, d\mathbb{P}(\omega),$$

the idea being to average the values of $X(\omega)$ over Ω, taking the probabilities into account. This level of abstraction is sometimes helpful. For example, the equality

$$\mathbb{E}(X+Y) = \mathbb{E}X + \mathbb{E}Y$$

follows directly from the linearity of integrals. By contrast, if we were to derive this fact using a joint density for X and Y, it would be a tedious, unenlightening computation.

On the other hand, the abstract space Ω is not a pleasant environment in which to actually compute integrals. For computations, we often need to rely on densities of the random variables under consideration, and we integrate these over the real numbers rather than over Ω. In this section, we develop the relationship between integrals over Ω and integrals over $\mathbb{R}$.

Recall that the distribution measure of X is the probability measure μ_X defined on $\mathbb{R}$ by

$$\mu_X(B) = \mathbb{P}\{X \in B\} \text{ for every Borel subset } B \text{ of } \mathbb{R}. \tag{1.5.1}$$

Because μ_X is a probability measure on $\mathbb{R}$, we can use it to integrate functions over $\mathbb{R}$. We have the following fundamental theorem relating integrals over $\mathbb{R}$ to integrals over Ω.

Theorem 1.5.1. *Let X be a random variable on a probability space $(\Omega, \mathcal{F}, \mathbb{P})$ and let g be a Borel-measurable function on R. Then*

$$\mathbb{E}|g(X)| = \int_{\mathbb{R}} |g(x)|\, d\mu_X(x), \tag{1.5.2}$$

and if this quantity is finite, then

$$\mathbb{E}g(X) = \int_{\mathbb{R}} g(x)\, d\mu_X(x). \tag{1.5.3}$$

PROOF: The proof proceeds by several steps, which collectively are called the *standard machine.*

Step 1. Indicator functions. Suppose the function $g(x) = \mathbb{I}_B(x)$ is the indicator of a Borel subset of $\mathbb{R}$. Since this function is nonnegative, (1.5.2) and (1.5.3) reduce to the same equation, namely

$$\mathbb{E}\mathbb{I}_B(X) = \int_{\mathbb{R}} \mathbb{I}_B(x)\, d\mu_X(x). \tag{1.5.4}$$

Since the random variable $\mathbb{I}_B(X)$ takes only the two values one and zero, its expectation is

$$\mathbb{E}\mathbb{I}_B(X) = 1 \cdot \mathbb{P}\{X \in B\} + 0 \cdot \mathbb{P}\{X \notin B\} = \mathbb{P}\{X \in B\}.$$

Similarly, the function $\mathbb{I}_B(x)$ of the dummy (not random!) variable x takes only the two values one and zero, so according to Theorem 1.3.1(i) with $\Omega = \mathbb{R}$, $X = \mathbb{I}_B$, and $\mathbb{P} = \mu_X$, its integral is

$$\int_{\mathbb{R}} \mathbb{I}_B(x)\, d\mu_X(x) = 1 \cdot \mu_X\{x; \mathbb{I}_B(x) = 1\} + 0 \cdot \mu_X\{x; \mathbb{I}_B(x) = 0\} = \mu_X(B).$$

In light of (1.5.1), we have gotten the same result in both cases, and (1.5.4) is proved.

Step 2. Nonnegative simple functions. A simple function is a finite sum of indicator functions times constants. In this step, we assume that

$$g(x) = \sum_{k=1}^{n} \alpha_k \mathbb{I}_{B_k}(x),$$

where $\alpha_1, \alpha_2, \ldots, \alpha_n$ are nonnegative constants and $B_1, B_2, \ldots, B_n$ are Borel subsets of $\mathbb{R}$. Because of linearity of integrals,

$$\mathbb{E}g(X) = \mathbb{E}\sum_{k=1}^{n} \alpha_k \mathbb{I}_{B_k}(X) = \sum_{k=1}^{n} \alpha_k \mathbb{E}\mathbb{I}_{B_k}(X) = \sum_{k=1}^{n} \alpha_k \int_R \mathbb{I}_{B_k}(x)\, d\mu_X(x),$$

where we have used (1.5.4) in the last step. But the linearity of integrals also implies that

$$\sum_{k=1}^{n} \alpha_k \int_{\mathbb{R}} \mathbb{I}_{B_k}(x)\, d\mu_X(x) = \int_{\mathbb{R}} \left(\sum_{k=1}^{n} \alpha_k \mathbb{I}_{B_k}(x) \right) d\mu_X(x) = \int_{\mathbb{R}} g(x)\, d\mu_X(x),$$

and we conclude that

$$\mathbb{E}g(X) = \int_{\mathbb{R}} g(x)\, d\mu_X(x)$$

when g is a nonnegative simple function.

Step 3. Nonnegative Borel-measurable functions. Let $g(x)$ be an arbitrary nonnegative Borel-measurable function defined on $\mathbb{R}$. For each positive integer n, define the sets

$$B_{k,n} = \left\{ x; \frac{k}{2^n} \le g(x) < \frac{k+1}{2^n} \right\}, \quad k = 0, 1, 2, \ldots 4^n - 1.$$

For each fixed n, the sets $B_{0,n}, B_{1,n}, \ldots, B_{4^n-1,n}$ correspond to the partition

$$0 < \frac{1}{2^n} < \frac{2}{2^n} < \cdots < \frac{4^n}{2^n} = 2^n.$$

At the next stage $n+1$, the partition points include all those at stage n and new partition points at the midpoints between the old ones. Because of this fact, the simple functions

$$g_n(x) = \sum_{k=0}^{4^n-1} \frac{k}{2^n} \mathbb{I}_{B_{k,n}}(x)$$

satisfy $0 \le g_1 \le g_2 \le \cdots \le g$. Furthermore, these functions become more and more accurate approximations of g as n becomes larger; indeed, $\lim_{n\to\infty} g_n(x) = g(x)$ for every $x \in \mathbb{R}$. From Step 2, we know that

$$\mathbb{E}g_n(X) = \int_{\mathbb{R}} g_n(x)\, d\mu_X(x)$$

for every n. Letting $n \to \infty$ and using the Monotone Convergence Theorem, Theorem 1.4.5, on both sides of the equation, we obtain

$$\mathbb{E}g(X) = \lim_{n\to\infty} \mathbb{E}g_n(X) = \lim_{n\to\infty} \int_{\mathbb{R}} g_n(x)\, d\mu_X(x) = \int_{\mathbb{R}} g(x)\, d\mu_X(x).$$

This proves (1.5.3) when g is a nonnegative Borel-measurable function.

Step 4. General Borel-measurable function. Let $g(x)$ be a general Borel-measurable function, which can take both positive and negative values. The functions

$$g^+(x) = \max\{g(x), 0\} \text{ and } g^-(x) = \max\{-g(x), 0\}$$

are both nonnegative, and from Step 3 we have

$$\mathbb{E}g^+(X) = \int_{\mathbb{R}} g^+(x)\, d\mu_X(x), \quad \mathbb{E}g^-(X) = \int_{\mathbb{R}} g^-(x)\, d\mu_X(x).$$

Adding these two equations, we obtain (1.5.2). If the quantity in (1.5.2) is finite, then

$$\begin{aligned} \mathbb{E}g^+(X) &= \textstyle\int_{\mathbb{R}} g^+(x)\, d\mu_X(x) < \infty, \\ \mathbb{E}g^-(X) &= \textstyle\int_{\mathbb{R}} g^-(x)\, d\mu_X(x) < \infty, \end{aligned}$$

and we can subtract these two equations because this is not an $\infty - \infty$ situation. The result of this subtraction is (1.5.3). □

Theorem 1.5.1 tells us that in order to compute the Lebesgue integral $\mathbb{E}X = \int_\Omega X(\omega)\, d\mathbb{P}(\omega)$ over the abstract space Ω, it suffices to compute the integral $\int_{\mathbb{R}} g(x)\, d\mu_X(x)$ over the set of real numbers. This is still a Lebesgue integral, and the integrator is the distribution measure μ_X rather than the Lebesgue measure. To actually perform a computation, we need to reduce this to something more familiar. Depending on the nature of the random variable X, the distribution measure μ_X on the right-hand side of (1.5.3) can have different forms. In the simplest case, X takes only finitely many values $x_0, x_1, x_2, \dots, x_n$, and then μ_X places a mass of size $p_k = \mathbb{P}\{X = x_k\}$ at each number x_k. In this case, formula (1.5.3) becomes

$$\mathbb{E}g(X) = \int_{\mathbb{R}} g(x)\mu_X(dx) = \sum_{k=0}^{n} g(x_k)p_k.$$

The most common case for continuous-time models in finance is when X has a density. This means that there is a nonnegative, Borel-measurable function f defined on $\mathbb{R}$ such that

$$\mu_X(B) = \int_B f(x)\,dx \text{ for every Borel subset } B \text{ of } \mathbb{R}. \tag{1.5.5}$$

This density allows us to compute the measure μ_X of a set B by computing an integral over B. In most cases, the density function f is bounded and continuous or almost everywhere continuous, so that the integral on the right-hand side of (1.5.5) can be computed as a Riemann integral.

If X has a density, we can use this density to compute expectations, as shown by the following theorem.

Theorem 1.5.2. *Let X be a random variable on a probability space $(\Omega, \mathcal{F}, \mathbb{P})$, and let g be a Borel-measurable function on $\mathbb{R}$. Suppose that X has a density f (i.e., f is a function satisfying (1.5.5)). Then*

$$\mathbb{E}|g(X)| = \int_{-\infty}^{\infty} |g(x)|f(x)\,dx. \tag{1.5.6}$$

If this quantity is finite, then

$$\mathbb{E}g(X) = \int_{-\infty}^{\infty} g(x)f(x)\,dx. \tag{1.5.7}$$

PROOF: The proof proceeds again by the standard machine.

Step 1. Indicator functions. If $g(x) = \mathbb{I}_B(x)$, then because g is nonnegative, equations (1.5.6) and (1.5.7) are the same and reduce to

$$\mathbb{E}\mathbb{I}_B(X) = \int_B f(x)\,dx.$$

The left-hand side is $\mathbb{P}\{X \in B\} = \mu_X(B)$, and (1.5.5) shows that the two sides are equal.

Step 2. Simple functions. If $g(x) = \sum_{k=1}^{n} \alpha_k \mathbb{I}_{B_k}(x)$, then

$$\mathbb{E}g(X) = \mathbb{E}\left(\sum_{k=1}^{n} \alpha_k \mathbb{I}_{B_k}(X)\right) = \sum_{k=1}^{n} \alpha_k \mathbb{E}\mathbb{I}_{B_k}(X)$$
$$= \sum_{k=1}^{n} \alpha_k \int_{-\infty}^{\infty} \mathbb{I}_{B_k}(x) f(x)\,dx = \int_{-\infty}^{\infty} \sum_{k=1}^{n} \alpha_k \mathbb{I}_{B_k}(x) f(x)\,dx$$
$$= \int_{-\infty}^{\infty} g(x) f(x)\,dx.$$

Step 3. Nonnegative Borel-measurable functions. Just as in the proof of Theorem 1.5.1 we construct a sequence of nonnegative simple functions $0 \leq g_1 \leq g_2 \leq \cdots \leq g$ such that $\lim_{n\to\infty} g_n(x) = g(x)$ for every $x \in R$. We have already shown that

$$\mathbb{E}g_n(X) = \int_{-\infty}^{\infty} g_n(x) f(x)\,dx$$

for every n. We let $n \to \infty$, using the Monotone Convergence Theorem, Theorem 1.4.5, on both sides of the equation, to obtain (1.5.7).

Step 4. General Borel-measurable functions. Let g be a general Borel-measurable function, which can take positive and negative values. We have just proved that

$$\mathbb{E}g^+(X) = \int_{-\infty}^{\infty} g^+(x) f(x)\,dx, \quad \mathbb{E}g^-(X) = \int_{-\infty}^{\infty} g^-(x) f(x)\,dx.$$

Adding these equations, we obtain (1.5.6). If the expression in (1.5.6) is finite, we can also subtract these equations to obtain (1.5.7). □

1.6 Change of Measure

We pick up the thread of Section 3.1 of Volume I, in which we used a positive random variable Z to change probability measures on a space Ω. We need to do this when we change from the actual probability measure $\mathbb{P}$ to the risk-neutral probability measure $\widetilde{\mathbb{P}}$ in models of financial markets. When Ω is uncountably infinite and $\mathbb{P}(\omega) = \widetilde{\mathbb{P}}(\omega) = 0$ for every $\omega \in \Omega$, it no longer makes sense to write (3.1.1) of Chapter 3 of Volume I,

$$Z(\omega) = \frac{\widetilde{\mathbb{P}}(\omega)}{\mathbb{P}(\omega)}, \tag{1.6.1}$$

because division by zero is undefined. We could rewrite this equation as

$$Z(\omega)\mathbb{P}(\omega) = \widetilde{\mathbb{P}}(\omega), \tag{1.6.2}$$

and now we have a meaningful equation, with both sides equal to zero, but the equation tells us nothing about the relationship among $\mathbb{P}$, $\widetilde{\mathbb{P}}$, and Z. Because

$\mathbb{P}(\omega) = \widetilde{\mathbb{P}}(\omega) = 0$, the value of $Z(\omega)$ could be anything and (1.6.2) would still hold.

However, (1.6.2) does capture the spirit of what we would like to accomplish. To change from $\mathbb{P}$ to $\widetilde{\mathbb{P}}$, we need to reassign probabilities in Ω using Z to tell us where in Ω we should revise the probability upward (where $Z > 1$) and where we should revise the probability downward (where $Z < 1$). However, we should do this set-by-set, rather than ω-by-ω. The process is described by the following theorem.

Theorem 1.6.1. *Let $(\Omega, \mathcal{F}, \mathbb{P})$ be a probability space and let Z be an almost surely nonnegative random variable with $\mathbb{E}Z = 1$. For $A \in \mathcal{F}$, define*

$$\widetilde{\mathbb{P}}(A) = \int_A Z(\omega)\, d\mathbb{P}(\omega). \tag{1.6.3}$$

Then $\widetilde{\mathbb{P}}$ is a probability measure. Furthermore, if X is a nonnegative random variable, then

$$\widetilde{\mathbb{E}}X = \mathbb{E}[XZ]. \tag{1.6.4}$$

If Z is almost surely strictly positive, we also have

$$\mathbb{E}Y = \widetilde{\mathbb{E}}\left[\frac{Y}{Z}\right] \tag{1.6.5}$$

for every nonnegative random variable Y.

The $\widetilde{\mathbb{E}}$ appearing in (1.6.4) is expectation under the probability measure $\widetilde{\mathbb{P}}$ (i.e., $\widetilde{\mathbb{E}}X = \int_\Omega X(\omega)\, d\widetilde{\mathbb{P}}(\omega)$).

Remark 1.6.2. Suppose X is a random variable that can take both positive and negative values. We may apply (1.6.4) to its positive and negative parts $X^+ = \max\{X, 0\}$ and $X^- = \max\{-X, 0\}$, and then subtract the resulting equations to see that (1.6.4) holds for this X as well, provided the subtraction does not result in an $\infty - \infty$ situation. The same remark applies to (1.6.5).

PROOF OF THEOREM 1.6.1: According to Definition 1.1.2, to check that $\widetilde{\mathbb{P}}$ is a probability measure, we must verify that $\widetilde{\mathbb{P}}(\Omega) = 1$ and that $\widetilde{\mathbb{P}}$ is countably additive. We have by assumption

$$\widetilde{\mathbb{P}}(\Omega) = \int_\Omega Z(\omega)\, d\mathbb{P}(\omega) = \mathbb{E}Z = 1.$$

For countable additivity, let $A_1, A_2, \ldots$ be a sequence of disjoint sets in $\mathcal{F}$, and define $B_n = \cup_{k=1}^n A_k$, $B_\infty = \cup_{k=1}^\infty A_k$. Because

$$\mathbb{I}_{B_1} \leq \mathbb{I}_{B_2} \leq \mathbb{I}_{B_3} \leq \ldots$$

and $\lim_{n\to\infty} \mathbb{I}_{B_n} = \mathbb{I}_{B_\infty}$, we may use the Monotone Convergence Theorem, Theorem 1.4.5, to write

$$\widetilde{\mathbb{P}}(B_\infty) = \int_\Omega \mathbb{I}_{B_\infty}(\omega) Z(\omega)\, d\mathbb{P}(\omega) = \lim_{n\to\infty} \int_\Omega \mathbb{I}_{B_n}(\omega) Z(\omega)\, d\mathbb{P}(\omega).$$

But $\mathbb{I}_{B_n}(\omega) = \sum_{k=1}^n \mathbb{I}_{A_k}(\omega)$, and so

$$\int_\Omega \mathbb{I}_{B_n}(\omega) Z(\omega)\, d\mathbb{P}(\omega) = \sum_{k=1}^n \int_\Omega \mathbb{I}_{A_k}(\omega) Z(\omega)\, d\mathbb{P}(\omega) = \sum_{k=1}^n \widetilde{\mathbb{P}}(A_k).$$

Putting these two equations together, we obtain the countable additivity property

$$\widetilde{\mathbb{P}}\left(\bigcup_{k=1}^\infty A_k\right) = \lim_{n\to\infty} \sum_{k=1}^n \widetilde{\mathbb{P}}(A_k) = \sum_{k=1}^\infty \widetilde{\mathbb{P}}(A_k).$$

Now suppose X is a nonnegative random variable. If X is an indicator function $X = \mathbb{I}_A$, then

$$\widetilde{\mathbb{E}}X = \widetilde{\mathbb{P}}(A) = \int_\Omega \mathbb{I}_A(\omega) Z(\omega)\, d\mathbb{P}(\omega) = \mathbb{E}\left[\mathbb{I}_A Z\right] = \mathbb{E}\left[XZ\right],$$

which is (1.6.4). We finish the proof of (1.6.4) using the standard machine developed in Theorem 1.5.1. When $Z > 0$ almost surely, $\frac{Y}{Z}$ is defined and we may replace X in (1.6.4) by $\frac{Y}{Z}$ to obtain (1.6.5). □

Definition 1.6.3. *Let Ω be a nonempty set and $\mathcal{F}$ a σ-algebra of subsets of Ω. Two probability measures $\mathbb{P}$ and $\widetilde{\mathbb{P}}$ on $(\Omega, \mathcal{F})$ are said to be* equivalent *if they agree which sets in $\mathcal{F}$ have probability zero.*

Under the assumptions of Theorem 1.6.1, including the assumption that $Z > 0$ almost surely, $\mathbb{P}$ and $\widetilde{\mathbb{P}}$ are equivalent. Suppose $A \in \mathcal{F}$ is given and $\mathbb{P}(A) = 0$. Then the random variable $\mathbb{I}_A Z$ is $\mathbb{P}$ almost surely zero, which implies

$$\widetilde{\mathbb{P}}(A) = \int_\Omega \mathbb{I}_A(\omega) Z(\omega)\, d\mathbb{P}(\omega) = 0.$$

On the other hand, suppose $B \in \mathcal{F}$ satisfies $\widetilde{\mathbb{P}}(B) = 0$. Then $\frac{1}{Z}\mathbb{I}_B = 0$ almost surely under $\widetilde{\mathbb{P}}$, so

$$\widetilde{\mathbb{E}}\left[\frac{1}{Z}\mathbb{I}_B\right] = 0.$$

Equation (1.6.5) implies $\mathbb{P}(B) = \mathbb{E}\mathbb{I}_B = 0$. This shows that $\mathbb{P}$ and $\widetilde{\mathbb{P}}$ agree which sets have probability zero. Because the sets with probability one are complements of the sets with probability zero, $\mathbb{P}$ and $\widetilde{\mathbb{P}}$ agree which sets have probability one as well. Because $\widetilde{\mathbb{P}}$ and $\mathbb{P}$ are equivalent, we do not need to specify which measure we have in mind when we say an event occurs *almost surely*.

In financial models, we will first set up a sample space Ω, which one can regard as the set of possible scenarios for the future. We imagine this set of possible scenarios has an actual probability measure $\mathbb{P}$. However, for purposes of pricing derivative securities, we will use a risk-neutral measure $\widetilde{\mathbb{P}}$. We will insist that these two measures are equivalent. They must agree on what is possible and what is impossible; they may disagree on how probable the possibilities are. This is the same situation we had in the binomial model; $\mathbb{P}$ and $\widetilde{\mathbb{P}}$ assigned different probabilities to the stock price paths, but they agreed which stock price paths were possible. In the continuous-time model, after we have $\mathbb{P}$ and $\widetilde{\mathbb{P}}$, we shall determine prices of derivative securities that allow us to set up hedges that work with $\widetilde{\mathbb{P}}$-probability one. These hedges then also work with $\mathbb{P}$-probability one. Although we have used the risk-neutral probability to compute prices, we will have obtained hedges that work with probability one under the actual (and the risk-neutral) probability measure.

It is common to refer to computations done under the actual measure as computations in the *real world* and computations done under the risk-neutral measure as computations in the *risk-neutral world*. This unfortunate terminology raises the question whether prices computed in the "risk-neutral world" are appropriate for the "real world" in which we live and have our profits and losses. Our answer to this question is that *there is only one world* in the models. There is a single sample space Ω representing all possible future states of the financial markets, and there is a single set of asset prices, modeled by random variables (i.e., functions of these future states of the market). We sometimes work in this world assuming that probabilities are given by an empirically estimated actual probability measure and sometimes assuming that they are given by risk-neutral probabilities, but we do not change our view of the world of possibilities. A hedge that works almost surely under one assumption of probabilities works almost surely under the other assumption as well, since the probability measures agree which events have probability one.

The change of measure discussed in Section 3.1 of Volume I is the special case of Theorem 1.6.1 for finite probability spaces, and Example 3.1.2 of Chapter 3 of Volume I provides a case with explicit formulas for $\mathbb{P}$, $\widetilde{\mathbb{P}}$, and Z when the expectations are sums. We give here two examples on uncountable probability spaces.

Example 1.6.4. Recall Example 1.2.4 in which $\Omega = [0,1]$, $\mathbb{P}$ is the uniform (i.e., Lebesgue) measure, and

$$\widetilde{\mathbb{P}}[a,b] = \int_a^b 2\omega\, d\omega = b^2 - a^2,\ 0 \le a \le b \le 1. \tag{1.2.2}$$

We may use the fact that $\mathbb{P}(d\omega) = d\omega$ to rewrite (1.2.2) as

$$\widetilde{\mathbb{P}}[a,b] = \int_{[a,b]} 2\omega\, d\mathbb{P}(\omega). \tag{1.2.2$'$}$$

Because $\mathcal{B}[0,1]$ is the σ-algebra generated by the closed intervals (i.e., begin with the closed intervals and put in all other sets necessary in order to have a

σ-algebra), the validity of (1.2.2)$'$ for all closed intervals $[a,b] \subset [0,1]$ implies its validity for all Borel subsets of $[0,1]$:

$$\widetilde{\mathbb{P}}(B) = \int_B 2\omega \, d\mathbb{P}(\omega) \text{ for every Borel set } B \subset \mathbb{R}.$$

This is (1.6.3) with $Z(\omega) = 2\omega$.

Note that $Z(\omega) = 2\omega$ is strictly positive almost surely ($\mathbb{P}\{0\} = 0$), and

$$\widetilde{\mathbb{E}} Z = \int_0^1 2\omega \, d\omega = 1.$$

According to (1.6.4), for every nonnegative random variable $X(\omega)$, we have the equation

$$\int_0^1 X(\omega) \, d\widetilde{\mathbb{P}}(\omega) = \int_0^1 X(\omega) \cdot 2\omega \, d\omega.$$

This suggests the notation

$$d\widetilde{\mathbb{P}}(\omega) = 2\omega \, d\omega = 2\omega \, d\mathbb{P}(\omega). \tag{1.6.6}$$

□

In general, when $\mathbb{P}$, $\widetilde{\mathbb{P}}$, and Z are related as in Theorem 1.6.1, we may rewrite the two equations (1.6.4) and (1.6.5) as

$$\int_\Omega X(\omega) \, d\widetilde{\mathbb{P}}(\omega) = \int_\Omega X(\omega) Z(\omega) \, d\mathbb{P}(\omega),$$

$$\int_\Omega Y(\omega) \, d\mathbb{P}(\omega) = \int_\Omega \frac{Y(\omega)}{Z(\omega)} \, d\widetilde{\mathbb{P}}(\omega).$$

A good way to remember these equations is to formally write $Z(\omega) = \frac{d\widetilde{\mathbb{P}}(\omega)}{d\mathbb{P}(\omega)}$. Equation (1.6.6) is a special case of this notation that captures the idea behind the nonsensical equation (1.6.1) that Z is somehow a "ratio of probabilities." In Example 1.6.4, $Z(\omega) = 2\omega$ is in fact a ratio of densities, with the denominator being the uniform density 1 for all $\omega \in [0,1]$.

Definition 1.6.5. *Let $(\Omega, \mathcal{F}, \mathbb{P})$ be a probability space, let $\widetilde{\mathbb{P}}$ be another probability measure on $(\Omega, \mathcal{F})$ that is equivalent to $\mathbb{P}$, and let Z be an almost surely positive random variable that relates $\mathbb{P}$ and $\widetilde{\mathbb{P}}$ via (1.6.3). Then Z is called the* Radon-Nikodým derivative *of $\widetilde{\mathbb{P}}$ with respect to $\mathbb{P}$, and we write*

$$Z = \frac{d\widetilde{\mathbb{P}}}{d\mathbb{P}}.$$

Example 1.6.6 (Change of measure for a normal random variable). Let X be a standard normal random variable defined on some probability space $(\Omega, \mathcal{F}, \mathbb{P})$.

Two ways of constructing X and $(\Omega, \mathcal{F}, \mathbb{P})$ were described in Example 1.2.6. For purposes of this example, we do not need to know the details about the probability space $(\Omega, \mathcal{F}, \mathbb{P})$, except we note that the set Ω is necessarily uncountably infinite and $\mathbb{P}(\omega) = 0$ for every $\omega \in \Omega$.

When we say X is a standard normal random variable, we mean that

$$\mu_X(B) = \mathbb{P}\{X \in B\} = \int_B \varphi(x)\, dx \text{ for every Borel subset } B \text{ of } \mathbb{R}, \quad (1.6.7)$$

where

$$\varphi(x) = \frac{1}{\sqrt{2\pi}} e^{-\frac{x^2}{2}}$$

is the standard normal density. If we take $B = (-\infty, b]$, this reduces to the more familiar condition

$$\mathbb{P}\{X \leq b\} = \int_{-\infty}^{b} \varphi(x) dx \text{ for every } b \in \mathbb{R}. \quad (1.6.8)$$

In fact, (1.6.8) is equivalent to the apparently stronger statement (1.6.7). Note that $\mathbb{E}X = 0$ and variance $\mathrm{Var}(X) = \mathbb{E}(X - \mathbb{E}X)^2 = 1$.

Let θ be a constant and define $Y = X + \theta$, so that under $\mathbb{P}$, the random variable Y is normal with $\mathbb{E}Y = \theta$ and variance $\mathrm{Var}(Y) = \mathbb{E}(Y - \mathbb{E}Y)^2 = 1$. Although it is not required by the formulas, we will assume θ is positive for the discussion below. We want to change to a new probability measure $\widetilde{\mathbb{P}}$ on Ω under which Y is a standard normal random variable. In other words, we want $\widetilde{\mathbb{E}}Y = 0$ and $\widetilde{\mathrm{Var}}(Y) = \widetilde{\mathbb{E}}(Y - \widetilde{\mathbb{E}}Y)^2 = 1$. We want to do this not by subtracting θ away from Y, but rather by assigning less probability to those ω for which $Y(\omega)$ is sufficiently positive and more probability to those ω for which $Y(\omega)$ is negative. *We want to change the distribution of Y without changing the random variable Y.* In finance, the change from the actual to the risk-neutral probability measure changes the distribution of asset prices without changing the asset prices themselves, and this example is a step in understanding that procedure.

We first define the random variable

$$Z(\omega) = \exp\left\{-\theta X(\omega) - \frac{1}{2}\theta^2\right\} \text{ for all } \omega \in \Omega.$$

This random variable has two important properties that allow it to serve as a Radon-Nikodým derivative for obtaining a probability measure $\widetilde{\mathbb{P}}$ equivalent to $\mathbb{P}$:

(i) $Z(\omega) > 0$ for all $\omega \in \Omega$ ($Z > 0$ almost surely would be good enough), and
(ii) $\mathbb{E}Z = 1$.

Property (i) is obvious because Z is defined as an exponential. Property (ii) follows from the integration

$$\begin{aligned}\mathbb{E}Z &= \int_{-\infty}^{\infty} \exp\left\{-\theta x - \frac{1}{2}\theta^2\right\} \varphi(x)dx \\ &= \frac{1}{\sqrt{2\pi}} \int_{-\infty}^{\infty} \exp\left\{-\frac{1}{2}(x^2 + 2\theta x + \theta^2)\right\} dx \\ &= \frac{1}{\sqrt{2\pi}} \int_{-\infty}^{\infty} \exp\left\{-\frac{1}{2}(x + \theta)^2\right\} dx \\ &= \frac{1}{\sqrt{2\pi}} \int_{-\infty}^{\infty} \exp\left\{-\frac{1}{2}y^2\right\} dy,\end{aligned}$$

where we have made the change of dummy variable $y = x + \theta$ in the last step. But $\frac{1}{\sqrt{2\pi}} \int_{-\infty}^{\infty} \exp\{-\frac{1}{2}y^2\}dy$, being the integral of the standard normal density, is equal to one.

We use the random variable Z to create a new probability measure $\widetilde{\mathbb{P}}$ by adjusting the probabilities of the events in Ω. We do this by defining

$$\widetilde{\mathbb{P}}(A) = \int_A Z(\omega)\, d\mathbb{P}(\omega) \text{ for all } A \in \mathcal{F}. \tag{1.6.9}$$

The random variable Z has the property that if $X(\omega)$ is positive, then $Z(\omega) < 1$ (we are still thinking of θ as a positive constant). This shows that $\widetilde{\mathbb{P}}$ assigns less probability than $\mathbb{P}$ to sets on which X is positive, a step in the right direction of statistically recentering Y. We claim not only that $\widetilde{\mathbb{E}}Y = 0$ but also that, under $\widetilde{\mathbb{P}}$, Y is a standard normal random variable. To see this, we compute

$$\begin{aligned}\widetilde{\mathbb{P}}\{Y \le b\} &= \int_{\{\omega; Y(\omega) \le b\}} Z(\omega)\, d\mathbb{P}(\omega) \\ &= \int_\Omega \mathbb{I}_{\{Y(\omega) \le b\}} Z(\omega)\, d\mathbb{P}(\omega) \\ &= \int_\Omega \mathbb{I}_{\{X(\omega) \le b - \theta\}} \exp\left\{-\theta X(\omega) - \frac{1}{2}\theta^2\right\} d\mathbb{P}(\omega).\end{aligned}$$

At this point, we have managed to write $\widetilde{\mathbb{P}}\{Y \le b\}$ in terms of a function of the random variable X, integrated with respect to the probability measure $\mathbb{P}$ under which X is standard normal. According to Theorem 1.5.2,

$$\begin{aligned}&\int_\Omega \mathbb{I}_{\{X(\omega) \le b - \theta\}} \exp\left\{-\theta X(\omega) - \frac{1}{2}\theta^2\right\} d\mathbb{P}(\omega) \\ &\qquad = \int_{-\infty}^{\infty} \mathbb{I}_{\{x \le b - \theta\}} e^{-\theta x - \frac{1}{2}\theta^2} \varphi(x)dx \\ &\qquad = \frac{1}{\sqrt{2\pi}} \int_{-\infty}^{b-\theta} e^{-\theta x - \frac{1}{2}\theta^2} e^{-\frac{x^2}{2}} dx\end{aligned}$$

$$= \frac{1}{\sqrt{2\pi}} \int_{-\infty}^{b-\theta} e^{-\frac{1}{2}(x+\theta)^2} dx$$
$$= \frac{1}{\sqrt{2\pi}} \int_{-\infty}^{b} e^{-\frac{1}{2}y^2} dy,$$

where we have made the change of dummy variable $y = x + \theta$ in the last step. We conclude that

$$\widetilde{\mathbb{P}}\{Y \le b\} = \frac{1}{\sqrt{2\pi}} \int_{-\infty}^{b} e^{-\frac{1}{2}y^2} dy,$$

which shows that Y is a standard normal random variable under the probability measure $\widetilde{\mathbb{P}}$. □

Following Corollary 2.4.6 of Chapter 2 of Volume I, we discussed how the existence of a risk-neutral measure guarantees that a financial model is free of arbitrage, the so-called *First Fundamental Theorem of Asset Pricing.* The same argument applies in continuous-time models and in fact underlies the Heath-Jarrow-Morton no-arbitrage condition for term-structure models. Consequently, we are interested in the existence of risk-neutral measures. As discussed earlier in this section, these must be equivalent to the actual probability measure. How can such probability measures $\widetilde{\mathbb{P}}$ arise? In Theorem 1.6.1, we began with the probability measure $\mathbb{P}$ and an almost surely positive random variable Z and constructed the equivalent probability measure $\widetilde{\mathbb{P}}$. It turns out that this is the only way to obtain a probability measure $\widetilde{\mathbb{P}}$ equivalent to $\mathbb{P}$. The proof of the following profound theorem is beyond the scope of this text.

Theorem 1.6.7 (Radon-Nikodým). *Let $\mathbb{P}$ and $\widetilde{\mathbb{P}}$ be equivalent probability measures defined on $(\Omega, \mathcal{F})$. Then there exists an almost surely positive random variable Z such that $\mathbb{E}Z = 1$ and*

$$\widetilde{\mathbb{P}}(A) = \int_A Z(\omega)\, d\mathbb{P}(\omega) \text{ for every } A \in \mathcal{F}.$$

1.7 Summary

Probability theory begins with a *probability space* $(\Omega, \mathcal{F}, \mathbb{P})$ (Definition 1.1.2). Here Ω is the set of all possible outcomes of a random experiment, $\mathcal{F}$ is the collection of subsets of Ω whose probability is defined, and $\mathbb{P}$ is a function mapping $\mathcal{F}$ to $[0,1]$. The two axioms of probability spaces are $\mathbb{P}(\Omega) = 1$ and *countable additivity*: the probability of a union of disjoint sets is the sum of the probabilities of the individual sets.

The collection of sets $\mathcal{F}$ in the preceding paragraph is a *σ-algebra*, which means that $\emptyset$ belongs to $\mathcal{F}$, the complement of every set in $\mathcal{F}$ is also in $\mathcal{F}$, and the union of any sequence of sets in $\mathcal{F}$ is also in $\mathcal{F}$. The Borel σ-algebra in $\mathbb{R}$, denoted $\mathcal{B}(\mathbb{R})$, is the smallest σ-algebra that contains all the closed interval

$[a, b]$ in $\mathbb{R}$. Every set encountered in practice is a Borel set (i.e., belongs to $\mathcal{B}(\mathbb{R})$).

A *random variable* X is a mapping from Ω to $\mathbb{R}$ (Definition 1.2.1). By definition, it has the property that, for every $B \in \mathcal{B}(\mathbb{R})$, the set $\{\omega \in \Omega; X(\omega) \in B\}$ is in the σ-algebra $\mathcal{F}$. A random variable X together with the probability measure $\mathbb{P}$ on Ω determines a *distribution* on $\mathbb{R}$. This distribution is not the random variable. Different random variables can have the same distribution, and the same random variable can have different distributions. We describe the distribution as a measure μ_X that assigns to each Borel subset B of $\mathbb{R}$ the mass $\mu_X(B) = \mathbb{P}\{X \in B\}$ (Definition 1.2.3). If X has a density $f(x)$, then $\mu_X(B) = \int_B f(x)\,dx$. If X is a discrete random variable, which means that it takes one of countably many values $x_1, x_2, \dots$, then we define $p_i = \mathbb{P}\{X = x_i\}$ and have $\mu_X(B) = \sum_{\{i; x_i \in B\}} p_i$.

The *expectation* of a random variable X is $\mathbb{E}X = \int_\Omega X(\omega)\,d\mathbb{P}(\omega)$, where the right-hand side is a Lebesgue integral over Ω. Lebesgue integrals are discussed in Section 1.3. They differ from Riemann integrals, which form approximating sums to the integral by partitioning the "x" (horizontal)-axis, because Lebesgue integrals form approximating sums to the integral by partitioning the "y" (vertical)-axis. Lebesgue integrals have the properties one would expect (Theorem 1.3.4):

Comparison. If $X \leq Y$ almost surely, then $\mathbb{E}X \leq \mathbb{E}Y$;

Linearity. $\mathbb{E}(\alpha X + \beta Y) = \alpha \mathbb{E}X + \beta \mathbb{E}Y$.

In addition, if φ is a convex function, we have *Jensen's inequality:* $\varphi(\mathbb{E}X) \leq \mathbb{E}\varphi(X)$.

If the random variable X has a density $f(x)$, then $\mathbb{E}X = \int_{-\infty}^{\infty} x f(x)\,dx$ and, more generally, $\mathbb{E}g(X) = \int_{-\infty}^{\infty} g(x) f(x)\,dx$ (Theorem 1.5.2). If the random variable is discrete with $p_i = \mathbb{P}\{X = x_i\}$, then $\mathbb{E}g(X) = \sum_i g(x_i) p_i$.

Suppose we have a sequence of random variables $X_1, X_2, X_3, \dots$ converging almost surely to a random variable X. It is not always true that

$$\mathbb{E}X = \lim_{n \to \infty} \mathbb{E}X_n. \tag{1.7.1}$$

However, if $0 \leq X_1 \leq X_2 \leq X_3 \leq \dots$ almost surely, then (1.7.1) holds (Monotone Convergence Theorem, Theorem 1.4.5). Alternatively, if there exists a random variable Y such that $\mathbb{E}Y < \infty$ and $|X_n| \leq Y$ almost surely for every n, then again (1.7.1) holds (Dominated Convergence Theorem, Theorem 1.4.9).

We may start with a probability space $(\Omega, \mathcal{F}, \mathbb{P})$ and change to a different measure $\widetilde{\mathbb{P}}$. Our motivation for considering two measures is that in finance there is both an actual probability measure and a risk-neutral probability measure. If $\mathbb{P}$ is a probability measure and Z is a nonnegative random variable satisfying $\mathbb{E}Z = 1$, then $\widetilde{\mathbb{P}}$ defined by

$$\widetilde{\mathbb{P}}(A) = \int_A Z(\omega)\,d\mathbb{P}(\omega) \text{ for all } A \in \mathcal{F}$$

is also a probability measure (Theorem 1.6.1). If Z is strictly positive almost surely, the two measures are *equivalent*: they agree about which sets have probability zero. For a random variable X, we have the change-of-expectation formula $\widetilde{\mathbb{E}}[X] = \mathbb{E}[XZ]$. If Z is strictly positive almost surely, there is a change-of-expectation formula in the other direction. Namely, if Y is a random variable, then $\mathbb{E}Y = \widetilde{\mathbb{E}}\left[\frac{Y}{Z}\right]$.

1.8 Notes

Probability theory is usually learned in two stages. In the first stage, one learns that a discrete random variable has a probability mass function and a continuous random variable has a density. These can be used to compute expectations and variances, and even conditional expectations, which are discussed in Chapter 2. Furthermore, one learns how transformations of continuous random variables change densities. A well-written book that contains all these things is DeGroot [48].

The second stage of probability theory, which is treated in this chapter, is measure-theoretic. In this stage, one views a random variable as a function from a sample space Ω to the set of real numbers $\mathbb{R}$. Certain subsets of Ω are called events, and the collection of all events forms a σ-algebra $\mathcal{F}$. Each set A in $\mathcal{F}$ has a probability $\mathbb{P}(A)$. This point of view handles both discrete and continuous random variables within the same unifying framework. It is necessary to adopt this point of view in order to understand the change from the actual to the risk-neutral measure in finance.

The measure-theoretic view of probability theory was begun by Kolmogorov [104]. A comprehensive book on measure-theoretic probability is Billingsley [10]. A succinct book on measure-theoretic probability and martingales is Williams [161]. A more detailed book is Chung [35]. All of these are at the level of a Ph.D. course in mathematics.

1.9 Exercises

Exercise 1.1. Using the properties of Definition 1.1.2 for a probability measure $\mathbb{P}$, show the following.

(i) If $A \in \mathcal{F}$, $B \in \mathcal{F}$, and $A \subset B$, then $\mathbb{P}(A) \le \mathbb{P}(B)$.
(ii) If $A \in \mathcal{F}$ and $\{A_n\}_{n=1}^{\infty}$ is a sequence of sets in $\mathcal{F}$ with $\lim_{n\to\infty} \mathbb{P}(A_n) = 0$ and $A \subset A_n$ for every n, then $\mathbb{P}(A) = 0$. (This property was used implicitly in Example 1.1.4 when we argued that the sequence of all heads, and indeed any particular sequence, must have probability zero.)

Exercise 1.2. The infinite coin-toss space Ω_∞ of Example 1.1.4 is *uncountably infinite*. In other words, we cannot list all its elements in a sequence.

To see that this is impossible, suppose there were such a sequential list of all elements of Ω_∞:

$$\begin{aligned}\omega^{(1)} &= \omega_1^{(1)}\omega_2^{(1)}\omega_3^{(1)}\omega_4^{(1)}\ldots,\\ \omega^{(2)} &= \omega_1^{(2)}\omega_2^{(2)}\omega_3^{(2)}\omega_4^{(2)}\ldots,\\ \omega^{(3)} &= \omega_1^{(3)}\omega_2^{(3)}\omega_3^{(3)}\omega_4^{(3)}\ldots,\\ &\vdots\end{aligned}$$

An element that does not appear in this list is the sequence whose first component is H if $\omega_1^{(1)}$ is T and is T if $\omega_1^{(1)}$ is H, whose second component is H if $\omega_2^{(2)}$ is T and is T if $\omega_2^{(2)}$ is H, whose third component is H if $\omega_3^{(3)}$ is T and is T if $\omega_3^{(3)}$ is H, etc. Thus, the list does not include every element of Ω_∞.

Now consider the set of sequences of coin tosses in which the outcome on each even-numbered toss matches the outcome of the toss preceding it, i.e.,

$$A = \{\omega = \omega_1\omega_2\omega_3\omega_4\omega_5\ldots;\omega_1 = \omega_2, \omega_3 = \omega_4, \ldots\}.$$

(i) Show that A is uncountably infinite.
(ii) Show that, when $0 < p < 1$, we have $\mathbb{P}(A) = 0$.

Uncountably infinite sets can have any probability between zero and one, including zero and one. The uncountability of the set does not help determine its probability.

Exercise 1.3. Consider the set function $\mathbb{P}$ defined for every subset of $[0,1]$ by the formula that $\mathbb{P}(A) = 0$ if A is a finite set and $\mathbb{P}(A) = \infty$ if A is an infinite set. Show that $\mathbb{P}$ satisfies (1.1.3)–(1.1.5), but $\mathbb{P}$ does not have the countable additivity property (1.1.2). We see then that the finite additivity property (1.1.5) does not imply the countable additivity property (1.1.2).

Exercise 1.4. (i) Construct a standard normal random variable Z on the probability space $(\Omega_\infty, \mathcal{F}_\infty, \mathbb{P})$ of Example 1.1.4 under the assumption that the probability for head is $p = \frac{1}{2}$. (Hint: Consider Examples 1.2.5 and 1.2.6.)
(ii) Define a sequence of random variables $\{Z_n\}_{n=1}^\infty$ on Ω_∞ such that

$$\lim_{n\to\infty} Z_n(\omega) = Z(\omega) \text{ for every } \omega \in \Omega_\infty$$

and, for each n, Z_n depends only on the first n coin tosses. (This gives us a procedure for approximating a standard normal random variable by random variables generated by a finite number of coin tosses, a useful algorithm for Monte Carlo simulation.)

Exercise 1.5. When dealing with double Lebesgue integrals, just as with double Riemann integrals, the order of integration can be reversed. The only assumption required is that the function being integrated be either nonnegative or integrable. Here is an application of this fact.

Let X be a nonnegative random variable with cumulative distribution function $F(x) = \mathbb{P}\{X \leq x\}$. Show that

$$\mathbb{E}X = \int_0^\infty (1 - F(x))\,dx$$

by showing that

$$\int_\Omega \int_0^\infty \mathbb{I}_{[0,X(\omega))}(x)\,dx\,d\mathbb{P}(\omega)$$

is equal to both $\mathbb{E}X$ and $\int_0^\infty (1 - F(x))\,dx$.

Exercise 1.6. Let u be a fixed number in $\mathbb{R}$, and define the convex function $\varphi(x) = e^{ux}$ for all $x \in \mathbb{R}$. Let X be a normal random variable with mean $\mu = \mathbb{E}X$ and standard deviation $\sigma = \left[\mathbb{E}(X-\mu)^2\right]^{\frac{1}{2}}$, i.e., with density

$$f(x) = \frac{1}{\sigma\sqrt{2\pi}} e^{-\frac{(x-\mu)^2}{2\sigma^2}}.$$

(i) Verify that

$$\mathbb{E}e^{uX} = e^{u\mu + \frac{1}{2}u^2\sigma^2}.$$

(ii) Verify that Jensen's inequality holds (as it must):

$$\mathbb{E}\varphi(X) \geq \varphi(\mathbb{E}X).$$

Exercise 1.7. For each positive integer n, define f_n to be the normal density with mean zero and variance n, i.e.,

$$f_n(x) = \frac{1}{\sqrt{2n\pi}} e^{-\frac{x^2}{2n}}.$$

(i) What is the function $f(x) = \lim_{n\to\infty} f_n(x)$?
(ii) What is $\lim_{n\to\infty} \int_{-\infty}^\infty f_n(x)\,dx$?
(iii) Note that

$$\lim_{n\to\infty} \int_{-\infty}^\infty f_n(x)\,dx \neq \int_{-\infty}^\infty f(x)\,dx.$$

Explain why this does not violate the Monotone Convergence Theorem, Theorem 1.4.5.

Exercise 1.8 (Moment-generating function). Let X be a nonnegative random variable, and assume that

$$\varphi(t) = \mathbb{E}e^{tX}$$

is finite for every $t \in \mathbb{R}$. Assume further that $\mathbb{E}\left[Xe^{tX}\right] < \infty$ for every $t \in \mathbb{R}$. The purpose of this exercise is to show that $\varphi'(t) = \mathbb{E}\left[Xe^{tX}\right]$ and, in particular, $\varphi'(0) = \mathbb{E}X$.

We recall the definition of derivative:

$$\varphi'(t) = \lim_{s \to t} \frac{\varphi(t) - \varphi(s)}{t - s} = \lim_{s \to t} \frac{\mathbb{E}e^{tX} - \mathbb{E}e^{sX}}{t - s} = \lim_{s \to t} \mathbb{E}\left[\frac{e^{tX} - e^{sX}}{t - s}\right].$$

The limit above is taken over a *continuous* variable s, but we can choose a sequence of numbers $\{s_n\}_{n=1}^{\infty}$ converging to t and compute

$$\lim_{s_n \to t} \mathbb{E}\left[\frac{e^{tX} - e^{s_n X}}{t - s_n}\right],$$

where now we are taking a limit of the expectations of the *sequence* of random variables

$$Y_n = \frac{e^{tX} - e^{s_n X}}{t - s_n}.$$

If this limit turns out to be the same, regardless of how we choose the sequence $\{s_n\}_{n=1}^{\infty}$ that converges to t, then this limit is also the same as $\lim_{s \to t} \mathbb{E}\left[\frac{e^{tX} - e^{sX}}{t - s}\right]$ and is $\varphi'(t)$.

The Mean Value Theorem from calculus states that if $f(t)$ is a differentiable function, then for any two numbers s and t, there is a number θ between s and t such that

$$f(t) - f(s) = f'(\theta)(t - s).$$

If we fix $\omega \in \Omega$ and define $f(t) = e^{tX(\omega)}$, then this becomes

$$e^{tX(\omega)} - e^{sX(\omega)} = (t - s)X(\omega)e^{\theta(\omega)X(\omega)}, \tag{1.9.1}$$

where $\theta(\omega)$ is a number depending on ω (i.e., a random variable lying between t and s).

(i) Use the Dominated Convergence Theorem (Theorem 1.4.9) and equation (1.9.1) to show that

$$\lim_{n \to \infty} \mathbb{E}Y_n = \mathbb{E}\left[\lim_{n \to \infty} Y_n\right] = \mathbb{E}\left[Xe^{tX}\right]. \tag{1.9.2}$$

This establishes the desired formula $\varphi'(t) = \mathbb{E}\left[Xe^{tX}\right]$.

(ii) Suppose the random variable X can take both positive and negative values and $\mathbb{E}e^{tX} < \infty$ and $E\left[|X|e^{tX}\right] < \infty$ for every $t \in \mathbb{R}$. Show that once again $\varphi'(t) = E\left[Xe^{tX}\right]$. (Hint: Use the notation (1.3.1) to write $X = X^+ - X^-$.)

Exercise 1.9. Suppose X is a random variable on some probability space $(\Omega, \mathcal{F}, \mathbb{P})$, A is a set in $\mathcal{F}$, and for every Borel subset B of $\mathbb{R}$, we have

$$\int_A \mathbb{I}_B(X(\omega))\, d\mathbb{P}(\omega) = \mathbb{P}(A) \cdot \mathbb{P}\{X \in B\}. \tag{1.9.3}$$

Then we say that X is *independent* of the event A.

Show that if X is independent of an event A, then

$$\int_A g(X(\omega))\,d\mathbb{P}(\omega) = \mathbb{P}(A) \cdot \mathbb{E}g(X)$$

for every nonnegative, Borel-measurable function g.

Exercise 1.10. Let $\mathbb{P}$ be the uniform (Lebesgue) measure on $\Omega = [0,1]$. Define

$$Z(\omega) = \begin{cases} 0 \text{ if } 0 \le \omega < \frac{1}{2}, \\ 2 \text{ if } \frac{1}{2} \le \omega \le 1. \end{cases}$$

For $A \in \mathcal{B}[0,1]$, define

$$\widetilde{\mathbb{P}}(A) = \int_A Z(\omega)\,d\mathbb{P}(\omega).$$

(i) Show that $\widetilde{\mathbb{P}}$ is a probability measure.
(ii) Show that if $\mathbb{P}(A) = 0$, then $\widetilde{\mathbb{P}}(A) = 0$. We say that $\widetilde{\mathbb{P}}$ is *absolutely continuous* with respect to $\mathbb{P}$.
(iii) Show that there is a set A for which $\widetilde{\mathbb{P}}(A) = 0$ but $\mathbb{P}(A) > 0$. In other words, $\widetilde{\mathbb{P}}$ and $\mathbb{P}$ are not equivalent.

Exercise 1.11. In Example 1.6.6, we began with a standard normal random variable X under a measure $\mathbb{P}$. According to Exercise 1.6, this random variable has the moment-generating function

$$\mathbb{E}e^{uX} = e^{\frac{1}{2}u^2} \text{ for all } u \in \mathbb{R}.$$

The moment-generating function of a random variable determines its distribution. In particular, any random variable that has moment-generating function $e^{\frac{1}{2}u^2}$ must be standard normal.

In Example 1.6.6, we also defined $Y = X + \theta$, where θ is a constant, we set $Z = e^{-\theta X - \frac{1}{2}\theta^2}$, and we defined $\widetilde{\mathbb{P}}$ by the formula (1.6.9):

$$\widetilde{\mathbb{P}}(A) = \int_A Z(\omega)\,d\mathbb{P}(\omega) \text{ for all } A \in \mathcal{F}.$$

We showed by considering its cumulative distribution function that Y is a standard normal random variable under $\widetilde{\mathbb{P}}$. Give another proof that Y is standard normal under $\widetilde{\mathbb{P}}$ by verifying the moment-generating function formula

$$\widetilde{\mathbb{E}}e^{uY} = e^{\frac{1}{2}u^2} \text{ for all } u \in \mathbb{R}.$$

Exercise 1.12. In Example 1.6.6, we began with a standard normal random variable X on a probability space $(\Omega, \mathcal{F}, \mathbb{P})$ and defined the random variable $Y = X + \theta$, where θ is a constant. We also defined $Z = e^{-\theta X - \frac{1}{2}\theta^2}$ and used Z as the Radon-Nikodým derivative to construct the probability measure $\widetilde{\mathbb{P}}$ by the formula (1.6.9):

$$\widetilde{\mathbb{P}}(A) = \int_A Z(\omega)\, d\mathbb{P}(\omega) \text{ for all } A \in \mathcal{F}.$$

Under $\widetilde{\mathbb{P}}$, the random variable Y was shown to be standard normal.

We now have a standard normal random variable Y on the probability space $(\Omega, \mathcal{F}, \widetilde{\mathbb{P}})$, and X is related to Y by $X = Y - \theta$. By what we have just stated, with X replaced by Y and θ replaced by $-\theta$, we could define $\widehat{Z} = e^{\theta Y - \frac{1}{2}\theta^2}$ and then use $\widehat{Z}$ as a Radon-Nikodým derivative to construct a probability measure $\widehat{\mathbb{P}}$ by the formula

$$\widehat{\mathbb{P}}(A) = \int_A \widehat{Z}(\omega)\, d\widetilde{\mathbb{P}}(\omega) \text{ for all } A \in \mathcal{F},$$

so that, under $\widehat{\mathbb{P}}$, the random variable X is standard normal. Show that $\widehat{Z} = \frac{1}{Z}$ and $\widehat{\mathbb{P}} = \mathbb{P}$.

Exercise 1.13 (Change of measure for a normal random variable). A nonrigorous but informative derivation of the formula for the Radon-Nikodým derivative $Z(\omega)$ in Example 1.6.6 is provided by this exercise. As in that example, let X be a standard normal random variable on some probability space $(\Omega, \mathcal{F}, \mathbb{P})$, and let $Y = X + \theta$. Our goal is to define a strictly positive random variable $Z(\omega)$ so that when we set

$$\widetilde{\mathbb{P}}(A) = \int_A Z(\omega)\, d\mathbb{P}(\omega) \text{ for all } A \in \mathcal{F}, \tag{1.9.4}$$

the random variable Y under $\widetilde{\mathbb{P}}$ is standard normal. If we fix $\overline{\omega} \in \Omega$ and choose a set A that contains $\overline{\omega}$ and is "small," then (1.9.4) gives

$$\widetilde{\mathbb{P}}(A) \approx Z(\overline{\omega})\mathbb{P}(A),$$

where the symbol $\approx$ means "is approximately equal to." Dividing by $\mathbb{P}(A)$, we see that

$$\frac{\widetilde{\mathbb{P}}(A)}{\mathbb{P}(A)} \approx Z(\overline{\omega})$$

for "small" sets A containing $\overline{\omega}$. We use this observation to identify $Z(\overline{\omega})$.

With $\overline{\omega}$ fixed, let $x = X(\overline{\omega})$. For $\epsilon > 0$, we define $B(x, \epsilon) = \left[x - \frac{\epsilon}{2}, x + \frac{\epsilon}{2}\right]$ to be the closed interval centered at x and having length ϵ. Let $y = x + \theta$ and $B(y, \epsilon) = \left[y - \frac{\epsilon}{2}, y + \frac{\epsilon}{2}\right]$.

(i) Show that

$$\frac{1}{\epsilon}\mathbb{P}\{X \in B(x, \epsilon)\} \approx \frac{1}{\sqrt{2\pi}} \exp\left\{-\frac{X^2(\overline{\omega})}{2}\right\}.$$

(ii) In order for Y to be a standard normal random variable under $\widetilde{\mathbb{P}}$, show that we must have

$$\frac{1}{\epsilon}\widetilde{\mathbb{P}}\{Y \in B(y, \epsilon)\} \approx \frac{1}{\sqrt{2\pi}} \exp\left\{-\frac{Y^2(\overline{\omega})}{2}\right\}.$$

(iii) Show that $\{X \in B(x,\epsilon)\}$ and $\{Y \in B(y,\epsilon)\}$ are the same set, which we call $A(\overline{\omega},\epsilon)$. This set contains $\overline{\omega}$ and is "small" when $\epsilon > 0$ is small.

(iv) Show that

$$\frac{\widetilde{\mathbb{P}}(A)}{\mathbb{P}(A)} \approx \exp\left\{-\theta X(\overline{\omega}) - \frac{1}{2}\theta^2\right\}.$$

The right-hand side is the value we obtained for $Z(\overline{\omega})$ in Example 1.6.6.

Exercise 1.14 (Change of measure for an exponential random variable). Let X be a nonnegative random variable defined on a probability space $(\Omega, \mathcal{F}, P)$ with the *exponential distribution*, which is

$$\mathbb{P}\{X \le a\} = 1 - e^{-\lambda a},\ a \ge 0,$$

where λ is a positive constant. Let $\tilde{\lambda}$ be another positive constant, and define

$$Z = \frac{\tilde{\lambda}}{\lambda} e^{-(\tilde{\lambda}-\lambda)X}.$$

Define $\widetilde{\mathbb{P}}$ by

$$\widetilde{\mathbb{P}}(A) = \int_A Z\, d\mathbb{P} \quad \text{for all } A \in \mathcal{F}.$$

(i) Show that $\widetilde{\mathbb{P}}(\Omega) = 1$.

(ii) Compute the cumulative distribution function

$$\widetilde{\mathbb{P}}\{X \le a\} \text{ for } a \ge 0$$

for the random variable X under the probability measure $\widetilde{\mathbb{P}}$.

Exercise 1.15 (Provided by Alexander Ng). Let X be a random variable on a probability space $(\Omega, \mathcal{F}, \mathbb{P})$, and assume X has a density function $f(x)$ that is positive for every $x \in \mathbb{R}$. Let g be a strictly increasing, differentiable function satisfying

$$\lim_{y\to-\infty} g(y) = -\infty, \quad \lim_{y\to\infty} g(y) = \infty,$$

and define the random variable $Y = g(X)$.

Let $h(y)$ be an arbitrary nonnegative function satisfying $\int_{-\infty}^{\infty} h(y)\,dy = 1$. We want to change the probability measure so that $h(y)$ is the density function for the random variable Y. To do this, we define

$$Z = \frac{h(g(X))g'(X)}{f(X)}.$$

(i) Show that Z is nonnegative and $\mathbb{E}Z = 1$.

Now define $\widetilde{\mathbb{P}}$ by

$$\widetilde{\mathbb{P}}(A) = \int_A Z\, d\mathbb{P} \quad \text{for all } A \in \mathcal{F}.$$

(ii) Show that Y has density h under $\widetilde{\mathbb{P}}$.

2

Information and Conditioning

2.1 Information and σ-algebras

The no-arbitrage theory of derivative security pricing is based on contingency plans. In order to price a derivative security, we determine the initial wealth we would need to set up a hedge of a short position in the derivative security. The hedge must specify what position we will take in the underlying security at each future time contingent on how the uncertainty between the present time and that future time is resolved. In order to make these contingency plans, we need a way to mathematically model the information on which our future decisions can be based. In the binomial model, that information was knowledge of the coin tosses between the initial time and the future time. For the continuous-time model, we need to develop somewhat more sophisticated machinery to capture this concept of information.

We imagine as always that some random experiment is performed, and the outcome is a particular ω in the set of all possible outcomes Ω. We might then be given some information—not enough to know the precise value of ω but enough to narrow down the possibilities. For example, the true ω might be the result of three coin tosses, and we are told only the first one. Or perhaps we are told the stock price at time two without being told any of the coin tosses. In such a situation, although we do not know the true ω precisely, we can make a list of sets that are sure to contain it and other sets that are sure not to contain it. These are the sets that are *resolved by the information.*

Indeed, suppose Ω is the set of eight possible outcomes of three coin tosses. If we are told the outcome of the first coin toss only, the sets

$$A_H = \{HHH, HHT, HTH, HTT\}, \quad A_T = \{THH, THT, TTH, TTT\} \tag{2.1.1}$$

are resolved. For each of these sets, once we are told the first coin toss, we know if the true ω is a member. The empty set $\emptyset$ and the whole space Ω are always resolved, even without any information; the true ω does not belong to $\emptyset$ and does belong to Ω. The four sets that are resolved by the first coin toss form the σ-algebra

$$\mathcal{F}_1 = \{\emptyset, \Omega, A_H, A_T\}.$$

We shall think of this σ-algebra as containing the information learned by observing the first coin toss. More precisely, if instead of being told the first coin toss, we are told, for each set in $\mathcal{F}_1$, whether or not the true ω belongs to the set, we know the outcome of the first coin toss and nothing more.

If we are told the first two coin tosses, we obtain a finer resolution. In particular, the four sets

$$\begin{aligned} A_{HH} &= \{HHH, HHT\}, \quad A_{HT} = \{HTH, HTT\}, \\ A_{TH} &= \{THH, THT\}, \quad A_{TT} = \{TTH, TTT\}, \end{aligned} \tag{2.1.2}$$

are resolved. Of course, the sets in $\mathcal{F}_1$ are still resolved. Whenever a set is resolved, so is its complement, which means that A_{HH}^c, A_{HT}^c, A_{TH}^c, and A_{TT}^c are resolved. Whenever two sets are resolved, so is their union, which means that $A_{HH} \cup A_{TH}$, $A_{HH} \cup A_{TT}$, $A_{HT} \cup A_{TH}$, and $A_{HT} \cup A_{TT}$ are resolved. We have already noted that the two other pairwise unions, $A_H = A_{HH} \cup A_{HT}$ and $A_T = A_{TH} \cup A_{TT}$, are resolved. The triple unions are also resolved, and these are the complements already mentioned, e.g.,

$$A_{HH} \cup A_{HT} \cup A_{TH} = A_{TT}^c.$$

In all, we have 16 resolved sets that together form a σ-algebra we call $\mathcal{F}_2$; i.e.,

$$\mathcal{F}_2 = \left\{ \begin{array}{l} \emptyset, \Omega, A_H, A_T, A_{HH}, A_{HT}, A_{TH}, A_{TT}, A_{HH}^c, A_{HT}^c, A_{TH}^c, A_{TT}^c, \\ A_{HH} \cup A_{TH}, \ A_{HH} \cup A_{TT}, \ A_{HT} \cup A_{TH}, \ A_{HT} \cup A_{TT} \end{array} \right\}. \tag{2.1.3}$$

We shall think of this σ-algebra as containing the information learned by observing the first two coin tosses.

If we are told all three coin tosses, we know the true ω and every subset of Ω is resolved. There are 256 subsets of Ω and, taken all together, they constitute the σ-algebra $\mathcal{F}_3$:

$$\mathcal{F}_3 = \text{ The set of all subsets of } \Omega.$$

If we are told nothing about the coin tosses, the only resolved sets are $\emptyset$ and Ω. We form the so-called *trivial σ-field* $\mathcal{F}_0$ with these two sets:

$$\mathcal{F}_0 = \{\emptyset, \Omega\}.$$

We have then four σ-algebras, $\mathcal{F}_0$, $\mathcal{F}_1$, $\mathcal{F}_2$, and $\mathcal{F}_3$, indexed by time. As time moves forward, we obtain finer resolution. In other words, if $n < m$, then $\mathcal{F}_m$ contains every set in $\mathcal{F}_n$ and even more. This means that $\mathcal{F}_m$ contains more information than $\mathcal{F}_n$. The collection of σ-algebras $\mathcal{F}_0$, $\mathcal{F}_1$, $\mathcal{F}_2$, $\mathcal{F}_3$ is an example of a *filtration*. We give the continuous-time formulation of this situation in the following definition.

Definition 2.1.1. *Let Ω be a nonempty set. Let T be a fixed positive number, and assume that for each $t \in [0,T]$ there is a σ-algebra $\mathcal{F}(t)$. Assume further that if $s \leq t$, then every set in $\mathcal{F}(s)$ is also in $\mathcal{F}(t)$. Then we call the collection of σ-algebras $\mathcal{F}(t)$, $0 \leq t \leq T$, a* filtration.

A filtration tells us the information we will have at future times. More precisely, when we get to time t, we will know for each set in $\mathcal{F}(t)$ whether the true ω lies in that set.

Example 2.1.2. Suppose our sample space is $\Omega = C_0[0,T]$, the set of continuous functions defined on $[0,T]$ taking the value zero at time zero. Suppose one of these functions $\overline{\omega}$ is chosen at random and we get to observe it up to time t, where $0 \leq t \leq T$. That is to say, we know the value of $\overline{\omega}(s)$ for $0 \leq s \leq t$, but we do not know the value of $\overline{\omega}(s)$ for $t < s \leq T$. Certain subsets of Ω are resolved. For example, the set $\{\omega \in \Omega; \max_{0 \leq s \leq t} \omega(s) \leq 1\}$ is resolved. We would put this in the σ-algebra $\mathcal{F}(t)$. Other subsets of Ω are not resolved by time t. For example, if $t < T$, the set $\{\omega \in \Omega; \omega(T) > 0\}$ is not resolved by time t. Indeed, the sets that are resolved by time t are just those sets that can be described in terms of the path of ω up to time t.[1] Every reasonable[2] subset of $\Omega = C_0[0,T]$ is resolved by time T. By contrast, at time zero we see only the value of $\overline{\omega}(0)$, which is equal to zero by the definition of Ω. We learn nothing about the outcome of the random experiment of choosing $\overline{\omega}$ by observing this. The only sets resolved at time zero are $\emptyset$ and Ω, and consequently $\mathcal{F}(0) = \{\emptyset, \Omega\}$. □

Example 2.1.2 provides the simplest setting in which we may construct a Brownian motion. It remains only to assign probability to the sets in $\mathcal{F} = \mathcal{F}(T)$, and then the paths $\omega \in C_0[0,T]$ will be the paths of the Brownian motion.

The discussion preceding Definition 2.1.1 suggests that the σ-algebras in a filtration can be built by taking unions and complements of certain fundamental sets in the way $\mathcal{F}_2$ was constructed from the four sets A_{HH}, A_{HT}, A_{TH}, and A_{TT}. If this were the case, it would be enough to work with these so-called *atoms* (indivisible sets in the σ-algebra) and not consider all the other sets. In uncountable sample spaces, however, there are sets that cannot be constructed as countable unions of atoms (and uncountable unions are forbidden because we cannot add up probabilities of such unions). For example, let us fix $t \in (0,T)$ in Example 2.1.2. Now choose a continuous function $f(u)$, defined only for $0 \leq u \leq t$ and satisfying $f(0) = 0$. The set of continuous functions $\omega \in C_0[0,T]$ that agree with f on $[0,t]$ and that are free to take any values on $(t,T]$ form an atom in $\mathcal{F}_t$. In symbols, this atom is

[1] For technical reasons, we would not include in $\mathcal{F}(t)$ sets such as $\{\omega \in \Omega; \max_{0 \leq s \leq t} \omega(s) \in B\}$ if B is a subset of $\mathbb{R}$ that is not Borel measurable. This technical issue can safely be ignored.

[2] Once again, there are pathological sets such as $\{\omega \in \Omega; \omega(T) \in B\}$, where B is a subset of $\mathbb{R}$ that is not Borel measurable. These are not included in $\mathcal{F}(T)$, but that shall not concern us.

$$\{\omega \in C_0[0,T]; \omega(u) = f(u) \text{ for all } u \in [0,t]\}.$$

Each time we choose a new function $f(u)$, defined for $0 \leq u \leq t$, we get a new atom. However, there is no way to obtain the important set $\{\omega \in \Omega; \omega(t) > 0\}$ by taking countable unions of these atoms. Moreover, it is usually the case that the atoms have zero probability. Consequently, in what follows we work with all the sets of $\mathcal{F}(t)$, especially those with positive probability, not with just the atoms.

Besides observing the evolution of an economy over time, which is the idea behind Example 2.1.2, there is a second way we might acquire information about the value of ω. Let X be a random variable. We assume throughout that there is a "formula" for X, and we know this formula even before the random experiment is performed. Because we already know this formula, we are waiting only to learn the value of ω to substitute into the formula so we can evaluate $X(\omega)$. But suppose that rather than being told the value of ω we are told only the value of $X(\omega)$. This resolves certain sets. For example, if we know the value of $X(\omega)$, then we know if ω is in the set $\{X \leq 1\}$ (yes if $X(\omega) \leq 1$ and no if $X(\omega) > 1$). Indeed, every set of the form $\{X \in B\}$, where B is a subset of $\mathbb{R}$, is resolved. Again, for technical reasons, we restrict attention to subsets B that are Borel measurable.

Definition 2.1.3. *Let X be a random variable defined on a nonempty sample space Ω. The σ-*algebra generated by X*, denoted $\sigma(X)$, is the collection of all subsets of Ω of the form $\{X \in B\}$,*[3] *where B ranges over the Borel subsets of $\mathbb{R}$.*

Example 2.1.4. We return to the three-period model of Example 1.2.1 of Chapter 1. In that model, Ω is the set of eight possible outcomes of three coin tosses, and

$$\begin{aligned}
&S_2(HHH) = S_2(HHT) = 16,\\
&S_2(HTH) = S_2(HTT) = S_2(THH) = S_2(THT) = 4,\\
&S_2(TTH) = S_2(TTT) = 1.
\end{aligned}$$

In Example 1.2.2 of Chapter 1, we wrote S_2 as a function of the first two coin tosses alone, but now we include the irrelevant third toss in the argument to get the full picture. If we take B to be the set containing the single number 16, then $\{S_2 \in B\} = \{HHH, HHT\} = A_{HH}$, where we are using the notation of (2.1.2). It follows that A_{HH} belongs to the σ-algebra $\sigma(S_2)$. Similarly, we can take B to contain the single number 4 and conclude that $A_{HT} \cup A_{TH}$ belongs to $\sigma(S_2)$, and we can take B to contain the single number 1 to see that A_{TT} belongs to $\sigma(S_2)$. Taking $B = \emptyset$, we obtain $\emptyset$. Taking $B = \mathbb{R}$, we obtain Ω. Taking $B = [4, 16]$, we obtain the set $A_{HH} \cup A_{HT} \cup A_{TH}$. In short, as B ranges over the Borel subsets of $\mathbb{R}$, we will obtain the list of sets

[3] We recall that $\{X \in B\}$ is shorthand notation for the subset $\{\omega \in \Omega; X(\omega) \in B\}$ of Ω.

$$\emptyset, \Omega, A_{HH}, A_{HT} \cup A_{TH}, A_{TT}$$

and all unions and complements of these. This is the σ-algebra $\sigma(S_2)$.

Every set in $\sigma(S_2)$ is in the σ-algebra $\mathcal{F}_2$ of (2.1.3), the information contained in the first two coin tosses. On the other hand, A_{HT} and A_{TH} appear separately in $\mathcal{F}_2$ and only their union appears in $\sigma(S_2)$. This is because seeing the first two coin tosses allows us to distinguish an initial head followed by a tail from an initial tail followed by a head, but knowing only the value of S_2 does not permit this. There is enough information in $\mathcal{F}_2$ to determine the value of S_2 and even more. We say that S_2 is *$\mathcal{F}_2$-measurable.* □

Definition 2.1.5. *Let X be a random variable defined on a nonempty sample space Ω. Let $\mathcal{G}$ be a σ-algebra of subsets of Ω. If every set in $\sigma(X)$ is also in $\mathcal{G}$, we say that X is* $\mathcal{G}$-measurable.

A random variable X is $\mathcal{G}$-measurable if and only if the information in $\mathcal{G}$ is sufficient to determine the value of X. If X is $\mathcal{G}$-measurable, then $f(X)$ is also $\mathcal{G}$-measurable for any Borel-measurable function f; if the information in $\mathcal{G}$ is sufficient to determine the value of X, it will also determine the value of $f(X)$. If X and Y are $\mathcal{G}$-measurable, then $f(X,Y)$ is $\mathcal{G}$-measurable for any Borel-measurable function $f(x,y)$ of two variables. In particular, $X+Y$ and XY are $\mathcal{G}$-measurable.

A portfolio position $\Delta(t)$ taken at time t must be $\mathcal{F}(t)$-measurable (i.e., must depend only on information available to the investor at time t). We revisit a concept first encountered in Definition 2.4.1 of Chapter 2 of Volume I.

Definition 2.1.6. *Let Ω be a nonempty sample space equipped with a filtration $\mathcal{F}(t)$, $0 \le t \le T$. Let $X(t)$ be a collection of random variables indexed by $t \in [0,T]$. We say this collection of random variables is an* adapted stochastic process *if, for each t, the random variable $X(t)$ is $\mathcal{F}(t)$-measurable.*

In the continuous-time models of this text, asset prices, portfolio processes (i.e., positions), and wealth processes (i.e., values of portfolio processes) will all be adapted to a filtration that we regard as a model of the flow of public information.

2.2 Independence

When a random variable is measurable with respect to a σ-algebra $\mathcal{G}$, the information contained in $\mathcal{G}$ is sufficient to determine the value of the random variable. The other extreme is when a random variable is independent of a σ-algebra. In this case, the information contained in the σ-algebra gives no clue about the value of the random variable. Independence is the subject of the present section. In the more common case, when we have a σ-algebra $\mathcal{G}$ and a random variable X that is neither measurable with respect to $\mathcal{G}$ nor independent of $\mathcal{G}$, the information in $\mathcal{G}$ is not sufficient to evaluate X, but we can estimate X based on the information in $\mathcal{G}$. We take up this case in the next section.

In contrast to the concept of measurability, we need a probability measure in order to talk about independence. Consequently, independence can be affected by changes of probability measure; measurability is not.

Let $(\Omega, \mathcal{F}, \mathbb{P})$ be a probability space. We say that two sets A and B in $\mathcal{F}$ are independent if

$$\mathbb{P}(A \cap B) = \mathbb{P}(A) \cdot \mathbb{P}(B).$$

For example, in $\Omega = \{HH, HT, TH, TT\}$ with $0 \leq p \leq 1$, $q = 1 - p$, and

$$\mathbb{P}(HH) = p^2,\ \mathbb{P}(HT) = pq,\ \mathbb{P}(TH) = pq,\ \mathbb{P}(TT) = q^2,$$

the sets

$$A = \{\text{head on first toss}\} = \{HH, HT\}$$

and

$$B = \{\text{head on the second toss}\} = \{HH, TH\}$$

are independent. Indeed,

$$\mathbb{P}(A \cap B) = \mathbb{P}(HH) = p^2 \text{ and } \mathbb{P}(A)\mathbb{P}(B) = (p^2 + pq)(p^2 + pq) = p^2.$$

Independence of sets A and B means that knowing that the outcome ω of a random experiment is in A does not change our estimation of the probability that it is in B. If we know the first toss results in head, we still have probability p for a head on the second toss.

In a similar way, we want to define independence of two random variables X and Y to mean that if ω occurs and we know the value of $X(\omega)$ (without actually knowing ω), then our estimation of the distribution of Y is the same as when we did not know the value of $X(\omega)$. The formal definitions are the following.

Definition 2.2.1. *Let $(\Omega, \mathcal{F}, \mathbb{P})$ be a probability space, and let $\mathcal{G}$ and $\mathcal{H}$ be sub-σ-algebras of $\mathcal{F}$ (i.e., the sets in $\mathcal{G}$ and the sets in $\mathcal{H}$ are also in $\mathcal{F}$). We say these two σ-*algebras are independent *if*

$$\mathbb{P}(A \cap B) = \mathbb{P}(A) \cdot \mathbb{P}(B) \textit{ for all } A \in \mathcal{G},\ B \in \mathcal{H}.$$

Let X and Y be random variables on $(\Omega, \mathcal{F}, \mathbb{P})$. We say these two random variables are independent *if the σ-algebras they generate, $\sigma(X)$ and $\sigma(Y)$, are independent. We say that the* random variable X is independent of the σ-algebra $\mathcal{G}$ *if $\sigma(X)$ and $\mathcal{G}$ are independent.*

Recall that $\sigma(X)$ is the collection of all sets of the form $\{X \in C\}$, where C ranges over the Borel subsets of $\mathbb{R}$. Similarly, every set in $\sigma(Y)$ is of the form $\{Y \in D\}$. Definition 2.2.1 says that X and Y are independent if and only if

$$\mathbb{P}\{X \in C \text{ and } Y \in D\} = \mathbb{P}\{X \in C\} \cdot \mathbb{P}\{Y \in D\}$$

for all Borel subsets C and D of $\mathbb{R}$.

Example 2.2.2. Recall the space Ω of three independent coin tosses on which the stock price random variables of Figure 1.2.2 of Chapter 1 are constructed. Let the probability measure $\mathbb{P}$ be given by

$$\begin{aligned}\mathbb{P}(HHH) = p^3,\ \ \mathbb{P}(HHT) = p^2q, \mathbb{P}(HTH) = p^2q, \mathbb{P}(HTT) = pq^2,\\ \mathbb{P}(THH) = p^2q,\ \mathbb{P}(THT) = pq^2,\ \mathbb{P}(TTH) = pq^2,\ \mathbb{P}(TTT) = q^3.\end{aligned}$$

Intuitively, the random variables S_2 and S_3 are not independent because if we know that S_2 takes the value 16, then we know that S_3 is either 8 or 32 and is not 2 or .50. To formalize this, we consider the sets $\{S_3 = 32\} = \{HHH\}$ and $\{S_2 = 16\} = \{HHH, HHT\}$, whose probabilities are $\mathbb{P}\{S_3 = 32\} = p^3$ and $\mathbb{P}\{S_2 = 16\} = p^2$. In order to have independence, we must have

$$\mathbb{P}\{S_2 = 16 \text{ and } S_3 = 32\} = \mathbb{P}\{S_2 = 16\} \cdot \mathbb{P}\{S_3 = 32\} = p^5.$$

But $\mathbb{P}\{S_2 = 16 \text{ and } S_3 = 32\} = \mathbb{P}\{HHH\} = p^3$, so independence requires $p = 1$ or $p = 0$. Indeed, if $p = 1$, then after learning that $S_2 = 16$, we do not revise our estimate of the distribution of S_3; we already knew it would be 32. If $p = 0$, then S_2 cannot be 16, and we do not have to worry about revising our estimate of the distribution of S_3 if this occurs because it will not occur.

As the previous discussion shows, in the interesting cases of $0 < p < 1$, the random variables S_2 and S_3 are not independent. However, the random variables S_2 and $\frac{S_3}{S_2}$ are independent. Intuitively, this is because S_2 depends on the first two tosses, and $\frac{S_3}{S_2}$ depends on the third toss only. The σ-algebra generated by S_2 comprises $\emptyset$, Ω_3, the atoms (fundamental sets)

$$\begin{aligned}\{S_2 = 16\} &= \{HHH, HHT\},\\ \{S_2 = 4\} &= \{HTH, HTT, THH, THT\},\\ \{S_2 = 1\} &= \{TTH, TTT\},\end{aligned}$$

and their unions. The σ-algebra generated by $\frac{S_3}{S_2}$ comprises $\emptyset$, Ω_3, and the atoms

$$\begin{aligned}\left\{\frac{S_3}{S_2} = 2\right\} &= \{HHH, HTH, THH, TTH\},\\ \left\{\frac{S_3}{S_2} = \frac{1}{2}\right\} &= \{HHT, HTT, THT, TTT\}.\end{aligned}$$

To verify independence, we can conduct a series of checks of the form

$$\mathbb{P}\left\{S_2 = 16 \text{ and } \frac{S_3}{S_2} = 2\right\} = \mathbb{P}\{S_2 = 16\} \cdot \mathbb{P}\left\{\frac{S_3}{S_2} = 2\right\}.$$

The left-hand side of this equality is

$$\mathbb{P}\left\{S_2 = 16 \text{ and } \frac{S_3}{S_2} = 2\right\} = \mathbb{P}\{HHH\} = p^3,$$

and the right-hand side is

$$\begin{aligned}\mathbb{P}\{S_2 = 16\} \cdot \mathbb{P}\left\{\frac{S_3}{S_2} = 2\right\} \\ &= \mathbb{P}\{HHH, HHT\} \cdot \mathbb{P}\{HHH, HTH, THH, TTH\} \\ &= p^2 \cdot p.\end{aligned}$$

Indeed, for every $A \in \sigma(S_2)$ and every $B \in \sigma\left(\frac{S_3}{S_2}\right)$, we have

$$\mathbb{P}(A \cap B) = \mathbb{P}(A) \cdot \mathbb{P}(B). \qquad \square$$

We shall often need independence of more than two random variables. We make the following definition.

Definition 2.2.3. *Let $(\Omega, \mathcal{F}, \mathbb{P})$ be a probability space and let $\mathcal{G}_1, \mathcal{G}_2, \mathcal{G}_3, \dots$ be a sequence of sub-σ-algebras of $\mathcal{F}$. For a fixed positive integer n, we say that the n* σ-algebras $\mathcal{G}_1, \mathcal{G}_2, \dots, \mathcal{G}_n$ are independent *if*

$$\mathbb{P}(A_1 \cap A_2 \cap \cdots \cap A_n) = \mathbb{P}(A_1) \cdot \mathbb{P}(A_2) \cdot \cdots \cdot \mathbb{P}(A_n) \\ \text{for all } A_1 \in \mathcal{G}_1, A_2 \in \mathcal{G}_2, \dots, A_n \in \mathcal{G}_n.$$

Let $X_1, X_2, X_3, \dots$ be a sequence of random variables on $(\Omega, \mathcal{F}, \mathbb{P})$. We say the n random variables $X_1, X_2, \dots, X_n$ are independent *if the σ-algebras $\sigma(X_1)$, $\sigma(X_2), \dots, \sigma(X_n)$ are independent. We say* the full sequence of σ-algebras $\mathcal{G}_1, \mathcal{G}_2, \mathcal{G}_3, \dots$ is independent *if, for every positive integer n, the n σ-algebras $\mathcal{G}_1, \mathcal{G}_2, \dots, \mathcal{G}_n$ are independent. We say the* full sequence of random variables $X_1, X_2, X_3, \dots$ is independent *if, for every positive integer n, the n random variables $X_1, X_2, \dots, X_n$ are independent.*

Example 2.2.4. The infinite independent coin-toss space $(\Omega_\infty, \mathcal{F}, \mathbb{P})$ of Example 1.1.4 of Chapter 1 exhibits the kind of independence described in Definition 2.2.3. Let $\mathcal{G}_k$ be the σ-algebra of information associated with the kth toss. In other words, $\mathcal{G}_k$ comprises the sets $\emptyset$, Ω_∞, and the atoms

$$\{\omega \in \Omega_\infty; \omega_k = H\} \text{ and } \{\omega \cdots \in \Omega_\infty; \omega_k = T\}.$$

Note that $\mathcal{G}_k$ is different from $\mathcal{F}_k$ in Example 1.1.4 of Chapter 1, the σ-algebra associated with the first k tosses. Under the probability measure constructed in Example 1.1.4 of Chapter 1, the full sequence of σ-algebras $\mathcal{G}_1, \mathcal{G}_2, \mathcal{G}_3, \dots$ is independent. Now recall the sequence of the random variables of (1.2.8) of Chapter 1:

$$Y_k(\omega) = \begin{cases} 1 \text{ if } \omega_k = H, \\ 0 \text{ if } \omega_k = T. \end{cases}$$

The full sequence of random variables $Y_1, Y_2, Y_3, \dots$ is likewise independent.
$\square$

The definition of independence of random variables, which was given in terms of independence of σ-algebras that they generate, is a strong condition that is conceptually useful but difficult to check in practice. We illustrate the first point with the following theorem and thereafter give a second theorem that simplifies the verification that two random variables are independent. Although this and the next section treat only the case of a pair of random variables, there are analogues of these results for n random variables.

Theorem 2.2.5. *Let X and Y be independent random variables, and let f and g be Borel-measurable functions on $\mathbb{R}$. Then $f(X)$ and $g(Y)$ are independent random variables.*

PROOF: Let A be in the σ-algebra generated by $f(X)$. This σ-algebra is a sub-σ-algebra of $\sigma(X)$. To see this, recall that, by definition, every set A in $\sigma(f(X))$ is of the form $\{\omega \in \Omega; f(X(\omega)) \in C\}$, where C is a Borel subset of $\mathbb{R}$. We define $D = \{x \in \mathbb{R}; f(x) \in C\}$ and then have

$$A = \{\omega \in \Omega; f(X(\omega)) \in C\} = \{\omega \in \Omega, X(\omega) \in D\}. \tag{2.2.1}$$

The set on the right-hand side of (2.2.1) is in $\sigma(X)$, so $A \in \sigma(X)$.

Let B be in the σ-algebra generated by $g(Y)$. This σ-algebra is a sub-σ-algebra of $\sigma(Y)$, so $B \in \sigma(Y)$. Since X and Y are independent, we have $\mathbb{P}(A \cap B) = \mathbb{P}(A) \cdot \mathbb{P}(B)$. □

Definition 2.2.6. *Let X and Y be random variables. The pair of random variables (X, Y) takes values in the plane $\mathbb{R}^2$, and the* joint distribution measure *of (X, Y) is given by*[4]

$$\mu_{X,Y}(C) = \mathbb{P}\{(X,Y) \in C\} \text{ for all Borel sets } C \subset \mathbb{R}^2. \tag{2.2.2}$$

This is a probability measure (i.e., a way of assigning measure between 0 and 1 to subsets of $\mathbb{R}^2$ so that $\mu_{X,Y}(\mathbb{R}^2) = 1$ and the countable additivity property is satisfied). The joint cumulative distribution function *of (X, Y) is*

$$F_{X,Y}(a,b) = \mu_{X,Y}\big((-\infty, a] \times (-\infty, b]\big) = \mathbb{P}\{X \le a, Y \le b\}, \; a \in \mathbb{R}, b \in \mathbb{R}. \tag{2.2.3}$$

We say that a nonnegative, Borel-measurable function $f_{X,Y}(x,y)$ is a joint density *for the pair of random variables (X, Y) if*

$$\mu_{X,Y}(C) = \int_{-\infty}^{\infty} \int_{-\infty}^{\infty} \mathbb{I}_C(x,y) f_{X,Y}(x,y)\, dy\, dx \text{ for all Borel sets } C \subset \mathbb{R}^2. \tag{2.2.4}$$

[4] One way to generate the σ-algebra of Borel subsets of $\mathbb{R}^2$ is to start with the collection of closed rectangles $[a_1, b_1] \times [a_2, b_2]$ and then add all other sets necessary in order to have a σ-algebra. Any set in this resulting σ-algebra is called a *Borel subset of* $\mathbb{R}^2$. All subsets of $\mathbb{R}^2$ normally encountered belong to this σ-algebra.

Condition (2.2.4) holds if and only if

$$F_{X,Y}(a,b) = \int_{-\infty}^{a}\int_{-\infty}^{b} f_{X,Y}(x,y)\,dy\,dx \text{ for all } a \in \mathbb{R}, b \in \mathbb{R}. \tag{2.2.5}$$

The *distribution measures* (generally called the *marginal distribution measures* in this context) of X and Y are

$$\begin{aligned} \mu_X(A) = \mathbb{P}\{X \in A\} &= \mu_{X,Y}(A \times \mathbb{R}) \text{ for all Borel subsets } A \subset \mathbb{R}, \\ \mu_Y(B) = \mathbb{P}\{Y \in B\} &= \mu_{X,Y}(\mathbb{R} \times B) \text{ for all Borel subsets } B \subset \mathbb{R}. \end{aligned}$$

The *(marginal) cumulative distribution functions* are

$$\begin{aligned} F_X(a) = \mu_X(-\infty, a] &= \mathbb{P}\{X \leq a\} \text{ for all } a \in \mathbb{R}, \\ F_Y(b) = \mu_Y(-\infty, b] &= \mathbb{P}\{Y \leq b\} \text{ for all } b \in \mathbb{R}. \end{aligned}$$

If the joint density $f_{X,Y}$ exists, then the marginal densities exist and are given by

$$f_X(x) = \int_{-\infty}^{\infty} f_{X,Y}(x,y)\,dy \text{ and } f_Y(y) = \int_{-\infty}^{\infty} f_{X,Y}(x,y)\,dx.$$

The *marginal densities*, if they exist, are nonnegative, Borel-measurable functions that satisfy

$$\mu_X(A) = \int_A f_X(x)\,dx \text{ for all Borel subsets } A \subset \mathbb{R},$$

$$\mu_Y(B) = \int_B f_Y(y)\,dy \text{ for all Borel subsets } B \subset \mathbb{R}.$$

These last conditions hold if and only if

$$F_X(a) = \int_{-\infty}^{a} f_X(x)\,dx \text{ for all } a \in \mathbb{R}, \tag{2.2.6}$$

$$F_Y(b) = \int_{-\infty}^{b} f_Y(y)\,dy \text{ for all } b \in \mathbb{R}. \tag{2.2.7}$$

Theorem 2.2.7. *Let X and Y be random variables. The following conditions are equivalent.*

(i) X and Y are independent.
(ii) The joint distribution measure factors:

$$\mu_{X,Y}(A \times B) = \mu_X(A) \cdot \mu_Y(B) \text{ for all Borel subsets } A \subset \mathbb{R}, B \subset \mathbb{R}. \tag{2.2.8}$$

(iii) The joint cumulative distribution function factors:

$$F_{X,Y}(a,b) = F_X(a) \cdot F_Y(b) \text{ for all } a \in \mathbb{R}, b \in \mathbb{R}. \tag{2.2.9}$$

(iv) The joint moment-generating function factors:

$$\mathbb{E}e^{uX+vY} = \mathbb{E}e^{uX} \cdot \mathbb{E}e^{vY} \tag{2.2.10}$$

for all $u \in \mathbb{R}$, $v \in \mathbb{R}$ for which the expectations are finite.

If there is a joint density, each of the conditions above is equivalent to the following.

(v) The joint density factors:

$$f_{X,Y}(x,y) = f_X(x) \cdot f_Y(y) \text{ for almost every } x \in \mathbb{R}, y \in \mathbb{R}. \tag{2.2.11}$$

The conditions above imply but are not equivalent to the following.

(vi) The expectation factors:

$$\mathbb{E}[XY] = \mathbb{E}X \cdot \mathbb{E}Y, \tag{2.2.12}$$

provided $\mathbb{E}|XY| < \infty$.

OUTLINE OF PROOF: We sketch the various steps that constitute the proof of this theorem.
(i)⇒(ii) Assume that X and Y are independent. Then

$$\begin{aligned}\mu_{X,Y}(A \times B) &= \mathbb{P}\{X \in A \text{ and } Y \in B\}\\ &= \mathbb{P}(\{X \in A\} \cap \{Y \in B\})\\ &= \mathbb{P}\{X \in A\} \cdot \mathbb{P}\{Y \in B\}\\ &= \mu_X(A) \cdot \mu_Y(B).\end{aligned}$$

(ii)⇒(i) A typical set in $\sigma(X)$ is of the form $\{X \in A\}$, and a typical set in $\sigma(Y)$ is of the form $\{Y \in B\}$. Assume (ii). Then

$$\begin{aligned}\mathbb{P}(\{X \in A\} \cap \{Y \in B\}) &= \mathbb{P}\{X \in A \text{ and } Y \in B\}\\ &= \mu_{X,Y}(A \times B)\\ &= \mu_X(A) \cdot \mu_Y(B)\\ &= \mathbb{P}\{X \in A\} \cdot \mathbb{P}\{Y \in B\}.\end{aligned}$$

This shows that every set in $\sigma(X)$ is independent of every set in $\sigma(Y)$.
(ii)⇒(iii) Assume (2.2.8). Then

$$\begin{aligned}F_{X,Y}(a,b) &= \mu_{X,Y}\big((-\infty,a] \times (-\infty,b]\big)\\ &= \mu_X(-\infty,a] \cdot \mu_Y(-\infty,b]\\ &= F_X(a) \cdot F_Y(b).\end{aligned}$$

(iii)⇒(ii) Equation (2.2.9) implies that (2.2.8) holds whenever A is of the form $A = (-\infty, a]$ and B is of the form $B = (-\infty, b]$. This is enough to

establish (2.2.8) for all Borel sets A and B, but the details of this are beyond the scope of the text.
(iii)⇒(v) If there is a joint density, then (iii) implies

$$\int_{-\infty}^{a}\int_{-\infty}^{b} f_{X,Y}(x,y)\,dy\,dx = \int_{-\infty}^{a} f_X(x)\,dx \cdot \int_{\infty}^{b} f_Y(y)\,dy.$$

Differentiating first with respect to a and then with respect to b, we obtain

$$f_{X,Y}(a,b) = f_X(a)\cdot f_Y(b),$$

which is just (2.2.11) with different dummy variables.
(v)⇒(iii) Assume there is a joint density. If we also assume (2.2.11), we can integrate both sides to get

$$\begin{aligned}
F_{X,Y}(a,b) &= \int_{-\infty}^{a}\int_{-\infty}^{b} f_{X,Y}(x,y)\,dy\,dx\\
&= \int_{-\infty}^{a}\int_{-\infty}^{b} f_X(x)\cdot f_Y(y)\,dy\,dx\\
&= \int_{-\infty}^{a} f_X(x)\,dx \cdot \int_{-\infty}^{b} f_Y(y)\,dy\\
&= F_X(a)\cdot F_Y(b).
\end{aligned}$$

(i)⇒(iv) We first use the "standard machine" as in the proof of Theorem 1.5.1 of Chapter 1, starting with the case when h is the indicator function of a Borel subset of $\mathbb{R}^2$, to show that, for every real-valued, Borel-measurable function $h(x,y)$ on $\mathbb{R}^2$, we have

$$\mathbb{E}|h(X,Y)| = \int_{\mathbb{R}^2} |h(x,y)|\,d\mu_{X,Y}(x,y),$$

and if this quantity is finite, then

$$\mathbb{E}h(X,Y) = \int_{\mathbb{R}^2} h(x,y)\,d\mu_{X,Y}(x,y). \tag{2.2.13}$$

This is true for any pair of random variables X and Y, whether or not they are independent. If X and Y are independent, then the joint distribution $\mu_{X,Y}$ is a product of marginal distributions, and this permits us to rewrite (2.2.13) as

$$\mathbb{E}h(X,Y) = \int_{-\infty}^{\infty}\int_{-\infty}^{\infty} h(x,y)\,d\mu_Y(y)\,d\mu_X(x). \tag{2.2.14}$$

We now fix numbers u and v and take $h(x,y) = e^{ux+vy}$. Equation (2.2.14) reduces to

$$\begin{aligned}\mathbb{E}e^{uX+vY} &= \int_{-\infty}^{\infty}\int_{-\infty}^{\infty} e^{ux+vy}\,d\mu_Y(y)\,d\mu_X(x)\\ &= \int_{-\infty}^{\infty} e^{ux}\,d\mu_X(x)\cdot\int_{-\infty}^{\infty} e^{vy}\,d\mu_Y(y)\\ &= \mathbb{E}e^{uX}\cdot\mathbb{E}e^{vY},\end{aligned}$$

where we have used Theorem 1.5.1 of Chapter 1 for the last step. The proof **(iv)⇒(i)** is beyond the scope of this text.
(i)⇒(vi) In the special case when $h(x,y) = xy$, (2.2.14) reduces to

$$\mathbb{E}[XY] = \int_{-\infty}^{\infty} x\,d\mu_X(x)\cdot\int_{-\infty}^{\infty} y\,d\mu_Y(y) = \mathbb{E}X\cdot\mathbb{E}Y,$$

where again we have used Theorem 1.5.1 of Chapter 1 for the last step. □

Example 2.2.8 (Independent normal random variables). Random variables X and Y are independent and standard normal if they have the joint density

$$f_{X,Y}(x,y) = \frac{1}{2\pi}e^{-\frac{1}{2}(x^2+y^2)} \text{ for all } x\in\mathbb{R}, y\in\mathbb{R}.$$

This is the product of the marginal densities

$$f_X(x) = \frac{1}{\sqrt{2\pi}}e^{-\frac{1}{2}x^2} \text{ and } f_Y(y) = \frac{1}{\sqrt{2\pi}}e^{-\frac{1}{2}y^2}.$$

We use the notation

$$N(a) = \frac{1}{\sqrt{2\pi}}\int_{-\infty}^{a} e^{-\frac{1}{2}x^2}\,dx \tag{2.2.15}$$

for the standard normal cumulative distribution function. The joint cumulative distribution function for (X,Y) factors:

$$\begin{aligned}F_{X,Y}(a,b) &= \int_{-\infty}^{a}\int_{-\infty}^{b} f_X(x)f_Y(y)\,dy\,dx\\ &= \int_{-\infty}^{a} f_X(x)\,dx\cdot\int_{-\infty}^{b} f_Y(y)\,dy\\ &= N(a)\cdot N(b).\end{aligned}$$

The joint distribution μ_X is the probability measure on $\mathbb{R}^2$ that assigns a measure to each Borel set $C\subset\mathbb{R}^2$ equal to the integral of $f_{X,Y}(x,y)$ over C. If $C = A\times B$, where $A\in\mathcal{B}(\mathbb{R})$ and $B\in\mathcal{B}(\mathbb{R})$, then $\mu_{X,Y}$ factors:

$$\begin{aligned}\mu_{X,Y}(A\times B) &= \int_A\int_B f_X(x)f_Y(y)\,dy\,dx\\ &= \int_A f_X(x)\,dx\cdot\int_B f_Y(y)\,dy\\ &= \mu_X(A)\cdot\mu_Y(B).\end{aligned}$$

□

We give an example to show that property (vi) of Theorem 2.2.7 does not imply independence. We precede this with a definition.

Definition 2.2.9. *Let X be a random variable whose expected value is defined. The* variance of X, *denoted* $\mathrm{Var}(X)$, *is*

$$\mathrm{Var}(X) = \mathbb{E}\left[(X - \mathbb{E}X)^2\right].$$

Because $(X - \mathbb{E}X)^2$ is nonnegative, $\mathrm{Var}(X)$ *is always defined, although it may be infinite. The* standard deviation of X *is* $\sqrt{\mathrm{Var}(X)}$. *The linearity of expectations shows that*

$$\mathrm{Var}(X) = \mathbb{E}\left[X^2\right] - (\mathbb{E}X)^2.$$

Let Y be another random variable and assume that $\mathbb{E}X$, $\mathrm{Var}(X)$, $\mathbb{E}Y$ and Var(Y) are all finite. The covariance of X and Y *is*

$$\mathrm{Cov}(X,Y) = \mathbb{E}\left[(X - \mathbb{E}X)(Y - \mathbb{E}Y)\right].$$

The linearity of expectations shows that

$$\mathrm{Cov}(X,Y) = \mathbb{E}[XY] - \mathbb{E}X \cdot \mathbb{E}Y.$$

In particular, $\mathbb{E}[XY] = \mathbb{E}X \cdot \mathbb{E}Y$ if and only if $\mathrm{Cov}(X,Y) = 0$. Assume, in addition to the finiteness of expectations and variances, that $\mathrm{Var}(X) > 0$ and Var(Y) > 0. The correlation coefficient of X and Y *is*

$$\rho(X,Y) = \frac{\mathrm{Cov}(X,Y)}{\sqrt{\mathrm{Var(X)Var}(Y)}}.$$

If $\rho(X,Y) = 0$ (or equivalently, $\mathrm{Cov}(X,Y) = 0$), we say that X and *Y* are uncorrelated.

Property (vi) of Theorem 2.2.7 implies that independent random variables are uncorrelated. The converse is not true, even for normal random variables, although it is true of *jointly normal* random variables (see Definition 2.2.11 below).

Example 2.2.10 (Uncorrelated, dependent normal random variables). Let X be a standard normal random variable and let Z be independent of X and satisfy[5]

[5] To construct such random variables, we can choose $\Omega = \{(\omega_1, \omega_2); 0 \leq \omega_1 \leq 1,\ 0 \leq \omega_2 \leq 1\}$ to be the unit square and choose $\mathbb{P}$ to be the two-dimensional Lebesgue measure according to which $\mathbb{P}(A)$ is equal to the area of A for every Borel subset of Ω. We then set $X(\omega_1, \omega_2) = N^{-1}(\omega_1)$, which is a standard normal random variable under $\mathbb{P}$ (see Example 1.2.6 for a discussion of this probability integral transform). We set $Z(\omega_1, \omega_2)$ to be -1 if $0 \leq \omega_2 \leq \frac{1}{2}$ and to be 1 if $\frac{1}{2} < \omega_2 \leq 1$.

$$\mathbb{P}\{Z = 1\} = \frac{1}{2} \text{ and } \mathbb{P}\{Z = -1\} = \frac{1}{2}. \tag{2.2.16}$$

Define $Y = ZX$. We show below that, like X, the random variable Y is standard normal. Furthermore, X and Y are uncorrelated, but they are not independent. The pair (X, Y) does not have a joint density.

Let us first determine the distribution of Y. We compute

$$\begin{aligned} F_Y(b) &= \mathbb{P}\{Y \leq b\} \\ &= \mathbb{P}\{Y \leq b \text{ and } Z = 1\} + \mathbb{P}\{Y \leq b \text{ and } Z = -1\} \\ &= \mathbb{P}\{X \leq b \text{ and } Z = 1\} + \mathbb{P}\{-X \leq b \text{ and } Z = -1\}. \end{aligned}$$

Because X and Z are independent, we have

$$\begin{aligned} &\mathbb{P}\{X \leq b \text{ and } Z = 1\} + \mathbb{P}\{-X \leq b \text{ and } Z = -1\} \\ &\quad = \mathbb{P}\{Z = 1\} \cdot \mathbb{P}\{X \leq b\} + \mathbb{P}\{Z = -1\} \cdot \mathbb{P}\{-X \leq b\} \\ &\quad = \frac{1}{2} \cdot \mathbb{P}\{X \leq b\} + \frac{1}{2} \cdot \mathbb{P}\{-X \leq b\}. \end{aligned}$$

Because X is a standard normal random variable, so is $-X$. Therefore, $\mathbb{P}\{X \leq b\} = \mathbb{P}\{-X \leq b\} = N(b)$. It follows that $F_Y(b) = N(b)$; in other words, Y is a standard normal random variable.

Since $\mathbb{E}X = \mathbb{E}Y = 0$, the covariance of X and Y is

$$\text{Cov}(X, Y) = \mathbb{E}[XY] = \mathbb{E}\big[ZX^2\big].$$

Because Z and X are independent, so are Z and X^2, and we may use Theorem 2.2.7(vi) to write

$$\mathbb{E}\big[ZX^2\big] = \mathbb{E}Z \cdot \mathbb{E}\big[X^2\big] = 0 \cdot 1 = 0.$$

Therefore, X and Y are uncorrelated.

The random variables X and Y cannot be independent for if they were, then $|X|$ and $|Y|$ would also be independent (Theorem 2.2.5). But $|X| = |Y|$. In particular,

$$\mathbb{P}\{|X| \leq 1, |Y| \leq 1\} = \mathbb{P}\{|X| \leq 1\} = N(1) - N(-1),$$

and

$$\mathbb{P}\{|X| \leq 1\} \cdot \mathbb{P}\{|Y| \leq 1\} = \big(N(1) - N(-1)\big)^2.$$

These two expressions are not equal, as they would be for independent random variables.

Finally, we want to examine the joint distribution measure $\mu_{X,Y}$ of (X, Y). Since $|X| = |Y|$, the pair (X, Y) takes values only in the set

$$C = \{(x, y); x = \pm y\}.$$

In other words, $\mu_{X,Y}(C) = 1$ and $\mu_{X,Y}(C^c) = 0$. But C has zero area. It follows that for any nonnegative function f, we must have

$$\int_{-\infty}^{\infty}\int_{-\infty}^{\infty} \mathbb{I}_C(x,y)f(x,y)\,dy\,dx = 0.$$

One way of thinking about this is to observe that if we want to integrate a function $\mathbb{I}_C(x,y)f(x,y)$ over the plane $\mathbb{R}^2$, we could first fix x and integrate out the y-variable, but since $f(x,y)\mathbb{I}_C(x,y)$ is zero except when $y = x$ and $y = -x$, we will get zero. When we next integrate out the x-variable, we will be integrating the zero function, and the end result will be zero. There cannot be a joint density for (X,Y) because with this choice of the set C, the left-hand side of (2.2.4) is one but the right-hand side is zero. Of course, X and Y have marginal densities because they are both standard normal. Moreover, the joint cumulative distribution function exists (as it always does). In this case, it is

$$\begin{aligned}
&F_{X,Y}(a,b)\\
&= \mathbb{P}\{X \le a \text{ and } Y \le b\}\\
&= \mathbb{P}\{X \le a, X \le b, \text{ and } Z = 1\} + \mathbb{P}\{X \le a, -X \le b, \text{ and } Z = -1\}\\
&= \mathbb{P}\{Z = 1\}\cdot\mathbb{P}\{X \le \min(a,b)\} + \mathbb{P}\{Z = -1\}\cdot\mathbb{P}\{-b \le X \le a\}\\
&= \frac{1}{2}N(\min(a,b)) + \frac{1}{2}\max\{N(a) - N(-b), 0\}.
\end{aligned}$$

There is no joint density $f_{X,Y}(x,y)$ that permits us to write this function in the form (2.2.5). □

Definition 2.2.11. *Two random variables X and Y are said to be* jointly normal *if they have the joint density*

$$\begin{aligned}
&f_{X,Y}(x,y)\\
&= \frac{1}{2\pi\sigma_1\sigma_2\sqrt{1-\rho^2}}\exp\left\{-\frac{1}{2(1-\rho^2)}\left[\frac{(x-\mu_1)^2}{\sigma_1^2} - \frac{2\rho(x-\mu_1)(y-\mu_2)}{\sigma_1\sigma_2}\right.\right.\\
&\qquad\left.\left. + \frac{(y-\mu_2)^2}{\sigma_2^2}\right]\right\}, \qquad (2.2.17)
\end{aligned}$$

where $\sigma_1 > 0$, $\sigma_2 > 0$, $|\rho| < 1$, and μ_1, μ_2 are real numbers. More generally, a random column vector $\mathbf{X} = (X_1, \dots, X_n)^{\mathrm{tr}}$, where the superscript tr *denotes transpose, is* jointly normal *if it has joint density*

$$f_{\mathbf{X}}(\mathbf{x}) = \frac{1}{\sqrt{(2\pi)^n \det(C)}}\exp\left\{-\frac{1}{2}(\mathbf{x}-\mu)C^{-1}(\mathbf{x}-\mu)^{\mathrm{tr}}\right\}. \qquad (2.2.18)$$

In equation (2.2.18), $\mathbf{x} = (x_1, \dots, x_n)$ is a row vector of dummy variables, $\mu = (\mu_1, \dots, \mu_n)$ is the row vector of expectations, and C is the positive definite matrix of covariances.

In the case of (2.2.17), X is normal with expectation μ_1 and variance σ_1^2, Y is normal with expectation μ_2 and variance σ_2^2, and the correlation between X and Y is ρ. The density factors (equivalently, X and Y are independent) if and only if $\rho = 0$. In the case (2.2.18), the density factors into the product of n normal densities (equivalently, the components of $\mathbf{X}$ are independent) if and only if C is a diagonal matrix (all the covariances are zero).

Linear combinations of jointly normal random variables (i.e., sums of constants times the random variables) are jointly normal. Since independent normal random variables are jointly normal, a general method for creating jointly normal random variables is to begin with a set of independent normal random variables and take linear combinations. Conversely, any set of jointly normal random variables can be reduced to linear combinations of independent normal random variables. We do this reduction for a pair of correlated normal random variables in Example 2.2.12 below.

Since the distribution of jointly normal random variables is characterized in terms of means and covariances, and joint normality is preserved under linear combinations, it is not necessary to deal directly with the density when making linear changes of variables. The following example illustrates this point.

Example 2.2.12. Let (X, Y) be jointly normal with the density (2.2.17). Define $W = Y - \frac{\rho\sigma_2}{\sigma_1}X$. Then X and W are independent. To verify this, it suffices to show that X and W have covariance zero since they are jointly normal. We compute

$$\begin{aligned}\operatorname{Cov}(X,W) &= \mathbb{E}\big[(X-\mathbb{E}X)(W-\mathbb{E}W)\big]\\ &= \mathbb{E}\big[(X-\mathbb{E}X)(Y-\mathbb{E}Y)\big] - \mathbb{E}\left[\frac{\rho\sigma_2}{\sigma_1}(X-\mathbb{E}X)^2\right]\\ &= \operatorname{Cov}(X,Y) - \frac{\rho\sigma_2}{\sigma_1}\sigma_1^2\\ &= 0.\end{aligned}$$

The expectation of W is $\mu_3 = \mu_2 - \frac{\rho\sigma_2\mu_1}{\sigma_1}$, and the variance is

$$\begin{aligned}\sigma_3^2 &= \mathbb{E}\big[(W-\mathbb{E}W)^2\big]\\ &= \mathbb{E}\big[(Y-\mathbb{E}Y)^2\big] - \frac{2\rho\sigma_2}{\sigma_1}\mathbb{E}\big[(X-\mathbb{E}X)(Y-\mathbb{E}Y)\big] + \frac{\rho^2\sigma_2^2}{\sigma_1^2}\mathbb{E}\big[(X-\mathbb{E}X)^2\big]\\ &= (1-\rho^2)\sigma_2^2.\end{aligned}$$

The joint density of X and W is

$$f_{X,W}(x,w) = \frac{1}{2\pi\sigma_1\sigma_3}\exp\left\{-\frac{(x-\mu_1)^2}{2\sigma_1^2} - \frac{(w-\mu_3)^2}{2\sigma_3^2}\right\}.$$

Note finally that we have decomposed Y into the linear combination

$$Y = \frac{\rho\sigma_2}{\sigma_1}X + W \tag{2.2.19}$$

of a pair of independent normal random variables X and W. □

2.3 General Conditional Expectations

We consider a random variable X defined on a probability space $(\Omega, \mathcal{F}, \mathbb{P})$ and a sub-σ-algebra $\mathcal{G}$ of $\mathcal{F}$. If X is $\mathcal{G}$-measurable, then the information in $\mathcal{G}$ is sufficient to determine the value of X. If X is independent of $\mathcal{G}$, then the information in $\mathcal{G}$ provides no help in determining the value of X. In the intermediate case, we can use the information in $\mathcal{G}$ to estimate but not precisely evaluate X. The *conditional expectation of X given $\mathcal{G}$* is such an estimate.

We have already discussed conditional expectations in the binomial model. Let Ω be the set of all possible outcomes of N coin tosses, and assume these coin tosses are independent with probability p for head and probability $q = 1-p$ for tail on each toss. Let $\mathbb{P}(\omega)$ denote the probability of a sequence of coin tosses under these assumptions. Let n be an integer, $1 \leq n \leq N-1$, and let X be a random variable. Then the conditional expectation of X under $\mathbb{P}$, based on the information at time n, is (see Definition 2.3.1 of Chapter 2)

$$\begin{aligned} &\mathbb{E}_n[X](\omega_1 \dots \omega_n) \\ &= \sum_{\omega_{n+1}\dots\omega_N} p^{\#H(\omega_{n+1}\dots\omega_N)} q^{\#T(\omega_{n+1}\dots\omega_N)} X(\omega_1 \dots \omega_n \omega_{n+1} \dots \omega_N). \quad (2.3.1) \end{aligned}$$

In the special cases $n = 0$ and $n = N$, we define

$$\mathbb{E}_0 X = \sum_{\omega_0\dots\omega_N} p^{\#H(\omega_0\dots\omega_N)} q^{\#T(\omega_0\dots\omega_N)} X(\omega_0 \dots \omega_N) = \mathbb{E}X, \quad (2.3.2)$$

$$E_N[X](\omega_0 \dots \omega_N) = X(\omega_0 \dots \omega_N). \quad (2.3.3)$$

In (2.3.2), we have the estimate of X based on no information, and in (2.3.3) we have the estimate based on full information.

We need to generalize (2.3.1)–(2.3.3) in a way suitable for a continuous-time model. Toward that end, we examine (2.3.1) within the context of a three-period example. Consider the general three-period model of Figure 2.3.1. We assume the probability of head on each toss is p and the probability of tail is $q = 1-p$, and we compute

$$\mathbb{E}_2[S_3](HH) = pS_3(HHH) + qS_3(HHT), \quad (2.3.4)$$
$$\mathbb{E}_2[S_3](HT) = pS_3(HTH) + qS_3(HTT), \quad (2.3.5)$$
$$\mathbb{E}_2[S_3](TH) = pS_3(THH) + qS_3(THT), \quad (2.3.6)$$
$$\mathbb{E}_2[S_3](TT) = pS_3(TTH) + qS_3(TTT). \quad (2.3.7)$$

Recall the σ-algebra $\mathcal{F}_2$ of (2.1.3), which is built up from the four fundamental sets (we call them *atoms* because they are indivisible within the σ-algebra) A_{HH}, A_{HT}, A_{TH}, and A_{TT} of (2.1.2). We multiply (2.3.4) by $\mathbb{P}(A_{HH}) = p^2$, multiply (2.3.5) by $\mathbb{P}(A_{HT}) = pq$, multiply (2.3.6) by $\mathbb{P}(A_{TH}) = pq$, and multiply (2.3.7) by $\mathbb{P}(A_{TT}) = q^2$. The resulting equations may be written as

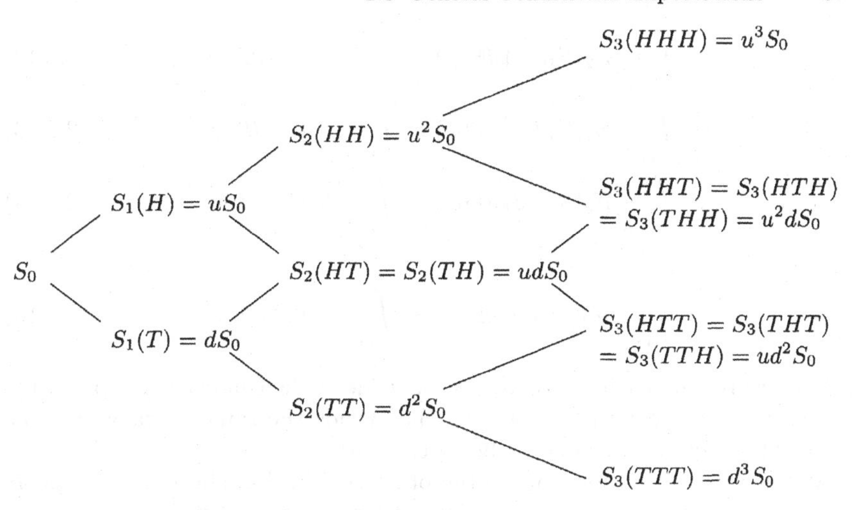

Fig. 2.3.1. General three-period model.

$$E_2[S_3](HH)\mathbb{P}(A_{HH}) = \sum_{\omega \in A_{HH}} S_3(\omega)\mathbb{P}(\omega), \tag{2.3.8}$$

$$E_2[S_3](HT)\mathbb{P}(A_{HT}) = \sum_{\omega \in A_{HT}} S_3(\omega)\mathbb{P}(\omega), \tag{2.3.9}$$

$$E_2[S_3](TH)\mathbb{P}(A_{TH}) = \sum_{\omega \in A_{TH}} S_3(\omega)\mathbb{P}(\omega), \tag{2.3.10}$$

$$E_2[S_3](TT)\mathbb{P}(A_{TT}) = \sum_{\omega \in A_{TT}} S_3(\omega)\mathbb{P}(\omega). \tag{2.3.11}$$

We could divide each of these equations by the probability of the atom appearing as the second factor on the left-hand sides and thereby recover the formulas (2.3.4)–(2.3.7) for the conditional expectations. However, in the continuous-time model, atoms typically have probability zero, and such a step cannot be performed. We therefore take an alternate route here to lay the groundwork for the continuous-time model.

On each of the atoms of $\mathcal{F}_2$, the conditional expectation $\mathbb{E}_2[S_3]$ is constant because the conditional expectation does not depend on the third toss and the atom is created by holding the first two tosses fixed. It follows that the left-hand sides of (2.3.8)–(2.3.11) may be written as integrals of the integrand $\mathbb{E}_2[S_3]$ over the atom. For this purpose, we shall write $\mathbb{E}_2[S_3](\omega) = \mathbb{E}_2[S_3](\omega_1\omega_2\omega_3)$, including the third toss in the argument, even though it is irrelevant. The right-hand sides of these equations are sums, which are Lebesgue integrals on a finite probability space. Using Lebesgue integral notation, we rewrite (2.3.8)–(2.3.11) as

$$\int_{A_{HH}} E_2[S_3](\omega)\,d\mathbb{P}(\omega) = \int_{A_{HH}} S_3(\omega)\,d\mathbb{P}(\omega), \tag{2.3.12}$$

$$\int_{A_{HT}} E_2[S_3](\omega)\,d\mathbb{P}(\omega) = \int_{A_{HT}} S_3(\omega)\,d\mathbb{P}(\omega), \tag{2.3.13}$$

$$\int_{A_{TH}} E_2[S_3](\omega)\,d\mathbb{P}(\omega) = \int_{A_{TH}} S_3(\omega)\,d\mathbb{P}(\omega), \tag{2.3.14}$$

$$\int_{A_{TT}} E_2[S_3](\omega)\,d\mathbb{P}(\omega) = \int_{A_{TT}} S_3(\omega)\,d\mathbb{P}(\omega). \tag{2.3.15}$$

In other words, on each of the atoms the value of the conditional expectation has been chosen to be that constant that yields the same average over the atom as the random variable S_3 being estimated.

We turn our attention now to the other sets in $\mathcal{F}_2$. The full list appears in (2.1.3), and every set on the list, except for the empty set, is a finite union of atoms. If we add equations (2.3.12) and (2.3.13), we obtain

$$\int_{A_H} E_2[S_3](\omega)\,d\mathbb{P}(\omega) = \int_{A_H} S_3(\omega)\,d\mathbb{P}(\omega).$$

Similarly, but adding various combinations of (2.3.12)–(2.3.15), we see that

$$\int_A \mathbb{E}_2[S_3](\omega)\,d\mathbb{P}(\omega) = \int_A S_3(\omega)\,d\mathbb{P}(\omega) \tag{2.3.16}$$

for every set $A \in \mathcal{F}_2$, except possibly for $A = \emptyset$. However, if $A = \emptyset$, equation (2.3.16) still holds, with both sides equal to zero. We call (2.3.16) the *partial-averaging property* of conditional expectations because it says that the conditional expectation and the random variable being estimated give the same value when averaged over "parts" of Ω (those "parts" that are sets in the conditioning σ-algebra $\mathcal{F}_2$).

We take (2.3.16) as the defining property of conditional expectations. The precise definition is the following.

Definition 2.3.1. *Let $(\Omega, \mathcal{F}, \mathbb{P})$ be a probability space, let $\mathcal{G}$ be a sub-σ-algebra of $\mathcal{F}$, and let X be a random variable that is either nonnegative or integrable. The* conditional expectation *of X given $\mathcal{G}$, denoted $\mathbb{E}[X|\mathcal{G}]$, is any random variable that satisfies*

(i) **(Measurability)** *$\mathbb{E}[X|\mathcal{G}]$ is $\mathcal{G}$-measurable, and*
(ii) **(Partial averaging)**

$$\int_A \mathbb{E}[X|\mathcal{G}](\omega)\,d\mathbb{P}(\omega) = \int_A X(\omega)\,d\mathbb{P}(\omega) \text{ for all } A \in \mathcal{G}. \tag{2.3.17}$$

If $\mathcal{G}$ is the σ-algebra generated by some other random variable W (i.e., $\mathcal{G} = \sigma(W)$), we generally write $\mathbb{E}[X|W]$ rather than $\mathbb{E}[X|\sigma(W)]$.

Property (i) in Definition 2.3.1 guarantees that, although the estimate of X based on the information in $\mathcal{G}$ is itself a random variable, the value of the estimate $\mathbb{E}[X|\mathcal{G}]$ can be determined from the information in $\mathcal{G}$. Property (i) captures the fact that the estimate $\mathbb{E}[X|\mathcal{G}]$ of X is *based on the information in* $\mathcal{G}$. Note in (2.3.4)–(2.3.7) that the conditional expectation $\mathbb{E}_2[S_3]$ is constant on the atoms of $\mathcal{F}_2$; this is property (i) for this case.

The second property ensures that $\mathbb{E}[X|\mathcal{G}]$ is indeed an estimate of X. It gives the same averages as X over all the sets in $\mathcal{G}$. If $\mathcal{G}$ has many sets, which provide a fine resolution of the uncertainty inherent in ω, then this partial-averaging property over the "small" sets in $\mathcal{G}$ says that $\mathbb{E}[X|\mathcal{G}]$ is a good estimator of X. If $\mathcal{G}$ has only a few sets, this partial-averaging property guarantees only that $\mathbb{E}[X|\mathcal{G}]$ is a crude estimate of X.

Definition 2.3.1 raises two immediate questions. First, does there always exist a random variable $\mathbb{E}[X|\mathcal{G}]$ satisfying properties (i) and (ii)? Second, if there is a random variable satisfying these properties, is it unique? The answer to the first question is yes, and the proof of the existence of $\mathbb{E}[X|\mathcal{G}]$ is based on the Radon-Nikodým Theorem, Theorem 1.6.7 (see Appendix B). The answer to the second question is a qualified yes, as we now explain. Suppose Y and Z both satisfy conditions (i) and (ii) of Definition 2.3.1. Because both Y and Z are $\mathcal{G}$-measurable, their difference $Y - Z$ is as well, and thus the set $A = \{Y - Z > 0\}$ is in $\mathcal{G}$. From (2.3.17), we have

$$\int_A Y(\omega)\,d\mathbb{P}(\omega) = \int_A X(\omega)\,d\mathbb{P}(\omega) = \int_A Z(\omega)\,d\mathbb{P}(\omega),$$

and thus

$$\int_A \big(Y(\omega) - Z(\omega)\big)\,d\mathbb{P}(\omega) = 0.$$

The integrand is strictly positive on the set A, so the only way this equation can hold is for A to have probability zero (i.e., $Y \leq Z$ almost surely). We can reverse the roles of Y and Z in this argument and conclude that $Z \leq Y$ almost surely. Hence $Y = Z$ almost surely. This means that although different procedures might result in different random variables when determining $\mathbb{E}[X|\mathcal{G}]$, these different random variables will agree almost surely. The set of ω for which the random variables are different has zero probability.

In this more general context, conditional expectations still have the five fundamental properties developed in Theorem 2.3.2 of Chapter 2 of Volume I. We restate them in the present context.

Theorem 2.3.2. *Let $(\Omega, \mathcal{F}, \mathbb{P})$ be a probability space and let $\mathcal{G}$ be a sub-σ-algebra of $\mathcal{F}$.*

(i) **(Linearity of conditional expectations)** *If X and Y are integrable random variables and c_1 and c_2 are constants, then*

$$\mathbb{E}[c_1X + c_2Y|\mathcal{G}] = c_1\mathbb{E}[X|\mathcal{G}] + c_2\mathbb{E}[Y|\mathcal{G}]. \tag{2.3.18}$$

This equation also holds if we assume that X and Y are nonnegative (rather than integrable) and c_1 and c_2 are positive, although both sides may be $+\infty$.

(ii) **(Taking out what is known)** *If X and Y are integrable random variables, Y and XY are integrable, and X is $\mathcal{G}$-measurable, then*

$$\mathbb{E}[XY|\mathcal{G}] = X\mathbb{E}[Y|\mathcal{G}]. \tag{2.3.19}$$

This equation also holds if we assume that X is positive and Y is nonnegative (rather than integrable), although both sides may be $+\infty$.

(iii) **(Iterated conditioning)** *If $\mathcal{H}$ is a sub-σ algebra of $\mathcal{G}$ ($\mathcal{H}$ contains less information than $\mathcal{G}$) and X is an integrable random variable, then*

$$\mathbb{E}\big[\mathbb{E}[X|\mathcal{G}]\big|\mathcal{H}\big] = \mathbb{E}\big[X\big|\mathcal{H}\big]. \tag{2.3.20}$$

This equation also holds if we assume that X is nonnegative (rather than integrable), although both sides may be $+\infty$.

(iv) **(Independence)** *If X is integrable and independent of $\mathcal{G}$, then*

$$\mathbb{E}[X|\mathcal{G}] = \mathbb{E}X. \tag{2.3.21}$$

This equation also holds if we assume that X is nonnegative (rather than integrable), although both sides may be $+\infty$.

(v) **(Conditional Jensen's inequality)** *If $\varphi(x)$ is a convex function of a dummy variable x and X is integrable, then*

$$\mathbb{E}\big[\varphi(X)|\mathcal{G}\big] \geq \varphi\big(E[X|\mathcal{G}]\big). \tag{2.3.22}$$

DISCUSSION AND SKETCH OF PROOF: We take each of these properties in turn.

(i) Linearity allows us to separate the estimation of random variables into estimation of separate pieces and then add the estimates of the pieces to estimate the whole. To verify that $\mathbb{E}[c_1X + c_2Y|\mathcal{G}]$ is given by the right-hand side of (2.3.18), we observe that the right-hand side is $\mathcal{G}$-measurable because $\mathbb{E}[X|\mathcal{G}]$ and $\mathbb{E}[Y|\mathcal{G}]$ are $\mathcal{G}$-measurable and then must check the partial-averaging property (ii) of Definition 2.3.1. Using the fact that $E[X|\mathcal{G}]$ and $E[Y|\mathcal{G}]$ themselves satisfy the partial-averaging property, we have for every $A \in \mathcal{G}$ that

$$\begin{aligned}
&\int_A \big(c_1\mathbb{E}[X|\mathcal{G}](\omega) + c_2\mathbb{E}[Y|\mathcal{G}](\omega)\big)\,d\mathbb{P}(\omega)\\
&= c_1\int_A \mathbb{E}[X|\mathcal{G}](\omega)\,d\mathbb{P}(\omega) + c_2\int_A \mathbb{E}[Y|\mathcal{G}](\omega)\,d\mathbb{P}(\omega)\\
&= c_1\int_A X(\omega)\,d\mathbb{P}(\omega) + c_2\int_A Y(\omega)\,d\mathbb{P}(\omega)\\
&= \int_A \big(c_1X(\omega) + c_2Y(\omega)\big)\,d\mathbb{P}(\omega),
\end{aligned}$$

which shows that $c_1\mathbb{E}[X|\mathcal{G}]+c_2\mathbb{E}[Y|\mathcal{G}]$ satisfies the partial-averaging property that characterizes $\mathbb{E}[c_1X+c_2Y|\mathcal{G}]$ and hence is $\mathbb{E}[c_1X+c_2Y|\mathcal{G}]$.

(ii) Taking out what is known permits us to remove X from the estimation problem if its value can be determined from the information in $\mathcal{G}$. To estimate XY, it suffices to estimate Y alone and then multiply the estimate by X. To prove (2.3.19), we observe first that $X\mathbb{E}[Y|\mathcal{G}]$ is $\mathcal{G}$-measurable because both X and $\mathbb{E}[Y|\mathcal{G}]$ are $\mathcal{G}$-measurable. We must check the partial-averaging property.

Let us first consider the case when X is a $\mathcal{G}$-measurable indicator random variable (i.e., $X=\mathbb{I}_B$, where B is a set in $\mathcal{G}$). Using the fact that $\mathbb{E}[Y|\mathcal{G}]$ itself satisfies the partial-averaging property, we have for every set $A\in\mathcal{G}$ that

$$\begin{aligned}\int_A X(\omega)\mathbb{E}[Y|\mathcal{G}](\omega)\,d\mathbb{P}(\omega) &= \int_{A\cap B}\mathbb{E}[Y|\mathcal{G}](\omega)\,d\mathbb{P}(\omega)\\ &= \int_{A\cap B} Y(\omega)\,d\mathbb{P}(\omega)\\ &= \int_A X(\omega)Y(\omega)\,d\mathbb{P}(\omega). \qquad (2.3.23)\end{aligned}$$

Having proved (2.3.23) for $\mathcal{G}$-measurable indicator random variables X, we may use the standard machine developed in the proof of Theorem 1.5.1 of Chapter 1 to obtain this equation for all $\mathcal{G}$-measurable random variables X for which XY is integrable. This shows that $X\mathbb{E}[Y|\mathcal{G}]$ satisfies the partial-averaging condition that characterizes $\mathbb{E}[XY|\mathcal{G}]$, and hence $X\mathbb{E}[Y|\mathcal{G}]$ is the conditional expectation $\mathbb{E}[XY|\mathcal{G}]$.

(iii) If we estimate X based on the information in $\mathcal{G}$ and then estimate the estimate based on the smaller amount of information in $\mathcal{H}$, we obtain the random variable we would have gotten by estimating X directly based on the smaller amount of information in $\mathcal{H}$. To prove this, we observe first that $\mathbb{E}[X|\mathcal{H}]$ is $\mathcal{H}$-measurable by definition. The partial-averaging property that characterizes $\mathbb{E}\big[\mathbb{E}[X|\mathcal{G}]\big|\mathcal{H}\big]$ is

$$\int_A \mathbb{E}\big[\mathbb{E}[X|\mathcal{G}]\big|\mathcal{H}\big](\omega)\,d\mathbb{P}(\omega) = \int_A \mathbb{E}[X|\mathcal{G}](\omega)\,\mathbb{P}(\omega) \text{ for all } A\in\mathcal{H}.$$

In order to prove (2.3.20), we must show that we can replace $\mathbb{E}\big[\mathbb{E}[X|\mathcal{G}]\big|\mathcal{H}\big]$ on the left-hand side of this equation by $\mathbb{E}[X|\mathcal{H}]$. But when $A\in\mathcal{H}$, it is also in $\mathcal{G}$, and the partial-averaging properties for $\mathbb{E}[X|\mathcal{H}]$ and $\mathbb{E}[X|\mathcal{G}]$ imply

$$\int_A \mathbb{E}[X|\mathcal{H}](\omega)\,d\mathbb{P}(\omega) = \int_A X(\omega)\,d\mathbb{P}(\omega) = \int_A \mathbb{E}[X|\mathcal{G}](\omega)\,d\mathbb{P}(\omega).$$

This shows that $\mathbb{E}[X|\mathcal{H}]$ satisfies the partial-averaging property that characterizes $\mathbb{E}\big[\mathbb{E}[X|\mathcal{G}]\big|\mathcal{H}\big]$, and hence $\mathbb{E}[X|\mathcal{H}]$ is $\mathbb{E}\big[\mathbb{E}[X|\mathcal{G}]\big|\mathcal{H}\big]$.

(iv) If X is independent of the information in $\mathcal{G}$, then the best estimate we can give of X is its expected value. This is also the estimate we would give based on no information. To prove this, we observe first that $\mathbb{E}X$ is $\mathcal{G}$-measurable.

Indeed, $\mathbb{E}X$ is not random and so is measurable with respect to every σ-algebra. We need to verify that $\mathbb{E}X$ satisfies the partial-averaging property that characterizes $\mathbb{E}[X|\mathcal{G}]$; i.e.,

$$\int_A \mathbb{E}X \, d\mathbb{P}(\omega) = \int_A X(\omega) \, d\mathbb{P}(\omega) \text{ for all } A \in \mathcal{G}. \tag{2.3.24}$$

Let us consider first the case when X is an indicator random variable independent of $\mathcal{G}$ (i.e., $X = \mathbb{I}_B$, where the set B is independent of $\mathcal{G}$). For all $A \in \mathcal{G}$, we have then

$$\int_A X(\omega) \, d\mathbb{P}(\omega) = \mathbb{P}(A \cap B) = \mathbb{P}(A) \cdot \mathbb{P}(B) = \mathbb{P}(A)\mathbb{E}X = \int_A \mathbb{E}X \, d\mathbb{P}(\omega),$$

and (2.3.24) holds. We complete the proof using the standard machine developed in the proof of Theorem 1.5.1 of Chapter 1.
(v) Using the linearity of conditional expectations, we can repeat the proof of Theorem 2.2.5 of Chapter 2 to prove the conditional Jensen's inequality. □

We note that $\mathbb{E}[X|\mathcal{G}]$ is an unbiased estimator of X:

$$\mathbb{E}\big(\mathbb{E}[X|\mathcal{G}]\big) = \mathbb{E}X. \tag{2.3.25}$$

This equality is just the partial-averaging property (2.3.17) with $A = \Omega$.

Example 2.3.3. Let X and Y be a pair of jointly normal random variables with joint density (2.2.17). As in Example 2.2.12, define $W = Y - \frac{\rho\sigma_2}{\sigma_1}X$ so that X and W are independent and (2.2.19) holds:

$$Y = \frac{\rho\sigma_2}{\sigma_1}X + W. \tag{2.2.19}$$

In Example 2.2.12, we saw that W is normal with mean $\mu_3 = \mu_2 - \frac{\rho\sigma_2\mu_1}{\sigma_1}$ and variance $\sigma_3^2 = (1-\rho^2)\sigma_2^2$. Let us take the conditioning σ-algebra to be $\mathcal{G} = \sigma(X)$. (When $\mathcal{G}$ is generated by a random variable X, it is customary to write $\mathbb{E}[\cdots|X]$ rather than $\mathbb{E}[\cdots|\sigma(X)]$.) We estimate Y, based on X, using (2.2.19) above and properties (i) (Linearity) and (iv) (Independence) from Theorem 2.3.2 to get the linear regression equation

$$\mathbb{E}[Y|X] = \frac{\rho\sigma_2}{\sigma_1}X + \mathbb{E}W = \frac{\rho\sigma_2}{\sigma_1}(X - \mu_1) + \mu_2. \tag{2.3.26}$$

Note that the right-hand side of (2.3.26) is random but is $\sigma(X)$-measurable (i.e., if we know the information in $\sigma(X)$, which is the same as knowing the value of X, then we can evaluate $\mathbb{E}[Y|X]$). Subtracting (2.3.26) from (2.2.19), we see that the error made by the estimator is

$$Y - \mathbb{E}[Y|X] = W - \mathbb{E}W.$$

The error is random, with expected value zero (the estimator is unbiased), and is independent of the estimate $\mathbb{E}[Y|X]$ (because $\mathbb{E}[Y|X]$ is $\sigma(X)$-measurable and W is independent of $\sigma(X)$). The independence between the error and the conditioning random variable X is a consequence of the joint normality in the example. In general, the error and the conditioning random variable are uncorrelated, but not necessarily independent; see Exercise 2.8. □

The Independence Lemma, Lemma 2.5.3 of Chapter 2 of Volume I, now takes the following more general form.

Lemma 2.3.4 (Independence). *Let $(\Omega, \mathcal{F}, \mathbb{P})$ be a probability space, and let $\mathcal{G}$ be a sub-σ-algebra of $\mathcal{F}$. Suppose the random variables $X_1, \dots, X_K$ are $\mathcal{G}$-measurable and the random variables $Y_1, \dots, Y_L$ are independent of $\mathcal{G}$. Let $f(x_1, \dots, x_K, y_1, \dots, y_L)$ be a function of the dummy variables $x_1, \dots, x_K$ and $y_1, \dots, y_L$, and define*

$$g(x_1, \dots, x_K) = \mathbb{E}f(x_1, \dots, x_K, Y_1, \dots, Y_L). \tag{2.3.27}$$

Then

$$\mathbb{E}\big[f(X_1, \dots, X_K, Y_1, \dots, Y_L)\big|\mathcal{G}\big] = g(X_1, \dots, X_K). \tag{2.3.28}$$

As with Lemma 2.5.3 of Volume I, the idea here is that since the information in $\mathcal{G}$ is sufficient to determine the values of $X_1, \dots, X_K$, we should hold these random variables constant when estimating $f(X_1, \dots, X_K, Y_1, \dots, Y_K)$. The other random variables, $Y_1, \dots, Y_L$, are independent of $\mathcal{G}$, and so we should integrate them out without regard to the information in $\mathcal{G}$. These two steps, holding $X_1, \dots, X_K$ constant and integrating out $Y_1, \dots, Y_L$, are accomplished by (2.3.27). We get an estimate that depends on the values of $X_1, \dots, X_K$ and, to capture this fact, we replaced the dummy (nonrandom) variables $x_1, \dots, x_K$ by the random variables $X_1, \dots, X_K$ at the last step. Although Lemma 2.5.3 of Volume I has a relatively simple proof, the proof of Lemma 2.3.4 requires some measure-theoretic ideas beyond the scope of this text, and will not be given.

Example 2.3.3 continued. Continuing with the notation of Example 2.3.3, suppose we want to estimate some function $f(x, y)$ of the random variables X and Y based on knowledge of X. We cannot use the Independence Lemma directly because X and Y are not independent. However, we can write Y as $Y = \frac{\rho\sigma_2}{\sigma_1}X + W$. Because X is $\sigma(X)$-measurable, W is independent of $\sigma(X)$ and W is normal with mean μ_3 and variance σ_3^2, the Independence Lemma tells us how to compute $\mathbb{E}[f(X, Y)|X]$. We should first replace the random variable X by a dummy variable x and then take the expectation (i.e., integrate with respect to the distribution of W). Thus, we define

$$\begin{aligned} g(x) &= \mathbb{E}f\left(x, \frac{\rho\sigma_1}{\sigma_1}x + W\right) \\ &= \frac{1}{\sigma_3\sqrt{2\pi}}\int_{-\infty}^{\infty} f\left(x, \frac{\rho\sigma_1}{\sigma_2}x + w\right)\exp\left\{-\frac{(w-\mu_3)^2}{2\sigma_3^2}\right\}dw. \end{aligned}$$

Then

$$\mathbb{E}[f(X,Y)|X] = g(X).$$

Our final answer is random but $\sigma(X)$-measurable, as it should be. □

We have all the tools required to introduce martingales and Markov processes in a continuous-time framework. The definitions are provided below. Examples will be given after we construct Brownian motion and Itô integrals in the next chapters.

Definition 2.3.5. *Let $(\Omega, \mathcal{F}, \mathbb{P})$ be a probability space, let T be a fixed positive number, and let $\mathcal{F}(t)$, $0 \le t \le T$, be a filtration of sub-σ-algebras of $\mathcal{F}$. Consider an adapted stochastic process $M(t)$, $0 \le t \le T$.*

(i) If

$$\mathbb{E}[M(t)|\mathcal{F}(s)] = M(s) \text{ for all } 0 \le s \le t \le T,$$

we say this process is a martingale. *It has no tendency to rise or fall.*

(ii) If

$$\mathbb{E}[M(t)|\mathcal{F}(s)] \ge M(s) \text{ for all } 0 \le s \le t \le T,$$

we say this process is a submartingale. *It has no tendency to fall; it may have a tendency to rise.*

(iii) If

$$\mathbb{E}[M(t)|\mathcal{F}(s)] \le M(s) \text{ for all } 0 \le s \le t \le T,$$

we say this process is a supermartingale. *It has no tendency to rise; it may have a tendency to fall.*

Definition 2.3.6. *Let $(\Omega, \mathcal{F}, \mathbb{P})$ be a probability space, let T be a fixed positive number, and let $\mathcal{F}(t)$, $0 \le t \le T$, be a filtration of sub-σ-algebras of $\mathcal{F}$. Consider an adapted stochastic process $X(t)$, $0 \le t \le T$. Assume that for all $0 \le s \le t \le T$ and for every nonnegative, Borel-measurable function f, there is another Borel-measurable function g such that*

$$\mathbb{E}[f(X(t))|\mathcal{F}(s)] = g(X(s)). \tag{2.3.29}$$

Then we say that the X is a Markov process.

Remark 2.3.7. In Definition 2.3.6, the function f is permitted to depend on t, and the function g will depend on s. These dependencies are not indicated in (2.3.29) because we wish there to emphasize how the dependence on the sample point ω works (i.e., the right-hand side depends on ω only through the random variable $X(s)$). If we indicate the dependence on time by writing $f(t,x)$ rather than $f(x)$, we can write $f(s,x)$ rather than $g(x)$ (we do not need different symbols f and g because the time variables t and s indicate we are dealing with different functions of x at the different times) and can rewrite (2.3.29) as

$$\mathbb{E}[f(t,X(t))|\mathcal{F}(s)] = f(s,X(s)),\ 0 \le s \le t \le T. \tag{2.3.30}$$

Ultimately, we shall see that when we regard $f(t,x)$ as a function of two variables this way, (2.3.30) implies that it satisfies a partial differential equation. This partial differential equation gives us a way to determine $f(s,x)$ if we know $f(t,x)$. The Black-Scholes-Merton partial differential equation is a special case of this. □

2.4 Summary

In measure-theoretic probability, information is modeled using σ-algebras. The information associated with a σ-algebra $\mathcal{G}$ can be thought of as follows. A random experiment is performed and an outcome ω is determined, but the value of ω is not revealed. Instead, for each set in the σ-algebra $\mathcal{G}$, we are told whether ω is in the set. The more sets there are on $\mathcal{G}$, the more information this provides. If $\mathcal{G}$ is the trivial σ-algebra containing only $\emptyset$ and Ω, this provides no information.

A random variable X is *$\mathcal{G}$-measurable* if and only if the set $\{X \in B\} = \{\omega \in \Omega; X(\omega) \in B\}$ is in $\mathcal{G}$ for every Borel subset of $\mathbb{R}$. In this case, the information in $\mathcal{G}$ is enough to determine the value of the random variable $X(\omega)$, even though it may not be enough to determine the value ω of the outcome of the random experiment.

At the other extreme, the information in a σ-algebra $\mathcal{G}$ may be irrelevant to the determination of the value of X. In this case, we say that $\mathcal{G}$ and X are *independent*. This idea is captured mathematically by Definition 2.2.3, which says that X and $\mathcal{G}$ are independent if, for every set $A \in \mathcal{G}$ and every Borel subset B of $\mathbb{R}$, we have

$$\mathbb{P}\{\omega \in \Omega; \omega \in A \text{ and } X(\omega) \in B\} = \mathbb{P}(A) \cdot \mathbb{P}\{\omega \in \Omega; X(\omega) \in B\}.$$

Two random variables X and Y are independent if and only if the *σ algebra generated by X*, defined to be the collection of sets of the form $\{X \in B\}$, is independent of the σ-algebra generated by Y. In other words, X and Y are independent if and only if

$$\mathbb{P}\{X \in B \text{ and } Y \in C\} = \mathbb{P}\{X \in B\} \cdot \mathbb{P}\{X \in C\} \text{ for all } B \in \mathcal{B}(\mathbb{R}), C \in \mathcal{B}(\mathbb{R}),$$

where $\mathcal{B}(\mathbb{R})$ denotes the σ-algebra of Borel subsets of $\mathbb{R}$. There are several equivalent ways to describe independence between two random variables having to do with factoring the joint cumulative distribution function, factoring the joint moment-generating function, and factoring the joint density (if there is a joint density). These are set out in Theorem 2.2.7. Independence implies uncorrelatedness, but uncorrelated random variables do not need to be independent. Jointly normally distributed random variables (Definition 2.2.11) are uncorrelated if and only if they are independent, but normally distributed random variables do not need to be *jointly* normal.

Often we find ourselves between the two extremes of random variables X that are $\mathcal{G}$-measurable and random variables X that are independent of $\mathcal{G}$. In such a case, the information in $\mathcal{G}$ is relevant to the determination of the value of X but is not sufficient to completely determine it. We then want to use the information in $\mathcal{G}$ to estimate X. We denote our estimate by $\mathbb{E}[X|\mathcal{G}]$ and call this the *conditional expectation of* X *given* $\mathcal{G}$. This is itself a random variable, but one that is $\mathcal{G}$-measurable (i.e., one that we can evaluate using only the information in $\mathcal{G}$). To be sure this is a good estimate of X, we require that it satisfy the *partial-averaging property* (see Definition 2.3.1(ii)):

$$\int_A \mathbb{E}[X|\mathcal{G}](\omega)\,d\mathbb{P}(\omega) = \int_A X(\omega)\,d\mathbb{P}(\omega) \text{ for every } A \in \mathcal{G}.$$

Conditional expectations behave in many ways like expectations, except that expectations do not depend on ω and conditional expectations do. The principal properties of conditional expectations are provided in Theorem 2.3.2, and these are reported briefly here.
Linearity: $\mathbb{E}[c_1X + c_2Y|\mathcal{G}] = c_1\mathbb{E}[X|\mathcal{G}] + c_2\mathbb{E}[Y|\mathcal{G}]$.
Taking out what is known: $\mathbb{E}[XY|\mathcal{G}] = X\mathbb{E}[Y|\mathcal{G}]$ if X is $\mathcal{G}$-measurable.
Iterated conditioning: $\mathbb{E}\big[\mathbb{E}[X|\mathcal{G}]\big|\mathcal{H}\big] = \mathbb{E}[X|\mathcal{H}]$ if $\mathcal{H}$ is a sub-σ-algebra of $\mathcal{G}$.
Independence: $\mathbb{E}[X|\mathcal{G}] = \mathbb{E}X$ if X is independent of $\mathcal{G}$.
Jensen's inequality: $\mathbb{E}[\varphi(X)|\mathcal{G}] \geq \varphi(\mathbb{E}[X|\mathcal{G}])$ if φ is convex.

In continuous-time finance, we work within the framework of a probability space $(\Omega, \mathcal{F}, \mathbb{P})$. We normally have a fixed final time T and then have a *filtration*, which is a collection of σ-algebras $\{\mathcal{F}(t); 0 \leq t \leq T\}$ indexed by the time variable t. We interpret $\mathcal{F}(t)$ as the information available at time t. For $0 \leq s \leq t \leq T$, every set in $\mathcal{F}(s)$ is also in $\mathcal{F}(t)$. In other words, information increases over time. Within this context, an *adapted stochastic process* is a collection of random variables $\{X(t); 0 \leq t \leq T\}$ also indexed by time such that, for every t, $X(t)$ is $\mathcal{F}(t)$-measurable; the information at time t is sufficient to evaluate the random variable $X(t)$. We think of $X(t)$ as the price of some asset at time t and $\mathcal{F}(t)$ as the information obtained by watching all the prices in the market up to time t.

Two important classes of adapted stochastic processes are *martingales* and *Markov processes*. These are defined in Definitions 2.3.5 and 2.3.6, respectively. A martingale has the property that

$$\mathbb{E}[M(t)|\mathcal{F}(s)] = M(s) \text{ for all } 0 \leq s \leq t \leq T.$$

If $\mathbb{E}[M(t)|\mathcal{F}(s)] \geq M(s)$ when $0 \leq s \leq t \leq T$, we have a *submartingale*. If the inequality is reversed, we have a *supermartingale*. A Markov process has the property that whenever $0 \leq s \leq t \leq T$ and we are given a function f, there is another function g such that

$$\mathbb{E}[f(X(t))|\mathcal{F}(s)] = g(X(s)).$$

The important feature here is that the estimate of $f(X(t))$ made at time s depends only on the process value $X(s)$ at time s and not on the path of the process before time s.

A useful tool for establishing that a process is Markov is the *Independence Lemma*, Lemma 2.3.4. The simplest version of this says that if X is a $\mathcal{G}$-measurable random variable and Y is independent of $\mathcal{G}$, then

$$\mathbb{E}[f(X,Y)|\mathcal{G}] = g(X),$$

where $g(x) = \mathbb{E}f(x,Y)$.

2.5 Notes

In the measure-theoretic view of probability theory, a conditional expectation is itself a random variable, measurable with respect to the conditioning σ-algebra. This point of view is indispensable for treating the rather complicated conditional expectations that arise in martingale theory. It was invented by Kolmogorov [104]. The term *martingale* was apparently first used by Ville [158], who assigned the name to a betting strategy. The concept dates back to 1934 work of Lévy. The first systematic treatment of martingales was provided by Doob [53].

2.6 Exercises

Exercise 2.1. Let $(\Omega, \mathcal{F}, \mathbb{P})$ be a general probability space, and suppose a random variable X on this space is measurable with respect to the trivial σ-algebra $\mathcal{F}_0 = \{\emptyset, \Omega\}$. Show that X is not random (i.e., there is a constant c such that $X(\omega) = c$ for all $\omega \in \Omega$). Such a random variable is called *degenerate*.

Exercise 2.2. Independence of random variables can be affected by changes of measure. To illustrate this point, consider the space of two coin tosses $\Omega_2 = \{HH, HT, TH, TT\}$, and let stock prices be given by

$$S_0 = 4,\ S_1(H) = 8,\ S_1(T) = 2,$$
$$S_2(HH) = 16,\ S_2(HT) = S_2(TH) = 4,\ S_2(TT) = 1.$$

Consider two probability measures given by

$$\widetilde{\mathbb{P}}(HH) = \tfrac{1}{4},\ \widetilde{\mathbb{P}}(HT) = \tfrac{1}{4},\ \widetilde{\mathbb{P}}(TH) = \tfrac{1}{4},\ \widetilde{\mathbb{P}}(TT) = \tfrac{1}{4},$$
$$\mathbb{P}(HH) = \tfrac{4}{9},\ \mathbb{P}(HT) = \tfrac{2}{9},\ \mathbb{P}(TH) = \tfrac{2}{9},\ \mathbb{P}(TT) = \tfrac{1}{9}.$$

Define the random variable

$$X = \begin{cases} 1 \text{ if } S_2 = 4, \\ 0 \text{ if } S_2 \neq 4. \end{cases}$$

(i) List all the sets in $\sigma(X)$.
(ii) List all the sets in $\sigma(S_1)$.
(iii) Show that $\sigma(X)$ and $\sigma(S_1)$ are independent under the probability measure $\widetilde{\mathbb{P}}$.
(iv) Show that $\sigma(X)$ and $\sigma(S_1)$ are not independent under the probability measure $\mathbb{P}$.
(v) Under $\mathbb{P}$, we have $\mathbb{P}\{S_1 = 8\} = \frac{2}{3}$ and $\mathbb{P}\{S_1 = 2\} = \frac{1}{3}$. Explain intuitively why, if you are told that $X = 1$, you would want to revise your estimate of the distribution of S_1.

Exercise 2.3 (Rotating the axes). Let X and Y be independent standard normal random variables. Let θ be a constant, and define random variables

$$V = X\cos\theta + Y\sin\theta \text{ and } W = -X\sin\theta + Y\cos\theta.$$

Show that V and W are independent standard normal random variables.

Exercise 2.4. In Example 2.2.8, X is a standard normal random variable and Z is an independent random variable satisfying

$$\mathbb{P}\{Z = 1\} = \mathbb{P}\{Z = -1\} = \frac{1}{2}.$$

We defined $Y = XZ$ and showed that Y is standard normal. We established that although X and Y are uncorrelated, they are not independent. In this exercise, we use moment-generating functions to show that Y is standard normal and X and Y are not independent.

(i) Establish the joint moment-generating function formula

$$\mathbb{E}e^{uX+vY} = e^{\frac{1}{2}(u^2+v^2)} \cdot \frac{e^{uv} + e^{-uv}}{2}.$$

(ii) Use the formula above to show that $\mathbb{E}e^{vY} = e^{\frac{1}{2}v^2}$. This is the moment-generating function for a standard normal random variable, and thus Y must be a standard normal random variable.
(iii) Use the formula in (i) and Theorem 2.2.7(iv) to show that X and Y are not independent.

Exercise 2.5. Let (X, Y) be a pair of random variables with joint density function

$$f_{X,Y}(x,y) = \begin{cases} \frac{2|x|+y}{\sqrt{2\pi}} \exp\left\{-\frac{(2|x|+y)^2}{2}\right\} & \text{if } y \geq -|x|, \\ 0 & \text{if } y < -|x|. \end{cases}$$

Show that X and Y are standard normal random variables and that they are uncorrelated but not independent.

Exercise 2.6. Consider a probability space Ω with four elements, which we call a, b, c, and d (i.e., $\Omega = \{a, b, c, d\}$). The σ-algebra $\mathcal{F}$ is the collection of all subsets of Ω; i.e., the sets in $\mathcal{F}$ are

$$\begin{aligned}&\Omega,\ \{a,b,c\},\ \{a,b,d\},\ \{a,c,d\},\ \{b,c,d\},\\&\{a,b\},\ \{a,c\},\ \{a,d\},\ \{b,c\},\ \{b,d\},\ \{c,d\},\\&\{a\},\ \{b\},\ \{c\},\ \{d\},\ \emptyset.\end{aligned}$$

We define a probability measure $\mathbb{P}$ by specifying that

$$\mathbb{P}\{a\} = \frac{1}{6},\ \mathbb{P}\{b\} = \frac{1}{3},\ \mathbb{P}\{c\} = \frac{1}{4},\ \mathbb{P}\{d\} = \frac{1}{4},$$

and, as usual, the probability of every other set in $\mathcal{F}$ is the sum of the probabilities of the elements in the set, e.g., $\mathbb{P}\{a,b,c\} = \mathbb{P}\{a\} + \mathbb{P}\{b\} + \mathbb{P}\{c\} = \frac{3}{4}$.

We next define two random variables, X and Y, by the formulas

$$\begin{aligned}&X(a) = 1, X(b) = 1, X(c) = -1, X(d) = -1,\\&Y(a) = 1, Y(b) = -1, Y(c) = 1, Y(d) = -1.\end{aligned}$$

We then define $Z = X + Y$.

(i) List the sets in $\sigma(X)$.
(ii) Determine $\mathbb{E}[Y|X]$ (i.e., specify the values of this random variable for a, b, c, and d). Verify that the partial-averaging property is satisfied.
(iii) Determine $\mathbb{E}[Z|X]$. Again, verify the partial-averaging property.
(iv) Compute $\mathbb{E}[Z|X] - \mathbb{E}[Y|X]$. Citing the appropriate properties of conditional expectation from Theorem 2.3.2, explain why you get X.

Exercise 2.7. Let Y be an integrable random variable on a probability space $(\Omega, \mathcal{F}, \mathbb{P})$ and let $\mathcal{G}$ be a sub-σ-algebra of $\mathcal{F}$. Based on the information in $\mathcal{G}$, we can form the estimate $\mathbb{E}[Y|\mathcal{G}]$ of Y and define the error of the estimation $\text{Err} = Y - \mathbb{E}[Y|\mathcal{G}]$. This is a random variable with expectation zero and some variance $\text{Var}(\text{Err})$. Let X be some other $\mathcal{G}$-measurable random variable, which we can regard as another estimate of Y. Show that

$$\text{Var}(\text{Err}) \leq \text{Var}(Y - X).$$

In other words, the estimate $\mathbb{E}[Y|\mathcal{G}]$ minimizes the variance of the error among all estimates based on the information in $\mathcal{G}$. (Hint: Let $\mu = \mathbb{E}(Y - X)$. Compute the variance of $Y - X$ as

$$\mathbb{E}\left[(Y - X - \mu)^2\right] = \mathbb{E}\left[\left((Y - \mathbb{E}[Y|\mathcal{G}]) + (\mathbb{E}[Y|\mathcal{G}] - X - \mu)\right)^2\right].$$

Multiply out the right-hand side and use iterated conditioning to show the cross-term is zero.)

Exercise 2.8. Let X and Y be integrable random variables on a probability space $(\Omega, \mathcal{F}, \mathbb{P})$. Then $Y = Y_1 + Y_2$, where $Y_1 = \mathbb{E}[Y|X]$ is $\sigma(X)$-measurable and $Y_2 = Y - E[Y|X]$. Show that Y_2 and X are uncorrelated. More generally, show that Y_2 is uncorrelated with every $\sigma(X)$-measurable random variable.

Exercise 2.9. Let X be a random variable.

(i) Give an example of a probability space $(\Omega, \mathcal{F}, \mathbb{P})$, a random variable X defined on this probability space, and a function f so that the σ-algebra generated by $f(X)$ is not the trivial σ-algebra $\{\emptyset, \Omega\}$ but is strictly smaller than the σ-algebra generated by X.
(ii) Can the σ-algebra generated by $f(X)$ ever be strictly larger than the σ-algebra generated by X?

Exercise 2.10. Let X and Y be random variables (on some unspecified probability space $(\Omega, \mathcal{F}, \mathbb{P})$), assume they have a joint density $f_{X,Y}(x,y)$, and assume $\mathbb{E}|Y| < \infty$. In particular, for every Borel subset C of $\mathbb{R}^2$, we have

$$\mathbb{P}\{(X,Y) \in C\} = \int_C f_{X,Y}(x,y)\, dx\, dy.$$

In elementary probability, one learns to compute $\mathbb{E}[Y|X = x]$, which is a *nonrandom* function of the *dummy variable* x, by the formula

$$\mathbb{E}[Y|X = x] = \int_{-\infty}^{\infty} y f_{Y|X}(y|x) dy, \tag{2.6.1}$$

where $f_{Y|X}(y|x)$ is the *conditional density* defined by

$$f_{Y|X}(y|x) = \frac{f_{X,Y}(x,y)}{f_X(x)}.$$

The denominator in this expression, $f_X(x) = \int_{-\infty}^{\infty} f_{X,Y}(x,\eta) d\eta$, is the *marginal density* of X, and we must assume it is strictly positive for every x. We introduce the symbol $g(x)$ for the function $\mathbb{E}[Y|X = x]$ defined by (2.6.1); i.e.,

$$g(x) = \int_{-\infty}^{\infty} y f_{Y|X}(y|x) dy = \int_{-\infty}^{\infty} \frac{y f_{X,Y}(x,y)}{f_X(x)}\, dy.$$

In measure-theoretic probability, conditional expectation is a *random variable* $\mathbb{E}[Y|X]$. This exercise is to show that when there is a joint density for (X,Y), this random variable can be obtained by substituting the random variable X in place of the dummy variable x in the function $g(x)$. In other words, this exercise is to show that

$$\mathbb{E}[Y|X] = g(X).$$

(We introduced the symbol $g(x)$ in order to avoid the mathematically confusing expression $E[Y|X = X]$.)

Since $g(X)$ is obviously $\sigma(X)$-measurable, to verify that $\mathbb{E}[Y|X] = g(X)$, we need only check that the partial-averaging property is satisfied. For every Borel-measurable function h mapping $\mathbb{R}$ to $\mathbb{R}$ and satisfying $\mathbb{E}|h(X)| < \infty$, we have

$$\mathbb{E}h(X) = \int_{-\infty}^{\infty} h(x) f_X(x)dx. \tag{2.6.2}$$

This is Theorem 1.5.2 in Chapter 1. Similarly, if h is a function of both x and y, then

$$\mathbb{E}h(X,Y) = \int_{-\infty}^{\infty}\int_{-\infty}^{\infty} h(x,y) f_{X,Y}(x,y)dxdy \tag{2.6.3}$$

whenever (X,Y) has a joint density $f_{X,Y}(x,y)$. You may use both (2.6.2) and (2.6.3) in your solution to this problem.

Let A be a set in $\sigma(X)$. By the definition of $\sigma(X)$, there is a Borel subset B of $\mathbb{R}$ such that $A = \{\omega \in \Omega; X(\omega) \in B\}$ or, more simply, $A = \{X \in B\}$. Show the partial-averaging property

$$\int_A g(X)d\mathbb{P} = \int_A Y d\mathbb{P}.$$

Exercise 2.11. (i) Let X be a random variable on a probability space $(\Omega, \mathcal{F}, \mathbb{P})$, and let W be a nonnegative $\sigma(X)$-measurable random variable. Show there exists a function g such that $W = g(X)$. (Hint: Recall that every set in $\sigma(X)$ is of the form $\{X \in B\}$ for some Borel set $B \subset \mathbb{R}$. Suppose first that W is the indicator of such a set, and then use the standard machine.)

(ii) Let X be a random variable on a probability space $(\Omega, \mathcal{F}, \mathbb{P})$, and let Y be a nonnegative random variable on this space. We do not assume that X and Y have a joint density. Nonetheless, show there is a function g such that $\mathbb{E}[Y|X] = g(X)$.

3

Brownian Motion

3.1 Introduction

In this chapter, we define Brownian motion and develop its basic properties. The definition of Brownian motion is provided in Section 3.3. Section 3.2 precedes it to give some intuition. For us, the most important properties of Brownian motion are that it is a martingale (Theorem 3.3.4) and that it accumulates quadratic variation at rate one per unit time (Theorem 3.4.3). The notion of quadratic variation is profound. It makes stochastic calculus different from ordinary calculus. For this reason, we begin already in Section 3.2 to talk about it.

Sections 3.5–3.7 develop properties of Brownian motion we shall need later but not in the development of stochastic calculus in Chapter 4. Therefore, the reader can go to Chapter 4 after completing Section 3.4. The Markov property is the concept used to relate stochastic calculus to partial differential equations. For Brownian motion, this property is presented in Section 3.5. The first passage time of Brownian motion to a level is presented in Section 3.6 and used in Chapter 8 to analyze a perpetual American put on a geometric Brownian motion. This is in the spirit of the perpetual American put analysis for the binomial model, which is given in Section 5.4 of Volume I. The reflection principle for Brownian motion developed in Section 3.7 is used in Chapter 7 to price exotic options.

3.2 Scaled Random Walks

3.2.1 Symmetric Random Walk

To create a Brownian motion, we begin with a symmetric random walk, one path of which is shown in Figure 3.2.1. To construct a symmetric random walk, we repeatedly toss a fair coin (p, the probability of H on each toss, and $q = 1 - p$, the probability of T on each toss, are both equal to $\frac{1}{2}$). We denote

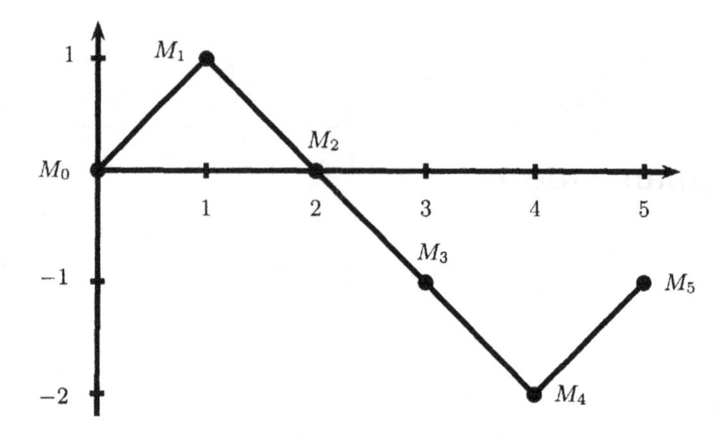

Fig. 3.2.1. Five steps of a random walk.

the successive outcomes of the tosses by $\omega = \omega_1\omega_2\omega_3 \ldots$. In other words, ω is the infinite sequence of tosses, and ω_n is the outcome of the nth toss. Let

$$X_j = \begin{cases} 1 & \text{if } \omega_j = H, \\ -1 & \text{if } \omega_j = T, \end{cases} \tag{3.2.1}$$

and define $M_0 = 0$,

$$M_k = \sum_{j=1}^{k} X_j, \quad k = 1, 2, \ldots. \tag{3.2.2}$$

The process M_k, $k = 0, 1, 2, \ldots$ is a *symmetric random walk*. With each toss, it either steps up one unit or down one unit, and each of the two possibilities is equally likely.

3.2.2 Increments of the Symmetric Random Walk

A random walk has *independent increments*. This means that if we choose nonnegative integers $0 = k_0 < k_1 < \cdots < k_m$, the random variables

$$M_{k_1} = (M_{k_1} - M_{k_0}),\ (M_{k_2} - M_{k_1}), \ldots, (M_{k_m} - M_{k_{m-1}})$$

are independent. Each of these random variables,

$$M_{k_{i+1}} - M_{k_i} = \sum_{j=k_i+1}^{k_{i+1}} X_j, \tag{3.2.3}$$

is called an *increment* of the random walk. It is the change in the position of the random walk between times k_i and k_{i+1}. Increments over nonoverlapping time intervals are independent because they depend on different coin tosses.

Moreover, each increment $M_{k_{i+1}} - M_{k_i}$ has expected value 0 and variance $k_{i+1} - k_i$. It is easy to see that the expected value is zero because the expected value of each X_j appearing on the right-hand side of (3.2.3) is zero. We also have $\mathrm{Var}(X_j) = \mathbb{E}X_j^2 = 1$, and since the different X_j are independent, we have from (3.2.3) that

$$\mathrm{Var}\big(M_{k_{i+1}} - M_{k_i}\big) = \sum_{j=k_i+1}^{k_{i+1}} \mathrm{Var}(X_j) = \sum_{j=k_i+1}^{k_{i+1}} 1 = k_{i+1} - k_i. \tag{3.2.4}$$

The variance of the symmetric random walk accumulates at rate one per unit time, so that the variance of the increment over any time interval k to ℓ for nonnegative integers $k < \ell$ is $\ell - k$.

3.2.3 Martingale Property for the Symmetric Random Walk

To see that the symmetric random walk is a martingale, we choose nonnegative integers $k < \ell$ and compute

$$\begin{aligned}
\mathbb{E}\big[M_\ell \big| \mathcal{F}_k\big] &= \mathbb{E}\big[(M_\ell - M_k) + M_k \big| \mathcal{F}_k\big] \\
&= \mathbb{E}\big[M_\ell - M_k \big| \mathcal{F}_k\big] + \mathbb{E}\big[M_k \big| \mathcal{F}_k\big] \\
&= \mathbb{E}\big[M_\ell - M_k \big| \mathcal{F}_k\big] + M_k \\
&= \mathbb{E}\big[M_\ell - M_k\big] + M_k \;=\; M_k.
\end{aligned} \tag{3.2.5}$$

Here we have used the notation $\mathbb{E}[\cdots|\mathcal{F}_k]$ of Chapter 2 to denote the conditional expectation based on the information at time k, which in this case is knowledge of the first k coin tosses. The second equality is a result of the linearity of conditional expectations (Theorem 2.3.2(i)). The third equality is because M_k depends only on the first k coin tosses (it is $\mathcal{F}_k$-measurable, where, in the language of Definition 2.1.5, $\mathcal{F}_k$ is the σ-algebra of information corresponding to the first k coin tosses). The fourth equality follows from independence (Theorem 2.3.2(iv)).

3.2.4 Quadratic Variation of the Symmetric Random Walk

Finally, we consider the *quadratic variation* of the symmetric random walk. The quadratic variation up to time k is defined to be

$$[M, M]_k = \sum_{j=1}^{k} \big(M_j - M_{j-1}\big)^2 = k. \tag{3.2.6}$$

Note that this is computed path-by-path. The quadratic variation up to time k along a path is computed by taking all the one-step increments $M_j - M_{j-1}$ along that path (these are equal to X_j, which is either 1 or -1, depending on the path), squaring these increments, and then summing them. Since $(M_j - M_{j-1})^2 = 1$, regardless of whether $M_j - M_{j-1}$ is 1 or -1, the sum in (3.2.6) is equal to $\sum_{j=1}^k 1 = k$, as reported in that equation.

We note that $[M, M]_k$ is the same as $\text{Var}(M_k)$ (set $k_{i+1} = k$ and $k_i = 0$ in (3.2.4)), but the computations of these two quantities are quite different. $\text{Var}(M_k)$ is computed by taking an average over all paths, taking their probabilities into account. If the random walk were not symmetric (i.e., if p were different from q), this would affect $\text{Var}(M_k)$. By contrast, $[M, M]_k$ is computed along a single path, and the probabilities of up and down steps do not enter the computation. One can compute the variance of a random walk only theoretically because it requires an average over all paths, realized and unrealized. However, from tick-by-tick price data, one can compute the quadratic variation along the realized path rather explicitly. For a random walk, there is the somewhat unusual feature that $[M, M]_k$ does not depend on the particular path chosen, but we shall see later that the quadratic variation for a random process generally does depend on the path along which it is computed.

3.2.5 Scaled Symmetric Random Walk

To approximate a Brownian motion, we speed up time and scale down the step size of a symmetric random walk. More precisely, we fix a positive integer n and define the *scaled symmetric random walk*

$$W^{(n)}(t) = \frac{1}{\sqrt{n}} M_{nt}, \tag{3.2.7}$$

provided nt is itself an integer. If nt is not an integer, we define $W^{(n)}(t)$ by linear interpolation between its values at the nearest points s and u to the left and right of t for which ns and nu are integers. We shall obtain a Brownian motion in the limit as $n \to \infty$. Figure 3.2.2 shows a simulated path of $W^{(100)}$ up to time 4; this was generated by 400 coin tosses with a step up or down of size $\frac{1}{10}$ on each coin toss.

Like the random walk, the scaled random walk has independent increments. If $0 = t_0 < t_1 < \cdots < t_m$ are such that each nt_j is an integer, then

$$\big(W^{(n)}(t_1) - W^{(n)}(t_0)\big),\ \big(W^{(n)}(t_2) - W^{(n)}(t_1)\big), \ldots, \big(W^{(n)}(t_m) - W^{(n)}(t_{m-1})\big)$$

are independent. These random variables depend on different coin tosses. For example, $W^{(100)}(0.20) - W^{(100)}(0)$ depends on the first 20 coin tosses and $W^{(100)}(0.70) - W^{(100)}(0.20)$ depends on the next 50 tosses. Furthermore, if $0 \leq s \leq t$ are such that ns and nt are integers, then

$$\mathbb{E}\big(W^{(n)}(t) - W^{(n)}(s)\big) = 0, \quad \text{Var}\big(W^{(n)}(t) - W^{(n)}(s)\big) = t - s. \tag{3.2.8}$$

This is because $W^{(n)}(t) - W^{(n)}(s)$ is the sum of $n(t-s)$ independent random variables, each with expected value zero and variance $\frac{1}{n}$. For example, $W^{(100)}(0.70) - W^{(100)}(0.20)$ is the sum of 50 independent random

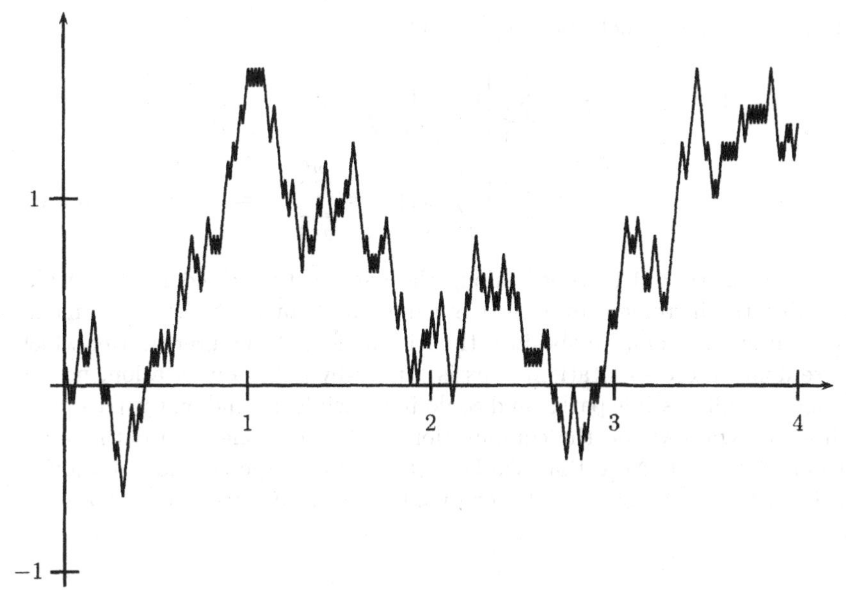

Fig. 3.2.2. A sample path of $W^{(100)}$.

variables, each of which takes the value $\frac{1}{10}$ or $-\frac{1}{10}$. Each of these random variables has expected value zero and variance $\frac{1}{100}$, so the variance of $W^{(100)}(0.70) - W^{(100)}(0.20)$ is $50 \cdot \frac{1}{100} = 0.50$.

Let $0 \leq s \leq t$ be given, and decompose $W^{(n)}(t)$ as

$$W^{(n)}(t) = \big(W^{(n)}(t) - W^{(n)}(s)\big) + W^{(n)}(s).$$

If s and t are chosen so that ns and nt are integers, then the first term on the right-hand side is independent of $\mathcal{F}(s)$, the σ-algebra of information available at time s (which is knowledge of the first ns coin tosses), and $W^{(n)}(s)$ is $\mathcal{F}(s)$-measurable (i.e., it depends only on the first ns coin tosses). We may prove the martingale property for the scaled random walk as we did for the random walk in (3.2.5):

$$\mathbb{E}\big[W^{(n)}(t)\big|\mathcal{F}(s)\big] = W^{(n)}(s) \tag{3.2.9}$$

for $0 \leq s \leq t$ such that ns and nt are integers.

Finally, we consider the *quadratic variation* of the scaled random walk. For $W^{(100)}$, the quadratic variation up to a time, say 1.37, is defined to be

$$\begin{aligned}
\big[W^{(100)}, W^{(100)}\big](1.37) &= \sum_{j=1}^{137}\left[W^{(100)}\left(\frac{j}{100}\right) - W^{(100)}\left(\frac{j-1}{100}\right)\right]^2 \\
&= \sum_{j=1}^{137}\left[\frac{1}{10}X_j\right]^2 = \sum_{j=1}^{137}\frac{1}{100} = 1.37.
\end{aligned}$$

In general, for $t \geq 0$ such that nt is an integer,

$$[W^{(n)}, W^{(n)}](t) = \sum_{j=1}^{nt} \left[W^{(n)}\left(\frac{j}{n}\right) - W^{(n)}\left(\frac{j-1}{n}\right) \right]^2$$

$$= \sum_{j=1}^{nt} \left[\frac{1}{\sqrt{n}} X_j \right]^2 = \sum_{j=1}^{nt} \frac{1}{n} = t. \tag{3.2.10}$$

If we go from time 0 to time t along the path of the scaled random walk, evaluating the increment over each time step and squaring these increments before summing them, we obtain t, the length of the time interval over which we are doing the computation. This is a path-by-path computation, not an average over all possible paths, and could in principle depend on the particular path along which we do the computation. However, along each path we get the same answer t. Note that $\operatorname{Var} W^{(n)}(t)$ is also t (the second equation in (3.2.8) with $s = 0$), but this latter quantity is an average over all possible paths.

3.2.6 Limiting Distribution of the Scaled Random Walk

In Figure 3.2.2 we see a single sample path of the scaled random walk. In other words, we have fixed a sequence of coin tosses $\omega = \omega_1\omega_2\ldots$ and drawn the path of the resulting process as time t varies. Another way to think about the scaled random walk is to fix the time t and consider the set of all possible paths evaluated at that time t. In other words, we can fix t and think about the scaled random walk corresponding to different values of ω, the sequence of coin tosses. For example, set $t = 0.25$ and consider the set of possible values of $W^{(100)}(0.25) = \frac{1}{10} M_{25}$. This random variable is generated by 25 coin tosses, and since the unscaled random walk M_{25} can take the value of any odd integer between -25 and 25, the scaled random walk $W^{(100)}(0.25)$ can take any of the values

$$-2.5, -2.3, -2.1, \ldots, -0.3, -0.1, 0.1, 0.3, \ldots, 2.1, 2.3, 2.5.$$

In order for $W^{(100)}(0.25)$ to take the value 0.1, we must get 13 heads and 12 tails in the 25 tosses. The probability of this is

$$\mathbb{P}\{W^{(100)}(0.25) = 0.1\} = \frac{25!}{13!\,12!}\left(\frac{1}{2}\right)^{25} = 0.1555. \tag{3.2.11}$$

We plot this information in Figure 3.2.3 by drawing a histogram bar centered at 0.1 with area 0.1555. Since this bar has width 0.2, its height must be $\frac{0.1555}{0.2} = 0.7775$. Figure 3.2.3 shows similar histogram bars for all possible values of $W^{(100)}(0.25)$ between -1.5 and 1.5.

The random variable $W^{(100)}(0.25)$ has expected value zero and variance 0.25. Superimposed on the histogram in Figure 3.2.3 is the normal density

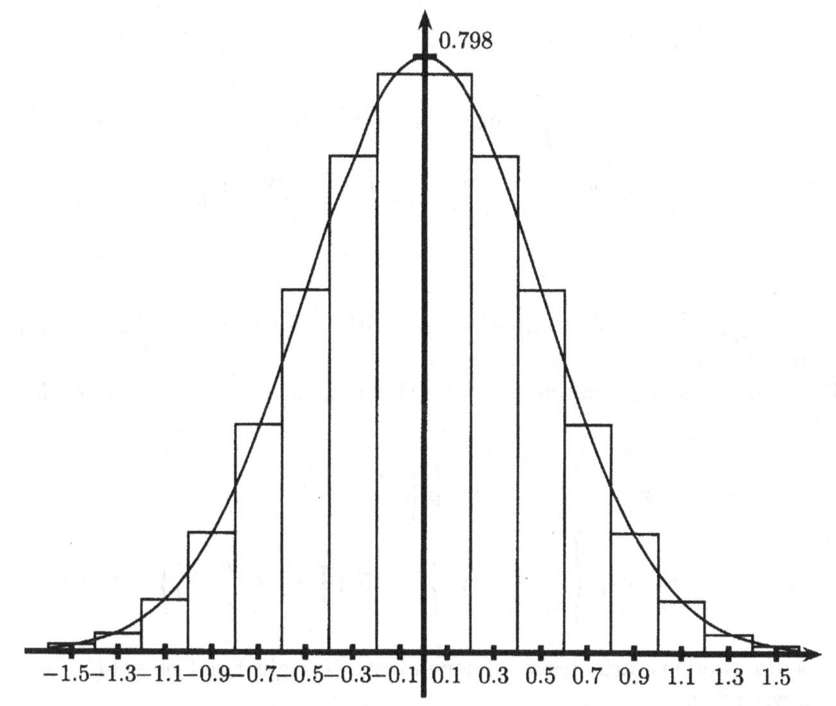

Fig. 3.2.3. Distribution of $W^{(100)}(0.25)$ and normal curve $y = \frac{2}{\sqrt{2\pi}}e^{-2x^2}$.

with this mean and variance. We see that the distribution of $W^{(100)}(0.25)$ is nearly normal. If we were given a continuous bounded function $g(x)$ and asked to compute $\mathbb{E}g\big(W^{(100)}(0.25)\big)$, a good approximation would be obtained by multiplying $g(x)$ by the normal density shown in Figure 3.2.3 and integrating:

$$\mathbb{E}g\big(W^{(100)}(0.25)\big) \approx \frac{2}{\sqrt{2\pi}} \int_{-\infty}^{\infty} g(x)e^{-2x^2}\,dx. \tag{3.2.12}$$

The Central Limit Theorem asserts that the approximation in (3.2.12) is valid. We provide the version of it that applies to our context.

Theorem 3.2.1 (Central limit). *Fix $t \geq 0$. As $n \to \infty$, the distribution of the scaled random walk $W^{(n)}(t)$ evaluated at time t converges to the normal distribution with mean zero and variance t.*

OUTLINE OF PROOF: One can identify distributions by identifying their moment-generating functions. For the normal density

$$f(x) = \frac{1}{\sqrt{2\pi t}}e^{-\frac{x^2}{2t}}$$

with mean zero and variance t, the moment-generating function is

$$\begin{aligned}\varphi(u) &= \int_{-\infty}^{\infty} e^{ux} f(x)\,dx \\ &= \frac{1}{\sqrt{2\pi t}} \int_{-\infty}^{\infty} \exp\left\{ux - \frac{x^2}{2t}\right\} dx \\ &= e^{\frac{1}{2}u^2 t} \cdot \frac{1}{\sqrt{2\pi t}} \int_{-\infty}^{\infty} \exp\left\{-\frac{(x-ut)^2}{2t}\right\} dx \\ &= e^{\frac{1}{2}u^2 t} \end{aligned} \tag{3.2.13}$$

because $\frac{1}{\sqrt{2\pi t}} e^{-\frac{(x-ut)^2}{2t}}$ is a normal density with mean ut and variance t and hence integrates to 1.

If t is such that nt is an integer, then the moment-generating function for $W^{(n)}(t)$ is

$$\begin{aligned}\varphi_n(u) = \mathbb{E}e^{uW^{(n)}(t)} &= \mathbb{E}\exp\left\{\frac{u}{\sqrt{n}} M_{nt}\right\} \\ = \mathbb{E}\exp\left\{\frac{u}{\sqrt{n}} \sum_{j=1}^{nt} X_j\right\} &= \mathbb{E}\prod_{j=1}^{nt} \exp\left\{\frac{u}{\sqrt{n}} X_j\right\}. \end{aligned} \tag{3.2.14}$$

Because the random variables are independent, the right-hand side of (3.2.14) may be written as

$$\prod_{j=1}^{nt} \mathbb{E}\exp\left\{\frac{u}{\sqrt{n}} X_j\right\} = \prod_{j=1}^{nt}\left(\frac{1}{2}e^{\frac{u}{\sqrt{n}}} + \frac{1}{2}e^{-\frac{u}{\sqrt{n}}}\right) = \left(\frac{1}{2}e^{\frac{u}{\sqrt{n}}} + \frac{1}{2}e^{-\frac{u}{\sqrt{n}}}\right)^{nt}.$$

We need to show that, as $n \to \infty$,

$$\varphi_n(u) = \left(\frac{1}{2}e^{\frac{u}{\sqrt{n}}} + \frac{1}{2}e^{-\frac{u}{\sqrt{n}}}\right)^{nt}$$

converges to the moment-generating function $\varphi(u) = e^{\frac{1}{2}u^2 t}$ in (3.2.13). To do this, it suffices to consider the logarithm of $\varphi_n(u)$ and show that

$$\log \varphi_n(u) = nt \log\left(\frac{1}{2}e^{\frac{u}{\sqrt{n}}} + \frac{1}{2}e^{-\frac{u}{\sqrt{n}}}\right)$$

converges to $\log \varphi(u) = \frac{1}{2}u^2 t$.

For this final computation, we make the change of variable $x = \frac{1}{\sqrt{n}}$ so that

$$\lim_{n\to\infty} \log \varphi_n(u) = t \lim_{x \downarrow 0} \frac{\log\left(\frac{1}{2}e^{ux} + \frac{1}{2}e^{-ux}\right)}{x^2}.$$

If we were to substitute $x = 0$ into the expression on the right-hand side, we would obtain $\frac{0}{0}$, and in this situation, we may use L'Hôpital's rule. The derivative of the numerator with respect to x is

$$\frac{\partial}{\partial x}\log\left(\frac{1}{2}e^{ux}+\frac{1}{2}e^{-ux}\right)=\frac{\frac{u}{2}e^{ux}-\frac{u}{2}e^{-ux}}{\frac{1}{2}e^{ux}+\frac{1}{2}e^{-ux}},$$

and the derivative of the denominator is

$$\frac{\partial}{\partial x}x^2=2x.$$

Therefore,

$$\lim_{n\to\infty}\log\varphi_n(u)=t\lim_{x\downarrow 0}\frac{\frac{u}{2}e^{ux}-\frac{u}{2}e^{-ux}}{2x\left(\frac{1}{2}e^{ux}+\frac{1}{2}e^{-ux}\right)}=\frac{t}{2}\lim_{x\downarrow 0}\frac{\frac{u}{2}e^{ux}-\frac{u}{2}e^{-ux}}{x},$$

where we have used the fact that $\lim_{x\downarrow 0}\left(\frac{1}{2}e^{ux}+\frac{1}{2}e^{-ux}\right)=1$. If we were to substitute $x=0$ into the expression on the right-hand side, we would again obtain $\frac{0}{0}$. In this situation, we apply L'Hôpital's rule again. The derivative of the numerator is

$$\frac{\partial}{\partial x}\left(\frac{u}{2}e^{ux}-\frac{u}{2}e^{-ux}\right)=\frac{u^2}{2}e^{ux}+\frac{u^2}{2}e^{-ux},$$

and the derivative of the denominator is $\frac{\partial}{\partial x}x=1$. Hence,

$$\lim_{n\to\infty}\log\varphi_n(u)=\frac{t}{2}\lim_{x\downarrow 0}\left(\frac{u^2}{2}e^{ux}+\frac{u^2}{2}e^{-ux}\right)=\frac{1}{2}u^2t,$$

as desired. □

3.2.7 Log-Normal Distribution as the Limit of the Binomial Model

The Central Limit Theorem, Theorem 3.2.1, can be used to show that the limit of a properly scaled binomial asset-pricing model leads to a stock price with a log-normal distribution. We present this limiting argument here under the assumption that the interest rate r is zero. The case of a nonzero interest rate is outlined in Exercise 3.8. These results show that the binomial model is a discrete-time version of the geometric Brownian motion model, which is the basis for the Black-Scholes-Merton option-pricing formula.

Let us build a model for a stock price on the time interval from 0 to t by choosing an integer n and constructing a binomial model for the stock price that takes n steps per unit time. We assume that n and t are chosen so that nt is an integer. We take the up factor to be $u_n=1+\frac{\sigma}{\sqrt{n}}$ and the down factor to be $d_n=1-\frac{\sigma}{\sqrt{n}}$. Here σ is a positive constant that will turn out to be the volatility of the limiting stock price process. The risk-neutral probabilities are then (see (1.1.8) of Chapter 1 of Volume I)

$$\tilde{p}=\frac{1+r-d_n}{u_n-d_n}=\frac{\sigma/\sqrt{n}}{2\sigma/\sqrt{n}}=\frac{1}{2},\quad \tilde{q}=\frac{u_n-1-r}{u_n-d_n}=\frac{\sigma/\sqrt{n}}{2\sigma/\sqrt{n}}=\frac{1}{2}.$$

The stock price at time t is determined by the initial stock price $S(0)$ and the result of the first nt coin tosses. The sum of the number of heads H_{nt} and number of tails T_{nt} in the first nt coin tosses is nt, a fact that we write as

$$nt = H_{nt} + T_{nt}.$$

The random walk M_{nt} is the number of heads minus the number of tails in these nt coin tosses:

$$M_{nt} = H_{nt} - T_{nt}.$$

Adding these two equations and dividing by 2, we see that

$$H_{nt} = \frac{1}{2}(nt + M_{nt}).$$

Subtracting them and dividing by 2, we see further that

$$T_{nt} = \frac{1}{2}(nt - M_{nt}).$$

In the model with up factor u_n and down factor d_n, the stock price at time t is

$$S_n(t) = S(0)u_n^{H_{nt}}d_n^{T_{nt}} = S(0)\left(1+\frac{\sigma}{\sqrt{n}}\right)^{\frac{1}{2}(nt+M_{nt})}\left(1-\frac{\sigma}{\sqrt{n}}\right)^{\frac{1}{2}(nt-M_{nt})}. \tag{3.2.15}$$

We wish to identify the distribution of this random variable as $n \to \infty$.

Theorem 3.2.2. *As $n \to \infty$, the distribution of $S_n(t)$ in (3.2.15) converges to the distribution of*

$$S(t) = S(0)\exp\left\{\sigma W(t) - \frac{1}{2}\sigma^2 t\right\}, \tag{3.2.16}$$

where $W(t)$ is a normal random variable with mean zero and variance t.

The distribution of $S(t)$ in (3.2.16) is called *log-normal.* More generally, any random variable of the form ce^X, where c is a constant and X is normally distributed, is said to have a log-normal distribution. In the case at hand, $X = \sigma W(t) - \frac{1}{2}\sigma^2 t$ is normal with mean $-\frac{1}{2}\sigma^2 t$ and variance $\sigma^2 t$.

PROOF OF THEOREM 3.2.2: It suffices to show that the distribution of

$$\log S_n(t) = \log S(0) + \frac{1}{2}(nt + M_{nt})\log\left(1+\frac{\sigma}{\sqrt{n}}\right) + \frac{1}{2}(nt - M_{nt})\log\left(1-\frac{\sigma}{\sqrt{n}}\right) \tag{3.2.17}$$

converges to the distribution of

$$\log S(t) = \log S(0) + \sigma W(t) - \frac{1}{2}\sigma^2 t,$$

where $W(t)$ is a normal random variable with mean zero and variance t. To do this, we need the Taylor series expansion of $f(x) = \log(1+x)$. We compute $f'(x) = (1+x)^{-1}$ and $f''(x) = -(1+x)^{-2}$ and evaluate them to obtain $f'(0) = 1$ and $f''(0) = -1$. According to Taylor's Theorem,

$$\log(1+x) = f(0) + f'(0)x + \frac{1}{2}f''(0)x^2 + O(x^3) = x - \frac{1}{2}x^2 + O(x^3),$$

where $O(x^3)$ indicates a term of order x^3. We apply this to (3.2.17) first with $x = \frac{\sigma}{\sqrt{n}}$ and then with $x = -\frac{\sigma}{\sqrt{n}}$. Our intention is to then let $n \to \infty$, and so we need to keep track of which terms have powers of n in the denominator and which terms do not. The former ones will have limit zero and the latter ones will not. We use the $O(\cdot)$ notation to do this. Not every term of the form $O(n^{-\frac{3}{2}})$ in the following equation is the same; their only common feature is that they have $n^{\frac{3}{2}}$ in their denominators. In particular, from (3.2.17) we have

$$\begin{aligned}
\log S(t) &= \log S(0) + \frac{1}{2}(nt + M_{nt})\left(\frac{\sigma}{\sqrt{n}} - \frac{\sigma^2}{2n} + O(n^{-\frac{3}{2}})\right) \\
&\quad + \frac{1}{2}(nt - M_{nt})\left(-\frac{\sigma}{\sqrt{n}} - \frac{\sigma^2}{2n} + O(n^{-\frac{3}{2}})\right) \\
&= \log S(0) + nt\left(-\frac{\sigma^2}{2n} + O(n^{-\frac{3}{2}})\right) + M_{nt}\left(\frac{\sigma}{\sqrt{n}} + O(n^{-\frac{3}{2}})\right) \\
&= \log S(0) - \frac{1}{2}\sigma^2 t + O(n^{-\frac{1}{2}}) + \sigma W^{(n)}(t) + O(n^{-1})W^{(n)}(t).
\end{aligned}$$

The term $W^{(n)}(t) = \frac{1}{\sqrt{n}}M_{nt}$ appears in two places in the last line. By the Central Limit Theorem, Theorem 3.2.1, its distribution converges to the distribution of a normal random variable with mean zero and variance t, a random variable we call $W(t)$. However, in one of its appearances, $W^{(n)}(t)$ is multiplied by a term that has n in the denominator, and this will have limit zero. The term $O(n^{-\frac{1}{2}})$ also has limit zero as $n \to \infty$. We conclude that as $n \to \infty$ the distribution of $\log S(t)$ approaches the distribution of $\log S(0) - \frac{1}{2}\sigma^2 t + \sigma W(t)$, which is what we set out to prove. □

3.3 Brownian Motion

3.3.1 Definition of Brownian Motion

We obtain Brownian motion as the limit of the scaled random walks $W^{(n)}(t)$ of (3.2.7) as $n \to \infty$. The Brownian motion inherits properties from these random walks. This leads to the following definition.

Definition 3.3.1. *Let $(\Omega, \mathcal{F}, \mathbb{P})$ be a probability space. For each $\omega \in \Omega$, suppose there is a continuous function $W(t)$ of $t \geq 0$ that satisfies $W(0) = 0$ and that depends on ω. Then $W(t)$, $t \geq 0$, is a* Brownian motion *if for all $0 = t_0 < t_1 < \cdots < t_m$ the increments*

$$W(t_1) = W(t_1) - W(t_0),\ W(t_2) - W(t_1),\ \ldots,\ W(t_m) - W(t_{m-1}) \quad (3.3.1)$$

are independent and each of these increments is normally distributed with

$$\mathbb{E}\big[W(t_{i+1}) - W(t_i)\big] = 0, \quad (3.3.2)$$

$$\mathrm{Var}\big[W(t_{i+1}) - W(t_i)\big] = t_{i+1} - t_i. \quad (3.3.3)$$

One difference between Brownian motion $W(t)$ and a scaled random walk, say $W^{(100)}(t)$, is that the scaled random walk has a natural time step $\frac{1}{100}$ and is linear between these time steps, whereas the Brownian motion has no linear pieces. The other difference is that, while the scaled random walk $W^{(100)}(t)$ is only approximately normal for each t (see Figure 3.2.3), the Brownian motion is exactly normal. This is a consequence of the Central Limit Theorem, Theorem 3.2.1. Not only is $W(t) = W(t) - W(0)$ normally distributed for each t, but the increments $W(t) - W(s)$ are normally distributed for all $0 \leq s < t$.

There are two ways to think of ω in Definition 3.3.1. One is to think of ω as the Brownian motion path. A random experiment is performed, and its outcome is the path of the Brownian motion. Then $W(t)$ is the value of this path at time t, and this value of course depends on which path resulted from the random experiment. Alternatively, one can think of ω as something more primitive than the path itself, akin to the outcome of a sequence of coin tosses, although now the coin is being tossed "infinitely fast." Once the sequence of coin tosses has been performed and the result ω obtained, then the path of the Brownian motion can be drawn. If the tossing is done again and a different ω is obtained, then a different path will be drawn.

In either case, the sample space Ω is the set of all possible outcomes of a random experiment, $\mathcal{F}$ is the σ-algebra of subsets of Ω whose probabilities are defined, and $\mathbb{P}$ is a probability measure. For each $A \in \mathcal{F}$, the probability of A is a number $\mathbb{P}(A)$ between zero and one. The distributional statements about Brownian motion pertain to $\mathbb{P}$.

For example, we might wish to determine the probability of the set A containing all $\omega \in \Omega$ that result in a Brownian motion path satisfying $0 \leq W(0.25) \leq 0.2$. Let us first consider this matter for the scaled random walk $W^{(100)}$. If we were asked to determine the set $\{\omega : 0 \leq W^{(100)}(0.25) \leq 0.2\}$, we would note that in order for the scaled random walk $W^{(100)}$ to fall between 0 and 0.2 at time 0.25, the unscaled random walk $M_{25} = 10W^{(100)}(0.25)$ must fall between 0 and 2 after 25 tosses. Since M_{25} can only be an odd number, it falls between 0 and 2 if and only if it is equal to 1 or, equivalently, if and only if $W^{(100)}(0.25) = 0.1$. To achieve this, the coin tossing must result in 13 heads and 12 tails in the first 25 tosses. Therefore, A is the set of all infinite sequences of coin tosses with the property that in the first 25 tosses there are 13 heads and 12 tails. The probability that one of these sequences occurs, given by (3.2.11), is $\mathbb{P}(A) = 0.1555$.

For the Brownian motion W, there is also a set of outcomes ω to the random experiment that results in a Brownian motion path satisfying $0 \leq W(0.25) \leq 0.2$. We choose not to describe this set as concretely as we just did for the scaled random walk $W^{(100)}$. Nonetheless, there is such a set of $\omega \in \Omega$, and the probability of this set is

$$\mathbb{P}\{0 \leq W(0.25) \leq 0.2\} = \frac{2}{\sqrt{2\pi}} \int_0^{0.2} e^{-2x^2}\, dx.$$

In place of the area in the histogram bar centered at 0.1 in Figure 3.2.3, which is 0.1555, we now have the area under the normal curve between 0 and 0.2 in that figure. These two areas are nearly the same.

3.3.2 Distribution of Brownian Motion

Because the increments

$$W(t_1) = W(t_1) - W(t_0), W(t_2) - W(t_1), \ldots, W(t_m) - W(t_{m-1})$$

of (3.3.1) are independent and normally distributed, the random variables $W(t_1), W(t_2), \ldots, W(t_m)$ are jointly normally distributed. The joint distribution of jointly normal random variables is determined by their means and covariances. Each of the random variables $W(t_i)$ has mean zero. For any two times, $0 \leq s < t$, the covariance of $W(s)$ and $W(t)$ is

$$\begin{aligned} \mathbb{E}\big[W(s)W(t)\big] &= \mathbb{E}\big[W(s)\big(W(t) - W(s)\big) + W^2(s)\big] \\ &= \mathbb{E}\big[W(s)\big] \cdot \mathbb{E}\big[W(t) - W(s)\big] + \mathbb{E}\big[W^2(s)\big] \\ &= 0 + \mathrm{Var}\big[W(s)\big] = s, \end{aligned}$$

where we have used the independence of $W(s)$ and $W(t) - W(s)$ in the second equality. Hence, the *covariance matrix for Brownian motion* (i.e., for the m-dimensional random vector $\big(W(t_1), W(t_2), \ldots, W(t_m)\big)$) is

$$\begin{bmatrix} \mathbb{E}\big[W^2(t_1)\big] & \mathbb{E}\big[W(t_1)W(t_2)\big] & \cdots & \mathbb{E}\big[W(t_1)W(t_m)\big] \\ \mathbb{E}\big[W(t_2)W(t_1)\big] & \mathbb{E}\big[W^2(t_2)\big] & \cdots & \mathbb{E}\big[W(t_2)W(t_m)\big] \\ \vdots & \vdots & & \vdots \\ \mathbb{E}\big[W(t_m)W(t_1)\big] & \mathbb{E}\big[W(t_m)W(t_2)\big] & \cdots & \mathbb{E}\big[W^2(t_m)\big] \end{bmatrix} = \begin{bmatrix} t_1 & t_1 & \cdots & t_1 \\ t_1 & t_2 & \cdots & t_2 \\ \vdots & \vdots & & \vdots \\ t_1 & t_2 & \cdots & t_m \end{bmatrix}. \quad (3.3.4)$$

The moment-generating function of this random vector can be computed using the moment-generating function (3.2.13) for a zero-mean normal random variable with variance t and the independence of the increments in (3.3.1). To assist in this computation, we note first that

$$\begin{aligned}u_3W(t_3)+u_2W(t_2)+u_1W(t_1)\\ =u_3\big(W(t_3)-W(t_2)\big)+(u_2+u_3)\big(W(t_2)-W(t_1)\big)\\ +(u_1+u_2+u_3)W(t_1)\end{aligned}$$

and more generally

$$\begin{aligned}&u_mW(t_m)+u_{m-1}W(t_{m-1})+u_{m-2}W(t_{m-2})+\cdots+u_1W(t_1)\\ &=u_m\big(W(t_m)-W(t_{m-1})\big)+(u_{m-1}+u_m)\big(W(t_{m-1})-W(t_{m-2})\big)\\ &\quad+(u_{m-2}+u_{m-1}+u_m)\big(W(t_{m-2})-W(t_{m-3})\big)+\dots\\ &\quad\cdots+(u_1+u_2+\cdots+u_m)W(t_1).\end{aligned}$$

We use these facts to compute the moment-generating function of the random vector $\big(W(t_1),W(t_2),\dots,W(t_m)\big)$:

$$\begin{aligned}&\varphi(u_1,u_2,\dots,u_m)\\ &=\mathbb{E}\exp\{u_mW(t_m)+u_{m-1}W(t_{m-1})+\cdots+u_1W(t_1)\}\\ &=\mathbb{E}\exp\{u_m\big(W(t_m)-W(t_{m-1})\big)+(u_{m-1}+u_m)\big(W(t_{m-1})-W(t_{m-2})\big)+\\ &\qquad\cdots+(u_1+u_2+\cdots+u_m)W(t_1)\}\\ &=\mathbb{E}\exp\big\{u_m\big(W(t_m)-W(t_{m-1})\big)\big\}\\ &\qquad\cdot\mathbb{E}\exp\big\{(u_{m-1}+u_m)\big(W(t_{m-1})-W(t_{m-2})\big)\big\}\\ &\qquad\cdots\mathbb{E}\exp\big\{(u_1+u_2+\cdots+u_m)W(t_1)\big\}\\ &=\exp\left\{\frac{1}{2}u_m^2(t_m-t_{m-1})\right\}\cdot\exp\left\{\frac{1}{2}(u_{m-1}+u_m)^2(t_{m-1}-t_{m-2})\right\}\\ &\qquad\cdots\exp\left\{\frac{1}{2}(u_1+u_2+\cdots+u_m)^2t_1\right\}.\end{aligned}$$

In conclusion, the *moment-generating function for Brownian motion* (i.e., for the m-dimensional random vector $\big(W(t_1),W(t_2),\dots,W(t_m)\big)$) is

$$\begin{aligned}&\varphi(u_1,u_2,\dots,u_m)\\ &=\mathbb{E}\exp\{u_mW(t_m)+u_{m-1}W(t_{m-1})+\cdots+u_1W(t_1)\}\\ &=\exp\left\{\frac{1}{2}(u_1+u_2+\cdots+u_m)^2t_1+\frac{1}{2}(u_2+u_3+\cdots+u_m)^2(t_2-t_1)+\right.\\ &\qquad\left.\cdots+\frac{1}{2}(u_{m-1}+u_m)^2(t_{m-1}-t_{m-2})+\frac{1}{2}u_m^2(t_m-t_{m-1})\right\}.\end{aligned}\tag{3.3.5}$$

The distribution of the Brownian increments in (3.3.1) can be specified by specifying the joint density or the joint moment-generating function of the random variables $W(t_1),W(t_2),\dots,W(t_m)$. This leads to the following theorem.

Theorem 3.3.2 (Alternative characterizations of Brownian motion). *Let $(\Omega,\mathcal{F},\mathbb{P})$ be a probability space. For each $\omega\in\Omega$, suppose there is a* *continuous function $W(t)$ of $t \geq 0$ that satisfies $W(0) = 0$ and that depends on ω. The following three properties are equivalent.*

(i) For all $0 = t_0 < t_1 < \cdots < t_m$, the increments

$$W(t_1) = W(t_1) - W(t_0),\ W(t_2) - W(t_1),\ \ldots,\ W(t_m) - W(t_{m-1})$$

are independent and each of these increments is normally distributed with mean and variance given by (3.3.2) and (3.3.3).

(ii) For all $0 = t_0 < t_1 < \cdots < t_m$, the random variables $W(t_1), W(t_2), \ldots, W(t_m)$ are jointly normally distributed with means equal to zero and covariance matrix (3.3.4).

(iii) For all $0 = t_0 < t_1 < \cdots < t_m$, the random variables $W(t_1), W(t_2), \ldots, W(t_m)$ have the joint moment-generating function (3.3.5).

If any of (i), (ii), or (iii) holds (and hence they all hold), then $W(t)$, $t \geq 0$, is a Brownian motion.

3.3.3 Filtration for Brownian Motion

In addition to the Brownian motion itself, we will need some notation for the amount of information available at each time. We do that with a filtration.

Definition 3.3.3. *Let $(\Omega, \mathcal{F}, \mathbb{P})$ be a probability space on which is defined a Brownian motion $W(t)$, $t \geq 0$. A* filtration for the Brownian motion *is a collection of σ-algebras $\mathcal{F}(t)$, $t \geq 0$, satisfying:*

(i) **(Information accumulates)** *For $0 \leq s < t$, every set in $\mathcal{F}(s)$ is also in $\mathcal{F}(t)$. In other words, there is at least as much information available at the later time $\mathcal{F}(t)$ as there is at the earlier time $\mathcal{F}(s)$.*

(ii) **(Adaptivity)** *For each $t \geq 0$, the Brownian motion $W(t)$ at time t is $\mathcal{F}(t)$-measurable. In other words, the information available at time t is sufficient to evaluate the Brownian motion $W(t)$ at that time.*

(iii) **(Independence of future increments)** *For $0 \leq t < u$, the increment $W(u) - W(t)$ is independent of $\mathcal{F}(t)$. In other words, any increment of the Brownian motion after time t is independent of the information available at time t.*

Let $\Delta(t)$, $t \geq 0$, be a stochastic process. We say that $\Delta(t)$ is adapted *to the filtration $\mathcal{F}(t)$ if for each $t \geq 0$ the random variable $\Delta(t)$ is $\mathcal{F}(t)$-measurable.*[1]

Properties (i) and (ii) in the definition above guarantee that the information available at each time t is at least as much as one would learn from observing the Brownian motion up to time t. Property (iii) says that this

[1] The adapted processes we encounter will serve as integrands, and for this one needs them to be jointly measurable in t and ω so that their integrals are defined and are themselves adapted processes. This is a technical requirement that we shall ignore in this text.

information is of no use in predicting future movements of the Brownian motion. In the asset-pricing models we build, property (iii) leads to the efficient market hypothesis.

There are two possibilities for the filtration $\mathcal{F}(t)$ for a Brownian motion. One is to let $\mathcal{F}(t)$ contain only the information obtained by observing the Brownian motion itself up to time t. The other is to include in $\mathcal{F}(t)$ information obtained by observing the Brownian motion and one or more other processes. However, if the information in $\mathcal{F}(t)$ includes observations of processes other than the Brownian motion W, this additional information is not allowed to give clues about the future increments of W because of property (iii).

3.3.4 Martingale Property for Brownian Motion

Theorem 3.3.4. *Brownian motion is a martingale.*

PROOF: Let $0 \leq s \leq t$ be given. Then

$$\begin{aligned} \mathbb{E}\big[W(t)\big|\mathcal{F}(s)\big] &= \mathbb{E}\big[\big(W(t) - W(s)\big) + W(s)\big|\mathcal{F}(s)\big] \\ &= \mathbb{E}\big[W(t) - W(s)\big|\mathcal{F}(s)\big] + \mathbb{E}\big[W(s)\big|\mathcal{F}(s)\big] \\ &= \mathbb{E}\big[W(t) - W(s)\big] + W(s) \\ &= W(s). \end{aligned}$$

The justifications for the steps in this equality are the same as the justifications for (3.2.5). □

3.4 Quadratic Variation

We computed the quadratic variation of the scaled random walk $W^{(n)}$ up to time T in (3.2.10), and this quadratic variation turned out to be T. This was computed by taking each of the steps of the scaled random walk between times 0 and T, squaring them, and summing them.

For Brownian motion, there is no natural step size. If we are given $T > 0$, we could simply choose a step size, say $\frac{T}{n}$ for some large n, and compute the quadratic variation up to time T with this step size. In other words, we could compute

$$\sum_{j=0}^{n-1} \left[W\left(\frac{(j+1)T}{n}\right) - W\left(\frac{jT}{n}\right) \right]^2. \tag{3.4.1}$$

We are interested in this quantity for small step sizes, and so as a last step we could evaluate the limit as $n \to \infty$. If we do this, we will get T, the same final answer as for the scaled random walk in (3.2.10). This is proved in Theorem 3.4.3 below.

The paths of Brownian motion are unusual in that their quadratic variation is not zero. This makes stochastic calculus different from ordinary calculus and is the source of the volatility term in the Black-Scholes-Merton partial differential equation. These matters will be discussed in the next chapter.

3.4.1 First-Order Variation

Before proving that Brownian motion accumulates T units of quadratic variation between times 0 and T, we digress slightly to discuss *first-order variation* (as opposed to *quadratic variation*, which is *second-order variation*). Consider the function $f(t)$ in Figure 3.4.1. We wish to compute the amount of up and down oscillation undergone by this function between times 0 and T, with the down moves adding to rather than subtracting from the up moves. We call this the *first-order variation* $\mathrm{FV}_T(f)$. For the function f shown, it is

$$\begin{aligned}\mathrm{FV}_T(f) &= \big[f(t_1) - f(0)\big] - \big[f(t_2) - f(t_1)\big] + \big[f(T) - f(t_2)\big]\\ &= \int_0^{t_1} f'(t)\,dt + \int_{t_1}^{t_2} (-f'(t))\,dt + \int_{t_2}^{T} f'(t)\,dt\\ &= \int_0^T |f'(t)|\,dt. \end{aligned} \tag{3.4.2}$$

The middle term

$$-[f(t_2) - f(t_1)] = f(t_1) - f(t_2)$$

is included in a way that guarantees that the magnitude of the down move of the function $f(t)$ between times t_1 and t_2 is added to rather than subtracted from the total.

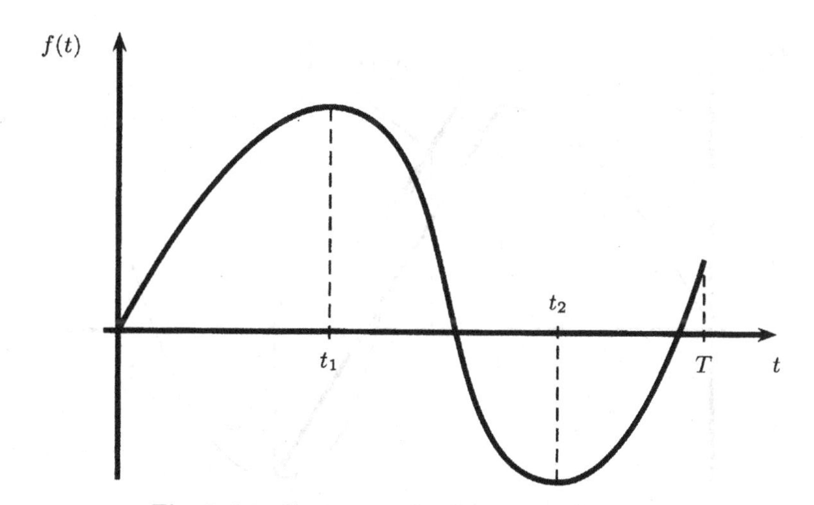

Fig. 3.4.1. Computing the first-order variation.

In general, to compute the first-order variation of a function up to time T, we first choose a *partition* $\Pi = \{t_0, t_1, \ldots, t_n\}$ of $[0, T]$, which is a set of times

$$0 = t_0 < t_1 < \cdots < t_n = T.$$

These will serve to determine the step size. We do not require the partition points $t_0 = 0, t_1, t_2, \ldots, t_n = T$ to be equally spaced, although they are allowed to be. The maximum step size of the partition will be denoted $\|\Pi\| = \max_{j=0,\ldots,n-1}(t_{j+1} - t_j)$. We then define

$$\mathrm{FV}_T(f) = \lim_{\|\Pi\|\to 0} \sum_{j=0}^{n-1} |f(t_{j+1}) - f(t_j)|. \tag{3.4.3}$$

The limit in (3.4.3) is taken as the number n of partition points goes to infinity and the length of the longest subinterval $t_{j+1} - t_j$ goes to zero.

Our first task is to verify that the definition (3.4.3) is consistent with the formula (3.4.2) for the function shown in Figure 3.4.1. To do this, we use the Mean Value Theorem, which applies to any function $f(t)$ whose derivative $f'(t)$ is defined everywhere. The Mean Value Theorem says that in each subinterval $[t_j, t_{j+1}]$ there is a point t_j^* such that

$$\frac{f(t_{j+1}) - f(t_j)}{t_{j+1} - t_j} = f'(t_j^*). \tag{3.4.4}$$

In other words, somewhere between t_j and t_{j+1}, the tangent line is parallel to the chord connecting the points $(t_j, f(t_j))$ and $(t_{j+1}, f(t_{j+1}))$ (see Figure 3.4.2).

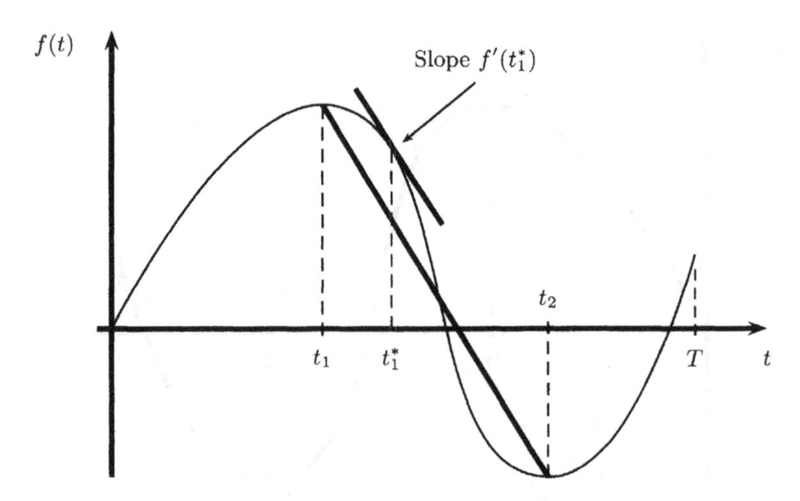

Fig. 3.4.2. Mean Value Theorem.

Multiplying (3.4.4) by $t_{j+1} - t_j$, we obtain

$$f(t_{j+1}) - f(t_j) = f'(t_j^*)(t_{j+1} - t_j).$$

The sum on the right-hand side of (3.4.3) may thus be written as

$$\sum_{j=0}^{n-1} |f'(t_j^*)|(t_{j+1} - t_j),$$

which is a Riemann sum for the integral of the function $|f'(t)|$. Therefore,

$$\mathrm{FV}_T(f) = \lim_{\|\Pi\| \to 0} \sum_{j=0}^{n-1} |f'(t_j^*)|(t_{j+1} - t_j) = \int_0^T |f'(t)|\, dt,$$

and we have rederived (3.4.2).

3.4.2 Quadratic Variation

Definition 3.4.1. *Let $f(t)$ be a function defined for $0 \le t \le T$. The* quadratic variation *of f up to time T is*

$$[f, f](T) = \lim_{\|\Pi\| \to 0} \sum_{j=0}^{n-1} [f(t_{j+1}) - f(t_j)]^2, \tag{3.4.5}$$

where $\Pi = \{t_0, t_1, \ldots, t_n\}$ and $0 = t_0 < t_1 < \cdots < t_n = T$.

Remark 3.4.2. Suppose the function f has a continuous derivative. Then

$$\sum_{j=0}^{n-1} [f(t_{j+1}) - f(t_j)]^2 = \sum_{j=0}^{n-1} |f'(t_j^*)|^2 (t_{j+1} - t_j)^2 \le \|\Pi\| \cdot \sum_{j=0}^{n-1} |f'(t_j^*)|^2 (t_{j+1} - t_j),$$

and thus

$$\begin{aligned} [f, f](T) &\le \lim_{\|\Pi\| \to 0} \left[\|\Pi\| \cdot \sum_{j=0}^{n-1} |f'(t_j^*)|^2 (t_{j+1} - t_j) \right] \\ &= \lim_{\|\Pi\| \to 0} \|\Pi\| \cdot \lim_{\|\Pi\| \to 0} \sum_{j=0}^{n-1} |f'(t_j^*)|^2 (t_{j+1} - t_j) \\ &= \lim_{\|\Pi\| \to 0} \|\Pi\| \cdot \int_0^T |f'(t)|^2\, dt \;=\; 0. \end{aligned}$$

In the last step of this argument, we use the fact that $f'(t)$ is continuous to ensure that $\int_0^T |f'(t)|^2 dt$ is finite. If $\int_0^T |f'(t)|^2 dt$ is infinite, then

$$\lim_{\|\Pi\| \to 0} \left[\|\Pi\| \cdot \sum_{j=0}^{n-1} |f'(t_j^*)|^2 (t_{j+1} - t_j) \right]$$

leads to a $0 \cdot \infty$ situation, which can be anything between 0 and ∞. □

Most functions have continuous derivatives, and hence their quadratic variations are zero. For this reason, one never considers quadratic variation in ordinary calculus. The paths of Brownian motion, on the other hand, cannot be differentiated with respect to the time variable. For functions that do not have derivatives, the Mean Value Theorem can fail and Remark 3.4.2 no longer applies. Consider, for example, the absolute value function $f(t) = |t|$ in Figure 3.4.3. The chord connecting $(t_1, f(t_1))$ and $(t_2, f(t_2))$ has slope $\frac{1}{5}$, but nowhere between t_1 and t_2 does the derivative of $f(t) = |t|$ equal $\frac{1}{5}$. Indeed, this derivative is always -1 for $t < 0$, is always 1 for $t > 0$, and is undefined at $t = 0$, where the the graph of the function $f(t) = |t|$ has a "point." Figure 3.2.2 suggests correctly that the paths of Brownian motion are very "pointy." Indeed, for a Brownian motion path $W(t)$, there is no value of t for which $\frac{d}{dt}W(t)$ is defined.

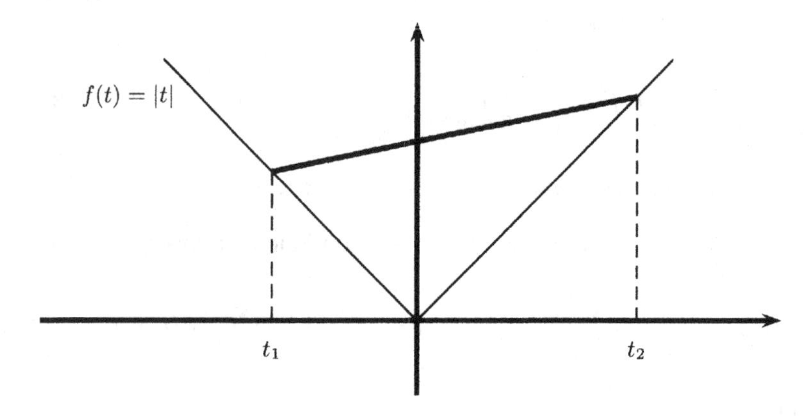

Fig. 3.4.3. Absolute value function.

Theorem 3.4.3. *Let W be a Brownian motion. Then $[W, W](T) = T$ for all $T \geq 0$ almost surely.*

We recall that the terminology *almost surely* means that there can be some paths of the Brownian motion for which the assertion $[W, W](T) = T$ is not true. However, the set of all such paths has zero probability. The set of paths for which the assertion of the theorem is true has probability one.

PROOF OF THEOREM 3.4.3: Let $\Pi = \{t_0, t_1, \ldots, t_n\}$ be a partition of $[0, T]$. Define the *sampled quadratic variation* corresponding to this partition to be

$$Q_\Pi = \sum_{j=0}^{n-1} \big(W(t_{j+1}) - W(t_j)\big)^2.$$

We must show that this sampled quadratic variation, which is a random variable (i.e., it depends on the path of the Brownian motion along which it is computed) converges to T as $\|\Pi\| \to 0$. We shall show that it has expected value T, and its variance converges to zero. Hence, it converges to its expected value T, regardless of the path along which we are doing the computation.[2]

The sampled quadratic variation is the sum of independent random variables. Therefore, its mean and variance are the sums of the means and variances of these random variables. We have

$$\mathbb{E}\left[\left(W(t_{j+1}) - W(t_j)\right)^2\right] = \mathrm{Var}\left[W(t_{j+1}) - W(t_j)\right] = t_{j+1} - t_j, \qquad (3.4.6)$$

which implies

$$\mathbb{E}Q_\Pi = \sum_{j=0}^{n-1} \mathbb{E}\left[\left(W(t_{j+1}) - W(t_j)\right)^2\right] = \sum_{j=0}^{n-1}(t_{j+1} - t_j) = T,$$

as desired. Moreover,

$$\begin{aligned}
&\mathrm{Var}\left[\left(W(t_{j+1}) - W(t_j)\right)^2\right] \\
&\quad = \mathbb{E}\left[\left(\left(W(t_{j+1}) - W(t_j)\right)^2 - (t_{j+1} - t_j)\right)^2\right] \\
&\quad = \mathbb{E}\left[\left(W(t_{j+1}) - W(t_j)\right)^4\right] - 2(t_{j+1} - t_j)\mathbb{E}\left[\left(W(t_{j+1}) - W(t_j)\right)^2\right] \\
&\qquad\quad +(t_{j+1} - t_j)^2.
\end{aligned}$$

The fourth moment of a normal random variable with zero mean is three times its variance squared (see Exercise 3.3). Therefore,

$$\mathbb{E}\left[\left(W(t_{j+1}) - W(t_j)\right)^4\right] = 3(t_{j+1} - t_j)^2,$$

$$\begin{aligned}
\mathrm{Var}\left[\left(W(t_{j+1}) - W(t_j)\right)^2\right] &= 3(t_{j+1} - t_j)^2 - 2(t_{j+1} - t_j)^2 + (t_{j+1} - t_j)^2 \\
&= 2(t_{j+1} - t_j)^2, \qquad (3.4.7)
\end{aligned}$$

and

$$\begin{aligned}
\mathrm{Var}(Q_\Pi) &= \sum_{j=0}^{n-1} \mathrm{Var}\left[\left(W(t_{j+1}) - W(t_j)\right)^2\right] = \sum_{j=0}^{n-1} 2(t_{j+1} - t_j)^2 \\
&\le \sum_{j=0}^{n-1} 2\|\Pi\|(t_{j+1} - t_j) = 2\|\Pi\|T.
\end{aligned}$$

In particular, $\lim_{\|\Pi\|\to 0} \mathrm{Var}(Q_\Pi) = 0$, and we conclude that $\lim_{\|\Pi\|\to 0} Q_\Pi = \mathbb{E}Q_\Pi = T$. □

[2] The convergence we prove is actually *convergence in mean square*, also called *L^2-convergence*. When this convergence takes place, there is a subsequence along which the convergence is *almost sure* (i.e., the convergence takes place for all paths except for a set of paths having probability zero). We shall not dwell on subtle differences among types of convergence of random variables.

Remark 3.4.4. In the proof above, we derived the equations (3.4.6) and (3.4.7):

$$\mathbb{E}\big[\big(W(t_{j+1}) - W(t_j)\big)^2\big] = t_{j+1} - t_j$$

and

$$\mathrm{Var}\big[\big(W(t_{j+1}) - W(t_j)\big)^2\big] = 2(t_{j+1} - t_j)^2.$$

It is tempting to argue that when $t_{j+1} - t_j$ is small, $(t_{j+1} - t_j)^2$ is *very* small, and therefore $\big(W(t_{j+1}) - W(t_j)\big)^2$, although random, is with high probability near its mean $t_{j+1} - t_j$. We could therefore claim that

$$\big(W(t_{j+1}) - W(t_j)\big)^2 \approx t_{j+1} - t_j. \tag{3.4.8}$$

This approximation is trivially true because, when $t_{j+1} - t_j$ is small, both sides are near zero. It would also be true if we squared the right-hand side, multiplied the right-hand side by 2, or made any of several other significant changes to the right-hand side. In other words, (3.4.8) really has no content. A better way to try to capture what we think is going on is to write

$$\frac{\big(W(t_{j+1}) - W(t_j)\big)^2}{t_{j+1} - t_j} \approx 1 \tag{3.4.9}$$

instead of (3.4.8). However,

$$\frac{\big(W(t_{j+1}) - W(t_j)\big)^2}{t_{j+1} - t_j}$$

is in fact not near 1, regardless of how small we make $t_{j+1} - t_j$. It is the square of the standard normal random variable

$$Y_{j+1} = \frac{W(t_{j+1}) - W(t_j)}{\sqrt{t_{j+1} - t_j}},$$

and its distribution is the same, no matter how small we make $t_{j+1} - t_j$.

To understand better the idea behind Theorem 3.4.3, we choose a large value of n and take $t_j = \frac{jT}{n}$, $j = 0, 1, \ldots, n$. Then $t_{j+1} - t_j = \frac{T}{n}$ for all j and

$$\big(W(t_{j+1}) - W(t_j)\big)^2 = T \cdot \frac{Y_{j+1}^2}{n}.$$

Since the random variables $Y_1, Y_2, \ldots, Y_n$ are independent and identically distributed, the Law of Large Numbers implies that $\sum_{j=0}^{n-1} \frac{Y_{j+1}^2}{n}$ converges to the common mean $\mathbb{E}Y_{j+1}^2$ as $n \to \infty$. This mean is 1, and hence $\sum_{j=0}^{n-1} \big(W(t_{j+1}) - W(t_j)\big)^2$ converges to T. Each of the terms $\big(W(t_{j+1}) - W(t_j)\big)^2$ in this sum can be quite different from its mean $t_{j+1} - t_j = \frac{T}{n}$, but when we sum many terms like this, the differences average out to zero.

We write informally

$$dW(t)\,dW(t) = dt, \tag{3.4.10}$$

but this should not be interpreted to mean either (3.4.8) or (3.4.9). It is only when we sum both sides of (3.4.9) and call upon the Law of Large Numbers to cancel errors that we get a correct statement. The statement is that on an interval $[0, T]$, Brownian motion accumulates T units of quadratic variation.

If we compute the quadratic variation of Brownian motion over the time interval $[0, T_1]$, we get $[W, W](T_1) = T_1$. If we compute the quadratic variation over $[0, T_2]$, where $0 < T_1 < T_2$, we get $[W, W](T_2) = T_2$. Therefore, if we partition the interval $[T_1, T_2]$, square the increments of Brownian motion for each of the subintervals in the partition, sum the squared increments, and take the limit as the maximal step size approaches zero, we will get the limit $[W, W](T_2) - [W, W](T_1) = T_2 - T_1$. Brownian motion accumulates $T_2 - T_1$ units of quadratic variation over the interval $[T_1, T_2]$. Since this is true for every interval of time, we conclude that

Brownian motion accumulates quadratic variation at rate one per unit time.

We write (3.4.10) to record this fact. In particular, the dt on the right-hand side of (3.4.10) is multiplied by an understood 1.

As mentioned earlier, the quadratic variation of Brownian motion is the source of volatility in asset prices driven by Brownian motion. We shall eventually scale Brownian motion, sometimes in time- and path-dependent ways, in order to vary the rate at which volatility enters these asset prices. □

Remark 3.4.5. Let $\Pi = \{t_0, t_1, \ldots, t_n\}$ be a partition of $[0, T]$ (i.e., $0 = t_0 < t_1 < \cdots < t_n = T$). In addition to computing the quadratic variation of Brownian motion

$$\lim_{\|\Pi\| \to 0} \sum_{j=0}^{n-1} \big(W(t_{j+1}) - W(t_j)\big)^2 = T, \tag{3.4.11}$$

we can compute the cross variation of $W(t)$ with t and the quadratic variation of t with itself, which are

$$\lim_{\|\Pi\| \to 0} \sum_{j=0}^{n-1} \big(W(t_{j+1}) - W(t_j)\big)(t_{j+1} - t_j) = 0, \tag{3.4.12}$$

$$\lim_{\|\Pi\| \to 0} \sum_{j=0}^{n-1} (t_{j+1} - t_j)^2 = 0. \tag{3.4.13}$$

To see that 0 is the limit in (3.4.12), we observe that

$$\Big|\big(W(t_{j+1}) - W(t_j)\big)(t_{j+1} - t_j)\Big| \le \max_{0 \le k \le n-1} \big|W(t_{k+1}) - W(k)\big|(t_{j+1} - t_j),$$

and so

$$\left|\sum_{j=0}^{n-1}(W(t_{j+1})-W(t_j))(t_{j+1}-t_j)\right| \le \max_{0\le k\le n-1}\left|W(t_{k+1})-W(t_k)\right|\cdot T.$$

Since W is continuous, $\max_{0\le k\le n-1}\left|W(t_{k+1})-W(k)\right|$ has limit zero as $\|\Pi\|$, the length of the longest subinterval, goes to zero. To see that 0 is the limit in (3.4.13), we observe that

$$\sum_{j=0}^{n-1}(t_{j+1}-t_j)^2 \le \max_{0\le k\le n-1}(t_{k+1}-t_k)\cdot\sum_{j=0}^{n-1}(t_{j+1}-t_j) = \|\Pi\|\cdot T,$$

which obviously has limit zero as $\|\Pi\|\to 0$.

Just as we capture (3.4.11) by writing (3.4.10), we capture (3.4.12) and (3.4.13) by writing

$$dW(t)\,dt = 0, \quad dt\,dt = 0. \tag{3.4.14}$$

□

3.4.3 Volatility of Geometric Brownian Motion

Let α and $\sigma > 0$ be constants, and define the *geometric Brownian motion*

$$S(t) = S(0)\exp\left\{\sigma W(t) + \left(\alpha - \frac{1}{2}\sigma^2\right)t\right\}.$$

This is the asset-price model used in the Black-Scholes-Merton option-pricing formula. Here we show how to use the quadratic variation of Brownian motion to identify the volatility σ from a path of this process.

Let $0 \le T_1 < T_2$ be given, and suppose we observe the geometric Brownian motion $S(t)$ for $T_1 \le t \le T_2$. We may then choose a partition of this interval, $T_1 = t_0 < t_2 < \cdots < t_m = T_2$, and observe "log returns"

$$\log\frac{S(t_{j+1})}{S(t_j)} = \sigma\big(W(t_{j+1})-W(t_j)\big) + \left(\alpha - \frac{1}{2}\sigma^2\right)(t_{j+1}-t_j)$$

over each of the subintervals $[t_j, t_{j+1}]$. The sum of the squares of the log returns, sometimes called the *realized volatility*, is

$$\begin{aligned}
&\sum_{j=0}^{m-1}\left(\log\frac{S(t_{j+1})}{S(t_j)}\right)^2\\
&= \sigma^2\sum_{j=0}^{m-1}\big(W(t_{j+1})-W(t_j)\big)^2 + \left(\alpha-\frac{1}{2}\sigma^2\right)^2\sum_{j=0}^{m-1}(t_{j+1}-t_j)^2\\
&\quad +2\sigma\left(\alpha-\frac{1}{2}\sigma^2\right)\sum_{j=0}^{m-1}\big(W(t_{j+1})-W(t_j)\big)(t_{j+1}-t_j).
\end{aligned} \tag{3.4.15}$$

When the maximum step size $\|\Pi\| = \max_{j=0,\ldots,m-1}(t_{j+1}-t_j)$ is small, then the first term on the right-hand side of (3.4.15) is approximately equal to its limit, which is σ^2 times the amount of quadratic variation accumulated by Brownian motion on the interval $[T_1, T_2]$, which is $T_2 - T_1$. The second term on the right-hand side of (3.4.15) is $\left(\alpha - \frac{1}{2}\sigma^2\right)^2$ times the quadratic variation of t, which was shown in Remark 3.4.5 to be zero. The third term on the right-hand side of (3.4.15) is $2\sigma\left(\alpha - \frac{1}{2}\sigma^2\right)$ times the cross variation of $W(t)$ and t, which was also shown in Remark 3.4.5 to be zero. We conclude that when the maximum step size $\|\Pi\|$ is small, the right-hand side of (3.4.15) is approximately equal to $\sigma^2(T_2 - T_1)$, and hence

$$\frac{1}{T_2 - T_1}\sum_{j=0}^{m-1}\left(\log\frac{S(t_{j+1})}{S(t_j)}\right)^2 \approx \sigma^2. \tag{3.4.16}$$

If the asset price $S(t)$ really is a geometric Brownian motion with constant volatility σ, then σ can be identified from price observations by computing the left-hand side of (3.4.16) and taking the square root. In theory, we can make this approximation as accurate as we like by decreasing the step size. In practice, there is a limit to how small the step size can be. Between trades, there is no information about prices, and when a trade takes place, it is sometimes at the bid price and sometimes at the ask price. On small time intervals, the difference in prices due to the bid–ask spread can be as large as the difference due to price fluctuations during the time interval.

3.5 Markov Property

In this section, we show that Brownian motion is a Markov process and discuss its *transition density*.

Theorem 3.5.1. *Let $W(t)$, $t \geq 0$, be a Brownian motion and let $\mathcal{F}(t)$, $t \geq 0$, be a filtration for this Brownian motion (see Definition 3.3.3). Then $W(t)$, $t \geq 0$, is a Markov process.*

PROOF: According to Definition 2.3.6, we must show that whenever $0 \leq s \leq t$ and f is a Borel-measurable function, there is another Borel-measurable function g such that

$$\mathbb{E}\big[f(W(t))\big|\mathcal{F}(s)\big] = g(W(s)). \tag{3.5.1}$$

To do this, we write

$$\mathbb{E}\big[f(W(t))\big|\mathcal{F}(s)\big] = \mathbb{E}\big[f((W(t) - W(s)) + W(s))\big|\mathcal{F}(s)\big]. \tag{3.5.2}$$

The random variable $W(t) - W(s)$ is independent of $\mathcal{F}(s)$, and the random variable $W(s)$ is $\mathcal{F}(s)$-measurable. This permits us to apply the Independence

Lemma, Lemma 2.3.4. In order to compute the expectation on the right-hand side of (3.5.2), we replace $W(s)$ by a dummy variable x to hold it constant and then take the unconditional expectation of the remaining random variable (i.e., we define $g(x) = \mathbb{E}f\big(W(t) - W(s) + x\big)$). But $W(t) - W(s)$ is normally distributed with mean zero and variance $t - s$. Therefore,

$$g(x) = \frac{1}{\sqrt{2\pi(t-s)}} \int_{-\infty}^{\infty} f(w+x) e^{-\frac{w^2}{2(t-s)}}\, dw. \tag{3.5.3}$$

The Independence Lemma states that if we now take the function $g(x)$ defined by (3.5.3) and replace the dummy variable x by the random variable $W(s)$, then equation (3.5.1) holds. □

We may make the change of variable $\tau = t - s$ and $y = w + x$ in (3.5.3) to obtain

$$g(x) = \frac{1}{\sqrt{2\pi\tau}} \int_{-\infty}^{\infty} f(y) e^{-\frac{(y-x)^2}{2\tau}}\, dy.$$

We define the *transition density* $p(\tau, x, y)$ for Brownian motion to be

$$p(\tau, x, y) = \frac{1}{\sqrt{2\pi\tau}} e^{-\frac{(y-x)^2}{2\tau}},$$

so that we may further rewrite (3.5.3) as

$$g(x) = \int_{-\infty}^{\infty} f(y) p(\tau, x, y)\, dy \tag{3.5.4}$$

and (3.5.1) as

$$\mathbb{E}\big[f\big(W(t)\big)\big|\mathcal{F}(s)\big] = \int_{-\infty}^{\infty} f(y) p(\tau, W(s), y)\, dy. \tag{3.5.5}$$

This equation has the following interpretation. Conditioned on the information in $\mathcal{F}(s)$ (which contains all the information obtained by observing the Brownian motion up to and including time s), the conditional density of $W(t)$ is $p(\tau, W(s), y)$. This is a density in the variable y. This density is normal with mean $W(s)$ and variance $\tau = t - s$. In particular, the only information from $\mathcal{F}(s)$ that is relevant is the value of $W(s)$. The fact that only $W(s)$ is relevant is the essence of the Markov property.

3.6 First Passage Time Distribution

In Chapter 5 of Volume I, we studied the first passage time for a random walk, first using the optional sampling theorem for martingales to obtain the distribution in Section 5.2 and then rederiving the distribution using the reflection principle in Section 5.3. Here we develop the first approach; the second is presented in the next section. In Sections 5.2 and 5.3 of Volume I, we observed after deriving the distribution of the first passage time for the symmetric random walk that our answer could easily be modified to obtain the first passage distribution for an asymmetric random walk. In this section, we work only with Brownian motion, the continuous-time counterpart of the symmetric random walk. The case of Brownian motion with drift, the continuous-time counterpart of an asymmetric random walk, is treated in Exercise 3.7. We revisit this problem in Chapter 7, where it is solved using Girsanov's Theorem. The resulting formulas often provide explicit pricing and hedging formulas for exotic options. Examples of the application of these formulas to such options are given in Chapter 7.

Just as we began in Section 5.2 of Volume I with a martingale that had the random walk in the exponential function, we must begin here with a martingale containing Brownian motion in the exponential function. We fix a constant σ. The so-called *exponential martingale* corresponding to σ, which is

$$Z(t) = \exp\left\{\sigma W(t) - \frac{1}{2}\sigma^2 t\right\}, \tag{3.6.1}$$

plays a key role in much of the remainder of this text.

Theorem 3.6.1 (Exponential martingale). *Let $W(t)$, $t \geq 0$, be a Brownian motion with a filtration $\mathcal{F}(t)$, $t \geq 0$, and let σ be a constant. The process $Z(t)$, $t \geq 0$, of (3.6.1) is a martingale.*

PROOF: For $0 \leq s \leq t$, we have

$$\begin{aligned}
&\mathbb{E}\big[Z(t)\big|\mathcal{F}(s)\big]\\
&= \mathbb{E}\left[\exp\left\{\sigma W(t) - \frac{1}{2}\sigma^2 t\right\}\Big|\mathcal{F}(s)\right]\\
&= \mathbb{E}\left[\exp\left\{\sigma\big(W(t) - W(s)\big)\right\}\cdot\exp\left\{\sigma W(s) - \frac{1}{2}\sigma^2 t\right\}\Big|\mathcal{F}(s)\right]\\
&= \exp\left\{\sigma W(s) - \frac{1}{2}\sigma^2 t\right\}\cdot\mathbb{E}\left[\exp\left\{\sigma\big(W(t) - W(s)\big)\right\}\big|\mathcal{F}(s)\right],
\end{aligned}\tag{3.6.2}$$

where we have used "taking out what is known" (Theorem 2.3.2(ii)) for the last step. We next use "independence" (Theorem 2.3.2(iv)) to write

$$\mathbb{E}\left[\exp\left\{\sigma\big(W(t) - W(s)\big)\right\}\big|\mathcal{F}(s)\right] = \mathbb{E}\left[\exp\left\{\sigma\big(W(t) - W(s)\big)\right\}\right].$$

Because $W(t) - W(s)$ is normally distributed with mean zero and variance $t - s$, this expected value is $\exp\left\{\frac{1}{2}\sigma^2(t - s)\right\}$ (see (3.2.13)). Substituting this into (3.6.2), we obtain the martingale property

$$\mathbb{E}\big[Z(t)\big|\mathcal{F}(s)\big] = \exp\left\{\sigma W(s) - \frac{1}{2}\sigma^2 s\right\} = Z(s). \qquad \square$$

Let m be a real number, and define the *first passage time* to level m

$$\tau_m = \min\{t \geq 0; W(t) = m\}. \tag{3.6.3}$$

This is the first time the Brownian motion W reaches the level m. If the Brownian motion never reaches the level m, we set $\tau_m = \infty$. A martingale that is stopped ("frozen" would be a more apt description) at a stopping time is still a martingale and thus must have constant expectation. (The text following Theorem 4.3.2 of Volume I discusses this in more detail.) Because of this fact,

$$1 = Z(0) = \mathbb{E}Z(t \wedge \tau_m) = \mathbb{E}\left[\exp\left\{\sigma W(t \wedge \tau_m) - \frac{1}{2}\sigma^2 (t \wedge \tau_m)\right\}\right], \tag{3.6.4}$$

where the notation $t \wedge \tau_m$ denotes the minimum of t and τ_m.

For the next step, we assume that $\sigma > 0$ and $m > 0$. In this case, the Brownian motion is always at or below level m for $t \leq \tau_m$ and so

$$0 \leq \exp\{\sigma W(t \wedge \tau_m)\} \leq e^{\sigma m}. \tag{3.6.5}$$

If $\tau_m < \infty$, the term $\exp\left\{-\frac{1}{2}\sigma^2(t \wedge \tau_m)\right\}$ is equal to $\exp\left\{-\frac{1}{2}\sigma^2 \tau_m\right\}$ for large enough t. On the other hand, if $\tau_m = \infty$, then the term $\exp\left\{-\frac{1}{2}\sigma^2(t \wedge \tau_m)\right\}$ is equal to $\exp\left\{-\frac{1}{2}\sigma^2 t\right\}$, and as $t \to \infty$, this converges to zero. We capture these two cases by writing

$$\lim_{t \to \infty} \exp\left\{-\frac{1}{2}\sigma^2(t \wedge \tau_m)\right\} = \mathbb{I}_{\{\tau_m < \infty\}} \exp\left\{-\frac{1}{2}\sigma^2 \tau_m\right\},$$

where the notation $\mathbb{I}_{\{\tau_m < \infty\}}$ is used to indicate the random variable that takes the value 1 if $\tau_m < \infty$ and otherwise takes the value zero. If $\tau_m < \infty$, then $\exp\{\sigma W(t \wedge \tau_m)\} = \exp\{\sigma W(\tau_m)\} = e^{\sigma m}$ when t becomes large enough. If $\tau_m = \infty$, then we do not know what happens to $\exp\{\sigma W(t \wedge \tau_m)\}$ as $t \to \infty$, but we at least know that this term is bounded because of (3.6.5). That is enough to ensure that the product of $\exp\{\sigma W(t \wedge \tau_m)\}$ and $\exp\left\{-\frac{1}{2}\sigma^2 \tau_m\right\}$ has limit zero in this case. In conclusion, we have

$$\lim_{t \to \infty} \exp\left\{\sigma W(t \wedge \tau_m) - \frac{1}{2}\sigma^2(t \wedge \tau_m)\right\} = \mathbb{I}_{\{\tau_m < \infty\}} \exp\left\{\sigma m - \frac{1}{2}\sigma^2 \tau_m\right\}.$$

We can now take the limit in (3.6.4)[3] to obtain

$$1 = \mathbb{E}\left[\mathbb{I}_{\{\tau_m < \infty\}} \exp\left\{\sigma m - \frac{1}{2}\sigma^2 \tau_m\right\}\right]$$

or, equivalently,

[3] The interchange of limit and expectation implicit in this step is justified by the Dominated Convergence Theorem, Theorem 1.4.9.

$$\mathbb{E}\left[\mathbb{I}_{\{\tau_m<\infty\}}\exp\left\{-\frac{1}{2}\sigma^2\tau_m\right\}\right]=e^{-\sigma m}. \tag{3.6.6}$$

Equation (3.6.6) holds when m and σ are positive. We may not substitute $\sigma=0$ into this equation, but since it holds for every positive σ, we may take the limit on both sides as $\sigma\downarrow 0$. This yields[4] $\mathbb{E}\left[\mathbb{I}_{\{\tau_m<\infty\}}\right]=1$ or, equivalently,

$$\mathbb{P}\{\tau_m<\infty\}=1. \tag{3.6.7}$$

Because τ_m is finite with probability one (we say τ_m is finite *almost surely*), we may drop the indicator of this event in (3.6.6) to obtain

$$\mathbb{E}\left[\exp\left\{-\frac{1}{2}\sigma^2\tau_m\right\}\right]=e^{-\sigma m}. \tag{3.6.8}$$

We have done the hard work in the proof of the following theorem.

Theorem 3.6.2. *For $m\in\mathbb{R}$, the first passage time of Brownian motion to level m is finite almost surely, and the Laplace transform of its distribution is given by*

$$\mathbb{E}e^{-\alpha\tau_m}=e^{-|m|\sqrt{2\alpha}}\ \textit{for all}\ \alpha>0. \tag{3.6.9}$$

PROOF: We consider first the case when m is positive. Let α be a positive constant, and set $\sigma=\sqrt{2\alpha}$, so that $\frac{1}{2}\sigma^2=\alpha$. Then (3.6.8) becomes (3.6.9). If m is negative, then because Brownian motion is symmetric, the first passage times τ_m and $\tau_{|m|}$ have the same distribution. Equation (3.6.9) for negative m follows. □

Remark 3.6.3. Differentiation of (3.6.9) with respect to α results in

$$\mathbb{E}\left[\tau_m e^{-\alpha\tau_m}\right]=\frac{|m|}{\sqrt{2\alpha}}e^{-|m|\sqrt{2\alpha}}\ \text{for all}\ \alpha>0.$$

Letting $\alpha\downarrow 0$, we obtain $\mathbb{E}\tau_m=\infty$ so long as $m\neq 0$.

3.7 Reflection Principle

3.7.1 Reflection Equality

In this section, we repeat for Brownian motion the reflection principle argument of Section 5.3 of Volume I for the random walk. The reader may wish to review that section before reading this one.

We fix a positive level m and a positive time t. We wish to "count" the Brownian motion paths that reach level m at or before time t (i.e., those paths for which the first passage time τ_m to level m is less than or equal to t). There are two types of such paths: those that reach level m prior to t but at time t are at some level w below m, and those that exceed level m at time t. There are also Brownian motion paths that are exactly at level m at time t, but unlike the case of the random walk in Section 5.3 of Volume I, the probability of this for Brownian motion is zero. We may thus ignore this possibility.

[4] Here we use the Monotone Convergence Theorem, Theorem 1.4.5.

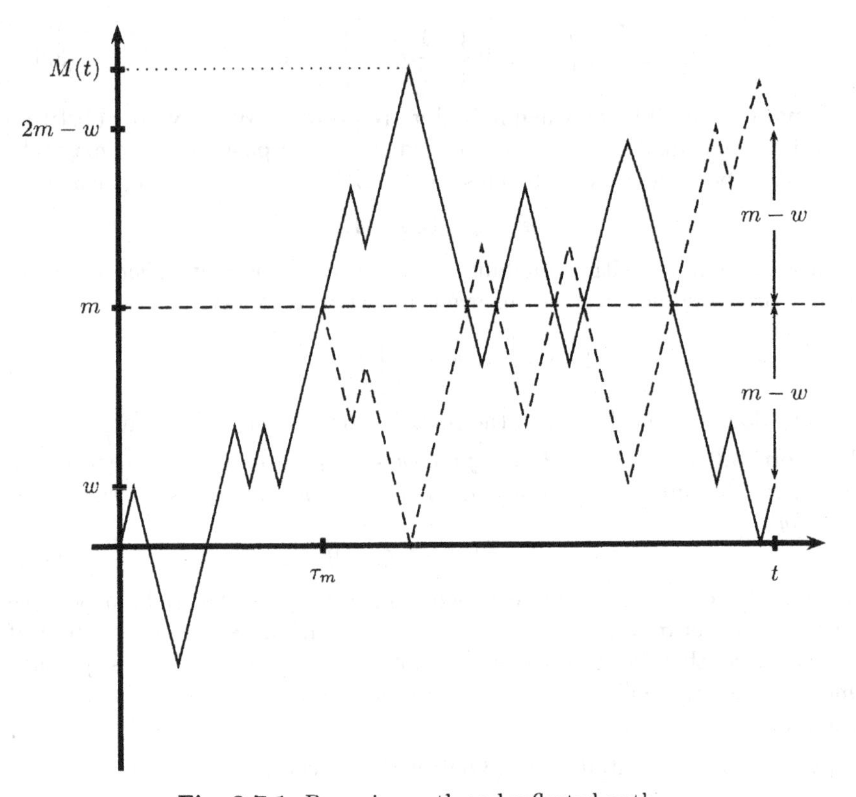

Fig. 3.7.1. Brownian path and reflected path.

As Figure 3.7.1 illustrates, for each Brownian motion path that reaches level m prior to time t but is at a level w below m at time t, there is a "reflected path" that is at level $2m - w$ at time t. This reflected path is constructed by switching the up and down moves of the Brownian motion from time τ_m onward. Of course, the probability that a Brownian motion path ends at exactly w or at exactly $2m - w$ is zero. In order to have nonzero probabilities, we consider the paths that reach level m prior to time t and are *at or below* level w at time t, and we consider their reflections, which are *at or above* $2m - w$ at time t. This leads to the key *reflection equality*

$$\mathbb{P}\{\tau_m \leq t, W(t) \leq w\} = \mathbb{P}\{W(t) \geq 2m - w\}, \quad w \leq m, m > 0. \tag{3.7.1}$$

3.7.2 First Passage Time Distribution

We draw two conclusions from (3.7.1). The first is the distribution for the random variable τ_m.

Theorem 3.7.1. *For all $m \neq 0$, the random variable τ_m has cumulative distribution function*

$$\mathbb{P}\{\tau_m \leq t\} = \frac{2}{\sqrt{2\pi}} \int_{\frac{|m|}{\sqrt{t}}}^{\infty} e^{-\frac{y^2}{2}}\, dy, \quad t \geq 0, \tag{3.7.2}$$

and density

$$f_{\tau_m}(t) = \frac{d}{dt}\mathbb{P}\{\tau_m \leq t\} = \frac{|m|}{t\sqrt{2\pi t}} e^{-\frac{m^2}{2t}}, \quad t \geq 0. \tag{3.7.3}$$

PROOF: We first consider the case $m > 0$. We substitute $w = m$ into the reflection formula (3.7.1) to obtain

$$\mathbb{P}\{\tau_m \leq t, W(t) \leq m\} = \mathbb{P}\{W(t) \geq m\}.$$

On the other hand, if $W(t) \geq m$, then we are guaranteed that $\tau_m \leq t$. In other words,

$$\mathbb{P}\{\tau_m \leq t, W(t) \geq m\} = \mathbb{P}\{W(t) \geq m\}.$$

Adding these two equations, we obtain the cumulative distribution function for τ_m:

$$\begin{aligned}\mathbb{P}\{\tau_m \leq t\} &= \mathbb{P}\{\tau_m \leq t, W(t) \leq m\} + \mathbb{P}\{\tau_m \leq t, W(t) \geq m\} \\ &= 2\mathbb{P}\{W(t) \geq m\} = \frac{2}{\sqrt{2\pi t}} \int_m^{\infty} e^{-\frac{x^2}{2t}}\, dx.\end{aligned}$$

We make the change of variable $y = \frac{x}{\sqrt{t}}$ in the integral, and this leads to (3.7.2) when m is positive. If m is negative, then τ_m and $\tau_{|m|}$ have the same distribution, and (3.7.2) provides the cumulative distribution function of the latter. Finally, (3.7.3) is obtained by differentiating (3.7.2) with respect to t. □

Remark 3.7.2. From (3.7.3), we see that

$$\mathbb{E}e^{-\alpha\tau_m} = \int_0^{\infty} e^{-\alpha m} f_{\tau_m}(t)\, dt = \int_0^{\infty} \frac{|m|}{t\sqrt{2\pi t}} e^{-\alpha m - \frac{m^2}{2t}}\, dt \text{ for all } \alpha > 0. \tag{3.7.4}$$

Theorem 3.6.2 provides the apparently different Laplace transform formula (3.6.9). These two formulas are in fact the same, and the steps needed to verify this are provided in Exercise 3.9. □

3.7.3 Distribution of Brownian Motion and Its Maximum

We define the *maximum to date* for Brownian motion to be

$$M(t) = \max_{0 \leq s \leq t} W(s). \tag{3.7.5}$$

This stochastic process is used in pricing barrier options. For the value of t in Figure 3.7.1, the random variable $M(t)$ is indicated. For positive m, we have

$M(t) \geq m$ if and only if $\tau_m \leq t$. This observation permits us to rewrite the reflection equality (3.7.1) as

$$\mathbb{P}\{M(t) \geq m, W(t) \leq w\} = \mathbb{P}\{W(t) \geq 2m - w\}, \quad w \leq m,\ m > 0. \quad (3.7.6)$$

From this, we can obtain the joint distribution of $W(t)$ and $M(t)$.

Theorem 3.7.3. *For $t > 0$, the joint density of $\big(M(t), W(t)\big)$ is*

$$f_{M(t),W(t)}(m, w) = \frac{2(2m - w)}{t\sqrt{2\pi t}} e^{-\frac{(2m-w)^2}{2t}}, \quad w \leq m,\ m > 0. \quad (3.7.7)$$

PROOF: Because

$$\mathbb{P}\{M(t) \geq m, W(t) \leq w\} = \int_m^\infty \int_{-\infty}^w f_{M(t),W(t)}(x, y)\, dy\, dx$$

and

$$\mathbb{P}\{W(t) \geq 2m - w\} = \frac{1}{\sqrt{2\pi t}} \int_{2m-w}^\infty e^{-\frac{z^2}{2t}}\, dz,$$

we have from (3.7.6) that

$$\int_m^\infty \int_{-\infty}^w f_{M(t),W(t)}(x, y)\, dy\, dx = \frac{1}{\sqrt{2\pi t}} \int_{2m-w}^\infty e^{-\frac{z^2}{2t}}\, dz.$$

We differentiate first with respect to m to obtain

$$-\int_{-\infty}^w f_{M(t),W(t)}(m, y)\, dy = -\frac{2}{\sqrt{2\pi t}} e^{-\frac{(2m-w)^2}{2t}}.$$

We next differentiate with respect to w to see that

$$-f_{M(t),W(t)}(m, w) = -\frac{2(2m - w)}{t\sqrt{2\pi t}} e^{-\frac{(2m-w)^2}{2t}}.$$

This is (3.7.7). □

When simulating Brownian motion to price exotic options, it is often convenient to first simulate the value of the Brownian motion at some time $T > 0$ and then simulate the maximum of the Brownian motion between times 0 and t. This second step requires that we know the distribution of the maximum of the Brownian motion $M(t)$ on $[0, t]$ conditioned on the value of $W(t)$. This conditional distribution is provided by the following corollary.

Corollary 3.7.4. *The conditional distribution of $M(t)$ given $W(t) = w$ is*

$$f_{M(t)|W(t)}(m|w) = \frac{2(2m - w)}{t} e^{-\frac{2m(m-w)}{t}}, \quad w \leq m, m > 0.$$

PROOF: The conditional density is the joint density divided by the marginal density of the conditioning random variable. The conditional density we seek here is

$$\begin{aligned} f_{M(t)|W(t)}(m|w) &= \frac{f_{M(t),W(t)}(m,w)}{f_{W(t)}(w)} \\ &= \frac{2(2m-w)}{t\sqrt{2\pi t}} \cdot \sqrt{2\pi t}\, e^{-\frac{(2m-w)^2}{2t}+\frac{w^2}{2t}} \\ &= \frac{2(2m-w)}{t} e^{-\frac{2m(m-w)}{t}}. \end{aligned}$$

□

3.8 Summary

Brownian motion is a continuous stochastic process $W(t)$, $t \geq 0$, that has independent, normally distributed increments. In this text, we adopt the convention that Brownian motion starts at zero at time zero, although one could add a constant a to our Brownian motion and obtain a "Brownian motion starting at a". For either Brownian motion starting at 0 or Brownian motion starting at a, if $0 = t_0 < t_1 < \cdots < t_m$, then the increments

$$W(t_1) - W(t_0),\ W(t_2) - W(t_1), \ldots, W(t_m) - W(t_{m-1})$$

are independent and normally distributed with

$$\mathbb{E}\big[W(t_{i+1}) - W(t_i)\big] = 0, \quad \mathrm{Var}\big[W(t_{i+1}) - W(t_i)\big] = t_{i+1} - t_i.$$

This is Definition 3.3.1. Associated with Brownian motion there is a filtration $\mathcal{F}(t)$, $t \geq 0$, such that for each $t \geq 0$ and $u \geq t$, $W(t)$ is $\mathcal{F}(t)$-measurable and $W(u) - W(t)$ is independent of $\mathcal{F}(t)$.

Brownian motion is both a martingale and a Markov process. Its transition density is

$$p(\tau, x, y) = \frac{1}{\sqrt{2\pi\tau}} e^{-\frac{(y-x)^2}{2\tau}}.$$

This is the density in the variable y for the random variable $W(s+\tau)$ given that $W(s) = x$.

A profound property of Brownian motion is that it accumulates quadratic variation at rate one per unit time (Theorem 3.4.3). If we choose a time interval $[T_1, T_2]$, choose partition points $T_1 = t_0 < t_1 < \cdots < t_m = T_2$, and compute $\sum_{j=0}^{m-1} \big(W(t_{j+1}) - W(t_j)\big)^2$, we get an answer that depends on the path along which the computation is done. However, if we let the number of partition points approach infinity and the length of the longest subinterval $t_{j+1} - t_j$ approach zero, this quantity has limit $T_2 - T_1$, the length of the interval over which the quadratic variation is being computed. We write $dW(t)\, dW(t) = dt$ to symbolize the fact that the amount of quadratic variation Brownian motion accumulates in an interval is equal to the length of the interval, *regardless of the path along which we do the computation.*

If we compute $\sum_{j=0}^{m-1}\left(W(t_{j+1})-W(t_j)\right)(t_{j+1}-t_j)$ or $\sum_{j=1}^{m-1}(t_{j+1}-t_j)^2$ and pass to the limit, we get zero (Remark 3.4.5). We symbolize this by writing $dW(t)\,dt = dt\,dt = 0$.

The first passage time of Brownian motion,

$$\tau_m = \min\{t \geq 0; W(t) = m\},$$

is the first time the Brownian motion reaches the level m. For $m \neq 0$, we have $\mathbb{P}\{\tau_m < \infty\} = 1$ (equation (3.6.7)) (i.e., the Brownian motion eventually reaches every nonzero level), but $\mathbb{E}\tau_m = \infty$ (Remark 3.6.3). The random variable τ_m is a stopping time, has density (Theorem 3.7.1)

$$f_{\tau_m}(t) = \frac{|m|}{t\sqrt{2\pi t}},$$

and this density has Laplace transform (Theorem 3.6.2; see also Exercise 3.9)

$$\mathbb{E}e^{-\alpha\tau_m} = e^{-|m|\sqrt{2\alpha}} \text{ for all } \alpha > 0.$$

The reflection principle used to determine the density $f_{\tau_m}(t)$ can also be used to determine the joint density of $W(t)$ and its maximum to date $M(t) = \max_{0\leq s\leq t} W(s)$. This joint density is (Theorem 3.7.3)

$$f_{M(t),W(t)}(m,w) = \frac{2(2m-w)}{t\sqrt{2\pi t}} e^{-\frac{(2m-w)^2}{2t}}, \quad w \leq m,\ m > 0.$$

3.9 Notes

In 1828, Robert Brown observed irregular movement of pollen suspended in water. This motion is now known to be caused by the buffeting of the pollen by water molecules, as explained by Einstein [62]. Bachelier [6] used Brownian motion (not geometric Brownian motion) as a model of stock prices, even though Brownian motion can take negative values. Lévy [107], [108] discovered many of the nonintuitive properties of Brownian motion. The first mathematically rigorous construction of Brownian motion is credited to Wiener [159], [160], and Brownian motion is sometimes called the *Wiener process.*

Brownian motion and its properties are presented in numerous texts, including Billingsley [10]. The development in these notes is a summary of that found in Karatzas and Shreve [101]. The properties of Brownian motion and many formulas useful for pricing exotic options are developed in Borodin and Salminen [18].

Convergence of discrete-time and/or discrete-state models to continuous-time models, a topic touched upon in Section 3.2.7, is treated by Amin and Khanna [3], Cox, Ross and Rubinstein [42], Duffie and Protter [60], and Willinger and Taqqu [162], among others.

3.10 Exercises

Exercise 3.1. According to Definition 3.3.3(iii), for $0 \leq t < u$, the Brownian motion increment $W(u) - W(t)$ is independent of the σ-algebra $\mathcal{F}(t)$. Use this property and property (i) of that definition to show that, for $0 \leq t < u_1 < u_2$, the increment $W(u_2) - W(u_1)$ is also independent of $\mathcal{F}(t)$.

Exercise 3.2. Let $W(t)$, $t \geq 0$, be a Brownian motion, and let $\mathcal{F}(t)$, $t \geq 0$, be a filtration for this Brownian motion. Show that $W^2(t) - t$ is a martingale. (Hint: For $0 \leq s \leq t$, write $W^2(t)$ as $(W(t) - W(s))^2 + 2W(t)W(s) - W^2(s)$.)

Exercise 3.3 (Normal kurtosis). The *kurtosis* of a random variable is defined to be the ratio of its fourth central moment to the square of its variance. For a normal random variable, the kurtosis is 3. This fact was used to obtain (3.4.7). This exercise verifies this fact.

Let X be a normal random variable with mean μ, so that $X - \mu$ has mean zero. Let the variance of X, which is also the variance of $X - \mu$, be σ^2. In (3.2.13), we computed the moment-generating function of $X - \mu$ to be $\varphi(u) = \mathbb{E}e^{u(X-\mu)} = e^{\frac{1}{2}u^2\sigma^2}$, where u is a real variable. Differentiating this function with respect to u, we obtain

$$\varphi'(u) = \mathbb{E}\left[(X - \mu)e^{u(X-\mu)}\right] = \sigma^2 u e^{\frac{1}{2}\sigma^2 u^2}$$

and, in particular, $\varphi'(0) = \mathbb{E}(X - \mu) = 0$. Differentiating again, we obtain

$$\varphi''(u) = \mathbb{E}\left[(X - \mu)^2 e^{u(X-\mu)}\right] = \left(\sigma^2 + \sigma^4 u^2\right) e^{\frac{1}{2}\sigma^2 u^2}$$

and, in particular, $\varphi''(0) = \mathbb{E}\left[(X - \mu)^2\right] = \sigma^2$. Differentiate two more times and obtain the normal kurtosis formula $\mathbb{E}\left[(X - \mu)^4\right] = 3\sigma^4$.

Exercise 3.4 (Other variations of Brownian motion). Theorem 3.4.3 asserts that if T is a positive number and we choose a partition Π with points $0 = t_0 < t_1 < t_2 < \cdots < t_n = T$, then as the number n of partition points approaches infinity and the length of the longest subinterval $\|\Pi\|$ approaches zero, the sample quadratic variation

$$\sum_{j=0}^{n-1} \left(W(t_{j+1}) - W(t_j)\right)^2$$

approaches T for almost every path of the Brownian motion W. In Remark 3.4.5, we further showed that $\sum_{j=0}^{n-1} \left(W(t_{j+1}) - W(t_j)\right)(t_{j+1} - t_j)$ and $\sum_{j=0}^{n-1} (t_{j+1} - t_j)^2$ have limit zero. We summarize these facts by the multiplication rules

$$dW(t)\,dW(t) = dt, \quad dW(t)\,dt = 0, \quad dt\,dt = 0. \tag{3.10.1}$$

(i) Show that as the number m of partition points approaches infinity and the length of the longest subinterval approaches zero, the sample first variation

$$\sum_{j=0}^{n-1} |W(t_{j+1}) - W(t_j)|$$

approaches ∞ for almost every path of the Brownian motion W. (Hint:

$$\sum_{j=0}^{n-1} \left(W(t_{j+1}) - W(t_j)\right)^2$$

$$\leq \max_{0 \leq k \leq n-1} |W(t_{k+1}) - W(t_k)| \cdot \sum_{j=0}^{n-1} |W(t_{j+1}) - W(t_j)|.)$$

(ii) Show that as the number n of partition points approaches infinity and the length of the longest subinterval approaches zero, the sample cubic variation

$$\sum_{j=0}^{n-1} |W(t_{j+1}) - W(t_j)|^3$$

approaches zero for almost every path of the Brownian motion W.

Exercise 3.5 (Black-Scholes-Merton formula). Let the interest rate r and the volatility $\sigma > 0$ be constant. Let

$$S(t) = S(0)e^{(r-\frac{1}{2}\sigma^2)t+\sigma W(t)}$$

be a geometric Brownian motion with mean rate of return r, where the initial stock price $S(0)$ is positive. Let K be a positive constant. Show that, for $T > 0$,

$$\mathbb{E}\left[e^{-rT}\left(S(T) - K\right)^+\right] = S(0)N\left(d_+(T, S(0))\right) - Ke^{-rT}N\left(d_-(T, S(0))\right),$$

where

$$d_\pm(T, S(0)) = \frac{1}{\sigma\sqrt{T}}\left[\log\frac{S(0)}{K} + \left(r \pm \frac{\sigma^2}{2}\right)T\right],$$

and N is the cumulative standard normal distribution function

$$N(y) = \frac{1}{\sqrt{2\pi}}\int_{-\infty}^{y} e^{-\frac{1}{2}z^2}\,dz = \frac{1}{\sqrt{2\pi}}\int_{-y}^{\infty} e^{-\frac{1}{2}z^2}\,dz.$$

Exercise 3.6. Let $W(t)$ be a Brownian motion and let $\mathcal{F}(t)$, $t \geq 0$, be an associated filtration.

(i) For $\mu \in \mathbb{R}$, consider the *Brownian motion with drift* μ:

$$X(t) = \mu t + W(t).$$

Show that for any Borel-measurable function $f(y)$, and for any $0 \le s < t$, the function

$$g(x) = \frac{1}{\sqrt{2\pi(t-s)}} \int_{-\infty}^{\infty} f(y) \exp\left\{-\frac{(y-x-\mu(t-s))^2}{2(t-s)}\right\} dy$$

satisfies $\mathbb{E}\big[f(X(t))\big|\mathcal{F}(s)\big] = g(X(s))$, and hence X has the Markov property. We may rewrite $g(x)$ as $g(x) = \int_{-\infty}^{\infty} f(y)p(\tau, x, y)\, dy$, where $\tau = t-s$ and

$$p(\tau, x, y) = \frac{1}{\sqrt{2\pi\tau}} \exp\left\{-\frac{(y-x-\mu\tau)^2}{2\tau}\right\}$$

is the *transition density* for Brownian motion with drift μ.

(ii) For $\nu \in \mathbb{R}$ and $\sigma > 0$, consider the *geometric Brownian motion*

$$S(t) = S(0)e^{\sigma W(t) + \nu t}.$$

Set $\tau = t - s$ and

$$p(\tau, x, y) = \frac{1}{\sigma y \sqrt{2\pi\tau}} \exp\left\{-\frac{\left(\log\frac{y}{x} - \nu\tau\right)^2}{2\sigma^2\tau}\right\}.$$

Show that for any Borel-measurable function $f(y)$ and for any $0 \le s < t$ the function $g(x) = \int_0^{\infty} h(y)p(\tau, x, y)\, dy$ satisfies $\mathbb{E}\big[f(S(t))\big|\mathcal{F}(s)\big] = g(S(s))$ and hence S has the Markov property and $p(\tau, x, y)$ is its transition density.

Exercise 3.7. Theorem 3.6.2 provides the Laplace transform of the density of the first passage time for Brownian motion. This problem derives the analogous formula for Brownian motions with drift. Let W be a Brownian motion. Fix $m > 0$ and $\mu \in \mathbb{R}$. For $0 \le t < \infty$, define

$$\begin{aligned} X(t) &= \mu t + W(t), \\ \tau_m &= \min\{t \ge 0; X(t) = m\}. \end{aligned}$$

As usual, we set $\tau_m = \infty$ if $X(t)$ never reaches the level m. Let σ be a positive number and set

$$Z(t) = \exp\left\{\sigma X(t) - \left(\sigma\mu + \frac{1}{2}\sigma^2\right)t\right\}.$$

(i) Show that $Z(t)$, $t \ge 0$, is a martingale.

(ii) Use (i) to conclude that

$$\mathbb{E}\left[\exp\left\{\sigma X(t\wedge\tau_m) - \left(\sigma\mu + \frac{1}{2}\sigma^2\right)(t\wedge\tau_m)\right\}\right] = 1, \quad t \ge 0.$$

(iii) Now suppose $\mu \geq 0$. Show that, for $\sigma > 0$,

$$\mathbb{E}\left[\exp\left\{\sigma m - \left(\sigma\mu + \frac{1}{2}\sigma^2\right)\tau_m\right\}\mathbb{I}_{\{\tau_m<\infty\}}\right] = 1.$$

Use this fact to show $\mathbb{P}\{\tau_m < \infty\} = 1$ and to obtain the Laplace transform

$$\mathbb{E}e^{-\alpha\tau_m} = e^{m\mu - m\sqrt{2\alpha+\mu^2}} \text{ for all } \alpha > 0.$$

(iv) Show that if $\mu > 0$, then $\mathbb{E}\tau_m < \infty$. Obtain a formula for $\mathbb{E}\tau_m$. (Hint: Differentiate the formula in (iii) with respect to α.)

(v) Now suppose $\mu < 0$. Show that, for $\sigma > -2\mu$,

$$\mathbb{E}\left[\exp\left\{\sigma m - \left(\sigma\mu + \frac{1}{2}\sigma^2\right)\tau_m\right\}\mathbb{I}_{\{\tau_m<\infty\}}\right] = 1.$$

Use this fact to show that $\mathbb{P}\{\tau_m < \infty\} = e^{-2x|\mu|}$, which is strictly less than one, and to obtain the Laplace transform

$$\mathbb{E}e^{-\alpha\tau_m} = e^{m\mu - m\sqrt{2\alpha+\mu^2}} \text{ for all } \alpha > 0.$$

Exercise 3.8. This problem presents the convergence of the distribution of stock prices in a sequence of binomial models to the distribution of geometric Brownian motion. In contrast to the analysis of Subsection 3.2.7, here we allow the interest rate to be different from zero.

Let $\sigma > 0$ and $r \geq 0$ be given. For each positive integer n, we consider a binomial model taking n steps per unit time. In this model, the interest rate per period is $\frac{r}{n}$, the up factor is $u_n = e^{\sigma/\sqrt{n}}$, and the down factor is $d_n = e^{-\sigma/\sqrt{n}}$. The risk-neutral probabilities are then

$$\tilde{p}_n = \frac{\frac{r}{n} + 1 - e^{-\sigma/\sqrt{n}}}{e^{\sigma/\sqrt{n}} - e^{-\sigma/\sqrt{n}}}, \quad \tilde{q}_n = \frac{e^{\sigma/\sqrt{n}} - \frac{r}{n} - 1}{e^{\sigma/\sqrt{n}} - e^{-\sigma/\sqrt{n}}}.$$

Let t be an arbitrary positive rational number, and for each positive integer n for which nt is an integer, define

$$M_{nt,n} = \sum_{k=1}^{nt} X_{k,n},$$

where $X_{1,n}, \ldots, X_{n,n}$ are independent, identically distributed random variables with

$$\widetilde{\mathbb{P}}\{X_{k,n} = 1\} = \tilde{p}_n, \quad \widetilde{\mathbb{P}}\{X_{k,n} = -1\} = \tilde{q}_n, \; k = 1, \ldots, n.$$

The stock price at time t in this binomial model, which is the result of nt steps from the initial time, is given by (see (3.2.15) for a similar equation)

$$
\begin{aligned}
S_n(t) &= S(0) u_n^{\frac{1}{2}(nt+M_{nt,n})} d_n^{\frac{1}{2}(nt-M_{nt,n})} \\
&= S(0) \exp\left\{ \frac{\sigma}{2\sqrt{n}} (nt + M_{nT,n}) \right\} \exp\left\{ -\frac{\sigma}{2\sqrt{n}} (nt - M_{nt,n}) \right\} \\
&= S(0) \exp\left\{ \frac{\sigma}{\sqrt{n}} M_{nt,n} \right\}.
\end{aligned}
$$

This problem shows that as $n \to \infty$, the distribution of the sequence of random variables $\frac{\sigma}{\sqrt{n}} M_{nt,n}$ appearing in the exponent above converges to the normal distribution with mean $(r - \frac{1}{2}\sigma^2)t$ and variance $\sigma^2 t$. Therefore, the limiting distribution of $S_n(t)$ is the same as the distribution of the geometric Brownian motion $S(0) \exp\{\sigma W(t) + (r - \frac{1}{2}\sigma)t\}$ at time t.

(i) Show that the moment-generating function $\varphi_n(u)$ of $\frac{1}{\sqrt{n}} M_{nt,n}$ is given by

$$
\varphi_n(u) = \left[e^{\frac{u}{\sqrt{n}}} \left(\frac{\frac{r}{n} + 1 - e^{-\sigma/\sqrt{n}}}{e^{\sigma/\sqrt{n}} - e^{-\sigma/\sqrt{n}}} \right) - e^{-\frac{u}{\sqrt{n}}} \left(\frac{\frac{r}{n} + 1 - e^{\sigma/\sqrt{n}}}{e^{\sigma/\sqrt{n}} - e^{-\sigma/\sqrt{n}}} \right) \right]^{nt}.
$$

(ii) We want to compute

$$
\lim_{n \to \infty} \varphi_n(u) = \lim_{x \downarrow 0} \varphi_{\frac{1}{x^2}}(u),
$$

where we have made the change of variable $x = \frac{1}{\sqrt{n}}$. To do this, we will compute $\log \varphi_{\frac{1}{x^2}}(u)$ and then take the limit as $x \downarrow 0$. Show that

$$
\log \varphi_{\frac{1}{x^2}}(u) = \frac{t}{x^2} \log \left[\frac{(rx^2 + 1)\sinh ux + \sinh(\sigma - u)x}{\sinh \sigma x} \right]
$$

(the definitions are $\sinh z = \frac{e^z - e^{-z}}{2}$, $\cosh z = \frac{e^z + e^{-z}}{2}$), and use the formula

$$
\sinh(A - B) = \sinh A \cosh B - \cosh A \sinh B
$$

to rewrite this as

$$
\log \varphi_{\frac{1}{x^2}}(u) = \frac{t}{x^2} \log \left[\cosh ux + \frac{(rx^2 + 1 - \cosh \sigma x)\sinh ux}{\sinh \sigma x} \right].
$$

(iii) Use the Taylor series expansions

$$
\cosh z = 1 + \frac{1}{2} z^2 + O(z^4), \ \sinh z = z + O(z^3),
$$

to show that

$$
\begin{aligned}
\cosh ux &+ \frac{(rx^2 + 1 - \cosh \sigma x)\sinh ux}{\sinh \sigma x} \\
&= 1 + \frac{1}{2} u^2 x^2 + \frac{rux^2}{\sigma} - \frac{1}{2} ux^2 \sigma + O(x^4). \qquad (3.10.2)
\end{aligned}
$$

The notation $O(x^j)$ is used to represent terms of the order x^j.

(iv) Use the Taylor series expansion $\log(1+x) = x + O(x^2)$ to compute $\lim_{x\downarrow 0}\log\varphi_{\frac{1}{x^2}}(u)$. Now explain how you know that the limiting distribution for $\frac{\sigma}{\sqrt{n}}M_{nt,n}$ is normal with mean $(r-\frac{1}{2}\sigma^2)t$ and variance $\sigma^2 t$.

Exercise 3.9 (Laplace transform of first passage density). The solution to this problem is long and technical. It is included for the sake of completeness, but the reader may safely skip it.

Let $m>0$ be given, and define

$$f(t,m) = \frac{m}{t\sqrt{2\pi t}}\exp\left\{-\frac{m^2}{2t}\right\}.$$

According to (3.7.3) in Theorem 3.7.1, $f(t,m)$ is the density in the variable t of the first passage time $\tau_m = \min\{t\geq 0; W(t)=m\}$, where W is a Brownian motion without drift. Let

$$g(\alpha,m) = \int_0^\infty e^{-\alpha t}f(t,m)\,dt, \quad \alpha>0,$$

be the Laplace transform of the density $f(t,m)$. This problem verifies that $g(\alpha,m) = e^{-m\sqrt{2\alpha}}$, which is the formula derived in Theorem 3.6.2.

(i) For $k\geq 1$, define

$$a_k(m) = \frac{1}{\sqrt{2\pi}}\int_0^\infty t^{-k/2}\exp\left\{-\alpha t - \frac{m^2}{2t}\right\}dt,$$

so $g(\alpha,m) = ma_3(m)$. Show that

$$\begin{aligned} g_m(\alpha,m) &= a_3(m) - m^2 a_5(m),\\ g_{mm}(\alpha,m) &= -3ma_5(m) + m^3 a_7(m). \end{aligned}$$

(ii) Use integration by parts to show that

$$a_5(m) = -\frac{2\alpha}{3}a_3(m) + \frac{m^2}{3}a_7(m).$$

(iii) Use (i) and (ii) to show that g satisfies the second-order ordinary differential equation

$$g_{mm}(\alpha,m) = 2\alpha g(\alpha,m).$$

(iv) The general solution to a second-order ordinary differential equation of the form

$$ay''(m) + by'(m) + cy(m) = 0$$

is

$$y(m) = A_1e^{\lambda_1 m} + A_2e^{\lambda_2 m},$$

where λ_1 and λ_2 are roots of the *characteristic equation*

$$a\lambda^2 + b\lambda + c = 0.$$

Here we are assuming that these roots are distinct. Find the general solution of the equation in (iii) when $\alpha > 0$. This solution has two undetermined parameters A_1 and A_2, and these may depend on α.

(v) Derive the bound

$$g(\alpha, m) \leq \frac{m}{\sqrt{2\pi}} \int_0^m \sqrt{\frac{m}{t}}\, t^{-3/2} \exp\left\{-\frac{m^2}{2t}\right\} dt + \frac{1}{\sqrt{2\pi m}} \int_m^\infty e^{-\alpha t}\, dt$$

and use it to show that, for every $\alpha > 0$,

$$\lim_{m \to \infty} g(\alpha, m) = 0.$$

Use this fact to determine one of the parameters in the general solution to the equation in (iii).

(vi) Using first the change of variable $s = t/m^2$ and then the change of variable $y = 1/\sqrt{s}$, show that

$$\lim_{m \downarrow 0} g(\alpha, m) = 1.$$

Use this fact to determine the other parameter in the general solution to the equation in (iii).

4

Stochastic Calculus

4.1 Introduction

This chapter defines Itô integrals and develops their properties. These are used to model the value of a portfolio that results from trading assets in continuous time. The calculus used to manipulate these integrals is based on the Itô-Doeblin formula of Section 4.4 and differs from ordinary calculus. This difference can be traced to the fact that Brownian motion has a nonzero quadratic variation and is the source of the volatility term in the Black-Scholes-Merton partial differential equation. The Black-Scholes-Merton equation is presented in Section 4.5. This is in the spirit of Sections 1.1 and 1.2 of Volume I in which we priced options by determining the portfolio that would hedge a short position. In particular, there is no discussion of risk-neutral pricing in this chapter. That topic is taken up in Chapter 5.

Section 4.6 extends stochastic calculus to multiple processes. Section 4.7 discusses the Brownian bridge, which plays a useful role in Monte Carlo methods for pricing. We do not treat Monte Carlo methods in this text; we include the Brownian bridge only because it is a natural application of the stochastic calculus developed in the earlier sections.

4.2 Itô's Integral for Simple Integrands

We fix a positive number T and seek to make sense of

$$\int_0^T \Delta(t)\, dW(t). \tag{4.2.1}$$

The basic ingredients here are a Brownian motion $W(t)$, $t \geq 0$, together with a filtration $\mathcal{F}(t)$, $t \geq 0$, for this Brownian motion. We will let the integrand $\Delta(t)$ be an adapted stochastic process. Our reason for doing this is that $\Delta(t)$ will eventually be the position we take in an asset at time t, and this typically depends on the price path of the asset up to time t. Anything that depends on the path of a random process is itself random. Requiring $\Delta(t)$ to be adapted means that we require $\Delta(t)$ to be $\mathcal{F}(t)$-measurable for each $t \geq 0$. In other words, the information available at time t is sufficient to evaluate $\Delta(t)$ at that time. When we are standing at time 0 and t is strictly positive, $\Delta(t)$ is unknown to us. It is a random variable. When we get to time t, we have sufficient information to evaluate $\Delta(t)$; its randomness has been resolved.

Recall that increments of the Brownian motion after time t are independent of $\mathcal{F}(t)$, and since $\Delta(t)$ is $\mathcal{F}(t)$-measurable, it must also be independent of these future Brownian increments. Positions we take in assets may depend on the price history of those assets, but they must be independent of the future increments of the Brownian motion that drives those prices.

The problem we face when trying to assign meaning to the Itô integral (4.2.1) is that Brownian motion paths cannot be differentiated with respect to time. If $g(t)$ is a differentiable function, then we can define

$$\int_0^T \Delta(t)\,dg(t) = \int_0^T \Delta(t)g'(t)\,dt,$$

where the right-hand side is an ordinary (Lebesgue) integral with respect to time. This will not work for Brownian motion.

4.2.1 Construction of the Integral

To define the integral (4.2.1), Itô devised the following way around the nondifferentiability of the Brownian paths. We first define the Itô integral for simple integrands $\Delta(t)$ and then extend it to nonsimple integrands as a limit of the integral of simple integrands. We describe this procedure.

Let $\Pi = \{t_0, t_1, \dots, t_n\}$ be a partition of $[0,T]$; i.e.,

$$0 = t_0 \leq t_1 \leq \cdots \leq t_n = T.$$

Assume that $\Delta(t)$ is constant in t on each subinterval $[t_j, t_{j+1})$. Such a process $\Delta(t)$ is a *simple process*.

Figure 4.2.1 shows a single path of a simple process $\Delta(t)$. We shall always choose these simple processes, as shown in this figure, to take a value at a partition time t_j and then hold it up to but not including the next partition time t_{j+1}. Although it is not apparent from Figure 4.2.1, the path shown depends on the same ω on which the path of the Brownian motion $W(t)$ (not shown) depends. If one were to choose a different ω, there would be a different path of the Brownian motion and possibly a different path of $\Delta(t)$. However, the value of $\Delta(t)$ can depend only on the information available at time t. Since there is no information at time 0, the value of $\Delta(0)$ must be the same for all paths, and hence the first piece of $\Delta(t)$, for $0 \leq t < t_1$, does not really depend on ω. The value of $\Delta(t)$ on the second interval, $[t_1, t_2)$, can depend on observations made during the first time interval $[0, t_1)$.

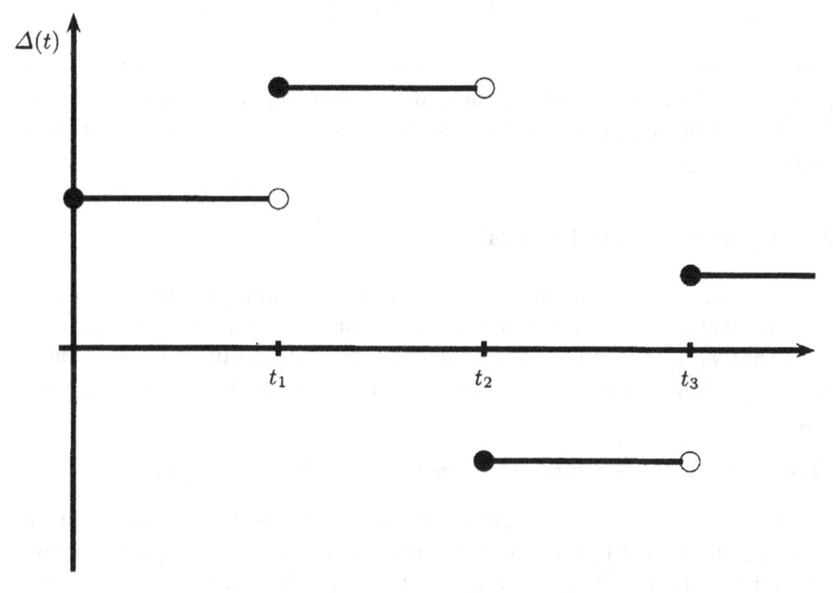

Fig. 4.2.1. A path of a simple process.

We shall think of the interplay between the simple process $\Delta(t)$ and the Brownian motion $W(t)$ in (4.2.1) in the following way. Regard $W(t)$ as the price per share of an asset at time t. (Since Brownian motion can take negative as well as positive values, it is not a good model of the price of a limited-liability asset such as a stock. For the sake of this illustration, we ignore that issue.) Think of $t_0, t_1, \ldots, t_{n-1}$ as the *trading dates* in the asset, and think of $\Delta(t_0), \Delta(t_1), \ldots, \Delta(t_{n-1})$ as the position (number of shares) taken in the asset at each trading date and held to the next trading date. The gain from trading at each time t is given by

$$
\begin{aligned}
I(t) &= \Delta(t_0)[W(t) - W(t_0)] = \Delta(0)W(t), \quad 0 \le t \le t_1, \\
I(t) &= \Delta(0)W(t_1) + \Delta(t_1)[W(t) - W(t_1)], \quad t_1 \le t \le t_2, \\
I(t) &= \Delta(0)W(t_1) + \Delta(t_1)[W(t_2) - W(t_1)] + \Delta(t_2)[W(t) - W(t_2)], \\
&\qquad\qquad t_2 \le t \le t_3,
\end{aligned}
$$

and so on. In general, if $t_k \le t \le t_{k+1}$, then

$$
I(t) = \sum_{j=0}^{k-1} \Delta(t_j)[W(t_{j+1}) - W(t_j)] + \Delta(t_k)[W(t) - W(t_k)]. \tag{4.2.2}
$$

The process $I(t)$ in (4.2.2) is the Itô integral of the simple process $\Delta(t)$, a fact that we write as

$$I(t) = \int_0^t \Delta(u)\, dW(u).$$

In particular, we can take $t = t_n = T$, and (4.2.2) provides a definition for the Itô integral (4.2.1). We have managed to define this integral not only for the upper limit of integration T but also for every upper limit of integration t between 0 and T.

4.2.2 Properties of the Integral

The Itô integral (4.2.2) is defined as the gain from trading in the martingale $W(t)$. A martingale has no tendency to rise or fall, and hence it is to be expected that $I(t)$, thought of as a process in its upper limit of integration t, also has no tendency to rise or fall. We formalize this observation by the next theorem and proof.

Theorem 4.2.1. *The Itô integral defined by (4.2.2) is a martingale.*

PROOF: Let $0 \leq s \leq t \leq T$ be given. We shall assume that s and t are in different subintervals of the partition Π (i.e., there are partition points t_ℓ and t_k such that $t_\ell < t_k$, $s \in [t_\ell, t_{\ell+1})$, and $t \in [t_k, t_{k+1})$). If s and t are in the same subinterval, the following proof simplifies. Equation (4.2.2) may be rewritten as

$$\begin{aligned} I(t) &= \sum_{j=0}^{\ell-1} \Delta(t_j)\big[W(t_{j+1}) - W(t_j)\big] + \Delta(t_\ell)\big[W(t_{\ell+1}) - W(t_\ell)\big] \\ &\quad + \sum_{j=\ell+1}^{k-1} \Delta(t_j)\big[W(t_{j+1}) - W(t_j)\big] + \Delta(t_k)\big[W(t) - W(t_k)\big]. \end{aligned} \tag{4.2.3}$$

We must show that $\mathbb{E}\big[I(t)\big|\mathcal{F}(s)\big] = I(s)$. We take the conditional expectation of each of the four terms on the right-hand side of (4.2.3). Every random variable in the first sum $\sum_{j=0}^{\ell-1} \Delta(t_j)\big[W(t_{j+1}) - W(t_j)\big]$ is $\mathcal{F}(s)$-measurable because the latest time appearing in this sum is t_ℓ and $t_\ell \leq s$. Therefore,

$$\mathbb{E}\left[\sum_{j=0}^{\ell-1} \Delta(t_j)\big[W(t_{j+1}) - W(t_j)\big]\,\middle|\, \mathcal{F}(s)\right] = \sum_{j=0}^{\ell-1} \Delta(t_j)\big[W(t_{j+1}) - W(t_j)\big]. \tag{4.2.4}$$

For the second term on the right-hand side of (4.2.3), we "take out what is known" (Theorem 2.3.2(ii)) and use the martingale property of W to write

$$\begin{aligned} \mathbb{E}\big[\Delta(t_\ell)\big(W(t_{\ell+1}) - W(t_\ell)\big)\big|\mathcal{F}(s)\big] &= \Delta(t_\ell)\big(\mathbb{E}\big[W(t_{\ell+1})\big|\mathcal{F}(s)\big] - W(t_\ell)\big) \\ &= \Delta(t_\ell)\big(W(s) - W(t_\ell)\big). \end{aligned} \tag{4.2.5}$$

Adding (4.2.4) and (4.2.5), we obtain $I(s)$.

It remains to show that the conditional expectations of the third and fourth terms on the right-hand side of (4.2.3) are zero. We will then have $\mathbb{E}\big[I(t)\big|\mathcal{F}(s)\big] = I(s)$.

The summands in the third term are of the form $\Delta(t_j)\big[W(t_{j+1}) - W(t_j)\big]$, where $t_j \geq t_{\ell+1} > s$. This permits us to use the following iterated conditioning trick, which is based on properties (iii) (iterated conditioning) and (ii) (taking out what is known) of Theorem 2.3.2:

$$\begin{aligned}
&\mathbb{E}\Big\{\Delta(t_j)\big(W(t_{j+1}) - W(t_j)\big)\Big|\mathcal{F}(s)\Big\} \\
&\quad = \mathbb{E}\Big\{\mathbb{E}\big[\Delta(t_j)\big(W(t_{j+1}) - W(t_j)\big)\big|\mathcal{F}(t_j)\big]\Big|\mathcal{F}(s)\Big\} \\
&\quad = \mathbb{E}\Big\{\Delta(t_j)\big(\mathbb{E}\big[W(t_{j+1})\big|\mathcal{F}(t_j)\big] - W(t_j)\big)\Big|\mathcal{F}(s)\Big\} \\
&\quad = \mathbb{E}\Big\{\Delta(t_j)\big(W(t_j) - W(t_j)\big)\Big|\mathcal{F}(s)\Big\} \;=\; 0.
\end{aligned}$$

At the end, we have used the fact that W is a martingale. Because the conditional expectation of each of the summands in the third term on the right-hand side of (4.2.3) is zero, the conditional expectation of the whole term is zero:

$$\mathbb{E}\left\{\sum_{j=\ell+1}^{k-1} \Delta(t_j)\big[W(t_{j+1}) - W(t_j)\big]\,\middle|\,\mathcal{F}(s)\right\} = 0.$$

The fourth term on the right-hand side of (4.2.3) is treated like the summands in the third term, with the result that

$$\begin{aligned}
&\mathbb{E}\Big\{\Delta(t_k)\big(W(t) - W(t_k)\big)\Big|\mathcal{F}(s)\Big\} \\
&\quad = \mathbb{E}\Big\{\mathbb{E}\big[\Delta(t_k)\big(W(t) - W(t_k)\big)\big|\mathcal{F}(t_k)\big]\Big|\mathcal{F}(s)\Big\} \\
&\quad = \mathbb{E}\Big\{\Delta(t_k)\big(\mathbb{E}\big[W(t)\big|\mathcal{F}(t_k)\big] - W(t_k)\big)\Big|\mathcal{F}(s)\Big\} \\
&\quad = \mathbb{E}\Big\{\Delta(t_k)\big(W(t_k) - W(t_k)\big)\Big|\mathcal{F}(s)\Big\} \;=\; 0.
\end{aligned}$$

This concludes the proof. □

Because $I(t)$ is a martingale and $I(0) = 0$, we have $\mathbb{E}I(t) = 0$ for all $t \geq 0$. It follows that $\operatorname{Var} I(t) = \mathbb{E}\, I^2(t)$, a quantity that can be evaluated by the formula in the next theorem.

Theorem 4.2.2 (Itô isometry). *The Itô integral defined by (4.2.2) satisfies*

$$\mathbb{E}\, I^2(t) = \mathbb{E}\int_0^t \Delta^2(u)\,du. \tag{4.2.6}$$

PROOF: To simplify the notation, we set $D_j = W(t_{j+1}) - W(t_j)$ for $j = 0, \dots, k-1$ and $D_k = W(t) - W(t_k)$ so that (4.2.2) may be written as $I(t) = \sum_{j=0}^{k} \Delta(t_j) D_j$ and

$$I^2(t) = \sum_{j=0}^{k} \Delta^2(t_j) D_j^2 + 2 \sum_{0 \le i < j \le k} \Delta(t_i)\Delta(t_j) D_i D_j.$$

We first show that the expected value of each of the cross terms is zero. For $i < j$, the random variable $\Delta(t_i)\Delta(t_j)D_i$ is $\mathcal{F}(t_j)$-measurable, while the Brownian increment D_j is independent of $\mathcal{F}(t_j)$. Furthermore, $\mathbb{E}D_j = 0$. Therefore,

$$\mathbb{E}\big[\Delta(t_i)\Delta(t_j)D_iD_j\big] = \mathbb{E}\big[\Delta(t_i)\Delta(t_j)D_i\big] \cdot \mathbb{E}D_j = \mathbb{E}\big[\Delta(t_i)\Delta(t_j)D_i\big] \cdot 0 = 0.$$

We next consider the square terms $\Delta^2(t_j)D_j^2$. The random variable $\Delta^2(t_j)$ is $\mathcal{F}(t_j)$-measurable, and the squared Brownian increment D_j^2 is independent of $\mathcal{F}(t_j)$. Furthermore, $\mathbb{E}D_j^2 = t_{j+1} - t_j$ for $j = 0, \dots, k-1$ and $\mathbb{E}D_k^2 = t - t_k$. Therefore,

$$\begin{aligned} \mathbb{E}I^2(t) &= \sum_{j=0}^{k} \mathbb{E}\big[\Delta^2(t_j)D_j^2\big] = \sum_{j=1}^{k} \mathbb{E}\Delta^2(t_j) \cdot \mathbb{E}D_j^2 \\ &= \sum_{j=0}^{k-1} \mathbb{E}\Delta^2(t_j)(t_{j+1} - t_j) + \mathbb{E}\Delta^2(t_k)(t - t_k). \end{aligned} \tag{4.2.7}$$

But $\Delta(t_j)$ is constant on the interval $[t_j, t_{j+1})$, and hence $\Delta^2(t_j)(t_{j+1} - t_j) = \int_{t_j}^{t_{j+1}} \Delta^2(u)\,du$. Similarly, $\Delta^2(t_k)(t - t_k) = \int_{t_k}^{t} \Delta^2(u)\,du$. We may thus continue (4.2.7) to obtain

$$\begin{aligned} \mathbb{E}I^2(t) &= \sum_{j=0}^{k-1} \mathbb{E} \int_{t_j}^{t_{j+1}} \Delta^2(u)\,du + \mathbb{E}\int_{t_k}^{t} \Delta^2(u)\,du \\ &= \mathbb{E}\left[\sum_{j=0}^{k-1} \int_{t_j}^{t_{j+1}} \Delta^2(u)\,du + \int_{t_k}^{t} \Delta^2(u)\,du\right] = \mathbb{E}\int_0^t \Delta^2(u)\,du. \qquad \square \end{aligned}$$

Finally, we turn to the quadratic variation of the Itô integral $I(t)$ thought of as a process in its upper limit of integration t. Brownian motion accumulates quadratic variation at rate one per unit time. However, Brownian motion is scaled in a time- and path-dependent way by the integrand $\Delta(u)$ as it enters the Itô integral $I(t) = \int_0^t \Delta(u)\,dB(u)$. Because increments are squared in the computation of quadratic variation, the quadratic variation of Brownian motion will be scaled by $\Delta^2(u)$ as it enters the Itô integral. The following theorem gives the precise statement.

Theorem 4.2.3. *The quadratic variation accumulated up to time t by the Itô integral (4.2.2) is*

$$[I,I](t) = \int_0^t \Delta^2(u)\,du. \tag{4.2.8}$$

PROOF: We first compute the quadratic variation accumulated by the Itô integral on one of the subintervals $[t_j, t_{j+1}]$ on which $\Delta(u)$ is constant. For this, we choose partition points

$$t_j = s_0 < s_1 < \cdots < s_m = t_{j+1}$$

and consider

$$\begin{aligned}\sum_{i=0}^{m-1} \big[I(s_{i+1}) - I(s_i)\big]^2 &= \sum_{i=0}^{m-1} \big[\Delta(t_j)\big(W(s_{i+1}) - W(s_i)\big)\big]^2 \\ &= \Delta^2(t_j) \sum_{i=0}^{m-1} \big(W(s_{i+1}) - W(s_i)\big)^2. \end{aligned} \tag{4.2.9}$$

As $m \to \infty$ and the step size $\max_{i=0,\dots,m-1}(s_{i+1} - s_i)$ approaches zero, the term $\sum_{i=0}^{m-1}\big(W(s_{i+1}) - W(s_i)\big)^2$ converges to the quadratic variation accumulated by Brownian motion between times t_j and t_{j+1}, which is $t_{j+1} - t_j$. Therefore, the limit of (4.2.9), which is the quadratic variation accumulated by the Itô integral between times t_j and t_{j+1}, is

$$\Delta^2(t_j)(t_{j+1} - t_j) = \int_{t_j}^{t_{j+1}} \Delta^2(u)\,du,$$

where again we have used the fact that $\Delta(u)$ is constant for $t_j \le u < t_{j+1}$. Analogously, the quadratic variation accumulated by the Itô integral between times t_k and t is $\int_{t_k}^t \Delta^2(u)\,du$. Adding up all these pieces, we obtain (4.2.8).
□

In Theorems 4.2.2 and 4.2.3, we finally see how the quadratic variation and the variance of a process can differ. The quadratic variation is computed path-by-path, and the result can depend on the path. If along one path of the Brownian motion we choose large positions $\Delta(u)$, the Itô integral will have a large quadratic variation. Along a different path, we could choose small positions $\Delta(u)$ and the Itô integral would have a small quadratic variation. The quadratic variation can be regarded as a measure of risk, and it depends on the size of the positions we take. The variance of $I(t)$ is an average over all possible paths of the quadratic variation. Because it is the expectation of something, it cannot be random. As an average over all possible paths, realized and unrealized, it is a more theoretical concept than quadratic variation. We emphasize here that what we are calling variance is not the empirical variance. Empirical (or sample) variance is computed from a realized path and is an estimator of the theoretical variance we are discussing. The empirical variance is sometimes carelessly called variance, which creates the possibility of confusion.

Finally, we recall the equation (3.4.10), $dW(t)\,dW(t) = dt$, of Remark 3.4.4. We interpret this equation as the statement that Brownian motion accumulates quadratic variation at rate one per unit time. It is another way of writing $[W,W](t) = t$, $t \geq 0$. The Itô integral formula $I(t) = \int_0^t \Delta(u)\,dW(u)$ can be written in differential form as $dI(t) = \Delta(t)\,dW(t)$, and we can then use (3.4.10) to square $dI(t)$:

$$dI(t)\,dI(t) = \Delta^2(t)\,dW(t)\,dW(t) = \Delta^2(t)\,dt. \tag{4.2.10}$$

This equation says that the Itô integral $I(t)$ accumulates quadratic variation at rate $\Delta^2(t)$ per unit time. The rate of accumulation is typically both time- and path-dependent. Equation (4.2.10) is another way of reporting the result of Theorem 4.2.3.

Remark 4.2.4 (on notation). The notations

$$I(t) = \int_0^t \Delta(u)\,dW(u) \tag{4.2.11}$$

and

$$dI(t) = \Delta(t)\,dW(t) \tag{4.2.12}$$

mean almost the same thing, although the second is probably more intuitive. Equation (4.2.11) has the precise meaning given by (4.2.2). Equation (4.2.12) has the imprecise meaning that when we move forward a little bit in time from time t, the change in the Itô integral I is $\Delta(t)$ times the change in the Brownian motion W. It also has a precise meaning, which one obtains by integrating both sides, remembering to put in a constant of integration $I(0)$:

$$I(t) = I(0) + \int_0^t \Delta(u)\,dW(u). \tag{4.2.13}$$

We say that (4.2.12) is the *differential form* of (4.2.13) and that (4.2.13) is the *integral form* of (4.2.12). These two equations mean exactly the same thing.

The only difference between (4.2.11) and (4.2.13), and hence the only difference between (4.2.11) and (4.2.12), is that (4.2.11) specifies the initial condition $I(0) = 0$, whereas (4.2.12) and (4.2.13) permit $I(0)$ to be any arbitrary constant. □

4.3 Itô's Integral for General Integrands

In this section, we define the Itô integral $\int_0^T \Delta(t)\,dW(t)$ for integrands $\Delta(t)$ that are allowed to vary continuously with time and also to jump. In particular, we no longer assume that $\Delta(t)$ is a simple process as shown in Figure

4.2.1. We do assume that $\Delta(t)$, $t \geq 0$, is adapted to the filtration $\mathcal{F}(t)$, $t \geq 0$. We also assume the square-integrability condition

$$\mathbb{E} \int_0^T \Delta^2(t)\,dt < \infty. \tag{4.3.1}$$

In order to define $\int_0^T \Delta(t)\,dW(t)$, we approximate $\Delta(t)$ by simple processes. Figure 4.3.1 suggests how this can be done. In that figure, the continuously varying $\Delta(t)$ is shown as a solid line and the approximating simple integrand is dashed. Notice that $\Delta(t)$ is allowed to jump. The approximating simple integrand is constructed by choosing a partition $0 = t_0 < t_1 < t_2 < t_3 < t_4$, setting the approximating simple process equal to $\Delta(t_j)$ at each t_j, and then holding the simple process constant over the subinterval $[t_j, t_{j+1})$. As the maximal step size of the partition approaches zero, the approximating integrand will become a better and better approximation of the continuously varying one.

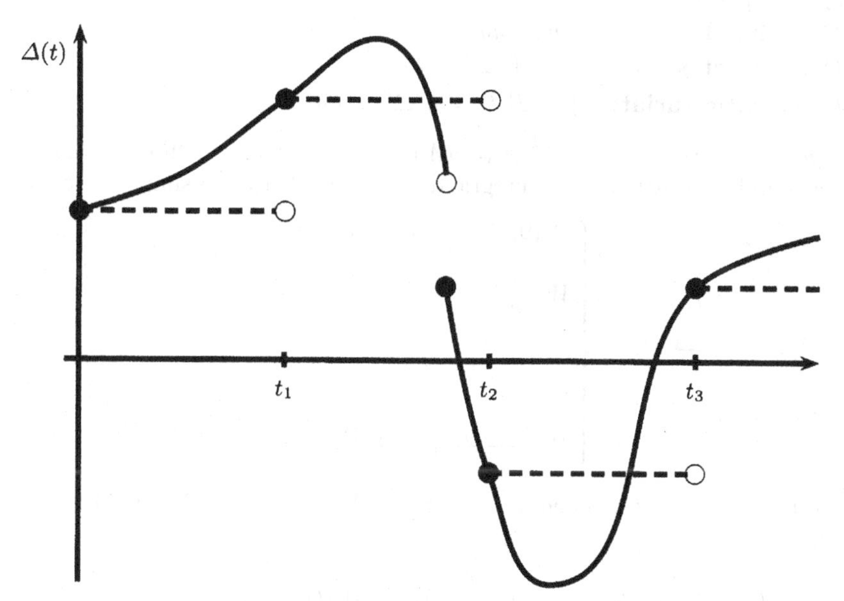

Fig. 4.3.1. Approximating a continuously varying integrand.

In general, then, it is possible to choose a sequence $\Delta_n(t)$ of simple processes such that as $n \to \infty$ these processes converge to the continuously varying $\Delta(t)$. By "converge," we mean that

$$\lim_{n\to\infty} \mathbb{E} \int_0^T |\Delta_n(t) - \Delta(t)|^2\,dt = 0. \tag{4.3.2}$$

For each $\Delta_n(t)$, the Itô integral $\int_0^t \Delta_n(u)\,dW(u)$ has already been defined for $0 \le t \le T$. We define the Itô integral for the continuously varying integrand $\Delta(t)$ by the formula[1]

$$\int_0^t \Delta(u)\,dW(u) = \lim_{n\to\infty} \int_0^t \Delta_n(u)\,dW(u), \quad 0 \le t \le T. \tag{4.3.3}$$

This integral inherits the properties of Itô integrals of simple processes. We summarize these in the next theorem.

Theorem 4.3.1. *Let T be a positive constant and let $\Delta(t)$, $0 \le t \le T$, be an adapted stochastic process that satisfies (4.3.1). Then $I(t) = \int_0^t \Delta(u)\,dW(u)$ defined by (4.3.3) has the following properties.*

(i) **(Continuity)** *As a function of the upper limit of integration t, the paths of $I(t)$ are continuous.*
(ii) **(Adaptivity)** *For each t, $I(t)$ is $\mathcal{F}(t)$-measurable.*
(iii) **(Linearity)** *If $I(t) = \int_0^t \Delta(u)\,dW(u)$ and $J(t) = \int_0^t \Gamma(u)\,dW(u)$, then $I(t) \pm J(t) = \int_0^t (\Delta(u) \pm \Gamma(u))\,dW(u)$; furthermore, for every constant c, $cI(t) = \int_0^t c\Delta(u)\,dW(u)$.*
(iv) **(Martingale)** *$I(t)$ is a martingale.*
(v) **(Itô isometry)** *$\mathbb{E}I^2(t) = \mathbb{E}\int_0^t \Delta^2(u)\,du$.*
(vi) **(Quadratic variation)** *$[I,I](t) = \int_0^t \Delta^2(u)\,du$.*

Example 4.3.2. We compute $\int_0^T W(t)\,dW(t)$. To do that, we choose a large integer n and approximate the integrand $\Delta(t) = W(t)$ by the simple process

$$\Delta_n(t) = \begin{cases} W(0) = 0 & \text{if } 0 \le t < \frac{T}{n}, \\ W\left(\frac{T}{n}\right) & \text{if } \frac{T}{n} \le t < \frac{2T}{n}, \\ \vdots & \\ W\left(\frac{(n-1)T}{n}\right) & \text{if } \frac{(n-1)T}{n} \le t < T, \end{cases}$$

as shown in Figure 4.3.2. Then $\lim_{n\to\infty} \mathbb{E}\int_0^T |\Delta_n(t) - W(t)|^2\,dt = 0$. By definition,

$$\begin{aligned} \int_0^T W(t)\,dW(t) &= \lim_{n\to\infty} \int_0^T \Delta_n(t)\,dW(t) \\ &= \lim_{n\to\infty} \sum_{j=0}^{n-1} W\left(\frac{jT}{n}\right)\left[W\left(\frac{(j+1)T}{n}\right) - W\left(\frac{jT}{n}\right)\right]. \end{aligned} \tag{4.3.4}$$

[1] For each t, the limit in (4.3.3) exists because $I_n(t) = \int_0^t \Delta_n(u)\,dW(u)$ is a Cauchy sequence in $L_2(\Omega, \mathcal{F}, \mathbb{P})$. This is because of Itô's isometry (Theorem 4.2.2), which yields $\mathbb{E}\big(I_n(t) - I_m(t)\big)^2 = \mathbb{E}\int_0^t \left|\Delta_n(u) - \Delta_m(u)\right|^2 du$. As a consequence of (4.3.2), the right-hand side has limit zero as n and m approach infinity.

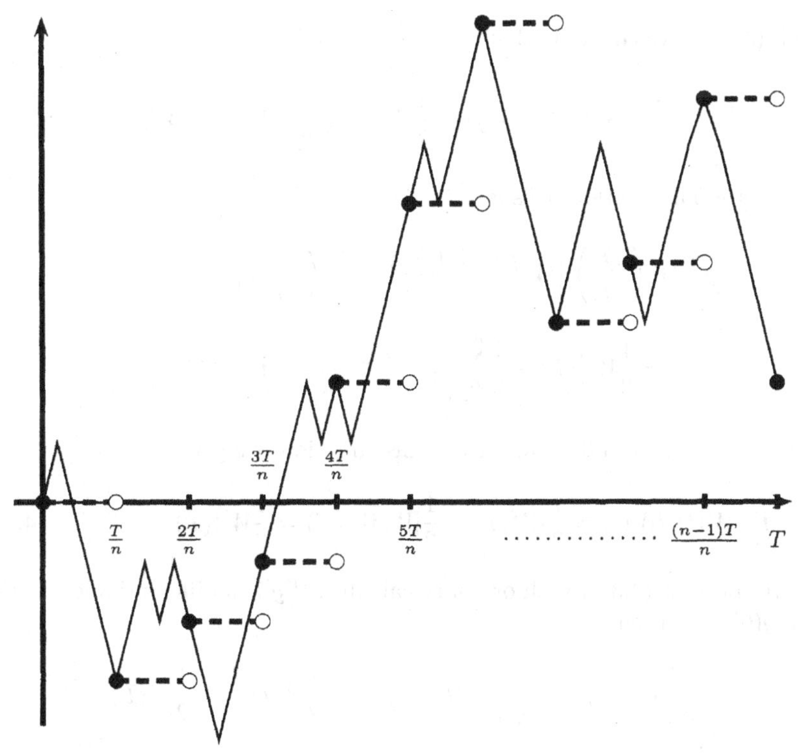

Fig. 4.3.2. Simple process approximating Brownian motion.

To simplify notation, we denote $W_j = W\left(\frac{jT}{n}\right)$. As a precursor to evaluating the limit in (4.3.4), we work out equation (4.3.5) below. The second equality in (4.3.5) is obtained by making the change of index $k = j + 1$ in the first sum. The third equality uses the fact that $W_0 = W(0) = 0$. We have

$$
\begin{aligned}
\frac{1}{2}\sum_{j=0}^{n-1}(W_{j+1} - W_j)^2 &= \frac{1}{2}\sum_{j=0}^{n-1} W_{j+1}^2 - \sum_{j=0}^{n-1} W_j W_{j+1} + \frac{1}{2}\sum_{j=0}^{n-1} W_j^2 \\
&= \frac{1}{2}\sum_{k=1}^{n} W_k^2 - \sum_{j=0}^{n-1} W_j W_{j+1} + \frac{1}{2}\sum_{j=0}^{n-1} W_j^2 \\
&= \frac{1}{2}W_n^2 + \frac{1}{2}\sum_{k=0}^{n-1} W_k^2 - \sum_{j=0}^{n-1} W_j W_{j+1} + \frac{1}{2}\sum_{j=0}^{n-1} W_j^2 \\
&= \frac{1}{2}W_n^2 + \sum_{j=0}^{n-1} W_j^2 - \sum_{j=0}^{n-1} W_j W_{j+1} \\
&= \frac{1}{2}W_n^2 + \sum_{j=0}^{n-1} W_j(W_j - W_{j+1}). \qquad (4.3.5)
\end{aligned}
$$

From (4.3.5), we conclude that

$$\sum_{j=0}^{n-1} W_j(W_{j+1} - W_j) = \frac{1}{2}W_n^2 - \frac{1}{2}\sum_{j=0}^{n-1}(W_{j+1} - W_j)^2.$$

In the original notation, this is

$$\sum_{j=0}^{n-1} W\left(\frac{jT}{n}\right)\left[W\left(\frac{(j+1)T}{n}\right) - W\left(\frac{jT}{n}\right)\right]$$
$$= \frac{1}{2}W^2(T) - \frac{1}{2}\sum_{j=0}^{n-1}\left[W\left(\frac{(j+1)T}{n}\right) - W\left(\frac{jT}{n}\right)\right]^2.$$

Letting $n \to \infty$ in (4.3.4) and using this equation, we get

$$\int_0^T W(t)\,dW(t) = \frac{1}{2}W^2(T) - \frac{1}{2}[W,W](T) = \frac{1}{2}W^2(T) - \frac{1}{2}T. \tag{4.3.6}$$

We contrast (4.3.6) with ordinary calculus. If g is a differentiable function with $g(0) = 0$, then

$$\int_0^T g(t)\,dg(t) = \int_0^T g(t)g'(t)\,dt = \frac{1}{2}g^2(t)\Big|_0^T = \frac{1}{2}g^2(T).$$

The extra term $-\frac{1}{2}T$ in (4.3.6) comes from the nonzero quadratic variation of Brownian motion and the way we constructed the Itô integral, always evaluating the integrand at the left-hand endpoint of the subinterval (see the right-hand side of (4.3.4)). If we were instead to evaluate at the midpoint, replacing the right-hand side of (4.3.4) by

$$\lim_{n\to\infty}\sum_{j=0}^{n-1} W\left(\frac{(j+\frac{1}{2})T}{n}\right)\left[W\left(\frac{(j+1)T}{n}\right) - W\left(\frac{jT}{n}\right)\right], \tag{4.3.7}$$

then we would not have gotten this term (see Exercise 4.4). The integral obtained by making this replacement is called the *Stratonovich integral*, and the ordinary rules of calculus apply to it. However, it is inappropriate for finance. In finance, the integrand represents a position in an asset and the integrator represents the price of that asset. We cannot decide at 1:00 p.m. which position we took at 9:00 a.m. We must decide the position at the beginning of each time interval, and the Itô integral is the limit of the gain achieved by that kind of trading as the time between trades approaches zero.

For functions $g(t)$ that have a derivative, integrals such as $\int_0^t g(t)\,dg(t)$ are not sensitive to this distinction (i.e., the Itô integral and Stratonovich integral approximations have the same limit, which is $\frac{1}{2}g^2(T)$). For functions that have a nonzero quadratic variation, integrals are sensitive to where in the subintervals the approximating integrands are evaluated.

The upper limit of integration T in (4.3.6) is arbitrary and can be replaced by any $t \geq 0$. In other words,

$$\int_0^t W(u)\,dW(u) = \frac{1}{2}W^2(t) - \frac{1}{2}t, \quad t \geq 0. \tag{4.3.8}$$

Theorem 4.3.1(iv) guarantees that $\int_0^t W(u)\,dW(u)$ is a martingale and hence has constant expectation. At $t = 0$, this martingale is 0, and hence its expectation must always be zero. This is indeed the case because $\mathbb{E}W^2(t) = t$. If the term $-\frac{1}{2}t$ were not present, we would not have a martingale. □

4.4 Itô-Doeblin Formula

The addition of Doeblin's name to what has traditionally been called the Itô formula is explained in the Notes, Section 4.9.

4.4.1 Formula for Brownian Motion

We want a rule to "differentiate" expressions of the form $f(W(t))$, where $f(x)$ is a differentiable function and $W(t)$ is a Brownian motion. If $W(t)$ were also differentiable, then the *chain rule* from ordinary calculus would give

$$\frac{d}{dt}f(W(t)) = f'(W(t))W'(t),$$

which could be written in differential notation as

$$df(W(t)) = f'(W(t))\,W'(t)\,dt = f'(W(t))\,dW(t).$$

Because W has nonzero quadratic variation, the correct formula has an extra term, namely,

$$df(W(t)) = f'(W(t))\,dW(t) + \frac{1}{2}f''(W(t))\,dt. \tag{4.4.1}$$

This is the *Itô-Doeblin formula in differential form*. Integrating this, we obtain the *Itô-Doeblin formula in integral form*:

$$f(W(t)) - f(W(0)) = \int_0^t f'(W(u))\,dW(u) + \frac{1}{2}\int_0^t f''(W(u))\,du. \tag{4.4.2}$$

The mathematically meaningful form of the Itô-Doeblin formula is the integral form (4.4.2). This is because we have precise definitions for both terms appearing on the right-hand side. The first, $\int_0^t f'(W(u))\,dW(u)$, is an Itô integral, defined in the previous section. The second, $\int_0^t f''(W(u))\,du$, is an ordinary (Lebesgue) integral with respect to the time variable.

For pencil and paper computations, the more convenient form of the Itô-Doeblin formula is the differential form (4.4.1). There is an intuitive meaning but no precise definition for the terms $df\big(W(t)\big)$, $dW(t)$, and dt appearing in this formula. The intuitive meaning is that $df\big(W(t)\big)$ is the change in $f\big(W(t)\big)$ when t changes a "little bit" dt, $dW(t)$ is the change in the Brownian motion when t changes a "little bit" dt, and the whole formula is exact only when the "little bit" is "infinitesimally small." Because there is no precise definition for "little bit" and "infinitesimally small," we rely on (4.4.2) to give precise meaning to (4.4.1).

The relationship between (4.4.1) and (4.4.2) is similar to that developed in ordinary calculus to assist in changing variables in an integral. If asked to compute the indefinite integral $\int f(u)f'(u)\,du$, we might make the change of variable $v = f(u)$ and write $dv = f'(u)\,du$, so that the indefinite integral becomes $\int v\,dv$, which is $\frac{1}{2}v^2 + C = \frac{1}{2}f^2(u) + C$, where C is a constant of integration. The final formula

$$\int f(u)f'(u)\,du = \frac{1}{2}f^2(u) + C$$

is correct, as can be verified by differentiating $\frac{1}{2}f^2(u)+C$ to get $f(u)f'(u)$. We do not attempt to give precise definitions to the terms dv and du appearing in the equation $dv = f'(u)\,du$ used in deriving it.

We formalize the preceding discussion with a theorem that provides a formula slightly more general than (4.4.2) in that it allows f to be a function of both t and x.

Theorem 4.4.1 (Itô-Doeblin formula for Brownian motion). *Let $f(t,x)$ be a function for which the partial derivatives $f_t(t,x)$, $f_x(t,x)$, and $f_{xx}(t,x)$ are defined and continuous, and let $W(t)$ be a Brownian motion. Then, for every $T \geq 0$,*

$$f(T,W(T)) = f(0,W(0)) + \int_0^T f_t\big(t,W(t)\big)\,dt + \int_0^T f_x\big(t,W(t)\big)\,dW(t) + \frac{1}{2}\int_0^T f_{xx}\big(t,W(t)\big)\,dt. \quad (4.4.3)$$

SKETCH OF PROOF: We first show why (4.4.3) holds when $f(x) = \frac{1}{2}x^2$. In this case, $f'(x) = x$ and $f''(x) = 1$. Let x_{j+1} and x_j be numbers. Taylor's formula implies

$$f(x_{j+1}) - f(x_j) = f'(x_j)(x_{j+1} - x_j) + \frac{1}{2}f''(x_j)(x_{j+1} - x_j)^2. \quad (4.4.4)$$

In this case, Taylor's formula to second order is exact (there is no remainder term) because f''' and all higher derivatives of f are zero. We return to this matter later.

Fix $T > 0$, and let $\Pi = \{t_0, t_1, \ldots, t_n\}$ be a partition of $[0, T]$ (i.e., $0 = t_0 < t_1 < \cdots < t_n = T$). We are interested in the difference between $f(W(0))$ and $f(W(T))$. This change in $f(W(t))$ between times $t = 0$ and $t = T$ can be written as the sum of the changes in $f(W(t))$ over each of the subintervals $[t_j, t_{j+1}]$. We do this and then use Taylor's formula (4.4.4) with $x_j = W(t_j)$ and $x_{j+1} = W(t_{j+1})$ to obtain

$$\begin{aligned} f(W(T)) - f(W(0)) &= \sum_{j=0}^{n-1} \left[f(W(t_{j+1})) - f(W(t_j)) \right] \\ &= \sum_{j=0}^{n-1} f'(W(t_j)) \left[W(t_{j+1}) - W(t_j) \right] \\ &\quad + \frac{1}{2} \sum_{j=0}^{n-1} f''(W(t_j)) \left[W(t_{j+1}) - W(t_j) \right]^2. \end{aligned} \tag{4.4.5}$$

For the function $f(x) = \frac{1}{2}x^2$, the right-hand side of (4.4.5) is

$$\sum_{j=0}^{n-1} W(t_j) \left[W(t_{j+1}) - W(t_j) \right] + \frac{1}{2} \sum_{j=0}^{n-1} \left[W(t_{j+1}) - W(t_j) \right]^2. \tag{4.4.6}$$

If we let $\|\Pi\| \to 0$, the left-hand side of (4.4.5) is unaffected and the terms on the right-hand side converge to an Itô integral and one-half of the quadratic variation of Brownian motion, respectively:

$$\begin{aligned} & f(W(T)) - f(W(0)) \\ &= \lim_{\|\Pi\| \to 0} \sum_{j=0}^{n-1} W(t_j) \left[W(t_{j+1}) - W(t_j) \right] + \lim_{\|\Pi\| \to 0} \frac{1}{2} \sum_{j=0}^{n-1} \left[W(t_{j+1}) - W(t_j) \right]^2 \\ &= \int_0^T W(t)\, dW(t) + \frac{1}{2} T \\ &= \int_0^T f'(W(t))\, dW(t) + \frac{1}{2} \int_0^T f''(W(t))\, dt. \end{aligned} \tag{4.4.7}$$

This is the Itô-Doeblin formula in integral form for the function $f(x) = \frac{1}{2}x^2$.

If instead of the quadratic function $f(x) = \frac{1}{2}x^2$ we had a general function $f(x)$, then in (4.4.5) we would have also gotten a sum of terms containing $\left[W(t_{j+1}) - W(t_j) \right]^3$. But according to Exercise 3.4 of Chapter 3, $\sum_{j=0}^{n-1} \left| W(t_{j+1}) - W(t_j) \right|^3$ has limit zero as $\|\Pi\| \to 0$. Therefore, this term would make no contribution to the final answer.

If we take a function $f(t, x)$ of both the time variable t and the variable x, then Taylor's Theorem says that

$$
\begin{aligned}
&f(t_{j+1}, x_{j+1}) - f(t_j, x_j) \\
&= f_t(t_j, x_j)(t_{j+1} - t_j) + f_x(t_j, x_j)(x_{j+1} - x_j) \\
&\quad + \frac{1}{2} f_{xx}(t_j, x_j)(x_{j+1} - x_j)^2 + f_{tx}(t_j, x_j)(t_{j+1} - t_j)(x_{j+1} - x_j) \\
&\quad + \frac{1}{2} f_{tt}(t_j, x_j)(t_{j+1} - t_j)^2 + \text{ higher-order terms.}
\end{aligned} \tag{4.4.8}
$$

We replace x_j by $W(t_j)$, replace x_{j+1} by $W(t_{j+1})$, and sum:

$$
\begin{aligned}
&f(T, W(T)) - f(0, W(0)) \\
&= \sum_{j=0}^{n-1} \left[f(t_{j+1}, W(t_{j+1})) - f(t_j, W(t_j)) \right] \\
&= \sum_{j=0}^{n-1} f_t(t_j, W(t_j))(t_{j+1} - t_j) + \sum_{j=0}^{n-1} f_x(t_j, W(t_j))\big(W(t_{j+1}) - W(t_j)\big) \\
&\quad + \frac{1}{2} \sum_{j=0}^{n-1} f_{xx}(t_j, W(t_j))\big(W(t_{j+1}) - W(t_j)\big)^2 \\
&\quad + \sum_{j=0}^{n-1} f_{tx}(t_j, W(t_j))(t_{j+1} - t_j)\big(W(t_{j+1}) - W(t_j)\big) \\
&\quad + \frac{1}{2} \sum_{j=0}^{n-1} f_{tt}(t_j, W(t_j))(t_{j+1} - t_j)^2 + \text{ higher-order terms.}
\end{aligned} \tag{4.4.9}
$$

When we take the limit as $\|\Pi\| \to 0$, the left-hand side of (4.4.9) is unaffected. The first term on the right-hand side of (4.4.9) contributes the ordinary (Lebesgue) integral

$$
\lim_{\|\Pi\| \to 0} \sum_{j=0}^{n-1} f_t(t_j, W(t_j))(t_{j+1} - t_j) = \int_0^T f_t(t, W(t))\, dt
$$

to the final answer. As $\|\Pi\| \to 0$, the second term contributes the Itô integral $\int_0^T f_x(t, W(t))\, dW(t)$. The third term contributes another ordinary (Lebesgue) integral, $\frac{1}{2} \int_0^T f_{xx}(t, W(t))\, dt$, similar to the way we obtained this integral in (4.4.7). In other words, in the third term we can replace $\big(W(t_{j+1}) - W(t_j)\big)^2$ by $t_{j+1} - t_j$. This is not an exact substitution, but when we sum the terms this substitution gives the correct limit as $\|\Pi\| \to 0$. See Remark 3.4.4 for more discussion of this point. With this substitution, the third term on the right-hand side of (4.4.9) contributes $\frac{1}{2} \int_0^T f_{xx}(t, W(t))\, dt$. These limits of the first three terms appear on the right-hand side of (4.4.3). The fourth and fifth terms contribute zero. Indeed, for the fourth term, we observe that

$$\lim_{\|\Pi\|\to 0}\left|\sum_{j=0}^{n-1} f_{tx}(t_j, W(t_j))(t_{j+1}-t_j)\big(W(t_{j+1})-W(t_j)\big)\right|$$

$$\leq \lim_{\|\Pi\|\to 0}\sum_{j=0}^{n-1}\big|f_{tx}(t_j, W(t_j))\big|\cdot(t_{j+1}-t_j)\cdot\big|W(t_{j+1})-W(t_j)\big|$$

$$\leq \lim_{\|\Pi\|\to 0}\max_{0\leq k\leq n-1}\big|W(t_{k+1})-W(t_k)\big|\cdot\lim_{\|\Pi\|\to 0}\sum_{j=0}^{n-1}\big|f_{tx}(t_j, W(t_j))\big|(t_{j+1}-t_j)$$

$$= 0\cdot\int_0^T \big|f_{tx}(t, W(t))\,dt \;=\; 0. \tag{4.4.10}$$

The fifth term is treated similarly:

$$\lim_{\|\Pi\|\to 0}\left|\frac{1}{2}\sum_{j=0}^{n-1} f_{tt}(t_j, W(t_j))(t_{j+1}-t_j)^2\right|$$

$$\leq \lim_{\|\Pi\|\to 0}\frac{1}{2}\sum_{j=0}^{n-1}\big|f_{tt}(t_j, W(t_j))\big|\cdot(t_{j+1}-t_j)^2$$

$$\leq \frac{1}{2}\lim_{\|\Pi\|\to 0}\max_{0\leq k\leq n-1}(t_{k+1}-t_k)\cdot\lim_{\|\Pi\|\to 0}\sum_{j=0}^{n-1}\big|f_{tt}(t_j, W(t_j))\big|(t_{j+1}-t_j)$$

$$= \frac{1}{2}\cdot 0\cdot\int_0^T f_{tt}(t, W(t))dt \;=\; 0. \tag{4.4.11}$$

The higher-order terms likewise contribute zero to the final answer. □

Remark 4.4.2. The fact that the sum (4.4.10) of terms containing the product $(t_{j+1}-t_j)\big(W(t_{j+1})-W(t_j)\big)$ has limit zero can be informally recorded by the formula $dt\,dW(t) = 0$. Similarly, the sum (4.4.11) of terms containing $(t_{j+1}-t_j)^2$ also has limit zero, and this can be recorded by the formula $dt\,dt = 0$. We can write these terms if we like in the Itô-Doeblin formula, so that in differential form it becomes

$$\begin{aligned} &df(t, W(t)) \\ &= f_t(t, W(t))\,dt + f_x(t, W(t))\,dW(t) + \frac{1}{2}f_{xx}(t, W(t))\,dW(t)\,dW(t) \\ &\qquad + f_{tx}(t, W(t))\,dt\,dW(t) + \frac{1}{2}f_{xx}(t, W(t))\,dt\,dt, \end{aligned}$$

but

$$dW(t)\,dW(t) = dt, \quad dt\,dW(t) = dW(t)\,dt = 0, \quad dt\,dt = 0, \tag{4.4.12}$$

and the Itô-Doeblin formula in differential form simplifies to

$$df(t, W(t)) = f_t(t, W(t))\,dt + f_x(t, W(t))\,dW(t) + \frac{1}{2}f_{xx}(t, W(t))\,dt. \tag{4.4.13}$$

In Figure 4.4.1, we illustrate the Taylor series approximation of the difference $f\big(W(t_{j+1})\big) - f\big(W(t_j)\big)$ for a function $f(x)$ that does not depend on t. The first-order approximation, which is $f'\big(W(t_j)\big)\big(W(t_{j+1}) - W(t_j)\big)$, has an error due to the convexity of the function $f(x)$. Most of this error is removed by adding in the second-order term $\frac{1}{2}f''\big(W(t_j)\big)\big(W(t_{j+1}) - W(t_j)\big)^2$, which captures the curvature of the function $f(x)$ at $x = W(t_j)$.

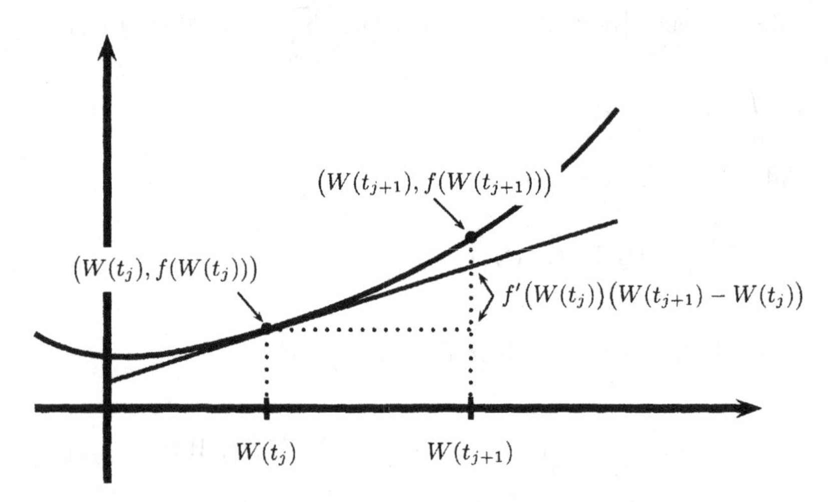

Fig. 4.4.1. Taylor approximation to $f\big(W(t_{j+1})\big) - f\big(W(t_j)\big)$.

In other words,

$$f\big(W(t_{j+1})\big) - f\big(W(t_j)\big) = f'\big(W(t_j)\big)\big(W(t_{j+1}) - W(t_j)\big) + \text{ small error,} \tag{4.4.14}$$

and

$$\begin{aligned} f\big(W(t_{j+1})\big) - f\big(W(t_j)\big) = &\, f'\big(W(t_j)\big)\big(W(t_{j+1}) - W(t_j)\big) \\ &+\frac{1}{2}f''\big(W(t_j)\big)\big(W(t_{j+1}) - W(t_j)\big)^2 \\ &+ \text{smaller error.} \end{aligned} \tag{4.4.15}$$

In both (4.4.14) and (4.4.15), as $\|\Pi\| \to 0$, the errors approach zero. However, before we let $\|\Pi\| \to 0$, we must first sum these equations over j, and the smaller we make $\|\Pi\|$, the more terms there are in the sum. When we sum both sides of (4.4.14), the errors accumulate, and although the error in each summand approaches zero as $\|\Pi\| \to 0$, the sum of the errors does not. When we use the more accurate approximation (4.4.15), this does not happen; the limit of the sum of the smaller errors is zero. We need the extra accuracy of (4.4.15) because the paths of Brownian motion are so volatile (i.e., they have nonzero quadratic variation). This extra term makes stochastic calculus different from ordinary calculus.

The Itô-Doeblin formula often simplifies the computation of Itô integrals. For example, with $f(x) = \frac{1}{2}x^2$, this formula says that

$$\begin{aligned}\frac{1}{2}W^2(T) &= f(W(T)) - f(W(0)) \\ &= \int_0^T f'(W(t))\,dW(t) + \frac{1}{2}\int_0^t f''(W(t))\,dt \\ &= \int_0^T W(t)\,dW(t) + \frac{1}{2}T.\end{aligned}$$

Rearranging terms, we have formula (4.3.6) and have obtained it without going through the approximation of the integrand by simple processes as we did in Example 4.3.2.

4.4.2 Formula for Itô Processes

We extend the Itô-Doeblin formula to stochastic processes more general than Brownian motion. The processes for which we develop stochastic calculus are the *Itô processes* defined below. Almost all stochastic processes, except those that have jumps, are Itô processes.

Definition 4.4.3. *Let $W(t)$, $t \geq 0$, be a Brownian motion, and let $\mathcal{F}(t)$, $t \geq 0$, be an associated filtration. An* Itô process *is a stochastic process of the form*

$$X(t) = X(0) + \int_0^t \Delta(u)\,dW(u) + \int_0^t \Theta(u)\,du, \tag{4.4.16}$$

where $X(0)$ is nonrandom and $\Delta(u)$ and $\Theta(u)$ are adapted stochastic processes.[2]

In order to understand the volatility associated with Itô processes, we must determine the rate at which they accumulate quadratic variation.

Lemma 4.4.4. *The quadratic variation of the Itô process (4.4.16) is*

$$[X,X](t) = \int_0^t \Delta^2(u)\,du. \tag{4.4.17}$$

PROOF: We introduce the notation $I(t) = \int_0^t \Delta(u)\,dW(u)$, $R(t) = \int_0^t \Theta(u)\,du$. Both these processes are continuous in their upper limit of integration t. To

[2] We assume that $\mathbb{E}\int_0^t \Delta^2(u)\,du$ and $\int_0^t |\Theta(u)|\,du$ are finite for every $t > 0$ so that the integrals on the right-hand side of (4.4.16) are defined and the Itô integral is a martingale. We shall always make such integrability assumptions, but we do not always explicitly state them.

determine the quadratic variation of X on $[0,t]$, we choose a partition $\Pi = \{t_0, t_1, \dots, t_n\}$ of $[0,t]$ (i.e., $0 = t_0 < t_1 < \cdots < t_n = t$) and we write the sampled quadratic variation

$$\sum_{j=0}^{n-1}\big[X(t_{j+1}) - X(t_j)\big]^2 = \sum_{j=0}^{n-1}\big[I(t_{j+1}) - I(t_j)\big]^2 + \sum_{j=0}^{n-1}\big[R(t_{j+1}) - R(t_j)\big]^2$$
$$+2\sum_{j=0}^{n-1}\big[I(t_{j+1}) - I(t_j)\big]\big[R(t_{j+1}) - R(t_j)\big].$$

As $\|\Pi\| \to 0$, the first term on the right-hand side, $\sum_{j=0}^{n-1}\big[I(t_{j+1}) - I(t_j)\big]^2$, converges to the quadratic variation of I on $[0,t]$, which according to Theorem 4.3.1(vi) is $[I,I](t) = \int_0^t \Delta^2(u)\,du$. The absolute value of the second term is bounded above by

$$\max_{0\le k\le n-1}\big|R(t_{k+1}) - R(t_k)\big| \cdot \sum_{j=0}^{n-1}\big|R(t_{j+1}) - R(t_j)\big|$$
$$= \max_{0\le k\le n-1}\big|R(t_{k+1}) - R(t_k)\big| \cdot \sum_{j=0}^{n-1}\left|\int_{t_j}^{t_{j+1}} \Theta(u)\,du\right|$$
$$\le \max_{0\le k\le n-1}\big|R(t_{k+1}) - R(t_k)\big| \cdot \sum_{j=0}^{n-1}\int_{t_j}^{t_{j+1}} |\Theta(u)|\,du$$
$$= \max_{0\le k\le n-1}\big|R(t_{k+1}) - R(t_k)\big| \cdot \int_0^t |\Theta(u)|\,du,$$

and as $\|\Pi\| \to 0$, this has limit $0 \cdot \int_0^t |\Theta(u)|\,du = 0$ because $R(t)$ is continuous. The absolute value of the third term is bounded above by

$$2\max_{0\le k\le n-1}\big|I(t_{k+1}) - I(t_k)\big| \cdot \sum_{j=0}^{n-1}\big|R(t_{j+1}) - R(t_j)\big|$$
$$\le 2\max_{0\le k\le n-1}\big|I(t_{k+1}) - I(t_k)\big| \cdot \int_0^t |\Theta(u)|\,du,$$

and this has limit $0 \cdot \int_0^t |\Theta(u)|^2\,du = 0$ as $\|\Pi\| \to 0$ because $I(t)$ is continuous. We conclude that $[X,X](t) = [I,I](t) = \int_0^t \Delta^2(u)\,du$. □

The conclusion of Lemma 4.4.4 is most easily remembered by first writing (4.4.16) in the differential notation

$$dX(t) = \Delta(t)\,dW(t) + \Theta(t)\,dt \tag{4.4.18}$$

and then using the differential multiplication table (4.4.12) to compute

$$\begin{aligned} dX(t)\,dX(t) &= \Delta^2(t)\,dW(t)\,dW(t) + 2\Delta(t)\Theta(t)\,dW(t)\,dt + \Theta^2(t)\,dt\,dt \\ &= \Delta^2(t)\,dt. \end{aligned} \tag{4.4.19}$$

This says that, at each time t, the process X is accumulating quadratic variation at rate $\Delta^2(t)$ per unit time, and hence the total quadratic variation accumulated on the time interval $[0,t]$ is $[X,X](t) = \int_0^t \Delta^2(u)\,du$. This quadratic variation is solely due to the quadratic variation of the Itô integral $I(t) = \int_0^t \Delta(u)\,dW(u)$. The ordinary integral $R(t) = \int_0^t \Theta(u)\,du$ has zero quadratic variation and thus contributes nothing to the quadratic variation of X.

Notice in this connection that having zero quadratic variation does not necessarily mean that $R(t)$ is nonrandom. Because $\Theta(u)$ can be random, $R(t)$ can also be random. However, $R(t)$ is not as volatile as $I(t)$. At each time t, we have a good estimate of the next increment of $R(t)$. For small time steps $h > 0$,

$$R(t+h) \approx R(t) + \Theta(t)h,$$

and we know both $R(t)$ and $\Theta(t)$ at time t. This is like investing in a money market account at a variable interest rate. At each time, we have a good estimate of the return over the near future because we know today's interest rate. Nonetheless, the return is random because the interest rate (Θ in this analogy) can change. In contrast, I is more volatile. At time t, one estimate of $I(t+h)$ is

$$I(t+h) \approx I(t) + \Delta(t)\big(W(t+h) - W(t)\big),$$

but we do not know $W(t+h) - W(t)$ at time t. In fact, $W(t+h) - W(t)$ is independent of the information available at time t. This is like investing in a stock.

So far we have discussed integrals with respect to time, such as $R(t) = \int_0^t \Theta(u)\,du$ appearing in (4.4.16) and Itô integrals (integrals with respect to Brownian motion) such as $I(t) = \int_0^t \Delta(u)\,dW(u)$, also appearing in (4.4.16). In addition, we shall need integrals with respect to Itô processes (i.e., integrals of the form $\int_0^t \Gamma(u)\,dX(u)$, where Γ is some adapted process). We define such an integral by separating $dX(t)$ into a $dW(t)$ term and a dt term as in (4.4.18).

Definition 4.4.5. *Let $X(t)$, $t \geq 0$, be an Itô process as described in Definition 4.4.3, and let $\Gamma(t)$, $t \geq 0$, be an adapted process. We define the* integral with respect to an Itô process[3]

$$\int_0^t \Gamma(u)\,dX(u) = \int_0^t \Gamma(u)\Delta(u)\,dW(u) + \int_0^t \Gamma(u)\Theta(u)\,du. \tag{4.4.20}$$

We again work through the sketch of the proof of Theorem 4.4.1, but with the Itô process $X(t)$ replacing the Brownian motion $W(t)$. In place of (4.4.9), we now have

[3] We assume that $E\int_0^t \Gamma^2(u)\Delta^2(u)\,du$ and $\int_0^t |\Gamma(u)\Theta(u)|\,du$ are finite for every $t > 0$ so that the integrals on the right-hand side of (4.4.20) are defined.

$$
\begin{aligned}
&f(T,X(T)) - f(0,X(0)) \\
&= \sum_{j=0}^{n-1} f_t(t_j,X(t_j))(t_{j+1}-t_j) + \sum_{j=0}^{n-1} f_x(t_j,X(t_j))\big(X(t_{j+1})-X(t_j)\big) \\
&\quad + \frac{1}{2}\sum_{j=0}^{n-1} f_{xx}(t_j,X(t_j))\big(X(t_{j+1})-X(t_j)\big)^2 \\
&\quad + \sum_{j=0}^{n-1} f_{tx}(t_j,X(t_j))(t_{j+1}-t_j)\big(X(t_{j+1})-X(t_j)\big) \\
&\quad + \frac{1}{2}\sum_{j=0}^{n-1} f_{tt}(t_j,X(t_j))(t_{j+1}-t_j)^2 + \text{ higher-order terms.} \quad (4.4.21)
\end{aligned}
$$

The last two sums on the right-hand side have zero limits as $\|\Pi\| \to 0$ for the same reasons the analogous terms have zero limits in the sketch of the proof of Theorem 4.4.1 (see (4.4.10) and (4.4.11)). The higher-order terms likewise have limit zero. The limit of the first term on the right-hand side of (4.4.21) is $\int_0^T f_t(t,X(t))dt$. The limit of the second term is

$$
\int_0^T f_x(t,X(t))\,dX(t) = \int_0^T f_x(t,X(t))\Delta(t)\,dW(t) + \int_0^T f_x(t,X(t))\Theta(t)\,dt.
$$

Finally, the limit of the third term on the right-hand side of (4.4.19) is

$$
\frac{1}{2}\int_0^T f_{xx}(t,X(t))\,d[X,X](t) = \frac{1}{2}\int_0^T f_{xx}(t,X(t))\Delta^2(t)\,dt
$$

because the Itô process $X(t)$ accumulates quadratic variation at rate $\Delta^2(t)$ per unit time (Lemma 4.4.4). These considerations lead to the following generalization of Theorem 4.4.1.

Theorem 4.4.6 (Itô-Doeblin formula for an Itô process). *Let $X(t)$, $t \geq 0$, be an Itô process as described in Definition 4.4.3, and let $f(t,x)$ be a function for which the partial derivatives $f_t(t,x)$, $f_x(t,x)$, and $f_{xx}(t,x)$ are defined and continuous. Then, for every $T \geq 0$,*

$$
\begin{aligned}
&f(T,X(T)) \\
&= f(0,X(0)) + \int_0^T f_t(t,X(t))\,dt + \int_0^T f_x(t,X(t))\,dX(t) \\
&\quad + \frac{1}{2}\int_0^T f_{xx}(t,X(t))\,d[X,X](t) \\
&= f(0,X(0)) + \int_0^T f_t(t,X(t))\,dt + \int_0^T f_x(t,X(t))\Delta(t)\,dW(t) \\
&\quad + \int_0^T f_x(t,X(t))\Theta(t)\,dt + \frac{1}{2}\int_0^T f_{xx}(t,X(t))\Delta^2(t)\,dt. \quad (4.4.22)
\end{aligned}
$$

Remark 4.4.7 (Summary of stochastic calculus). Theorem 4.4.6 is stated in mathematically precise language. Every term on the right-hand side has a solid definition, and in the end the right-hand side reduces to a sum of a nonrandom quantity $f\big(0, X(0)\big)$, three ordinary (Lebesgue) integrals with respect to time, and an Itô integral.

However, it is easier to remember and use the result of this theorem if we recast it in differential notation. We may rewrite (4.4.22) as

$$df\big(t, X(t)\big) = f_t\big(t, X(t)\big)\,dt + f_x\big(t, X(t)\big)\,dX(t) + \frac{1}{2}f_{xx}\big(t, X(t)\big)\,dX(t)\,dX(t). \tag{4.4.23}$$

The guiding principle here is that we write out the Taylor series expansion of $f\big(t, X(t)\big)$ with respect to all its arguments, which in this case are t and $X(t)$. We take this Taylor series expansion out to first order for every argument that has zero quadratic variation, which in this case is t, and we take the expansion out to second order for every argument that has nonzero quadratic variation, which in this case is $X(t)$.

We may reduce (4.4.23) to an expression that involves only dt and $dW(t)$ by using the differential form (4.4.18) of the Itô process (i.e., $dX(t) = \Delta(t)\,dW(t) + \Theta(t)\,dt$) and the formula (4.4.19) for the rate at which $X(t)$ accumulates quadratic variation (i.e., $dX(t)\,dX(t) = \Delta^2(t)\,dt$). This is obtained by squaring the formula for $dX(t)$ and using the multiplication table (4.4.12). Making these substitutions in (4.4.23), we obtain

$$\begin{aligned} df\big(t, X(t)\big) &= f_t\big(t, X(t)\big)\,dt + f_x\big(t, X(t)\big)\Delta(t)\,dW(t) \\ &\quad + f_x\big(t, X(t)\big)\Theta(t)\,dt + \frac{1}{2}f_{xx}\big(t, X(t)\big)\Delta^2(t)\,dt. \end{aligned} \tag{4.4.24}$$

Itô calculus is little more than repeated use of this formula in a variety of situations. □

4.4.3 Examples

We conclude this section with three examples illustrating Remark 4.4.7. Many more examples are developed in subsequent sections and in the exercises.

Example 4.4.8 (Generalized geometric Brownian motion). Let $W(t)$, $t \geq 0$, be a Brownian motion, let $\mathcal{F}(t)$, $t \geq 0$, be an associated filtration, and let $\alpha(t)$ and $\sigma(t)$ be adapted processes. Define the Itô process

$$X(t) = \int_0^t \sigma(s)\,dW(s) + \int_0^t \Big(\alpha(s) - \frac{1}{2}\sigma^2(s)\Big)ds. \tag{4.4.25}$$

Then

$$dX(t) = \sigma(t)\,dW(t) + \Big(\alpha(t) - \frac{1}{2}\sigma^2(t)\Big)\,dt,$$

and

$$dX(t)\,dX(t) = \sigma^2(t)\,dW(t)\,dW(t) = \sigma^2(t)\,dt.$$

Consider an asset price process given by

$$S(t) = S(0)e^{X(t)} = S(0)\exp\left\{\int_0^t \sigma(s)\,dW(s) + \int_0^t \left(\alpha(s) - \frac{1}{2}\sigma^2(s)\right)ds\right\}, \tag{4.4.26}$$

where $S(0)$ is nonrandom and positive. We may write $S(t) = f\big(X(t)\big)$, where $f(x) = S(0)e^x$, $f'(x) = S(0)e^x$, and $f''(x) = S(0)e^x$. According to the Itô-Doeblin formula

$$\begin{aligned} dS(t) &= df\big(X(t)\big) \\ &= f'\big(X(t)\big)\,dX(t) + \frac{1}{2}f''\big(X(t)\big)\,dX(t)\,dX(t) \\ &= S(0)e^{X(t)}\,dX(t) + \frac{1}{2}S(0)e^{X(t)}\,dX(t)\,dX(t) \\ &= S(t)\,dX(t) + \frac{1}{2}S(t)\,dX(t)\,dX(t) \\ &= \alpha(t)S(t)\,dt + \sigma(t)S(t)\,dW(t). \end{aligned} \tag{4.4.27}$$

The asset price $S(t)$ has instantaneous mean rate of return $\alpha(t)$ and volatility $\sigma(t)$. Both the instantaneous mean rate of return and the volatility are allowed to be time-varying and random.

This example includes all possible models of an asset price process that is always positive, has no jumps, and is driven by a single Brownian motion. Although the model is driven by a Brownian motion, the distribution of $S(t)$ does not need to be log-normal because $\alpha(t)$ and $\sigma(t)$ are allowed to be time-varying and random. If α and σ are constant, we have the usual geometric Brownian motion model, and the distribution of $S(t)$ is log-normal.

In the case of constant α and σ, (4.4.26) becomes

$$S(t) = S(0)\exp\left\{\sigma W(t) + \left(\alpha - \frac{1}{2}\sigma^2\right)t\right\}. \tag{4.4.28}$$

One can incorrectly argue from this formula that since Brownian motion is a martingale (i.e., it has no overall tendency to rise or fall), the mean rate of return for $S(t)$ must be $\alpha - \frac{1}{2}\sigma^2$. The error in this argument is that although $W(t)$ is a martingale, $S(0)e^{\sigma W(t)}$ is not. The convexity of the function $e^{\sigma x}$ imparts an upward drift to $S(0)e^{\sigma W(t)}$. In order to correct for this, one must subtract $\frac{1}{2}\sigma^2 t$ in the exponential; the process $S(0)\exp\left\{\sigma W(t) - \frac{1}{2}\sigma^2 t\right\}$ is a martingale (see Theorem 3.6.1). If we now add αt in the exponential, we get $S(t)$, a process with mean rate of return α.

The Itô-Doeblin formula automatically keeps track of these effects, even when α and σ are time-varying and random. If $\alpha = 0$, then (4.4.27) yields

$$dS(t) = \sigma(t)S(t)\,dW(t).$$

Integration of both sides yields

$$S(t) = S(0) + \int_0^t \sigma(s)S(s)\,dW(s).$$

The right-hand side is the nonrandom constant $S(0)$ plus an Itô integral, which is a martingale, and hence (in the case $\alpha = 0$)

$$S(t) = S(0)\exp\left\{\int_0^t \sigma(s)\,dW(s) - \frac{1}{2}\int_0^t \sigma^2(s)\,ds\right\} \tag{4.4.29}$$

is a martingale. In other words, the term $\sigma(t)S(t)\,dW(t)$ on the right-hand side of (4.4.27) contributes no drift, just pure volatility, to the asset price.

When $\alpha(t)$ is a nonzero random process, (4.4.27) shows that it plays the role of the mean rate of return. In the case of time-varying and random $\alpha(t)$, we will call this the *instantaneous* mean rate of return since it depends on the time (and the sample path) where it is evaluated. □

The preceding example supplies the heart of the proof of the following theorem.

Theorem 4.4.9 (Itô integral of a deterministic integrand). *Let $W(s)$, $s \geq 0$, be a Brownian motion, and let $\Delta(s)$ be a* nonrandom *function of time. Define $I(t) = \int_0^t \Delta(s)\,dW(s)$. For each $t \geq 0$, the random variable $I(t)$ is normally distributed with expected value zero and variance $\int_0^t \Delta^2(s)\,ds$.*

PROOF: The mean and variance of $I(t)$ are easy to determine. Since $I(t)$ is a martingale and $I(0) = 0$, we must have $\mathbb{E}I(t) = I(0) = 0$. Itô's isometry (Theorem 4.3.1(v)) implies that

$$\mathrm{Var}I(t) = \mathbb{E}I^2(t) = \int_0^t \Delta^2(s)\,ds.$$

We do not need to take the expected value of $\int_0^t \Delta^2(s)\,ds$ on the right-hand side of this formula because $\Delta(s)$ is not random.

The challenge is to show that $I(t)$ is normally distributed. We shall do this by establishing that $I(t)$ has the moment-generating function of a normal random variable with mean zero and variance $\int_0^t \Delta^2(s)\,ds$, which is (see (3.2.13))

$$\mathbb{E}e^{uI(t)} = \exp\left\{\frac{1}{2}u^2\int_0^t \Delta^2(s)\,ds\right\} \text{ for all } u \in \mathbb{R}. \tag{4.4.30}$$

Because $\Delta(s)$ is not random, (4.4.30) is equivalent to

$$\mathbb{E}\exp\left\{uI(t) - \frac{1}{2}u^2\int_0^t \Delta^2(s)\,ds\right\} = 1,$$

which may be rewritten as

$$\mathbb{E}\exp\left\{\int_0^t u\Delta(s)\,dW(s) - \frac{1}{2}\int_0^t \big(u\Delta(s)\big)^2\,ds\right\} = 1. \tag{4.4.31}$$

But the process

$$\exp\left\{\int_0^t u\Delta(s)\,dW(s) - \frac{1}{2}\int_0^t \big(u\Delta(s)\big)^2\,ds\right\}$$

is a martingale. Indeed, it is a generalized geometric Brownian motion with mean rate of return $\alpha = 0$; see (4.4.29) with $\sigma(s) = u\Delta(s)$. Furthermore, this process takes the value 1 at $t = 0$, and hence its expectation is always 1. This gives us (4.4.31). □

Note that (4.4.31) always holds, regardless of whether $\Delta(s)$ is random. However, we need to assume that $\Delta(s)$ is nonrandom in order to obtain the moment-generating function formula (4.4.30) from (4.4.31). When $\Delta(s)$ is random, there is no reason that the distribution of $\int_0^t \Delta(s)\,dW(s)$ should be normal.

Example 4.4.10 (Vasicek interest rate model). Let $W(t), t \geq 0$, be a Brownian motion. The Vasicek model for the interest rate process $R(t)$ is

$$dR(t) = \big(\alpha - \beta R(t)\big)\,dt + \sigma\,dW(t), \tag{4.4.32}$$

where α, β, and σ are positive constants. Equation (4.4.32) is an example of a *stochastic differential equation.* It defines a random process, $R(t)$ in this case, by giving a formula for its differential, and the formula involves the random process itself and the differential of a Brownian motion.

The solution to the stochastic differential equation (4.4.32) can be determined in closed form and is

$$R(t) = e^{-\beta t}R(0) + \frac{\alpha}{\beta}\big(1 - e^{-\beta t}\big) + \sigma e^{-\beta t}\int_0^t e^{\beta s}\,dW(s), \tag{4.4.33}$$

a claim that we now verify. In particular, we compute the differential of the right-hand side of (4.4.33). To do this, we use the Itô-Doeblin formula with

$$f(t,x) = e^{-\beta t}R(0) + \frac{\alpha}{\beta}\big(1 - e^{-\beta t}\big) + \sigma e^{-\beta t}x$$

and $X(t) = \int_0^t e^{\beta s}\,dW(s)$. Then the right-hand side of (4.4.33) is $f\big(t, X(t)\big)$. The technique we are using is to separate the right-hand side into two parts: an ordinary function of two variables t and x, which has no randomness in it, and an Itô process $X(t)$, which contains all the randomness. For the Itô-Doeblin formula, we shall need the following partial derivatives of $f(t,x)$:

$$f_t(t,x) = -\beta e^{-\beta t}R(0) + \alpha e^{-\beta t} - \sigma\beta e^{-\beta t}x \;=\; \alpha - \beta f(t,x),$$
$$f_x(t,x) = \sigma e^{-\beta t},$$
$$f_{xx}(t,x) = 0.$$

We shall also need the differential of $X(t)$, which is $dX(t) = e^{\beta t}\,dW(t)$. We shall not need $dX(t)\,dX(t) = e^{2\beta t}\,dt$ because $f_{xx}(t,x) = 0$. The Itô-Doeblin formula states that

$$\begin{aligned} &df\big(t,X(t)\big)\\ &= f_t\big(t,X(t)\big)\,dt + f_x\big(t,X(t)\big)\,dX(t) + \frac{1}{2}f_{xx}\big(t,X(t)\big)\,dX(t)\,dX(t)\\ &= \Big(\alpha - \beta f\big(t,X(t)\big)\Big)\,dt + \sigma\,dW(t). \end{aligned}$$

This shows that $f\big(t,X(t)\big)$ satisfies the stochastic differential equation (4.4.32) that defines $R(t)$. Moreover, $f\big(0,X(0)\big) = R(0)$. Because $f\big(t,X(t)\big)$ satisfies the equation defining $R(t)$ and has the same initial condition as $R(t)$, it must be the case that $f(t,X(t)) = R(t)$ for all $t \geq 0$.

Theorem 4.4.9 implies that the random variable $\int_0^t e^{\beta s}\,dW(s)$ appearing on the right-hand side of (4.4.33) is normally distributed with mean zero and variance

$$\int_0^t e^{2\beta s}\,ds = \frac{1}{2\beta}\big(e^{2\beta t} - 1\big).$$

Therefore, $R(t)$ is normally distributed with mean $e^{-\beta t}R(0) + \frac{\alpha}{\beta}\big(1 - e^{-\beta t}\big)$ and variance $\frac{\sigma^2}{2\beta}\big(1 - e^{-2\beta t}\big)$. In particular, no matter how the parameters $\alpha > 0$, $\beta > 0$, and $\sigma > 0$ are chosen, there is positive probability that $R(t)$ is negative, an undesirable property for an interest rate model.

The Vasicek model has the desirable property that the interest rate is *mean-reverting*. When $R(t) = \frac{\alpha}{\beta}$, the drift term (the dt term) in (4.4.32) is zero. When $R(t) > \frac{\alpha}{\beta}$, this term is negative, which pushes $R(t)$ back toward $\frac{\alpha}{\beta}$. When $R(t) < \frac{\alpha}{\beta}$, this term is positive, which again pushes $R(t)$ back toward $\frac{\alpha}{\beta}$. If $R(0) = \frac{\alpha}{\beta}$, then $\mathbb{E}R(t) = \frac{\alpha}{\beta}$ for all $t \geq 0$. If $R(0) \neq \frac{\alpha}{\beta}$, then $\lim_{t\to\infty}\mathbb{E}R(t) = \frac{\alpha}{\beta}$. □

Example 4.4.11 (Cox-Ingersoll-Ross (CIR) interest rate model). Let $W(t)$, $t \geq 0$, be a Brownian motion. The Cox-Ingersoll-Ross model for the interest rate process $R(t)$ is

$$dR(t) = \big(\alpha - \beta R(t)\big)\,dt + \sigma\sqrt{R(t)}\,dW(t), \tag{4.4.34}$$

where α, β, and σ are positive constants. Unlike the Vasicek equation (4.4.32), the CIR equation (4.4.34) does not have a closed-form solution. The advantage of (4.4.34) over the Vasicek model is that the interest rate in the CIR model does not become negative. If $R(t)$ reaches zero, the term multiplying $dW(t)$

vanishes and the positive drift term $\alpha\,dt$ in equation (4.4.34) drives the interest rate back into positive territory. Like the Vasicek model, the CIR model is mean-reverting.

Although one cannot derive a closed-form solution for (4.4.34), the distribution of $R(t)$ for each positive t can be determined. That computation would take us too far afield. We instead content ourselves with the derivation of the expected value and variance of $R(t)$. To do this, we use the function $f(t,x) = e^{\beta t}x$ and the Itô-Doeblin formula to compute

$$\begin{aligned}
&d\big(e^{\beta t}R(t)\big)\\
&= df\big(t, R(t)\big)\\
&= f_t\big(t, R(t)\big)\,dt + f_x\big(t, R(t)\big)\,dR(t) + \frac{1}{2}f_{xx}\big(t, R(t)\big)\,dR(t)\,dR(t)\\
&= \beta e^{\beta t}R(t)\,dt + e^{\beta t}\big(\alpha - \beta R(t)\big)\,dt + e^{\beta t}\sigma\sqrt{R(t)}\,dW(t)\\
&= \alpha e^{\beta t}\,dt + \sigma e^{\beta t}\sqrt{R(t)}\,dW(t).
\end{aligned} \tag{4.4.35}$$

Integration of both sides of (4.4.35) yields

$$\begin{aligned}
e^{\beta t}R(t) &= R(0) + \alpha\int_0^t e^{\beta u}\,du + \sigma\int_0^t e^{\beta u}\sqrt{R(u)}\,dW(u)\\
&= R(0) + \frac{\alpha}{\beta}\big(e^{\beta t} - 1\big) + \sigma\int_0^t e^{\beta u}\sqrt{R(u)}\,dW(u).
\end{aligned}$$

Recalling that the expectation of an Itô integral is zero, we obtain

$$e^{\beta t}\mathbb{E}R(t) = R(0) + \frac{\alpha}{\beta}\big(e^{\beta t} - 1\big)$$

or, equivalently,

$$\mathbb{E}R(t) = e^{-\beta t}R(0) + \frac{\alpha}{\beta}\big(1 - e^{-\beta t}\big). \tag{4.4.36}$$

This is the same expectation as in the Vasicek model.

To compute the variance of $R(t)$, we set $X(t) = e^{\beta t}R(t)$, for which we have already computed

$$dX(t) = \alpha e^{\beta t}\,dt + \sigma e^{\beta t}\sqrt{R(t)}\,dW(t) = \alpha e^{\beta t}\,dt + \sigma e^{\frac{\beta t}{2}}\sqrt{X(t)}\,dW(t)$$

and $\mathbb{E}X(t) = R(0) + \frac{\alpha}{\beta}\big(e^{\beta t} - 1\big)$. According to the Itô-Doeblin formula (with $f(x) = x^2$, $f'(x) = 2x$, and $f''(x) = 2$),

$$\begin{aligned}
d\big(X^2(t)\big) &= 2X(t)\,dX(t) + dX(t)\,dX(t)\\
&= 2\alpha e^{\beta t}X(t)\,dt + 2\sigma e^{\frac{\beta t}{2}}X^{\frac{3}{2}}(t)\,dW(t) + \sigma^2 e^{\beta t}X(t)\,dt.
\end{aligned} \tag{4.4.37}$$

Integration of (4.4.37) yields

$$X^2(t) = X^2(0) + (2\alpha + \sigma^2)\int_0^t e^{\beta u}X(u)\,du + 2\sigma\int_0^t e^{\frac{\beta u}{2}}X^{\frac{3}{2}}(u)\,dW(u).$$

Taking expectations, using the fact that the expectation of an Itô integral is zero and the formula already derived for $\mathbb{E}X(t)$, we obtain

$$\begin{aligned}
\mathbb{E}X^2(t) &= X^2(0) + (2\alpha + \sigma^2)\int_0^t e^{\beta u}\mathbb{E}X(u)\,du \\
&= R^2(0) + (2\alpha + \sigma^2)\int_0^t e^{\beta u}\Big(R(0) + \frac{\alpha}{\beta}(e^{\beta u} - 1)\Big)\,du \\
&= R^2(0) + \frac{2\alpha + \sigma^2}{\beta}\Big(R(0) - \frac{\alpha}{\beta}\Big)(e^{\beta t} - 1) + \frac{2\alpha + \sigma^2}{2\beta}\cdot\frac{\alpha}{\beta}(e^{2\beta t} - 1).
\end{aligned}$$

Therefore,

$$\begin{aligned}
\mathbb{E}R^2(t) &= e^{-2\beta t}\mathbb{E}X^2(t) \\
&= e^{-2\beta t}R^2(0) + \frac{2\alpha + \sigma^2}{\beta}\Big(R(0) - \frac{\alpha}{\beta}\Big)(e^{-\beta t} - e^{-2\beta t}) \\
&\quad + \frac{\alpha(2\alpha + \sigma^2)}{2\beta^2}(1 - e^{-2\beta t}).
\end{aligned}$$

Finally,

$$\begin{aligned}
\mathrm{Var}\big(R(t)\big) &= \mathbb{E}R^2(t) - \big(\mathbb{E}R(t)\big)^2 \\
&= e^{-2\beta t}R^2(0) + \frac{2\alpha + \sigma^2}{\beta}\Big(R(0) - \frac{\alpha}{\beta}\Big)(e^{-\beta t} - e^{-2\beta t}) \\
&\quad + \frac{\alpha(2\alpha + \sigma^2)}{2\beta^2}(1 - e^{-2\beta t}) - e^{-2\beta t}R^2(0) \\
&\quad - \frac{2\alpha}{\beta}R(0)(e^{-\beta t} - e^{-2\beta t}) - \frac{\alpha^2}{\beta^2}(1 - e^{-\beta t})^2 \\
&= \frac{\sigma^2}{\beta}R(0)\Big(e^{-\beta t} - e^{-2\beta t}\Big) + \frac{\alpha\sigma^2}{2\beta^2}\Big(1 - 2e^{-\beta t} + e^{-2\beta t}\Big).
\end{aligned} \tag{4.4.38}$$

In particular,

$$\lim_{t\to\infty}\mathrm{Var}\big(R(t)\big) = \frac{\alpha\sigma^2}{2\beta^2}.$$

4.5 Black-Scholes-Merton Equation

The addition of Merton's name to what has traditionally been called the Black-Scholes equation is explained in the Notes, Section 4.9.

In this section, we derive the Black-Scholes-Merton partial differential equation for the price of an option on an asset modeled as a geometric Brownian motion. The idea behind this derivation is the same as in the binomial model of Chapter 1 of Volume I, which is to determine the initial capital required to perfectly hedge a short position in the option.

4.5.1 Evolution of Portfolio Value

Consider an agent who at each time t has a portfolio valued at $X(t)$. This portfolio invests in a money market account paying a constant rate of interest r and in a stock modeled by the geometric Brownian motion

$$dS(t) = \alpha S(t)\,dt + \sigma S(t)\,dW(t). \tag{4.5.1}$$

Suppose at each time t, the investor holds $\Delta(t)$ shares of stock. The position $\Delta(t)$ can be random but must be adapted to the filtration associated with the Brownian motion $W(t)$, $t \geq 0$. The remainder of the portfolio value, $X(t) - \Delta(t)S(t)$, is invested in the money market account.

The differential $dX(t)$ for the investor's portfolio value at each time t is due to two factors, the capital gain $\Delta(t)\,dS(t)$ on the stock position and the interest earnings $r\big(X(t) - \Delta(t)S(t)\big)\,dt$ on the cash position. In other words,

$$\begin{aligned} dX(t) &= \Delta(t)\,dS(t) + r\big(X(t) - \Delta(t)S(t)\big)\,dt \\ &= \Delta(t)\big(\alpha S(t)\,dt + \sigma S(t)\,dW(t)\big) + r\big(X(t) - \Delta(t)S(t)\big)\,dt \\ &= rX(t)\,dt + \Delta(t)(\alpha - r)S(t)\,dt + \Delta(t)\sigma S(t)\,dW(t). \end{aligned} \tag{4.5.2}$$

The three terms appearing in the last line of (4.5.2) can be understood as follows:

(i) an average underlying rate of return r on the portfolio, which is reflected by the term $rX(t)\,dt$,
(ii) a risk premium $\alpha - r$ for investing in the stock, which is reflected by the term $\Delta(t)(\alpha - r)S(t)\,dt$, and
(iii) a volatility term proportional to the size of the stock investment, which is the term $\Delta(t)\sigma S(t)\,dW(t)$.

The discrete-time analogue of equation (4.5.2) appears in Chapter 1 of Volume I as (1.2.12):

$$X_{n+1} = \Delta_n S_{n+1} + (1+r)(X_n - \Delta_n S_n).$$

We may rearrange terms in this equation to obtain

$$X_{n+1} - X_n = \Delta_n(S_{n+1} - S_n) + r(X_n - \Delta_n S_n), \tag{4.5.3}$$

which is analogous to the first line of (4.5.2), except in (4.5.3) time steps forward one unit at a time, whereas in (4.5.2) time moves forward continuously.

See Exercise 4.10 for additional discussion of the rationale for equation (4.5.2) in option pricing.

We shall often consider the discounted stock price $e^{-rt}S(t)$ and the discounted portfolio value of an agent, $e^{-rt}X(t)$. According to the Itô-Doeblin formula with $f(t,x) = e^{-rt}x$, the differential of the discounted stock price is

$$\begin{aligned} & d\big(e^{-rt}S(t)\big) \\ & = df\big(t, S(t)\big) \\ & = f_t\big(t, S(t)\big)\,dt + f_x\big(t, S(t)\big)\,dS(t) + \frac{1}{2}f_{xx}\big(t, S(t)\big)\,dS(t)\,dS(t) \\ & = -re^{-rt}S(t)\,dt + e^{-rt}\,dS(t) \\ & = (\alpha - r)e^{-rt}S(t)\,dt + \sigma e^{-rt}S(t)\,dW(t), \end{aligned} \tag{4.5.4}$$

and the differential of the discounted portfolio value is

$$\begin{aligned} & d\big(e^{-rt}X(t)\big) \\ & = df\big(t, X(t)\big) \\ & = f_t\big(t, X(t)\big)\,dt + f_x\big(t, X(t)\big)\,dX(t) + \frac{1}{2}f_{xx}\big(t, X(t)\big)\,dX(t)\,dX(t) \\ & = -re^{-rt}X(t)\,dt + e^{-rt}\,dX(t) \\ & = \Delta(t)(\alpha - r)e^{-rt}S(t)\,dt + \Delta(t)\sigma e^{-rt}S(t)\,dW(t) \\ & = \Delta(t)\,d\big(e^{-rt}S(t)\big). \end{aligned} \tag{4.5.5}$$

Discounting the stock price reduces the mean rate of return from α, the term multiplying $S(t)\,dt$ in (4.5.1), to $\alpha - r$, the term multiplying $e^{-rt}S(t)\,dt$ in (4.5.4). Discounting the portfolio value removes the underlying rate of return r; compare the last line of (4.5.2) to the next-to-last line of (4.5.5). The last line of (4.5.5) shows that change in the discounted portfolio value is solely due to change in the discounted stock price.

4.5.2 Evolution of Option Value

Consider a European call option that pays $\big(S(T) - K\big)^+$ at time T. The strike price K is some nonnegative constant. Black, Scholes, and Merton argued that the value of this call at any time should depend on the time (more precisely, on the time to expiration) and on the value of the stock price at that time, and of course it should also depend on the model parameters r and σ and the contractual strike price K. Only two of these quantities, time and stock price, are variable. Following this reasoning, we let $c(t,x)$ denote the value of the call at time t if the stock price at that time is $S(t) = x$. There is nothing random about the function $c(t,x)$. However, the value of the option is random; it is the stochastic process $c\big(t, S(t)\big)$ obtained by replacing the dummy variable x by the random stock price $S(t)$ in this function. At the initial time, we do not know the future stock prices $S(t)$ and hence do not know the future option values $c\big(t, S(t)\big)$. Our goal is to determine the function $c(t, x)$ so we at least have a formula for the future option values in terms of the future stock prices.

We begin by computing the differential of $c\big(t, S(t)\big)$. According to the Itô-Doeblin formula, it is

$$\begin{aligned}
&dc\big(t, S(t)\big)\\
&= c_t\big(t, S(t)\big)\,dt + c_x\big(t, S(t)\big)\,dS(t) + \frac{1}{2}c_{xx}\big(t, S(t)\big)\,dS(t)\,dS(t)\\
&= c_t\big(t, S(t)\big)\,dt + c_x\big(t, S(t)\big)\big(\alpha S(t)\,dt + \sigma S(t)\,dW(t)\big)\\
&\qquad + \frac{1}{2}c_{xx}\big(t, S(t)\big)\sigma^2 S^2(t)\,dt\\
&= \left[c_t\big(t, S(t)\big) + \alpha S(t)c_x\big(t, S(t)\big) + \frac{1}{2}\sigma^2 S^2(t)c_{xx}\big(t, S(t)\big)\right]dt\\
&\qquad + \sigma S(t)c_x\big(t, S(t)\big)\,dW(t).
\end{aligned} \tag{4.5.6}$$

We next compute the differential of the discounted option price $e^{-rt}c\big(t, S(t)\big)$. Let $f(t, x) = e^{-rt}x$. According to the Itô-Doeblin formula,

$$\begin{aligned}
&d\big(e^{-rt}c(t, S(t))\big)\\
&\quad= df\big(t, c(t, S(t))\big)\\
&\quad= f_t\big(t, c(t, S(t))\big)\,dt + f_x\big(t, c(t, S(t))\big)\,dc(t, S(t))\\
&\qquad\quad + \frac{1}{2}f_{xx}\big(t, c(t, S(t))\big)\,dc(t, S(t))\,dc(t, S(t))\\
&\quad= -re^{-rt}c(t, S(t))\,dt + e^{-rt}\,dc(t, S(t))\\
&\quad= e^{-rt}\left[-rc(t, S(t)) + c_t(t, S(t)) + \alpha S(t)c_x(t, S(t))\right.\\
&\qquad\quad \left. + \frac{1}{2}\sigma^2 S^2(t)c_{xx}(t, S(t))\right]dt + e^{-rt}\sigma S(t)c_x(t, S(t))dW(t).
\end{aligned} \tag{4.5.7}$$

4.5.3 Equating the Evolutions

A (short option) hedging portfolio starts with some initial capital $X(0)$ and invests in the stock and money market account so that the portfolio value $X(t)$ at each time $t \in [0, T]$ agrees with $c(t, S(t))$. This happens if and only if $e^{-rt}X(t) = e^{-rt}c(t, S(t))$ for all t. One way to ensure this equality is to make sure that

$$d\big(e^{-rt}X(t)\big) = d\big(e^{-rt}c(t, S(t))\big) \text{ for all } t \in [0, T) \tag{4.5.8}$$

and $X(0) = c(0, S(0))$. Integration of (4.5.8) from 0 to t then yields

$$e^{-rt}X(t) - X(0) = e^{-rt}c(t, S(t)) - c(0, S(0)) \text{ for all } t \in [0, T). \tag{4.5.9}$$

If $X(0) = c(0, S(0))$, then we can cancel this term in (4.5.9) and get the desired equality.

Comparing (4.5.5) and (4.5.7), we see that (4.5.8) holds if and only if

$$\begin{aligned}
&\Delta(t)(\alpha - r)S(t)\,dt + \Delta(t)\sigma S(t)\,dW(t)\\
&= \left[-rc(t,S(t)) + c_t(t,S(t)) + \alpha S(t)c_x(t,S(t)) + \frac{1}{2}\sigma^2 S^2(t)c_{xx}(t,S(t))\right]dt\\
&\qquad + \sigma S(t)c_x(t,S(t))\,dW(t). \qquad (4.5.10)
\end{aligned}$$

We examine what is required in order for (4.5.10) to hold.

We first equate the $dW(t)$ terms in (4.5.10), which gives

$$\Delta(t) = c_x(t,S(t)) \text{ for all } t \in [0,T). \qquad (4.5.11)$$

This is called the *delta-hedging rule*. At each time t prior to expiration, the number of shares held by the hedge of the short option position is the partial derivative with respect to the stock price of the option value at that time. This quantity, $c_x(t,S(t))$, is called the *delta* of the option.

We next equate the dt terms in (4.5.10), using (4.5.11), to obtain

$$\begin{aligned}
&(\alpha - r)S(t)c_x(t,S(t))\\
&= -rc(t,S(t)) + c_t(t,S(t)) + \alpha S(t)c_x(t,S(t)) + \frac{1}{2}\sigma^2 S^2(t)c_{xx}(t,S(t))\\
&\qquad \text{for all } t \in [0,T). \qquad (4.5.12)
\end{aligned}$$

The term $\alpha S(t)c_x(t,S(t))$ appears on both sides of (4.5.12), and after canceling it, we obtain

$$\begin{aligned}
rc(t,S(t)) = c_t(t,S(t)) + rS(t)c_x(t,S(t)) + \frac{1}{2}\sigma^2 S^2(t)c_{xx}(t,S(t))&\\
\text{for all } t \in [0,T). \qquad (4.5.13)&
\end{aligned}$$

In conclusion, we should seek a continuous function $c(t,x)$ that is a solution to the *Black-Scholes-Merton partial differential equation*

$$c_t(t,x) + rxc_x(t,x) + \frac{1}{2}\sigma^2 x^2 c_{xx}(t,x) = rc(t,x) \text{ for all } t \in [0,T),\ x \geq 0, \qquad (4.5.14)$$

and that satisfies the *terminal condition*

$$c(T,x) = (x-K)^+. \qquad (4.5.15)$$

Suppose we have found this function. If an investor starts with initial capital $X(0) = c(0,S(0))$ and uses the hedge $\Delta(t) = c_x(t,S(t))$, then (4.5.10) will hold for all $t \in [0,T)$. Indeed, the $dW(t)$ terms on the left and right sides of (4.5.10) agree because $\Delta(t) = c_x(t,S(t))$, and the dt terms agree because (4.5.14) guarantees (4.5.13). Equality in (4.5.10) gives us (4.5.9). Canceling $X(0) = c(0,S(0))$ and e^{-rt} in this equation, we see that $X(t) = c(t,S(t))$ for all $t \in [0,T)$. Taking the limit as $t \uparrow T$ and using the fact that both $X(t)$ and

$c(t, S(t))$ are continuous, we conclude that $X(T) = c(T, S(T)) = \big(S(T)-K\big)^+$. This means that the short position has been successfully hedged. No matter which of its possible paths the stock price follows, when the option expires, the agent hedging the short position has a portfolio whose value agrees with the option payoff.

4.5.4 Solution to the Black-Scholes-Merton Equation

The Black-Scholes-Merton equation (4.5.14) does not involve probability. It is a partial differential equation, and the arguments t and x are dummy variables, not random variables. One can solve it by partial differential equation methods. In this section, however, rather than showing how to solve the equation, we shall simply present the solution and check that it works. In Subsection 5.2.5, we present a derivation of this solution based on probability theory.

We want the Black-Scholes-Merton equation to hold for all $x \geq 0$ and $t \in [0, T)$ so that (4.5.14) will hold regardless of which of its possible paths the stock price follows. If the initial stock price is positive, then the stock price is always positive, and it can take any positive value. If the initial stock price is zero, then subsequent stock prices are all zero. We cover both of these cases by asking (4.5.14) to hold for all $x \geq 0$. We do not need (4.5.14) to hold at $t = T$, although we need the function $c(t, x)$ to be continuous at $t = T$. If the hedge works at all times strictly prior to T, it also works at time T because of continuity.

Equation (4.5.14) is a partial differential equation of the type called *backward parabolic*. For such an equation, in addition to the terminal condition (4.5.15), one needs boundary conditions at $x = 0$ and $x = \infty$ in order to determine the solution. The boundary condition at $x = 0$ is obtained by substituting $x = 0$ into (4.5.14), which then becomes

$$c_t(t, 0) = rc(t, 0). \tag{4.5.16}$$

This is an ordinary differential equation for the function $c(t, 0)$ of t, and the solution is

$$c(t, 0) = e^{rt} c(0, 0).$$

Substituting $t = T$ into this equation and using the fact that $c(T, 0) = (0 - K)^+ = 0$, we see that $c(0, 0) = 0$ and hence

$$c(t, 0) = 0 \text{ for all } t \in [0, T]. \tag{4.5.17}$$

This is the *boundary condition at* $x = 0$.

As $x \to \infty$, the function $c(t, x)$ grows without bound. In such a case, we give the boundary condition at $x = \infty$ by specifying the rate of growth. One way to specify a *boundary condition at* $x = \infty$ for the European call is

$$\lim_{x\to\infty} \left[c(t, x) - \left(x - e^{-r(T-t)}K\right)\right] = 0 \text{ for all } t \in [0, T]. \tag{4.5.18}$$

In particular, $c(t,x)$ grows at the same rate as x as $x \to \infty$. Recall that $c(t,x)$ is the value at time t of a call on a stock whose price at time t is x. For large x, this call is deep in the money and very likely to end in the money. In this case, the price of the call is almost as much as the price of the forward contract discussed in Subsection 4.5.6 below (see (4.5.26)). This is the assertion of (4.5.18).

The solution to the Black-Scholes-Merton equation (4.5.14) with terminal condition (4.5.15) and boundary conditions (4.5.17) and (4.5.18) is

$$c(t,x) = xN\big(d_+(T-t,x)\big) - Ke^{-r(T-t)}N\big(d_-(T-t,x)\big), \quad 0 \le t < T, \; x > 0, \tag{4.5.19}$$

where

$$d_\pm(\tau,x) = \frac{1}{\sigma\sqrt{\tau}}\left[\log\frac{x}{K} + \left(r \pm \frac{\sigma^2}{2}\right)\tau\right], \tag{4.5.20}$$

and N is the cumulative standard normal distribution

$$N(y) = \frac{1}{\sqrt{2\pi}}\int_{-\infty}^{y} e^{-\frac{z^2}{2}}\,dz = \frac{1}{\sqrt{2\pi}}\int_{-y}^{\infty} e^{-\frac{z^2}{2}}\,dz. \tag{4.5.21}$$

We shall sometimes use the notation

$$\mathrm{BSM}(\tau,x;K,r,\sigma) = xN\big(d_+(\tau,x)\big) - Ke^{-r\tau}N\big(d_-(\tau,x)\big), \tag{4.5.22}$$

and call $\mathrm{BSM}(\tau,x;K,r,\sigma)$ the *Black-Scholes-Merton function.* In this formula, τ and x denote the time to expiration and the current stock price, respectively. The parameters K, r, and σ are the strike price, the interest rate, and the stock volatility, respectively.

Formula (4.5.19) does not define $c(t,x)$ when $t = T$ (because then $\tau = T - t = 0$ and this appears in the denominator in (4.5.20)), nor does it define $c(t,x)$ when $x = 0$ (because $\log x$ appears in (4.5.20), and $\log 0$ is not a real number). However, (4.5.19) defines $c(t,x)$ in such a way that $\lim_{t\to T} c(t,x) = (x-K)^+$ and $\lim_{x\downarrow 0} c(t,x) = 0$. Verification of all of these claims is given as Exercise 4.9.

4.5.5 The Greeks

The derivatives of the function $c(t,x)$ of (4.5.19) with respect to various variables are called the *Greeks.* Two of these are derived in Exercise 4.9, namely *delta,* which is

$$c_x(t,x) = N\big(d_+(T-t,x)\big), \tag{4.5.23}$$

and *theta,* which is

$$c_t(t,x) = -rKe^{-r(T-t)}N\big(d_-(T-t,x)\big) - \frac{\sigma x}{2\sqrt{T-t}}N'\big(d_+(T-t,x)\big). \tag{4.5.24}$$

Because both N and N' are always positive, delta is always positive and theta is always negative. Another of the Greeks is *gamma,* which is

$$c_{xx}(t,x) = N'\big(d_+(T-t,x)\big)\frac{\partial}{\partial x}d_+(T-t,x) = \frac{1}{\sigma x\sqrt{T-t}}N'\big(d_+(T-t,x)\big). \tag{4.5.25}$$

Like delta, gamma is always positive.

In order to simplify notation in the following discussion, we sometimes suppress the arguments (t,x) of $c(t,x)$ and $(T-t,x)$ of $d_\pm(T-t,x)$. If at time t the stock price is x, then the short option hedge of (4.5.11) calls for holding $c_x(t,x)$ shares of stock, a position whose value is $xc_x = xN(d_+)$. The hedging portfolio value is $c = xN(d_+) - Ke^{-r(T-t)}N(d_-)$, and since $xc_x(t,x)$ of this value is invested in stock, the amount invested in the money market must be

$$c(t,x) - xc_x(t,x) = -Ke^{-r(T-t)}N(d_-),$$

a negative number. To hedge a short position in a call option, one must borrow money. To hedge a long position in a call option, one does the opposite. In other words, to hedge a long call position one should hold $-c_x$ shares of stock (i.e., have a short position in stock) and invest $Ke^{-r(T-t)}N(d_-)$ in the money market account.

Because delta and gamma are positive, for fixed t, the function $c(t,x)$ is increasing and convex in the variable x, as shown in Figure 4.5.1. Suppose at time t the stock price is x_1 and we wish to take a long position in the option and hedge it. We do this by purchasing the option for $c(t,x_1)$, shorting $c_x(t,x_1)$ shares of stock, which generates income $x_1c_x(t,x_1)$, and investing the difference,

$$M = x_1c_x(t,x_1) - c(t,x_1),$$

in the money market account. We wish to consider the sensitivity to stock price changes of the portfolio that has these three components: long option, short stock, and long money market account. The initial portfolio value

$$c(t,x_1) - x_1c_x(t,x_1) + M$$

is zero at the moment t when we set up these positions.

If the stock price were to instantaneously fall to x_0 as shown in Figure 4.5.1 and we do not change our positions in the stock or money market account, then the value of the option we hold would fall to $c(t,x_0)$ and the liability due to our short position in stock would decrease to $x_0c_x(t,x_1)$. Our total portfolio value, including M in the money market account, would be

$$c(t,x_0) - x_0c_x(t,x_1) + M = c(t,x_0) - c_x(t,x_1)(x_0-x_1) - c(t,x_1).$$

This is the difference at x_0 between the curve $y = c(t,x)$ and the straight line $y = c_x(t,x_1)(x-x_1) + c(t,x_1)$ in Figure 4.5.1. Because this difference is positive, our portfolio benefits from an instantaneous drop in the stock price.

On the other hand, if the stock price were to instantaneously rise to x_2 and we do not change our positions in the stock or money market account, then the value of the option would rise to $c(t,x_2)$ and the liability due to our short position in stock would increase to $x_2 c_x(t, x_1)$. Our total portfolio value, including M in the money market account, would be

$$c(t, x_2) - x_2 c_x(t, x_1) + M = c(t, x_2) - c_x(t, x_1)(x_2 - x_1) - c(t, x_1).$$

This is the difference at x_2 between the curve $y = c(t, x)$ and the straight line $y = c_x(t, x_1)(x - x_1) + c(t, x_1)$ in Figure 4.5.1. This difference is positive, so our portfolio benefits from an instantaneous rise in the stock price.

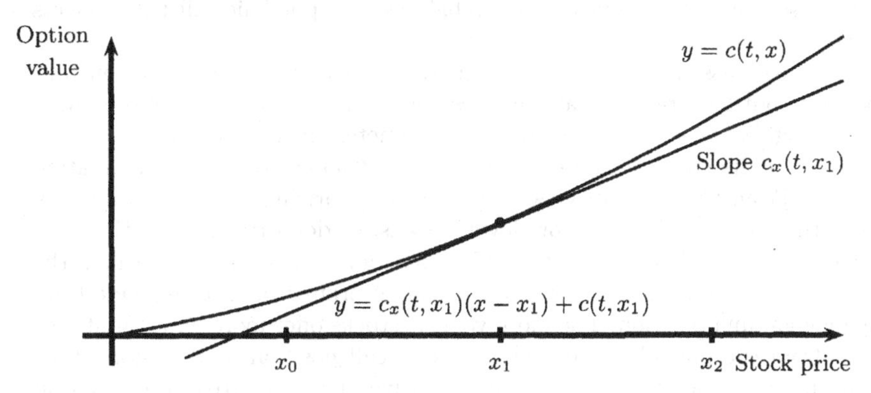

Fig. 4.5.1. Delta-neutral position.

The portfolio we have set up is said to be *delta-neutral* and *long gamma.* The portfolio is long gamma because it benefits from the convexity of $c(t, x)$ as described above. If there is an instantaneous rise or an instantaneous fall in the stock price, the value of the portfolio increases. A long gamma portfolio is profitable in times of high stock volatility.

"Delta-neutral" refers to the fact that the line in Figure 4.5.1 is tangent to the curve $y = c(t, x)$. Therefore, when the stock price makes a small move, the change of portfolio value due to the corresponding change in option price is nearly offset by the change in the value of our short position in the stock. The straight line is a good approximation to the option price *for small stock price moves.* If the straight line were steeper than the option price curve at the starting point x_1, then we would be *short delta*; an upward move in the stock price would hurt the portfolio because the liability from the short position in stock would rise faster than the value of the option. On the other hand, a downward move would increase the portfolio value because the option price would fall more slowly than the rate of decrease in the liability from the short stock position. Unless a trader has a view on the market, he tries to set up portfolios that are delta-neutral. If he expects high volatility, he would at the same time try to choose the portfolio to be long gamma.

The portfolio described above may at first appear to offer an arbitrage opportunity. When we let time move forward, not only does the long gamma position offer an opportunity for profit, but the positive investment in the money market account enhances this opportunity. The drawback is that theta, the derivative of $c(t,x)$ with respect to time, is negative. As we move forward in time, the curve $y = c(t,x)$ is shifting downward. Figure 4.5.1 is misleading because it is drawn with t fixed. In principle, the portfolio can lose money because the curve $c(t,x)$ shifts downward more rapidly than the money market investment and the long gamma position generate income. The essence of the hedging argument in Subsection 4.5.3 is that if the stock really is a geometric Brownian motion and we have determined the right value of the volatility σ, then so long as we continuously rebalance our portfolio, all these effects exactly cancel!

Of course, assets are not really geometric Brownian motions with constant volatility, but the argument above gives a good first approximation to reality. It also highlights volatility as the key parameter. In fact, the mean rate of return α of the stock does not appear in the Black-Scholes-Merton equation (4.5.14). From the point of view of no-arbitrage pricing, it is irrelevant how likely the stock is to go up or down because a delta-neutral position is a hedge against both possibilities. What matters is how much volatility the stock has, for we need to know the amount of profit that can be made from the long gamma position. The more volatile stocks offer more opportunity for profit from the portfolio that hedges a long call position with a short stock position, and hence the call is more expensive. The derivative of the option price with respect to the volatility σ is called *vega*, and it is positive. As volatility increases, so do option prices in the Black-Scholes-Merton model.

4.5.6 Put–Call Parity

A *forward contract* with delivery price K obligates its holder to buy one share of the stock at expiration time T in exchange for payment K. At expiration, the value of the forward contract is $S(T) - K$. Let $f(t,x)$ denote the value of the forward contract at earlier times $t \in [0,T]$ if the stock price at time t is $S(t) = x$.

We argue that the value of a forward contract is given by

$$f(t,x) = x - e^{-r(T-t)}K. \tag{4.5.26}$$

If an agent sells this forward contract at time zero for $f(t,S(0)) = S(0) - e^{-rT}K$, he can set up a *static hedge*, a hedge that does not trade except at the initial time, in order to protect himself. Specifically, the agent should purchase one share of stock. Since he has initial capital $S(0) - e^{-rT}K$ from the sale of the forward contract, this requires that he borrow $e^{-rT}K$ from the money market account. The agent makes no further trades. At expiration of the forward contract, he owns one share of stock and his debt to the money market account has grown to K, so his portfolio value is $S(T)-K$, exactly the value of the forward contract. Because the agent has been able to replicate the payoff of the forward contract with a portfolio whose value at each time t is

$S(t)-e^{-r(T-t)}K$, this must be the value at each time of the forward contract. This is $f(t,S(t))$, where $f(t,x)$ is defined by (4.5.26).

The *forward price* of a stock at time t is defined to be the value of K that causes the forward contract at time t to have value zero (i.e., it is the value of K that satisfies the equation $S(t)-e^{-r(T-t)}K=0$). Hence, we see that in a model with a constant interest rate, the forward price at time t is

$$\text{For}(t)=e^{r(T-t)}S(t). \tag{4.5.27}$$

Note that the forward price is not the price (or value) of a forward contract. For $0\leq t\leq T$, the forward price at time t is the price one can lock in at time t for the purchase of one share of stock at time T, paying the price (*settling*) at time T. No money changes hands at the time the price is locked in.

Let us consider this situation at time $t=0$. At that time, one can lock in a price $\text{For}(0)=e^{rT}S(0)$ for purchase of the stock at time T. Let us do this, which means we set $K=e^{rT}S(0)$ in (4.5.26). The value of this forward contract is zero at time $t=0$, but as soon as time begins to move forward, the value of the forward contract changes. Indeed, its value at time t is

$$f(t,S(t))=S(t)-e^{rt}S(0).$$

Finally, let us consider a *European put*, which pays off $(K-S(T))^+$ at time T. We observe that for any number x, the equation

$$x-K=(x-K)^+-(K-x)^+ \tag{4.5.28}$$

holds. Indeed, if $x\geq K$, then $(x-K)^+=x-K$ and $(K-x)^+=0$. On the other hand, if $x\leq K$, then $(x-K)^+=0$ and $-(K-x)^+=-(K-x)=x-K$. In either case, the right-hand side of (4.5.28) equals the left-hand side. We denote by $p(t,x)$ the value of the European put at time t if the time-t stock price is $S(t)=x$. Similarly, we denote by $c(t,x)$ the value of the European call expiring at time T with strike price K and by $f(t,x)$ the value of the forward contract for the purchase of one share of stock at time T in exchange for payment K. Equation (4.5.28) implies

$$f(T,S(T))=c(T,S(T))-p(T,S(T));$$

the payoff of the forward contract agrees with the payoff of a portfolio that is long one call and short one put. Since the value at time T of the forward contract agrees with the value of the portfolio that is long one call and short one put, these values must agree at all previous times:

$$f(t,x)=c(t,x)-p(t,x),\quad x\geq 0,\ 0\leq t\leq T. \tag{4.5.29}$$

If this were not the case, one could at some time t either sell or buy the portfolio that is long the forward, short the call, and long the put, realizing an instant profit, and have no liability upon expiration of the contracts. The relationship (4.5.29) is called *put–call parity*.

Note that we have derived the put–call parity formula (4.5.29) without appealing to the Black-Scholes-Merton model of a geometric Brownian motion for the stock price and a constant interest rate. Indeed, without any assumptions on the prices except sufficient liquidity that permits one to form the portfolio that is long one call and short one put, we have put–call parity. If we make the assumption of a constant interest rate r, then $f(t,x)$ is given by (4.5.26). If we make the additional assumption that the stock is a geometric Brownian motion with constant volatility $\sigma > 0$, then we have also the Black-Scholes-Merton call formula (4.5.19). We can then solve (4.5.29) to obtain the Black-Scholes-Merton put formula

$$\begin{aligned} p(t,x) &= x\big(N(d_+(T-t,x)) - 1\big) - Ke^{-r(T-t)}\big(N(d_-(T-t,x)) - 1\big) \\ &= Ke^{-r(T-t)}N(-d_-(T-t,x)) - xN(-d_+(T-t,x)), \end{aligned} \tag{4.5.30}$$

where $d_\pm(T-t,x)$ is given by (4.5.20).

4.6 Multivariable Stochastic Calculus

4.6.1 Multiple Brownian Motions

Definition 4.6.1. *A* d-dimensional Brownian motion *is a process*

$$W(t) = (W_1(t), \dots, W_d(t))$$

with the following properties.

(i) Each $W_i(t)$ is a one-dimensional Brownian motion.
(ii) If $i \neq j$, then the processes $W_i(t)$ and $W_j(t)$ are independent.

Associated with a d-dimensional Brownian motion, we have a filtration $\mathcal{F}(t)$, $t \geq 0$, such that the following holds.

(iii) **(Information accumulates)** *For $0 \leq s < t$, every set in $\mathcal{F}(s)$ is also in $\mathcal{F}(t)$.*
(iv) **(Adaptivity)** *For each $t \geq 0$, the random vector $W(t)$ is $\mathcal{F}(t)$-measurable.*
(v) **(Independence of future increments)** *For $0 \leq t < u$, the vector of increments $W(u) - W(t)$ is independent of $\mathcal{F}(t)$.*

Although we have defined a multidimensional Brownian motion to be a vector of *independent* one-dimensional Brownian motions, we shall see in Example 4.6.6 how to build correlated Brownian motions from this.

Because each component W_i of a d-dimensional Brownian motion is a one-dimensional Brownian motion, we have the quadratic variation formula $[W_i, W_i](t) = t$, which we write informally as

$$dW_i(t)\,dW_i(t) = dt.$$

However, if $i \neq j$, we shall see that independence of W_i and W_j implies $[W_i, W_j](t) = 0$, which we write informally as

$$dW_i(t)\,dW_j(t) = 0, \quad i \neq j.$$

We justify this claim.

Let $\Pi = \{t_0, \ldots, t_n\}$ be a partition of $[0, T]$. For $i \neq j$, define the *sampled cross variation* of W_i and W_j on $[0, T]$ to be

$$C_\Pi = \sum_{k=0}^{n-1} \big[W_i(t_{k+1}) - W_i(t_k)\big]\big[W_j(t_{k+1}) - W_j(t_k)\big].$$

The increments appearing on the right-hand side of the equation above are all independent of one another and all have mean zero. Therefore, $\mathbb{E}C_\Pi = 0$.

We compute $\mathrm{Var}(C_\Pi)$. Note first that

$$\begin{aligned} C_\Pi^2 = &\sum_{k=0}^{n-1} \big[W_i(t_{k+1}) - W_i(t_k)\big]^2\big[W_j(t_{k+1}) - W_j(t_k)\big]^2 \\ &+2\sum_{\ell<k}^{n-1} \big[W_i(t_{\ell+1}) - W_i(t_\ell)\big]\big[W_j(t_{\ell+1}) - W_j(t_\ell)\big] \\ &\qquad\cdot\big[W_i(t_{k+1}) - W_i(t_k)\big]\big[W_j(t_{k+1}) - W_j(t_k)\big]. \end{aligned}$$

All the increments appearing in the sum of cross-terms are independent of one another and all have mean zero. Therefore,

$$\mathrm{Var}(C_\Pi) = \mathbb{E}C_\Pi^2 = \mathbb{E}\sum_{k=0}^{n-1} \big[W_i(t_{k+1}) - W_i(t_k)\big]^2\big[W_j(t_{k+1}) - W_j(t_k)\big]^2.$$

But $\big[W_i(t_{k+1}) - W_i(t_k)\big]^2$ and $\big[W_j(t_{k+1}) - W_j(t_k)\big]^2$ are independent of one another, and each has expectation $(t_{k+1} - t_k)$. It follows that

$$\mathrm{Var}(C_\Pi) = \sum_{k=0}^{n-1}(t_{k+1} - t_k)^2 \leq \|\Pi\| \cdot \sum_{k=0}^{n-1}(t_{k+1} - t_k) = \|\Pi\| \cdot T.$$

As $\|\Pi\| \to 0$, we have $\mathrm{Var}(C_\Pi) \to 0$, so C_Π converges to the constant $\mathbb{E}C_\Pi = 0$.

4.6.2 Itô-Doeblin Formula for Multiple Processes

To keep the notation as simple as possible, we write the Itô formula for two processes driven by a two-dimensional Brownian motion. In the obvious way, the formula generalizes to any number of processes driven by a Brownian motion of any number (not necessarily the same number) of dimensions.

Let $X(t)$ and $Y(t)$ be Itô processes, which means they are processes of the form

$$X(t) = X(0) + \int_0^t \Theta_1(u)\,du + \int_0^t \sigma_{11}(u)\,dW_1(u) + \int_0^t \sigma_{12}(u)\,dW_2(u),$$

$$Y(t) = Y(0) + \int_0^t \Theta_2(u)\,du + \int_0^t \sigma_{21}(u)\,dW_1(u) + \int_0^t \sigma_{22}(u)\,dW_2(u).$$

The integrands $\Theta_i(u)$ and $\sigma_{ij}(u)$ are assumed to be adapted processes. In differential notation, we write

$$dX(t) = \Theta_1(t)\,dt + \sigma_{11}(t)\,dW_1(t) + \sigma_{12}(t)\,dW_2(t), \tag{4.6.1}$$

$$dY(t) = \Theta_2(t)\,dt + \sigma_{21}(t)\,dW_1(t) + \sigma_{22}(t)\,dW_2(t). \tag{4.6.2}$$

The Itô integral $\int_0^t \sigma_{11}(u)\,dW_1(u)$ accumulates quadratic variation at rate $\sigma_{11}^2(t)$ per unit time, and the Itô integral $\int_0^t \sigma_{12}(u)\,dW_2(u)$ accumulates quadratic variation at rate $\sigma_{12}^2(t)$ per unit time. Because both of these integrals appear in $X(t)$, the process $X(t)$ accumulates quadratic variation at rate $\sigma_{11}^2(t) + \sigma_{12}^2(t)$ per unit time:

$$[X,X](t) = \int_0^t \big(\sigma_{11}^2(u) + \sigma_{12}^2(u)\big)\,du.$$

We may write this equation in differential form as

$$dX(t)\,dX(t) = \big(\sigma_{11}^2(t) + \sigma_{12}^2(t)\big)\,dt. \tag{4.6.3}$$

One can informally derive (4.6.3) by squaring (4.6.1) and using the multiplication rules

$$dt\,dt = 0,\ \ dt\,dW_i(t) = 0,\ \ dW_i(t)\,dW_i(t) = dt,\ \ dW_i(t)\,dW_j(t) = 0 \text{ for } i \neq j.$$

In a similar way, we may derive the differential formulas

$$dY(t)\,dY(t) = \big(\sigma_{21}^2(t) + \sigma_{22}^2(t)\big)\,dt, \tag{4.6.4}$$

$$dX(t)\,dY(t) = \big(\sigma_{11}(t)\sigma_{21}(t) + \sigma_{12}(t)\sigma_{22}(t)\big)\,dt. \tag{4.6.5}$$

Equation (4.6.5) says that, for every $T \geq 0$,

$$[X,Y](T) = \int_0^T \big(\sigma_{11}(t)\sigma_{21}(t) + \sigma_{12}(t)\sigma_{22}(t)\big)\,dt. \tag{4.6.6}$$

The term $[X,Y](T)$ on the left-hand side is defined as follows. Let $\Pi = \{t_0, t_1, \ldots, t_n\}$ be a partition of $[0,T]$ (i.e., $0 = t_0 < t_1 < \cdots < t_n = T$) and set up the sampled cross variation

$$\sum_{k=0}^{n-1} \big[X(t_{k+1}) - X(t_k)\big]\big[Y(t_{k+1}) - Y(t_k)\big]. \tag{4.6.7}$$

Now let the number of partition points n go to infinity as the length of the longest subinterval $\|\Pi\| = \max_{0\leq k\leq n-1}(t_{k+1} - t_k)$ goes to zero. The limit of the sum in (4.6.7) is $[X,Y](T)$. This limit is given by the right-hand side of (4.6.6). The proof of this assertion is similar to the proof of Lemma 4.4.4, with the additional feature that we must use the fact that $[W_1, W_2](t) = 0$. We omit the details.

The following theorem generalizes the Itô-Doeblin formula of Theorem 4.4.6. The justification, which we omit, is similar to that of Theorem 4.4.6.

Theorem 4.6.2 (Two-dimensional Itô-Doeblin formula). *Let $f(t,x,y)$ be a function whose partial derivatives f_t, f_x, f_y, f_{xx}, f_{xy}, f_{yx}, and f_{yy} are defined and are continuous. Let $X(t)$ and $Y(t)$ be Itô processes as discussed above. The* two-dimensional Itô-Doeblin formula in differential form *is*

$$\begin{aligned}
&df\big(t,X(t),Y(t)\big)\\
&=f_t\big(t,X(t),Y(t)\big)\,dt + f_x\big(t,X(t),Y(t)\big)\,dX(t) + f_y\big(t,X(t),Y(t)\big)\,dY(t)\\
&\quad+\frac{1}{2}f_{xx}\big(t,X(t),Y(t)\big)\,dX(t)\,dX(t) + f_{xy}\big(t,X(t),Y(t)\big)\,dX(t)\,dY(t)\\
&\quad+\frac{1}{2}f_{yy}\big(t,X(t),Y(t)\big)\,dY(t)\,dY(t). \qquad (4.6.8)
\end{aligned}$$

Before discussing formula (4.6.8), we rewrite it, leaving out t wherever possible, to obtain the same formula in the more compact notation

$$\begin{aligned}
df(t,X,Y) &= f_t\,dt + f_x\,dX + f_y\,dY\\
&\quad+\frac{1}{2}f_{xx}\,dX\,dX + f_{xy}\,dX\,dY + \frac{1}{2}f_{yy}\,dY\,dY. \qquad (4.6.9)
\end{aligned}$$

The right-hand side of (4.6.9) is the Taylor series expansion of f out to second order. The full expansion would have the additional second-order terms $f_{tt}\,dt\,dt$, $\frac{1}{2}f_{tx}\,dt\,dX$, and $\frac{1}{2}f_{ty}\,dt\,dY$, but $dt\,dt$, $dt\,dX$, and $dt\,dY$ are zero. The Taylor series expansion actually has two mixed partial terms, $\frac{1}{2}f_{xy}\,dX\,dY$ and $\frac{1}{2}f_{yx}\,dY\,dX$. For functions f whose second partial derivatives exist and are continuous, $f_{xy} = f_{yx}$, and so we have combined these terms into the single term $f_{xy}\,dX\,dY$ in (4.6.9).

The differentials dX, dY, $dX\,dX$, $dX\,dY$, and $dY\,dY$ appearing in (4.6.9) are given by (4.6.1)–(4.6.5). Making these substitutions and then integrating (4.6.9), we obtain the Itô-Doeblin formula in integral form:

$$\begin{aligned}
&f(t,X(t),Y(t)) - f(0,X(0),Y(0))\\
&= \int_0^t \Big[\sigma_{11}(u)f_x(u,X(u),Y(u)) + \sigma_{21}(u)f_y(u,X(u),Y(u))\Big]\,dW_1(u)\\
&\quad+\int_0^t \Big[\sigma_{12}(u)f_x(u,X(u),Y(u)) + \sigma_{22}(u)f_y(u,X(u),Y(u))\Big]\,dW_2(u)
\end{aligned}$$

$$
\begin{aligned}
&+\int_0^t \Big[f_t(u,X(u),Y(u))\\
&\quad+\Theta_1(u)f_x(u,X(u),Y(u))+\Theta_2(u)f_y(u,X(u),Y(u))\\
&\quad+\frac{1}{2}(\sigma_{11}^2(u)+\sigma_{12}^2(u))f_{xx}(u,X(u),Y(u))\\
&\quad+(\sigma_{11}(u)\sigma_{21}(u)+\sigma_{12}(u)\sigma_{22}(u))f_{xy}(u,X(u),Y(u))\\
&\quad+\frac{1}{2}(\sigma_{21}^2(u)+\sigma_{22}^2(u))f_{yy}(u,X(u),Y(u))\Big]\,du. \qquad (4.6.10)
\end{aligned}
$$

The right-hand side of this equation has one ordinary (Lebesgue) integral with respect to du and two Itô integrals, one with respect to $dW_1(u)$ and the other with respect to $dW_2(u)$. All terms have precise mathematical meanings. This equation demonstrates why it is preferable to work with differential notation, such as in (4.6.9).

Corollary 4.6.3 (Itô product rule). *Let $X(t)$ and $Y(t)$ be Itô processes. Then*

$$d\big(X(t)Y(t)\big) = X(t)\,dY(t) + Y(t)\,dX(t) + dX(t)\,dY(t).$$

PROOF: In (4.6.9), take $f(t,x,y) = xy$, so that $f_t = 0$, $f_x = y$, $f_y = x$, $f_{xx} = 0$, $f_{xy} = 1$, and $f_{yy} = 0$. □

4.6.3 Recognizing a Brownian Motion

A Brownian motion $W(t)$ is a martingale with continuous paths whose quadratic variation is $[W,W](t) = t$. It turns out that these conditions characterize Brownian motion in the sense of the following theorem.

Theorem 4.6.4 (Lévy, one dimension). *Let $M(t)$, $t \geq 0$, be a martingale relative to a filtration $\mathcal{F}(t)$, $t \geq 0$. Assume that $M(0) = 0$, $M(t)$ has continuous paths, and $[M,M](t) = t$ for all $t \geq 0$. Then $M(t)$ is a Brownian motion.*

IDEA OF THE PROOF: A Brownian motion is a martingale whose increments are normally distributed. The surprising feature of Lévy's Theorem is that the assumptions do not say anything about normality, and yet implicit in the conclusion is the assertion that $M(t)$ is normally distributed.

The method used to establish normality is to first check that in the derivation of the Itô-Doeblin formula, Theorem 4.4.1, for Brownian motion, the only properties of Brownian motion that were used are assumed in this theorem: a continuous process with quadratic variation $[M,M](t) = t$. Therefore, the Itô-Doeblin formula may be applied to M with the result that, for any function $f(t,x)$ whose derivatives exist and are continuous,

$$df(t,M(t)) = f_t(t,M(t))\,dt + f_x(t,M(t))\,dM(t) + \frac{1}{2}f_{xx}(t,M(t))\,dt. \quad (4.6.11)$$

The last term uses the fact that $dM(t)\,dM(t) = dt$. In integrated form, (4.6.11) is

$$f(t, M(t)) = f(0, M(0)) + \int_0^t \left[f_t(s, M(s)) + \frac{1}{2} f_{xx}(s, M(s)) \right] ds + \int_0^t f_x(s, M(s))\, dM(s). \quad (4.6.12)$$

Because $M(t)$ is a martingale, the stochastic integral $\int_0^t f_x(s, M(s))\, dM(s)$ is also. (See Exercise 4.1 for the case of a simple integrand; the general case follows from this exercise upon passage to the limit.) At $t = 0$, this stochastic integral takes the value zero, and so its expectation is always zero. Taking expectations in (4.6.12), we obtain

$$\mathbb{E} f(t, M(t)) = f(0, M(0)) + \mathbb{E} \int_0^t \left[f_t(s, M(s)) + \frac{1}{2} f_{xx}(s, M(s)) \right] ds. \quad (4.6.13)$$

We fix a number u and define

$$f(t, x) = \exp\left\{ ux - \frac{1}{2} u^2 t \right\}.$$

Then $f_t(t, x) = -\frac{1}{2} u^2 f(t, x)$, $f_x(t, x) = u f(t, x)$, and $f_{xx}(t, x) = u^2 f(t, x)$. In particular,

$$f_t(t, x) + \frac{1}{2} f_{xx}(t, x) = 0.$$

For this function $f(t, x)$, the second term on the right-hand side of (4.6.13) is zero, and that equation becomes

$$\mathbb{E} \exp\left\{ uM(t) - \frac{1}{2} u^2 t \right\} = 1.$$

In other words, we have the moment-generating function formula

$$\mathbb{E} e^{uM(t)} = e^{\frac{1}{2} u^2 t}.$$

This is the moment-generating function for the normal distribution with mean zero and variance t (see (3.2.13)). Hence, that is the distribution that $M(t)$ must have. □

The idea used to justify Theorem 4.6.4 can be combined with the two-dimensional Itô-Doeblin formula used to show independence. In particular, we have the following two-dimensional version of Lévy's Theorem.

Theorem 4.6.5 (Lévy, two dimensions). *Let $M_1(t)$ and $M_2(t)$, $t \geq 0$, be martingales relative to a filtration $\mathcal{F}(t)$, $t \geq 0$. Assume that for $i = 1, 2$, we have $M_i(0) = 0$, $M_i(t)$ has continuous paths, and $[M_i, M_i](t) = t$ for all $t \geq 0$. If, in addition, $[M_1, M_2](t) = 0$ for all $t \geq 0$, then $M_1(t)$ and $M_2(t)$ are independent Brownian motions.*

IDEA OF THE PROOF: The one-dimensional Lévy Theorem, Theorem 4.6.4, implies that M_1 and M_2 are Brownian motions. To show independence, we examine the joint moment-generating function.

Let $f(t,x,y)$ be a function whose derivatives are defined and continuous. The two-dimensional Itô-Doeblin formula implies that

$$\begin{aligned} df(t, M_1, M_2) &= f_t\,dt + f_x\,dM_1 + f_y\,dM_2 \\ &\quad + \frac{1}{2} f_{xx}\,dM_1\,dM_1 + f_{xy}\,dM_1\,dM_2 + f_{yy}\,dM_2\,dM_2 \\ &= f_t\,dt + f_x\,dM_1 + f_y\,dM_2 + \frac{1}{2} f_{xx}\,dt + \frac{1}{2} f_{yy}\,dt, \end{aligned}$$

where we have used the assumptions $[M_1, M_1](t) = t$, $[M_2, M_2](t) = t$, and $[M_1, M_2](t) =$ in their equivalent form $dM_1(t)\,dM_1(t) = dt$, $dM_2(t)\,dM_2(t) = dt$, and $dM_1(t)\,dM_2(t) = 0$. We integrate both sides to obtain

$$\begin{aligned} & f(t, M_1(t), M_2(t)) \\ &= f(0, M_1(0), M_2(0)) + \int_0^t \Big[f_t(s, M_1(s), M_2(s)) + \frac{1}{2} f_{xx}(s, M_1(s), M_2(s)) \\ &\qquad\qquad + \frac{1}{2} f_{yy}(s, M_1(s), M_2(s)) \Big]\,ds \\ &\quad + \int_0^t f_x(s, M_1(s), M_2(s))\,dM_1(s) + \int_0^t f_y(x, M_1(s), M_2(s))\,dM_2(s). \end{aligned}$$

The last two terms on the right-hand side are martingales, starting at zero at time zero, and hence having expectation zero. Therefore,

$$\begin{aligned} & \mathbb{E} f(t, M_1(t), M_2(t)) \\ &= f(0, M_1(0), M_2(0)) + \mathbb{E} \int_0^t \Big[f_t(s, M_1(s), M_2(s)) + \frac{1}{2} f_{xx}(s, M_1(s), M_2(s)) \\ &\qquad\qquad + \frac{1}{2} f_{yy}(s, M_1(s), M_2(s)) \Big]\,ds. \qquad (4.6.14) \end{aligned}$$

We now fix numbers u_1 and u_2 and define

$$f(t, x, y) = \exp\left\{ u_1 x + u_2 y - \frac{1}{2}(u_1^2 + u_2^2) t \right\}.$$

Then $f_t(t,x,y) = -\frac{1}{2}(u_1^2 + u_2^2) f(t,x,y)$, $f_x(t,x,y) = u_1 f(t,x,y)$, $f_y(t,x,y) = u_2 f(t,x,y)$. $f_{xx}(t,x,y) = u_1^2 f(t,x,y)$, and $f_{yy}(t,x,y) = u_2^2 f(t,x,y)$. For this function $f(t,x,y)$, the second term on the right-hand side of (4.6.14) is zero. We conclude that

$$\mathbb{E} \exp\left\{ u_1 M_1(t) + u_2 M_2(t) - \frac{1}{2}(u_1^2 + u_2^2) t \right\} = 1,$$

which gives us the moment-generating function formula

$$\mathbb{E}e^{u_1 M_1(t)+u_2 M_2(t)} = e^{\frac{1}{2}u_1^2 t} \cdot e^{\frac{1}{2}u_2^2 t}.$$

Because the joint moment-generating function factors into the product of moment-generating functions, $M_1(t)$ and $M_2(t)$ must be independent. □

Example 4.6.6 (Correlated stock prices). Suppose

$$\frac{dS_1(t)}{S_1(t)} = \alpha_1\,dt + \sigma_1\,dW_1(t),$$
$$\frac{dS_2(t)}{S_2(t)} = \alpha_2\,dt + \sigma_2\Big[\rho\,dW_1(t) + \sqrt{1-\rho^2}\,dW_2(t)\Big],$$

where $W_1(t)$ and $W_2(t)$ are *independent* Brownian motions and $\sigma_1 > 0$, $\sigma_2 > 0$ and $-1 \le \rho \le 1$ are constant. To analyze the second stock price process, we define

$$W_3(t) = \rho W_1(t) + \sqrt{1-\rho^2}\,W_2(t).$$

Then $W_3(t)$ is a continuous martingale with $W_3(0) = 0$, and

$$\begin{aligned} dW_3(t)\,dW_3(t) &= \rho^2\,dW_1(t)\,dW_1(t) + 2\rho\sqrt{1-\rho^2}\,dW_1(t)\,dW_2(t) \\ &\quad + (1-\rho^2)\,dW_2(t)\,dW_2(t) \\ &= \rho^2\,dt + (1-\rho^2)\,dt \;=\; dt. \end{aligned}$$

In other words, $[W_3, W_3](t) = t$. According to the one-dimensional Lévy Theorem, Theorem 4.6.4, $W_3(t)$ is a Brownian motion. Because we can write the differential of $S_2(t)$ as

$$\frac{dS_2(t)}{S_2(t)} = \alpha_2\,dt + \sigma_2\,dW_3(t),$$

we see that $S_2(t)$ is a geometric Brownian motion with mean rate of return α_2 and volatility σ_2.

The Brownian motions $W_1(t)$ and $W_3(t)$ are correlated. According to Itô's product rule (Corollary 4.6.3),

$$\begin{aligned} d\big(W_1(t)W_3(t)\big) &= W_1(t)\,dW_3(t) + W_3(t)\,dW_1(t) + dW_1(t)\,dW_3(t) \\ &= W_1(t)\,dW_3(t) + W_3(t)\,dW_1(t) + \rho\,dt. \end{aligned}$$

Integrating, we obtain

$$W_1(t)W_3(t) = \int_0^t W_1(s)\,dW_3(s) + \int_0^t W_3(s)\,dW_1(s) + \rho t.$$

The Itô integrals on the right-hand side have expectation zero, so the covariance of $W_1(t)$ and $W_3(t)$ is

$$\mathbb{E}\big[W_1(t)W_3(t)\big] = \rho t.$$

Because both $W_1(t)$ and $W_3(t)$ have standard deviation $\sqrt{t}$, the number ρ is the correlation between $W_1(t)$ and $W_3(t)$. The case of nonconstant correlation ρ is presented in Exercise 4.17. □

4.7 Brownian Bridge

We conclude this chapter with a the discussion of the Brownian bridge. This is a stochastic process that is like a Brownian motion except that with probability one it reaches a specified point at a specified positive time. We first discuss Gaussian processes in general, the class to which the Brownian bridge belongs, and we then define the Brownian bridge and present its properties. The primary use for the Brownian bridge in finance is as an aid to Monte Carlo simulation. We make no use of it in this text.

4.7.1 Gaussian Processes

Definition 4.7.1. *A* Gaussian process $X(t)$, $t \geq 0$, *is a stochastic process that has the property that, for arbitrary times* $0 < t_1 < t_2 < \cdots < t_n$, *the random variables* $X(t_1), X(t_2), \dots X(t_n)$ *are jointly normally distributed.*

The joint normal distribution of a set of vectors is determined by their means and covariances. Therefore, for a Gaussian process, the joint distribution of $X(t_1), X(t_2), \dots, X(t_n)$ is determined by the means and covariances of these random variables. We denote the mean of $X(t)$ by $m(t)$, and, for $s \geq 0$, $t \geq 0$, we denote the covariance of $X(s)$ and $X(t)$ by $c(s,t)$; i.e.,

$$m(t) = \mathbb{E}X(t), \quad c(s,t) = \mathbb{E}\big[\big(X(s) - m(s)\big)\big(X(t) - m(t)\big)\big].$$

Example 4.7.2 (Brownian motion). Brownian motion $W(t)$ is a Gaussian process. For $0 < t_1 < t_2 < \cdots < t_n$, the increments

$$I_1 = W(t_1), \; I_2 = W(t_2) - W(t_1), \; \dots, \; I_n = W(t_n) - W(t_{n-1})$$

are independent and normally distributed. Writing

$$W(t_1) = I_1, \; W(t_2) = \sum_{j=1}^{2} I_j, \; \dots, \; W(t_n) = \sum_{j=1}^{n} I_j,$$

we see that the random variables $W(t_1), W(t_2), \dots, W(t_n)$ are jointly normally distributed. These random variables are *not* independent. It is the *increments* of Brownian motion that are independent. Of course, the mean function for Brownian motion is

$$m(t) = \mathbb{E}W(t) = 0.$$

We may compute the covariance by letting $0 \leq s \leq t$ be given and noting that

$$\begin{aligned} c(s,t) &= \mathbb{E}\big[W(s)W(t)\big] \\ &= \mathbb{E}\big[W(s)\big(W(t) - W(s) + W(s)\big)\big] \\ &= \mathbb{E}\big[W(s)\big(W(t) - W(s)\big)\big] + \mathbb{E}\big[W^2(s)\big]. \end{aligned}$$

Because $W(s)$ and $W(t)-W(s)$ are independent and both have mean zero, we see that $\mathbb{E}\big[W(s)\big(W(t)-W(s)\big)\big]=0$. The other term, $\mathbb{E}\big[W^2(s)\big]$, is the variance of $W(s)$, which is s. We conclude that $c(s,t)=s$ when $0\le s\le t$. Reversing the roles of s and t, we conclude that $c(s,t)=t$ when $0\le t\le s$. In general, the covariance function for Brownian motion is then

$$c(s,t)=s\wedge t,$$

where $s\wedge t$ denotes the minimum of s and t. □

Example 4.7.3 (Itô integral of a deterministic integrand). Let $\Delta(t)$ be a nonrandom function of time, and define

$$I(t)=\int_0^t \Delta(s)\,dW(s),$$

where $W(t)$ is a Brownian motion. Then $I(t)$ is a Gaussian process, as we now show.

In the proof of Theorem 4.4.9, we showed that, for fixed $u\in\mathbb{R}$, the process

$$M_u(t)=\exp\left\{uI(t)-\frac{1}{2}u^2\int_0^t \Delta^2(s)\,ds\right\}$$

is a martingale. We used this fact to argue that

$$1=M_u(0)=\mathbb{E}M_u(t)=e^{-\frac{1}{2}u^2\int_0^t \Delta^2(s)\,ds}\cdot\mathbb{E}e^{uI(t)},$$

and we thus obtained the moment-generating function formula

$$\mathbb{E}e^{uI(t)}=e^{\frac{1}{2}u^2\int_0^t \Delta^2(s)\,ds}. \tag{4.7.1}$$

The right-hand side is the moment generating function for a normal random variable with mean zero and variance $\int_0^t \Delta^2(s)\,ds$. Therefore, this is the distribution of $I(t)$.

Although we have shown that $I(t)$ is normally distributed, verification that the process is Gaussian requires more. We must verify that, for $0<t_1<t_2<\cdots<t_n$, the random variables $I(t_1),I(t_2),\dots,I(t_n)$ are *jointly* normally distributed. It turns out that the increments

$$I(t_1)-I(0)=I(t_1),\ I(t_2)-I(t_1),\dots,I(t_n)-I(t_{n-1})$$

are normally distributed and independent, and from this the joint normality of $I(t_1),I(t_2),\dots,I(t_n)$ follows by the same argument as used in Example 4.7.2 for Brownian motion.

We show that, for $0<t_1<t_2$, the two random increments $I(t_1)-I(0)=I(t_1)$ and $I(t_2)-I(t_1)$ are normally distributed and independent. The argument we provide can be iterated to prove this result for any number of increments. For fixed $u_2\in\mathbb{R}$, the martingale property of M_{u_2} implies that

$$M_{u_2}(t_1) = \mathbb{E}[M_{u_2}(t_2)|\mathcal{F}(t_1)].$$

Now let $u_1 \in \mathbb{R}$ be fixed. Because $\frac{M_{u_1}(t_1)}{M_{u_2}(t_1)}$ is $\mathcal{F}(t_1)$-measurable, we may multiply the equation above by this quotient to obtain

$$\begin{aligned} M_{u_1}(t_1) &= \mathbb{E}\Big[\frac{M_{u_1}(t_1)M_{u_2}(t_2)}{M_{u_2}(t_1)}\Big|\mathcal{F}(t_1)\Big] \\ &= \mathbb{E}\left[\exp\left\{u_1 I(t_1) + u_2\big(I(t_2) - I(t_1)\big) - \frac{1}{2}u_1^2\int_0^{t_1}\Delta^2(s)\,ds \right.\right. \\ &\qquad \left.\left. -\frac{1}{2}u_2^2\int_{t_1}^{t_2}\Delta^2(s)\,ds\right\}\right|\mathcal{F}(t_1)\Bigg]. \end{aligned}$$

We now take expectations

$$\begin{aligned} 1 &= M_{u_1}(0) \\ &= \mathbb{E}M_{u_1}(t_1) \\ &= \mathbb{E}\left[\exp\left\{u_1 I(t_1) + u_2\big(I(t_2) - I(t_1)\big) - \frac{1}{2}u_1^2\int_0^{t_1}\Delta^2(s)\,ds \right.\right. \\ &\qquad \left.\left. -\frac{1}{2}u_2^2\int_{t_1}^{t_2}\Delta^2(s)\,ds\right\}\right] \\ &= \mathbb{E}\left[\exp\left\{u_1 I(t_1) + u_2\big(I(t_2) - I(t_1)\big)\right\}\right] \\ &\qquad \cdot\exp\left\{-\frac{1}{2}u_1^2\int_0^{t_1}\Delta^2(s)\,ds - \frac{1}{2}u_2^2\int_{t_1}^{t_2}\Delta^2(s)\,ds\right\}, \end{aligned}$$

where we have used the fact that $\Delta^2(s)$ is nonrandom to take the integrals of $\Delta^2(s)$ outside the expectation on the right-hand side. This leads to the moment-generating function formula

$$\begin{aligned} &\mathbb{E}\left[\exp\left\{u_1 I(t_1) + u_2\big(I(t_2) - I(t_1)\big)\right\}\right] \\ &\qquad = \exp\left\{\frac{1}{2}u_1^2\int_0^{t_1}\Delta^2(s)\,ds\right\}\cdot\exp\left\{\frac{1}{2}u_2^2\int_{t_1}^{t_2}\Delta^2(s)\,ds\right\}. \end{aligned}$$

The right-hand side is the product of the moment-generating function for a normal random variable with mean zero and variance $\int_0^{t_1}\Delta^2(s)\,ds$ and the moment-generating function for a normal random variable with mean zero and variance $\int_{t_1}^{t_2}\Delta^2(s)\,ds$. It follows that $I(t_1)$ and $I(t_2) - I(t_1)$ must have these distributions, and because their joint moment-generating function factors into this product of moment-generating functions, they must be independent.

The covariance of $I(t_1)$ and $I(t_2)$ can be computed using the same trick as in Example 4.7.2 for the covariance of Brownian motion. We have

$$\begin{aligned}
c(t_1,t_2) &= \mathbb{E}\big[I(t_1)I(t_2)\big] \\
&= \mathbb{E}\big[I(t_1)\big(I(t_2)-I(t_1)+I(t_1)\big)\big] \\
&= \mathbb{E}\big[I(t_1)\big(I(t_2)-I(t_1)\big)\big]+\mathbb{E}I^2(t_1) \\
&= \mathbb{E}I(t_1)\cdot\mathbb{E}\big[I(t_2)-I(t_1)\big]+\int_0^{t_1}\Delta^2(s)\,ds \\
&= \int_0^{t_1}\Delta^2(s)\,ds.
\end{aligned}$$

For the general case where $s \geq 0$ and $t \geq 0$ and we do not know the relationship between s and t, we have the covariance formula

$$c(s,t) = \int_0^{s\wedge t}\Delta^2(u)\,du.$$

□

4.7.2 Brownian Bridge as a Gaussian Process

Definition 4.7.4. *Let $W(t)$ be a Brownian motion. Fix $T > 0$. We define the* Brownian bridge from 0 to 0 on $[0,T]$ *to be the process*

$$X(t) = W(t) - \frac{t}{T}W(T),\ 0 \leq t \leq T. \tag{4.7.2}$$

Note that $\frac{t}{T}W(T)$ as a function of t is the line from $(0,0)$ to $\big(T, W(T)\big)$. In (4.7.2), we have subtracted this line away from the Brownian motion $W(t)$, so that the resulting process $X(t)$ satisfies

$$X(0) = X(T) = 0.$$

Because $W(T)$ enters the definition of $X(t)$ for $0 \leq t \leq T$, the Brownian bridge $X(t)$ is not adapted to the filtration $\mathcal{F}(t)$ generated by $W(t)$. We shall later obtain a different process that has the same distribution as the process $X(t)$ but is adapted to this filtration.

For $0 < t_1 < t_2 < \cdots < t_n < T$, the random variables

$$X(t_1) = W(t_1) - \frac{t_1}{T}W(T), \ldots, X(t_n) = W(t_n) - \frac{t_n}{T}W(T)$$

are jointly normal because $W(t_1), \ldots, W(t_n), W(T)$ are jointly normal. Hence, the Brownian bridge from 0 to 0 is a Gaussian process. Its mean function is easily seen to be

$$m(t) = \mathbb{E}X(t) = \mathbb{E}\Big[W(t) - \frac{t}{T}W(T)\Big] = 0.$$

For $s, t \in (0,T)$, we compute the covariance function

$$\begin{aligned}
&c(s,t)\\
&= \mathbb{E}\left[\left(W(s) - \frac{s}{T}W(T)\right)\left(W(t) - \frac{t}{T}W(T)\right)\right]\\
&= \mathbb{E}\big[W(s)W(t)\big] - \frac{t}{T}\mathbb{E}\big[W(s)W(T)\big] - \frac{s}{T}\mathbb{E}\big[W(t)W(T)\big] + \frac{st}{T^2}\mathbb{E}W^2(T)\\
&= s \wedge t - \frac{2st}{T} + \frac{st}{T} \ = \ s \wedge t - \frac{st}{T}.
\end{aligned} \tag{4.7.3}$$

Definition 4.7.5. *Let $W(t)$ be a Brownian motion. Fix $T > 0$, $a \in \mathbb{R}$, and $b \in \mathbb{R}$. We define the* Brownian bridge from a to b on $[0, T]$ *to be the process*

$$X^{a \to b}(t) = a + \frac{(b-a)t}{T} + X(t), \ 0 \le t \le T,$$

where $X(t) = X^{0 \to 0}$ is the Brownian bridge from 0 to 0 of Definition 4.7.4.

The function $a + \frac{(b-a)t}{T}$, as a function of t, is the line from $(0, a)$ to (T, b). When we add this line to the Brownian bridge from 0 to 0 on $[0, T]$, we obtain a process that begins at a at time 0 and ends at b at time T. Adding a nonrandom function to a Gaussian process gives us another Gaussian process. The mean function is affected:

$$m^{a \to b}(t) = \mathbb{E}X^{a \to b}(t) = a + \frac{(b-a)t}{T}.$$

However, the covariance function is not affected:

$$c^{a \to b}(s,t) = \mathbb{E}\left[\left(X^{a \to b}(s) - m^{a \to b}(s)\right)\left(X^{a \to b}(t) - m^{a \to b}(t)\right)\right] = s \wedge t - \frac{st}{T}.$$

4.7.3 Brownian Bridge as a Scaled Stochastic Integral

We cannot write the Brownian bridge as a stochastic integral of a deterministic integrand because the variance of the Brownian bridge,

$$\mathbb{E}X^2(t) = c(t,t) = t - \frac{t^2}{T} = \frac{t(T-t)}{T},$$

increases for $0 \le t \le \frac{T}{2}$ and then decreases for $\frac{T}{2} \le t \le T$. In Example 4.7.3, the variance of $I(t) = \int_0^t \Delta(u)\, dW(u)$ is $\int_0^t \Delta^2(u)\, du$, which is nondecreasing in t. However, we can obtain a process with the same distribution as the Brownian bridge from 0 to 0 as a scaled stochastic integral. In particular, consider

$$Y(t) = (T-t)\int_0^t \frac{1}{T-u}\, dW(u), \ 0 \le t < T. \tag{4.7.4}$$

The integral

$$I(t) = \int_0^t \frac{1}{T-u}\, dW(u)$$

is a Gaussian process of the type discussed in Example 4.7.3, provided $t < T$ so the integrand is defined. For $0 < t_1 < t_2 < \cdots < t_n < T$, the random variables

$$Y(t_1) = (T - t_1)I(t_1),\ Y(t_2) = (T - t_2)I(t_2), \dots, Y(t_n) = (T - t_n)I(t_n)$$

are jointly normal because $I(t_1), I(t_2), \dots, I(t_n)$ are jointly normal. In particular, Y is a Gaussian process.

The mean and covariance functions of I are

$$m^I(t) = 0,$$
$$c^I(s,t) = \int_0^{s \wedge t} \frac{1}{(T-u)^2}\,du = \frac{1}{T - s \wedge t} - \frac{1}{T} \text{ for all } s,t \in [0,T).$$

This means that the mean function for Y is $m^Y(t) = 0$. To compute the covariance function for Y, we assume for the moment that $0 \le s \le t < T$ so that

$$c^I(s,t) = \frac{1}{T-s} - \frac{1}{T} = \frac{s}{T(T-s)}.$$

Then

$$\begin{aligned} c^Y(s,t) &= \mathbb{E}\left[(T-s)(T-t)I(s)I(t)\right] \\ &= (T-s)(T-t)\frac{s}{T(T-s)} \\ &= \frac{(T-t)s}{T} \\ &= s - \frac{st}{T}. \end{aligned}$$

If we had taken $0 \le t \le s < T$, the roles of s and t would have been reversed. In general,

$$c^Y(s,t) = s \wedge t - \frac{st}{T} \text{ for all } s,t \in [0,T). \tag{4.7.5}$$

This is the same covariance formula (4.7.3) we obtained for the Brownian bridge. Because the mean and covariance functions for a Gaussian process completely determine the distribution of the process, we conclude that the process Y has the same distribution as the Brownian bridge from 0 to 0 on $[0,T]$.

We now consider the variance

$$\mathbb{E}Y^2(t) = c^Y(t,t) = \frac{t(T-t)}{T},\quad 0 < t < T.$$

Note that, as $t \uparrow T$, this variance converges to 0. In other words, as $t \uparrow T$, the random process $Y(t)$, which always has mean zero, has a variance that converges to zero. We did not initially define $Y(T)$, but this observation suggests that it makes sense to define $Y(T) = 0$. If we do that, then $Y(t)$ is continuous at $t = T$. We summarize this discussion with the following theorem.

Theorem 4.7.6. *Define the process*

$$Y(t) = \begin{cases} (T-t)\int_0^t \frac{1}{T-u}\,dW(u) & \text{for } 0 \le t < T, \\ 0 & \text{for } t = T. \end{cases}$$

Then $Y(t)$ is a continuous Gaussian process on $[0,T]$ and has mean and covariance functions

$$m^Y(t) = 0, \ t \in [0,T],$$
$$c^Y(s,t) = s \wedge t - \frac{st}{T} \text{ for all } s,t \in [0,T].$$

In particular, the process $Y(t)$ has the same distribution as the Brownian bridge from 0 to 0 on $[0,T]$ (Definition 4.7.5).

We note that the process $Y(t)$ is adapted to the filtration generated by the Brownian motion $W(t)$. It is interesting to compute the stochastic differential of $Y(t)$, which is

$$\begin{aligned} dY(t) &= \int_0^t \frac{1}{T-u}\,dW(u)\cdot d(T-t) + (T-t)\cdot d\int_0^t \frac{1}{T-u}\,dW(u) \\ &= -\int_0^t \frac{1}{T-u}\,dW(u)\cdot dt + dW(t) \\ &= -\frac{Y(t)}{T-t}dt + dW(t). \end{aligned}$$

If $Y(t)$ is positive as t approaches T, the drift term $-\frac{Y(t)}{T-t}dt$ becomes large in absolute value and is negative. This drives $Y(t)$ toward zero. On the other hand, if $Y(t)$ is negative, the drift term becomes large and positive, and this again drives $Y(t)$ toward zero. This strongly suggests, and it is indeed true, that as $t \uparrow T$ the process $Y(t)$ converges to zero almost surely.

4.7.4 Multidimensional Distribution of the Brownian Bridge

We fix $a \in \mathbb{R}$ and $b \in \mathbb{R}$ and let $X^{a\to b}(t)$ denote the Brownian bridge from a to b on $[0,T]$. We also fix $0 = t_0 < t_1 < t_2 < \cdots < t_n < T$. In this section, we compute the joint density of $X^{a\to b}(t_1), \ldots, X^{a\to b}(t_n)$.

We recall that the Brownian bridge from a to b has the mean function

$$m^{a\to b}(t) = a + \frac{(b-a)t}{T} = \frac{(T-t)a}{T} + \frac{bt}{T}$$

and covariance function

$$c(s,t) = s \wedge t - \frac{st}{T}.$$

When $s \le t$, we may write this as

$$c(s,t) = s - \frac{st}{T} = \frac{s(T-t)}{T}, \quad 0 \le s \le t \le T.$$

To simplify notation, we set $\tau_j = T - t_j$ so that $\tau_0 = T$. We define random variables

$$Z_j = \frac{X^{a\to b}(t_j)}{\tau_j} - \frac{X^{a\to b}(t_{j-1})}{\tau_{j-1}}.$$

Because $X^{a\to b}(t_1), \ldots, X^{a\to b}(t_n)$ are jointly normal, so are $Z(t_1), \ldots, Z(t_n)$. We compute

$$\begin{aligned}
\mathbb{E}Z_j &= \frac{1}{\tau_j}\mathbb{E}X^{a\to b}(t_j) - \frac{1}{\tau_j}\mathbb{E}X^{a\to b}(t_{j+1}) \\
&= \frac{a}{T} + \frac{bt_j}{T\tau_j} - \frac{a}{T} - \frac{bt_{j-1}}{T\tau_{j-1}} \\
&= \frac{bt_j(T - t_{j-1}) - bt_{j-1}(T - t_j)}{T\tau_j\tau_{j-1}} \\
&= \frac{b(t_j - t_{j-1})}{\tau_j\tau_{j-1}}.
\end{aligned}$$

Furthermore,

$$\begin{aligned}
\mathrm{Var}(Z_j) &= \frac{1}{\tau_j^2}\mathrm{Var}\big(X^{a\to b}(t_j)\big) - \frac{2}{\tau_j\tau_{j-1}}\mathrm{Cov}\big(X^{a\to b}(t_j), X^{a\to b}(t_{j-1})\big) \\
&\quad + \frac{1}{\tau_{j-1}^2}\mathrm{Var}\big(X^{a\to b}(t_{j-1})\big) \\
&= \frac{1}{\tau_j^2}c(t_j,t_j) - \frac{2}{\tau_j\tau_{j-1}}c(t_j,t_{j-1}) + \frac{1}{\tau_{j-1}^2}c(t_{j-1},t_{j-1}) \\
&= \frac{t_j}{T\tau_j} - \frac{2t_{j-1}}{T\tau_{j-1}} + \frac{t_{j-1}}{T\tau_{j-1}} \\
&= \frac{t_j(T - t_{j-1}) - 2t_{j-1}(T - t_j) + t_{j-1}(T - t_j)}{T\tau_j\tau_{j-1}} \\
&= \frac{t_j - t_{j-1}}{\tau_j\tau_{j-1}}.
\end{aligned}$$

Finally, we compute the covariance of Z_i and Z_j when $i < j$. We obtain

$$\begin{aligned}
\mathrm{Cov}(Z_i, Z_j) &= \frac{1}{\tau_i\tau_j}c(t_i,t_j) - \frac{1}{\tau_i\tau_{j-1}}c(t_i,t_{j-1}) - \frac{1}{\tau_{i-1}\tau_j}c(t_{i-1},t_j) \\
&\quad + \frac{1}{\tau_{i-1}\tau_{j-1}}c(t_{i-1},t_{j-1}) \\
&= \frac{t_i(T-t_j)}{T\tau_i\tau_j} - \frac{t_i(T-t_{j-1})}{T\tau_i\tau_{j-1}} - \frac{t_{i-1}(T-t_j)}{T\tau_{i-1}\tau_j} + \frac{t_{i-1}(T-t_{j-1})}{T\tau_{i-1}\tau_{j-1}} \\
&= 0.
\end{aligned}$$

We conclude that the normal random variables $Z_1, \dots, Z_n$ are independent, and we can write down their joint density, which is

$$
\begin{aligned}
&f_{Z(t_1),\dots,Z(t_n)}(z_1,\dots,z_n)\\
&= \prod_{j=1}^n \frac{1}{\sqrt{2\pi \frac{t_j-t_{j-1}}{\tau_j\tau_{j-1}}}} \exp\left\{ -\frac{1}{2} \cdot \frac{\left(z_j - \frac{b(t_j-t_{j-1})}{\tau_j\tau_{j-1}}\right)^2}{\frac{t_j-t_{j-1}}{\tau_j\tau_{j-1}}} \right\}\\
&= \exp\left\{ -\frac{1}{2} \sum_{j=1}^n \frac{\left(z_j - \frac{b(t_j-t_{j-1})}{\tau_j\tau_{j-1}}\right)^2}{\frac{t_j-t_{j-1}}{\tau_j\tau_{j-1}}} \right\} \cdot \prod_{j=1}^n \frac{1}{\sqrt{2\pi \frac{t_j-t_{j-1}}{\tau_j\tau_{j-1}}}}.
\end{aligned}
$$

We make the change of variables

$$
z_j = \frac{x_j}{\tau_j} - \frac{x_{j-1}}{\tau_{j-1}}, \quad j = 1, \dots, n,
$$

where $x_0 = a$, to find the joint density for $X^{a\to b}(t_1), \dots, X^{a\to b}(t_n)$. We work first on the sum in the exponent to see the effect of this change of variables. We have

$$
\begin{aligned}
&\sum_{j=1}^n \frac{\left(z_j - \frac{b(t_j-t_{j-1})}{\tau_j\tau_{j-1}}\right)^2}{\frac{t_j-t_{j-1}}{\tau_j\tau_{j-1}}}\\
&= \sum_{j=1}^n \frac{\tau_j\tau_{j-1}}{t_j - t_{j-1}} \left(\frac{x_j}{\tau_j} - \frac{x_{j-1}}{\tau_{j-1}} - \frac{b(t_j - t_{j-1})}{\tau_j\tau_{j-1}} \right)^2\\
&= \sum_{j=1}^n \frac{\tau_j\tau_{j-1}}{t_j - t_{j-1}} \left(\frac{x_j^2}{\tau_j^2} + \frac{x_{j-1}^2}{\tau_{j-1}^2} + \frac{b^2(t_j - t_{j-1})^2}{\tau_j^2\tau_{j-1}^2} - \frac{2x_jx_{j-1}}{\tau_j\tau_{j-1}} \right.\\
&\qquad\qquad \left. - \frac{2x_jb(t_j - t_{j-1})}{\tau_j^2\tau_{j-1}} + \frac{2x_{j-1}b(t_j - t_{j-1})}{\tau_j\tau_{j-1}^2} \right)\\
&= \sum_{j=1}^n \left(\frac{\tau_{j-1}x_j^2}{\tau_j(t_j - t_{j-1})} + \frac{\tau_jx_{j-1}^2}{\tau_{j-1}(t_j - t_{j-1})} + \frac{b^2(t_j - t_{j-1})}{\tau_j\tau_{j-1}} - \frac{2x_jx_{j-1}}{t_j - t_{j-1}} \right.\\
&\qquad\qquad \left. - \frac{2x_jb}{\tau_j} + \frac{2x_{j-1}b}{\tau_{j-1}} \right)\\
&= \sum_{j=1}^n \left[\frac{x_j^2}{t_j - t_{j-1}} \left(1 + \frac{\tau_{j-1} - \tau_j}{\tau_j}\right) + \frac{x_{j-1}^2}{t_j - t_{j-1}} \left(1 - \frac{\tau_{j-1} - \tau_j}{\tau_{j-1}}\right) \right.\\
&\qquad \left. - \frac{2x_jx_{j-1}}{t_j - t_{j-1}} \right] + b^2 \sum_{j=1}^n \left(\frac{1}{\tau_j} - \frac{1}{\tau_{j-1}} \right) - 2b \sum_{j=1}^n \left(\frac{x_j}{\tau_j} - \frac{x_{j-1}}{\tau_{j-1}} \right).
\end{aligned}
$$

Now

$$\tau_{j-1} - \tau_j = (T - t_{j-1}) - (T - t_j) = t_j - t_{j-1},$$

and so this last expression is equal to

$$\sum_{j=1}^{n} \left[\frac{x_j^2 - 2x_j x_{j-1} + x_{j-1}^2}{t_j - t_{j-1}} \right] + \sum_{j=1}^{n} \left(\frac{x_j^2}{\tau_j} - \frac{x_{j-1}^2}{\tau_{j-1}} \right)$$
$$+b^2 \sum_{j=1}^{n} \left(\frac{1}{\tau_j} - \frac{1}{\tau_{j-1}} \right) - 2b \sum_{j=1}^{n} \left(\frac{x_j}{\tau_j} - \frac{x_{j-1}}{\tau_{j-1}} \right)$$
$$= \sum_{j=1}^{n} \frac{(x_j - x_{j-1})^2}{t_j - t_{j-1}} + \frac{x_n^2}{T - t_n} - \frac{a^2}{T} + b^2 \left(\frac{1}{T - t_n} - \frac{1}{T} \right)$$
$$-2b \left(\frac{x_n}{T - t_n} - \frac{a}{T} \right)$$
$$= \sum_{j=1}^{n} \frac{(x_j - x_{j-1})^2}{t_j - t_{j-1}} + \frac{(b - x_n)^2}{T - t_n} - \frac{(b-a)^2}{T}.$$

In conclusion, when we change variables from z_j to x_j, we have the equation

$$\exp\left\{ -\frac{1}{2} \sum_{j=1}^{n} \frac{\left(z_j - \frac{b(t_j - t_j - 1)}{\tau_j \tau_{j-1}} \right)^2}{\frac{t_j - t_{j-1}}{\tau_j \tau_{j-1}}} \right\}$$
$$= \exp\left\{ -\frac{1}{2} \sum_{j=1}^{n} \frac{(x_j - x_{j-1})^2}{t_j - t_{j-1}} - \frac{(b - x_n)^2}{2(T - t_n)} + \frac{(b-a)^2}{2T} \right\}.$$

To change a density, we also need to account for the Jacobian of the change of variables. In this case, we have

$$\frac{\partial z_j}{\partial x_j} = \frac{1}{\tau_j}, \; j = 1, \ldots, n,$$
$$\frac{\partial z_j}{\partial x_{j-1}} = -\frac{1}{\tau_{j-1}}, j = 2, \ldots, n,$$

and all other partial derivatives are zero. This leads to the Jacobian matrix

$$J = \begin{bmatrix} \frac{1}{\tau_1} & 0 & \cdots & 0 \\ -\frac{1}{\tau_1} & \frac{1}{\tau_2} & \cdots & 0 \\ \vdots & \vdots & & \vdots \\ 0 & 0 & \cdots & \frac{1}{\tau_n} \end{bmatrix},$$

whose determinant is $\prod_{j=1}^{n} \frac{1}{\tau_j}$. Multiplying $f_{Z(t_1),\ldots,Z(t_n)}(z_1, \ldots, z_n)$ by this determinant and using the change of variables worked out above, we obtain the density for $X^{a \to b}(t_1), \ldots, X^{a \to b}(t_n)$,

$$
\begin{aligned}
&f_{X^{a\to b}(t_1),\ldots,X^{a\to b}(t_n)}(x_1,\ldots,x_n)\\
&=\prod_{j=1}^{n}\frac{1}{\sqrt{2\pi(t_j-t_{j-1})}}\sqrt{\frac{\tau_{j-1}}{\tau_j}}\\
&\qquad\cdot\exp\left\{-\frac{1}{2}\sum_{j=1}^{n}\frac{(x_j-x_{j-1})^2}{t_j-t_{j-1}}-\frac{(b-x_n)^2}{2(T-t_n)}+\frac{(b-a)^2}{2T}\right\}\\
&=\sqrt{\frac{T}{T-t_n}}\cdot\prod_{j=1}^{n}\frac{1}{\sqrt{2\pi(t_j-t_{j-1})}}\\
&\qquad\cdot\exp\left\{-\frac{1}{2}\sum_{j=1}^{n}\frac{(x_j-x_{j-1})^2}{t_j-t_{j-1}}-\frac{(b-x_n)^2}{2(T-t_n)}+\frac{(b-a)^2}{2T}\right\}\\
&=\frac{p(T-t_n,x_n,b)}{p(T,a,b)}\prod_{j=1}^{n}p(t_j-t_{j-1},x_{j-1},x_j),
\end{aligned}
\tag{4.7.6}
$$

where

$$p(\tau,x,y)=\frac{1}{\sqrt{2\pi\tau}}\exp\left\{-\frac{(y-x)^2}{2\tau}\right\}$$

is the transition density for Brownian motion.

4.7.5 Brownian Bridge as a Conditioned Brownian Motion

The joint density (4.7.6) for $X^{a\to b}(t_1),\ldots,X^{a\to b}(t_n)$ permits us to give one more interpretation for the Brownian bridge from a to b on $[0,T]$. It is a Brownian motion $W(t)$ on this time interval, starting at $W(0)=a$ and conditioned to arrive at b at time T (i.e., conditioned on $W(T)=b$). Let $0=t_0<t_1<t_2<\cdots<t_n<T$ be given. The joint density of $W(t_1),\ldots,W(t_n),W(T)$ is

$$f_{W(t_1),\ldots,W(t_n),W(T)}(x_1,\ldots,x_n,b)=p(T-t_n,x_n,b)\prod_{j=1}^{n}p(t_j-t_{j-1},x_{j-1},x_j),\tag{4.7.7}$$

where $W(0)=x_0=a$. This is because $p(t_1-t_0,x_0,x_1)=p(t_1,a,x_1)$ is the density for the Brownian motion going from $W(0)=a$ to $W(t_1)=x_1$ in the time between $t=0$ and $t=t_1$. Similarly, $p(t_2-t_1,x_1,x_2)$ is the density for going from $W(t_1)=x_1$ to $W(t_2)=x_2$ between time $t=t_1$ and $t=t_2$. The joint density for $W(t_1)$ and $W(t_2)$ is then the product

$$p(t_1,a,x_1)p(t_2-t_1,x_1,x_2).$$

Continuing in this way, we obtain the joint density (4.7.7). The marginal density of $W(T)$ is $p(T,a,b)$. The density of $W(t_1),\ldots,W(t_n)$ conditioned on $W(T)=b$ is thus the quotient

$$\frac{p(T-t_n,x_n,b)}{p(T,a,b)}\prod_{j=1}^{n}p(t_j-t_{j-1},x_{j-1},x_j),$$

and this is $f_{X^{a\to b}(t_1),\dots,X^{a\to b}(t_n)}(x_1,\dots,x_n)$ of (4.7.6).

Finally, let us define

$$M^{a\to b}(T)=\max_{0\le t\le T}X^{a\to b}(t)$$

to be the maximum value obtained by the Brownian bridge from a to b on $[0,T]$. This random variable has the following distribution.

Corollary 4.7.7. *The density of $M^{a\to b}(T)$ is*

$$f_{M^{a\to b}(T)}(y)=\frac{2(2y-b-a)}{T}e^{-\frac{2}{T}(y-a)(y-b)},\quad y>\max\{a,b\}. \tag{4.7.8}$$

PROOF: Because the Brownian bridge from 0 to w on $[0,T]$ is a Brownian motion conditioned on $W(T)=w$, the maximum of $X^{0\to w}$ on $[0,T]$ is the maximum of W on $[0,T]$ conditioned on $W(T)=w$. Therefore, the density of $M^{0\to w}(T)$ was computed in Corollary 3.7.4 and is

$$f_{M^{0\to w}(T)}(m)=\frac{2(2m-w)}{T}e^{-\frac{2m(m-w)}{T}},\quad w<m,m>0. \tag{4.7.9}$$

The density of $f_{M^{a\to b}(T)}(y)$ can be obtained by translating from the initial condition $W(0)=a$ to $W(0)=0$ and using (4.7.9). In particular, in (4.7.9) we replace m by $y-a$ and replace w by $b-a$. This results in (4.7.8). □

4.8 Summary

Let $W(t)$ be a Brownian motion and $\Delta(t)$ a stochastic process adapted to the filtration of the Brownian motion. The Itô integral

$$I(t)=\int_0^t\Delta(u)\,dW(u) \tag{4.8.1}$$

is a martingale. Because it is zero at time $t=0$, its expectation is zero for all t. Its variance is given by *Itô's isometry*

$$\mathbb{E}I^2(t)=\mathbb{E}\int_0^t\Delta^2(u)\,du. \tag{4.8.2}$$

The quadratic variation accumulated by the Itô integral up to time t is

$$[I,I](t)=\int_0^t\Delta^2(u)\,du. \tag{4.8.3}$$

These assertions appear in Theorem 4.3.1. Note that the quadratic variation (4.8.3) is computed path-by-path and the result may depend on the path, whereas the variance (4.8.2) is an average over all paths. In differential notation, we write (4.8.1) as

$$dI(t) = \Delta(t)\, dW(t)$$

and (4.8.3) as

$$dI(t)\, dI(t) = \Delta^2(t)\, dW(t)\, dW(t) = \Delta^2(t)\, dt.$$

An *Itô process* (Definition 4.4.3) is a process of the form

$$X(t) = X(0) + \int_0^t \Delta(u)\, dW(u) + \int_0^t \Theta(u)\, du, \tag{4.8.4}$$

where $X(0)$ is nonrandom and $\Delta(u)$ and $\Theta(u)$ are adapted stochastic processes. According to Lemma 4.4.4, the quadratic variation accumulated by X up to time t is

$$[X,X](t) = \int_0^t \Delta^2(u)\, du. \tag{4.8.5}$$

In differential notation, we write (4.8.4) as

$$dX(t) = \Delta(t)\, dW(t) + \Theta(t)\, dt$$

and (4.8.5) as

$$\begin{aligned}
dX(t)\, dX(t) &= \big(\Delta(t)\, dW(t) + \Theta(t)\, dt\big)^2 \\
&= \Delta^2(t)\, dW(t)\, dW(t) + 2\Delta(t)\,\Theta(t)\, dW(t)\, dt + \Theta^2(t)\, dt\, dt \\
&= \Delta^2(t)\, dt,
\end{aligned}$$

where we have used the multiplication table

$$dW(t)\, dW(t) = dt, \quad dW(t)\, dt = dt\, dW(t) = 0, \quad dt\, dt = 0.$$

Suppose X and Y are Itô processes with differentials

$$dX(t) = \Theta_1(t)\, dt + \sigma_{11}(t)\, dW_1(t) + \sigma_{12}(t)\, dW_2(t), \tag{4.8.6}$$

$$dY(t) = \Theta_2(t)\, dt + \sigma_{21}(t)\, dW_1(t) + \sigma_{22}(t)\, dW_2(t), \tag{4.8.7}$$

where W_1 and W_2 are independent Brownian motions. Then

$$dX(t)\, dX(t) = \big(\sigma_{11}^2(t) + \sigma_{12}^2(t)\big)\, dt, \tag{4.8.8}$$

$$dX(t)\, dY(t) = \big(\sigma_{11}(t)\sigma_{21}(t) + \sigma_{12}(t)\sigma_{22}(t)\big)\, dt, \tag{4.8.9}$$

$$dY(t)\, dY(t) = \big(\sigma_{21}^2(t) + \sigma_{22}^2(t)\big)\, dt. \tag{4.8.10}$$

Equations (4.8.8)–(4.8.10) can be obtained by multiplying the equations (4.8.6) and (4.8.7) for $dX(t)$ and $dY(t)$ and using the multiplication table

$$dW_i(t)\,dW_i(t) = dt, \;\; dW_i(t)\,dt = dt\,dW_i(t) = 0, \;\; dt\,dt = 0,$$

and

$$dW_1(t)\,dW_2(t) = 0. \tag{4.8.11}$$

Equation (4.8.11) holds for *independent* Brownian motions. If instead we had

$$dW_1(t)\,dW_2(t) = \rho\,dt,$$

for a constant $\rho \in [-1, 1]$, then ρ would be the correlation between $W_1(t)$ and $W_2(t)$ (i.e., $\mathbb{E}[W_1(t)W_2(t)] = \rho t$).

Now suppose $f(t, x, y)$ is a function of the time variable t and two dummy variables x and y. The multidimensional Itô-Doeblin formula (Theorem 4.6.2) says

$$\begin{aligned}
&df\big(t, X(t), Y(t)\big) \\
&= f_t\big(t, X(t), Y(t)\big)\,dt + f_x\big(t, X(t), Y(t)\big)\,dX(t) + f_y\big(t, X(t), Y(t)\big)\,dY(t) \\
&\quad \frac{1}{2} f_{xx}\big(t, X(t), Y(t)\big)\,dX(t)\,dX(t) + f_{xy}\big(t, X(t), Y(t)\big)\,dX(t)\,dY(t) \\
&\quad\quad + \frac{1}{2} f_{yy}\big(t, X(t), Y(t)\big)\,dY(t)\,dY(t).
\end{aligned} \tag{4.8.12}$$

Replacing all the differentials on the right-hand side of (4.8.12) by their formulas (4.8.6)–(4.8.10) and integrating, one obtains a formula for the stochastic process $f(t, X(t), Y(t))$ as the sum of $f(0, X(0), Y(0))$, an ordinary integral with respect to time, an Itô integral with respect to dW_1, and an Itô integral with respect to dW_2.

There are two important special cases of (4.8.12). If the second process Y is not present, (4.8.12) reduces to the Itô-Doeblin formula for one process (Theorem 4.4.6):

$$df\big(t, X(t)\big) = f_t\big(t, X(t)\big)\,dt + f_x\big(t, X(t)\big)\,dX(t) + \frac{1}{2} f_{xx}\big(t, X(t)\big)\,dX(t)\,dX(t).$$

If both X and Y are present and $f(t, x, y) = xy$, then (4.8.12) gives us *Itô's product rule* (Corollary 4.6.3):

$$d\big(X(t)Y(t)\big) = X(t)\,dY(t) + Y(t)\,dX(t) + dX(t)\,dY(t).$$

Using the Itô-Doeblin formula, we can derive the Black-Scholes-Merton partial differential equation. This was done in Section 4.5, and that section is summarized here. Let the stock price $S(t)$ be a geometric Brownian motion:

$$dS(t) = \alpha S(t)\,dt + \sigma S(t)\,dW(t).$$

Let $c(t, S(t))$ be the price at time $t \in [0, T]$ of a European call paying $(S(T) - K)^+$ at expiration time T. Suppose we sell this call for $X(0) = c(0, S(0))$ at time zero and, starting with initial capital $X(0)$, invest in a stock and a money market account paying a constant rate of interest r. If $\Delta(t)$ is the number of shares of stock held by the portfolio at time t, then

$$dX(t) = \Delta(t)\,dS(t) + r\big(X(t) - \Delta(t)S(t)\big)\,dt.$$

We compute the differential of the discounted portfolio value $e^{-rt}X(t)$, the differential of the discounted call price $e^{-rt}c(t,S(t))$, and set these two equal. This results in the *delta-hedging rule* (4.5.11),

$$\Delta(t) = c_x(t, S(t)), \tag{4.8.13}$$

and the Black-Scholes-Merton partial differential equation (4.5.14),

$$c_t(t,x) + rxc_x(t,x) + \frac{1}{2}\sigma^2 x^2 c_{xx}(t,x) = rc(t,x).$$

In addition to satisfying this partial differential equation, the function $c(t,x)$ must satisfy the boundary conditions

$$c(T,x) = (x-K)^+,\ c(t,0) = 0,\ \lim_{x\to\infty}\big[c(t,x) - \big(x - e^{-r(T-t)}K\big)\big] = 0.$$

The function satisfying these conditions is (see (4.5.19))

$$c(t,x) = xN\big(d_+(T-t,x)\big) - Ke^{-r(T-t)}N\big(d_-(T-t,x)\big), \tag{4.8.14}$$

where

$$d_\pm(\tau,x) = \frac{1}{\sigma\sqrt{\tau}}\left[\log\frac{x}{K} + \left(r \pm \frac{\sigma^2}{2}\right)\tau\right].$$

Using the function given by (4.8.14), if one starts with initial capital $X(0) = c(0, S(0))$ and uses the delta-hedging rule (4.8.13), then at every time t, $X(t) = c(t, S(t))$. In particular, at the final time, the value of the hedging portfolio is $X(T) = c(T, S(T)) = (S(T) - K)^+$ almost surely. The short position in the European call has been hedged.

Lévy's Theorem, Theorem 4.6.4, says that if $M(t)$ is a continuous martingale starting at $M(0) = 0$ and if $[M, M](t) = t$ (i.e., $dM(t)\,dM(t) = dt$), then $M(t)$ is a Brownian motion. If $M_1(t)$ and $M_2(t)$ are two such processes and $[M_1, M_2](t) = 0$ (i.e., $dM_1(t)\,dM_2(t) = 0$), then $M_1(t)$ and $M_2(t)$ are independent Brownian motions (Theorem 4.6.5). One can use this theorem to construct independent Brownian motions from correlated Brownian motions and vice versa (see Exercise 4.13).

A Gaussian process $X(t)$ is one for which $X(t_1), X(t_2), \dots X(t_n)$ are jointly normally distributed whenever $0 < t_1 < t_2 < \cdots < t_n$ (Definition 4.7.1). Because the joint distribution of jointly normal random variables is determined by means, variances, and covariances, the distribution of a Gaussian process is determined by its mean function $m(t) = \mathbb{E}X(t)$ and covariance function $c(s,t) = \mathrm{Cov}(X(s), X(t))$. Brownian motion is a Gaussian process with $m(t) = 0$ and $c(s,t) = s \wedge t$ (Example 4.7.2). If $\Delta(u)$ is nonrandom, then $I(t) = \int_0^t \Delta(u)\,dW(u)$ is a Gaussian process with $m(t) = 0$ and $c(s,t) = \int_0^{s\wedge t} \Delta^2(u)\,du$ (Example 4.7.3). The Brownian bridge from a to b on $[0,T]$ is a Gaussian process with $m(t) = \frac{(T-t)a+bt}{T}$ for $t \in [0,T]$ and $c(s,t) = s \wedge t - \frac{st}{T}$ for $s,t \in [0,T]$ (see Subsection 10.7.2). The Brownian bridge from a to b on $[0,T]$ is the process one obtains by starting a Brownian motion at a at time $t = 0$ and conditioning on $W(T) = b$ (see Subsection 10.7.5).

4.9 Notes

The modern theory of stochastic calculus developed from the work of Itô [92]. Not only did Itô define the integral with respect to Brownian motion, but he also developed the change-of-variable formula commonly called *Itô's rule* or *Itô's formula*. As demonstrated in this chapter, this formula is at the heart of a wide range of useful calculations. An amazing twist to the story of stochastic calculus has recently emerged. In February 1940, the French National Academy of Sciences received a document from W. Doeblin, a French soldier on the German front. Doeblin died shortly thereafter, and the document remained sealed until May 2000. When it was opened, the document was found to contain a construction of the stochastic integral slightly different from Itô's and a clear statement of the change-of-variable formula. Doeblin's work [52], Yor's [166] analysis of the work, and a detailed history by Bru [24] of the context of the work appeared in the December 2000 issue of *Comptes Rendus de L'Académie des Sciences.* An English translation of this material is [25]. Because of this remarkable development, in this text the change-of-variable formula is called the Itô-Doeblin formula.

We have defined the Itô integral $\int_0^T \Delta^2(t)\,dW(t)$ under the condition

$$\mathbb{E}\int_0^T \Delta^2(t)\,dt < \infty. \tag{4.3.1}$$

The integral can be defined under the weaker condition

$$\int_0^T \Delta^2(t) < \infty \text{ almost surely}$$

but then is not guaranteed to be a martingale. It is still a *local martingale*, a topic discussed in advanced books on stochastic calculus (e.g., [101]). In this text, we do not consider local martingales. We work only under the condition (4.3.1), and every Itô integral we encounter is a martingale.

Brownian motion was introduced to finance by Bachelier [6]. Samuelson [143], [145] presents the argument that geometric Brownian motion is a good model for stock prices. The application of stochastic calculus to finance began with the work of Merton [121]. (The paper [121] and many other papers by Merton that use stochastic calculus in finance are collected in Merton [124].) The Black-Scholes-Merton formula is based on the geometric Brownian motion model for stock prices. However, no-arbitrage pricing theory has now moved far beyond this assumption. As seen in this and subsequent chapters, this theory and the accompanying risk-neutral pricing formula can be applied in the presence of a time-varying random volatility, a time-varying random mean rate of return, and a time-varying random interest rate.

Many finance books, including (in order of increasing mathematical difficulty) Hull [87], Dothan [54], and Duffie [56], include sections on Itô's integral and the Itô-Doeblin formula. Some other books on dynamic models in finance are Cox and Rubinstein [43], Huang and Litzenberger [86], Ingersoll [91], and Jarrow [97]. A comprehensive text is Wilmott [164]. Some good references for practitioners are Baxter and Rennie [8] (reviewed in [134]), Björk [11] (reviewed in [135]), and Musiela and Rutkowski [126] (reviewed in [134]). More mathematical texts on stochastic calculus with applications to finance are Lamberton and Lapeyre [105] (reviewed in [134]) and Steele [150] (reviewed in [136]). Other texts on stochastic calculus are Chung and Williams [36], Karatzas and Shreve [101], Øksendal [129], and Protter [133]. Karatzas and Shreve [102] is a sequel to [101] that focuses on finance. Protter [133] is the easiest place to learn about stochastic calculus for processes with jumps, and this is not at all easy. We introduce this topic in Chapter 11.

No-arbitrage pricing theory and the accompanying risk-neutral pricing formula is predicated on the assumption that there is no arbitrage in the market. An arbitrage is defined to be a trading strategy which begins with zero capital and at a later time has positive capital with positive probability without having any risk of loss. Absence of arbitrage is similar to but different from the *efficient market hypothesis*, which asserts that technical analysis of stock prices is of no value. This hypothesis asserts that patterns in stock prices may be useful to estimate the parameters of the distribution of future returns, but they do not provide clues to whether the next price movement will be up or down. In particular, technical analysis does not permit one to outperform the market. This hypothesis could be violated in a way which permits one to outperform the market with high probability without actually admitting arbitrage because there is still a nonzero probability of underperforming the market. This is sometimes called *statistical arbitrage.* An empirical study supporting the efficient market hypothesis is Fama [64], which also discusses distributions that fit stock prices better than geometric Brownian motion. A criticism of the efficient market hypothesis is provided by LeRoy [106], and a recent paper that finds long-range dependence (but not much) in stock price data is Willinger, Taqqu, and Teverovsky [163]. A provocative article on the source of stock price movements is Black [14].

Geometric fractional Brownian motion has been proposed as an alternative model for stock prices because it has fatter tails than geometric Brownian motion. One can assume such a model and compute the prices of derivative securities as their expected discounted payoffs, but the model is inconsistent with the usual delta-hedging formula. Indeed, geometric fractional Brownian motion violates the efficient market hypothesis so strongly that it admits arbitrage (not just "statistical arbitrage" but actual arbitrage). An example of this is provided by Rogers [138]. Further examples of arbitrage and a market-trading restriction that prevents arbitrage in such markets are provided by Cheridito [33].

The Vasicek model of Example 4.4.10 is taken from [154]. The Cox-Ingersoll-Ross model of Example 4.4.11 is due to [41], where the distribution of the interest rate process in the model is provided.

The derivation of the Black-Scholes-Merton formula in Section 4.5 is similar to that originally given by Black and Scholes [17] but also relies heavily on the no-arbitrage idea appearing in Merton [122]. It is well-documented that the three men cooperated on development of the option-pricing formula, and in recognition of this the 1997 Nobel Prize in Economics was awarded to Scholes and Merton. (Black died in 1995, and the prize is not awarded posthumously). In this text, the role of all three men is acknowledged by the terminology *Black-Scholes-Merton* option-pricing formula. Even though geometric Brownian motion is a less than perfect model for stock prices, the Black-Scholes-Merton pricing formula for vanilla options (i.e., European calls and puts) seems not to be terribly sensitive to deficiencies in the model.

The passage from discrete to continuous time in the model of evolution of the portfolio value, which is touched upon in Subsection 4.5.1, is given a more detailed treatment by Duffie and Protter [60]; see also Exercise 4.10.

4.10 Exercises

Exercise 4.1. Suppose $M(t)$, $0 \le t \le T$, is a martingale with respect to some filtration $\mathcal{F}(t)$, $0 \le t \le T$. Let $\Delta(t)$, $0 \le t \le T$, be a simple process adapted to $\mathcal{F}(t)$ (i.e., there is a partition $\Pi = \{t_0, t_1, \dots, t_n\}$ of $[0, T]$ such that, for every j, $\Delta(t_j)$ is $\mathcal{F}(t_j)$-measurable and $\Delta(t)$ is constant in t on each subinterval $[t_j, t_{j+1})$). For $t \in [t_k, t_{k+1})$, define the stochastic integral

$$I(t) = \sum_{j=0}^{k-1} \Delta(t_j)[M(t_{j+1}) - M(t_j)] + \Delta(t_k)[M(t) - M(t_k)].$$

We think of $M(t)$ as the price of an asset at time t and $\Delta(t_j)$ as the number of shares of the asset held by an investor between times t_j and t_{j+1}. Then $I(t)$ is the capital gains that accrue to the investor between times 0 and t. Show that $I(t)$, $0 \le t \le T$, is a martingale.

Exercise 4.2. Let $W(t)$, $0 \le t \le T$, be a Brownian motion, and let $\mathcal{F}(t)$, $0 \le t \le T$, be an associated filtration. Let $\Delta(t)$, $0 \le t \le T$, be a *nonrandom* simple process (i.e., there is a partition $\Pi = \{t_0, t_1, \dots, t_n\}$ of $[0, T]$ such that for every j, $\Delta(t_j)$ is a nonrandom quantity and $\Delta(t) = \Delta(t_j)$ is constant in t on the subinterval $[t_j, t_{j+1})$). For $t \in [t_k, t_{k+1}]$, define the stochastic integral

$$I(t) = \sum_{j=0}^{k-1} \Delta(t_j)[W(t_{j+1}) - W(t_j)] + \Delta(t_k)[W(t) - W(t_k)].$$

(i) Show that whenever $0 \le s < t \le T$, the increment $I(t) - I(s)$ is independent of $\mathcal{F}(s)$. (Simplification: If s is between two partition points, we can always insert s as an extra partition point. Then we can relabel the partition points so that they are still called $t_0, t_1, \ldots, t_n$, but with a larger value of n and now with $s = t_k$ for some value of k. Of course, we must set $\Delta(s) = \Delta(t_{k-1})$ so that Δ takes the same value on the interval $[s, t_{k+1})$ as on the interval $[t_{k-1}, s)$. Similarly, we can insert t as an extra partition point if it is not already one. Consequently, to show that $I(t) - I(s)$ is independent of $\mathcal{F}(s)$ for all $0 \le s < t \le T$, it suffices to show that $I(t_k) - I(t_\ell)$ is independent of $\mathcal{F}(t_\ell)$ whenever t_k and t_ℓ are two partition points with $t_\ell < t_k$. This is all you need to do.)

(ii) Show that whenever $0 \le s < t \le T$, the increment $I(t) - I(s)$ is a normally distributed random variable with mean zero and variance $\int_s^t \Delta^2(u)\,du$.

(iii) Use (i) and (ii) to show that $I(t)$, $0 \le t \le T$, is a martingale.

(iv) Show that $I^2(t) - \int_0^t \Delta^2(u)\,du$, $0 \le t \le T$, is a martingale.

Exercise 4.3. We now consider a case in which $\Delta(t)$ in Exercise 4.2 is simple but random. In particular, let $t_0 = 0$, $t_1 = s$, and $t_2 = t$, and let $\Delta(0)$ be nonrandom and $\Delta(s) = W(s)$. Which of the following assertions is true? Justify your answers.

(i) $I(t) - I(s)$ is independent of $\mathcal{F}(s)$.

(ii) $I(t) - I(s)$ is normally distributed. (Hint: Check if the fourth moment is three times the square of the variance; see Exercise 3.3 of Chapter 3.)

(iii) $\mathbb{E}[I(t)|\mathcal{F}(s)] = I(s)$.

(iv) $\mathbb{E}\left[I^2(t) - \int_0^t \Delta^2(u)\,du \middle| \mathcal{F}(s)\right] = I^2(s) - \int_0^s \Delta^2(u)\,du$.

Exercise 4.4 (Stratonovich integral). Let $W(t)$, $t \ge 0$, be a Brownian motion. Let T be a fixed positive number and let $\Pi = \{t_0, t_1, \ldots, t_n\}$ be a partition of $[0, T]$ (i.e., $0 = t_0 < t_1 < \cdots < t_n = T$). For each j, define $t_j^* = \frac{t_j + t_{j+1}}{2}$ to be the midpoint of the interval $[t_j, t_{j+1}]$.

(i) Define the *half-sample quadratic variation* corresponding to Π to be

$$Q_{\Pi/2} = \sum_{j=0}^{n-1} \left(W(t_j^*) - W(t_j)\right)^2.$$

Show that $Q_{\Pi/2}$ has limit $\frac{1}{2}T$ as $\|\Pi\| \to 0$. (Hint: It suffices to show that $\mathbb{E}Q_{\Pi/2} = \frac{1}{2}T$ and $\lim_{\|\Pi\| \to 0} \mathrm{Var}(Q_{\Pi/2}) = 0$.)

(ii) Define the Stratonovich integral of $W(t)$ with respect to $W(t)$ to be

$$\int_0^T W(t) \circ dW(t) = \lim_{\|\Pi\| \to 0} \sum_{j=0}^{n-1} W(t_j^*)\big(W(t_{j+1}) - W(t_j)\big). \tag{4.10.1}$$

In contrast to the Itô integral $\int_0^T W(t)\,dW(t) = \frac{1}{2}W^2(T) - \frac{1}{2}T$ of (4.3.4), which evaluates the integrand at the left endpoint of each subinterval $[t_j, t_{j+1}]$, here we evaluate the integrand at the midpoint t_j^*. Show that

$$\int_0^T W(t) \circ dW(t) = \frac{1}{2}W^2(T).$$

(Hint: Write the approximating sum in (4.10.1) as the sum of an approximating sum for the Itô integral $\int_0^T W(t)\,dW(t)$ and $Q_{\Pi/2}$. The approximating sum for the Itô integral is the one corresponding to the partition $0 = t_0 < t_0^* < t_1 < t_1^* < \cdots < t_{n-1}^* < t_n = T$, not the partition Π.)

Exercise 4.5 (Solving the generalized geometric Brownian motion equation). Let $S(t)$ be a positive stochastic process that satisfies the generalized geometric Brownian motion differential equation (see Example 4.4.8)

$$dS(t) = \alpha(t)S(t)\,dt + \sigma(t)S(t)\,dW(t), \tag{4.10.2}$$

where $\alpha(t)$ and $\sigma(t)$ are processes adapted to the filtration $\mathcal{F}(t)$, $t \geq 0$, associated with the Brownian motion $W(t)$, $t \geq 0$. In this exercise, we show that $S(t)$ must be given by formula (4.4.26) (i.e., that formula provides the only solution to the stochastic differential equation (4.10.2)). In the process, we provide a method for solving this equation.

(i) Using (4.10.2) and the Itô-Doeblin formula, compute $d\log S(t)$. Simplify so that you have a formula for $d\log S(t)$ that does not involve $S(t)$.
(ii) Integrate the formula you obtained in (i), and then exponentiate the answer to obtain (4.4.26).

Exercise 4.6. Let $S(t) = S(0)\exp\left\{\sigma W(t) + \left(\alpha - \frac{1}{2}\sigma^2\right)t\right\}$ be a geometric Brownian motion. Let p be a positive constant. Compute $d\left(S^p(t)\right)$, the differential of $S(t)$ raised to the power p.

Exercise 4.7. (i) Compute $dW^4(t)$ and then write $W^4(T)$ as the sum of an ordinary (Lebesgue) integral with respect to time and an Itô integral.
(ii) Take expectations on both sides of the formula you obtained in (i), use the fact that $\mathbb{E}W^2(t) = t$, and derive the formula $\mathbb{E}W^4(T) = 3T^2$.
(iii) Use the method of (i) and (ii) to derive a formula for $\mathbb{E}W^6(T)$.

Exercise 4.8 (Solving the Vasicek equation). The Vasicek interest rate stochastic differential equation (4.4.32) is

$$dR(t) = \big(\alpha - \beta R(t)\big)\,dt + \sigma\,dW(t),$$

where α, β, and σ are positive constants. The solution to this equation is given in Example 4.4.10. This exercise shows how to derive this solution.

(i) Use (4.4.32) and the Itô-Doeblin formula to compute $d\big(e^{\beta t}R(t)\big)$. Simplify it so that you have a formula for $d\big(e^{\beta t}R(t)\big)$ that does not involve $R(t)$.
(ii) Integrate the equation you obtained in (i) and solve for $R(t)$ to obtain (4.4.33).

Exercise 4.9. For a European call expiring at time T with strike price K, the Black-Scholes-Merton price at time t, if the time-t stock price is x, is

$$c(t,x) = xN\big(d_+(T-t,x)\big) - Ke^{-r(T-t)}N\big(d_-(T-t,x)\big),$$

where

$$d_+(\tau,x) = \frac{1}{\sigma\sqrt{\tau}}\left[\log\frac{x}{K} + \left(r + \frac{1}{2}\sigma^2\right)\tau\right],$$
$$d_-(\tau,x) = d_+(\tau,x) - \sigma\sqrt{\tau},$$

and $N(y)$ is the cumulative standard normal distribution

$$N(y) = \frac{1}{\sqrt{2\pi}}\int_{-\infty}^{y} e^{-\frac{z^2}{2}}\,dz = \frac{1}{\sqrt{2\pi}}\int_{-y}^{\infty} e^{-\frac{z^2}{2}}\,dz.$$

The purpose of this exercise is to show that the function c satisfies the Black-Scholes-Merton partial differential equation

$$c_t(t,x) + rxc_x(t,x) + \frac{1}{2}\sigma^2x^2c_{xx}(t,x) = rc(t,x),\ 0 \le t < T, x > 0, \quad (4.10.3)$$

the *terminal condition*

$$\lim_{t\uparrow T} c(t,x) = (x-K)^+, \quad x > 0, x \ne K, \quad (4.10.4)$$

and the *boundary conditions*

$$\lim_{x\downarrow 0} c(t,x) = 0, \quad \lim_{x\to\infty}\left[c(t,x) - \big(x - e^{-r(T-t)}K\big)\right] = 0, \quad 0 \le t < T. \quad (4.10.5)$$

Equation (4.10.4) and the first part of (4.10.5) are usually written more simply but less precisely as

$$c(T,x) = (x-K)^+,\ x \ge 0$$

and

$$c(t,0) = 0,\ 0 \le t \le T.$$

For this exercise, we abbreviate $c(t,x)$ as simply c and $d_\pm(T-t,x)$ as simply $d_\pm$.

(i) Verify first the equation

$$Ke^{-r(T-t)}N'(d_-) = xN'(d_+). \tag{4.10.6}$$

(ii) Show that $c_x = N(d_+)$. This is the *delta* of the option. (Be careful! Remember that d_+ is a function of x.)

(iii) Show that

$$c_t = -rKe^{-r(T-t)}N(d_-) - \frac{\sigma x}{2\sqrt{T-t}}N'(d_+).$$

This is the *theta* of the option.

(iv) Use the formulas above to show that c satisfies (4.10.3).

(v) Show that for $x > K$, $\lim_{t \uparrow T} d_\pm = \infty$, but for $0 < x < K$, $\lim_{t \uparrow T} d_\pm = -\infty$. Use these equalities to derive the terminal condition (4.10.4).

(vi) Show that for $0 \le t < T$, $\lim_{x \downarrow 0} d_\pm = -\infty$. Use this fact to verify the first part of boundary condition (4.10.5) as $x \downarrow 0$.

(vii) Show that for $0 \le t < T$, $\lim_{x \to \infty} d_\pm = \infty$. Use this fact to verify the second part of boundary condition (4.10.5) as $x \to \infty$. In this verification, you will need to show that

$$\lim_{x \to \infty} \frac{N(d_+) - 1}{x^{-1}} = 0.$$

This is an indeterminate form $\frac{0}{0}$, and L'Hôpital's rule implies that this limit is

$$\lim_{x \to \infty} \frac{\frac{d}{dx}\left[N(d_+) - 1\right]}{\frac{d}{dx}x^{-1}}.$$

Work out this expression and use the fact that

$$x = K \exp\left\{\sigma\sqrt{T-t}\, d_+ - (T-t)\left(r + \frac{1}{2}\sigma^2\right)\right\}$$

to write this expression solely in terms of d_+ (i.e., without the appearance of any x except the x in the argument of $d_+(T-t,x)$). Then argue that the limit is zero as $d_+ \to \infty$.

Exercise 4.10 (Self-financing trading). The fundamental idea behind no-arbitrage pricing is to reproduce the payoff of a derivative security by trading in the underlying asset (which we call a stock) and the money market account. In discrete time, we let X_k denote the value of the hedging portfolio at time k and let Δ_k denote the number of shares of stock held between times k and $k+1$. Then, at time k, after rebalancing (i.e., moving from a position of Δ_{k-1} to a position Δ_k in the stock), the amount in the money market account is $X_k - S_k\Delta_k$. The value of the portfolio at time $k+1$ is

$$X_{k+1} = \Delta_k S_{k+1} + (1+r)\big(X_k - \Delta_k S_k\big). \tag{4.10.7}$$

This formula can be rearranged to become

$$X_{k+1} - X_k = \Delta_k\big(S_{k+1} - S_k\big) + r\big(X_k - \Delta_k S_k\big), \tag{4.10.8}$$

which says that the gain between time k and time $k+1$ is the sum of the capital gain on the stock holdings, $\Delta_k(S_{k+1} - S_k)$, and the interest earnings on the money market account, $r(X_k - \Delta_k S_k)$. The continuous-time analogue of (4.10.8) is

$$dX(t) = \Delta(t)\,dS(t) + r\big(X(t) - \Delta(t)S(t)\big)\,dt. \tag{4.10.9}$$

Alternatively, one could define the value of a share of the money market account at time k to be

$$M_k = (1+r)^k$$

and formulate the discrete-time model with two processes, Δ_k as before and Γ_k denoting the number of shares of the money market account held at time k after rebalancing. Then

$$X_k = \Delta_k S_k + \Gamma_k M_k, \tag{4.10.10}$$

so that (4.10.7) becomes

$$X_{k+1} = \Delta_k S_{k+1} + (1+r)\Gamma_k M_k = \Delta_k S_{k+1} + \Gamma_k M_{k+1}. \tag{4.10.11}$$

Subtracting (4.10.10) from (4.10.11), we obtain in place of (4.10.8) the equation

$$X_{k+1} - X_k = \Delta_k\big(S_{k+1} - S_k\big) + \Gamma_k(M_{k+1} - M_k), \tag{4.10.12}$$

which says that the gain between time k and time $k+1$ is the sum of the capital gain on stock holdings, $\Delta_k(S_{k+1} - S_k)$, and the earnings from the money market investment, $\Gamma_k(M_{k+1} - M_k)$.

But Δ_k and Γ_k cannot be chosen arbitrarily. The agent arrives at time $k+1$ with some portfolio of Δ_k shares of stock and Γ_k shares of the money market account and then rebalances. In terms of Δ_k and Γ_k, the value of the portfolio upon arrival at time $k+1$ is given by (4.10.11). After rebalancing, it is

$$X_{k+1} = \Delta_{k+1} S_{k+1} + \Gamma_{k+1} M_{k+1}.$$

Setting these two values equal, we obtain the *discrete-time self-financing condition*

$$S_{k+1}(\Delta_{k+1} - \Delta_k) + M_{k+1}(\Gamma_{k+1} - \Gamma_k) = 0. \tag{4.10.13}$$

The first term is the cost of rebalancing in the stock, and the second is the cost of rebalancing in the money market account. If the sum of these two terms is not zero, then money must either be put into the position or can be taken out as a by-product of rebalancing. The point is that when the two processes Δ_k and Γ_k are used to describe the evolution of the portfolio value

X_k, then two equations, (4.10.12) and (4.10.13), are required rather than the single equation (4.10.8) when only the process Δ_k is used.

Finally, we note that we may rewrite the *discrete-time self-financing condition* (4.10.13) as

$$\begin{aligned} S_k(\Delta_{k+1} - \Delta_k) + (S_{k+1} - S_k)(\Delta_{k+1} - \Delta_k) \\ + M_k(\Gamma_{k+1} - \Gamma_k) + (M_{k+1} - M_k)(\Gamma_{k+1} - \Gamma_k) = 0. \end{aligned} \quad (4.10.14)$$

This is suggestive of the *continuous-time self-financing condition*

$$S(t)\,d\Delta(t) + dS(t)\,d\Delta(t) + M(t)\,d\Gamma(t) + dM(t)\,d\Gamma(t) = 0, \quad (4.10.15)$$

which we derive below.

(i) In continuous time, let $M(t) = e^{rt}$ be the price of a share of the money market account at time t, let $\Delta(t)$ denote the number of shares of stock held at time t, and let $\Gamma(t)$ denote the number of shares of the money market account held at time t, so that the total portfolio value at time t is

$$X(t) = \Delta(t)S(t) + \Gamma(t)M(t). \quad (4.10.16)$$

Using (4.10.16) and (4.10.9), derive the continuous-time self-financing condition (4.10.15).

A common argument used to derive the Black-Scholes-Merton partial differential equation and delta-hedging formula goes like this. Let $c(t, x)$ be the price of a call at some time t if the stock price at that time is $S(t) = x$. Form a portfolio that is long the call and short $\Delta(t)$ shares of stock, so that the value of the portfolio at time t is $N(t) = c\big(t, S(t)\big) - \Delta(t)S(t)$. We want to choose $\Delta(t)$ so this is "instantaneously riskless," in which case its value would have to grow at the interest rate. Otherwise, according to this argument, we could arbitrage this portfolio against the money market account. This means we should have $dN(t) = rN(t)\,dt$. We compute the differential of $N(t)$ and get

$$\begin{aligned} dN(t) &= c_t\big(t, S(t)\big)\,dt + c_x\big(t, S(t)\big)\,dS(t) \\ &\quad + \frac{1}{2}c_{xx}\big(t, S(t)\big)\,dS(t)\,dS(t) - \Delta(t)\,dS(t) \\ &= \big[c_x\big(t, S(t)\big) - \Delta(t)\big]\,dS(t) \\ &\quad + \left[c_t\big(t, S(t)\big) + \frac{1}{2}\sigma^2 S^2(t) c_{xx}\big(t, S(t)\big)\right] dt. \end{aligned} \quad (4.10.17)$$

In order for this to be instantaneously riskless, we must cancel out the $dS(t)$ term, which contains the risk. This gives us the delta-hedging formula $\Delta(t) = c_x\big(t, S(t)\big)$. Having chosen $\Delta(t)$ this way, we recall that we expect to have $dN(t) = rN(t)\,dt$, and this yields

$$rN(t)\,dt = \left[c_t\big(t,S(t)\big) + \frac{1}{2}\sigma^2 S^2(t)c_{xx}\big(t,S(t)\big)\right]dt. \tag{4.10.18}$$

But

$$N(t) = c\big(t,S(t)\big) - \Delta(t)S(t) = c\big(t,S(t)\big) - S(t)c_x\big(t,S(t)\big), \tag{4.10.19}$$

and substitution of (4.10.19) into (4.10.18) yields the Black-Scholes-Merton partial differential equation

$$c_t\big(t,S(t)\big) + rS(t)c_s\big(t,S(t)\big) + \frac{1}{2}\sigma^2 S^2(t)c_{ss}\big(t,S(t)\big) = rc\big(t,S(t)\big). \tag{4.10.20}$$

One can question the first step of this argument, where we failed to use Itô's product rule (Corollary 4.6.3) on the term $\Delta(t)S(t)$ when we differentiated $N(t)$ in (4.10.17). In discrete time, we hold Δ_k fixed for a period and let S move, computing the capital gain according to the formula $\Delta_k(S_{k+1} - S_k)$, and in (4.10.17) we are attempting something analogous to that in continuous time. However, as soon as we set $\Delta(t) = c_x\big(t,S(t)\big)$, then $\Delta(t)$ moves continuously in time and the differential of $N(t)$ is really

$$\begin{aligned} dN(t) &= c_t\big(t,S(t)\big)\,dt + c_x\big(t,S(t)\big)\,dS(t) + \frac{1}{2}c_{xx}\big(t,S(t)\big)\,dS(t)\,dS(t) \\ &\quad -\Delta(t)\,dS(t) - S(t)\,d\Delta(t) - d\Delta(t)\,dS(t) \end{aligned} \tag{4.10.21}$$

rather than the expression in (4.10.17).

This exercise shows that the argument is correct after all. At least, equation (4.10.18) is correct, and from that the Black-Scholes-Merton partial differential equation (4.10.20) follows.

Recall from Subsection 4.5.3 that if we take $X(0) = c\big(0,S(0)\big)$ and at each time t hold $\Delta(t) = c_x\big(t,S(t)\big)$ shares of stock, borrowing or investing in the money market as necessary to finance this, then at each time t we have a portfolio of stock and a money market account valued at $X(t) = c(t,S(t))$. The amount invested in the money market account at each time t is

$$X(t) - \Delta(t)S(t) = c\big(t,S(t)\big) - \Delta(t)S(t) = N(t),$$

and so the number of money market account shares held is

$$\Gamma(t) = \frac{N(t)}{M(t)}.$$

(ii) Now replace (4.10.17) by its corrected version (4.10.21) and use the continuous-time self-financing condition you derived in part (i) to derive (4.10.18).

Exercise 4.11. Let

$$c(t,x) = xN(d_+(T-t,x)) - Ke^{-r(T-t)}N(d_-(T-t,x))$$

denote the price for a European call, expiring at time T with strike price K, where

$$d_{\pm}(T-t,x) = \frac{1}{\sigma_1\sqrt{T-t}}\left[\log\frac{x}{K} + \left(r \pm \frac{\sigma_1^2}{2}\right)(T-t)\right].$$

This option price assumes the underlying stock is a geometric Brownian motion with volatility $\sigma_1 > 0$. For this problem, we take this to be the market price of the option.

Suppose, however, that the underlying asset is really a geometric Brownian motion with volatility $\sigma_2 > \sigma_1$, i.e.,

$$dS(t) = \alpha S(t)\,dt + \sigma_2 S(t)\,dW(t).$$

Consequently, the market price of the call is incorrect.

We set up a portfolio whose value at each time t we denote by $X(t)$. We begin with $X(0) = 0$. At each time t, the portfolio is long one European call, is short $c_x(t, S(t))$ shares of stock, and thus has a cash position

$$X(t) - c(t, S(t)) + S(t)c_x(t, S(t)),$$

which is invested at the constant interest rate r. We also remove cash from this portfolio at a rate $\frac{1}{2}(\sigma_2^2 - \sigma_1^2)S^2(t)c_{xx}(t, S(t))$. Therefore, the differential of the portfolio value is

$$\begin{aligned} dX(t) &= dc(t, S(t)) - c_x(t, S(t))\,dS(t) \\ &\quad + r\big[X(t) - c(t, S(t)) + S(t)c_x(t, S(t))\big]\,dt \\ &\quad - \frac{1}{2}(\sigma_2^2 - \sigma_1^2)S^2(t)c_{xx}(t, S(t))\,dt,\ 0 \le t \le T. \end{aligned}$$

Show that $X(t) = 0$ for all $t \in [0, T]$. In particular, because $c_{xx}(t, S(t)) > 0$ and $\sigma_2 > \sigma_1$, we have an arbitrage opportunity; we can start with zero initial capital, remove cash at a positive rate between times 0 and T, and at time T have zero liability. (Hint: Compute $d\big(e^{-rt}X(t)\big)$.)

Exercise 4.12. (i) Use formulas (4.5.23)–(4.5.25), (4.5.26), and (4.5.29) to determine the delta $p_x(t,x)$, the gamma $p_{xx}(t,x)$, and the theta $p_t(t,x)$ of a European put.

(ii) Show that an agent hedging a short position in the put should have a short position in the underlying stock and a long position in the money market account.

(iii) Show that $f(t,x)$ of (4.5.26) and $p(t,x)$ satisfy the same Black-Scholes-Merton partial differential equation (4.5.14) satisfied by $c(t,x)$.

Exercise 4.13 (Decomposition of correlated Brownian motions into independent Brownian motions). Suppose $B_1(t)$ and $B_2(t)$ are Brownian motions and

$$dB_1(t)\,dB_2(t) = \rho(t)\,dt,$$

where ρ is a stochastic process taking values strictly between -1 and 1. Define processes $W_1(t)$ and $W_2(t)$ such that

$$B_1(t) = W_1(t),$$
$$B_2(t) = \int_0^t \rho(s)\, dW_1(s) + \int_0^t \sqrt{1-\rho^2(s)}\, dW_2(s),$$

and show that $W_1(t)$ and $W_2(t)$ are independent Brownian motions.

Exercise 4.14. In the derivation of the Itô-Doeblin formula, Theorem 4.4.1, we considered only the case of the function $f(x) = \frac{1}{2}x^2$, for which $f''(x) = 1$. This made it easy to determine the limit of the last term,

$$\frac{1}{2}\sum_{j=0}^{n-1} f''(W(t_j))\big[W(t_{j+1}) - W(t_j)\big]^2,$$

appearing in (4.4.5). Indeed,

$$\begin{aligned}\lim_{\|\Pi\|\to 0}\sum_{j=0}^{n-1} f''(W(t_j))\big[W(t_{j+1}) - W(t_j)\big]^2 &= \lim_{\|\Pi\|\to 0}\sum_{j=0}^{n-1}\big[W(t_{j+1}) - W(t_j)\big]^2\\ &= [W,W](T) \;=\; T\\ &= \int_0^T f''(W(t))\,dt.\end{aligned}$$

If we had been working with an arbitrary function $f(x)$, we could not replace $f''(W(t_j))$ by 1 in the argument above. It is tempting in this case to just argue that $\big[W(t_{j+1}) - W(t_j)\big]^2$ is approximately equal to $(t_{j+1} - t_j)$, so that

$$\sum_{j=0}^{n-1} f''(W(t_j))\big[W(t_{j+1}) - W(t_j)\big]^2$$

is approximately equal to

$$\sum_{j=0}^{n-1} f''(W(t_j))(t_{j+1} - t_j),$$

and this has limit $\int_0^T f''(W(t))\,dt$ as $\|\Pi\| \to 0$. However, as discussed in Remark 3.4.4, it does not make sense to say that $\big[W(t_{j+1}) - W(t_j)\big]^2$ is approximately equal to $(t_{j+1} - t_j)$. In this exercise, we develop a correct explanation for the equation

$$\lim_{\|\Pi\|\to 0}\sum_{j=0}^{n-1} f''(W(t_j))\big[W(t_{j+1}) - W(t_j)\big]^2 = \int_0^T f''(W(t))\,dt. \tag{4.10.22}$$

Define

$$Z_j = f''(W(t_j))\big[\big(W(t_{j+1}) - W(t_j)\big)^2 - (t_{j+1} - t_j)\big]$$

so that

$$\sum_{j=0}^{n-1} f''(W(t_j))\big[W(t_{j+1}) - W(t_j)\big]^2 = \sum_{j=0}^{n-1} Z_j + \sum_{j=0}^{n-1} f''(W(t_j))(t_{j+1} - t_j). \tag{4.10.23}$$

(i) Show that Z_j is $\mathcal{F}(t_{j+1})$-measurable and

$$\mathbb{E}[Z_j|\mathcal{F}(t_j)] = 0, \quad \mathbb{E}[Z_j^2|\mathcal{F}(t_j)] = 2\big[f''(W(t_j))\big]^2(t_{j+1} - t_j)^2.$$

It remains to show that

$$\lim_{\|\Pi\|\to 0} \sum_{j=0}^{n-1} Z_j = 0. \tag{4.10.24}$$

This will cause us to obtain (4.10.22) when we take the limit in (4.10.23). Prove (4.10.24) in the following steps.

(ii) Show that $\mathbb{E}\sum_{j=0}^{n-1} Z_j = 0$.

(iii) Under the assumption that $\mathbb{E}\int_0^T [f''(W(t))]^2\,dt$ is finite, show that

$$\lim_{\|\Pi\|\to 0} \operatorname{Var}\left[\sum_{j=0}^{n-1} Z_j\right] = 0.$$

(Warning: The random variables $Z_1, Z_2, \ldots, Z_{n-1}$ are not independent.)

From (iii), we conclude that $\sum_{j=0}^{n-1} Z_j$ converges to its mean, which by (ii) is zero. This establishes (4.10.24).

Exercise 4.15 (Creating correlated Brownian motions from independent ones). Let $(W_1(t), \ldots, W_d(t))$ be a d-dimensional Brownian motion. In particular, these Brownian motions are independent of one another. Let $(\sigma_{ij}(t))_{i=1,\ldots,m;j=1,\ldots,d}$ be an $m \times d$ matrix-valued process adapted to the filtration associated with the d-dimensional Brownian motion. For $i = 1, \ldots, m$, define

$$\sigma_i(t) = \left[\sum_{j=1}^{d} \sigma_{ij}^2(t)\right]^{\frac{1}{2}},$$

and assume this is never zero. Define also

$$B_i(t) = \sum_{j=1}^{d} \int_0^t \frac{\sigma_{ij}(u)}{\sigma_i(u)}\,dW_j(u).$$

(i) Show that, for each i, B_i is a Brownian motion.

(ii) Show that $dB_i(t)\,dB_k(t) = \rho_{ik}(t)$, where

$$\rho_{ik}(t) = \frac{1}{\sigma_i(t)\sigma_k(t)} \sum_{j=1}^{d} \sigma_{ij}(t)\sigma_{kj}(t).$$

Exercise 4.16 (Creating independent Brownian motions to represent correlated ones). Let $B_1(t), \ldots, B_m(t)$ be m one-dimensional Brownian motions with

$$dB_i(t)\,dB_k(t) = \rho_{ik}(t)\,dt \text{ for all } i, k = 1, \ldots, m,$$

where $\rho_{ik}(t)$ are adapted processes taking values in $(-1, 1)$ for $i \neq k$ and $\rho_{ik}(t) = 1$ for $i = k$. Assume that the symmetric matrix

$$C(t) = \begin{bmatrix} \rho_{11}(t) & \rho_{12}(t) & \cdots & \rho_{1m}(t) \\ \rho_{21}(t) & \rho_{22}(t) & \cdots & \rho_{2m}(t) \\ \vdots & \vdots & & \vdots \\ \rho_{m1}(t) & \rho_{m2}(t) & \cdots & \rho_{mm}(t) \end{bmatrix}$$

is positive definite for all t almost surely. Because the matrix $C(t)$ is symmetric and positive definite, it has a *matrix square root*. In other words, there is a matrix

$$A(t) = \begin{bmatrix} a_{11}(t) & a_{12}(t) & \cdots & a_{1m}(t) \\ a_{21}(t) & a_{22}(t) & \cdots & a_{2m}(t) \\ \vdots & \vdots & & \vdots \\ a_{m1}(t) & a_{m2}(t) & \cdots & a_{mm}(t) \end{bmatrix}$$

such that $C(t) = A(t)A^{\mathrm{tr}}(t)$, which when written componentwise is

$$\rho_{ik}(t) = \sum_{j=1}^{m} a_{ij}(t)a_{kj}(t) \text{ for all } i, k = 1, \ldots, m. \tag{4.10.25}$$

This matrix can be chosen so that its components $a_{ik}(t)$ are adapted processes. Furthermore, the matrix $A(t)$ has an inverse

$$A^{-1}(t) = \begin{bmatrix} \alpha_{11}(t) & \alpha_{12}(t) & \cdots & \alpha_{1m}(t) \\ \alpha_{21}(t) & \alpha_{22}(t) & \cdots & \alpha_{2m}(t) \\ \vdots & \vdots & & \vdots \\ \alpha_{m1}(t) & \alpha_{m2}(t) & \cdots & \alpha_{mm}(t) \end{bmatrix},$$

which means that

$$\sum_{j=1}^{m} a_{ij}(t)\alpha_{jk}(t) = \sum_{j=1}^{m} \alpha_{ij}(t)a_{jk}(t) = \delta_{ik}, \tag{4.10.26}$$

where we define

$$\delta_{ik} = \begin{cases} 1 & \text{if } i = k, \\ 0 & \text{if } i \neq k, \end{cases}$$

to be the so-called *Kronecker delta*. Show that there exist m independent Brownian motions $W_1(t), \dots, W_m(t)$ such that

$$B_i(t) = \sum_{j=1}^{m} \int_0^t a_{ij}(u)\, dW_j(u) \text{ for all } i = 1, \dots, m. \tag{4.10.27}$$

Exercise 4.17 (Instantaneous correlation). Let

$$X_1(t) = X_1(0) + \int_0^t \Theta_1(u)\, du + \int_0^t \sigma_1(u)\, dB_1(u),$$
$$X_2(t) = X_2(0) + \int_0^t \Theta_2(u)\, du + \int_0^t \sigma_2(u)\, dB_2(u),$$

where $B_1(t)$ and $B_2(t)$ are Brownian motions satisfying $dB_1(t)\, dB_2(t) = \rho(t)$ and $\rho(t)$, $\Theta_1(t)$, $\Theta_2(t)$, $\sigma_1(t)$, and $\sigma_2(t)$ are adapted processes. Then

$$dX_1(t)\, dX_2(t) = \sigma_1(t)\sigma_2(t)\, dB_1(t)\, dB_2(t) = \rho(t)\sigma_1(t)\sigma_2(t)\, dt.$$

We call $\rho(t)$ the *instantaneous correlation* between $X_1(t)$ and $X_2(t)$ for the reason explained by this exercise.

We first consider the case when ρ, Θ_1, Θ_2, σ_1, and σ_2 are constant. Then

$$X_1(t) = X_1(0) + \Theta_1 t + \sigma_1 B_1(t),$$
$$X_2(t) = X_2(0) + \Theta_2 t + \sigma_2 B_2(t).$$

Fix $t_0 > 0$, and let $\epsilon > 0$ be given.

(i) Use Itô's product rule to show that

$$\mathbb{E}\left[\big(B_1(t_0+\epsilon) - B_1(t_0)\big)\big(B_2(t_0+\epsilon) - B_2(t_0)\big) \big| \mathcal{F}(t_0) \right] = \rho\epsilon.$$

(ii) Show that, conditioned on $\mathcal{F}(t_0)$, the pair of random variables

$$\big(X_1(t_0+\epsilon) - X_1(t_0), X_2(t_0+\epsilon) - X_2(t_0)\big)$$

has means, variances, and covariance

$$M_i(\epsilon) = \mathbb{E}\left[X_i(t_0+\epsilon) - X_i(t_0) \big| \mathcal{F}(t_0) \right] = \Theta_i \epsilon \text{ for } i = 1, 2, \tag{4.10.28}$$

$$\begin{aligned} V_i(\epsilon) &= \mathbb{E}\left[\big(X_i(t_0+\epsilon) - X_i(t_0)\big)^2 \Big| \mathcal{F}(t_0) \right] - M_i^2(\epsilon) \\ &= \sigma_i^2 \epsilon \text{ for } i = 1, 2, \end{aligned} \tag{4.10.29}$$

$$\begin{aligned} C(\epsilon) = \mathbb{E}&\left[\big(X_1(t_0+\epsilon) - X_1(t_0)\big)\big(X_2(t_0+\epsilon) - X_2(t_0)\big) \big| \mathcal{F}(t_0) \right] \\ &- M_1(\epsilon) M_2(\epsilon) = \rho\sigma_1\sigma_2\epsilon. \end{aligned} \tag{4.10.30}$$

In particular, conditioned on $\mathcal{F}(t_0)$, the correlation between the increments $X_1(t_0+\epsilon) - X_1(t_0)$ and $X_2(t_0+\epsilon) - X_2(t_0)$ is

$$\frac{C(\epsilon)}{\sqrt{V_1(\epsilon)V_2(\epsilon)}} = \rho.$$

We now allow $\rho(t)$, $\Theta_1(t)$, $\Theta_2(t)$, $\sigma_1(t)$, and $\sigma_2(t)$ to be continuous adapted processes, assuming only that there is a constant M such that

$$|\Theta_1(t)| \leq M, \quad |\sigma_1(t)| \leq M, \quad |\Theta_2(t)| \leq M, \quad |\sigma_2(T)| \leq M, \quad |\rho(t)| \leq M \tag{4.10.31}$$

for all $t \geq 0$ almost surely. We again fix $t_0 \geq 0$.

(iii) Show that, conditioned on $\mathcal{F}(t_0)$, we have the conditional mean formulas

$$M_i(\epsilon) = E\left[X_i(t_0+\epsilon) - X_i(t_0) \middle| \mathcal{F}(t_0)\right] = \Theta_i(t_0)\epsilon + o(\epsilon) \text{ for } i = 1, 2, \tag{4.10.32}$$

where we denote by $o(\epsilon)$ any quantity that is so small that $\lim_{\epsilon\downarrow 0} \frac{o(\epsilon)}{\epsilon} = 0$. In other words, show that

$$\lim_{\epsilon\downarrow 0} \frac{1}{\epsilon} M_i(\epsilon) = \Theta_i(t_0) \text{ for } i = 1, 2. \tag{4.10.33}$$

(Hint: First show that

$$M_i(\epsilon) = \mathbb{E}\left[\int_{t_0}^{t_0+\epsilon} \Theta_i(u)\,du \middle| \mathcal{F}(t_0)\right]. \tag{4.10.34}$$

The Dominated Convergence Theorem, Theorem 1.4.9, works for conditional expectations as well as for expectations in the following sense. Let X be a random variable. Suppose for every $\epsilon > 0$ we have a random variable $X(\epsilon)$ such that $\lim_{\epsilon\downarrow 0} X(\epsilon) = X$ almost surely. Finally, suppose there is another random variable Y such that $\mathbb{E}Y < \infty$ and $|X(\epsilon)| \leq Y$ almost surely for every $\epsilon > 0$. Then

$$\lim_{\epsilon\downarrow 0} \mathbb{E}[X(\epsilon)|\mathcal{F}(t_0)] = \mathbb{E}[X|\mathcal{F}(t_0)].$$

Use this to obtain (4.10.33) from (4.10.34).)

(iv) Show that $D_{ij}(\epsilon)$ defined by

$$\begin{aligned} D_{ij}(\epsilon) = \mathbb{E}\big[&\big(X_i(t_0+\epsilon) - X_i(t_0)\big)\big(X_j(t_0+\epsilon) - X_j(t_0)\big)\big| \mathcal{F}(t_0)\big] \\ &- M_i(\epsilon)M_j(\epsilon) \end{aligned}$$

for $i = 1, 2$ and $j = 1, 2$ satisfies

$$D_{ij}(\epsilon) = \rho_{ij}(t_0)\sigma_i(t_0)\sigma_j(t_0)\epsilon + o(\epsilon), \tag{4.10.35}$$

where we set $\rho_{11}(t) = \rho_{22}(t) = 1$ and $\rho_{12}(t) = \rho_{21}(t) = \rho(t)$. (Hint: You should define the martingales

$$Y_i(t) = \int_0^t \sigma_i(u)\, dB_i(u) \text{ for } i = 1, 2,$$

so you can write

$$D_{ij}(\epsilon) = \mathbb{E}\Bigg[\bigg(Y_i(t_0+\epsilon) - Y_i(t_0) + \int_{t_0}^{t_0+\epsilon} \Theta_i(u)\, du\bigg) \cdot \bigg(Y_j(t_0+\epsilon) - Y_j(t_0) + \int_{t_0}^{t_0+\epsilon} \Theta_j(u)\, du\bigg)\bigg|\, \mathcal{F}(t_0)\Bigg] - M_i(\epsilon)M_j(\epsilon). \tag{4.10.36}$$

Then expand the expression on the right-hand side of (4.10.36). You should use Itô's product rule to show that the first term in the expansion is

$$\mathbb{E}\left[\big(Y_i(t_0+\epsilon) - Y_i(t_0)\big)\big(Y_j(t_0+\epsilon) - Y_j(t_0)\big) \big|\, \mathcal{F}(t_0)\right] = \mathbb{E}\left[\int_{t_0}^{t_0+\epsilon} \rho_{ij}(u)\sigma_i(u)\sigma_j(u)\, du \bigg|\, \mathcal{F}(t_0)\right].$$

This equation is similar to (4.10.34), and you can use the Dominated Convergence Theorem as stated in the hint for (iii) to conclude that

$$\lim_{\epsilon \downarrow 0} \frac{1}{\epsilon} \mathbb{E}\left[\big(Y_i(t_0+\epsilon) - Y_i(t_0)\big)\big(Y_j(t_0+\epsilon) - Y_j(t_0)\big) \big|\, \mathcal{F}(t_0)\right] = \rho_{ij}(t_0)\sigma_i(t_0)\sigma_j(t_0).$$

To handle the other terms that arise from expanding (4.10.36), you will need (4.10.31) and the fact that

$$\lim_{\epsilon \downarrow 0} \mathbb{E}\big[|Y_i(t_0+\epsilon) - Y_1(t_0)| \big| \mathcal{F}(t_0)\big] = 0. \tag{4.10.37}$$

You may use (4.10.37) without proving it.

(v) Show that, conditioned on $\mathcal{F}(t_0)$, the pair of random variables

$$\big(X_1(t_0+\epsilon) - X_1(t_0), X_2(t_0+\epsilon) - X_2(t_0)\big)$$

has variances and covariance

$$V_i(\epsilon) = \mathbb{E}\left[\big(X_i(t_0+\epsilon) - X_i(t_0)\big)^2 \Big|\, \mathcal{F}(t_0)\right] - M_i^2(\epsilon) = \sigma_i^2(t_0)\epsilon + o(\epsilon) \text{ for } i = 1, 2, \tag{4.10.38}$$

$$C(\epsilon) = \mathbb{E}\left[\big(X_1(t_0+\epsilon) - X_1(t_0)\big)\big(X_2(t_0+\epsilon) - X_2(t_0)\big) \big|\, \mathcal{F}(t_0)\right] - M_1(\epsilon)M_2(\epsilon) \tag{4.10.39}$$
$$= \rho(t_0)\sigma_1(t_0)\sigma_2(t_0)\epsilon + o(\epsilon).$$

(vi) Show that

$$\lim_{\epsilon \downarrow 0} \frac{C(\epsilon)}{\sqrt{V_1(\epsilon)V_2(\epsilon)}} = \rho(t_0). \tag{4.10.40}$$

In other words, for small values of $\epsilon > 0$, conditioned on $\mathcal{F}(t_0)$, the correlation between the increments $X_1(t_0+\epsilon) - X_1(t_0)$ and $X_2(t_0+\epsilon) - X_2(t_0)$ is approximately equal to $\rho(t_0)$, and this approximation becomes exact as $\epsilon \downarrow 0$.

Exercise 4.18. Let a stock price be a geometric Brownian motion

$$dS(t) = \alpha S(t)\,dt + \sigma S(t)\,dW(t),$$

and let r denote the interest rate. We define the *market price of risk* to be

$$\theta = \frac{\alpha - r}{\sigma}$$

and the *state price density process* to be

$$\zeta(t) = \exp\left\{ -\theta W(t) - \left(r + \frac{1}{2}\theta^2\right)t \right\}.$$

(i) Show that

$$d\zeta(t) = -\theta\zeta(t)\,dW(t) - r\zeta(t)\,dt.$$

(ii) Let X denote the value of an investor's portfolio when he uses a portfolio process $\Delta(t)$. From (4.5.2), we have

$$dX(t) = rX(t)\,dt + \Delta(t)(\alpha - r)S(t)\,dt + \Delta(t)\sigma S(t)\,dW(t).$$

Show that $\zeta(t)X(t)$ is a martingale. (Hint: Show that the differential $d\big(\zeta(t)X(t)\big)$ has no dt term.)

(iii) Let $T > 0$ be a fixed terminal time. Show that if an investor wants to begin with some initial capital $X(0)$ and invest in order to have portfolio value $V(T)$ at time T, where $V(T)$ is a given $\mathcal{F}(T)$-measurable random variable, then he must begin with initial capital

$$X(0) = \mathbb{E}[\zeta(T)V(T)].$$

In other words, the *present value* at time zero of the random payment $V(T)$ at time T is $\mathbb{E}[\zeta(T)V(T)]$. This justifies calling $\zeta(t)$ the state price density process.

Exercise 4.19. Let $W(t)$ be a Brownian motion, and define

$$B(t) = \int_0^t \operatorname{sign}(W(s))\,dW(s),$$

where

$$\operatorname{sign}(x) = \begin{cases} 1 & \text{if } x \geq 0, \\ -1 & \text{if } x < 0. \end{cases}$$

(i) Show that $B(t)$ is a Brownian motion.
(ii) Use Itô's product rule to compute $d\big[B(t)W(t)\big]$. Integrate both sides of the resulting equation and take expectations. Show that $\mathbb{E}\big[B(t)W(t)\big] = 0$ (i.e., $B(t)$ and $W(t)$ are uncorrelated).
(iii) Verify that

$$dW^2(t) = 2W(t)\,dW(t) + dt.$$

(iv) Use Itô's product rule to compute $d\big[B(t)W^2(t)\big]$. Integrate both sides of the resulting equation and take expectations to conclude that

$$\mathbb{E}\big[B(t)W^2(t)\big] \neq \mathbb{E}B(t) \cdot \mathbb{E}W^2(t).$$

Explain why this shows that, although they are uncorrelated normal random variables, $B(t)$ and $W(t)$ are not independent.

Exercise 4.20 (Local time). Let $W(t)$ be a Brownian motion. The Itô-Doeblin formula in differential form says that

$$df\big(W(t)\big) = f'\big(W(t)\big)\,dW(t) + \frac{1}{2}f''\big(W(t)\big)\,dt. \tag{4.10.41}$$

In integrated form, this formula is

$$f\big(W(T)\big) = f\big(W(0)\big) + \int_0^T f'\big(W(t)\big)\,dW(t) + \frac{1}{2}\int_0^T f''\big(W(t)\big)\,dt. \tag{4.10.42}$$

The usual statement of this formula assumes that the function $f''(x)$ is defined for every $x \in \mathbb{R}$ and is a continuous function of x. In fact, the formula still holds if there are finitely many points x where $f''(x)$ is undefined, provided that $f'(x)$ is defined for every $x \in \mathbb{R}$ and is a continuous function of x (and provided that $|f''(x)|$ is bounded so that the integral $\int_0^T f''\big(W(t)\big)\,dt$ is defined). However, if $f'(x)$ is not defined at some point, naive application of the Itô-Doeblin formula can give wrong answers, as this problem demonstrates.

(i) Let K be a positive constant, and define $f(x) = (x-K)^+$. Compute $f'(x)$ and $f''(x)$. Note that there is a point x where $f'(x)$ is not defined, and note also that $f''(x)$ is zero everywhere except at this point, where $f''(x)$ is also undefined.
(ii) Substitute the function $f(x) = (x-K)^+$ into (4.10.42), replacing the term $\frac{1}{2}\int_0^T f''\big(W(t)\big)\,dt$ by zero since f'' is zero everywhere except at one point, where it is not defined. Show that the two sides of this equation cannot be equal by computing their expected values.
(iii) To get some idea of what is going on here, define a sequence of functions $\{f_n\}_{n=1}^{\infty}$ by the formula

$$f_n(x) = \begin{cases} 0 & \text{if } x \leq K - \frac{1}{2n}, \\ \frac{n}{2}(x-K)^2 + \frac{1}{2}(x-K) + \frac{1}{8n} & \text{if } K - \frac{1}{2n} \leq x \leq K + \frac{1}{2n}, \\ x - K & \text{if } x \geq K + \frac{1}{2n}. \end{cases}$$

Show that

$$f_n'(x) = \begin{cases} 0 & \text{if } x \le K - \frac{1}{2n}, \\ n(x-K) + \frac{1}{2} & \text{if } K - \frac{1}{2n} \le x \le K + \frac{1}{2n}, \\ 1 & \text{if } x \ge K + \frac{1}{2n}. \end{cases}$$

In particular, because we get the same value for $f_n'\left(K - \frac{1}{2n}\right)$ regardless of whether we use the formula for $x \le K - \frac{1}{2n}$ or the formula for $K - \frac{1}{2n} \le x \le K + \frac{1}{2n}$, the derivative $f'\left(K - \frac{1}{2n}\right)$ is defined. By the same argument, $f_n'\left(K + \frac{1}{2n}\right)$ is also defined. Verify that

$$f_n''(x) = \begin{cases} 0 \text{ if } x < K - \frac{1}{2n}, \\ n \text{ if } K - \frac{1}{2n} < x < K + \frac{1}{2n}, \\ 0 \text{ if } x < K + \frac{1}{2n}. \end{cases}$$

The second derivative $f''(x)$ is not defined when $x = K \pm \frac{1}{2n}$ because the formulas above disagree at those points.

(iv) Show that

$$\lim_{n \to \infty} f_n(x) = (x - K)^+$$

for every $x \in \mathbb{R}$ and

$$\lim_{n \to \infty} f_n'(x) = \begin{cases} 0 \text{ if } x < K, \\ \frac{1}{2} \text{ if } x = K, \\ 1 \text{ if } x > K. \end{cases}$$

The value of $\lim_{n \to \infty} f_n'(x)$ at a single point will not matter when we integrate in part (v) below, so instead of using the formula just derived, we will replace $\lim_{n \to \infty} f_n'(x)$ by

$$\mathbb{I}_{(K,\infty)}(x) = \begin{cases} 0 \text{ if } x \le K, \\ 1 \text{ if } x > K, \end{cases}$$

in (4.10.44) below. The two functions $\lim_{n \to \infty} f_n'(x)$ and $\mathbb{I}_{(K,\infty)}(x)$ agree except at the single point $x = K$.

For each n, the Itô-Doeblin formula applies to the function f_n because $f_n'(x)$ is defined for every x and is a continuous function of x, $f_n''(x)$ is defined for every x except the two points $x = K \pm \frac{1}{2n}$, and $|f''(x)|$ is bounded above by n. In integrated form, the Itô-Doeblin formula applied to f_n gives

$$f_n(W(T)) = f_n(W(0)) + \int_0^T f_n'(W(t))\, dW(t) + \int_0^T f_n''(W(t))\, dt. \quad (4.10.43)$$

If we now let $n \to \infty$ in this formula, we obtain

$$\begin{aligned}(W(T) - K)^+ &= (W(0) - K)^+ + \int_0^T \mathbb{I}_{(K,\infty)}(W(t))\, dW(t) \\ &\quad + \lim_{n \to \infty} n \int_0^T \mathbb{I}_{\left(K - \frac{1}{2n}, K + \frac{1}{2n}\right)}(W(t))\, dt. \quad (4.10.44)\end{aligned}$$

Let us define the *local time of the Brownian motion at* K to be

$$L_K(T) = \lim_{n\to\infty} n \int_0^T \mathbb{I}_{(K-\frac{1}{2n}, K+\frac{1}{2n})}(W(t))\, dt.$$

(This formula is sometimes written as

$$L_K(T) = \int_0^T \delta_K(W(t))\, dt,$$

where δ_K is the so-called "Dirac delta function" at K.) For a fixed n, the expression $\int_0^T \mathbb{I}_{(K-\frac{1}{2n}, K+\frac{1}{2n})}(W(t))\, dt$ measures how much time between time 0 and time T the Brownian motion spends in the band of length $\frac{1}{n}$ centered at K. As $n \to \infty$, this has limit zero because the width of the band is approaching zero. However, before taking the limit, we multiply by n, and now it is not clear whether the limit will be zero, $+\infty$, or something in between. The limit will, of course, be random; it depends on the path of the Brownian motion.

(v) Show that if the path of the Brownian motion stays strictly below K on the time interval $[0, T]$, we have $L_K(T) = 0$.
(vi) We may solve (4.10.44) for $L_K(T)$, using the fact that $W(0) = 0$ and $K > 0$, to obtain

$$L_K(T) = \big(W(T) - K\big)^+ - \int_0^T \mathbb{I}_{(K,\infty)}(W(t))\, dW(t). \tag{4.10.45}$$

From this, we see that $L_K(T)$ is never $+\infty$. Show that we cannot have $L_K(T) = 0$ almost surely. In other words, for some paths of the Brownian motion, we must have $L_K(T) > 0$. (It turns out that the paths that reach level K are those for which $L_K(T) > 0$.)

Exercise 4.21 (Stop-loss start-gain paradox). Let $S(t)$ be a geometric Brownian motion with mean rate of return zero. In other words,

$$dS(t) = \sigma S(t)\, dW(t),$$

where the volatility σ is constant. We assume the interest rate is $r = 0$.

Suppose we want to hedge a short position in a European call with strike price K and expiration date T. We assume that the call is initially out of the money (i.e., $S(0) < K$). Starting with zero capital ($X(0) = 0$), we could try the following portfolio strategy: own one share of the stock whenever its price strictly exceeds K, and own zero shares whenever its price is K or less. In other words, we use the hedging portfolio process

$$\Delta(t) = \mathbb{I}_{(K,\infty)}\big(S(t)\big).$$

The value of this hedge follows the stochastic differential equation

$$dX(t) = \Delta(t)\,dS(t) + r\big(X(t) - \Delta(t)X(t)\big)\,dt,$$

and since $r = 0$ and $X(0) = 0$, we have

$$X(T) = \sigma \int_0^T \mathbb{I}_{(K,\infty)}(S(t))S(t)\,dW(t). \tag{4.10.46}$$

Executing this hedge requires us to borrow from the money market to buy a share of stock whenever the stock price rises across level K and sell the share, repaying the money market debt, when it falls back across level K. (Recall that we have taken the interest rate to be zero. The situation we are describing can also occur with a nonzero interest rate, but it is more complicated to set up.) At expiration, if the stock price $S(T)$ is below K, there would appear to have been an even number of crossings of the level K, half in the up direction and half in the down direction, so that we would have bought and sold the stock repeatedly, each time at the same price K, and at the final time have no stock and zero debt to the money market. In other words, if $S(T) < K$, then $X(T) = 0$. On the other hand, if at the final time $S(T)$ is above K, we have bought the stock one more time than we sold it, so that we end with one share of stock and a debt of K to the money market. Hence, if $S(T) > K$, we have $X(T) = S(T) - K$. If at the final time $S(T) = K$, then we either own a share of stock valued at K and have a money market debt K or we have sold the stock and have zero money market debt. In either case, $X(T) = 0$. According to this argument, regardless of the final stock price, we have $X(T) = \big(S(T) - K\big)^+$. This kind of hedging is called a *stop-loss start-gain strategy*.

(i) Discuss the practical aspects of implementing the stop-loss start-gain strategy described above. Can it be done?

(ii) Apart from the practical aspects, does the mathematics of continuous-time stochastic calculus suggest that the stop-loss start-gain strategy can be implemented? In other words, with $X(T)$ defined by (4.10.46), is it really true that $X(T) = \big(S(T) - K\big)^+$?

5

Risk-Neutral Pricing

5.1 Introduction

In the binomial asset pricing model of Chapter 1 of Volume I, we showed how to price a derivative security by determining the initial capital required to hedge a short position in the derivative security. In a two-period model, this method led to the six equations (1.2.2), (1.2.3), and (1.2.5)–(1.2.8) in six unknowns in Volume I. Three of these unknowns were the position the hedge should take in the underlying asset at time zero, the position taken by the hedge at time one if the first coin toss results in H, and the position taken by the hedge at time one if the first coin toss results in T. The three other unknowns were the value of the derivative security at time zero, the value of the derivative security at time one if the first coin toss results in H, and the value of the derivative security at time one if the first coin toss results in T. The solution to these six equations provides both the value of the derivative security at all times and the hedge for the short position at all times, regardless of the outcome of the first coin toss.

In Theorem 1.2.2 of Volume I, we discovered a clever way to solve these six equations in six unknowns by first solving for the derivative security values V_n using the risk-neutral probabilities $\tilde{p}$ and $\tilde{q}$ in (1.2.16) and then computing the hedge positions from (1.2.17). Equation (1.2.16) says that under the risk-neutral probabilities, the discounted derivative security value is a martingale.

In Section 4.5 of this volume, we repeated the first part of this program. To determine the value of a European call, we determined the initial capital required to set up a portfolio that with probability one hedges a short position in the derivative security. Subsection 4.5.3, in which we equated the evolution of the discounted portfolio value with the evolution of the discounted option value, provides the continuous-time analogue of solving the six equations (1.2.2), (1.2.3), and (1.2.5)–(1.2.8) of Volume I. From that process, we obtained the delta-hedging rule (4.5.11) and we obtained the Black-Scholes-Merton partial differential equation (4.5.14) for the value of the call.

Now we execute the second part of the program. In this chapter, we discover a clever way to solve the partial differential equation (4.5.14) using a risk-neutral probability measure. After solving this equation, we can then compute the short option hedge using (4.5.11).

To accomplish this second part of the program, we show in Section 5.2 how to construct the risk-neutral measure in a model with a single underlying security. This step relies on Girsanov's Theorem, which is presented in Section 5.2. Risk-neutral pricing is a powerful method for computing prices of derivative securities, but it is fully justified only when it is accompanied by a hedge for a short position in the security being priced. In Section 5.3, we provide the conditions under which such a hedge exists in a model with a single underlying security. Section 5.4 generalizes the ideas of Sections 5.2 and 5.3 to models with multiple underlying securities. Furthermore, Section 5.4 provides conditions that guarantee that such a model does not admit arbitrage and that every derivative security in the model can be hedged.

5.2 Risk-Neutral Measure

5.2.1 Girsanov's Theorem for a Single Brownian Motion

In Theorem 1.6.1, we began with a probability space $(\Omega, \mathcal{F}, \mathbb{P})$ and a nonnegative random variable Z satisfying $\mathbb{E}Z = 1$. We then defined a new probability measure $\widetilde{\mathbb{P}}$ by the formula

$$\widetilde{\mathbb{P}}(A) = \int_A Z(\omega)\, dP(\omega) \text{ for all } A \in \mathcal{F}. \tag{5.2.1}$$

Any random variable X now has two expectations, one under the original probability measure $\mathbb{P}$, which we denote $\mathbb{E}X$, and the other under the new probability measure $\widetilde{\mathbb{P}}$, which we denote $\widetilde{\mathbb{E}}X$. These are related by the formula

$$\widetilde{\mathbb{E}}X = \mathbb{E}[XZ]. \tag{5.2.2}$$

If $\mathbb{P}\{Z > 0\} = 1$, then $\mathbb{P}$ and $\widetilde{\mathbb{P}}$ agree which sets have probability zero and (5.2.2) has the companion formula

$$\mathbb{E}X = \widetilde{\mathbb{E}}\left[\frac{X}{Z}\right]. \tag{5.2.3}$$

We say Z is the *Radon-Nikodým derivative* of $\widetilde{\mathbb{P}}$ with respect to $\mathbb{P}$, and we write

$$Z = \frac{d\widetilde{\mathbb{P}}}{d\mathbb{P}}.$$

This is supposed to remind us that Z is like a ratio of these two probability measures. The reader may wish to review Section 3.1 of Volume I, where this concept is discussed in a finite probability model. In the case of a finite probability model, we actually have

$$Z(\omega) = \frac{\widetilde{\mathbb{P}}(\omega)}{\mathbb{P}(\omega)}. \tag{5.2.4}$$

If we multiply both sides of (5.2.4) by $\mathbb{P}(\omega)$ and then sum over ω in a set A, we obtain

$$\widetilde{\mathbb{P}}(A) = \sum_{\omega \in A} Z(\omega)P(\omega) \text{ for all } A \subset \Omega. \tag{5.2.5}$$

In a general probability model, we cannot write (5.2.4) because $\mathbb{P}(\omega)$ is typically zero for each individual ω, but we can write an analogue of (5.2.5). This is (5.2.1).

Example 1.6.6 shows how we can use this change-of-measure idea to move the mean of a normal random variable. In particular, if X is a standard normal random variable on a probability space $(\Omega, \mathcal{F}, \mathbb{P})$, θ is a constant, and we define

$$Z = \exp\left\{-\theta X - \frac{1}{2}\theta^2\right\},$$

then under the probability measure $\widetilde{\mathbb{P}}$ given by (5.2.1), the random variable $Y = X + \theta$ is standard normal. In particular, $\widetilde{\mathbb{E}}Y = 0$, whereas $\mathbb{E}Y = \mathbb{E}X + \theta = \theta$. By changing the probability measure, we have changed the expectation of Y.

In this section, we perform a similar change of measure in order to change a mean, but this time for a whole process rather than for a single random variable. To set the stage, suppose we have a probability space $(\Omega, \mathcal{F}, \mathbb{P})$ and a filtration $\mathcal{F}(t)$, defined for $0 \le t \le T$, where T is a fixed final time. Suppose further that Z is an almost surely positive random variable satisfying $\mathbb{E}Z = 1$, and we define $\widetilde{\mathbb{P}}$ by (5.2.1). We can then define the *Radon-Nikodým derivative process*

$$Z(t) = \mathbb{E}[Z|\mathcal{F}(t)], \quad 0 \le t \le T. \tag{5.2.6}$$

This process in discrete time is discussed in Section 3.2 of Volume I. The Radon-Nikodým derivative process (5.2.6) is a martingale because of iterated conditioning (Theorem 2.3.2(iii)): for $0 \le s \le t \le T$,

$$\mathbb{E}[Z(t)|\mathcal{F}(s)] = \mathbb{E}\big[\mathbb{E}[Z|\mathcal{F}(t)]\big|\mathcal{F}(s)\big] = \mathbb{E}[Z|\mathcal{F}(s)] = Z(s). \tag{5.2.7}$$

Furthermore, it has the properties presented in the following two lemmas, which are continuous-time analogues of Lemmas 3.2.5 and 3.2.6 of Volume I.

Lemma 5.2.1. *Let t satisfying $0 \le t \le T$ be given and let Y be an $\mathcal{F}(t)$-measurable random variable. Then*

$$\widetilde{\mathbb{E}}Y = \mathbb{E}[YZ(t)]. \tag{5.2.8}$$

PROOF: We use (5.2.2), the unbiasedness of conditional expectations (2.3.25), the property "taking out what is known" (Theorem 2.3.2(ii)), and the definition of $Z(t)$ to write

$$\widetilde{\mathbb{E}}Y = \mathbb{E}[YZ] = \mathbb{E}\big[\mathbb{E}[YZ|\mathcal{F}(t)]\big] = \mathbb{E}\big[Y\mathbb{E}[Z|\mathcal{F}(t)]\big] = \mathbb{E}[YZ(t)]. \qquad \square$$

Lemma 5.2.2. *Let s and t satisfying $0 \le s \le t \le T$ be given and let Y be an $\mathcal{F}(t)$-measurable random variable. Then*

$$\widetilde{\mathbb{E}}[Y|\mathcal{F}(s)] = \frac{1}{Z(s)}\mathbb{E}[YZ(t)|\mathcal{F}(s)]. \tag{5.2.9}$$

PROOF: It is clear that $\frac{1}{Z(s)}\mathbb{E}[YZ(t)|\mathcal{F}(s)]$ is $\mathcal{F}(s)$-measurable. We must check the partial-averaging property (Definition 2.3.1(ii)), which in this case is

$$\int_A \frac{1}{Z(s)}\mathbb{E}[YZ(t)|\mathcal{F}(s)]\,d\widetilde{\mathbb{P}} = \int_A Y\,d\widetilde{\mathbb{P}} \text{ for all } A \in \mathcal{F}(s). \tag{5.2.10}$$

Note that because we are claiming that the right-hand side of (5.2.9) is the conditional expectation of Y under the $\widetilde{\mathbb{P}}$ probability measure, we must integrate with respect to the measure $\widetilde{\mathbb{P}}$ in the statement of the partial-averaging property (5.2.10). We may write the left-hand side of (5.2.10) as

$$\widetilde{\mathbb{E}}\left[\mathbb{I}_A \frac{1}{Z(s)}\mathbb{E}[YZ(t)|\mathcal{F}(s)]\right]$$

and then use Lemma 5.2.1 for $\mathcal{F}(s)$-measurable random variables, use "taking out what is known," use the unbiasedness of conditional expectations (2.3.25), and finally use Lemma 5.2.1 for $\mathcal{F}(t)$-measurable random variables to write

$$\begin{aligned}
\widetilde{\mathbb{E}}\left[\mathbb{I}_A \frac{1}{Z(s)}\mathbb{E}[YZ(t)|\mathcal{F}(s)]\right] &= \mathbb{E}\big[\mathbb{I}_A\mathbb{E}[YZ(t)|\mathcal{F}(s)]\big]\\
&= \mathbb{E}\big[\mathbb{E}[\mathbb{I}_AYZ(t)|\mathcal{F}(s)]\big]\\
&= \mathbb{E}[\mathbb{I}_AYZ(t)]\\
&= \widetilde{\mathbb{E}}[\mathbb{I}_AY]\\
&= \int_A Y\,d\widetilde{\mathbb{P}}.
\end{aligned}$$

This verifies (5.2.10), which in turn proves (5.2.9). $\square$

Theorem 5.2.3 (Girsanov, one dimension). *Let $W(t)$, $0 \le t \le T$, be a Brownian motion on a probability space $(\Omega, \mathcal{F}, \mathbb{P})$, and let $\mathcal{F}(t)$, $0 \le t \le T$, be a filtration for this Brownian motion. Let $\Theta(t)$, $0 \le t \le T$, be an adapted process. Define*

$$Z(t) = \exp\left\{-\int_0^t \Theta(u)\,dW(u) - \frac{1}{2}\int_0^t \Theta^2(u)\,du\right\}, \tag{5.2.11}$$

$$\widetilde{W}(t) = W(t) + \int_0^t \Theta(u)\,du, \tag{5.2.12}$$

and assume that[1]

$$\mathbb{E}\int_0^T \Theta^2(u)Z^2(u)\,du < \infty. \tag{5.2.13}$$

Set $Z = Z(T)$. Then $\mathbb{E}Z = 1$ and under the probability measure $\widetilde{\mathbb{P}}$ given by (5.2.1), the process $\widetilde{W}(t)$, $0 \le t \le T$, is a Brownian motion.

PROOF: We use Lévy's Theorem, Theorem 4.6.4, which says that a martingale starting at zero at time zero, with continuous paths and with quadratic variation equal to t at each time t, is a Brownian motion. The process $\widetilde{W}$ starts at zero at time zero and is continuous. Furthermore, $[\widetilde{W},\widetilde{W}](t) = [W,W](t) = t$ because the term $\int_0^t \Theta(u)\,du$ in the definition of $\widetilde{W}(t)$ contributes zero quadratic variation. In other words,

$$d\widetilde{W}(t)\,d\widetilde{W}(t) = \big(dW(t) + \Theta(t)\,dt\big)^2 = dW(t)\,dW(t) = dt.$$

It remains to show that $\widetilde{W}(t)$ is a martingale under $\widetilde{\mathbb{P}}$. We first observe that $Z(t)$ is a martingale under $\mathbb{P}$. With

$$X(t) = -\int_0^t \Theta(u)\,dW(u) - \frac{1}{2}\int_0^t \Theta^2(u)\,du$$

and $f(x) = e^x$ so that $f'(x) = e^x$ and $f''(x) = e^x$, we have

$$\begin{aligned}
dZ(t) &= df\big(X(t)\big)\\
&= f'\big(X(t)\big)\,dX(t) + \frac{1}{2}f''\big(X(t)\big)\,dX(t)\,dX(t)\\
&= e^{X(t)}\Big(-\Theta(t)\,dW(t) - \frac{1}{2}\Theta^2(t)\,dt\Big) + \frac{1}{2}e^{X(t)}\Theta^2(t)\,dt\\
&= -\Theta(t)Z(t)\,dW(t).
\end{aligned}$$

Integrating both sides of the equation above, we see that

$$Z(t) = Z(0) - \int_0^t \Theta(u)Z(u)\,dW(u). \tag{5.2.14}$$

Because Itô integrals are martingales, $Z(t)$ is a martingale. In particular, $\mathbb{E}Z = \mathbb{E}Z(T) = Z(0) = 1$.

Because $Z(t)$ is a martingale and $Z = Z(T)$, we have

$$Z(t) = \mathbb{E}[Z(T)|\mathcal{F}(t)] = \mathbb{E}[Z|\mathcal{F}(t)], \quad 0 \le t \le T.$$

This shows that $Z(t)$, $0 \le t \le T$, is a Radon-Nikodým derivative process as defined in (5.2.6), and Lemmas 5.2.1 and 5.2.2 apply to this situation.

[1] Condition (5.2.13) is imposed to ensure that the Itô integral in (5.2.14) is defined and is a martingale. This is (4.3.1) imposed in the construction of Itô integrals.

We next show that $\widetilde{W}(t)Z(t)$ is a martingale under $\mathbb{P}$. To see this, we compute the differential using Itô's product rule (Corollary 4.6.3):

$$\begin{aligned} d\big(\widetilde{W}(t)Z(t)\big) &= \widetilde{W}(t)\,dZ(t) + Z(t)\,d\widetilde{W}(t) + d\widetilde{W}(t)\,dZ(t) \\ &= -\widetilde{W}(t)\Theta(t)Z(t)\,dW(t) + Z(t)\,dW(t) + Z(t)\,\Theta(t)\,dt \\ &\qquad + \big(dW(t) + \Theta(t)\,dt\big)\big(-\Theta(t)Z(t)\,dW(t)\big) \\ &= \big(-\widetilde{W}(t)\Theta(t) + 1\big)Z(t)\,dW(t). \end{aligned}$$

Because the final expression has no dt term, the process $\widetilde{W}(t)Z(t)$ is a martingale under $\mathbb{P}$.

Now let $0 \le s \le t \le T$ be given. Lemma 5.2.2 and the martingale property for $\widetilde{W}(t)Z(t)$ under $\mathbb{P}$ imply

$$\widetilde{\mathbb{E}}[\widetilde{W}(t)|\mathcal{F}(s)] = \frac{1}{Z(s)}\mathbb{E}[\widetilde{W}(t)Z(t)|\mathcal{F}(s)] = \frac{1}{Z(s)}\widetilde{W}(s)Z(s) = \widetilde{W}(s).$$

This shows that $\widetilde{W}(t)$ is a martingale under $\widetilde{\mathbb{P}}$. The proof is complete. □

The probability measures $\mathbb{P}$ and $\widetilde{\mathbb{P}}$ in Girsanov's Theorem are equivalent (i.e., they agree about which sets have probability zero and hence about which sets have probability one). This is because $\mathbb{P}\{Z > 0\} = 1$; see Definition 1.6.3 and the discussion following it. In the remainder of this section, we set up an asset price model in which $\mathbb{P}$ is the actual probability measure and $\widetilde{\mathbb{P}}$ is the risk-neutral measure. We want these probabilities to agree about what is possible and what is impossible, and they do. In the discrete-time binomial model of Volume I, the actual and risk-neutral probability measures agree about which moves are possible (i.e., they both give positive probability to an up move, positive probability to a down move, and the sizes (but not the probabilities) of the up and down moves are the same whether we are working under the actual probability measure or the risk-neutral probability measure). The set of possible asset price paths is a tree in the binomial model, and both the actual probability measure and the risk-neutral probability measure are based on the same tree. In the continuous-time model, there are infinitely many possible paths, and this agreement about what is possible and what is not possible is the equivalence of Definition 1.6.3.

5.2.2 Stock Under the Risk-Neutral Measure

Let $W(t)$, $0 \le t \le T$, be a Brownian motion on a probability space $(\Omega, \mathcal{F}, \mathbb{P})$, and let $\mathcal{F}(t)$, $0 \le t \le T$, be a filtration for this Brownian motion. Here T is a fixed final time. Consider a stock price process whose differential is

$$dS(t) = \alpha(t)S(t)\,dt + \sigma(t)S(t)\,dW(t), \quad 0 \le t \le T. \tag{5.2.15}$$

The mean rate of return $\alpha(t)$ and the volatility $\sigma(t)$ are allowed to be adapted processes. We assume that, for all $t \in [0, T]$, $\sigma(t)$ is almost surely not zero.

This stock price is a generalized geometric Brownian motion (see Example 4.4.8, in particular, (4.4.27)), and an equivalent way of writing (5.2.15) is (see (4.4.26))

$$S(t) = S(0)\exp\left\{\int_0^t \sigma(s)\,dW(s) + \int_0^t \left(\alpha(s) - \frac{1}{2}\sigma^2(s)\right)ds\right\}. \tag{5.2.16}$$

In addition, suppose we have an adapted interest rate process $R(t)$. We define the *discount process*

$$D(t) = e^{-\int_0^t R(s)\,ds} \tag{5.2.17}$$

and note that

$$dD(t) = -R(t)D(t)\,dt. \tag{5.2.18}$$

To obtain (5.2.18) from (5.2.17), we can define $I(t) = \int_0^t R(s)\,ds$ so that $dI(t) = R(t)\,dt$ and $dI(t)\,dI(t) = 0$. We introduce the function $f(x) = e^{-x}$, for which $f'(x) = -f(x)$, $f''(x) = f(x)$, and then use the Itô-Doeblin formula to write

$$\begin{aligned}
dD(t) &= df\big(I(t)\big)\\
&= f'\big(I(t)\big)\,dI(t) + \frac{1}{2}f''\big(I(t)\big)\,dI(t)\,dI(t)\\
&= -f\big(I(t)\big)R(t)\,dt\\
&= -R(t)D(t)\,dt.
\end{aligned}$$

Observe that although $D(t)$ is random, it has zero quadratic variation. This is because it is "smooth." It has a derivative, namely $D'(t) = -R(t)D(t)$, and one does not need stochastic calculus to do this computation. The stock price $S(t)$ is random and has nonzero quadratic variation. It is "more random" than $D(t)$. If we invest in the stock, we have no way of knowing whether the next move of the driving Brownian motion will be up or down, and this move directly affects the stock price. Hence, we face a high degree of uncertainty. On the other hand, consider a money market account with variable interest rate $R(t)$, where money is rolled over at this interest rate. If the price of a share of this money market account at time zero is 1, then the price of a share of this money market account at time t is $e^{\int_0^t R(s)\,ds} = \frac{1}{D(t)}$. If we invest in this account, we know the interest rate at the time of the investment and hence have a high degree of certainty about what the return will be over a short period of time. Over longer periods, we are less certain because the interest rate is variable, and at the time of investment, we do not know the future interest rates that will be applied. However, the randomness in the model affects the money market account only indirectly by affecting the interest rate. Changes in the interest rate do not affect the money market account instantaneously but only when they act over time. (Warning: The money market account is not a bond. For a bond, a change in the interest rate can have an instantaneous effect on price.) Unlike the price of the money market account, the stock price is susceptible to instantaneous unpredictable changes and is, in this sense, "more random" than $D(t)$. Our mathematical model captures this effect because $S(t)$ has nonzero quadratic variation, while $D(t)$ has zero quadratic variation.

The discounted stock price process is

$$D(t)S(t) = S(0)\exp\left\{\int_0^t \sigma(s)dW(s) + \int_0^t \left(\alpha(s) - R(s) - \frac{1}{2}\sigma^2(s)\right) ds\right\}, \tag{5.2.19}$$

and its differential is

$$\begin{aligned} d\big(D(t)S(t)\big) &= \big(\alpha(t) - R(t)\big)D(t)S(t)\,dt + \sigma(t)D(t)S(t)\,dW(t) \\ &= \sigma(t)D(t)S(t)\big[\Theta(t)\,dt + dW(t)\big], \end{aligned} \tag{5.2.20}$$

where we define the *market price of risk* to be

$$\Theta(t) = \frac{\alpha(t) - R(t)}{\sigma(t)}. \tag{5.2.21}$$

One can derive (5.2.20) either by applying the Itô-Doeblin formula to the right-hand side of (5.2.19) or by using Itô's product rule and the formulas (5.2.15) and (5.2.18). The first line of (5.2.20), compared with (5.2.15), shows that the mean rate of return of the discounted stock price is $\alpha(t) - R(t)$, which is the mean rate $\alpha(t)$ of the undiscounted stock price, reduced by the interest rate $R(t)$. The volatility of the discounted stock price is the same as the volatility of the undiscounted stock price.

We introduce the probability measure $\widetilde{\mathbb{P}}$ defined in Girsanov's Theorem, Theorem 5.2.3, which uses the market price of risk $\Theta(t)$ given by (5.2.21). In terms of the Brownian motion $\widetilde{W}(t)$ of that theorem, we may rewrite (5.2.20) as

$$d\big(D(t)S(t)\big) = \sigma(t)D(t)S(t)\,d\widetilde{W}(t). \tag{5.2.22}$$

We call $\widetilde{\mathbb{P}}$, the measure defined in Girsanov's Theorem, the *risk-neutral measure* because it is equivalent to the original measure $\mathbb{P}$ and it renders the discounted stock price $D(t)S(t)$ into a martingale. Indeed, according to (5.2.22),

$$D(t)S(t) = S(0) + \int_0^t \sigma(u)D(u)S(u)\,d\widetilde{W}(u),$$

and under $\widetilde{\mathbb{P}}$ the process $\int_0^t \sigma(u)D(u)S(u)\,d\widetilde{W}(u)$ is an Itô integral and hence a martingale.

The undiscounted stock price $S(t)$ has mean rate of return equal to the interest rate under $\widetilde{\mathbb{P}}$, as one can verify by making the replacement $dW(t) = -\Theta(t)\,dt + d\widetilde{W}(t)$ in (5.2.15). With this substitution, (5.2.15) becomes

$$dS(t) = R(t)S(t)\,dt + \sigma(t)S(t)\,d\widetilde{W}(t). \tag{5.2.23}$$

We can either solve this equation for $S(t)$ or simply replace the Itô integral $\int_0^t \sigma(s)\,dW(s)$ by its equivalent $\int_0^t \sigma(s)\,d\widetilde{W}(s) - \int_0^t \big(\alpha(s) - R(s)\big)\,ds$ in (5.2.16) to obtain the formula

$$S(t) = S(0)\exp\left\{\int_0^t \sigma(s)\,d\widetilde{W}(s) + \int_0^t \left(R(s) - \frac{1}{2}\sigma^2(s)\right) ds\right\}. \qquad (5.2.24)$$

In discrete time, the change of measure does not change the binomial tree, only the probabilities on the branches of the tree. In continuous time, the change from the actual measure $\mathbb{P}$ to the risk-neutral measure $\widetilde{\mathbb{P}}$ changes the mean rate of return of the stock but not the volatility. The volatility tells us which stock price paths are possible—namely those for which the log of the stock price accumulates quadratic variation at rate $\sigma^2(t)$ per unit time. After the change of measure, we are still considering the same set of stock price paths, but we have shifted the probability on them. If $\alpha(t) > R(t)$, as it normally is, then the change of measure puts more probability on the paths with lower return so that the overall mean rate of return is reduced from $\alpha(t)$ to $R(t)$.

5.2.3 Value of Portfolio Process Under the Risk-Neutral Measure

Consider an agent who begins with initial capital $X(0)$ and at each time t, $0 \leq t \leq T$, holds $\Delta(t)$ shares of stock, investing or borrowing at the interest rate $R(t)$ as necessary to finance this. The differential of this agent's portfolio value is given by the analogue of (4.5.2) for this case of random $\alpha(t)$, $\sigma(t)$, and $R(t)$, and this works out to be

$$\begin{aligned}
dX(t) &= \Delta(t)\,dS(t) + R(t)\big(X(t) - \Delta(t)S(t)\big)\,dt \\
&= \Delta(t)\big(\alpha(t)S(t)\,dt + \sigma(t)S(t)\,dW(t)\big) + R(t)\big(X(t) - \Delta(t)S(t)\big)dt \\
&= R(t)X(t)\,dt + \Delta(t)\big(\alpha(t) - R(t)\big)S(t)\,dt + \Delta(t)\sigma(t)S(t)\,dW(t) \\
&= R(t)X(t)\,dt + \Delta(t)\sigma(t)S(t)\big[\Theta(t)\,dt + dW(t)\big]. \qquad (5.2.25)
\end{aligned}$$

Itô's product rule, (5.2.18), and (5.2.20) imply

$$\begin{aligned}
d\big(D(t)X(t)\big) &= \Delta(t)\sigma(t)D(t)S(t)\big[\Theta(t)\,dt + dW(t)\big] \\
&= \Delta(t)\,d\big(D(t)S(t)\big). \qquad (5.2.26)
\end{aligned}$$

Changes in the discounted value of an agent's portfolio are entirely due to fluctuations in the discounted stock price. We may use (5.2.22) to rewrite (5.2.26) as

$$d\big(D(t)X(t)\big) = \Delta(t)\sigma(t)D(t)S(t)\,d\widetilde{W}(t). \qquad (5.2.27)$$

Our agent has two investment options: (1) a money market account with rate of return $R(t)$, and (2) a stock with mean rate of return $R(t)$ under $\widetilde{\mathbb{P}}$. Regardless of how the agent invests, the mean rate of return for his portfolio will be $R(t)$ under $\widetilde{\mathbb{P}}$, and hence the discounted value of his portfolio, $D(t)X(t)$, will be a martingale. This is the content of (5.2.27).

5.2.4 Pricing Under the Risk-Neutral Measure

In Section 4.5, we derived the Black-Scholes-Merton equation for the value of a European call by asking what initial capital $X(0)$ and portfolio process $\Delta(t)$ an agent would need in order to hedge a short position in the call (i.e., in order to have $X(T) = (S(T) - K)^+$ almost surely). In this section, we generalize the question. Let $V(T)$ be an $\mathcal{F}(T)$-measurable random variable. This represents the payoff at time T of a derivative security. We allow this payoff to be path-dependent (i.e., to depend on anything that occurs between times 0 and T), which is what $\mathcal{F}(T)$-measurability means. We wish to know what initial capital $X(0)$ and portfolio process $\Delta(t)$, $0 \le t \le T$, an agent would need in order to hedge a short position in this derivative security, i.e., in order to have

$$X(T) = V(T) \text{ almost surely.} \tag{5.2.28}$$

In Section 4.5, the mean rate of return, volatility, and interest rate were constant. In this section, we do not assume a constant mean rate of return, volatility, and interest rate.

Our agent wishes to choose initial capital $X(0)$ and portfolio strategy $\Delta(t)$, $0 \le t \le T$, such that (5.2.28) holds. We shall see in the next section that this can be done. Once it has been done, the fact that $D(t)X(t)$ is a martingale under $\widetilde{\mathbb{P}}$ implies

$$D(t)X(t) = \widetilde{\mathbb{E}}\big[D(T)X(T)|\mathcal{F}(t)\big] = \widetilde{\mathbb{E}}\big[D(T)V(T)|\mathcal{F}(t)\big]. \tag{5.2.29}$$

The value $X(t)$ of the hedging portfolio in (5.2.29) is the capital needed at time t in order to successfully complete the hedge of the short position in the derivative security with payoff $V(T)$. Hence, we can call this the *price* $V(t)$ of the derivative security at time t, and (5.2.29) becomes

$$D(t)V(t) = \widetilde{\mathbb{E}}\big[D(T)V(T)|\mathcal{F}(t)\big],\ 0 \le t \le T. \tag{5.2.30}$$

This is the continuous-time analogue of the risk-neutral pricing formula (2.4.10) in the binomial model of Volume I. Dividing (5.2.30) by $D(t)$, which is $\mathcal{F}(t)$-measurable and hence can be moved inside the conditional expectation on the right-hand side of (5.2.30), and recalling the definition of $D(t)$, we may write (5.2.30) as

$$V(t) = \widetilde{\mathbb{E}}\left[e^{-\int_t^T R(u)\,du} V(T)\Big|\,\mathcal{F}(t)\right],\ 0 \le t \le T. \tag{5.2.31}$$

This is the continuous-time analogue of (2.4.11) of Volume I. We shall refer to both (5.2.30) and (5.2.31) as the *risk-neutral pricing formula* for the continuous-time model.

5.2.5 Deriving the Black-Scholes-Merton Formula

The addition of Merton's name to what has traditionally been called the Black-Scholes formula is explained in the Notes to Chapter 4, Section 4.9.

To obtain the Black-Scholes-Merton price of a European call, we assume a constant volatility σ, constant interest rate r, and take the derivative security payoff to be $V(T) = \big(S(T) - K\big)^+$. The right-hand side of (5.2.31) becomes

$$\widetilde{\mathbb{E}}\left[e^{-r(T-t)}\big(S(T)-K\big)^+\Big|\mathcal{F}(t)\right].$$

Because geometric Brownian motion is a Markov process, this expression depends on the stock price $S(t)$ and of course on the time t at which the conditional expectation is computed, but not on the stock price prior to time t. In other words, there is a function $c(t,x)$ such that

$$c\big(t,S(t)\big) = \widetilde{\mathbb{E}}\left[e^{-r(T-t)}\big(S(T)-K\big)^+\Big|\mathcal{F}(t)\right]. \tag{5.2.32}$$

We can compute $c(t,x)$ using the Independence Lemma, Lemma 2.3.4. With constant σ and r, equation (5.2.24) becomes

$$S(t) = S(0)\exp\left\{\sigma\widetilde{W}(t) + \left(r - \frac{1}{2}\sigma^2\right)t\right\},$$

and we may thus write

$$\begin{aligned} S(T) &= S(t)\exp\left\{\sigma\big(\widetilde{W}(T)-\widetilde{W}(t)\big) + \left(r-\frac{1}{2}\sigma^2\right)\tau\right\} \\ &= S(t)\exp\left\{-\sigma\sqrt{\tau}\,Y + \left(r-\frac{1}{2}\sigma^2\right)\tau\right\}, \end{aligned}$$

where Y is the standard normal random variable

$$Y = -\frac{\widetilde{W}(T)-\widetilde{W}(t)}{\sqrt{T-t}},$$

and τ is the "time to expiration" $T-t$. We see that $S(T)$ is the product of the $\mathcal{F}(t)$-measurable random variable $S(t)$ and the random variable

$$\exp\left\{-\sigma\sqrt{\tau}\,Y + \left(r-\frac{1}{2}\sigma^2\right)\tau\right\},$$

which is independent of $\mathcal{F}(t)$. Therefore, (5.2.32) holds with

$$\begin{aligned} c(t,x) &= \widetilde{\mathbb{E}}\left[e^{-r\tau}\left(x\exp\left\{-\sigma\sqrt{\tau}\,Y + \left(r-\frac{1}{2}\sigma^2\right)\tau\right\} - K\right)^+\right] \\ &= \frac{1}{\sqrt{2\pi}}\int_{-\infty}^{\infty} e^{-r\tau}\left(x\exp\left\{-\sigma\sqrt{\tau}\,y + \left(r-\frac{1}{2}\sigma^2\right)\tau\right\} - K\right)^+ e^{-\frac{1}{2}y^2}\,dy. \end{aligned}$$

The integrand

$$\left(x\exp\left\{-\sigma\sqrt{\tau}\,y+\left(r-\frac{1}{2}\sigma^2\right)\tau\right\}-K\right)^+$$

is positive if and only if

$$y<d_-(\tau,x)=\frac{1}{\sigma\sqrt{\tau}}\left[\log\frac{x}{K}+\left(r-\frac{1}{2}\sigma^2\right)\tau\right]. \tag{5.2.33}$$

Therefore,

$$\begin{aligned}
&c(t,x)\\
&=\frac{1}{\sqrt{2\pi}}\int_{-\infty}^{d_-(\tau,x)}e^{-r\tau}\left(x\exp\left\{-\sigma\sqrt{\tau}\,y+\left(r-\frac{1}{2}\sigma^2\right)\tau\right\}-K\right)e^{-\frac{1}{2}y^2}\,dy\\
&=\frac{1}{\sqrt{2\pi}}\int_{-\infty}^{d_-(\tau,x)}x\exp\left\{-\frac{y^2}{2}-\sigma\sqrt{\tau}\,y-\frac{\sigma^2\tau}{2}\right\}dy\\
&\qquad-\frac{1}{\sqrt{2\pi}}\int_{-\infty}^{d_-(\tau,x)}e^{-r\tau}Ke^{-\frac{1}{2}y^2}\,dy\\
&=\frac{x}{\sqrt{2\pi}}\int_{-\infty}^{d_-(\tau,x)}\exp\left\{-\frac{1}{2}\left(y+\sigma\sqrt{\tau}\right)^2\right\}dy-e^{-r\tau}KN\big(d_-(\tau,x)\big)\\
&=\frac{x}{\sqrt{2\pi}}\int_{-\infty}^{d_-(\tau,x)+\sigma\sqrt{\tau}}\exp\left\{-\frac{z^2}{2}\right\}dz-e^{-r\tau}KN\big(d_-(\tau,x)\big)\\
&=xN\big(d_+(\tau,x)\big)-e^{-r\tau}KN\big(d_-(\tau,x)\big),
\end{aligned}$$

where

$$d_+(\tau,x)=d_-(\tau,x)+\sigma\sqrt{\tau}=\frac{1}{\sigma\sqrt{\tau}}\left[\log\frac{x}{K}+\left(r+\frac{1}{2}\sigma^2\right)\tau\right]. \tag{5.2.34}$$

For future reference, we introduce the notation

$$\mathrm{BSM}(\tau,x;K,r,\sigma)=\widetilde{\mathbb{E}}\left[e^{-r\tau}\left(x\exp\left\{-\sigma\sqrt{\tau}\,Y+\left(r-\frac{1}{2}\sigma^2\right)\tau\right\}-K\right)^+\right], \tag{5.2.35}$$

where Y is a standard normal random variable under $\widetilde{\mathbb{P}}$. We have just shown that

$$\mathrm{BSM}(\tau,x;K,r,\sigma)=xN\big(d_+(\tau,x)\big)-e^{-r\tau}KN\big(d_-(\tau,x)\big). \tag{5.2.36}$$

In Section 4.5, we derived the Black-Scholes-Merton partial differential equation (4.5.14) and then provided the solution in equation (4.5.19) without explaining how one obtains this solution (although one can verify after the fact that (4.5.19) does indeed solve (4.5.14); see Exercise 4.9 of Chapter 4). Here we have derived the solution by the device of switching to the risk-neutral measure.

5.3 Martingale Representation Theorem

The risk-neutral pricing formula for the price (value) at time t of a derivative security paying $V(T)$ at time T, equation (5.2.31), was derived under the assumption that if an agent begins with the correct initial capital, there is a portfolio process $\Delta(t)$, $0 \leq t \leq T$, such that the agent's portfolio value at the final time T will be $V(T)$ almost surely. Under this assumption, we determined the "correct initial capital" to be (set $t = 0$ in (5.2.31))

$$V(0) = \widetilde{\mathbb{E}}\big[D(T)V(T)\big],$$

and the value of the hedging portfolio at every time t, $0 \leq t \leq T$, to be $V(t)$ given by (5.2.31). In this section, in the model with one stock driven by one Brownian motion, we verify the assumption on which the risk-neutral pricing formula (5.2.31) is based. We take up the case of multiple Brownian motions and multiple stocks in Section 5.4.

5.3.1 Martingale Representation with One Brownian Motion

The existence of a hedging portfolio in the model with one stock and one Brownian motion depends on the following theorem, which we state without proof.

Theorem 5.3.1 (Martingale representation, one dimension). *Let $W(t)$, $0 \leq t \leq T$, be a Brownian motion on a probability space $(\Omega, \mathcal{F}, \mathbb{P})$, and let $\mathcal{F}(t)$, $0 \leq t \leq T$, be the filtration generated by this Brownian motion. Let $M(t)$, $0 \leq t \leq T$, be a martingale with respect to this filtration (i.e., for every t, $M(t)$ is $\mathcal{F}(t)$-measurable and for $0 \leq s \leq t \leq T$, $\mathbb{E}[M(t)|\mathcal{F}(s)] = M(s)$). Then there is an adapted process $\Gamma(u)$, $0 \leq u \leq T$, such that*

$$M(t) = M(0) + \int_0^t \Gamma(u)\, dW(u), \ 0 \leq t \leq T. \tag{5.3.1}$$

The Martingale Representation Theorem asserts that when the filtration is the one generated by a Brownian motion (i.e., the only information in $\mathcal{F}(t)$ is that gained from observing the Brownian motion up to time t), then every martingale with respect to this filtration is an initial condition plus an Itô integral with respect to the Brownian motion. The relevance to hedging of this is that the *only* source of uncertainty in the model is the Brownian motion appearing in Theorem 5.3.1, and hence there is only one source of uncertainty to be removed by hedging. This assumption implies that the martingale cannot have jumps because Itô integrals are continuous. If we want to have a martingale with jumps, we will need to build a model that includes sources of uncertainty different from (or in addition to) Brownian motion.

The assumption that the filtration in Theorem 5.3.1 is the one generated by the Brownian motion is more restrictive than the assumption of Girsanov's

Theorem, Theorem 5.2.3, in which the filtration can be larger than the one generated by the Brownian motion. If we include this extra restriction in Girsanov's Theorem, then we obtain the following corollary. The first paragraph of this corollary is just a repeat of Girsanov's Theorem; the second part contains the new assertion.

Corollary 5.3.2. *Let $W(t)$, $0 \le t \le T$, be a Brownian motion on a probability space $(\Omega, \mathcal{F}, \mathbb{P})$, and let $\mathcal{F}(t)$, $0 \le t \le T$, be the filtration generated by this Brownian motion. Let $\Theta(t)$, $0 \le t \le T$, be an adapted process, define*

$$Z(t) = \exp\left\{ -\int_0^t \Theta(u)\, dW(u) - \frac{1}{2}\int_0^t \Theta^2(u)\, du \right\},$$

$$\widetilde{W}(t) = W(t) + \int_0^t \Theta(u)\, du,$$

and assume that $\widetilde{\mathbb{E}} \int_0^T \Theta^2(u) Z^2(u)\, du < \infty$. Set $Z = Z(T)$. Then $\mathbb{E}Z = 1$, and under the probability measure $\widetilde{\mathbb{P}}$ given by (5.2.1), the process $\widetilde{W}(t)$, $0 \le t \le T$, is a Brownian motion.

Now let $\widetilde{M}(t)$, $0 \le t \le T$, be a martingale under $\widetilde{\mathbb{P}}$. Then there is an adapted process $\widetilde{\Gamma}(u)$, $0 \le u \le T$, such that

$$\widetilde{M}(t) = \widetilde{M}(0) + \int_0^t \widetilde{\Gamma}(u)\, d\widetilde{W}(u),\ 0 \le t \le T. \tag{5.3.2}$$

Corollary 5.3.2 is not a trivial consequence of the Martingale Representation Theorem, Theorem 5.3.1, with $\widetilde{W}(t)$ replacing $W(t)$ because the filtration $\mathcal{F}(t)$ in this corollary is generated by the process $W(t)$, not the $\widetilde{\mathbb{P}}$-Brownian motion $\widetilde{W}(t)$. However, the proof is not difficult and is left to the reader as Exercise 5.5.

5.3.2 Hedging with One Stock

We now return to the hedging problem. We begin with the model of Subsection 5.2.2, which has the stock price process (5.2.15) and an interest rate process $R(t)$ that generates the discount process (5.2.17). Recall the assumption that, for all $t \in [0, T]$, the volatility $\sigma(t)$ is almost surely not zero. We make the additional assumption that the filtration $\mathcal{F}(t)$, $0 \le t \le T$, is generated by the Brownian motion $W(t)$, $0 \le t \le T$.

Let $V(T)$ be an $\mathcal{F}(T)$-measurable random variable and, for $0 \le t \le T$, define $V(t)$ by the risk-neutral pricing formula (5.2.31). Then, according to (5.2.30),

$$D(t)V(t) = \widetilde{\mathbb{E}}\big[D(T)V(T)\big|\mathcal{F}(t)\big].$$

This is a $\widetilde{\mathbb{P}}$-martingale; indeed, iterated conditioning implies that, for $0 \le s \le t \le T$,

$$\begin{aligned}\widetilde{\mathbb{E}}\big[D(t)V(t)\big|\mathcal{F}(s)\big] &= \widetilde{\mathbb{E}}\big[\widetilde{\mathbb{E}}[D(T)V(T)|\mathcal{F}(t)]\big|\mathcal{F}(s)\big]\\ &= \widetilde{\mathbb{E}}\big[D(T)V(T)\big|\mathcal{F}(s)\big]\\ &= D(s)V(s).\end{aligned} \tag{5.3.3}$$

Therefore, $D(t)V(t)$ has a representation as (recall that $D(0)V(0) = V(0)$)

$$D(t)V(t) = V(0) + \int_0^t \widetilde{\Gamma}(u)\, d\widetilde{W}(u),\ 0 \le t \le T. \tag{5.3.4}$$

On the other hand, for any portfolio process $\Delta(t)$, the differential of the discounted portfolio value is given by (5.2.27), and hence

$$D(t)X(t) = X(0) + \int_0^t \Delta(u)\sigma(u)D(u)S(u)\, d\widetilde{W}(u),\ 0 \le t \le T. \tag{5.3.5}$$

In order to have $X(t) = V(t)$ for all t, we should choose

$$X(0) = V(0) \tag{5.3.6}$$

and choose $\Delta(t)$ to satisfy

$$\Delta(t)\sigma(t)D(t)S(t) = \widetilde{\Gamma}(t),\ 0 \le t \le T, \tag{5.3.7}$$

which is equivalent to

$$\Delta(t) = \frac{\widetilde{\Gamma}(t)}{\sigma(t)D(t)S(t)},\ 0 \le t \le T. \tag{5.3.8}$$

With these choices, we have a hedge for a short position in the derivative security with payoff $V(T)$ at time T.

There are two key assumptions that make the hedge possible. The first is that the volatility $\sigma(t)$ is not zero, so equation (5.3.7) can be solved for $\Delta(t)$. If the volatility vanishes, then the randomness of the Brownian motion does not enter the stock, although it may still enter the payoff $V(T)$ of the derivative security. In this case, the stock is no longer an effective hedging instrument. The other key assumption is that $\mathcal{F}(t)$ is generated by the underlying Brownian motion (i.e., there is no randomness in the derivative security apart from the Brownian motion randomness, which can be hedged by trading the stock). Under these two assumptions, every $\mathcal{F}(T)$-measurable derivative security can be hedged. Such a model is said to be *complete*.

The Martingale Representation Theorem argument of this section justifies the risk-neutral pricing formulas (5.2.30) and (5.2.31), but it does not provide a practical method of finding the hedging portfolio $\Delta(t)$. The final formula (5.3.8) for $\Delta(t)$ involves the integrand $\widetilde{\Gamma}(t)$ in the martingale representation (5.3.4) of the discounted derivative security price. While the Martingale Representation Theorem guarantees that such a process $\widetilde{\Gamma}$ exists and hence a hedge $\Delta(t)$ exists, it does not provide a method for finding $\widetilde{\Gamma}(t)$. We return to this point in Chapter 6.

5.4 Fundamental Theorems of Asset Pricing

In this section, we extend the discussions of Sections 5.2 and 5.3 to the case of multiple stocks driven by multiple Brownian motions. In the process, we develop and illustrate the two fundamental theorems of asset pricing. In addition to providing these theorems, in this section we give precise definitions of some of the basic concepts of derivative security pricing in continuous-time models

5.4.1 Girsanov and Martingale Representation Theorems

The two theorems on which this section is based are the multidimensional Girsanov Theorem and the multidimensional Martingale Representation Theorem. We state them here.

Throughout this section,

$$W(t) = \big(W_1(t), \dots, W_d(t)\big)$$

is a multidimensional Brownian motion on a probability space $(\Omega, \mathcal{F}, \mathbb{P})$. We interpret $\mathbb{P}$ to be the *actual probability measure*, the one that would be observed from empirical studies of price data. Associated with this Brownian motion, we have a filtration $\mathcal{F}(t)$ (see Definition 3.3.3). We shall have a fixed final time T, and we shall assume that $\mathcal{F} = \mathcal{F}(T)$. We do not always assume that the filtration is the one generated by the Brownian motion. When that is assumed, we say so explicitly.

Theorem 5.4.1 (Girsanov, multiple dimensions). *Let T be a fixed positive time, and let $\Theta(t) = \big(\Theta_1(t), \dots, \Theta_d(t)\big)$ be a d-dimensional adapted process. Define*

$$Z(t) = \exp\left\{-\int_0^t \Theta(u) \cdot dW(u) - \frac{1}{2}\int_0^t \|\Theta(u)\|^2 du\right\}, \tag{5.4.1}$$

$$\widetilde{W}(t) = W(t) + \int_0^t \Theta(u)\, du, \tag{5.4.2}$$

and assume that

$$\mathbb{E}\int_0^T \|\Theta(u)\|^2 Z^2(u)\, du < \infty. \tag{5.4.3}$$

Set $Z = Z(T)$. Then $\mathbb{E}Z = 1$, and under the probability measure $\widetilde{\mathbb{P}}$ given by

$$\widetilde{\mathbb{P}}(A) = \int_A Z(\omega)\, d\mathbb{P}(\omega) \text{ for all } A \in \mathcal{F},$$

the process $\widetilde{W}(t)$ is a d-dimensional Brownian motion.

The Itô integral in (5.4.1) is

$$\int_0^t \Theta(u) \cdot dW(u) = \int_0^t \sum_{j=1}^d \Theta_j(u)\, dW_j(u) = \sum_{j=1}^d \int_0^t \Theta_j(u)\, dW_j(u).$$

Also, in (5.4.1), $\|\Theta(u)\|$ denotes the Euclidean norm

$$\|\Theta(u)\| = \left(\sum_{j=1}^d \Theta_j^2(u) \right)^{\frac{1}{2}},$$

and (5.4.2) is shorthand notation for $\widetilde{W}(t) = \big(\widetilde{W}_1(t), \dots, \widetilde{W}_d(t)\big)$ with

$$\widetilde{W}_j(t) = W_j(t) + \int_0^t \Theta_j(u)\, du, \quad j = 1, \dots, d.$$

The remarkable thing about the conclusion of the multidimensional Girsanov Theorem is that the component processes of $\widetilde{W}(t)$ are independent under $\widetilde{\mathbb{P}}$. This is part of what it means to be a d-dimensional Brownian motion. The component processes of $W(t)$ are independent under $\mathbb{P}$, but each of the $\Theta_j(t)$ processes can depend in a path-dependent, adapted way on all of the Brownian motions $W_1(t), \dots, W_d(t)$. Therefore, under $\mathbb{P}$, the components of $\widetilde{W}(t)$ can be far from independent. Yet, after the change to the measure $\widetilde{\mathbb{P}}$, these components are independent. The proof of Theorem 5.4.1 is like that of the one-dimensional Girsanov Theorem 5.2.3, except it uses a d-dimensional version of Lévy's Theorem. The proof for $d = 2$ based on the two-dimensional Lévy Theorem, Theorem 4.6.5, is left to the reader as Exercise 5.6.

Theorem 5.4.2 (Martingale representation, multiple dimensions). *Let T be a fixed positive time, and assume that $\mathcal{F}(t)$, $0 \le t \le T$, is the filtration generated by the d-dimensional Brownian motion $W(t)$, $0 \le t \le T$. Let $M(t)$, $0 \le t \le T$, be a martingale with respect to this filtration under $\mathbb{P}$. Then there is an adapted, d-dimensional process $\Gamma(u) = \big(\Gamma_1(u), \dots, \Gamma_d(u)\big)$, $0 \le u \le T$, such that*

$$M(t) = M(0) + \int_0^t \Gamma(u) \cdot dW(u),\ 0 \le t \le T. \tag{5.4.4}$$

If, in addition, we assume the notation and assumptions of Theorem 5.4.1 and if $\widetilde{M}(t)$, $0 \le t \le T$, is a $\widetilde{\mathbb{P}}$-martingale, then there is an adapted, d-dimensional process $\widetilde{\Gamma}(u) = \big(\widetilde{\Gamma}_1(u), \dots, \widetilde{\Gamma}_d(u)\big)$ such that

$$\widetilde{M}(t) = \widetilde{M}(0) + \int_0^t \widetilde{\Gamma}(u) \cdot d\widetilde{W}(u),\ 0 \le t \le T. \tag{5.4.5}$$

5.4.2 Multidimensional Market Model

We assume there are m stocks, each with stochastic differential

$$dS_i(t) = \alpha_i(t)S_i(t)\,dt + S_i(t)\sum_{j=1}^{d}\sigma_{ij}(t)\,dW_j(t), \quad i = 1,\ldots,m. \tag{5.4.6}$$

We assume that the mean rate of return vector $(\alpha_i(t))_{i=1,\ldots,m}$ and the volatility matrix $(\sigma_{ij}(t))_{i=1,\ldots,m;j=1,\ldots,d}$ are adapted processes. These stocks are typically correlated. To see the nature of this correlation, we set $\sigma_i(t) = \sqrt{\sum_{j=1}^{d}\sigma_{ij}^2(t)}$, which we assume is never zero, and we define processes

$$B_i(t) = \sum_{j=1}^{d}\int_0^t \frac{\sigma_{ij}(u)}{\sigma_i(u)}\,dW_j(u), \quad i = 1,\ldots,m. \tag{5.4.7}$$

Being a sum of stochastic integrals, each $B_i(t)$ is a continuous martingale. Furthermore,

$$dB_i(t)\,dB_i(t) = \sum_{j=1}^{d}\frac{\sigma_{ij}^2(t)}{\sigma_i^2(t)}\,dt = dt.$$

According to Lévy's Theorem, Theorem 4.6.4, $B_i(t)$ is a Brownian motion. We may rewrite (5.4.6) in terms of the Brownian motion $B_i(t)$ as

$$dS_i(t) = \alpha_i(t)S_i(t)\,dt + \sigma_i(t)S_i(t)\,dB_i(t). \tag{5.4.8}$$

From this formula, we see that $\sigma_i(t)$ is the volatility of $S_i(t)$.

For $i \neq k$, the Brownian motions $B_i(t)$ and $B_k(t)$ are typically not independent. To see this, we first note that

$$dB_i(t)\,dB_k(t) = \sum_{j=1}^{d}\frac{\sigma_{ij}(t)\sigma_{kj}(t)}{\sigma_i(t)\sigma_k(t)}\,dt = \rho_{ik}(t)\,dt, \tag{5.4.9}$$

where

$$\rho_{ik}(t) = \frac{1}{\sigma_i(t)\sigma_k(t)}\sum_{j=1}^{d}\sigma_{ij}(t)\sigma_{kj}(t). \tag{5.4.10}$$

Itô's product rule implies

$$d\big(B_i(t)B_k(t)\big) = B_i(t)\,dB_k(t) + B_k(t)\,dB_i(t) + dB_i(t)\,dB_k(t),$$

and integration of this equation yields

$$B_i(t)B_k(t) = \int_0^t B_i(u)\,dB_k(u) + \int_0^t B_k(u)\,dB_i(u) + \int_0^t \rho_{ik}(u)\,du. \tag{5.4.11}$$

Taking expectations and using the fact that the expectation of an Itô integral is zero, we obtain the covariance formula

$$\mathrm{Cov}\big[B_i(t), B_k(t)\big] = \mathbb{E}\int_0^t \rho_{ik}(u)\,du. \tag{5.4.12}$$

If the processes $\sigma_{ij}(t)$ and $\sigma_{kj}(t)$ are constant (i.e., independent of t and not random), then so are $\sigma_i(t)$, $\sigma_k(t)$, and $\rho_{ik}(t)$. In this case, (5.4.12) reduces to $\mathrm{Cov}\big[B_i(t), B_k(t)\big] = \rho_{ik}t$. Because both $B_i(t)$ and $B_k(t)$ have standard deviation $\sqrt{t}$, the correlation between $B_i(t)$ and $B_j(t)$ is simply ρ_{ik}. When $\sigma_{ij}(t)$ and $\sigma_{kj}(t)$ are themselves random processes, we call $\rho_{ik}(t)$ the *instantaneous correlation* between $B_i(t)$ and $B_k(t)$. At a fixed time t along a particular path, $\rho_{ik}(t)$ is the conditional correlation between the next increments of B_i and B_k over a "small" time interval following time t (see Exercise 4.17 of Chapter 4 with $\Theta_1 = \Theta_2 = 0$, $\sigma_1 = \sigma_2 = 1$).

Finally, we note from (5.4.8) and (5.4.9) that

$$\begin{aligned} dS_i(t)\,dS_k(t) &= \sigma_i(t)\sigma_k(t)S_i(t)S_k(t)\,dB_i(t)\,dB_k(t) \\ &= \rho_{ik}(t)\sigma_i(t)\sigma_k(t)S_i(t)S_k(t)\,dt. \end{aligned} \tag{5.4.13}$$

Rewriting (5.4.13) in terms of "relative differentials," we obtain

$$\frac{dS_i(t)}{S_i(t)} \cdot \frac{dS_k(t)}{S_k(t)} = \rho_{ik}(t)\sigma_i(t)\sigma_k(t)\,dt.$$

The volatility processes $\sigma_i(t)$ and $\sigma_k(t)$ are the respective *instantaneous standard deviations* of the relative changes in S_i and S_k at time t, and the process $\rho_{ik}(t)$ is the *instantaneous correlation* between these relative changes.

Mean rates of return are affected by the change to a risk-neutral measure in the next subsection. Instantaneous standard deviations and correlations are unaffected (Exercise 5.12(ii) and (iii)). If the instantaneous standard deviations and correlations are not random, then (noninstantaneous) standard deviations and correlations are unaffected by the change of measure (see Exercise 5.12(iv) for the case of correlations). However, (noninstantaneous) standard deviations and correlations can be affected by a change of measure when the instantaneous standard deviations and correlations are random (see Exercises 5.12(v) and 5.13 for the case of correlations).

We define a *discount process*

$$D(t) = e^{-\int_0^t R(u)\,du}. \tag{5.4.14}$$

We assume that the interest rate process $R(t)$ is adapted. In addition to stock prices, we shall often work with discounted stock prices. Their differentials are

$$\begin{aligned} d\big(D(t)S_i(t)\big) &= D(t)\big[dS_i(t) - R(t)S_i(t)\,dt\big] \\ &= D(t)S_i(t)\Big[\big(\alpha_i(t) - R(t)\big)\,dt + \sum_{j=1}^{d}\sigma_{ij}(t)\,dW_j(t)\Big] \\ &= D(t)S_i(t)\big[\big(\alpha_i(t) - R(t)\big)\,dt + \sigma_i(t)\,dB_i(t)\big],\ i = 1,\dots,m. \end{aligned} \tag{5.4.15}$$

5.4.3 Existence of the Risk-Neutral Measure

Definition 5.4.3. *A probability measure* $\widetilde{\mathbb{P}}$ *is said to be* risk-neutral *if*

(i) $\widetilde{\mathbb{P}}$ *and* $\mathbb{P}$ *are equivalent (i.e., for every* $A \in \mathcal{F}$, $\mathbb{P}(A) = 0$ *if and only if* $\widetilde{\mathbb{P}}(A) = 0$*), and*

(ii) under $\widetilde{\mathbb{P}}$*, the discounted stock price* $D(t)S_i(t)$ *is a martingale for every* $i = 1,\dots,m$.

In order to make discounted stock prices be martingales, we would like to rewrite (5.4.15) as

$$d\big(D(t)S_i(t)\big) = D(t)S_i(t)\sum_{j=1}^{d}\sigma_{ij}(t)\big[\Theta_j(t)\,dt + dW_j(t)\big]. \tag{5.4.16}$$

If we can find the market price of risk processes $\Theta_j(t)$ that make (5.4.16) hold, with one such process for each source of uncertainty $W_j(t)$, we can then use the multidimensional Girsanov Theorem to construct an equivalent probability measure $\widetilde{\mathbb{P}}$ under which $\widetilde{W}(t)$ given by (5.4.2) is a d-dimensional Brownian motion. This permits us to reduce (5.4.16) to

$$d\big(D(t)S_i(t)\big) = D(t)S_i(t)\sum_{j=1}^{d}\sigma_{ij}(t)\,d\widetilde{W}_j(t), \tag{5.4.17}$$

and hence $D(t)S_i(t)$ is a martingale under $\widetilde{\mathbb{P}}$. The problem of finding a risk-neutral measure is simply one of finding processes $\Theta_j(t)$ that make (5.4.15) and (5.4.16) agree. Since these equations have the same coefficient multiplying each $dW_j(t)$, they agree if and only if the coefficient multiplying dt is the same in both cases, which means that

$$\alpha_i(t) - R(t) = \sum_{j=1}^{d}\sigma_{ij}(t)\Theta_j(t), \quad i = 1,\dots,m. \tag{5.4.18}$$

We call these the *market price of risk equations.* These are m equations in the d unknown processes $\Theta_1(t),\dots,\Theta_d(t)$.

If one cannot solve the market price of risk equations, then there is an arbitrage lurking in the model; the model is bad and should not be used for pricing. We do not give the detailed proof of this fact. Instead, we give a simple example to illustrate it.

Example 5.4.4. Suppose there are two stocks ($m = 2$) and one Brownian motion ($d = 1$), and suppose further that all coefficient processes are constant. Then, the market price of risk equations are

$$\alpha_1 - r = \sigma_1 \theta, \tag{5.4.19}$$

$$\alpha_2 - r = \sigma_2 \theta. \tag{5.4.20}$$

These equations have a solution θ if and only if

$$\frac{\alpha_1 - r}{\sigma_1} = \frac{\alpha_2 - r}{\sigma_2}.$$

If this equation does not hold, then one can arbitrage one stock against the other. Suppose, for example, that

$$\frac{\alpha_1 - r}{\sigma_1} > \frac{\alpha_2 - r}{\sigma_2}$$

and define

$$\mu = \frac{\alpha_1 - r}{\sigma_1} - \frac{\alpha_2 - r}{\sigma_2} > 0.$$

Suppose that at each time an agent holds $\Delta_1(t) = \frac{1}{S_1(t)\sigma_1}$ shares of stock one and $\Delta_2(t) = -\frac{1}{S_2(t))\sigma_2}$ shares of stock two, borrowing or investing as necessary at the interest rate r to set up and maintain this portfolio. The initial capital required to take the stock positions is $\frac{1}{\sigma_1} - \frac{1}{\sigma_2}$, but if this is positive we borrow from the money market account, and if it is negative we invest in the money market account, so the initial capital required to set up the whole portfolio, including the money market position, is $X(0) = 0$. The differential of the portfolio value $X(t)$ is

$$\begin{aligned}
&dX(t) \\
&= \Delta_1(t)\, dS_1(t) + \Delta_2(t)\, dS_2(t) + r\big(X(t) - \Delta_1(t)S_1(t) - \Delta_2(t)S_2(t)\big)\, dt \\
&= \frac{\alpha_1 - r}{\sigma_1}\, dt + dW(t) - \frac{\alpha_2 - r}{\sigma_2}\, dt - dW(t) + rX(t)\, dt \\
&= \mu\, dt + rX(t)\, dt.
\end{aligned}$$

The differential of the discounted portfolio value is

$$d\big(D(t)X(t)\big) = D(t)\big(dX(t) - rX(t)\, dt\big) = \mu D(t)\, dt.$$

The right-hand side $\mu D(t)$ is strictly positive and nonrandom. Therefore, this portfolio will make money for sure and do so faster than the interest rate r because the discounted portfolio value has a nonrandom positive derivative. We have managed to synthesize a second money market account with rate of return higher than r, and now the arbitrage opportunities are limitless. □

When there is no solution to the market price of risk equations, the arbitrage in the model may not be as obvious as in Example 5.4.4, but it does exist. If there is a solution to the market price of risk equations, then there is no arbitrage. To show this, we need to introduce some notation and terminology. In the market with stock prices $S_i(t)$ given by (5.4.6) and interest rate process $R(t)$, an agent can begin with initial capital $X(0)$ and choose adapted portfolio processes $\Delta_i(t)$, one for each stock $S_i(t)$. The differential of the agent's portfolio value will then be

$$\begin{aligned} dX(t) &= \sum_{i=1}^{m} \Delta_i(t)\, dS_i(t) + R(t)\left(X(t) - \sum_{i=1}^{m} \Delta_i(t) S_i(t)\right) dt \\ &= R(t)X(t)\, dt + \sum_{i=1}^{m} \Delta_i(t)\big(dS_i(t) - R(t)S_i(t)\, dt\big) \\ &= R(t)X(t)\, dt + \sum_{i=1}^{m} \frac{\Delta_i(t)}{D(t)}\, d\big(D(t)S_i(t)\big). \end{aligned} \tag{5.4.21}$$

The differential of the discounted portfolio value is

$$\begin{aligned} d\big(D(t)X(t)\big) &= D(t)\big(dX(t) - R(t)X(t)\, dt\big) \\ &= \sum_{i=1}^{m} \Delta_i(t)\, d\big(D(t)S_i(t)\big). \end{aligned} \tag{5.4.22}$$

If $\widetilde{\mathbb{P}}$ is a risk-neutral measure, then under $\widetilde{\mathbb{P}}$ the processes $D(t)S_i(t)$ are martingales, and hence the process $D(t)X(t)$ must also be a martingale. Put another way, under $\widetilde{\mathbb{P}}$ each of the stocks has mean rate of return $R(t)$, the same as the rate of return of the money market account. Hence, no matter how an agent invests, the mean rate of return of his portfolio value under $\widetilde{\mathbb{P}}$ must also be $R(t)$, and the discounted portfolio value must then be a martingale. We have proved the following result.

Lemma 5.4.5. *Let $\widetilde{\mathbb{P}}$ be a risk-neutral measure, and let $X(t)$ be the value of a portfolio. Under $\widetilde{\mathbb{P}}$, the discounted portfolio value $D(t)X(t)$ is a martingale.*

Definition 5.4.6. *An* arbitrage *is a portfolio value process $X(t)$ satisfying $X(0) = 0$ and also satisfying for some time $T > 0$*

$$\mathbb{P}\{X(T) \geq 0\} = 1, \quad \mathbb{P}\{X(T) > 0\} > 0. \tag{5.4.23}$$

An arbitrage is a way of trading so that one starts with zero capital and at some later time T is sure not to have lost money and furthermore has a positive probability of having made money. Such an opportunity exists if and only if there is a way to start with positive capital $X(0)$ and to beat the money market account. In other words, there exists an arbitrage if and only if there is a way to start with $X(0)$ and at a later time T have a portfolio value satisfying

$$\mathbb{P}\left\{X(T) \geq \frac{X(0)}{D(T)}\right\} = 1, \quad \mathbb{P}\left\{X(T) > \frac{X(0)}{D(T)}\right\} > 0 \tag{5.4.24}$$

(see Exercise 5.7).

Theorem 5.4.7 (First fundamental theorem of asset pricing). *If a market model has a risk-neutral probability measure, then it does not admit arbitrage.*

PROOF: If a market model has a risk-neutral probability measure $\widetilde{\mathbb{P}}$, then every discounted portfolio value process is a martingale under $\widetilde{\mathbb{P}}$. In particular, every portfolio value process satisfies $\widetilde{\mathbb{E}}\big[D(T)X(T)\big] = X(0)$. Let $X(t)$ be a portfolio value process with $X(0) = 0$. Then we have

$$\widetilde{\mathbb{E}}\big[D(T)X(T)\big] = 0. \tag{5.4.25}$$

Suppose $X(T)$ satisfies the first part of (5.4.23) (i.e., $\mathbb{P}\{X(T) < 0\} = 0$). Since $\widetilde{\mathbb{P}}$ is equivalent to $\mathbb{P}$, we have also $\widetilde{\mathbb{P}}\{X(T) < 0\} = 0$. This, coupled with (5.4.25), implies $\widetilde{\mathbb{P}}\{X(T) > 0\} = 0$, for otherwise we would have $\widetilde{\mathbb{P}}\{D(T)X(T) > 0\} > 0$, which would imply $\widetilde{\mathbb{E}}\big[D(T)X(T)\big] > 0$. Because $\mathbb{P}$ and $\widetilde{\mathbb{P}}$ are equivalent, we have also $\mathbb{P}\{X(T) > 0\} = 0$. Hence, $X(t)$ is not an arbitrage. In fact, there cannot exist an arbitrage since every portfolio value process $X(t)$ satisfying $X(0) = 0$ cannot be an arbitrage. □

One should never offer prices derived from a model that admits arbitrage, and the First Fundamental Theorem provides a simple condition one can apply to check that the model one is using does not have this fatal flaw. In our model with d Brownian motions and m stocks, this amounts to producing a solution to the market price of risk equations (5.4.18). In models of the term structure of interest rates (i.e., models that provide prices for bonds of every maturity), there are many instruments available for trading, and possible arbitrages in the model prices are a real concern. An application of the First Fundamental Theorem of Asset Pricing in such a model leads directly to the Heath-Jarrow-Morton condition for no arbitrage in term-structure models.

5.4.4 Uniqueness of the Risk-Neutral Measure

Definition 5.4.8. *A market model is* complete *if every derivative security can be hedged.*

Let us suppose we have a market model with a filtration generated by a d-dimensional Brownian motion and with a risk-neutral measure $\widetilde{\mathbb{P}}$ (i.e., we have solved the market price of risk equations (5.4.18), used the resulting market prices of risk $\Theta_1(t), \dots, \Theta_d(t)$ to define the Radon-Nikodým derivative process $Z(t)$, and have changed to the measure $\widetilde{\mathbb{P}}$ under which $\widetilde{W}(t)$ defined by (5.4.2) is a d-dimensional Brownian motion). Suppose further that we are given an $\mathcal{F}(T)$-measurable random variable $V(T)$, which is the payoff of some derivative security.

We would like to be sure we can hedge a short position in the derivative security whose payoff at time T is $V(T)$. We can define $V(t)$ by (5.2.31), so that $D(t)V(t)$ satisfies (5.2.30), and just as in (5.3.3), we see that $D(t)V(t)$ is a martingale under $\widetilde{\mathbb{P}}$. According to the Martingale Representation Theorem 5.4.2, there are processes $\widetilde{\Gamma}_1(u), \dots, \widetilde{\Gamma}_d(u)$ such that

$$D(t)V(t) = V(0) + \sum_{j=1}^{d} \int_0^t \widetilde{\Gamma}_j(u)\, d\widetilde{W}_j(u), \quad 0 \le t \le T. \tag{5.4.26}$$

Consider a portfolio value process that begins at $X(0)$. According to (5.4.22) and (5.4.17),

$$\begin{aligned} d\big(D(t)X(t)\big) &= \sum_{i=1}^{m} \Delta_i(t)\, d\big(D(t)S_i(t)\big) \\ &= \sum_{j=1}^{d} \sum_{i=1}^{m} \Delta_i(t) D(t) S_i(t) \sigma_{ij}(t)\, d\widetilde{W}_j(t) \end{aligned} \tag{5.4.27}$$

or, equivalently,

$$D(t)X(t) = X(0) + \sum_{j=1}^{d} \int_0^t \sum_{i=1}^{m} \Delta_i(u) D(u) S_i(u) \sigma_{ij}(u)\, d\widetilde{W}_j(u). \tag{5.4.28}$$

Comparing (5.4.26) and (5.4.28), we see that to hedge the short position, we should take $X(0) = V(0)$ and choose the portfolio processes $\Delta_1(t), \dots, \Delta_m(t)$ so that the *hedging equations*

$$\frac{\widetilde{\Gamma}_j(t)}{D(t)} = \sum_{i=1}^{m} \Delta_i(t) S_i(t) \sigma_{ij}(t), \quad j = 1, \dots, d, \tag{5.4.29}$$

are satisfied. These are d equations in m unknown processes $\Delta_1(t), \dots, \Delta_m(t)$.

Theorem 5.4.9 (Second fundamental theorem of asset pricing). *Consider a market model that has a risk-neutral probability measure. The model is complete if and only if the risk-neutral probability measure is unique.*

SKETCH OF PROOF: We first assume that the model is complete. We wish to show that there can be only one risk-neutral measure. Suppose the model has two risk-neutral measures, $\widetilde{\mathbb{P}}_1$ and $\widetilde{\mathbb{P}}_2$. Let A be a set in $\mathcal{F}$, which we assumed at the beginning of this section is the same as $\mathcal{F}(T)$. Consider the derivative security with payoff $V(T) = \mathbb{I}_A \frac{1}{D(T)}$. Because the model is complete, a short position in this derivative security can be hedged (i.e., there is a portfolio value process with some initial condition $X(0)$ that satisfies $X(T) = V(T)$). Since both $\widetilde{\mathbb{P}}_1$ and $\widetilde{\mathbb{P}}_2$ are risk-neutral, the discounted portfolio value process $D(t)X(t)$ is a martingale under both $\widetilde{\mathbb{P}}_1$ and $\widetilde{\mathbb{P}}_2$. It follows that

$$\begin{aligned} \widetilde{\mathbb{P}}_1(A) = \widetilde{\mathbb{E}}_1\big[D(T)V(T)\big] &= \widetilde{\mathbb{E}}_1\big[D(T)X(T)\big] = X(0) \\ = \widetilde{\mathbb{E}}_2\big[D(T)X(T)\big] &= \widetilde{\mathbb{E}}_2\big[D(T)V(T)\big] = \widetilde{\mathbb{P}}_2(A). \end{aligned}$$

Since A is an arbitrary set in $\mathcal{F}$ and $\widetilde{\mathbb{P}}_1(A) = \widetilde{\mathbb{P}}_2(A)$, these two risk-neutral measures are really the same.

For the converse, suppose there is only one risk-neutral measure. This means first of all that the filtration for the model is generated by the d-dimensional Brownian motion driving the assets. If that were not the case (i.e., if there were other sources of uncertainty in the model besides the driving Brownian motions), then we could assign arbitrary probabilities to those sources of uncertainty without changing the distributions of the driving Brownian motions and hence without changing the distributions of the assets. This would permit us to create multiple risk-neutral measures. Because the driving Brownian motions are the only sources of uncertainty, the only way multiple risk-neutral measures can arise is via multiple solutions to the market price of risk equations (5.4.18). Hence, uniqueness of the risk-neutral measure implies that the market price of risk equations (5.4.18) have only one solution $\big(\Theta_1(t), \dots, \Theta_d(t)\big)$. For fixed t and ω, these equations are of the form

$$Ax = b, \tag{5.4.30}$$

where A is the $m \times d$-dimensional matrix

$$A = \begin{bmatrix} \sigma_{11}(t), & \sigma_{12}(t), & \dots, & \sigma_{1d}(t) \\ \sigma_{21}(t), & \sigma_{22}(t), & \dots, & \sigma_{2d}(t) \\ \vdots & \vdots & & \vdots \\ \sigma_{m1}(t), & \sigma_{m2}(t), & \dots, & \sigma_{md}(t) \end{bmatrix}, \tag{5.4.31}$$

x is the d-dimensional column vector

$$x = \begin{bmatrix} \Theta_1(t) \\ \Theta_2(t) \\ \vdots \\ \Theta_d(t) \end{bmatrix},$$

and b is the m-dimensional column vector

$$b = \begin{bmatrix} \alpha_1(t) - R(t) \\ \alpha_2(t) - R(t) \\ \vdots \\ \alpha_m(t) - R(t) \end{bmatrix}.$$

Our assumption that there is only one risk-neutral measure means that the system of equations (5.4.30) has a unique solution x.

In order to be assured that every derivative security can be hedged, we must be able to solve the hedging equations (5.4.29) for $\Delta_1(t), \dots, \Delta_m(t)$ no matter what values of $\frac{\widetilde{\Gamma}_j(t)}{D(t)}$ appear on the left-hand side. For fixed t and ω, the hedging equations are of the form

$$A^{tr} y = c, \tag{5.4.32}$$

where A^{tr} is the transpose of the matrix in (5.4.31), y is the m-dimensional vector

$$y = \begin{bmatrix} y_1 \\ y_2 \\ \vdots \\ y_m \end{bmatrix} = \begin{bmatrix} \Delta_1(t)S_1(t) \\ \Delta_2(t)S_2(t) \\ \vdots \\ \Delta_m(t)S_m(t) \end{bmatrix},$$

and c is the d-dimensional vector

$$c = \begin{bmatrix} \frac{\widetilde{\Gamma}_1(t)}{D(t)} \\ \frac{\widetilde{\Gamma}_2(t)}{D(t)} \\ \vdots \\ \frac{\widetilde{\Gamma}_d(t)}{D(t)} \end{bmatrix}.$$

In order to be assured that the market is complete, there must be a solution y to the system of equations (5.4.32), no matter what vector c appears on the right-hand side. If there is always a solution $y_1, \dots, y_m$, then there are portfolio processes $\Delta_i(t) = \frac{y_i}{S_i(t)}$ satisfying the hedging equations (5.4.29), no matter what processes appear on the left-hand side of those equations. We could then conclude that a short position in an arbitrary derivative security can be hedged.

The uniqueness of the solution x to (5.4.30) implies the existence of a solution y to (5.4.32). We give a proof of this fact in Appendix C. Consequently, uniqueness of the risk-neutral measure implies that the market model is complete. □

5.5 Dividend-Paying Stocks

According to Definition 5.4.3, discounted stock prices are martingales under the risk-neutral measure. This is the case provided the stock pays no dividend. The key feature of a risk-neutral measure is that it causes discounted portfolio values to be martingales (see Lemma 5.4.5), and that ensures the absence of arbitrage (First Fundamental Theorem of Asset Pricing, Theorem 5.4.7). In order for the discounted value of a portfolio that invests in a dividend-paying stock to be a martingale, the discounted value of the stock *with the dividends reinvested* must be a martingale, but the discounted stock price itself is not a martingale. This section works out the details of this situation. We consider a single stock price driven by a single Brownian motion, although the results we obtain here also apply when there are multiple stocks and multiple Brownian motions.

5.5.1 Continuously Paying Dividend

Consider a stock, modeled as a generalized geometric Brownian motion, that pays dividends continuously over time at a rate $A(t)$ per unit time. Here $A(t)$, $0 \leq t \leq T$, is a nonnegative adapted process. A continuously paid dividend is not a bad model for a mutual fund, which collects lump sum dividends at a variety of times on a variety of stocks. In the case of a single stock, it is more reasonable to assume there are periodic lump sum dividend payments. We consider that case in Subsections 5.5.3 and 5.5.4.

Dividends paid by a stock reduce its value, and so we shall take as our model of the stock price

$$dS(t) = \alpha(t)S(t)\,dt + \sigma(t)\,S(t)\,dW(t) - A(t)S(t)\,dt. \tag{5.5.1}$$

If the stock were to withhold dividends, its mean rate of return would be $\alpha(t)$. Equivalently, if an agent holding the stock were to reinvest the dividends, the mean rate of return on his investment would be $\alpha(t)$. The mean rate of return $\alpha(t)$, the volatility $\sigma(t)$, and the interest rate $R(t)$ appearing in (5.5.2) below are all assumed to be adapted processes.

An agent who holds the stock receives both the capital gain or loss due to stock price movements and the continuously paying dividend. Thus, if $\Delta(t)$ is the number of shares held at time t, then the portfolio value $X(t)$ satisfies

$$\begin{aligned} dX(t) &= \Delta(t)\,dS(t) + \Delta(t)A(t)S(t)\,dt + R(t)\big[X(t) - \Delta(t)S(t)\big]\,dt \\ &= R(t)X(t)\,dt + \big(\alpha(t) - R(t)\big)\Delta(t)S(t)\,dt + \sigma(t)\Delta(t)S(t)\,dW(t) \\ &= R(t)X(t)\,dt + \Delta(t)S(t)\sigma(t)\big[\Theta(t)\,dt + dW(t)\big], \end{aligned} \tag{5.5.2}$$

where

$$\Theta(t) = \frac{\alpha(t) - R(t)}{\sigma(t)} \tag{5.5.3}$$

is the usual market price of risk.

We define

$$\widetilde{W}(t) = W(t) + \int_0^t \Theta(u)\,du \tag{5.5.4}$$

and use Girsanov's Theorem to change to a measure $\widetilde{\mathbb{P}}$ under which $\widetilde{W}$ is a Brownian motion, so we may rewrite (5.5.2) as

$$dX(t) = R(t)X(t)\,dt + \Delta(t)S(t)\sigma(t)\,d\widetilde{W}(t).$$

The discounted portfolio value satisfies

$$d\big[D(t)X(t)\big] = \Delta(t)D(t)S(t)\sigma(t)\,d\widetilde{W}(t).$$

In particular, under the risk-neutral measure $\widetilde{\mathbb{P}}$, the discounted portfolio process is a martingale. Here we denote by $D(t) = e^{-\int_0^t R(u)du}$ the usual discount process.

If we now wish to hedge a short position in a derivative security paying $V(T)$ at time T, where $V(T)$ is an $\mathcal{F}(T)$-measurable random variable, we will need to choose the initial capital $X(0)$ and the portfolio process $\Delta(t)$, $0 \le t \le T$, so that $X(T) = V(T)$. Because $D(t)X(t)$ is a martingale under $\widetilde{\mathbb{P}}$, we must have

$$D(t)X(t) = \widetilde{\mathbb{E}}\big[D(T)V(T)\big|\mathcal{F}(t)\big],\ 0 \le t \le T.$$

The value $X(t)$ of this portfolio at each time t is the value (price) of the derivative security at that time, which we denote by $V(t)$. Making this replacement in the formula above, we obtain the risk-neutral pricing formula

$$D(t)V(t) = \widetilde{\mathbb{E}}\big[D(T)V(T)\big|\mathcal{F}(t)\big],\ 0 \le t \le T. \tag{5.5.5}$$

We have obtained the same risk-neutral pricing formula (5.2.30) as in the case of no dividends. Furthermore, conditions that guarantee that a short position can be hedged, and hence risk-neutral pricing is fully justified, are the same as in the no-dividend case; see Section 5.3.

The difference between the dividend and no-dividend cases is in the evolution of the underlying stock under the risk-neutral measure. From (5.5.1) and the definition of $\widetilde{W}(t)$, we see that

$$dS(t) = \big[R(t) - A(t)\big]S(t)\,dt + \sigma(t)S(t)\,d\widetilde{W}(t). \tag{5.5.6}$$

Under the risk-neutral measure, the stock does not have mean rate of return $R(t)$, and consequently the discounted stock price is not a martingale. Indeed,

$$S(t) = S(0)\exp\left\{\int_0^t \sigma(u)\,d\widetilde{W}(u) + \int_0^t \Big[R(u) - A(u) - \frac{1}{2}\sigma^2(u)\Big]\,du\right\}. \tag{5.5.7}$$

The process

$$e^{\int_0^t A(u)}D(t)S(t) = \exp\left\{\int_0^t \sigma(u)\,d\widetilde{W}(u) - \frac{1}{2}\int_0^t \sigma^2(u)\,du\right\}$$

is a martingale. This is the interest-rate-discounted value at time t of an account that initially purchases one share of the stock and continuously reinvests the dividends in the stock.

5.5.2 Continuously Paying Dividend with Constant Coefficients

In the event that the volatility σ, the interest rate r, and the dividend rate a are constant, the stock price at time t, given by (5.5.7), is

$$S(t) = S(0)\exp\left\{\sigma\widetilde{W}(t) + \left(r - a - \frac{1}{2}\sigma^2\right)t\right\}. \tag{5.5.8}$$

For $0 \leq t \leq T$, we have

$$S(T) = S(t)\exp\left\{\sigma\big(\widetilde{W}(T) - \widetilde{W}(t)\big) + \left(r - a - \frac{1}{2}\sigma^2\right)(T-t)\right\}.$$

According to the risk-neutral pricing formula, the price at time t of a European call expiring at time T with strike K is

$$V(t) = \widetilde{\mathbb{E}}\left[e^{-r(T-t)}(S(T)-K)^+ \big| \mathcal{F}(t)\right]. \tag{5.5.9}$$

To evaluate this, we first compute

$$\begin{aligned}
&c(t,x)\\
&= \widetilde{\mathbb{E}}\left[e^{-r(T-t)}\left(x\exp\left\{\sigma\big(\widetilde{W}(T) - \widetilde{W}(t)\big) + \left(r - a - \frac{1}{2}\sigma^2\right)(T-t)\right\} - K\right)^+\right]\\
&= \widetilde{\mathbb{E}}\left[e^{-r\tau}\left(x\exp\left\{-\sigma\sqrt{\tau}Y + \left(r - a - \frac{1}{2}\sigma^2\right)\tau\right\} - K\right)^+\right],
\end{aligned} \tag{5.5.10}$$

where $\tau = T - t$ and

$$Y = -\frac{\widetilde{W}(T) - \widetilde{W}(t)}{\sqrt{T-t}}$$

is a standard normal random variable under $\widetilde{\mathbb{P}}$. We define

$$d_{\pm}(\tau,x) = \frac{1}{\sigma\sqrt{\tau}}\left[\log\frac{x}{K} + \left(r - a \pm \frac{1}{2}\sigma^2\right)\tau\right]. \tag{5.5.11}$$

We note that the random variable whose expectation we are computing in (5.5.10) is nonzero (the call expires in the money) if and only if $Y < d_-(\tau,x)$. Therefore,

$$\begin{aligned}
&c(t,x)\\
&= \frac{1}{\sqrt{2\pi}}\int_{-\infty}^{d_-(\tau,x)} e^{-r\tau}\left(x\exp\left\{-\sigma\sqrt{\tau}\,y + \left(r - a - \frac{1}{2}\sigma^2\right)\tau\right\} - K\right)e^{-\frac{1}{2}y^2}\,dy\\
&= \frac{1}{\sqrt{2\pi}}\int_{-\infty}^{d_-(\tau,x)} x\exp\left\{-\sigma\sqrt{\tau}\,y - \left(a + \frac{1}{2}\sigma^2\right)\tau - \frac{1}{2}y^2\right\}dy\\
&\qquad - \frac{1}{\sqrt{2\pi}}\int_{-\infty}^{d_-(\tau,x)} e^{-r\tau}Ke^{-\frac{1}{2}y^2}\,dy\\
&= \frac{1}{\sqrt{2\pi}}\int_{-\infty}^{d_-(\tau,x)} xe^{-a\tau}\exp\left\{-\frac{1}{2}(y + \sigma\sqrt{\tau})^2\right\}dy - e^{-r\tau}KN(d_-(\tau,x)).
\end{aligned}$$

We make the change of variable $z = y + \sigma\sqrt{\tau}$ in the integral, which leads us to the formula

$$\begin{aligned} c(t,x) &= \frac{1}{\sqrt{2\pi}} \int_{-\infty}^{d_+(\tau,x)} x e^{-a\tau} e^{-\frac{z^2}{2}}\, dz - e^{-r\tau} K N(d_-(\tau,x)) \\ &= x e^{-a\tau} N(d_+(\tau,x)) - e^{-r\tau} K N(d_-(\tau,x)). \end{aligned} \tag{5.5.12}$$

According to the Independence Lemma, Lemma 2.3.4, the option price $V(t)$ in (5.5.9) is $c(t, S(t))$. The only differences between this formula and the one for a non-dividend-paying stock is in the definition (5.5.11) of $d_\pm(\tau,x)$ (see (5.2.33) and (5.2.34)) and in the presence of $e^{-a\tau}$ in the first term on the right-hand side of (5.5.12).

5.5.3 Lump Payments of Dividends

Finally, let us consider the case when the dividend is paid in lumps. That is to say there are times $0 < t_1 < t_2 < t_n < T$ and, at each time t_j, the dividend paid is $a_j S(t_j-)$, where $S(t_j-)$ denotes the stock price just prior to the dividend payment. The stock price after the dividend payment is the stock price before the dividend payment less the dividend payment:

$$S(t_j) = S(t_j-) - a_j S(t_j-) = (1-a_j) S(t_j-). \tag{5.5.13}$$

We assume that each a_j is an $\mathcal{F}(t_j)$-measurable random variable taking values in $[0,1]$. If $a_j = 0$, no dividend is paid at time t_j. If $a_j = 1$, the full value of the stock is paid as a dividend at time t_j, and the stock value is zero thereafter. To simplify the notation, we set $t_0 = 0$ and $t_{n+1} = T$. However, neither $t_0 = 0$ nor $t_{n+1} = T$ is a dividend payment date (i.e., $a_0 = 0$ and $a_{n+1} = 0$). We assume that, between dividend payment dates, the stock price follows a generalized geometric Brownian motion:

$$dS(t) = \alpha(t) S(t)\, dt + \sigma(t) S(t)\, dW(t), \quad t_j \le t < t_{j+1}, \; j = 0, 1, \dots, n. \tag{5.5.14}$$

Equations (5.5.13) and (5.5.14) fully determine the evolution of the stock price.

Between dividend payment dates, the differential of the portfolio value corresponding to a portfolio process $\Delta(t)$, $0 \le t \le T$, is

$$\begin{aligned} dX(t) &= \Delta(t)\, dS(t) + R(t)\big[X(t) - \Delta(t)S(t)\big]\, dt \\ &= R(t)X(t)\, dt + \big(\alpha(t) - R(t)\big)\Delta(t)S(t)\, dt + \sigma(t)\Delta(t)S(t)\, dW(t) \\ &= R(t)X(t)\, dt + \Delta(t)\sigma(t)S(t)\big[\Theta(t)\, dt + dW(t)\big], \end{aligned}$$

where the market price of risk $\Theta(t)$ is again defined by (5.5.3). At the dividend payment dates, the value of the portfolio stock holdings drops by $a_j\Delta(t_j)S(t_j-)$, but the portfolio collects the dividend $a_j\Delta(t_j)S(t_j-)$, and so the portfolio value does not jump. It follows that

$$dX(t) = R(t)X(t)\,dt + \Delta(t)\sigma(t)S(t)\big[\Theta(t)\,dt + dW(t)\big] \tag{5.5.15}$$

is the correct formula for the evolution of the portfolio value at all times t. We again define $\widetilde{W}$ by (5.5.4), change to a measure $\widetilde{\mathbb{P}}$ under which $\widetilde{W}$ is a Brownian motion, and obtain the risk-neutral pricing formula (5.5.5).

5.5.4 Lump Payments of Dividends with Constant Coefficients

We price a European call under the assumption that σ, r, and each a_j are constant. From (5.5.14) and the definition of $\widetilde{W}$, we have

$$dS(t) = rS(t)\,dt + \sigma S(t)\,d\widetilde{W}(t),\ t_j \le t < t_{j+1},\ j = 0, 1, \dots, n.$$

Therefore,

$$S(t_{j+1}-) = S(t_j)\exp\left\{\sigma\big(\widetilde{W}(t_{j+1}) - \widetilde{W}(t_j)\big) + \Big(r - \frac{1}{2}\sigma^2\Big)(t_{j+1} - t_j)\right\}. \tag{5.5.16}$$

From (5.5.13), we see that

$$\begin{aligned} &S(t_{j+1}) \\ &= (1 - a_{j+1})S(t_j)\exp\left\{\sigma\big(\widetilde{W}(t_{j+1}) - \widetilde{W}(t_j)\big) + \Big(r - \frac{1}{2}\sigma^2\Big)(t_{j+1} - t_j)\right\} \end{aligned}$$

or, equivalently, for $j = 0, 1, \dots, n$,

$$\frac{S(t_{j+1})}{S(t_j)} = (1 - a_{j+1})\exp\left\{\sigma\big(\widetilde{W}(t_{j+1}) - \widetilde{W}(t_j)\big) + \Big(r - \frac{1}{2}\sigma^2\Big)(t_{j+1} - t_j)\right\}.$$

It follows that

$$\begin{aligned} \frac{S(T)}{S(0)} &= \frac{S(t_{n+1})}{S(t_0)} \\ &= \prod_{j=0}^{n} \frac{S(t_{j+1})}{S(t_j)} \\ &= \prod_{j=0}^{n-1}(1 - a_{j+1}) \cdot \exp\left\{\sigma\widetilde{W}(T) + \Big(r - \frac{1}{2}\sigma^2\Big)T\right\}. \end{aligned}$$

In other words,

$$S(T) = S(0)\prod_{j=0}^{n-1}(1 - a_{j+1}) \cdot \exp\left\{\sigma\widetilde{W}(T) + \Big(r - \frac{1}{2}\sigma^2\Big)T\right\}. \tag{5.5.17}$$

This is the same formula we would have for the price at time T of a geometric Brownian motion not paying dividends if the initial stock price were

$S(0)\prod_{j=0}^{n-1}(1-a_{j+1})$ rather than $S(0)$. Therefore, the price at time zero of a European call on this dividend-paying asset, a call that expires at time T with strike price K, is obtained by replacing the initial stock price by $S(0)\prod_{j=0}^{n-1}(1-a_{j+1})$ in the classical Black-Scholes-Merton formula. This results in the call price

$$S(0)\prod_{j=0}^{n-1}(1-a_{j+1})N(d_+) - e^{-rT}KN(d_-),$$

where

$$d_\pm = \frac{1}{\sigma\sqrt{T}}\left[\log\frac{S(0)}{K} + \sum_{j=0}^{n-1}\log(1-a_{j+1}) + \left(r \pm \frac{1}{2}\sigma^2\right)T\right].$$

A similar formula holds for the call price at times t between 0 and T. In those cases, one includes only the terms $(1-a_{j+1})$ corresponding to the dividend dates between times t and T.

5.6 Forwards and Futures

In this section, we assume there is a unique risk-neutral measure $\widetilde{\mathbb{P}}$, and all assets satisfy the risk-neutral pricing formula. Under this assumption, we study forward and futures prices and the relationship between them. The formulas we develop apply to any tradable, non-dividend-paying asset, not just to a stock. In a binomial model, these topics were addressed in Sections 6.3 and 6.5 of Volume I.

5.6.1 Forward Contracts

Let $S(t)$, $0 \le t \le \overline{T}$, be an asset price process, and let $R(t)$, $0 \le t \le \overline{T}$, be an interest rate process. We choose here some large time $\overline{T}$, and all bonds and derivative securities we consider will mature or expire at or before time $\overline{T}$. As usual, we define the discount process $D(t) = e^{-\int_0^t R(u)du}$. According to the risk-neutral pricing formula (5.2.30), the price at time t of a zero-coupon bond paying 1 at time T is

$$B(t,T) = \frac{1}{D(t)}\widetilde{\mathbb{E}}[D(T)|\mathcal{F}(t)], \quad 0 \le t \le T \le \overline{T}. \tag{5.6.1}$$

This pricing formula guarantees that no arbitrage can be found by trading in these bonds because any such portfolio, when discounted, will be a martingale under the risk-neutral measure. The details of this argument in the binomial model are presented in Theorem 6.2.6 and Remark 6.2.7 of Volume I.

Definition 5.6.1. *A* forward contract *is an agreement to pay a specified* delivery price K *at a* delivery date T, *where* $0 \leq T \leq \overline{T}$, *for the asset whose price at time* t *is* $S(t)$. *The* T-forward price $\mathrm{For}_S(t,T)$ *of this asset at time* t, *where* $0 \leq t \leq T \leq \overline{T}$, *is the value of* K *that makes the forward contract have no-arbitrage price zero at time* t.

Theorem 5.6.2. *Assume that zero-coupon bonds of all maturities can be traded. Then*

$$\mathrm{For}_S(t,T) = \frac{S(t)}{B(t,T)}, \quad 0 \leq t \leq T \leq \overline{T}. \tag{5.6.2}$$

PROOF: Suppose that at time t an agent sells the forward contract with delivery date T and delivery price K. Suppose further that the value K is chosen so that the forward contract has price zero at time t. Then selling the forward contract generates no income. Having sold the forward contract at time t, suppose the agent immediately shorts $\frac{S(t)}{B(t,T)}$ zero-coupon bonds and uses the income $S(t)$ generated to buy one share of the asset. The agent then does no further trading until time T, at which time she owns one share of the asset, which she delivers according to the forward contract. In exchange, she receives K. After covering the short bond position, she is left with $K - \frac{S(t)}{B(t,T)}$. If this is positive, the agent has found an arbitrage. If it is negative, the agent could instead have taken the opposite position, going long the forward, long the T-maturity bond, and short the asset, to again achieve an arbitrage. In order to preclude arbitrage, K must be given by (5.6.2). □

Remark 5.6.3. The proof of Theorem 5.6.2 does not use the notion of risk-neutral pricing. It shows that the forward price must be given by (5.6.2) in order to preclude arbitrage. Because we have assumed the existence of a risk-neutral measure and are pricing all assets by the risk-neutral pricing formula, we must be able to obtain (5.6.2) from the risk-neutral pricing formula as well. Indeed, using (5.2.30), (5.6.1), and the fact that the discounted asset price is a martingale under $\widetilde{\mathbb{P}}$, we compute the price at time t of the forward contract to be

$$\begin{aligned}
&\frac{1}{D(t)}\widetilde{\mathbb{E}}\big[D(T)\big(S(T)-K\big)\big|\mathcal{F}(t)\big] \\
&= \frac{1}{D(t)}\widetilde{\mathbb{E}}\big[D(T)S(T)\big|\mathcal{F}(t)\big] - \frac{K}{D(t)}\widetilde{\mathbb{E}}[D(T)|\mathcal{F}(t)] \\
&= S(t) - KB(t,T).
\end{aligned}$$

In order for this to be zero, K must be given by (5.6.2).

5.6.2 Futures Contracts

Consider a time interval $[0,T]$, which we divide into subintervals using the partition points $0 = t_0 < t_1 < t_2 < \cdots < t_n = T$. We shall refer to each subinterval $[t_k, t_{k+1})$ as a "day."

Suppose the interest rate is constant within each day. Then the discount process is given by $D(0) = 1$ and, for $k = 0, 1, \ldots, n-1$,

$$D(t_{k+1}) = \exp\Big\{-\int_0^{t_{k+1}} R(u)\,du\Big\} = \exp\Big\{-\sum_{j=0}^{k} R(t_j)(t_{j+1} - t_j)\Big\},$$

which is $\mathcal{F}(t_k)$-measurable. According to the risk-neutral pricing formula (5.6.1), the zero-coupon bond paying 1 at maturity T has time-t_k price

$$B(t_k, T) = \frac{1}{D(t_k)} \widetilde{\mathbb{E}}[D(T)|\mathcal{F}(t_k)].$$

An asset whose price at time t is $S(t)$ has time-t_k forward price

$$\mathrm{For}_S(t_k, T) = \frac{S(t_k)}{B(t_k, T)}, \tag{5.6.3}$$

an $\mathcal{F}(t_k)$-measurable quantity. Suppose we take a long position in the forward contract at time t_k (i.e., agree to receive $S(T)$ and pay $\mathrm{For}_S(t_k, T)$ at time T). The value of this position at time $t_j \geq t_k$ is

$$\begin{aligned}
V_{k,j} &= \frac{1}{D(t_j)} \widetilde{\mathbb{E}}\left[D(T)\Big(S(T) - \frac{S(t_k)}{B(t_k,T)}\Big)\Big|\, \mathcal{F}(t_j)\right] \\
&= \frac{1}{D(t_j)} \widetilde{\mathbb{E}}[D(T)S(T)|\mathcal{F}(t_j)] - \frac{S(t_k)}{B(t_k,T)} \cdot \frac{1}{D(t_j)} \widetilde{\mathbb{E}}[D(T)|\mathcal{F}(t_j)] \\
&= S(t_j) - S(t_k) \cdot \frac{B(t_j, T)}{B(t_k, T)}.
\end{aligned}$$

If $t_j = t_k$, this is zero, as it should be. However, for $t_j > t_k$, it is generally different from zero. For example, if the interest rate is a constant r so that $B(t, T) = e^{-r(T-t)}$, then

$$V_{k,j} = S(t_j) - e^{r(t_j - t_k)} S(t_k).$$

If the asset grows faster than the interest rate, the forward contract takes on a positive value. Otherwise, it takes on a negative value. In either case, one of the parties to the forward contract could become concerned about default by the other party.

To alleviate the problem of default risk, parties to a forward contract could agree to settle one day after the contract is entered. The original forward contract purchaser could then seek to purchase a new forward contract one day later than the initial purchase. By repeating this process, the forward contract purchaser could generate the cash flow

$$\begin{aligned}
V_{0,1} &= S(t_1) - S(t_0) \cdot \frac{B(t_1, T)}{B(t_0, T)} = S(t_1) - S(0) \cdot \frac{B(t_1, T)}{B(0, T)}, \\
V_{1,2} &= S(t_2) - S(t_1) \cdot \frac{B(t_2, T)}{B(t_1, T)}, \\
&\vdots
\end{aligned}$$

$$V_{n-1,n} = S(t_n) - S(t_{n-1}) \cdot \frac{B(t_n,T)}{B(t_{n-1},T)} = S(T) - \frac{S(t_{n-1})}{B(t_{n-1},T)}.$$

There are two problems with this. First of all, the purchaser of the forward contract was presumably motivated by a desire to hedge against a price increase in the underlying asset. It is not clear the extent to which receiving this cash flow provides such a hedge. Second, this daily buying and selling of forward contracts requires that there be a liquid market each day for forward contracts initiated that day and forward contracts initiated one day before. This is too much to expect.

A better idea than daily repurchase of forward contracts is to create a *futures price* $\mathrm{Fut}_S(t,T)$, and use it as described below. If an agent holds a long futures position between times t_k and t_{k+1}, then at time t_{k+1} he receives a payment

$$\mathrm{Fut}_S(t_{k+1},T) - \mathrm{Fut}_S(t_k,T).$$

This is called *marking to margin.* The stochastic process $\mathrm{Fut}_S(t,T)$ is constructed so that $\mathrm{Fut}_S(t,T)$ is $\mathcal{F}(t)$-measurable for every t and

$$\mathrm{Fut}_S(T,T) = S(T).$$

Therefore, the sum of payments received by an agent who purchases a futures contract at time zero and holds it until delivery date T is

$$\begin{aligned}\big(\mathrm{Fut}_S(t_1,T) - \mathrm{Fut}_S(t_0,T)\big) + \big(\mathrm{Fut}_S(t_2,T) - \mathrm{Fut}_S(t_1,T)\big) + \dots \\ \dots + \big(\mathrm{Fut}_S(t_n,T) - \mathrm{Fut}_S(t_{n-1},T) &= \mathrm{Fut}_S(T,T) - \mathrm{Fut}_S(0,T) \\ &= S(T) - \mathrm{Fut}_S(0,T).\end{aligned}$$

If the agent takes delivery of the asset at time T, paying market price $S(T)$ for it, his total income from the futures contract and the delivery payment is $-\mathrm{Fut}_S(0,T)$. Ignoring the time value of money, he has effectively paid the price $\mathrm{Fut}_S(0,T)$ for the asset, a price that was locked in at time zero.

In contrast to the case of a forward contract, the payment from holding a futures contract is distributed over the life of the contract rather than coming solely at the end. The mechanism for these payments is the margin account, which the owner of the futures contract must open at the time of purchase of the contract and to which he must contribute or from which he may withdraw money, depending on the trajectory of the futures price. Whereas the owner of a forward contract is exposed to counterparty default risk, the owner of a futures contract is exposed to the risk that some of the intermediate payments (margin calls) will force him to close out his position prematurely.

In addition to satisfying $\mathrm{Fut}_S(T,T) = S(T)$, the futures price process is chosen so that at each time t_k the value of the payment to be received at time t_{k+1}, and indeed at all future times $t_j > t_k$, is zero. This means that at any time one may enter or close out a position in the contract without incurring any cost other than payments already made. The condition that the value at time t_k of the payment to be received at time t_{k+1} be zero may be written as

$$\begin{aligned}0 &= \frac{1}{D(t_k)}\widetilde{\mathbb{E}}\big[D(t_{k+1})\big(\mathrm{Fut}_S(t_{k+1},T) - \mathrm{Fut}_S(t_k,T)\big)\big|\mathcal{F}(t_k)\big] \\ &= \frac{D(t_{k+1})}{D(t_k)}\big\{\widetilde{\mathbb{E}}[\mathrm{Fut}_S(t_{k+1},T)|\mathcal{F}(t_k)] - \mathrm{Fut}_S(t_k,T)\big\},\end{aligned}$$

where we have used the fact that $D(t_{k+1})$ is $\mathcal{F}(t_k)$-measurable to take $D(t_{k+1})$ out of the conditional expectation. From the equation above, we see that

$$\widetilde{\mathbb{E}}[\mathrm{Fut}_S(t_{k+1},T)|\mathcal{F}(t_k)] = \mathrm{Fut}_S(t_k,T), \quad k = 0,1,\dots,n-1. \tag{5.6.4}$$

This shows that $\mathrm{Fut}_S(t_k,T)$ must be a discrete-time martingale under $\widetilde{\mathbb{P}}$. But we also require that $\mathrm{Fut}_S(T,T) = S(T)$, from which we conclude that the futures prices must be given by the formula

$$\mathrm{Fut}_S(t_k,T) = \widetilde{\mathbb{E}}[S(T)|\mathcal{F}(t_k)], \ k = 0,1,\dots,n. \tag{5.6.5}$$

Indeed, under the condition that $\mathrm{Fut}_S(T,T) = S(T)$, equations (5.6.4) and (5.6.5) are equivalent.

We note finally that with $\mathrm{Fut}_S(t,T)$ given by (5.6.5), the value at time t_k of the payment to be received at time t_j is zero for every $j \geq k+1$. Indeed, using the $\mathcal{F}(t_{j-1})$-measurability of $D(t_j)$ and the martingale property for $\mathrm{Fut}_S(t,T)$, we have

$$\begin{aligned}&\frac{1}{D(t_k)}\widetilde{\mathbb{E}}\big[D(t_j)\big(\mathrm{Fut}_S(t_j,T) - \mathrm{Fut}_S(t_{j-1},T)\big)\big|\mathcal{F}(t_k)\big] \\ &= \frac{1}{D(t_k)}\widetilde{\mathbb{E}}\Big[\widetilde{\mathbb{E}}\big[D(t_j)\big(\mathrm{Fut}_S(t_j,T) - \mathrm{Fut}_S(t_{j-1},T)\big)\big|\mathcal{F}(t_{j-1})\big]\Big|\mathcal{F}(t_k)\Big] \\ &= \frac{1}{D(t_k)}\widetilde{\mathbb{E}}\Big[D(t_j)\widetilde{\mathbb{E}}\big[\mathrm{Fut}_S(t_j,T)\big|\mathcal{F}(t_{j-1})\big] - D(t_j)\mathrm{Fut}_S(t_{j-1},T)\Big|\mathcal{F}(t_k)\Big] \\ &= \frac{1}{D(t_k)}\widetilde{\mathbb{E}}\Big[D(t_j)\mathrm{Fut}_S(t_{j-1},T) - D(t_j)\mathrm{Fut}_S(t_{j-1},T)\Big|\mathcal{F}(t_k)\Big] = 0.\end{aligned}$$

These considerations lead us to make the following definition for the fully continuous case (i.e., the case when $R(t)$ is assumed only to be an adapted stochastic process, not necessarily constant on time intervals of the form $[t_k, t_{k+1})$).

Definition 5.6.4. *The* futures price *of an asset whose value at time T is $S(T)$ is given by the formula*

$$\mathrm{Fut}_S(t,T) = \widetilde{\mathbb{E}}[S(T)|\mathcal{F}(t)], \quad 0 \leq t \leq T. \tag{5.6.6}$$

A long position in the futures contract *is an agreement to receive as a cash flow the changes in the futures price (which may be negative as well as positive) during the time the position is held. A short position in the futures contract receives the opposite cash flow.*

Theorem 5.6.5. *The futures price is a martingale under the risk-neutral measure* $\widetilde{\mathbb{P}}$, *it satisfies* $\mathrm{Fut}_S(T,T) = S(T)$, *and the value of a long (or short) futures position to be held over an interval of time is always zero.*

OUTLINE OF PROOF: The usual iterated conditioning argument shows that $\mathrm{Fut}_S(t,T)$ given by (5.6.6) is a $\widetilde{\mathbb{P}}$-martingale satisfying the terminal condition $\mathrm{Fut}_S(T,T) = S(T)$. In fact, this is the only $\widetilde{\mathbb{P}}$-martingale satisfying this terminal condition.

If the filtration $\mathcal{F}(t)$, $0 \le t \le T$, is generated by a Brownian motion $W(t)$, $0 \le t \le T$, then Corollary 5.3.2 of the Martingale Representation Theorem implies that

$$\mathrm{Fut}_S(t,T) = \mathrm{Fut}_S(0,T) + \int_0^t \widetilde{\Gamma}(u)\, d\widetilde{W}(u), \quad 0 \le t \le T,$$

for some adapted integrand process $\widetilde{\Gamma}$ (i.e., $d\mathrm{Fut}_S(t,T) = \widetilde{\Gamma}(t)\, d\widetilde{W}(t)$). Let $0 \le t_0 < t_1 \le T$ be given and consider an agent who at times t between times t_0 and t_1 holds $\Delta(t)$ futures contracts. It costs nothing to change the position in futures contracts, but because the futures contracts generate cash flow, the agent may have cash to invest or need to borrow in order to execute this strategy. He does this investing and/or borrowing at the interest rate $R(t)$ prevailing at the time of the investing or borrowing. The agent's profit $X(t)$ from this trading satisfies

$$dX(t) = \Delta(t)\, d\mathrm{Fut}_S(t,T) + R(t)X(t)\, dt = \Delta(t)\widetilde{\Gamma}(t)\, d\widetilde{W}(t) + R(t)X(t)\, dt,$$

and thus

$$d\big(D(t)X(t)\big) = D(t)\Delta(t)\widetilde{\Gamma}(t)\, d\widetilde{W}(t).$$

Assume that at time t_0 the agent's profit is $X(t_0) = 0$. At time t_1, the agent's profit $X(t_1)$ will satisfy

$$D(t_1)X(t_1) = \int_{t_0}^{t_1} D(u)\Delta(u)\widetilde{\Gamma}(u)\, d\widetilde{W}(u). \tag{5.6.7}$$

Because Itô integrals are martingales, we have

$$\begin{aligned}
&\widetilde{\mathbb{E}}[D(t_1)X(t_1)|\mathcal{F}(t_0)] \\
&= \widetilde{\mathbb{E}}\left[\int_0^{t_1} D(u)\Delta(u)\widetilde{\Gamma}(u)\, d\widetilde{W}(u) - \int_0^{t_0} D(u)\Delta(u)\widetilde{\Gamma}(u)\, d\widetilde{W}(u)\,\Big|\, \mathcal{F}(t_0)\right] \\
&= \widetilde{\mathbb{E}}\left[\int_0^{t_1} D(u)\Delta(u)\widetilde{\Gamma}(u)\, d\widetilde{W}(u)\,\Big|\, \mathcal{F}(t_0)\right] - \int_0^{t_0} D(u)\Delta(u)\widetilde{\Gamma}(u)\, d\widetilde{W}(u) \\
&= 0.
\end{aligned} \tag{5.6.8}$$

According to the risk-neutral pricing formula, the value at time t_0 of a payment of $X(t_1)$ at time t_1 is $\frac{1}{D(t_0)}\widetilde{\mathbb{E}}\big[D(t_1)X(t_1)\big|\mathcal{F}(t_0)\big]$, and we have just shown that this is zero. The value of owning a long futures position over the interval t_0 to t_1 is obtained by setting $\Delta(u) = 1$ for all u; the value of holding a short position is obtained by setting $\Delta(u) = -1$ for all u. In both cases, we see that this value is zero.

If the filtration $\mathcal{F}(t)$, $0 \leq t \leq T$, is not generated by a Brownian motion, so that we cannot use Corollary 5.3.2, then we must write (5.6.7) as

$$D(t_1)X(t_1) = \int_{t_0}^{t_1} D(u)\Delta(u)\, d\mathrm{Fut}_S(u,T). \tag{5.6.9}$$

This integral can be defined and it will be a martingale. We will again have

$$\widetilde{\mathbb{E}}[D(t_1)X(t_1)|\mathcal{F}(t_0)] = 0. \qquad \square$$

Remark 5.6.6 (Risk-neutral valuation of a cash flow). Suppose an asset generates a cash flow so that between times 0 and u a total of $C(u)$ is paid, where $C(u)$ is $\mathcal{F}(u)$-measurable. Then a portfolio that begins with one share of this asset at time t and holds this asset between times t and T, investing or borrowing at the interest rate r as necessary, satisfies

$$dX(u) = dC(u) + R(u)X(u)\, du,$$

or equivalently

$$d\big(D(u)X(u)\big) = D(u)\, dC(u).$$

Suppose $X(t) = 0$. Then integration shows that

$$D(T)X(T) = \int_t^T D(u)\, dC(u).$$

The risk-neutral value at time t of $X(T)$, which is the risk-neutral value at time t of the cash flow received between times t and T, is thus

$$\frac{1}{D(t)}\widetilde{\mathbb{E}}[D(T)X(T)] = \frac{1}{D(t)}\widetilde{\mathbb{E}}\left[\int_t^T D(u)\, dC(u)\middle| \mathcal{F}(t)\right], \quad 0 \leq t \leq T. \tag{5.6.10}$$

Formula (5.6.10) generalizes the risk-neutral pricing formula (5.2.30) to allow for a cash flow rather than payment at the single time T. In (5.6.10), the process $C(u)$ can represent a succession of lump sum payments $A_1, A_2, \dots, A_n$ at times $t_1 < t_2 < \cdots < t_n$, where each A_i is an $\mathcal{F}(t_i)$-measurable random variable. The formula for this is

$$C(u) = \sum_{i=1}^{n} A_i \mathbb{I}_{[0,u]}(t_i).$$

In this case,

$$\int_t^T D(u)\,dC(u) = \sum_{i=1}^n D(t_i)A_i \mathbb{I}_{(t,T]}(t_i).$$

Only payments made strictly later than time t appear in this sum. Equation (5.6.10) says that the value at time t of the string of payments to be made strictly later than time t is

$$\frac{1}{D(t)}\widetilde{\mathbb{E}}\left[\sum_{i=1}^n D(t_i)A_i\mathbb{I}_{(t,T]}(t_i)\,\middle|\,\mathcal{F}(t)\right] = \sum_{i=1}^n \mathbb{I}_{(t,T]}(t_i)\frac{1}{D(t)}\widetilde{\mathbb{E}}[D(t_i)A_i|\mathcal{F}(t)],$$

which is the sum of the time-t values of the payments made strictly later than time t.

The process $C(u)$ can also be continuous, as in (5.6.9). The process $C(u)$ may decrease as well as increase (i.e., the cash flow may be negative as well as positive). □

5.6.3 Forward–Futures Spread

We conclude with a comparison of forward and futures prices. We have defined these prices to be

$$\begin{aligned}\mathrm{For}_S(t,T) &= \frac{S(t)}{B(t,T)},\\ \mathrm{Fut}_S(t,T) &= \widetilde{\mathbb{E}}[S(T)|\mathcal{F}(t)].\end{aligned}$$

If the interest rate is a constant r, then $B(t,T) = e^{-r(T-t)}$ and

$$\begin{aligned}&\mathrm{For}_S(t,T) = e^{r(T-t)}S(t),\\ &\mathrm{Fut}_S(t,T) = e^{rT}\widetilde{\mathbb{E}}\big[e^{-rT}S(T)\big|\mathcal{F}(t)\big] \;=\; e^{rT}e^{-rt}S(t) \;=\; e^{r(T-t)}S(t).\end{aligned}$$

In this case, the forward and futures prices agree.

We compare $\mathrm{For}_S(0,T)$ and $\mathrm{Fut}_S(0,T)$ in the case of a random interest rate. In this case, $B(0,T) = \widetilde{\mathbb{E}}D(T)$, and the so-called *forward–futures spread* is

$$\begin{aligned}\mathrm{For}_S(0,T) - \mathrm{Fut}_S(0,T) &= \frac{S(0)}{\widetilde{\mathbb{E}}D(T)} - \widetilde{\mathbb{E}}S(T)\\ &= \frac{1}{\widetilde{\mathbb{E}}D(T)}\{\widetilde{\mathbb{E}}[D(T)S(T)] - \widetilde{\mathbb{E}}D(T)\cdot\widetilde{\mathbb{E}}S(T)\}\\ &= \frac{1}{B(0,T)}\widetilde{\mathrm{Cov}}\big(D(T),S(T)\big), \qquad (5.6.11)\end{aligned}$$

where $\widetilde{\mathrm{Cov}}\big(D(T),S(T)\big)$ denotes the covariance of $D(T)$ and $S(T)$ under the risk-neutral measure. If the interest rate is nonrandom, this covariance is zero and the futures price agrees with the forward price.

One can explain this last formula as follows. If $D(T)$ and $S(T)$ are positively correlated, then higher asset prices tend to correspond to higher discount levels, which tend to correspond to lower interest rates. But when the asset goes up, the long position in the futures contract receives a payment (because the futures price is positively correlated with the underlying asset price). The long position in the futures contract thus receives money when the interest rate for investing is unfavorable (low) and conversely must pay money when the interest rate at which money can be borrowed is also unfavorable (high). The owner of the futures contract would have rather owned the forward contract, in which all payments are postponed until the end. Therefore, to make the futures contract attractive, the futures price must be lower than the forward price. (Recall that this price is what the investor ultimately pays for the asset.) This creates a positive forward–futures spread when the discount factor $D(T)$ and the asset price $S(T)$ are positively correlated. Note that all correlations in this argument are computed under the risk-neutral measure, not the actual probability measure. In a Brownian-motion-driven model, in which the multidimensional Girsanov Theorem, Theorem 5.4.1, is used to change to the risk-neutral measure, instantaneous asset correlations are the same under both measures (see Exercise 5.12). However, correlations between random variables (as opposed to instantaneous correlations between stochastic processes) can be affected by changes of measure (see Exercise 5.13).

5.7 Summary

This chapter treats the application to finance of two major theorems, Girsanov (Theorem 5.4.1) and Martingale Representation (Theorem 5.4.2). These lead to the two *Fundamental Theorems of Asset Pricing*, Theorem 5.4.7 and Theorem 5.4.9. Both of these are stated for models with multiple assets whose prices are driven by multiple Brownian motions.

According to the Fundamental Theorems of Asset Pricing, there are three possible situations when we build a mathematical model of a multiasset market.

Case 1. There is no risk-neutral measure (i.e., the market price of risk equations (5.4.18) cannot be solved for $\Theta_1(t), \dots, \Theta_d(t)$). This is a bad model. There must be some way to form an arbitrage by trading at the prices given by this model. Do not use this model.

Case 2. There are multiple risk-neutral measures (i.e., the market price of risk equations (5.4.18) have more than one solution). The different risk-neutral measures lead to different prices for derivative securities in the model. Any derivative security that has more than one price cannot be synthesized by trading in the model (i.e., a position in this derivative security cannot be hedged). (If the derivative security could be hedged, this would determine a unique price; see the proof of Theorem 5.4.9.) It may still be possible to calibrate the model (i.e., determine its parameters by getting it to match market prices, and the model might then give reasonable prices for nontraded instruments). However, it cannot be used to fully hedge the exposure associated with derivative positions.

At the present time, credit derivative models fall into Case 2. They are used for pricing, but are incomplete because the derivatives in question pay off contingent upon the default of some party and it is impossible to perfectly hedge default risk by trading in primary assets. These models have multiple risk-neutral measures, all of which can be consistent with market prices of the primary assets but give different prices for derivatives. In practical applications, one of these risk-neutral measures is singled out and used for pricing. Which of the risk-neutral measures is chosen for this purpose depends on the way the model is specified and calibrated.

Case 3. There is one and only one set of processes $\Theta_1(t), \dots, \Theta_d(t)$ that solve the market price of risk equations (5.4.18). There is a unique risk-neutral measure, and risk-neutral pricing is justified. In other words,the price (value) at time t of any security that pays $V(T)$ at time T is

$$V(t) = \frac{1}{D(t)} \widetilde{\mathbb{E}}\big[D(T)V(T)\big|\mathcal{F}(t)\big]. \tag{5.7.1}$$

In particular, the price at time zero of the security is its risk-neutral expected discounted payoff. The risk-neutral price of a derivative security is the initial capital that permits an agent to set up a perfect hedge for a short position in that derivative security. These perfect hedges are the solutions $\Delta_1(t), \dots, \Delta_m(t)$ of the hedging equations (5.4.29), and these solutions are guaranteed to exist (by the second part of the proof of Theorem 5.4.9). However, we do not generally attempt to determine the hedging positions $\Delta_1(t), \dots, \Delta_m(t)$ by solving (5.4.29). Instead, we determine hedges by the technique presented in Chapter 6.

When assets pay *dividends*, their discounted prices are no longer martingales under the risk-neutral measure. Instead, the martingale under the risk-neutral measure is the discounted value of any portfolio that trades in the assets and receives dividends in proportion to its position in the assets at the time of dividend payment. For the case of a continuous payment of dividends at a constant rate, the Black-Scholes-Merton formula is given by (5.5.12). If dividend payments are made in lump sums, the necessary modification to the classical Black-Scholes-Merton formula is presented in Subsection 11.5.4.

The *forward price* of an asset is defined to be that price that one can agree today to pay at a future delivery date so that the present value of the forward contract is zero. For assets that pay no dividends (and, unlike most commodities, cost nothing to hold), the forward price is the asset price divided by the price of a zero-coupon bond maturing on the delivery date and having face value 1:

$$\mathrm{For}_S(t,T) = \frac{S(t)}{B(t,T)}, \quad 0 \le t \le T.$$

The *futures price* of an asset is an adapted stochastic process $\text{Fut}_S(t,T)$ with two properties.

(i) The futures price agrees with the asset price on the delivery date (i.e., $\text{Fut}_S(T,T) = S(T)$).
(ii) The value of holding the futures contract over a period of time and receiving the cash flows associated with this position is zero:

$$\frac{1}{D(t_0)}\widetilde{\mathbb{E}}\left[\int_{t_0}^{t_1} D(u)d\text{Fut}_S(u,T)\bigg|\mathcal{F}(t)\right] = 0, \quad 0 \le t_0 < t_1 \le T.$$

The unique process having these two properties is

$$\text{Fut}_S(t,T) = \widetilde{\mathbb{E}}[S(T)|\mathcal{F}(t)], \quad 0 \le t \le T.$$

When the interest rate process is nonrandom, forward and futures prices agree. When interest rates are random, the difference between forward and futures prices is proportional to the covariance under the risk-neutral measure between the discount factor $D(T)$ and the underlying asset price $S(T)$ (see (5.6.11)).

5.8 Notes

The idea of risk-neutral pricing is implicit in the classical papers by Black and Scholes [17] and Merton [122] but was not fully developed and appreciated until the work of Ross [140], Harrison and Kreps [77], and Harrison and Pliska [78], [79]. Ross [140] treats a one-period model, Harrison and Kreps [77] treat a continuous-time model with trading at discrete dates, and Harrison and Pliska [78], [79] treat a continuous-time model with continuous trading. The closely related concept of state price density (see Exercise 5.2) is due to Arrow and Debreu [5].

Girsanov's Theorem, Theorem 5.2.3, in the generality stated here is due to Girsanov [72], although the result for constant θ was established much earlier by Cameron and Martin [26]. The theorem requires a technical condition to ensure that $\mathbb{E}Z(T) = 1$ so that $\widetilde{\mathbb{P}}$ is a probability measure. For this purpose, we imposed (5.2.13). An easier condition to verify, due to Novikov [128], is

$$\mathbb{E}\left\{\frac{1}{2}\int_0^T \Theta^2(u)\,du\right\} < \infty;$$

see Karatzas and Shreve [101], page 198. The multidimensional version of both Girsanov's Theorem and the Martingale Representation Theorem (Theorems 5.4.1 and 5.4.2) can be found in Karatzas and Shreve [101] as Theorems 5.1 and 4.15 of Chapter 3. A mathematically rigorous application of these theorems to Brownian-motion-driven models in finance is provided by Karatzas and Shreve [102].

The application of the Girsanov Theorem to risk-neutral pricing is due to Harrison and Pliska [78]. This methodology frees the Brownian-motion-driven model from the assumption of a constant interest rate and volatility. When both of these are stochastic, the Brownian-motion-driven model is mathematically the most general possible for continuous stock prices that do not admit arbitrage. In particular, *the log-normal model for asset prices is just one special case of the Brownian-motion-driven model.*

The Fundamental Theorems of Asset Pricing, Theorems 5.4.7 and 5.4.9, can be found in Harrison and Pliska [78], [79]. It is tempting to believe the converse of Theorem 5.4.7 (i.e., that the absence of arbitrage implies the existence of a risk-neutral measure). This is true in discrete-time models (see Dalang, Morton, and Willinger [45]), but in continuous-time models a slightly stronger condition is needed to guarantee existence of a risk-neutral measure. See Delbaen and Schachermayer [49] for a summary of relevant results.

The distinction between forward contracts and futures was pointed out by Margrabe [118] and Black [13]. No-arbitrage pricing of futures in a discrete-time model was developed by Cox, Ingersoll, and Ross [40] and Jarrow and Oldfield [98].

5.9 Exercises

Exercise 5.1. Consider the discounted stock price $D(t)S(t)$ of (5.2.19). In this problem, we derive the formula (5.2.20) for $d\big(D(t)S(t)\big)$ by two methods.

(i) Define $f(x) = S(0)e^x$ and set

$$X(t) = \int_0^t \sigma(s)\,dW(s) + \int_0^t \Big(\alpha(s) - R(s) - \frac{1}{2}\sigma^2(s)\Big)\,ds$$

so that $D(t)S(t) = f(X(t))$. Use the Itô-Doeblin formula to compute $df(X(t))$.

(ii) According to Itô's product rule,

$$d\big(D(t)S(t)\big) = S(t)\,dD(t) + D(t)\,dS(t) + dD(t)\,dS(t).$$

Use (5.2.15) and (5.2.18) to work out the right-hand side of this equation.

Exercise 5.2 (State price density process). Show that the risk-neutral pricing formula (5.2.30) may be rewritten as

$$D(t)Z(t)V(t) = \mathbb{E}\big[D(T)Z(T)V(T)\big|\mathcal{F}(t)\big]. \tag{5.9.1}$$

Here $Z(t)$ is the Radon-Nikodým derivative process (5.2.11) when the market price of risk process $\Theta(t)$ is given by (5.2.21) and the conditional expectation on the right-hand side of (5.9.1) is taken under the actual probability measure

$\mathbb{P}$, not the risk-neutral measure $\widetilde{\mathbb{P}}$. In particular, if for some $A \in \mathcal{F}(T)$ a derivative security pays off $\mathbb{I}_A$ (i.e., pays 1 if A occurs and 0 if A does not occur), then the value of this derivative security at time zero is $\mathbb{E}[D(T)Z(T)\mathbb{I}_A]$. The process $D(t)Z(t)$ appearing in (5.9.1) is called the *state price density process.*

Exercise 5.3. According to the Black-Scholes-Merton formula, the value at time zero of a European call on a stock whose initial price is $S(0) = x$ is given by

$$c(0,x) = xN(d_+(T,x)) - Ke^{-rT}N(d_-(T,x)),$$

where

$$d_+(T,x) = \frac{1}{\sigma\sqrt{T}}\left[\log\frac{x}{K} + \left(r + \frac{1}{2}\sigma^2\right)T\right],$$
$$d_-(T,x) = d_+(T,x) - \sigma\sqrt{T}.$$

The stock is modeled as a geometric Brownian motion with constant volatility $\sigma > 0$, the interest rate is constant r, the call strike is K, and the call expiration time is T. This formula is obtained by computing the discounted expected payoff of the call under the risk-neutral measure,

$$\begin{aligned} c(0,x) &= \widetilde{\mathbb{E}}\left[e^{-rT}(S(T) - K)^+\right] \\ &= \widetilde{\mathbb{E}}\left[e^{-rT}\left(x\exp\left\{\sigma\widetilde{W}(T) + \left(r - \frac{1}{2}\sigma^2\right)T\right\} - K\right)^+\right], \end{aligned} \tag{5.9.2}$$

where $\widetilde{W}$ is a Brownian motion under the risk-neutral measure $\widetilde{\mathbb{P}}$. In Exercise 4.9(ii), the *delta* of this option is computed to be $c_x(0,x) = N(d_+(T,x))$. This problem provides an alternate way to compute $c_x(0,x)$.

(i) We begin with the observation that if $h(s) = (s - K)^+$, then

$$h'(s) = \begin{cases} 0 \text{ if } s < K, \\ 1 \text{ if } s > K. \end{cases}$$

If $s = K$, then $h'(s)$ is undefined, but that will not matter in what follows because $S(T)$ has zero probability of taking the value K. Using the formula for $h'(s)$, differentiate inside the expected value in (5.9.2) to obtain a formula for $c_x(0,x)$.

(ii) Show that the formula you obtained in (i) can be rewritten as

$$c_x(0,x) = \widehat{\mathbb{P}}(S(T) > K),$$

where $\widehat{\mathbb{P}}$ is a probability measure equivalent to $\widetilde{\mathbb{P}}$. Show that

$$\widehat{W}(t) = \widetilde{W}(t) - \sigma t$$

is a Brownian motion under $\widehat{\mathbb{P}}$.

(iii) Rewrite $S(T)$ in terms of $\widehat{W}(T)$, and then show that

$$\widehat{\mathbb{P}}\{S(T) > K\} = \widehat{\mathbb{P}}\left\{-\frac{\widehat{W}(T)}{\sqrt{T}} < d_+(T,x)\right\} = N(d_+(T,x)).$$

Exercise 5.4 (Black-Scholes-Merton formula for time-varying, nonrandom interest rate and volatility). Consider a stock whose price differential is

$$dS(t) = r(t)S(t)\,dt + \sigma(t)\,d\widetilde{W}(t),$$

where $r(t)$ and $\sigma(t)$ are nonrandom functions of t and $\widetilde{W}$ is a Brownian motion under the risk-neutral measure $\widetilde{\mathbb{P}}$. Let $T > 0$ be given, and consider a European call, whose value at time zero is

$$c(0, S(0)) = \mathbb{E}\left[\exp\left\{-\int_0^T r(t)\,dt\right\}(S(T) - K)^+\right].$$

(i) Show that $S(T)$ is of the form $S(0)e^X$, where X is a normal random variable, and determine the mean and variance of X.

(ii) Let

$$\begin{aligned}\mathrm{BSM}(T, x; K, R, \Sigma) &= xN\left(\frac{1}{\Sigma\sqrt{T}}\left[\log\frac{x}{K} + (R + \Sigma^2/2)T\right]\right)\\ &\quad - e^{-RT}KN\left(\frac{1}{\Sigma\sqrt{T}}\left[\log\frac{x}{K} + (R - \Sigma^2/2)T\right]\right)\end{aligned}$$

denote the value at time zero of a European call expiring at time T when the underlying stock has constant volatility Σ and the interest rate R is constant. Show that

$$c(0, S(0)) = \mathrm{BSM}\left(S(0), T, \frac{1}{T}\int_0^T r(t)dt, \sqrt{\frac{1}{T}\int_0^T \sigma^2(t)dt}\right).$$

Exercise 5.5. Prove Corollary 5.3.2 by the following steps.

(i) Compute the differential of $\frac{1}{Z(t)}$, where $Z(t)$ is given in Corollary 5.3.2.

(ii) Let $\widetilde{M}(t)$, $0 \le t \le T$, be a martingale under $\widetilde{\mathbb{P}}$. Show that $M(t) = Z(t)\widetilde{M}(t)$ is a martingale under $\mathbb{P}$.

(iii) According to Theorem 5.3.1, there is an adapted process $\Gamma(u)$, $0 \le u \le T$, such that

$$M(t) = M(0) + \int_0^T \Gamma(u)\,dW(u),\ 0 \le t \le T.$$

Write $\widetilde{M}(t) = M(t) \cdot \frac{1}{Z(t)}$ and take its differential using Itô's product rule.

(iv) Show that the differential of $\widetilde{M}(t)$ is the sum of an adapted process, which we call $\widetilde{\Gamma}(t)$, times $d\widetilde{W}(t)$, and zero times dt. Integrate to obtain (5.3.2).

Exercise 5.6. Use the two-dimensional Lévy Theorem, Theorem 4.6.5, to prove the two-dimensional Girsanov Theorem (i.e., Theorem 5.4.1 with $d = 2$).

Exercise 5.7. (i) Suppose a multidimensional market model as described in Section 5.4.2 has an arbitrage. In other words, suppose there is a portfolio value process satisfying $X_1(0) = 0$ and

$$\mathbb{P}\{X_1(T) \geq 0\} = 1, \quad \mathbb{P}\{X_1(T) > 0\} > 0, \tag{5.4.23}$$

for some positive T. Show that if $X_2(0)$ is positive, then there exists a portfolio value process $X_2(t)$ starting at $X_2(0)$ and satisfying

$$\mathbb{P}\left\{X_2(T) \geq \frac{X_2(0)}{D(T)}\right\} = 1, \quad \mathbb{P}\left\{X_2(T) > \frac{X_2(0)}{D(T)}\right\} > 0. \tag{5.4.24}$$

(ii) Show that if a multidimensional market model has a portfolio value process $X_2(t)$ such that $X_2(0)$ is positive and (5.4.24) holds, then the model has a portfolio value process $X_1(0)$ such that $X_1(0) = 0$ and (5.4.23) holds.

Exercise 5.8 (Every strictly positive asset is a generalized geometric Brownian motion). Let $(\Omega, \mathcal{F}, \mathbb{P})$ be a probability space on which is defined a Brownian motion $W(t)$, $0 \leq t \leq T$. Let $\mathcal{F}(t)$, $0 \leq t \leq T$, be the filtration generated by this Brownian motion. Assume there is a unique risk-neutral measure $\widetilde{\mathbb{P}}$, and let $\widetilde{W}(t)$, $0 \leq t \leq T$, be the Brownian motion under $\widetilde{\mathbb{P}}$ obtained by an application of Girsanov's Theorem, Theorem 5.2.3.

Corollary 5.3.2 of the Martingale Representation Theorem asserts that every martingale $\widetilde{M}(t)$, $0 \leq t \leq T$, under $\widetilde{\mathbb{P}}$ can be written as a stochastic integral with respect to $\widetilde{W}(t)$, $0 \leq t \leq T$. In other words, there exists an adapted process $\widetilde{\Gamma}(t)$, $0 \leq t \leq T$, such that

$$\widetilde{M}(t) = \widetilde{M}(0) + \int_0^t \widetilde{\Gamma}(u)\, d\widetilde{B}(u), \quad 0 \leq t \leq T.$$

Now let $V(T)$ be an almost surely positive ("almost surely" means with probability one under both $\mathbb{P}$ and $\widetilde{\mathbb{P}}$ since these two measures are equivalent), $\mathcal{F}(T)$-measurable random variable. According to the risk-neutral pricing formula (5.2.31), the price at time t of a security paying $V(T)$ at time T is

$$V(t) = \widetilde{\mathbb{E}}\left[e^{-\int_t^T R(u)du} V(T) \middle| \mathcal{F}(t)\right], \quad 0 \leq t \leq T.$$

(i) Show that there exists an adapted process $\widetilde{\Gamma}(t)$, $0 \leq t \leq T$, such that

$$dV(t) = R(t)V(t)\, dt + \frac{\widetilde{\Gamma}(t)}{D(t)}\, d\widetilde{W}(t), \quad 0 \leq t \leq T.$$

(ii) Show that, for each $t \in [0, T]$, the price of the derivative security $V(t)$ at time t is almost surely positive.

(iii) Conclude from (i) and (ii) that there exists an adapted process $\sigma(t)$, $0 \leq t \leq T$, such that

$$dV(t) = R(t)V(t)\,dt + \sigma(t)V(t)\,d\widetilde{W}(t), \quad 0 \leq t \leq T.$$

In other words, prior to time T, the price of every asset with almost surely positive price at time T follows a generalized (because the volatility may be random) geometric Brownian motion.

Exercise 5.9 (Implying the risk-neutral distribution). Let $S(t)$ be the price of an underlying asset, which is not necessarily a geometric Brownian motion (i.e., does not necessarily have constant volatility). With $S(0) = x$, the risk-neutral pricing formula for the price at time zero of a European call on this asset, paying $\big(S(T) - K\big)^+$ at time T, is

$$c(0, T, x, K) = \widetilde{\mathbb{E}}\left[e^{-rT}\big(S(T) - K\big)^+\right].$$

(Normally we consider this as a function of the current time 0 and the current stock price x, but in this exercise we shall also treat the expiration time T and the strike price K as variables, and for that reason we include them as arguments of c.) We denote by $\tilde{p}(0, T, x, y)$ the risk-neutral density in the y variable of the distribution of $S(T)$ when $S(0) = x$. Then we may rewrite the risk-neutral pricing formula as

$$c(0, T, x, K) = e^{-rT} \int_K^\infty (y - K)\tilde{p}(0, T, x, y)\,dy. \tag{5.9.3}$$

Suppose we know the market prices for calls of all strikes (i.e., we know $c(0, T, x, K)$ for all $K > 0$).[2] We can then compute $c_K(0, T, x, K)$ and $c_{KK}(0, T, x, K)$, the first and second derivatives of the option price with respect to the strike. Differentiate (5.9.3) twice with respect to K to obtain the equations

$$c_K(0, T, x, K) = -e^{-rT} \int_K^\infty \tilde{p}(0, T, x, y)dy = -e^{-rT}\widetilde{\mathbb{P}}\{S(T) > K\},$$
$$c_{KK}(0, T, x, K) = e^{-rT}\tilde{p}(0, T, x, K).$$

The second of these equations provides a formula for the risk-neutral distribution of $S(T)$ in terms of call prices:

$$\tilde{p}(0, T, x, K) = e^{rT} c_{KK}(0, T, x, K) \text{ for all } K > 0.$$

[2] In practice, we do not have this many prices. We have the prices of calls at some strikes, and we can infer the prices of calls at other strikes by knowing the prices of puts and using put–call parity. We must create prices for the calls of other strikes by interpolation of the prices we do have.

Exercise 5.10 (Chooser option). Consider a model with a unique risk-neutral measure $\widetilde{\mathbb{P}}$ and constant interest rate r. According to the risk-neutral pricing formula, for $0 \le t \le T$, the price at time t of a European call expiring at time T is

$$C(t) = \widetilde{\mathbb{E}}\left[e^{-r(T-t)}(S(T)-K)^{+}\Big|\mathcal{F}(t)\right],$$

where $S(T)$ is the underlying asset price at time T and K is the strike price of the call. Similarly, the price at time t of a European put expiring at time T is

$$P(t) = \widetilde{\mathbb{E}}\left[e^{-r(T-t)}(K-S(T))^{+}\Big|\mathcal{F}(t)\right].$$

Finally, because $e^{-rt}S(t)$ is a martingale under $\widetilde{\mathbb{P}}$, the price at time t of a forward contract for delivery of one share of stock at time T in exchange for a payment of K at time T is

$$\begin{aligned}F(t) &= \widetilde{\mathbb{E}}\left[e^{-r(T-t)}(S(T)-K)\Big|\mathcal{F}(t)\right]\\ &= e^{rt}\widetilde{\mathbb{E}}\left[e^{-rT}S(T)\big|\mathcal{F}(t)\right] - e^{-r(T-t)}K\\ &= S(t) - e^{-r(T-t)}K.\end{aligned}$$

Because

$$(S(T)-K)^{+} - (K-S(T))^{+} = S(T)-K,$$

we have the *put–call parity* relationship

$$\begin{aligned}C(t)-P(t) &= \widetilde{\mathbb{E}}\left[e^{-r(T-t)}(S(T)-K)^{+} - e^{-r(T-t)}(K-S(T))^{+}\Big|\mathcal{F}(t)\right]\\ &= \widetilde{\mathbb{E}}\left[e^{-r(T-t)}(S(T)-K)\Big|\mathcal{F}(t)\right] \;=\; F(t).\end{aligned}$$

Now consider a date t_0 between 0 and T, and consider a *chooser option*, which gives the right at time t_0 to choose to own either the call or the put.

(i) Show that at time t_0 the value of the chooser option is

$$C(t_0) + \max\{0, -F(t_0)\} = C(t_0) + \left(e^{-r(T-t_0)}K - S(t_0)\right)^{+}.$$

(ii) Show that the value of the chooser option at time 0 is the sum of the value of a call expiring at time T with strike price K and the value of a put expiring at time t_0 with strike price $e^{-r(T-t_0)}K$.

Exercise 5.11 (Hedging a cash flow). Let $W(t)$, $0 \le t \le T$, be a Brownian motion on a probability space $(\Omega, \mathcal{F}, \mathbb{P})$, and let $\mathcal{F}(t)$, $0 \le t \le T$, be the filtration generated by this Brownian motion. Let the mean rate of return $\alpha(t)$, the interest rate $R(t)$, and the volatility $\sigma(t)$ be adapted processes, and assume that $\sigma(t)$ is never zero. Consider a stock price process whose differential is given by (5.2.15):

$$dS(t) = \alpha(t)S(t)\,dt + \sigma(t)S(t)\,dW(t), \quad 0 \le t \le T.$$

Suppose an agent must pay a cash flow at rate $C(t)$ at each time t, where $C(t)$, $0 \le t \le T$, is an adapted process. If the agent holds $\Delta(t)$ shares of stock at each time t, then the differential of her portfolio value will be

$$dX(t) = \Delta(t)\,dS(t) + R(t)\big(X(t) - \Delta(t)S(t)\big)\,dt - C(t)\,dt. \tag{5.9.4}$$

Show that there is a nonrandom value of $X(0)$ and a portfolio process $\Delta(t)$, $0 \le t \le T$, such that $X(T) = 0$ almost surely. (Hint: Define the risk-neutral measure and apply Corollary 5.3.2 of the Martingale Representation Theorem to the process

$$\widetilde{M}(t) = \widetilde{\mathbb{E}}\left[\int_0^T D(u)C(u)\,du \,\middle|\, \mathcal{F}(t)\right], \quad 0 \le t \le T, \tag{5.9.5}$$

where $D(t)$ is the discount process (5.2.17).)

Exercise 5.12 (Correlation under change of measure). Consider the multidimensional market model of Subsection 5.4.2, and let $B_i(t)$ be defined by (5.4.7). Assume that the market price of risk equations (5.4.18) have a solution $\Theta_1(t), \dots, \Theta_d(t)$, and let $\widetilde{\mathbb{P}}$ be the corresponding risk-neutral measure under which

$$\widetilde{W}_j(t) = W_j(t) + \int_0^t \Theta_j(u)\,du, \quad j = 1, \dots, d,$$

are independent Brownian motions.

(i) For $i = 1, \dots, d$, define $\gamma_i(t) = \sum_{j=1}^d \frac{\sigma_{ij}(t)\theta_j(t)}{\sigma_i(t)}$. Show that

$$\widetilde{B}_i(t) = B_i(t) + \int_0^t \gamma_i(u)\,du$$

is a Brownian motion under $\widetilde{\mathbb{P}}$.

(ii) We saw in (5.4.8) that

$$dS_i(t) = \alpha_i(t)S_i(t)\,dt + \sigma_i(t)S_i(t)\,dB_i(t), \quad i = 1, \dots, m.$$

Show that

$$dS_i(t) = R(t)S_i(t)\,dt + \sigma_i S_i(t)\,d\widetilde{B}_i(t), \quad i = 1, \dots, m.$$

(iii) We saw in (5.4.9) that $dB_i(t)\,dB_k(t) = \rho_{ik}(t)$. This is the *instantaneous correlation* between $B_i(t)$ and $B_k(t)$. Because (5.4.9) makes no reference to the probability measure, Exercise 4.17 of Chapter 4 implies that under both $\mathbb{P}$ and $\widetilde{\mathbb{P}}$, the correlation between the pair of increments $B_1(t_0+\epsilon) - B_1(t_0)$ and $B_2(t_0+\epsilon) - B_2(t_0)$ is approximately $\rho_{ik}(t_0)$. Show that

$$d\widetilde{B}_i(t)\,d\widetilde{B}_k(t) = \rho_{ik}(t).$$

This formula means that, conditioned on $\mathcal{F}(t_0)$, under both $\mathbb{P}$ and $\widetilde{\mathbb{P}}$ the correlation between the pair of increments $\widetilde{B}_1(t_0+\epsilon) - \widetilde{B}_1(t_0)$ and $\widetilde{B}_2(t_0+\epsilon) - \widetilde{B}_2(t_0)$ is approximately $\rho_{ik}(t_0)$.

(iv) Show that if $\rho_{ik}(t)$ is not random (although it may still depend on t), then for every $t \geq 0$,

$$\mathbb{E}\big[B_i(t)B_k(t)\big] = \widetilde{\mathbb{E}}\big[\widetilde{B}_i(t)\widetilde{B}_k(t)\big] = \int_0^t \rho_{ik}(u)\,du.$$

Since $B_i(t)$ and $B_k(t)$ both have variance t under $\mathbb{P}$ and $\widetilde{B}_i(t)$ and $\widetilde{B}_k(t)$ both have variance t under $\widetilde{\mathbb{P}}$, this shows that the correlation between $B_i(t)$ and $B_k(t)$ under $\mathbb{P}$ is the same as the correlation between $\widetilde{B}_i(t)$ and $\widetilde{B}_k(t)$ under $\widetilde{\mathbb{P}}$. In both cases, this correlation is $\frac{1}{t}\int_0^t \rho_{ik}(u)\,du$. If ρ_{ik} is constant, then the correlation is simply ρ_{ik}.

(v) When $\rho_{ik}(t)$ is random, we can have

$$\mathbb{E}\big[B_i(t)B_k(t)\big] \neq \widetilde{\mathbb{E}}\big[\widetilde{B}_i(t)\widetilde{B}_k(t)\big].$$

Even though instantaneous correlations are unaffected by a change of measure, correlations can be. To see this, we take $m = d = 2$ and let $W_1(t)$ and $W_2(t)$ be independent Brownian motions under $\mathbb{P}$. Take $\sigma_{11}(t) = \sigma_{21}(t) = 0$, $\sigma_{12}(t) = 1$, and $\sigma_{22}(t) = \operatorname{sign}(W_1(t))$, where

$$\operatorname{sign}(x) = \begin{cases} 1 & \text{if } x \geq 0, \\ -1 & \text{if } x < 0. \end{cases}$$

Then $\sigma_1(t) = 1$, $\sigma_2(t) = 1$, $\rho_{11}(t) = 1$, $\rho_{22}(t) = 1$ and $\rho_{12}(t) = \rho_{21}(t) = \operatorname{sign}(W_1(t))$. Take $\Theta_1(t) = 1$ and $\Theta_2(t) = 0$, so that $\widetilde{W}_1(t) = W_1(t) + t$ and $\widetilde{W}_2(t) = W_2(t)$. Then $\gamma_1(t) = \gamma_2(t) = 0$. We have

$$B_1(t) = W_2(t), \quad B_2(t) = \int_0^t \operatorname{sign}(W_1(u))\,dW_2(u),$$
$$\widetilde{B}_1(t) = B_1(t), \quad \widetilde{B}_2(t) = B_2(t).$$

Show that

$$\mathbb{E}\big[B_1(t)B_2(t)\big] \neq \widetilde{\mathbb{E}}\big[\widetilde{B}_1(t)\widetilde{B}_2(t)\big] \text{ for all } t > 0.$$

Exercise 5.13. In part (v) of Exercise 5.12, we saw that when we change measures and change Brownian motions, correlations can change if the instantaneous correlations are random. This exercise shows that a change of measure without a change of Brownian motions can change correlations if the market prices of risk are random.

Let $W_1(t)$ and $W_2(t)$ be independent Brownian motions under a probability measure $\widetilde{\mathbb{P}}$. Take $\Theta_1(t) = 0$ and $\Theta_2(t) = W_1(t)$ in the multidimensional Girsanov Theorem, Theorem 5.4.1. Then $\widetilde{W}_1(t) = W_1(t)$ and $\widetilde{W}_2(t) = W_2(t) + \int_0^t W_1(u)\,du$.

(i) Because $\widetilde{W}_1(t)$ and $\widetilde{W}_2(t)$ are Brownian motions under $\widetilde{\mathbb{P}}$, the equation $\widetilde{\mathbb{E}}\widetilde{W}_1(t) = \widetilde{\mathbb{E}}\widetilde{W}_2(t) = 0$ must hold for all $t \in [0,T]$. Use this equation to conclude that

$$\widetilde{\mathbb{E}}W_1(t) = \widetilde{\mathbb{E}}W_2(t) = 0 \text{ for all } t \in [0,T].$$

(ii) From Itô's product rule, we have

$$d\big(W_1(t)W_2(t)\big) = W_1(t)\,dW_2(t) + W_2(t)\,dW_1(t).$$

Use this equation to show that

$$\widetilde{\mathrm{Cov}}[W_1(T), W_2(T)] = \widetilde{\mathbb{E}}[W_1(T)W_2(T)] = -\frac{1}{2}T^2.$$

This is different from

$$\mathrm{Cov}[W_1(T), W_2(T)] = \mathbb{E}[W_1(T)W_2(T)] = 0.$$

Exercise 5.14 (Cost of carry). Consider a commodity whose unit price at time t is $S(t)$. Ownership of a unit of this commodity requires payment at a rate a per unit time (*cost of carry*) for storage. Note that this payment is per unit of commodity, not a fraction of the price of the commodity. Thus, the value of a portfolio that holds $\Delta(t)$ units of the commodity at time t and also invests in a money market account with constant rate of interest r has differential

$$dX(t) = \Delta(t)\,dS(t) - a\Delta(t)\,dt + r\big(X(t) - \Delta(t)S(t)\big)\,dt. \tag{5.9.6}$$

As with the dividend-paying stock in Section 5.5, we must choose the risk-neutral measure so that the discounted portfolio value $e^{-rt}X(t)$ is a martingale. We shall assume a constant volatility, so in place of (5.5.6) we have

$$dS(t) = rS(t)\,dt + \sigma S(t)\,d\widetilde{W}(t) + a\,dt, \tag{5.9.7}$$

where $\widetilde{W}(t)$ is a Brownian motion under the risk-neutral measure $\widetilde{\mathbb{P}}$.

(i) Show that when $dS(t)$ is given by (5.9.7), then under $\widetilde{\mathbb{P}}$ the discounted portfolio value process $e^{-rt}X(t)$, where $X(t)$ is given by (5.9.6), is a martingale.

(ii) Define

$$Y(t) = \exp\left\{\sigma\widetilde{W}(t) + \left(r - \frac{1}{2}\sigma^2\right)t\right\}.$$

Verify that, for $0 \le t \le T$,

$$dY(t) = rY(t)\,dt + \sigma Y(t)\,d\widetilde{W}(t),$$

that $e^{-rt}Y(t)$ is a martingale under $\widetilde{\mathbb{P}}$, and that

$$S(t) = S(0)Y(t) + Y(t)\int_0^t \frac{a}{Y(s)}\,ds \tag{5.9.8}$$

satisfies (5.9.7).

(iii) For $0 \le t \le T$, derive a formula for $\widetilde{\mathbb{E}}[S(T)|\mathcal{F}(t)]$ in terms of $S(t)$ by writing

$$\begin{aligned}\widetilde{\mathbb{E}}[S(T)|\mathcal{F}(t)] = S(0)\widetilde{\mathbb{E}}[Y(T)|\mathcal{F}(t)] + \widetilde{\mathbb{E}}[Y(T)|\mathcal{F}(t)]\int_0^t \frac{a}{Y(s)}\,ds \\ +a\int_t^T \widetilde{\mathbb{E}}\left[\frac{Y(T)}{Y(s)}\middle|\mathcal{F}(t)\right]ds\end{aligned} \tag{5.9.9}$$

and then simplifying the right-hand side of this equation.

(iv) The process $\widetilde{\mathbb{E}}[S(T)|\mathcal{F}(t)]$ is the futures price process for the commodity (i.e., $\mathrm{Fut}_S(t,T) = \widetilde{\mathbb{E}}[S(T)|\mathcal{F}(t)]$). This must be a martingale under $\widetilde{\mathbb{P}}$. To check the formula you obtained in (iii), differentiate it and verify that $\widetilde{\mathbb{E}}[S(T)|\mathcal{F}(t)]$ is a martingale under $\widetilde{\mathbb{P}}$.

(v) Let $0 \le t \le T$ be given. Consider a forward contract entered at time t to purchase one unit of the commodity at time T for price K paid at time T. The value of this contract at time t when it is entered is

$$\widetilde{\mathbb{E}}\big[e^{-r(T-t)}(S(T) - K)\big|\mathcal{F}(t)\big]. \tag{5.9.10}$$

The forward price $\mathrm{For}_S(t,T)$ is the value of K that makes the contract value (5.9.10) equal to zero. Show that $\mathrm{For}_S(t,T) = \mathrm{Fut}_S(t,T)$.

(vi) Consider an agent who takes a short position in a forward contract at time zero. This costs nothing and generates no income at time zero. The agent hedges this position by borrowing $S(0)$ from the money market account and purchasing one unit of the commodity, which she holds until time T. At time T, the agent delivers the commodity under the forward contract and receives the forward price $\mathrm{For}_S(0,T)$ set at time zero. Show that this is exactly what the agent needs to cover her debt to the money market account, which has two parts. First of all, at time zero, the agent borrows $S(0)$ from the money market account in order to purchase the unit of the commodity. Second, between times zero and T, the agent pays the cost of carry a per unit time, borrowing from the money market account to finance this. (Hint: The value of the agent's portfolio of commodity and money market account begins at $X(0) = 0$ (one unit of the commodity and a money market position of $-S(0)$) and is governed by (5.9.6) with $\Delta(t) = 1$. Write this equation, determine $d(e^{-rt}X(t))$, integrate both sides from zero to T, and solve for $X(T)$. You will need the fact that $e^{-rt}(dS(t) - rS(t)\,dt) = d(e^{-rt}S(t))$. You should get $X(T) = S(T) - \text{For}_S(0,T)$.)

6

Connections with Partial Differential Equations

6.1 Introduction

There are two ways to compute a derivative security price: (1) use Monte Carlo simulation to generate paths of the underlying security or securities under the risk-neutral measure and use these paths to estimate the risk-neutral expected discounted payoff; or (2) numerically solve a partial differential equation. This chapter addresses the second of these methods by showing how to connect the risk-neutral pricing problem to partial differential equations. Section 6.2 explains the concept of stochastic differential equations, which is used to model asset prices. Solutions to stochastic differential equations have the Markov property, as is discussed in Section 6.3. Because of this, related to each stochastic differential equation there are two partial differential equations, one that includes discounting and one that does not. These partial differential equations and their derivations are the subject of Section 6.4. Section 6.5 shows how these ideas can be applied to interest rate models to compute bond prices and the prices of derivatives on bonds. The discussion of Sections 6.2–6.5 concerns one-dimensional processes. The multidimensional theory is outlined in Section 6.6, and a representative example that uses this theory, pricing and hedging an Asian option, is presented in that section.

6.2 Stochastic Differential Equations

A *stochastic differential equation* is an equation of the form

$$dX(u) = \beta(u, X(u))\, du + \gamma(u, X(u))\, dW(u). \tag{6.2.1}$$

Here $\beta(u, x)$ and $\gamma(u, x)$ are given functions, called the *drift* and *diffusion*, respectively. In addition to this equation, an *initial condition* of the form $X(t) = x$, where $t \geq 0$ and $x \in \mathbb{R}$, is specified. The problem is then to find a stochastic process $X(T)$, defined for $T \geq t$, such that

$$X(t) = x, \tag{6.2.2}$$

$$X(T) = X(t) + \int_t^T \beta(u, X(u))\,du + \int_t^T \gamma(u, X(u))\,dW(u). \tag{6.2.3}$$

Under mild conditions on the functions $\beta(u,x)$ and $\gamma(u,x)$, there exists a unique process $X(T)$, $T \geq t$, satisfying (6.2.2) and (6.2.3). However, this process can be difficult to determine explicitly because it appears on both the left- and right-hand sides of equation (6.2.3).

The solution $X(T)$ at time T will be $\mathcal{F}(T)$-measurable (i.e., $X(T)$ only depends on the path of the Brownian motion up to time T.) In fact, since the initial condition $X(t) = x$ is specified, all that is really needed to determine $X(T)$ is the path of the Brownian motion between times t and T.

Although stochastic differential equations are, in general, difficult to solve, a *one-dimensional linear stochastic differential equation* can be solved explicitly. This is a stochastic differential equation of the form

$$dX(u) = \big(a(u) + b(u)X(u)\big)\,du + \big(\gamma(u) + \sigma(u)X(u)\big)\,dW(u), \tag{6.2.4}$$

where $a(u)$, $b(u)$, $\sigma(u)$, and $\gamma(u)$ are nonrandom functions of time. Indeed, this equation can even be solved when $a(u)$, $b(u)$, $\gamma(u)$, and $\sigma(u)$ are adapted random processes (see Exercise 6.1), although it is then no longer of the form (6.2.1). In order to guarantee that the solution to (6.2.1) has the Markov property discussed in Section 6.3 below, the only randomness we permit on the right-hand side of (6.2.1) is the randomness inherent in the solution $X(u)$ and in the driving Brownian motions $W(u)$. There cannot be additional randomness such as would occur if any of the processes $a(u)$, $b(u)$, $\gamma(u)$, and $\sigma(u)$ appearing in (6.2.4) were themselves random. The next two examples are special cases of (6.2.4) in which $a(u)$, $b(u)$, $\gamma(u)$, and $\sigma(u)$ are nonrandom.

Example 6.2.1 (Geometric Brownian motion). The stochastic differential equation for geometric Brownian motion is

$$dS(u) = \alpha S(u)\,du + \sigma S(u)\,dW(u).$$

In the notation of (6.2.1), $\beta(u,x) = \alpha x$ and $\gamma(u,x) = \sigma x$. We know the formula for the solution to this stochastic differential equation when the initial time is zero and the initial position is $S(0)$, namely

$$S(t) = S(0)\exp\left\{\sigma W(t) + \left(\alpha - \frac{1}{2}\sigma^2\right)t\right\}.$$

Similarly, for $T \geq t$,

$$S(T) = S(0)\exp\left\{\sigma W(T) + \left(\alpha - \frac{1}{2}\sigma^2\right)T\right\}.$$

Dividing $S(T)$ by $S(t)$, we obtain

$$\frac{S(T)}{S(t)} = \exp\left\{\sigma\big(W(T) - W(t)\big) + \left(\alpha - \frac{1}{2}\sigma^2\right)(T-t)\right\}.$$

If the initial condition is given at time t rather than at time zero and is $S(t) = x$, then this last equation becomes

$$S(T) = x\exp\left\{\sigma\big(W(T) - W(t)\big) + \left(\alpha - \frac{1}{2}\sigma^2\right)(T-t)\right\}.$$

As expected, when we use the initial condition $S(t) = x$, then $S(T)$ depends only on the path of the Brownian motion between times t and T.

Example 6.2.2 (Hull-White interest rate model). Consider the stochastic differential equation

$$dR(u) = \big(a(u) - b(u)R(u)\big)\,du + \sigma(u)\,d\widetilde{W}(u),$$

where $a(u)$, $b(u)$, and $\sigma(u)$ are nonrandom positive functions of the time variable u and $\widetilde{W}(u)$ is a Brownian motion under a risk-neutral measure $\widetilde{\mathbb{P}}$. In this case, we use the dummy variable r rather than x, and $\beta(u,r) = a(u) - b(u)r$, $\gamma(u,r) = \sigma(u)$. Let us take the initial condition $R(t) = r$. We can solve the stochastic differential equation by first using the stochastic differential equation to compute

$$\begin{aligned} d\left(e^{\int_0^u b(v)dv}R(u)\right) &= e^{\int_0^u b(v)dv}\big(b(u)R(u)\,du + dR(u)\big) \\ &= e^{\int_0^u b(v)dv}\big(\alpha(u)\,du + \sigma(u)\,d\widetilde{W}(u)\big). \end{aligned}$$

Integrating both sides from t to T and using the initial condition $R(t) = r$, we obtain the formula

$$e^{\int_0^T b(v)dv}R(T) = re^{\int_0^t b(v)dv} + \int_t^T e^{\int_0^u b(v)\,dv}\alpha(u)\,du + \int_t^T e^{\int_0^u b(v)\,dv}\sigma(u)\,d\widetilde{W}(u),$$

which we can solve for $R(T)$:

$$R(T) = re^{-\int_t^T b(v)dv} + \int_t^T e^{-\int_u^T b(v)\,dv}\alpha(u)\,du + \int_t^T e^{-\int_u^T b(v)\,dv}\sigma(u)\,d\widetilde{W}(u).$$

This is an explicit formula for the solution $R(T)$. The right-hand side of the final equation does not involve the interest rate process $R(u)$ apart from the initial condition $R(t) = r$; it contains only this initial condition, an integral with respect to time, and an Itô integral of given functions. Note also that the Brownian motion path between times t and T only enters this formula.

Recall from Theorem 4.4.9 that the Itô integral $\int_t^T e^{-\int_u^T b(v)dv}\sigma(u)\,d\widetilde{W}(u)$ of the nonrandom integrand $e^{-\int_u^T b(v)dv}\sigma(u)$ is normally distributed with mean zero and variance $\int_t^T e^{-2\int_u^T b(v)dv}\sigma^2(u)\,du$. The other terms appearing in the formula above for $R(T)$ are nonrandom. Therefore, under the risk-neutral measure $\widetilde{\mathbb{P}}$, $R(T)$ is normally distributed with mean

$$re^{-\int_t^T b(v)dv} + \int_t^T e^{-\int_u^T b(v)\,dv}\alpha(u)\,du$$

and variance

$$\int_t^T e^{-2\int_u^T b(v)dv}\sigma^2(u)\,du.$$

In particular, there is a positive probability that $R(T)$ is negative. This is one of the principal objections to the Hull-White model. □

Example 6.2.3 (Cox-Ingersoll-Ross interest rate model). In the Cox-Ingersoll-Ross (CIR) model, the interest rate is given by the stochastic differential equation

$$dR(u) = \big(a - bR(u)\big)\,du + \sigma\sqrt{R(u)}\,d\widetilde{W}(u), \tag{6.2.5}$$

where a, b, and σ are positive constants. Suppose an initial condition $R(t) = r$ is given. Although there is no formula for $R(T)$, there is one and only one solution to this differential equation starting from the given initial condition. This solution can be approximated by Monte Carlo simulation, and many of its properties can be determined, even though we do not have an explicit formula for it. For instance, in Example 4.4.11, the mean and variance of $R(T)$ were computed when the initial time is $t = 0$ and the initial interest rate is $R(0)$.

Unlike the interest rate in the Hull-White model, the interest rate in the Cox-Ingersoll-Ross model cannot take negative values. When the interest rate approaches zero, the term $\sigma\sqrt{R(u)}\,d\widetilde{W}(u)$ also approaches zero. With the volatility disappearing, the behavior of the interest rate near zero depends on the drift term $a - bR(u)$, and this is $a > 0$ when $R(u) = 0$. The positive drift prevents the interest rate from crossing zero into negative territory.

More information about the solution to (6.2.5) is provided in Exercise 6.6 and Remark 6.9.1 following that exercise. □

6.3 The Markov Property

Consider the stochastic differential equation (6.2.1). Let $0 \le t \le T$ be given, and let $h(y)$ be a Borel-measurable function. Denote by

$$g(t,x) = \mathbb{E}^{t,x}h(X(T)) \tag{6.3.1}$$

the expectation of $h(X(T))$, where $X(T)$ is the solution to (6.2.1) with initial condition $X(t) = x$. (We assume that $\mathbb{E}^{t,x}|h(X(T))| < \infty$.) Note that there is nothing random about $g(t,x)$; it is an ordinary (actually, Borel-measurable) function of the two dummy variables t and x.

If we do not have an explicit formula for the distribution of $X(T)$, we could compute $g(t,x)$ numerically by beginning at $X(t) = x$ and simulating the stochastic differential equation. One way to do this would be to use the *Euler method*, a particular type of Monte Carlo method: choose a small positive step size δ, and then set

$$X(t+\delta) = x + \beta(t,x)\delta + \gamma(t,x)\sqrt{\delta}\,\epsilon_1,$$

where ϵ_1 is a standard normal random variable. Then set

$$X(t+2\delta) = X(t+\delta) + \beta\big(t+\delta, X(t+\delta)\big)\delta + \gamma\big(t+\delta, X(t+\delta)\big)\sqrt{\delta}\,\epsilon_2,$$

where ϵ_2 is a standard normal random variable independent of ϵ_1. By this device, one eventually determines a value for $X(T)$ (assuming δ is chosen so that $\frac{T-t}{\delta}$ is an integer). This gives one realization of $X(T)$ (corresponding to one ω). Now repeat this process many times and compute the average of $h(X(T))$ over all these simulations to get an approximate value for $g(t,x)$. Note that if one were to begin with a different time t and initial value x, one would get a different answer (i.e., the answer is a function of t and x). This dependence on t and x is emphasized by the notation $E^{t,x}$ in (6.3.1).

Theorem 6.3.1. *Let $X(u)$, $u \geq 0$, be a solution to the stochastic differential equation (6.2.1) with initial condition given at time 0. Then, for $0 \leq t \leq T$,*

$$\mathbb{E}\big[h(X(T))\big|\mathcal{F}(t)\big] = g\big(t, X(t)\big). \tag{6.3.2}$$

While the details of the proof of Theorem 6.3.1 are quite technical and will not be given, the intuitive content is clear. Suppose the process $X(u)$ begins at time zero, being generated by the stochastic differential equation (6.2.1), and one watches it up to time t. Suppose now that one is asked, based on this information, to compute the conditional expectation of $h(X(T))$, where $T \geq t$. Then one should pretend that the process is starting at time t at its current position, generate the solution to the stochastic differential equation corresponding to this initial condition, and compute the expected value of $h(X(T))$ generated in this way. In other words, replace $X(t)$ by a dummy x in order to hold it constant, compute $g(t,x) = \mathbb{E}^{t,x}h(X(T))$, and after computing this function put the random variable $X(t)$ back in place of the dummy x. This is the procedure set forth in the Independence Lemma, Lemma 2.3.4, and it is applicable here because the value of $X(T)$ is determined by the value of $X(t)$, which is $\mathcal{F}(t)$-measurable, and the increments of the Brownian motion between times t and T, which are independent of $\mathcal{F}(t)$.

Notice in the discussion above that although one watches the stochastic process $X(u)$ for $0 \leq u \leq t$, the only relevant piece of information when computing $\mathbb{E}\big[h(X(T))\big|\mathcal{F}(t)\big]$ is the value of $X(t)$. This means that $X(t)$ is a Markov process (see Definition 2.3.6). We highlight this fact as a corollary.

Corollary 6.3.2. *Solutions to stochastic differential equations are Markov processes.*

6.4 Partial Differential Equations

The Feynman-Kac Theorem below relates stochastic differential equations and partial differential equations. When this partial differential equation is solved (usually numerically), it produces the function $g(t,x)$ of (6.3.1). The Euler method described in the previous section for determining this function converges slowly and gives the function value for only one pair (t,x). Numerical algorithms for solving equation (6.4.1) below converge quickly in the case of one-dimensional x being considered here and give the function $g(t,x)$ for all values of (t,x) simultaneously. The relationship between geometric Brownian motion and the Black-Scholes-Merton partial differential equation is a special case of the relationship between stochastic differential equations and partial differential equations developed in the following theorems.

Theorem 6.4.1 (Feynman-Kac). *Consider the stochastic differential equation*

$$dX(u) = \beta(u, X(u))\,du + \gamma(u, X(u))\,dW(u). \tag{6.2.1}$$

Let $h(y)$ be a Borel-measurable function. Fix $T > 0$, and let $t \in [0,T]$ be given. Define the function

$$g(t,x) = \mathbb{E}^{t,x} h(X(T)). \tag{6.3.1}$$

(We assume that $\mathbb{E}^{t,x}|h(X(T))| < \infty$ for all t and x.) Then $g(t,x)$ satisfies the partial differential equation

$$g_t(t,x) + \beta(t,x)g_x(t,x) + \frac{1}{2}\gamma^2(t,x)g_{xx}(t,x) = 0 \tag{6.4.1}$$

and the terminal condition

$$g(T,x) = h(x) \text{ for all } x. \tag{6.4.2}$$

The proof of the Feynman-Kac Theorem depends on the following lemma.

Lemma 6.4.2. *Let $X(u)$ be a solution to the stochastic differential equation (6.2.1) with initial condition given at time 0. Let $h(y)$ be a Borel-measurable function, fix $T > 0$, and let $g(t,x)$ be given by (6.3.1). Then the stochastic process*

$$g(t, X(t)), \quad 0 \leq t \leq T,$$

is a martingale.

PROOF: Let $0 \leq s \leq t \leq T$ be given. Theorem 6.3.1 implies

$$\mathbb{E}\big[h(X(T))\big|\mathcal{F}(s)\big] = g(s, X(s)),$$

$$\mathbb{E}\big[h(X(T))\big|\mathcal{F}(t)\big] = g(t, X(t)).$$

Take conditional expectations of the second equation, using iterated conditioning and the first equation, to obtain

$$\begin{aligned}\mathbb{E}\big[g(t,X(t))\big|\mathcal{F}(s)\big] &= \mathbb{E}\big[\mathbb{E}\big[h(X(T))\big|\mathcal{F}(t)\big]\big|\mathcal{F}(s)\big]\\ &= \mathbb{E}\big[h(X(T))\big|\mathcal{F}(s)\big]\\ &= g(s,X(s)).\end{aligned}$$ □

Outline of Proof of Theorem 6.4.1: Let $X(t)$ be the solution to the stochastic differential equation (6.2.1) starting at time zero. Since $g(t,X(t))$ is a martingale, the net dt term in the differential $dg(t,X(t))$ must be zero. If it were positive at any time, then $g(t,X(t))$ would have a tendency to rise at that time; if it were negative, $g(t,X(t))$ would have a tendency to fall. Omitting the argument $(t,X(t))$ in several places below, we compute

$$\begin{aligned}dg(t,X(t)) &= g_t\,dt + g_x\,dX + \frac{1}{2}g_{xx}\,dX\,dX\\ &= g_t\,dt + \beta g_x\,dt + \gamma g_x\,dW + \frac{1}{2}\gamma^2 g_{xx}\,dt\\ &= \left[g_t + \beta g_x + \frac{1}{2}\gamma^2 g_{xx}\right]dt + \gamma g_x\,dW.\end{aligned}$$

Setting the dt term to zero and putting back the argument $(t,X(t))$, we obtain

$$g_t(t,X(t)) + \beta(t,X(t))g_x(t,X(t)) + \frac{1}{2}\gamma^2(t,X(t))g_{xx}(t,X(t)) = 0$$

along every path of X. Therefore,

$$g_t(t,x) + \beta(t,x)g_x(t,x) + \frac{1}{2}\gamma^2(t,x)g_{xx}(t,x) = 0$$

at every point (t,x) that can be reached by $(t,X(t))$. For example, if $X(t)$ is a geometric Brownian motion, then (6.4.1) must hold for every $t \in [0,T)$ and every $x > 0$. On the other hand, if $X(t)$ is a Hull-White interest rate process, which can take any positive or negative value, then (6.4.1) must hold for every $t \in [0,T)$ and every $x \in \mathbb{R}$. □

The general principle behind the proof of the Feynman-Kac theorem is:

1. find the martingale,
2. take the differential, and
3. set the dt term equal to zero.

This gives a partial differential equation, which can then be solved numerically. We illustrate this three-step procedure in the following theorem and subsequent examples.

Theorem 6.4.3 (Discounted Feynman-Kac). *Consider the stochastic differential equation*

$$dX(u) = \beta(u,X(u))\,du + \gamma(u,X(u))\,dW(u). \tag{6.2.1}$$

Let $h(y)$ be a Borel-measurable function and let r be constant. Fix $T > 0$, and let $t \in [0, T]$ be given. Define the function

$$f(t,x) = \mathbb{E}^{t,x}\big[e^{-r(T-t)}h(X(T))\big]. \tag{6.4.3}$$

(We assume that $\mathbb{E}^{t,x}|h(X(T))| < \infty$ for all t and x.) Then $f(t,x)$ satisfies the partial differential equation

$$f_t(t,x) + \beta(t,x)f_x(t,x) + \frac{1}{2}\gamma^2(t,x)f_{xx}(t,x) = rf(t,x) \tag{6.4.4}$$

and the terminal condition

$$f(T,x) = h(x) \text{ for all } x. \tag{6.4.5}$$

OUTLINE OF PROOF: Let $X(t)$ be the solution to the stochastic differential equation (6.2.1) starting at time zero. Then

$$f(t, X(t)) = \mathbb{E}\big[e^{-r(T-t)}h(X(T))\big|\mathcal{F}(t)\big].$$

However, it is not the case that $f(t, X(t))$ is a martingale. Indeed, if $0 \le s \le t \le T$, then

$$\begin{aligned}\mathbb{E}\big[f(t,X(t))\big|\mathcal{F}(s)\big] &= \mathbb{E}\big[\mathbb{E}\big[e^{-r(T-t)}h(X(T))\big|\mathcal{F}(t)\big]\big|\mathcal{F}(s)\big]\\ &= \mathbb{E}\big[e^{-r(T-t)}h(X(T))\big|\mathcal{F}(s)\big],\end{aligned}$$

which is not the same as

$$f(s, X(s)) = \mathbb{E}\big[e^{-r(T-s)}h(X(T))\big|\mathcal{F}(s)\big]$$

because of the differing discount terms. The difficulty here is that in order to get the martingale property from iterated conditioning, we need the random variable being estimated not to depend on t, the time of the conditioning. To achieve this, we "complete the discounting," observing that

$$e^{-rt}f(t, X(t)) = \mathbb{E}\big[e^{-rT}h(X(T))\big|\mathcal{F}(t)\big].$$

We may now apply iterated conditioning to show that $e^{-rt}f(t, X(t))$ is a martingale. The differential of this martingale is

$$\begin{aligned}d\big(e^{-rt}f(t,X(t))\big) &= e^{-rt}\Big[-rf\,dt + f_t\,dt + f_x\,dX + \frac{1}{2}f_{xx}\,dX\,dX\Big]\\ &= e^{-rt}\Big[-rf + f_t + \beta f_x + \frac{1}{2}\gamma^2 f_{xx}\Big]\,dt + e^{-rt}\gamma f_x\,dW.\end{aligned}$$

Setting the dt term equal to zero, we obtain (6.4.4). □

Example 6.4.4 (Options on a geometric Brownian motion). Let $h(S(T))$ be the payoff at time T of a derivative security whose underlying asset is the geometric Brownian motion

$$dS(u) = \alpha S(u)\,du + \sigma S(u)\,dW(u). \tag{6.4.6}$$

We may rewrite this as

$$dS(u) = rS(u)\,du + \sigma S(u)\,d\widetilde{W}(u), \tag{6.4.7}$$

where $\widetilde{W}(u)$ is a Brownian motion under the risk-neutral probability measure $\widetilde{\mathbb{P}}$. Here we assume that σ and the interest rate r are constant. According to the risk-neutral pricing formula (5.2.31), the price of the derivative security at time t is

$$V(t) = \widetilde{\mathbb{E}}\big[e^{-r(T-t)}h(S(T))\big|\mathcal{F}(t)\big]. \tag{6.4.8}$$

Because the stock price is Markov and the payoff is a function of the stock price alone, there is a function $v(t,x)$ such that $V(t) = v(t, S(t))$. Moreover, the function $v(t,x)$ must satisfy the discounted partial differential equation (6.4.4). This is the Black-Scholes-Merton equation

$$v_t(t,x) + rxv_x(t,x) + \frac{1}{2}\sigma^2x^2v_{xx}(t,x) = rv(t,x). \tag{6.4.9}$$

When the underlying asset is a geometric Brownian motion, this is the right pricing equation for a European call, a European put, a forward contract, and any other option that pays off some function of $S(T)$ at time T.

Note that to derive (6.4.9) we use the discounted partial differential equation (6.4.4) when the stochastic differential equation for the underlying process is (6.4.7) rather than (6.4.6) (i.e., we have $rxv_x(t,x)$ in (6.4.9) rather than $\alpha xv_x(t,x)$). This is because we are computing the conditional expectation in (6.4.8) under the risk-neutral measure $\widetilde{\mathbb{P}}$ and hence must use the differential equation that represents $S(u)$ in terms of $\widetilde{W}(u)$, the Brownian motion under $\widetilde{\mathbb{P}}$. In other words, we are using the Discounted Feynman-Kac Theorem with $\widetilde{W}(u)$ replacing $W(u)$ and $\widetilde{\mathbb{P}}$ replacing $\mathbb{P}$. □

In the previous example, if σ were a function of time and stock price (i.e., $\sigma(t,x)$), then the stock price would no longer be a geometric Brownian motion and the Black-Scholes-Merton formula would no longer apply. However, one can still solve for the option price by solving the partial differential equation (6.4.9), where now the constant σ^2 is replaced by $\sigma^2(t,x)$:

$$v_t(t,x) + rxv_x(t,x) + \frac{1}{2}\sigma^2(t,x)x^2v_{xx}(t,x) = rv(t,x). \tag{6.4.10}$$

This equation is not difficult to solve numerically.

It has been observed in markets that if one assumes a constant volatility, the parameter σ that makes the theoretical option price given by (6.4.9) agree with the market price, the so-called *implied volatility*, is different for options having different strikes. In fact, this implied volatility is generally a convex function of the strike price. One refers to this phenomenon as the *volatility smile*.

One simple model with nonconstant volatility is the *constant elasticity of variance* (CEV) model, in which $\sigma(t,x) = \sigma x^{\delta-1}$ depends on x but not t. The parameter $\delta \in (0,1)$ is chosen so that the model gives a good fit to option prices across different strikes at a single expiration date. For this model, the stock price is governed by the stochastic differential equation

$$dS(t) = rS(t)\,dt + \sigma S^{\delta}(t)\,d\widetilde{W}(t).$$

The volatility $\sigma S^{\delta-1}(t)$ is a decreasing function of the stock price.

When one wishes to account for different volatilities implied by options expiring at different dates as well as different strikes, one needs to allow σ to depend on t as well as x. This function $\sigma(t,x)$ is called the *volatility surface* (see Exercise 6.10).

6.5 Interest Rate Models

The simplest models for fixed income markets begin with a stochastic differential equation for the interest rate, e.g.,

$$dR(t) = \beta(t, R(t))\,dt + \gamma(t, R(t))\,d\widetilde{W}(t), \tag{6.5.1}$$

where $\widetilde{W}(t)$ is a Brownian motion under a risk-neutral probability measure $\widetilde{\mathbb{P}}$. In these models, one begins with a risk-neutral measure $\widetilde{\mathbb{P}}$ and uses the risk-neutral pricing formula to price all assets. This guarantees that discounted asset prices are martingales under the risk-neutral measure, and hence there is no arbitrage. The issue of *calibration* of these models (i.e., choosing the model and the model parameters so that they give a good fit to market prices) is not discussed in this text.

Models for the interest rate $R(t)$ are sometimes called *short-rate models* because $R(t)$ is the interest rate for short-term borrowing. When the interest rate is determined by only one stochastic differential equation, as is the case in this section, the model is said to have *one factor.* The primary shortcoming of one-factor models is that they cannot capture complicated yield curve behavior; they tend to produce parallel shifts in the yield curve but not changes in its slope or curvature.

The *discount process* is as given in (5.2.17),

$$D(t) = e^{-\int_0^t R(s)ds},$$

and we denote the *money market account price process* to be

$$\frac{1}{D(t)} = e^{\int_0^t R(s)ds}.$$

This is the value at time t of one unit of currency invested in the money market account at time zero and continuously rolled over at the short-term interest rate $R(s)$, $s \geq 0$. As discussed following (5.2.18), we have the differential formulas

$$dD(t) = -R(t)D(t)\,dt, \quad d\Big(\frac{1}{D(t)}\Big) = \frac{R(t)}{D(t)}\,dt.$$

A *zero-coupon bond* is a contract promising to pay a certain "face" amount, which we take to be 1, at a fixed maturity date T. Prior to that, the bond makes no payments. The risk-neutral pricing formula (5.2.30) says that the discounted price of this bond should be a martingale under the risk-neutral measure. In other words, for $0 \leq t \leq T$, the price of the bond $B(t,T)$ should satisfy

$$D(t)B(t,T) = \widetilde{\mathbb{E}}\big[D(T)\big|\mathcal{F}(t)\big]. \tag{6.5.2}$$

(Note that $B(T,T) = 1$.) This gives us the *zero-coupon bond pricing formula*

$$B(t,T) = \widetilde{\mathbb{E}}\left[e^{-\int_t^T R(s)ds}\middle|\,\mathcal{F}(t)\right], \tag{6.5.3}$$

which we take as a definition. Once zero-coupon bond prices have been computed, we can define the *yield* between times t and T to be

$$Y(t,T) = -\frac{1}{T-t}\log B(t,T)$$

or, equivalently,

$$B(t,T) = e^{-Y(t,T)(T-t)}.$$

The yield $Y(t,T)$ is the constant rate of continuously compounding interest between times t and T that is consistent with the bond price $B(t,T)$. The 30-year rate at time t is $Y(t,30+t)$; this is an example of a long rate. Notice that once we adopt a model (6.5.1) for the short rate, the long rate is determined by the formulas above; we may not model the long rate separately.

Since R is given by a stochastic differential equation, it is a Markov process and we must have

$$B(t,T) = f(t,R(t))$$

for some function $f(t,r)$ of the dummy variables t and r. This is a slight step beyond the way we have used the Markov property previously because the random variable $e^{-\int_t^T R(s)ds}$ being estimated in (6.5.3) depends on the path segment $R(s)$, $t \leq s \leq T$, not just on $R(T)$. However, the only relevant part of the path of R before time t is its value at time t, and so the bond price $B(t,T)$ must be a function of time t and $R(t)$.

To find the partial differential equation for the unknown function $f(t,r)$, we find a martingale, take its differential, and set the dt term equal to zero. The martingale in this case is $D(t)B(t,T) = D(t)f(t,R(t))$. Its differential is

$$\begin{aligned}
d\big(D(t)f(t,R(t))\big) &= f(t,R(t))\,dD(t) + D(t)\,df(t,R(t))\\
&= D(t)\left[-Rf\,dt + f_t\,dt + f_r dR + \frac{1}{2}f_{rr}\,dR\,dR\right]\\
&= D(t)\left[-Rf + f_t + \beta f_r + \frac{1}{2}\gamma^2 f_{rr}\right]dt + D(t)\gamma f_r\,d\widetilde{W}.
\end{aligned}$$

Setting the dt term equal to zero, we obtain the partial differential equation

$$f_t(t,r) + \beta(t,r)f_r(t,r) + \frac{1}{2}\gamma^2(t,r)f_{rr}(t,r) = rf(t,r). \tag{6.5.4}$$

We also have the terminal condition

$$f(T,r) = 1 \text{ for all } r \tag{6.5.5}$$

because the value of the bond at maturity is its face value 1.

Example 6.5.1 (Hull-White interest rate model). In the Hull-White model, the evolution of the interest rate is given by

$$dR(t) = \big(a(t) - b(t)R(t)\big)\,dt + \sigma(t)\,d\widetilde{W}(t),$$

where $a(t)$, $b(t)$, and $\sigma(t)$ are nonrandom positive functions of time. The partial differential equation (6.5.4) for the zero-coupon bond price becomes

$$f_t(t,r) + \big(a(t) - b(t)r\big)f_r(t,r) + \frac{1}{2}\sigma^2(t)f_{rr}(t,r) = rf(t,r). \tag{6.5.6}$$

We initially guess and subsequently verify that the solution has the form

$$f(t,r) = e^{-rC(t,T)-A(t,T)}$$

for some nonrandom functions $C(t,T)$ and $A(t,T)$ to be determined. These are functions of $t \in [0,T]$; the maturity T is fixed. In this case, the yield

$$Y(t,T) = -\frac{1}{T-t}\log f(t,r) = \frac{1}{T-t}\Big(rC(t,T) + A(t,T)\Big)$$

is an *affine* function of r (i.e., a number times r plus another number). The Hull-White model is a special case of a class of models called *affine yield models.*

Furthermore,

$$\begin{aligned}
f_t(t,r) &= \big(-rC'(t,T) - A'(t,T)\big)f(t,r),\\
f_r(t,r) &= -C(t,T)f(t,r),\\
f_{rr}(t,r) &= C^2(t,T)f(t,r),
\end{aligned}$$

where $C'(t,T) = \frac{\partial}{\partial t}C(t,T)$ and $A'(t,T) = \frac{\partial}{\partial t}A(t,T)$. Substitution into the partial differential equation (6.5.6) gives

$$\begin{aligned}
&\Big[\big(-C'(t,T) + b(t)C(t,T) - 1\big)r\\
&\quad -A'(t,T) - a(t)C(t,T) + \frac{1}{2}\sigma^2(t)C^2(t,T)\Big]\,f(t,r) = 0.
\end{aligned} \tag{6.5.7}$$

Because this equation must hold for all r, the term that multiplies r in this equation must be zero. Otherwise, changing the value of r would change the value of the left-hand side of (6.5.7), and hence it could not always be equal to zero. This gives us an ordinary differential equation in t:

$$C'(t,T) = b(t)C(t,T) - 1. \tag{6.5.8}$$

Setting this term equal to zero in (6.5.7), we now see that

$$A'(t,T) = -a(t)C(t,T) + \frac{1}{2}\sigma^2(t)C^2(t,T). \tag{6.5.9}$$

The terminal condition (6.5.5) must hold for all r, and this implies that $C(T,T) = A(T,T) = 0$. Equations (6.5.8) and (6.5.9) and these terminal conditions provide enough information to determine the functions $A(t,T)$ and $C(t,T)$ for $0 \le t \le T$. They are

$$C(t,T) = \int_t^T e^{-\int_t^s b(v)dv} ds, \tag{6.5.10}$$

$$A(t,T) = \int_t^T \left(a(s)C(s,T) - \frac{1}{2}\sigma^2(s)C^2(s,T) \right) ds. \tag{6.5.11}$$

It is clear that these formulas give functions that satisfy $C(T,T) = A(T,T) = 0$. The verification that these formulas provide the unique solutions to (6.5.8) and (6.5.9) is Exercise 6.3.

In conclusion, we have derived an explicit formula for the price of a zero-coupon bond as a function of the interest rate in the Hull-White model. It is

$$B(t,T) = e^{-R(t)C(t,T)-A(t,T)}, \quad 0 \le t \le T,$$

where $C(t,T)$ and $A(t,T)$ are given by (6.5.10) and (6.5.11). □

Example 6.5.2 (Cox-Ingersoll-Ross interest rate model). In the CIR model, the evolution of the interest rate is given by

$$dR(t) = \big(a - bR(t)\big)\,dt + \sigma\sqrt{R(t)}\,d\widetilde{W}(t),$$

where a, b, and σ are positive constants. The partial differential equation (6.5.4) for the bond price becomes

$$f_t(t,r) + (a - br)f_r(t,r) + \frac{1}{2}\sigma^2 r f_{rr}(t,r) = rf(t,r). \tag{6.5.12}$$

Again, we initially guess and subsequently verify that the solution has the form

$$f(t,r) = e^{-rC(t,T)-A(t,T)}.$$

The Cox-Ingersoll-Ross model is another example of an affine yield model. Substitution into the differential equation (6.5.12) gives

$$\Big[\big(-C'(t,T)+bC(t,T)+\frac{1}{2}\sigma^2C^2(t,T)-1\big)r \\ -A'(t,T)-aC(t,T)\Big]f(t,r)=0. \qquad (6.5.13)$$

We can again conclude that the term multiplying r must be zero and then conclude that the other term must also be zero, thereby obtaining two ordinary differential equations in t:

$$C'(t,T)=bC(t,T)+\frac{1}{2}\sigma^2C^2(t,T)-1, \qquad (6.5.14)$$

$$A'(t,T)=-aC(t,T). \qquad (6.5.15)$$

The solutions to these equations satisfying the terminal conditions $C(T,T)=A(T,T)=0$ are

$$C(t,T)=\frac{\sinh(\gamma(T-t))}{\gamma\cosh(\gamma(T-t))+\frac{1}{2}b\sinh(\gamma(T-t))}, \qquad (6.5.16)$$

$$A(t,T)=-\frac{2a}{\sigma^2}\log\left[\frac{\gamma e^{\frac{1}{2}b(T-t)}}{\gamma\cosh(\gamma(T-t))+\frac{1}{2}b\sinh(\gamma(T-t))}\right], \qquad (6.5.17)$$

where $\gamma=\frac{1}{2}\sqrt{b^2+2\sigma^2}$, $\sinh u=\frac{e^u-e^{-u}}{2}$, and $\cosh u=\frac{e^u+e^{-u}}{2}$. The verification of this assertion is Exercise 6.4. □

Example 6.5.3 (Option on a bond). Consider the general short-rate model (6.5.1). Let $0\le t\le T_1<T_2$ be given. In this example, the fixed time T_2 is the maturity date for a zero-coupon bond. The fixed time T_1 is the expiration date for a European call on this bond. We wish to determine the value of this call at time t.

Suppose we have solved for the function $f(t,r)$ satisfying the partial differential equation (6.5.4) together with the terminal condition (6.5.5). This gives us the price of the zero-coupon bond as a function of time and the underlying interest rate.

According to the risk-neutral pricing formula (5.2.31) and the Markov property, the value of the call at time t is

$$\begin{aligned} c(t,R(t)) &= \widetilde{\mathbb{E}}\left[e^{-\int_t^{T_1}R(s)ds}\Big(f(T_1,R(T_1))-K\Big)^+\Big|\mathcal{F}(t)\right] \\ &= \frac{1}{D(t)}\widetilde{\mathbb{E}}\left[D(T_1)\Big(f(T_1,R(T_1))-K\Big)^+\Big|\mathcal{F}(t)\right] \end{aligned}$$

for some function $c(t,r)$ of the dummy variables t and r. The discounted call price

$$D(t)c(t,R(t))=\widetilde{\mathbb{E}}\left[D(T_1)\Big(f(T_1,R(T_1))-K\Big)^+\Big|\mathcal{F}(t)\right],\ 0\le t\le T_1,$$

is a martingale. The differential of the discounted call price is

$$\begin{aligned}
d\big(D(t)c(t,R(t))\big) &= c(t,R(t))\,dD(t) + D(t)\,dc(t,R(t)) \\
&= D\left[-Rc\,dt + c_t\,dt + c_r\,dR + \frac{1}{2}c_{rr}\,dR\,dR\right] \\
&= D\left[-Rc + c_t + \beta c_r + \frac{1}{2}\gamma^2 c_{rr}\right] dt + D\gamma c_r\,d\widetilde{W}.
\end{aligned}$$

Setting the dt term to zero, we obtain the partial differential equation

$$c_t(t,r) + \beta(t,r)c_r(t,r) + \frac{1}{2}\gamma^2(t,r)c_{rr}(t,r) = rc(t,r).$$

This is the same partial differential equation that governs $f(t,r)$. However, $c(t,r)$ and $f(t,r)$ have different terminal conditions. The terminal condition for $c(t,r)$ is

$$c(T_1,r) = (f(T_1,r) - K)^+ \text{ for all } r.$$

One can use these conditions to numerically determine the call price function $c(t,r)$. □

6.6 Multidimensional Feynman-Kac Theorems

The Feynman-Kac and Discounted Feynman-Kac Theorems, Theorems 6.4.1 and 6.4.3, have multidimensional versions. The number of differential equations and the number of Brownian motions entering those differential equations can both be larger than one and do not need to be the same. We illustrate the general situation by working out the details for two stochastic differential equations driven by two Brownian motions.

Let $W(t) = (W_1(t), W_2(t))$ be a two-dimensional Brownian motion (i.e., a vector of two independent, one-dimensional Brownian motions). Consider two stochastic differential equations

$$\begin{aligned}
dX_1(u) &= \beta_1(u,X_1(u),X_2(u))\,du + \gamma_{11}(u,X_1(u),X_2(u))\,dW_1(u) \\
&\quad + \gamma_{12}(u,X_1(u),X_2(u))\,dW_2(u), \\
dX_2(u) &= \beta_2(u,X_1(u),X_2(u))\,du + \gamma_{21}(u,X_1(u),X_2(u))\,dW_1(u) \\
&\quad + \gamma_{22}(u,X_1(u),X_2(u))\,dW_2(u).
\end{aligned}$$

The solution to this pair of stochastic differential equations, starting at $X_1(t) = x_1$ and $X_2(t) = x_2$, depends on the specified initial time t and the initial positions x_1 and x_2. Regardless of the initial condition, the solution is a Markov process.

Let a Borel-measurable function $h(y_1, y_2)$ be given. Corresponding to the initial condition t, x_1, x_2, where $0 \le t \le T$, we define

$$g(t,x_1,x_2) = \mathbb{E}^{t,x_1,x_2} h(X_1(T),X_2(T)), \tag{6.6.1}$$

$$f(t,x_1,x_2) = \mathbb{E}^{t,x_1,x_2}\left[e^{-r(T-t)}h(X_1(T),X_2(T))\right]. \tag{6.6.2}$$

Then

$$g_t + \beta_1 g_{x_1} + \beta_2 g_{x_2} \\ + \frac{1}{2}(\gamma_{11}^2 + \gamma_{12}^2) g_{x_1 x_1} + (\gamma_{11}\gamma_{21} + \gamma_{12}\gamma_{22}) g_{x_1 x_2} + \frac{1}{2}(\gamma_{21}^2 + \gamma_{22}^2) g_{x_2 x_2} = 0, \tag{6.6.3}$$

$$f_t + \beta_1 f_{x_1} + \beta_2 f_{x_2} \\ + \frac{1}{2}(\gamma_{11}^2 + \gamma_{12}^2) f_{x_1 x_1} + (\gamma_{11}\gamma_{21} + \gamma_{12}\gamma_{22}) f_{x_1 x_2} + \frac{1}{2}(\gamma_{21}^2 + \gamma_{22}^2) f_{x_2 x_2} = rf. \tag{6.6.4}$$

Of course, these functions also satisfy the terminal conditions

$$g(T, x_1, x_2) = f(T, x_1, x_2) = h(x_1, x_2) \text{ for all } x_1 \text{ and } x_2.$$

Equations (6.6.3) and (6.6.4) are derived by starting the pair of processes X_1, X_2 at time zero, observing that the processes $g(t, X_1(t), X_2(t))$ and $e^{-rt} f(t, X_1(t), X_2(t))$ are martingales, taking their differentials, and setting the dt terms equal to zero. When taking the differentials, one uses the fact that W_1 and W_2 are independent. We leave the details to the reader in Exercise 6.5. This exercise also provides the counterparts of (6.6.3) and (6.6.4) when W_1 and W_2 are correlated Brownian motions.

Example 6.6.1 (Asian option). We show by example how the Discounted Feynman-Kac Theorem can be used to find prices and hedges, even for path-dependent options. The option we choose for this example is an Asian option. A more detailed discussion of this option is presented in Section 7.5. The payoff we consider is

$$V(T) = \left(\frac{1}{T} \int_0^T S(u)\, du - K \right)^+,$$

where $S(u)$ is a geometric Brownian motion, the expiration time T is fixed and positive, and K is a positive strike price. In terms of the Brownian motion $\widetilde{W}(u)$ under the risk-neutral measure $\widetilde{\mathbb{P}}$, we may write the stochastic differential equation for $S(u)$ as

$$dS(u) = rS(u)\, du + \sigma S(u)\, d\widetilde{W}(u). \tag{6.6.5}$$

Because the payoff depends on the whole path of the stock price via its integral, at each time t prior to expiration it is not enough to know just the stock price in order to determine the value of the option. We must also know the integral of the stock price,

$$Y(t) = \int_0^t S(u)\, du,$$

up to the current time t. Similarly, it is not enough to know just the integral $Y(t)$. We must also know the current stock price $S(t)$. Indeed, for the same value of $Y(t)$, the Asian option is worth more for high values of $S(t)$ than for low values because the high values of $S(t)$ make it more likely that the option will have a high payoff.

For the process $Y(u)$, we have the stochastic differential equation

$$dY(u) = S(u)\,du. \tag{6.6.6}$$

Because the pair of processes $(S(u), Y(u))$ is given by the pair of stochastic differential equations (6.6.5) and (6.6.6), the pair of processes $(S(u), Y(u))$ is a two-dimensional Markov process.

Note that $Y(u)$ alone is not a Markov process because its stochastic differential equation involves the process $S(u)$. However, the pair $(S(u), Y(u))$ is Markov because the pair of stochastic differential equations for these processes involves only these processes (and, of course, the driving Brownian motion $\widetilde{W}(u)$).

If we use (6.6.5) and (6.6.6) to generate the processes $S(u)$ and $Y(u)$ starting with initial values $S(0) > 0$ and $Y(0) = 0$ at time zero, then the payoff of the Asian option at expiration time T is $V(T) = (\frac{1}{T}Y(T) - K)^+$. According to the risk-neutral pricing formula (5.2.31), the value of the Asian option at times prior to expiration is

$$V(t) = \widetilde{\mathbb{E}}\left[e^{-r(T-t)}\left(\frac{1}{T}Y(T) - K\right)^+ \middle| \mathcal{F}(t)\right], \quad 0 \le t \le T.$$

Because the pair of processes $(S(u), Y(u))$ is Markov, this can be written as some function of the time variable t and the values at time t of these processes. In other words, there is a function $v(t, x, y)$ such that

$$v\big(t, S(t), Y(t)\big) = V(t) = \widetilde{\mathbb{E}}\left[e^{-r(T-t)}\left(\frac{1}{T}Y(T) - K\right)^+ \middle| \mathcal{F}(t)\right].$$

Note that this function must satisfy the terminal condition

$$v(T, x, y) = \left(\frac{y}{T} - K\right)^+ \text{ for all } x \text{ and } y. \tag{6.6.7}$$

Using iterated conditioning, it is easy to see that the discounted option value $e^{-rt}v(t, S(t), Y(t))$ is martingale. Its differential is

$$\begin{aligned}
&d\big(e^{-rt}v(t, S(t), Y(t))\big) \\
&= e^{-rt}\bigg[-rv\,dt + v_t\,dt + v_x\,dS + v_y\,dY + \frac{1}{2}v_{xx}\,dS\,dS + v_{xy}\,dS\,dY \\
&\qquad + \frac{1}{2}v_{yy}\,dY\,dY\bigg]
\end{aligned}$$

$$= e^{-rt}\left[-rv\,dt + v_t\,dt + v_x(rS\,dt + \sigma S\,d\widetilde{W}) + v_y S\,dt + \frac{1}{2}\sigma^2 S^2 v_{xx}\,dt\right]$$

$$= e^{-rt}\Bigg[-rv(t,S(t),Y(t)) + v_t(t,S(t),Y(t)) + rS(t)v_x(t,S(t),Y(t))$$

$$+S(t)v_y(t,S(t),Y(t)) + \frac{1}{2}\sigma^2 S^2(t)v_{xx}(t,S(t),Y(T))\Bigg]\,dt$$

$$+e^{-rt}\sigma S(t)v_x(t,S(t),Y(t))\,d\widetilde{W}(t). \tag{6.6.8}$$

Because the discounted option price is a martingale, the dt term in this differential must be zero. We obtain the partial differential equation

$$v_t(t,x,y)+rxv_x(t,x,y)+xv_y(t,x,y)+\frac{1}{2}\sigma^2x^2v_{xx}(t,x,y) = rv(t,x,y). \tag{6.6.9}$$

This is an example of the Discounted Feynman-Kac Theorem, a special case of equation (6.6.4). In particular, (6.6.8) simplifies to

$$d\big(e^{-rt}v(t,S(t),Y(t))\big) = e^{-rt}\sigma S(t)v_x(t,S(t),Y(t))\,d\widetilde{W}(t). \tag{6.6.10}$$

Recall from (5.2.27) that the discounted value of a portfolio satisfies the equation

$$d\big(e^{-rt}X(t)\big) = e^{-rt}\sigma S(t)\Delta(t)\,d\widetilde{W}(t). \tag{6.6.11}$$

If we sell the Asian option at time zero for $v(0,S(0),0)$ and use this as the initial capital for a hedging portfolio (i.e., take $X(0) = v(0,S(0),0)$), and at each time t use the portfolio process $\Delta(t) = v_x(t,S(t),Y(t))$, then we will have

$$d\big(e^{-rt}X(t)\big) = d\big(e^{-rt}v(t,S(t),Y(t))\big)$$

for all times t, and hence

$$X(T) = v(T,S(T),Y(T)) = \Big(\frac{1}{T}Y(T) - K\Big)^+.$$

This procedure hedges a short position in the Asian option. We have obtained the usual formula that the number of shares held to hedge a short position in the option is the derivative of the option value with respect to the underlying stock price. However, the Asian option price is the solution to a partial differential equation that contains a term $xv_y(t,x,y)$ that does not appear in the partial differential equation for the price of a European option. □

6.7 Summary

When the underlying price of an asset is given by a stochastic differential equation, the asset price is Markov and the price of any non-path-dependent derivative security based on that asset is given by a partial differential equation. In order to price path-dependent securities, one first seeks to determine the variables on which the path-dependent payoff depends and then introduce one or more additional stochastic differential equations in order to have a system of such equations that describes the relevant variables. If this can be done, then again the price of the derivative security is given by a partial differential equation.

This leads to the following four-step procedure for finding the pricing differential equation and for constructing a hedge for a derivative security.

1. Determine the variables on which the derivative security price depends. In addition to time t, these are the underlying asset price $S(t)$ and possibly other stochastic processes. We call these stochastic processes the *state processes*. One must be able to represent the derivative security payoff in terms of these state processes.
2. Write down a system of stochastic differential equations for the state processes. Be sure that, except for the driving Brownian motions, the only random processes appearing on the right-hand sides of these equations are the state processes themselves. This ensures that the vector of state processes is Markov.
3. The Markov property guarantees that the derivative security price at each time is a function of time and the state processes at that time. The discounted option price is a martingale under the risk-neutral measure. Compute the differential of the discounted option price, set the dt term equal to zero, and obtain thereby a partial differential equation.
4. The terms multiplying the Brownian motion differentials in the discounted derivative security price differential must be matched by the terms multiplying the Brownian motion differentials in the evolution of the hedging portfolio value; see (5.4.27). Matching these terms determines the hedge for a short position in the derivative security.

6.8 Notes

Conditions for the existence and uniqueness of solutions to stochastic differential equations are provided by Karatzas and Shreve [101], Chapter 5, Section 2, who also show in Chapter 5, Section 4, that solutions to stochastic differential equations have the Markov property. This is based on work of Stroock and Varadhan [151]. The ideas behind the Feynman-Kac Theorem, although not the presentation we give here, trace back to Feynman [65] and Kac [99].

Hull and White presented their interest rate model in [88], in which they generalized a model of Vasicek [154] to allow time-varying coefficients. The origin of the Cox-Ingersoll-Ross model is [41], where one can find a closed-form formula for the distribution of the interest rate in the model. These are examples of *affine-yield models*, a class identified by Duffie and Kan [58]. They are sometimes called *multifactor CIR models*.

Example 6.6.1 obtains a partial differential equation for the price of an Asian option but does not address computational issues. In the form given here, the equation is difficult to handle numerically. Večer [156] and Rogers and Shi [139] present transformations of this equation that are numerically more stable. See also Andreasen [4] for an application of the change-of-numéraire idea of Chapter 9 to discretely sampled Asian options. The transformation of Večer and its use for both continuously sampled and discretely sampled Asian options is presented in Section 7.5.

The Heston stochastic volatility model of Exercise 6.7 is taken from Heston [84]. Exercise 6.10 on implying the volatility surface comes from Dupire [61]. The same idea for binomial trees was worked out by Derman et al. [50], [51].

6.9 Exercises

Exercise 6.1. Consider the stochastic differential equation

$$dX(u) = \big(a(u) + b(u)X(u)\big)\,du + \big(\gamma(u) + \sigma(u)X(u)\big)\,dW(u), \tag{6.2.4}$$

where $W(u)$ is a Brownian motion relative to a filtration $\mathcal{F}(u)$, $u \geq 0$, and we allow $a(u)$, $b(u)$, $\gamma(u)$, and $\sigma(u)$ to be processes adapted to this filtration. Fix an initial time $t \geq 0$ and an initial position $x \in \mathbb{R}$. Define

$$Z(u) = \exp\left\{\int_t^u \sigma(v)\,dW(v) + \int_t^u \left(b(v) - \frac{1}{2}\sigma^2(v)\right)dv\right\},$$

$$Y(u) = x + \int_t^u \frac{a(v) - \sigma(v)\gamma(v)}{Z(v)}\,dv + \int_t^u \frac{\gamma(v)}{Z(v)}\,dW(v).$$

(i) Show that $Z(t) = 1$ and

$$dZ(u) = b(u)Z(u)\,du + \sigma(u)Z(u)\,dW(u),\ u \geq t.$$

(ii) By its very definition, $Y(u)$ satisfies $Y(t) = x$ and

$$dY(u) = \frac{a(u) - \sigma(u)\gamma(u)}{Z(u)}\,du + \frac{\gamma(u)}{Z(u)}\,dW(u),\ u \geq t.$$

Show that $X(u) = Y(u)Z(u)$ solves the stochastic differential equation (6.2.4) and satisfies the initial condition $X(t) = x$.

Exercise 6.2 (No-arbitrage derivation of bond-pricing equation). In Section 6.5, we began with the stochastic differential equation (6.5.1) for the interest rate under the risk-neutral measure $\widetilde{\mathbb{P}}$, used the risk-neutral pricing formula (6.5.3) to consider a zero-coupon bond maturing at time T whose price $B(t,T)$ at time t before maturity is a function $f(t, R(t))$ of the time and the interest rate, and derived the partial differential equation (6.5.4) for the function $f(t,r)$. In this exercise, we show how to derive this partial differential equation from no-arbitrage considerations rather than by using the risk-neutral pricing formula.

Suppose the interest rate is given by a stochastic differential equation

$$dR(t) = \alpha(t, R(t))\,dt + \gamma(t, R(t))\,dW(t), \tag{6.9.1}$$

where $W(t)$ is a Brownian motion under a probability measure $\mathbb{P}$ not assumed to be risk-neutral. Assume further that, for each T, the T-maturity zero-coupon bond price is a function $f(t, R(t), T)$ of the current time t, the current interest rate $R(t)$, and the maturity of the bond T. We do *not* assume that this bond price is given by the risk-neutral pricing formula (6.5.3).

Assume for the moment that $f_r(t, r, T) \neq 0$ for all values of r and $0 \leq t \leq T$, so we can define

$$\beta(t, r, T) = -\frac{1}{f_r(t, r, T)}\left[-rf(t, r, T) + f_t(t, r, T) + \frac{1}{2}\gamma^2(t, r)f_{rr}(t, r, T)\right], \tag{6.9.2}$$

and then have

$$f_t(t, r, T) + \beta(t, r, T)f_r(t, r, T) + \frac{1}{2}\gamma^2(t, r)f_{rr}(t, r, T) = rf(t, r, T). \tag{6.9.3}$$

Equation (6.9.3) will reduce to (6.5.4) for the function $f(t, r, T)$ if we can show that $\beta(t, r, T)$ does not depend on T.

(i) Consider two maturities $0 < T_1 < T_2$, and consider a portfolio that at each time $t \leq T_1$ holds $\Delta_1(t)$ bonds maturing at time T_1 and $\Delta_2(t)$ bonds maturing at time T_2, financing this by investing or borrowing at the interest rate $R(t)$. Show that the value of this portfolio satisfies

$$\begin{aligned}
&d\big(D(t)X(t)\big)\\
&= \Delta_1(t)D(t)\Big[-R(t)f(t, R(t), T_1) + f_t(t, R(t), T_1)\\
&\qquad +\alpha(t, R(t))f_r(t, R(t), T_1) + \frac{1}{2}\gamma^2(t, R(t))f_{rr}(t, R(t), T_1)\Big]\,dt\\
&\quad +\Delta_2(t)D(t)\Big[-R(t)f(t, R(t), T_2) + f_t(t, R(t), T_2)\\
&\qquad +\alpha(t, R(t))f_r(t, R(t), T_2) + \frac{1}{2}\gamma^2(t, R(t))f_{rr}(t, R(t), T_2)\Big]\,dt\\
&\quad +D(t)\gamma(t, R(t))\big[\Delta_1(t)f_r(t, R(t), T_1) + \Delta_2(t)f_r(t, R(t), T_2)\big]\,dW(t)\\
&= \Delta_1(t)D(t)\big[\alpha(t, R(t)) - \beta(t, R(t), T_1)\big]f_r(t, R(t), T_1)\,dt\\
&\quad +\Delta_2(t)D(t)\big[\alpha(t, R(t)) - \beta(t, R(t), T_2)\big]f_r(t, R(t), T_2)\,dt\\
&\quad +D(t)\gamma(t, R(t))\big[\Delta_1(t)f_r(t, R(t), T_1) + \Delta_2(t)f_r(t, R(t), T_2)\big]\,dW(t).
\end{aligned} \tag{6.9.4}$$

(ii) Denote

$$\text{sign}(x) = \begin{cases} 1 & \text{if } x > 0, \\ 0 & \text{if } x = 0, \\ -1 & \text{if } x < 0, \end{cases}$$

and

$$S(t) = \text{sign}\left\{\left[\beta(t, R(t), T_2) - \beta(t, R(t), T_1)\right] f_r(t, R(t), T_1) f_r(t, R(t), T_2)\right\}.$$

Show that the portfolio processes $\Delta_1(t) = S(t) f_r(t, R(t), T_2)$ and $\Delta_2(t) = -S(t) f_r(t, R(t), T_1)$ result in arbitrage unless $\beta(t, R(t), T_1) = \beta(t, R(t), T_2)$. Since T_1 and T_2 are arbitrary, we conclude that $\beta(t, r, T)$ does not depend on T.

(iii) Now let a maturity $T > 0$ be given and consider a portfolio $\Delta(t)$ that invests only in the bond of maturity T, financing this by investing or borrowing at the interest rate $R(t)$. Show that the value of this portfolio satisfies

$$\begin{aligned} & d\big(D(t)X(t)\big) \\ & = \Delta(t)D(t)\big[-R(t)f(t, R(t), T) + f_t(t, R(t), T) \\ & \qquad + \alpha(t, R(t)) f_r(t, R(t), T) + \frac{1}{2}\gamma^2(t, R(t)) f_{rr}(t, R(t), T)\big]\, dt \\ & \qquad + D(t)\Delta(t)\gamma(t, R(t)) f_r(t, R(t), T)\, dW(t). \end{aligned} \tag{6.9.5}$$

Show that if $f_r(t, r, T) = 0$, then there is an arbitrage unless

$$f_t(t, r, T) + \frac{1}{2}\gamma^2(t, r) f_{rr}(t, r, T) = r f(t, r, T). \tag{6.9.6}$$

In other words, if $f_r(t, r, T) = 0$, then (6.9.3) must hold no matter how we choose $\beta(t, r, T)$.

In conclusion, we have shown that if trading in the zero-coupon bonds presents no arbitrage opportunity, then for all t, r, and T such that $f_r(t, r, T) \neq 0$, we can define $\beta(t, r)$ by (6.9.2) because the right-hand side of (6.9.2) does not depend on T. We then have

$$f_t(t, r, T) + \beta(t, r) f_r(t, r, T) + \frac{1}{2}\gamma^2(t, r) f_{rr}(t, r, T) = r f(t, r, T), \tag{6.9.7}$$

which is (6.5.4) for the T-maturity bond. If $f_r(t, r, T) = 0$, then (6.9.6) holds, so (6.9.7) must still hold, no matter how $\beta(t, r)$ is defined. If we now change to a measure $\widetilde{\mathbb{P}}$ under which

$$\widetilde{W}(t) = W(t) + \int_0^t \frac{1}{\gamma(u, R(u))}\big[\alpha(u, R(u)) - \beta(u, R(u))\big]\, du$$

is a Brownian motion, then (6.9.1) can be rewritten as (6.5.1). The probability measure $\widetilde{\mathbb{P}}$ is risk-neutral.

Exercise 6.3 (Solution of Hull-White model). This exercise solves the ordinary differential equations (6.5.8) and (6.5.9) to produce the solutions $C(t,T)$ and $A(t,T)$ given in (6.5.10) and (6.5.11).

(i) Use equation (6.5.8) with s replacing t to show that

$$\frac{d}{ds}\left[e^{-\int_0^s b(v)dv}C(s,T)\right] = -e^{-\int_0^s b(v)dv}.$$

(ii) Integrate the equation in (i) from $s = t$ to $s = T$, and use the terminal condition $C(T,T)$ to obtain (6.5.10).

(iii) Replace t by s in (6.5.9), integrate the resulting equation from $s = t$ to $s = T$, use the terminal condition $A(T,T) = 0$, and obtain (6.5.11).

Exercise 6.4 (Solution of Cox-Ingersoll-Ross model). This exercise solves the ordinary differential equations (6.5.14) and (6.5.15) to produce the solutions $C(t,T)$ and $A(t,T)$ given in (6.5.16) and (6.5.17).

(i) Define the function

$$\varphi(t) = \exp\left\{\frac{1}{2}\sigma^2 \int_t^T C(u,T)\,du\right\}.$$

Show that

$$C(t,T) = -\frac{2\varphi'(t)}{\sigma^2\varphi(t)}, \tag{6.9.8}$$

$$C'(t,T) = -\frac{2\varphi''(t)}{\sigma^2\varphi(t)} + \frac{1}{2}\sigma^2 C^2(t,T). \tag{6.9.9}$$

(ii) Use the equation (6.5.14) to show that

$$\varphi''(t) - b\varphi'(t) - \frac{1}{2}\sigma^2\varphi(t) = 0. \tag{6.9.10}$$

This is a constant-coefficient linear ordinary differential equation. All solutions are of the form

$$\varphi(t) = a_1 e^{\lambda_1 t} + a_2 e^{\lambda_2 t},$$

where λ_1 and λ_2 are solutions of the so-called *characteristic equation* $\lambda^2 - b\lambda - \frac{1}{2}\sigma^2 = 0$, and a_1 and a_2 are constants.

(iii) Show that $\varphi(t)$ must be of the form

$$\varphi(t) = \frac{c_1}{\frac{1}{2}b+\gamma}e^{-(\frac{1}{2}b+\gamma)(T-t)} - \frac{c_2}{\frac{1}{2}b-\gamma}e^{-(\frac{1}{2}b-\gamma)(T-t)} \tag{6.9.11}$$

for some constants c_1 and c_2, where $\gamma = \frac{1}{2}\sqrt{b^2+2\sigma^2}$.

(iv) Show that

$$\varphi'(t) = c_1 e^{-(\frac{1}{2}b+\gamma)(T-t)} - c_2 e^{-(\frac{1}{2}b-\gamma)(T-t)}. \tag{6.9.12}$$

Use the fact that $C(T,T) = 0$ to show that $c_1 = c_2$.

(v) Show that

$$\begin{aligned}\varphi(t) &= c_1 e^{-\frac{1}{2}b(T-t)} \left[\frac{\frac{1}{2}b-\gamma}{\frac{1}{4}b^2-\gamma^2} e^{-\gamma(T-t)} - \frac{\frac{1}{2}b+\gamma}{\frac{1}{4}b^2-\gamma^2} e^{\gamma(T-t)} \right] \\ &= \frac{2c_1}{\sigma^2} e^{-\frac{1}{2}b(T-t)} \big[b\sinh(\gamma(T-t)) + 2\gamma\cosh(\gamma(T-t)) \big], \\ \varphi'(t) &= -2c_1 e^{-\frac{1}{2}b(T-t)} \sinh(\gamma(T-t)).\end{aligned}$$

Conclude that $C(t,T)$ is given by (6.5.16).

(vi) From (6.5.15) and (6.9.8), we have

$$A'(t,T) = \frac{2a\varphi'(t)}{\sigma^2\varphi(t)}.$$

Replace t by s in this equation, integrate from $s = t$ to $s = T$, and show that $A(t,T)$ is given by (6.5.17).

Exercise 6.5 (Two-dimensional Feynman-Kac).

(i) With $g(t,x_1,x_2)$ and $f(t,x_1,x_2)$ defined by (6.6.1) and (6.6.2), show that $g(t,X_1(t),X_2(t))$ and $e^{-rt}f(t,X_1(t),X_2(t))$ are martingales.

(ii) Assuming that W_1 and W_2 are independent Brownian motions, use the Itô-Doeblin formula to compute the differentials of $g(t,X_1(t),X_2(t))$ and $e^{-rt}f(t,X_1(t),X_2(t))$, set the dt term to zero, and thereby obtain the partial differential equations (6.6.3) and (6.6.4).

(iii) Now consider the case that $dW_1(t)\,dW_2(t) = \rho\,dt$, where ρ is a constant. Compute the differentials of $g(t,X_1(t),X_2(t))$ and $e^{-rt}f(t,X_1(t),X_2(t))$, set the dt term to zero, and obtain the partial differential equations

$$\begin{aligned} g_t + \beta_1 g_{x_1} + \beta_2 g_{x_2} &+ \left(\frac{1}{2}\gamma_{11}^2 + \rho\gamma_{11}\gamma_{12} + \frac{1}{2}\gamma_{12}^2\right) g_{x_1x_1} \\ &+ (\gamma_{11}\gamma_{21} + \rho\gamma_{11}\gamma_{22} + \rho\gamma_{12}\gamma_{21} + \gamma_{12}\gamma_{22}) g_{x_1x_2} \\ &+ \left(\frac{1}{2}\gamma_{21}^2 + \rho\gamma_{21}\gamma_{22} + \frac{1}{2}\gamma_{22}^2\right) g_{x_2x_2} = 0, \end{aligned} \tag{6.9.13}$$

$$\begin{aligned} f_t + \beta_1 f_{x_1} + \beta_2 f_{x_2} &+ \left(\frac{1}{2}\gamma_{11}^2 + \rho\gamma_{11}\gamma_{12} + \frac{1}{2}\gamma_{12}^2\right) f_{x_1x_1} \\ &+ (\gamma_{11}\gamma_{21} + \rho\gamma_{11}\gamma_{22} + \rho\gamma_{12}\gamma_{21} + \gamma_{12}\gamma_{22}) f_{x_1x_2} \\ &+ \left(\frac{1}{2}\gamma_{21}^2 + \rho\gamma_{21}\gamma_{22} + \frac{1}{2}\gamma_{22}^2\right) f_{x_2x_2} = rf. \end{aligned} \tag{6.9.14}$$

Exercise 6.6 (Moment-generating function for Cox-Ingersoll-Ross process).

(i) Let $W_1, \dots, W_d$ be independent Brownian motions and let a and σ be positive constants. For $j = 1, \dots, d$, let $X_j(t)$ be the solution of the *Ornstein-Uhlenbeck* stochastic differential equation

$$dX_j(t) = -\frac{b}{2}X_j(t)\,dt + \frac{1}{2}\sigma\,dW_j(t). \tag{6.9.15}$$

Show that

$$X_j(t) = e^{-\frac{1}{2}bt}\left[X_j(0) + \frac{\sigma}{2}\int_0^t e^{\frac{1}{2}bu}\,dW_j(u)\right]. \tag{6.9.16}$$

Show further that for fixed t, the random variable $X_j(t)$ is normal with

$$\mathbb{E}X_j(t) = e^{-\frac{1}{2}bt}X_j(0), \quad \mathrm{Var}(X_j(t)) = \frac{\sigma^2}{4b}\left[1 - e^{-bt}\right]. \tag{6.9.17}$$

(Hint: Use Theorem 4.4.9.)

(ii) Define

$$R(t) = \sum_{j=1}^{d} X_j^2(t), \tag{6.9.18}$$

and show that

$$dR(t) = \big(a - bR(t)\big)\,dt + \sigma\sqrt{R(t)}\,dB(t), \tag{6.9.19}$$

where $a = \frac{d\sigma^2}{4}$ and

$$B(t) = \sum_{j=1}^{d}\int_0^t \frac{X_j(s)}{\sqrt{R(s)}}\,dW_j(s) \tag{6.9.20}$$

is a Brownian motion. In other words, $R(t)$ is a Cox-Ingersoll-Ross interest rate process (Example 6.5.2). (Hint: Use Lévy's Theorem, Theorem 4.6.4, to show that $B(t)$ is a Brownian motion.)

(iii) Suppose $R(0) > 0$ is given, and define

$$X_j(0) = \sqrt{\frac{R(0)}{d}}.$$

Show then that $X_1(t), \dots, X_d(t)$ are independent, identically distributed, normal random variables, each having expectation

$$\mu(t) = e^{-\frac{1}{2}bt}\sqrt{\frac{R(0)}{d}}$$

and variance

$$v(t) = \frac{\sigma^2}{4b}\left[1 - e^{-bt}\right].$$

(iv) Part (iii) shows that $R(t)$ given by (6.9.18) is the sum of squares of independent, identically distributed, normal random variables and hence has a *noncentral* χ^2 *distribution*, the term "noncentral" referring to the fact that $\mu(t) = \mathbb{E}X_j(t)$ is not zero. To compute the moment-generating function of $R(t)$, first compute the moment-generating function

$$\mathbb{E}\exp\left\{uX_j^2(t)\right\} = \frac{1}{\sqrt{1-2v(t)u}}\exp\left\{\frac{u\mu^2(t)}{1-2v(t)u}\right\} \text{ for all } u < \frac{1}{2v(t)}. \tag{6.9.21}$$

(Hint: You will need to complete a square, first deriving and then using the equation

$$ux^2 - \frac{1}{2v(t)}\left(x-\mu(t)\right)^2 = -\frac{1-2v(t)u}{2v(t)}\left(x - \frac{\mu(t)}{1-2v(t)u}\right)^2 + \frac{u\mu^2(t)}{1-2v(t)u}.$$

The integral from $-\infty$ to ∞ of the normal density with mean $\mu(t)/(1-2v(t)u)$ and variance $v(t)/(1-2v(t)u)$,

$$\sqrt{\frac{1-2v(t)u}{2\pi v(t)}}\exp\left\{-\frac{1-2v(t)u}{2v(t)}\left(x-\frac{\mu(t)}{1-2v(t)u}\right)^2\right\},$$

is equal to 1.)

(v) Show that $R(t)$ given by (6.9.19) has moment-generating function

$$\begin{aligned}\mathbb{E}e^{uR(t)} &= \left(\frac{1}{1-2v(t)u}\right)^{d/2}\exp\left\{\frac{e^{-bt}uR(0)}{1-2v(t)u}\right\}\\ &= \left(\frac{1}{1-2v(t)u}\right)^{2a/\sigma^2}\exp\left\{\frac{e^{-bt}uR(0)}{1-2v(t)u}\right\} \text{ for all } u < \frac{1}{2v(t)}.\end{aligned} \tag{6.9.22}$$

Remark 6.9.1 (Cox-Ingersoll-Ross process hitting zero). Although we have derived (6.9.22) under the assumption that d is a positive integer, the second line of (6.9.22) is expressed in terms of only the parameters a, b, and σ entering (6.9.19), and this formula is valid for all $a > 0$, $b > 0$, and $\sigma > 0$. When $d \geq 2$ (i.e., $a \geq \frac{1}{2}\sigma^2$), the multidimensional process $\left(X_1(t), \dots, X_d(t)\right)$ never hits the origin in $\mathbb{R}^d$, and hence $R(t)$ is never zero. In fact, $R(t)$ is never zero if and only if $a \geq \frac{1}{2}\sigma^2$. If $0 < a < \frac{1}{2}\sigma^2$, then $R(t)$ hits zero repeatedly but after each hit becomes positive again.

Exercise 6.7 (Heston stochastic volatility model). Suppose that under a risk-neutral measure $\widetilde{\mathbb{P}}$ a stock price is governed by

$$dS(t) = rS(t)\,dt + \sqrt{V(t)}\,S(t)\,d\widetilde{W}_1(t), \tag{6.9.23}$$

where the interest rate r is constant and the volatility $\sqrt{V(t)}$ is itself a stochastic process governed by the equation

$$dV(t) = \left(a - bV(t)\right)dt + \sigma\sqrt{V(t)}\,d\widetilde{W}_2(t). \tag{6.9.24}$$

The parameters a, b, and σ are positive constants, and $\widetilde{W}_1(t)$ and $\widetilde{W}_2(t)$ are correlated Brownian motions under $\widetilde{\mathbb{P}}$ with

$$d\widetilde{W}_1(t)\,d\widetilde{W}_2(t) = \rho\,dt$$

for some $\rho \in (-1, 1)$. Because the two-dimensional process $(S(t), V(t))$ is governed by the pair of stochastic differential equations (6.9.23) and (6.9.24), it is a two-dimensional Markov process.

So long as trading takes place only in the stock and money market account, this model is incomplete. One can create a one-parameter family of risk-neutral measures by changing the dt term in (6.9.24) without affecting (6.9.23).

At time t, the risk-neutral price of a call expiring at time $T \geq t$ in this stochastic volatility model is $\widetilde{\mathbb{E}}[e^{-r(T-t)}(S(T) - K)^+ | \mathcal{F}(t)]$. Because of the Markov property, there is a function $c(t, s, v)$ such that

$$c(t, S(t), V(t)) = \widetilde{\mathbb{E}}\left[e^{-r(T-t)}\left(S(T) - K\right)^+ \middle| \mathcal{F}(t)\right], \quad 0 \leq t \leq T. \qquad (6.9.25)$$

This problem shows that the function $c(t, s, v)$ satisfies the partial differential equation

$$c_t + rsc_s + (a - bv)c_v + \frac{1}{2}s^2 vc_{ss} + \rho\sigma svc_{sv} + \frac{1}{2}\sigma^2 vc_{vv} = rc \qquad (6.9.26)$$

in the region $0 \leq t < T$, $s \geq 0$, and $v \geq 0$. The function $c(t, s, v)$ also satisfies the boundary conditions

$$c(T, s, v) = (s - K)^+ \text{ for all } s \geq 0, v \geq 0, \qquad (6.9.27)$$
$$c(t, 0, v) = 0 \text{ for all } 0 \leq t \leq T, v \geq 0, \qquad (6.9.28)$$
$$c(t, s, 0) = \left(s - e^{-r(T-t)}K\right)^+ \text{ for all } 0 \leq t \leq T, s \geq 0, \qquad (6.9.29)$$
$$\lim_{s \to \infty} \frac{c(t, s, v)}{s - K} = 1 \text{ for all } 0 \leq t \leq T, v \geq 0, \qquad (6.9.30)$$
$$\lim_{v \to \infty} c(t, s, v) = s \text{ for all } 0 \leq t \leq T, s \geq 0. \qquad (6.9.31)$$

In this problem, we shall be concerned only with (6.9.27).

(i) Show that $e^{-rt}c(t, S(t), V(t))$ is a martingale under $\widetilde{\mathbb{P}}$, and use this fact to obtain (6.9.26).

(ii) Suppose there are functions $f(t, x, v)$ and $g(t, x, v)$ satisfying

$$f_t + \left(r + \frac{1}{2}v\right) f_x + (a - bv + \rho\sigma v)f_v + \frac{1}{2}vf_{xx} + \rho\sigma vf_{xv} + \frac{1}{2}\sigma^2 vf_{vv} = 0, \qquad (6.9.32)$$

$$g_t + \left(r - \frac{1}{2}v\right) g_x + (a - bv)g_v + \frac{1}{2}vg_{xx} + \rho\sigma vg_{xv} + \frac{1}{2}\sigma^2 vg_{vv} = 0, \qquad (6.9.33)$$

in the region $0 \le t < T$, $-\infty < x < \infty$, and $v \ge 0$. Show that if we define

$$c(t,s,v) = sf(t,\log s,v) - e^{-r(T-t)}Kg(t,\log s,v), \tag{6.9.34}$$

then $c(t,s,v)$ satisfies the partial differential equation (6.9.26).

(iii) Suppose a pair of processes $(X(t),V(t))$ is governed by the stochastic differential equations

$$dX(t) = \left(r + \frac{1}{2}V(t)\right)dt + \sqrt{V(t)}\,dW_1(t), \tag{6.9.35}$$

$$dV(t) = \big(a - bV(t) + \rho\sigma V(t)\big)\,dt + \sigma\sqrt{V(t)}\,dW_2(t), \tag{6.9.36}$$

where $W_1(t)$ and $W_2(t)$ are Brownian motions under some probability measure $\mathbb{P}$ with $dW_1(t)\,dW_2(t) = \rho\,dt$. Define

$$f(t,x,v) = \mathbb{E}^{t,x,v}\mathbb{I}_{\{X(T)\ge\log K\}}. \tag{6.9.37}$$

Show that $f(t,x,v)$ satisfies the partial differential equation (6.9.32) and the boundary condition

$$f(T,x,v) = \mathbb{I}_{\{x\ge\log K\}} \text{ for all } x\in\mathbb{R}, v\ge 0. \tag{6.9.38}$$

(iv) Suppose a pair of processes $(X(t),V(t))$ is governed by the stochastic differential equations

$$dX(t) = \left(r - \frac{1}{2}V(t)\right)dt + \sqrt{V(t)}\,dW_1(t), \tag{6.9.39}$$

$$dV(t) = \big(a - bV(t)\big)\,dt + \sigma\sqrt{V(t)}\,dW_2(t), \tag{6.9.40}$$

where $W_1(t)$ and $W_2(t)$ are Brownian motions under some probability measure $\mathbb{P}$ with $dW_1(t)\,dW_2(t) = \rho\,dt$. Define

$$g(t,x,v) = \mathbb{E}^{t,x,v}\mathbb{I}_{\{X(T)\ge\log K\}}. \tag{6.9.41}$$

Show that $g(t,x,v)$ satisfies the partial differential equation (6.9.33) and the boundary condition

$$g(T,x,v) = \mathbb{I}_{\{x\ge\log K\}} \text{ for all } x\in\mathbb{R}, v\ge 0. \tag{6.9.42}$$

(v) Show that with $f(t,x,v)$ and $g(t,x,v)$ as in (iii) and (iv), the function $c(t,x,v)$ of (6.9.34) satisfies the boundary condition (6.9.27).

Remark 6.9.2. In fact, with $f(t,x,v)$ and $g(t,x,v)$ as in (iii) and (iv), the function $c(t,x,v)$ of (6.9.34) satisfies all the boundary conditions (6.9.27)–(6.9.31) and is the function appearing on the left-hand side of (6.9.25).

Exercise 6.8 (Kolmogorov backward equation). Consider the stochastic differential equation

$$dX(u) = \beta(u, X(u))\,du + \gamma(u, X(u))\,dW(u).$$

We assume that, just as with a geometric Brownian motion, if we begin a process at an arbitrary initial positive value $X(t) = x$ at an arbitrary initial time t and evolve it forward using this equation, its value at each time $T > t$ could be any positive number but cannot be less than or equal to zero. For $0 \leq t < T$, let $p(t, T, x, y)$ be the transition density for the solution to this equation (i.e., if we solve the equation with the initial condition $X(t) = x$, then the random variable $X(T)$ has density $p(t, T, x, y)$ in the y variable). We are assuming that $p(t, T, x, y) = 0$ for $0 \leq t < T$ and $y \leq 0$.

Show that $p(t, T; x, y)$ satisfies the *Kolmogorov backward equation*

$$-p_t(t, T, x, y) = \beta(t, x)p_x(t, T, x, y) + \frac{1}{2}\gamma^2(t, x)p_{xx}(t, T, x, y). \tag{6.9.43}$$

(Hint: We know from the Feynman-Kac Theorem, Theorem 6.4.1, that, for any function $h(y)$, the function

$$g(t, x) = \mathbb{E}^{t,x}h\big(X(T)\big) = \int_0^\infty h(y)p(t, T, x, y)dy \tag{6.9.44}$$

satisfies the partial differential equation

$$g_t(t, x) + \beta(t, x)g_x(t, x) + \frac{1}{2}\gamma^2(t, x)g_{xx}(t, x) = 0. \tag{6.9.45}$$

Use (6.9.44) to compute g_t, g_x, and g_{xx}, and then argue that the only way (6.9.45) can hold regardless of the choice of the function $h(y)$ is for $p(t, T, x, y)$ to satisfy the Kolmogorov backward equation.)

Exercise 6.9 (Kolmogorov forward equation). (Also called the *Fokker-Planck equation*). We begin with the same stochastic differential equation,

$$dX(u) = \beta\big(u, X(u)\big)\,du + \gamma\big(u, X(u)\big)\,dW(u), \tag{6.9.46}$$

as in Exercise 6.8, use the same notation $p(t, T, x, y)$ for the transition density, and again assume that $p(t, T, x, y) = 0$ for $0 \leq t < T$ and $y \leq 0$. In this problem, we show that $p(t, T, x, y)$ satisfies the *Kolmogorov forward equation*

$$\frac{\partial}{\partial T}p(t, T, x, y) = -\frac{\partial}{\partial y}\big(\beta(t, y)p(t, T, x, y)\big) + \frac{1}{2}\frac{\partial^2}{\partial y^2}\big(\gamma^2(T, y)p(t, T, x, y)\big). \tag{6.9.47}$$

In contrast to the Kolmogorov backward equation, in which T and y were held constant and the variables were t and x, here t and x are held constant and the variables are y and T. The variables t and x are sometimes called the *backward variables*, and T and y are called the *forward variables*.

(i) Let b be a positive constant and let $h_b(y)$ be a function with continuous first and second derivatives such that $h_b(x) = 0$ for all $x \leq 0$, $h_b'(x) = 0$ for all $x \geq b$, and $h_b(b) = h_b'(b) = 0$. Let $X(u)$ be the solution to the stochastic differential equation with initial condition $X(t) = x \in (0, b)$, and use Itô's formula to compute $dh_b(X(u))$.

(ii) Let $0 \leq t < T$ be given, and integrate the equation you obtained in (i) from t to T. Take expectations and use the fact that $X(u)$ has density $p(t, u, x, y)$ in the y-variable to obtain

$$\int_0^b h_b(y)p(t,T,x,y)dy = h_b(x) + \int_t^T \int_0^b \beta(u,y)p(t,u,x,y)h_b'(y)dy\,du + \frac{1}{2}\int_t^T \int_0^b \gamma^2(u,y)p(t,u,x,y)h_b''(y)dy. \tag{6.9.48}$$

(iii) Integrate the integrals $\int_0^b \cdots dy$ on the right-hand side of (6.9.48) by parts to obtain

$$\begin{aligned}\int_0^b & h_b(y)p(t,T,x,y)dy \\ &= h_b(x) - \int_t^T \int_0^b \frac{\partial}{\partial y}\left[\beta(u,y)p(t,u,x,y)\right] h_b(y)dy\,du \\ &\quad + \frac{1}{2}\int_t^T \int_0^b \frac{\partial^2}{\partial y^2}\left[\gamma^2(u,y)p(t,u,x,y)\right] h_b(y)dy\,du.\end{aligned} \tag{6.9.49}$$

(iv) Differentiate (6.9.49) with respect to T to obtain

$$\int_0^b h_b(y)\left[\frac{\partial}{\partial T}p(t,T,x,y) + \frac{\partial}{\partial y}\left(\beta(T,y)p(t,T,x,y)\right) - \frac{1}{2}\frac{\partial^2}{\partial y^2}\left(\gamma^2(T,y)p(t,T,x,y)\right)\right]dy = 0. \tag{6.9.50}$$

(v) Use (6.9.50) to show that there cannot be numbers $0 < y_1 < y_2$ such that

$$\frac{\partial}{\partial T}p(t,T,x,y) + \frac{\partial}{\partial y}\left(\beta(T,y)p(t,T,x,y)\right) - \frac{1}{2}\frac{\partial^2}{\partial y^2}\left(\gamma^2(T,y)p(t,T,x,y)\right) > 0 \text{ for all } y \in (y_1, y_2).$$

Similarly, there cannot be numbers $0 < y_1 < y_2$ such that

$$\frac{\partial}{\partial T}p(t,T,x,y) + \frac{\partial}{\partial y}\left(\beta(T,y)p(t,T,x,y)\right) - \frac{1}{2}\frac{\partial^2}{\partial y^2}\left(\gamma^2(T,y)p(t,T,x,y)\right) < 0 \text{ for all } y \in [y_1, y_2].$$

This is as much as you need to do for this problem. It is now obvious that if

$$\frac{\partial}{\partial T}p(t,T,x,y) + \frac{\partial}{\partial y}\big(\beta(T,y)p(t,T,x,y)\big) - \frac{1}{2}\frac{\partial}{\partial y^2}\big(\gamma^2(T,y)p(t,T,x,y)\big)$$

is a continuous function of y, then this expression must be zero for every $y > 0$, and hence $p(t,T,x,y)$ satisfies the Kolmogorov forward equation stated at the beginning of this problem.

Exercise 6.10 (Implying the volatility surface). Assume that a stock price evolves according to the stochastic differential equation

$$dS(u) = rS(u)\,dt + \sigma\big(u,S(u)\big)S(u)\,d\widetilde{W}(u),$$

where the interest rate r is constant, the volatility $\sigma(u,x)$ is a function of time and the underlying stock price, and $\widetilde{W}$ is a Brownian motion under the risk-neutral measure $\widetilde{\mathbb{P}}$. This is a special case of the stochastic differential equation (6.9.46) with $\beta(u,x) = rx$ and $\gamma(u,x) = \sigma(u,x)x$. Let $\tilde{p}(t,T,x,y)$ denote the transition density.

According to Exercise 6.9, the transition density $\tilde{p}(t,T,x,y)$ satisfies the Kolmogorov forward equation

$$\frac{\partial}{\partial T}\tilde{p}(t,T,x,y) = -\frac{\partial}{\partial y}\left(ry\tilde{p}(t,T,x,y)\right) + \frac{1}{2}\frac{\partial^2}{\partial y^2}\left(\sigma^2(T,y)y^2\tilde{p}(t,T,x,y)\right). \tag{6.9.51}$$

Let

$$c(0,T,x,K) = e^{-rT}\int_K^\infty (y-K)\tilde{p}(0,T,x,y)dy \tag{6.9.52}$$

denote the time-zero price of a call expiring at time T, struck at K, when the initial stock price is $S(0) = x$. Note that

$$c_T(0,T,x,K) = -rc(0,T,x,K) + e^{-rT}\int_K^\infty (y-K)\tilde{p}_T(0,T,x,y)dy. \tag{6.9.53}$$

(i) Integrate once by parts to show that

$$-\int_K^\infty (y-K)\frac{\partial}{\partial y}\left(ry\tilde{p}(0,T,x,y)\right)dy = \int_K^\infty ry\tilde{p}(0,T,x,y)dy. \tag{6.9.54}$$

You may assume that

$$\lim_{y\to\infty}(y-K)ry\tilde{p}(0,T,x,y) = 0. \tag{6.9.55}$$

(ii) Integrate by parts and then integrate again to show that

$$\frac{1}{2}\int_K^\infty (y-K)\frac{\partial^2}{\partial y^2}\left(\sigma^2(T,y)y^2\tilde{p}(0,T,x,y)\right)dy = \frac{1}{2}\sigma^2(T,K)K^2\tilde{p}(0,T,x,K). \tag{6.9.56}$$

You may assume that

$$\lim_{y\to\infty}(y-K)\frac{\partial}{\partial y}\left(\sigma^2(T,y)y^2\tilde{p}(0,T,x,y)\right)=0, \tag{6.9.57}$$

$$\lim_{y\to\infty}\sigma^2(T,y)y^2\tilde{p}(0,T,x,y)=0. \tag{6.9.58}$$

(iii) Now use (6.9.53), (6.9.52), (6.9.51), (6.9.54), (6.9.56), and Exercise 5.9 of Chapter 5 in that order to obtain the equation

$$\begin{aligned}
&c_T(0,T,x,K)\\
&=e^{-rT}rK\int_K^\infty \tilde{p}(0,T,x,y)dy+\frac{1}{2}e^{-rT}\sigma^2(T,K)K^2\tilde{p}(0,T,x,K)\\
&=-rKc_K(0,T,x,K)+\frac{1}{2}\sigma^2(T,K)K^2c_{KK}(0,T,x,K).
\end{aligned} \tag{6.9.59}$$

This is the end of the problem. Note that under the assumption that $c_{KK}(0,T,x,K)\neq 0$, (6.9.59) can be solved for the volatility term $\sigma^2(T,K)$ in terms of the quantities $c_T(0,T,x,K)$, $c_K(0,T,x,K)$, and $c_{KK}(0,T,x,K)$, which can be inferred from market prices.

7

Exotic Options

7.1 Introduction

The European calls and puts considered thus far in this text are sometimes called *vanilla* or even *plain vanilla* options. Their payoffs depend only on the final value of the underlying asset. Options whose payoffs depend on the path of the underlying asset are called *path-dependent* or *exotic.*

In this chapter, we present three types of exotic options on a geometric Brownian motion asset and work out a detailed analysis for one option of each type. The types considered are *barrier options*, *lookback options*, and *Asian options.* In each case, we work out the standard partial differential equation governing the option price. The first two options have explicit pricing formulas, which are based on the reflection principle for Brownian motion. Such a formula for Asian options is not known. However, for the Asian option there is a change-of-numéraire argument that reduces the pricing partial differential equation to a simple form that can easily be solved numerically. We present this argument in Subsection 7.5.3.

7.2 Maximum of Brownian Motion with Drift

In this section, we derive the joint density for a Brownian motion with drift and its maximum to date. This density is used in Sections 7.3 and 7.4 to obtain explicit pricing formulas for a barrier option and a lookback option. To derive this formula, we begin with a Brownian motion $\widetilde{W}(t)$, $0 \leq t \leq T$, defined on a probability space $(\Omega, \mathcal{F}, \widetilde{\mathbb{P}})$. Under $\widetilde{\mathbb{P}}$, the Brownian motion $\widetilde{W}(t)$ has zero drift (i.e., it is a martingale). Let α be a given number, and define

$$\widehat{W}(t) = \alpha t + \widetilde{W}(t), \quad 0 \leq t \leq T. \tag{7.2.1}$$

This Brownian motion $\widehat{W}(t)$ has drift α under $\widetilde{\mathbb{P}}$. We further define

$$\widehat{M}(T) = \max_{0 \le t \le T} \widehat{W}(t). \tag{7.2.2}$$

Because $\widehat{W}(0) = 0$, we have $\widehat{M}(T) \ge 0$. We also have $\widehat{W}(T) \le \widehat{M}(T)$. Therefore, the pair of random variables $(\widehat{M}(t), \widehat{W}(T))$ takes values in the set $\{(m, w); w \le m, m \ge 0\}$ shown in Figure 7.2.1.

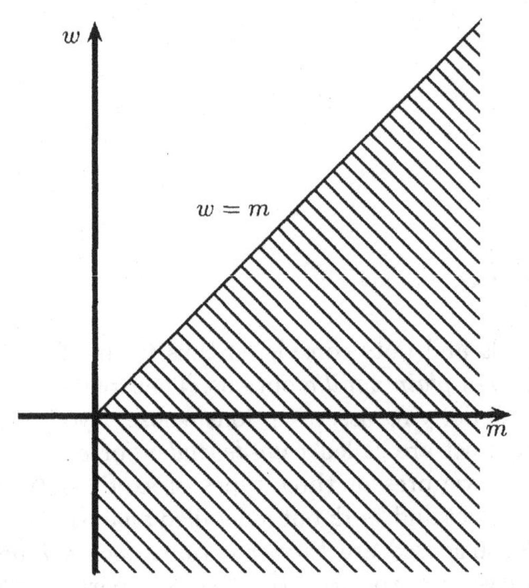

Fig. 7.2.1. Range of $(\widehat{M}(T), \widehat{W}(T))$.

Theorem 7.2.1. *The joint density under* $\widetilde{\mathbb{P}}$ *of the pair* $(\widehat{M}(T), \widehat{W}(T))$ *is*

$$\widetilde{f}_{\widehat{M}(T),\widehat{W}(T)}(m, w) = \frac{2(2m - w)}{T\sqrt{2\pi T}} e^{\alpha w - \frac{1}{2}\alpha^2 T - \frac{1}{2T}(2m-w)^2}, \quad w \le m,\ m \ge 0, \tag{7.2.3}$$

and is zero for other values of m *and* w.

PROOF: We define the exponential martingale

$$\widehat{Z}(t) = e^{-\alpha \widetilde{W}(t) - \frac{1}{2}\alpha^2 t} = e^{-\alpha \widehat{W}(t) + \frac{1}{2}\alpha^2 t}, \quad 0 \le t \le T,$$

and use $\widehat{Z}(T)$ to define a new probability measure $\widehat{\mathbb{P}}$ by

$$\widehat{\mathbb{P}}(A) = \int_A Z(T)\, d\widetilde{\mathbb{P}} \text{ for all } A \in \mathcal{F}.$$

According to Girsanov's Theorem, Theorem5.2.3, $\widehat{W}(t)$ is a Brownian motion (with zero drift) under $\widehat{\mathbb{P}}$. Theorem 3.7.3 gives us the joint density of $(\widehat{M}(T), \widehat{W}(T))$ under $\widehat{\mathbb{P}}$, which is

$$\widehat{f}_{\widehat{M}(T),\widehat{W}(T)}(m,w) = \frac{2(2m-w)}{T\sqrt{2\pi T}} e^{-\frac{1}{2T}(2m-w)^2}, \quad w \le m,\ m \ge 0, \tag{7.2.4}$$

and is zero for other values of m and w. To work out the density of $(\widehat{M}(T), \widehat{W}(T))$ under $\widetilde{\mathbb{P}}$, we use Lemma 5.2.1, which implies

$$\begin{aligned}
&\widetilde{\mathbb{P}}\{\widehat{M}(T) \le m, \widehat{W}(T) \le w\} \\
&\qquad = \widetilde{\mathbb{E}}\big[\mathbb{I}_{\{\widehat{M}(T)\le m, \widehat{W}(T)\le w\}}\big] \\
&\qquad = \widehat{\mathbb{E}}\left[\frac{1}{\widehat{Z}(T)}\mathbb{I}_{\{\widehat{M}(T)\le m, \widehat{W}(T)\le w\}}\right] \\
&\qquad = \widehat{\mathbb{E}}\left[e^{\alpha\widehat{W}(T)-\frac{1}{2}\alpha^2 T}\mathbb{I}_{\{\widehat{M}(T)\le m, \widehat{W}(T)\le w\}}\right] \\
&\qquad = \int_{-\infty}^{w}\int_{-\infty}^{m} e^{\alpha y-\frac{1}{2}\alpha^2 T}\widehat{f}_{\widehat{M}(T),\widehat{W}(T)}(x,y)\,dx\,dy.
\end{aligned}$$

Therefore, the density of $(\widehat{M}(T), \widehat{W}(T))$ under $\widetilde{\mathbb{P}}$ is

$$\frac{\partial^2}{\partial m\,\partial w}\widetilde{\mathbb{P}}\{\widehat{M}(T) \le m, \widehat{W}(T) \le w\} = e^{\alpha w-\frac{1}{2}\alpha^2 T}\widehat{f}_{\widehat{M}(T),\widehat{W}(T)}(m,w). \tag{7.2.5}$$

When $w \le m$ and $m \ge 0$, this is formula (7.2.3). For other values of m and w, we obtain zero because $\widehat{f}_{\widehat{M}(T),\widehat{W}(T)}(m,w)$ is zero. □

Corollary 7.2.2. *We have*

$$\widetilde{\mathbb{P}}\{\widehat{M}(T) \le m\} = N\left(\frac{m-\alpha T}{\sqrt{T}}\right) - e^{2\alpha m}N\left(\frac{-m-\alpha T}{\sqrt{T}}\right), \quad m \ge 0, \tag{7.2.6}$$

and the density under $\widetilde{\mathbb{P}}$ *of the random variable* $\widehat{M}(T)$ *is*

$$\widetilde{f}_{\widehat{M}(T)}(m) = \frac{2}{\sqrt{2\pi T}}e^{-\frac{1}{2T}(m-\alpha T)^2} - 2\alpha e^{2\alpha m}N\left(\frac{-m-\alpha T}{\sqrt{T}}\right), \quad m \ge 0, \tag{7.2.7}$$

and is zero for $m < 0$.

PROOF: We integrate the density (7.2.3) over the region in Figure 7.2.2 to compute

$$\begin{aligned}
&\widetilde{\mathbb{P}}\{\widehat{M}(T) \le m\} \\
&= \int_0^m \int_w^m \frac{2(2\mu-w)}{T\sqrt{2\pi T}} e^{\alpha w-\frac{1}{2}\alpha^2 T-\frac{1}{2T}(2\mu-w)^2}\,d\mu\,dw \\
&\quad + \int_{-\infty}^0 \int_0^m \frac{2(2\mu-w)}{T\sqrt{2\pi T}} e^{\alpha w-\frac{1}{2}\alpha^2 T-\frac{1}{2T}(2\mu-w)^2}\,d\mu\,dw
\end{aligned}$$

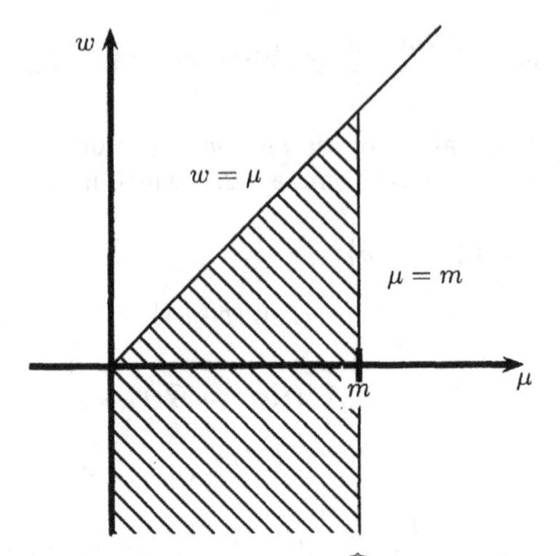

Fig. 7.2.2. The region $\widehat{M}(T) \le m$.

$$
\begin{aligned}
&= -\int_0^m \frac{1}{\sqrt{2\pi T}} e^{\alpha w - \frac{1}{2}\alpha^2 T - \frac{1}{2T}(2\mu - w)^2}\Big|_{\mu=w}^{\mu=m} dw \\
&\quad - \int_{-\infty}^0 \frac{1}{\sqrt{2\pi T}} e^{\alpha w - \frac{1}{2}\alpha^2 T - \frac{1}{2T}(2\mu - w)^2}\Big|_{\mu=0}^{\mu=m} dw \\
&= -\frac{1}{\sqrt{2\pi T}} \int_0^m e^{\alpha w - \frac{1}{2}\alpha^2 T - \frac{1}{2T}(2m-w)^2} dw + \frac{1}{\sqrt{2\pi T}} \int_0^m e^{\alpha w - \frac{1}{2}\alpha^2 T - \frac{1}{2T} w^2} dw \\
&\quad - \frac{1}{\sqrt{2\pi T}} \int_{-\infty}^0 e^{\alpha w - \frac{1}{2}\alpha^2 T - \frac{1}{2T}(2m-w)^2} dw + \frac{1}{\sqrt{2\pi T}} \int_{-\infty}^0 e^{\alpha w - \frac{1}{2}\alpha^2 T - \frac{1}{2T} w^2} dw \\
&= -\frac{1}{\sqrt{2\pi T}} \int_{-\infty}^m e^{\alpha w - \frac{1}{2}\alpha^2 T - \frac{1}{2T}(2m-w)^2} dw + \frac{1}{\sqrt{2\pi T}} \int_{-\infty}^m e^{\alpha w - \frac{1}{2}\alpha^2 T - \frac{1}{2T} w^2} dw.
\end{aligned}
$$

We complete the squares. Observe that

$$
\begin{aligned}
-\frac{1}{2T}(w - 2m - \alpha T)^2 &= -\frac{(2m-w)^2}{2T} + \alpha w - 2\alpha m - \frac{1}{2}\alpha^2 T, \\
-\frac{1}{2T}(w - \alpha T)^2 &= -\frac{w^2}{2T} + \alpha w - \frac{1}{2}\alpha^2 T.
\end{aligned}
$$

Therefore,

$$
\begin{aligned}
&\widetilde{\mathbb{P}}\{\widehat{M}(T) \le m\} \\
&= -\frac{e^{2\alpha m}}{\sqrt{2\pi T}} \int_{-\infty}^m e^{-\frac{1}{2T}(w - 2m - \alpha T)^2} dw + \frac{1}{\sqrt{2\pi T}} \int_{-\infty}^m e^{-\frac{1}{2T}(w - \alpha T)^2} dw.
\end{aligned}
$$

We make the change of variable $y = \frac{w - 2m - \alpha T}{\sqrt{T}}$ in the first integral and $y = \frac{w - \alpha T}{\sqrt{T}}$ in the second, thereby obtaining

$$\widetilde{\mathbb{P}}\{\widehat{M}(T) \le m\} = -\frac{e^{2\alpha m}}{\sqrt{2\pi}} \int_{-\infty}^{\frac{-m-\alpha T}{\sqrt{T}}} e^{-\frac{1}{2}y^2}\,dy + \frac{1}{\sqrt{2\pi}} \int_{-\infty}^{\frac{m-\alpha T}{\sqrt{T}}} e^{-\frac{1}{2}y^2}\,du$$

$$= -e^{2\alpha m} N\left(\frac{-m-\alpha T}{\sqrt{T}}\right) + N\left(\frac{m-\alpha T}{\sqrt{T}}\right).$$

This establishes (7.2.6).

To obtain the density (7.2.7), we differentiate (7.2.6) with respect to m:

$$\frac{d}{dm}\widetilde{\mathbb{P}}\{\widehat{M}(T) \le m\}$$

$$= N'\left(\frac{m-\alpha T}{\sqrt{T}}\right)\left(\frac{1}{\sqrt{T}}\right) - 2\alpha e^{2\alpha m} N\left(\frac{-m-\alpha T}{\sqrt{T}}\right)$$

$$-e^{2\alpha m} N'\left(\frac{-m-\alpha T}{\sqrt{T}}\right)\left(-\frac{1}{\sqrt{T}}\right)$$

$$= \frac{1}{\sqrt{2\pi T}} e^{-\frac{1}{2T}(m-\alpha T)^2} - 2\alpha e^{2\alpha m} N\left(\frac{-m-\alpha T}{\sqrt{T}}\right) + \frac{e^{2\alpha m}}{\sqrt{2\pi T}} e^{-\frac{1}{2T}(-m-\alpha T)^2}.$$

The exponent in the third term is

$$2\alpha m - \frac{(-m-\alpha T)^2}{2T} = \frac{4\alpha m}{2T} - \frac{m^2 + 2\alpha m T + \alpha^2 T^2}{2T}$$

$$= -\frac{m^2 - 2\alpha m T + \alpha^2 T^2}{2T}$$

$$= -\frac{(m-\alpha T)^2}{2T},$$

which is the exponent in the first term. Combining the first and third terms, we obtain (7.2.7). □

7.3 Knock-out Barrier Options

There are several types of barrier options. Some "knock out" when the underlying asset price crosses a barrier (i.e., they become worthless). If the underlying asset price begins below the barrier and must cross above it to cause the knock-out, the option is said to be *up-and-out.* A *down-and-out* option has the barrier below the initial asset price and knocks out if the asset price falls below the barrier. Other options "knock in" at a barrier (i.e., they pay off zero unless they cross a barrier). Knock-in options also fall into two classes, *up-and-in* and *down-and-in.* The payoff at expiration for barrier options is typically either that of a put or a call. More complex barrier options require the asset price to not only cross a barrier but spend a certain amount of time across the barrier in order to knock in or knock out.

In this section, we treat an up-and-out call on a geometric Brownian motion. The methodology we develop works equally well for up-and-in, down-and-out, and down-and-in puts and calls.

7.3.1 Up-and-Out Call

Our underlying risky asset is geometric Brownian motion

$$dS(t) = rS(t)\,dt + \sigma S(t)\,d\widetilde{W}(t),$$

where $\widetilde{W}(t)$, $0 \le t \le T$, is a Brownian motion under the risk-neutral measure $\widetilde{\mathbb{P}}$. Consider a European call, expiring at time T, with strike price K and up-and-out barrier B. We assume $K < B$; otherwise, the option must knock out in order to be in the money and hence could only pay off zero. The solution to the stochastic differential equation for the asset price is

$$S(t) = S(0)e^{\sigma\widetilde{W}(t)+(r-\frac{1}{2}\sigma^2)t} = S(0)e^{\sigma\widehat{W}(t)}, \tag{7.3.1}$$

where $\widehat{W}(t) = \alpha t + \widetilde{W}(t)$, and

$$\alpha = \frac{1}{\sigma}\left(r - \frac{1}{2}\sigma^2\right).$$

We define $\widehat{M}(T) = \max_{0\le t\le T}\widehat{W}(t)$, so

$$\max_{0\le t\le T} S(t) = S(0)e^{\sigma\widehat{M}(T)}.$$

The option knocks out if and only if $S(0)e^{\sigma\widehat{M}(T)} > B$; if $S(0)e^{\sigma\widehat{M}(T)} \le B$, the option pays off

$$\big(S(T) - K\big)^+ = \left(S(0)e^{\sigma\widehat{W}(T)} - K\right)^+.$$

In other words, the payoff of the option is

$$\begin{aligned} V(T) &= \left(S(0)e^{\sigma\widehat{W}(T)} - K\right)^+ \mathbb{I}_{\{S(0)e^{\sigma\widehat{M}(T)}\le B\}} \\ &= \left(S(0)e^{\sigma\widehat{W}(T)} - K\right)\mathbb{I}_{\{S(0)e^{\sigma\widehat{W}(T)}\ge K,\,S(0)e^{\sigma\widehat{M}(T)}\le B\}} \\ &= \left(S(0)e^{\sigma\widehat{W}(T)} - K\right)\mathbb{I}_{\{\widehat{W}(T)\ge k,\,\widehat{M}(T)\le b\}}, \end{aligned} \tag{7.3.2}$$

where

$$k = \frac{1}{\sigma}\log\frac{K}{S(0)}, \quad b = \frac{1}{\sigma}\log\frac{B}{S(0)}. \tag{7.3.3}$$

7.3.2 Black-Scholes-Merton Equation

The price of an up-and-out call satisfies a Black-Scholes-Merton equation that has been modified to account for the barrier. This equation can be used to solve for the price. In this particular case, we do not need to find the price this way because it can be computed analytically (see Subsection 7.3.3). However, we provide the equation and its derivation because this methodology works in situations where analytical solutions cannot be obtained.

Theorem 7.3.1. *Let $v(t,x)$ denote the price at time t of the up-and-out call under the assumption that the call has not knocked out prior to time t and $S(t)=x$. Then $v(t,x)$ satisfies the Black-Scholes-Merton partial differential equation*

$$v_t(t,x)+rxv_x(t,x)+\frac{1}{2}\sigma^2x^2v_{xx}(t,x)=rv(t,x) \tag{7.3.4}$$

in the rectangle $\{(t,x); 0\le t<T, 0\le x\le B\}$ and satisfies the boundary conditions

$$v(t,0)=0,\quad 0\le t\le T, \tag{7.3.5}$$
$$v(t,B)=0,\quad 0\le t<T, \tag{7.3.6}$$
$$v(T,x)=(x-K)^+,\quad 0\le x\le B. \tag{7.3.7}$$

The lower boundary condition (7.3.5) follows as in the usual Black-Scholes-Merton framework: If the asset price begins at zero, it stays there and the option expires out of the money. The upper boundary condition follows from the fact that when the geometric Brownian $S(t)$ hits the level B, it immediately rises above B. In fact, because it has nonzero quadratic variation, the asset price $S(t)$ oscillates, rising and falling across the level B infinitely many times immediately after hitting it. The option price is zero when the asset price hits B because the option is on the verge of knocking out. The only exception to this is if the level B is first reached at the expiration time T, for then there is no time left for the knock-out. In this case, the option price is given by the terminal condition (7.3.7). In particular, the function $v(t,x)$ is not continuous at the corner of its domain where $t=T$ and $x=B$. It is continuous everywhere else in the rectangle $\{(t,x); 0\le t\le T, 0\le x\le B\}$.

Exercise 7.8 outlines the steps to verify the Black-Scholes-Merton equation by direct computation, starting with the analytical formula (7.3.20) obtained in Subsection 7.3.3. Here we derive this partial differential equation (7.3.4) by the simpler but more generally applicable argument used previously: (1) find the martingale, (2) take the differential, and (3) set the dt term equal to zero.

Let us begin with an initial asset price $S(0)\in(0,B)$. We then define the option payoff $V(T)$ by (7.3.2). The price of the option at time t between initiation and expiration is given by the risk-neutral pricing formula

$$V(t)=\widetilde{\mathbb{E}}\left[e^{-r(T-t)}V(T)\middle|\mathcal{F}(t)\right],\quad 0\le t\le T. \tag{7.3.8}$$

The usual iterated conditioning argument (e.g., (5.3.3)) shows that

$$e^{-rt}V(t)=\widetilde{\mathbb{E}}\left[e^{-rT}V(T)\middle|\mathcal{F}(t)\right],\quad 0\le t\le T, \tag{7.3.9}$$

is a martingale. We would like to use the Markov property as we did in Example 6.4.4 to say that $V(t)=v(t,S(t))$, where $v(t,x)$ is the function in Theorem 7.3.1. However, this equation does not hold for all values of t along all paths. Recall that $v(t,S(t))$ is the value of the option under the assumption that it has not knocked out prior to t, whereas $V(t)$ is the value of the option without any assumption. In particular, if the underlying asset price rises above the barrier B and then returns below the barrier by time t, then $V(t)$ will be zero because the option has knocked out, but $v\big(t, S(t)\big)$ will be strictly positive because $v(t, x)$ given by (7.3.20) is strictly positive for all values of $0 \le t < T$ and $0 < x < B$. The process $V(t)$ is path-dependent and remembers that the option has knocked out. The process $v\big(t, S(t)\big)$ is not path-dependent, and when $S(t) < B$, it gives the price of the option under the assumption that it has not knocked out, *even if that assumption is incorrect.*

We resolve this annoyance by defining ρ to be the first time t at which the asset price reaches the barrier B. In other words, ρ is chosen in a path-dependent way so that $S(t) < B$ for $0 \le t \le \rho$ and $S(\rho) = B$. Since the asset price almost surely exceeds the barrier immediately after reaching it, we may regard ρ as the time of knock-out. If the asset price does not reach the barrier before expiration, we set $\rho = \infty$. If the asset price first reaches the barrier at time T, then $\rho = T$ but knock-out does not occur because there is no time left for the asset price to exceed the barrier. However, the probability that the asset price first reaches the barrier at time T is zero, so this anomaly does not matter.

The random variable ρ is a *stopping time* because it chooses its value based on the path of the asset price up to time ρ. Stopping times in the binomial model were defined in Definition 4.3.1 of Volume I. The Optional Sampling Theorem, Theorem 4.3.2 of Volume I, asserts that a martingale stopped at a stopping time is still a martingale. The same is true in continuous time. In particular, the process

$$e^{-r(t\wedge\rho)}V(t\wedge\rho) = \begin{cases} e^{-rt}V(t) & \text{if } 0 \le t \le \rho, \\ e^{-r\rho}V(\rho) & \text{if } \rho < t \le T, \end{cases} \tag{7.3.10}$$

is a $\widetilde{\mathbb{P}}$-martingale. Before t gets to ρ, this is just the martingale $e^{-rt}V(t)$. Once t gets to ρ, although the time parameter t can march on, the value of the process is frozen at $e^{-r\rho}V(\rho)$. A process that does not move is trivially a martingale. The only way the martingale property could be ruined would be if ρ "looked ahead" when deciding to stop the process. If ρ stopped at a time because the process was about to go up and let the process continue if it was about to go down, the stopped process would have a downward tendency. So long as ρ makes the decision to stop at the current time based only on the path up to and perhaps including the current time, the act of stopping a martingale at time ρ preserves the martingale property.

Lemma 7.3.2. *We have*

$$V(t) = v\big(t, S(t)\big), \quad 0 \le t \le \rho. \tag{7.3.11}$$

In particular, $e^{-rt}v\big(t, S(t)\big)$ is a $\widetilde{\mathbb{P}}$-martingale up to time ρ, or, put another way, the stopped process

$$e^{-r(t\wedge\rho)}v\big(t\wedge\rho, S(t\wedge\rho)\big), \quad 0\le t\le T, \tag{7.3.12}$$

is a martingale under $\widetilde{\mathbb{P}}$.

SKETCH OF PROOF: Because $v\big(t,S(t)\big)$ is the value of the up-and-out call under the assumption that it has not knocked out before time t, and for $t\le\rho$ this assumption is correct, we have (7.3.11) for $t\le\rho$. From (7.3.11), we conclude that the process in (7.3.12) is the $\widetilde{\mathbb{P}}$-martingale (7.3.10). □

PROOF OF THEOREM 7.3.1: We compute the differential

$$\begin{aligned}
d\big(e^{-rt}v\big(t,S(t)\big)\big) &= e^{-rt}\Big[-rv\big(t,S(t)\big)\,dt + v_t\big(t,S(t)\big)\,dt + v_x\big(t,S(t)\big)\,dS(t)\\
&\qquad + \frac{1}{2}v_{xx}\big(t,S(t)\big)\,dS(t)\,dS(t)\Big]\\
&= e^{-rt}\Big[-rv\big(t,S(t)\big) + v_t\big(t,S(t)\big) + rS(t)v_x\big(t,S(t)\big)\\
&\qquad + \frac{1}{2}\sigma^2S^2(t)v_{xx}\big(t,S(t)\big)\Big]\,dt\\
&\quad + e^{-rt}\sigma S(t)v_x\big(t,S(t)\big)\,d\widetilde{W}(t).
\end{aligned} \tag{7.3.13}$$

The dt term must be zero for $0\le t\le\rho$, (i.e., before the option knocks out). But since $\big(t,S(t)\big)$ can reach any point in $\{(t,x); 0\le t<T, 0\le x\le B\}$ before the option knocks out, the Black-Scholes-Merton equation (7.3.4) must hold for every $t\in[0,T)$ and $x\in[0,B]$. □

Remark 7.3.3. From Theorem 7.3.1 and its proof, we see how to construct a hedge, at least theoretically. Setting the dt term in (7.3.13) equal to zero, we obtain

$$d\big(e^{-rt}v\big(t,S(t)\big)\big) = e^{-rt}\sigma S(t)v_x\big(t,S(t)\big)\,d\widetilde{W}(t), \quad 0\le t\le\rho. \tag{7.3.14}$$

The discounted value of a portfolio that at each time t holds $\Delta(t)$ shares of the underlying asset is given by (see (5.2.27))

$$d\big(e^{-rt}X(t)\big) = e^{-rt}\sigma S(t)\Delta(t)\,d\widetilde{W}(t).$$

At least theoretically, if an agent begins with a short position in the up-and-out call and with initial capital $X(0) = v\big(0,S(0)\big)$, then the usual delta-hedging formula

$$\Delta(t) = v_x\big(t,S(t)\big) \tag{7.3.15}$$

will cause her portfolio value $X(t)$ to track the option value $v\big(t,S(t)\big)$ up to the time ρ of knock-out or up to expiration T, whichever comes first.

In practice, the delta hedge is impossible to implement if the option has not knocked out and the underlying asset price approaches the barrier near expiration of the option. The function $v(T,x)$ is discontinuous at $x = B$, jumping from $B - K$ to 0 at that point. For t near T and x just below B, the function $v(t,x)$ is approaching a discontinuity and has large negative delta $v_x(t,x)$ and large negative gamma $v_{xx}(t,x)$ values. Near expiration near the barrier, the delta-hedging formula (7.3.15) requires the agent to take a large short position in the underlying asset and to make large adjustments in the position (because of the large negative gamma) whenever the asset price moves. The Black-Scholes-Merton model assumes the bid–ask spread is zero, and here that assumption is a poor model of reality. The delta-hedging formula calls for such a large amount of trading that the bid–ask spread becomes significant. The common industry practice is to price and hedge the up-and-out call as if the barrier were at a level slightly higher than B. In this way, the large delta and gamma values of the option occur in the region above the contractual barrier B, and the hedging position will be closed out upon knock-out at the contractual barrier before the asset price reaches this region. □

7.3.3 Computation of the Price of the Up-and-Out Call

The risk-neutral price at time zero of the up-and-out call with payoff $V(T)$ given by (7.3.2) is $V(0) = \widetilde{\mathbb{E}}[e^{-rT}V(T)]$. We use the density formula (7.2.3) to compute this. If $k \geq 0$, we must integrate over the region $\{(m,w); k \leq w \leq m \leq b\}$. On the other hand, if $k < 0$, we integrate over the region $\{(m,w); k \leq w \leq m, 0 \leq m \leq b\}$. In both cases, the region can be described as $\{(m,w); k \leq w \leq b, w^+ \leq m \leq b\}$; see Figure 7.3.1. We assume here that $S(0) \leq B$ so that $b > 0$. Otherwise, the region over which we integrate has zero area, and the time-zero value of the call is zero rather than the integral computed below. We also assume $S(0) > 0$ so that b and k are finite.

When $0 < S(0) \leq B$, the time-zero value of the up-and-out call is

$$\begin{aligned}
V(0) &= \int_k^b \int_{w^+}^b e^{-rT}\big(S(0)e^{\sigma w} - K\big)\frac{2(2m-w)}{T\sqrt{2\pi T}}e^{\alpha w - \frac{1}{2}\alpha^2 T - \frac{1}{2T}(2m-w)^2}\,dm\,dw \\
&= -\int_k^b e^{-rT}\big(S(0)e^{\sigma w} - K\big)\frac{1}{\sqrt{2\pi T}}\,e^{\alpha w - \frac{1}{2}\alpha^2 T - \frac{1}{2T}(2m-w)^2}\Big|_{m=w^+}^{m=b}\,dw \\
&= \frac{1}{\sqrt{2\pi T}}\int_k^b \big(S(0)e^{\sigma w} - K\big)e^{-rT + \alpha w - \frac{1}{2}\alpha^2 T - \frac{1}{2T}w^2}\,dw \\
&\quad - \frac{1}{\sqrt{2\pi T}}\int_k^b \big(S(0)e^{\sigma w} - K\big)e^{-rT + \alpha w - \frac{1}{2}\alpha^2 T - \frac{1}{2T}(2b-w)^2}\,dw \\
&= S(0)I_1 - KI_2 - S(0)I_3 + KI_4,
\end{aligned}$$

where

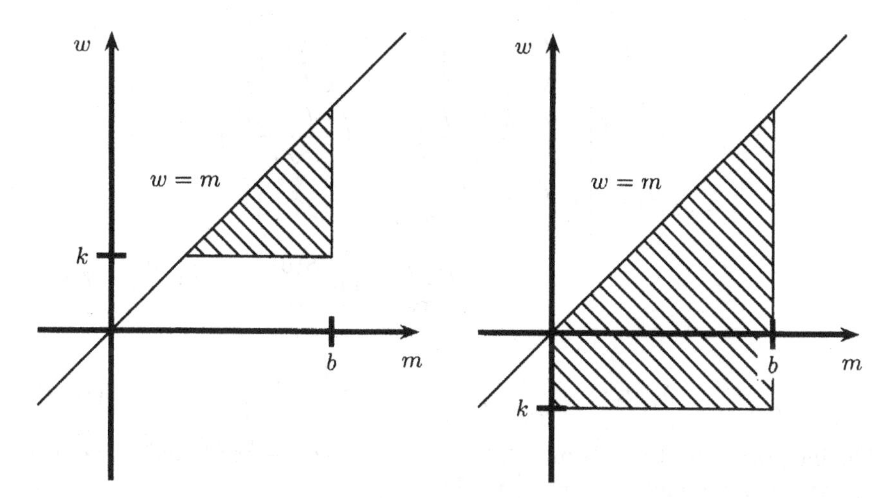

Fig. 7.3.1. Regions of integration for $k \geq 0$ and $k < 0$.

$$
\begin{aligned}
I_1 &= \frac{1}{\sqrt{2\pi T}} \int_k^b e^{\sigma w - rT + \alpha w - \frac{1}{2}\alpha^2 T - \frac{1}{2T} w^2}\, dw, \\
I_2 &= \frac{1}{\sqrt{2\pi T}} \int_k^b e^{-rT + \alpha w - \frac{1}{2}\alpha^2 T - \frac{1}{2T} w^2}\, dw, \\
I_3 &= \frac{1}{\sqrt{2\pi T}} \int_k^b e^{\sigma w - rT + \alpha w - \frac{1}{2}\alpha^2 T - \frac{1}{2T}(2b-w)^2}\, dw, \\
&= \frac{1}{\sqrt{2\pi T}} \int_k^b e^{\sigma w - rT + \alpha w - \frac{1}{2}\alpha^2 T - \frac{2}{T}b^2 + \frac{2}{T}bw - \frac{1}{2T} w^2}\, dw, \\
I_4 &= \frac{1}{\sqrt{2\pi T}} \int_k^b e^{-rT + \alpha w - \frac{1}{2}\alpha^2 T - \frac{1}{2T}(2b-w)^2}\, dw \\
&= \frac{1}{\sqrt{2\pi T}} \int_k^b e^{-rT + \alpha w - \frac{1}{2}\alpha^2 T - \frac{2}{T}b^2 + \frac{2}{T}bw - \frac{1}{2T} w^2}\, dw.
\end{aligned}
$$

Each of these integrals is of the form

$$
\begin{aligned}
\frac{1}{\sqrt{2\pi T}} \int_k^b e^{\beta + \gamma w - \frac{1}{2T} w^2}\, dw &= \frac{1}{\sqrt{2\pi T}} \int_k^b e^{-\frac{1}{2T}(w - \gamma T)^2 + \frac{1}{2}\gamma^2 T + \beta}\, dw \\
&= e^{\frac{1}{2}\gamma^2 T + \beta} \frac{1}{\sqrt{2\pi}} \int_{\frac{1}{\sqrt{T}}(k - \gamma T)}^{\frac{1}{\sqrt{T}}(b - \gamma T)} e^{-\frac{1}{2}y^2}\, dy, \quad (7.3.16)
\end{aligned}
$$

where we have made the change of variable $y = \frac{w - \gamma T}{\sqrt{T}}$. Using the standard cumulative normal distribution property $N(z) = 1 - N(-z)$ and (7.3.3), we continue, writing

$$
\begin{aligned}
\frac{1}{\sqrt{2\pi T}} \int_k^b & e^{\beta+\gamma w-\frac{w^2}{2T}}\,dw \\
&= e^{\frac{1}{2}\gamma^2 T+\beta}\left[N\left(\frac{b-\gamma T}{\sqrt{T}}\right)-N\left(\frac{k-\gamma T}{\sqrt{T}}\right)\right] \\
&= e^{\frac{1}{2}\gamma^2 T+\beta}\left[N\left(\frac{-k+\gamma T}{\sqrt{T}}\right)-N\left(\frac{-b+\gamma T}{\sqrt{T}}\right)\right] \\
&= e^{\frac{1}{2}\gamma^2 T+\beta}\left[N\left(\frac{1}{\sigma\sqrt{T}}\left[\log\frac{S(0)}{K}+\gamma\sigma T\right]\right)\right. \\
&\qquad\left. -N\left(\frac{1}{\sigma\sqrt{T}}\left[\log\frac{S(0)}{B}+\gamma\sigma T\right]\right)\right]. \qquad (7.3.17)
\end{aligned}
$$

Set

$$
\delta_\pm(\tau,s)=\frac{1}{\sigma\sqrt{\tau}}\left[\log s+\left(r\pm\frac{1}{2}\sigma^2\right)\tau\right]. \qquad (7.3.18)
$$

The integral I_1 is of the form (7.3.17) with $\beta=-rT-\frac{1}{2}\alpha^2 T$ and $\gamma=\alpha+\sigma$, so $\frac{1}{2}\gamma^2 T+\beta=0$ and $\gamma\sigma=r+\frac{1}{2}\sigma^2$. Therefore,

$$
I_1=N\left(\delta_+\left(T,\frac{S(0)}{K}\right)\right)-N\left(\delta_+\left(T,\frac{S(0)}{B}\right)\right).
$$

The integral I_2 is of the form (7.3.17) with $\beta=-rT-\frac{1}{2}\alpha^2 T$ and $\gamma=\alpha$, so $\frac{1}{2}\gamma^2 T+\beta=-rT$ and $\gamma\sigma=r-\frac{1}{2}\sigma^2$. Therefore,

$$
I_2=e^{-rT}\left[N\left(\delta_-\left(T,\frac{S(0)}{K}\right)\right)-N\left(d_-\left(T,\frac{S(0)}{B}\right)\right)\right].
$$

For I_3, we have $\beta=-rT-\frac{1}{2}\alpha^2 T-\frac{2b^2}{T}$ and $\gamma=\alpha+\sigma+\frac{2b}{T}$, so

$$
\begin{aligned}
\frac{1}{2}\gamma^2 T+\beta &= \log\left(\frac{S(0)}{B}\right)^{-\frac{2r}{\sigma^2}-1}, \\
\gamma\sigma T &= \left(r+\frac{1}{2}\sigma^2\right)T+\log\left(\frac{B}{S(0)}\right)^2.
\end{aligned}
$$

Therefore,

$$
I_3=\left(\frac{S(0)}{B}\right)^{-\frac{2r}{\sigma^2}-1}\left[N\left(\delta_+\left(T,\frac{B^2}{KS(0)}\right)\right)-N\left(\delta_+\left(T,\frac{B}{S(0)}\right)\right)\right].
$$

Finally, for I_4, we have $\beta=-rT-\frac{1}{2}\alpha^2 T-\frac{2b^2}{T}$ and $\gamma=\alpha+\frac{2b}{T}$, so

$$
\begin{aligned}
\frac{1}{2}\gamma^2 T+\beta &= -rT+\log\left(\frac{S(0)}{B}\right)^{-\frac{2r}{\sigma^2}+1}, \\
\gamma\sigma T &= \left(r-\frac{1}{2}\sigma^2\right)T+\log\left(\frac{B}{S(0)}\right)^2.
\end{aligned}
$$

Therefore,

$$I_4 = e^{-rT}\left(\frac{S(0)}{B}\right)^{-\frac{2r}{\sigma^2}+1}\left[N\left(\delta_-\left(T,\frac{B^2}{KS(0)}\right)\right) - N\left(\delta_-\left(T,\frac{B}{S(0)}\right)\right)\right].$$

Putting all this together, under the assumption $0 < S(0) \leq B$, we have the up-and-out call price formula

$$\begin{aligned}
&V(0)\\
&= S(0)\left[N\left(\delta_+\left(T,\frac{S(0)}{K}\right)\right) - N\left(\delta_+\left(T,\frac{S(0)}{B}\right)\right)\right]\\
&\quad -e^{-rT}K\left[N\left(\delta_-\left(T,\frac{S(0)}{K}\right)\right) - N\left(\delta_-\left(T,\frac{S(0)}{B}\right)\right)\right]\\
&\quad -B\left(\frac{S(0)}{B}\right)^{-\frac{2r}{\sigma^2}}\left[N\left(\delta_+\left(T,\frac{B^2}{KS(0)}\right)\right) - N\left(\delta_+\left(T,\frac{B}{S(0)}\right)\right)\right]\\
&\quad +e^{-rT}K\left(\frac{S(0)}{B}\right)^{-\frac{2r}{\sigma^2}+1}\left[N\left(\delta_-\left(T,\frac{B^2}{KS(0)}\right)\right) - N\left(\delta_-\left(T,\frac{B}{S(0)}\right)\right)\right].
\end{aligned} \tag{7.3.19}$$

Now let $t \in [0,T)$ be given, and assume the underlying asset price at time t is $S(t) = x$. As above, we assume $0 < x \leq B$. If the call has not knocked out prior to time t, its price at time t is obtained by replacing T by the time to expiration $\tau = T - t$ and replacing $S(0)$ by x in (7.3.19). This gives us the call price as a function $v(t,x)$ of the two variables t and x:

$$\begin{aligned}
v(t,x) &= x\left[N\left(\delta_+\left(\tau,\frac{x}{K}\right)\right) - N\left(\delta_+\left(\tau,\frac{x}{B}\right)\right)\right]\\
&\quad -e^{-r\tau}K\left[N\left(\delta_-\left(\tau,\frac{x}{K}\right)\right) - N\left(\delta_-\left(\tau,\frac{x}{B}\right)\right)\right]\\
&\quad -B\left(\frac{x}{B}\right)^{-\frac{2r}{\sigma^2}}\left[N\left(\delta_+\left(\tau,\frac{B^2}{Kx}\right)\right) - N\left(\delta_+\left(\tau,\frac{B}{x}\right)\right)\right]\\
&\quad +e^{-r\tau}K\left(\frac{x}{B}\right)^{-\frac{2r}{\sigma^2}+1}\left[N\left(\delta_-\left(\tau,\frac{B^2}{Kx}\right)\right) - N\left(\delta_-\left(\tau,\frac{B}{x}\right)\right)\right],\\
&\qquad\qquad 0 \leq t < T,\ 0 < x \leq B.
\end{aligned} \tag{7.3.20}$$

Formula (7.3.20) was derived under the assumption that $\tau > 0$ (i.e., $t < T$) and $0 < x \leq B$. For $0 \leq t \leq T$ and $x > B$, we have $v(t,x) = 0$ because the option knocks out when the asset price exceeds the barrier B. Indeed, if the asset price reaches the barrier before expiration, then it will immediately exceed the barrier almost surely, and so $v(t,B) = 0$ for $0 \leq t < T$. However, $v(T,B) = B - K$. We also have $v(t,0) = 0$ because geometric Brownian motion starting at 0 stays at zero, and hence the call expires out of the money. Finally, if the option does not knock out prior to expiration, then its payoff is that of a European call (i.e., $v(T,x) = (x-K)^+$). In summary, $v(t,x)$ satisfies the boundary conditions (7.3.5)–(7.3.7). Formula (7.3.6) can be obtained by substitution of $x = B$ in (7.3.20), but for $x > B$, the right-hand side of (7.3.20) is not $v(t,x) = 0$. Formula (7.3.20) was derived under the assumption $0 < x \leq B$, and it is incorrect if $x > B$. Formulas (7.3.5) and (7.3.7) cannot be obtained by substitution of $x = 0$ and $t = T$ ($\tau = 0$) into (7.3.20) because this leads to zeroes in denominators, but it can be shown that (7.3.20) gives these formulas as limits as $x \downarrow 0$ and $\tau \downarrow 0$; see Exercise 7.2.

7.4 Lookback Options

An option whose payoff is based on the maximum that the underlying asset price attains over some interval of time prior to expiration is called a *lookback option*. In this section we price a *floating strike lookback option*. The payoff of this option is the difference between the maximum asset price over the time between initiation and expiration and the asset price at expiration. The discussion of this option introduces a new type of differential, a differential that is neither dt nor $d\widetilde{W}(t)$.

7.4.1 Floating Strike Lookback Option

We begin with a geometric Brownian motion asset price, which may be written as in (7.3.1) as

$$S(t) = S(0)e^{\sigma \widehat{W}(t)}, \tag{7.4.1}$$

where, as in Subsection 7.3.1, $\widehat{W}(t) = \alpha t + \widetilde{W}(t)$ and

$$\alpha = \frac{1}{\sigma}\left(r - \frac{1}{2}\sigma^2\right).$$

With

$$\widehat{M}(t) = \max_{0 \leq u \leq t} \widehat{W}(u), \quad 0 \leq t \leq T, \tag{7.4.2}$$

we may write the maximum of the asset price up to time t as

$$Y(t) = \max_{0 \leq u \leq t} S(u) = S(0)e^{\sigma \widehat{M}(t)}. \tag{7.4.3}$$

The lookback option considered in this section pays off

$$V(T) = Y(T) - S(T) \tag{7.4.4}$$

at expiration time T. This payoff is nonnegative because $Y(T) \geq S(T)$.

Let $t \in [0,T]$ be given. At time t, the risk-neutral price of the lookback option is

$$V(t) = \widetilde{\mathbb{E}}\left[e^{-r(T-t)}(Y(T) - S(T))\Big|\mathcal{F}(t)\right]. \tag{7.4.5}$$

Because the pair of processes $\big(S(t), Y(t)\big)$ has the Markov property (see Exercise 7.3), there must exist a function $v(t,x,y)$ such that

$$V(t) = v\big(t, S(t), Y(t)\big).$$

In Subsection 7.4.2, we characterize this function by the Black-Scholes-Merton equation. In Subsection 7.4.3, we compute it explicitly.

7.4.2 Black-Scholes-Merton Equation

Theorem 7.4.1. *Let $v(t,x,y)$ denote the price at time t of the floating strike lookback option under the assumption that $S(t) = x$ and $Y(t) = y$. Then $v(t,x,y)$ satisfies the Black-Scholes-Merton partial differential equation*

$$v_t(t,x,y) + rxv_x(t,x,y) + \frac{1}{2}\sigma^2 x^2 v_{xx}(t,x,y) = rv(t,x,y) \tag{7.4.6}$$

in the region $\{(t,x,y); 0 \le t < T,\ 0 \le x \le y\}$ and satisfies the boundary conditions

$$v(t,0,y) = e^{-r(T-t)}y, \quad 0 \le t \le T,\ y \ge 0, \tag{7.4.7}$$
$$v_y(t,y,y) = 0, \quad 0 \le t \le T,\ y > 0, \tag{7.4.8}$$
$$v(T,x,y) = y - x, \quad 0 \le x \le y. \tag{7.4.9}$$

Iterated conditioning implies that $e^{-rt}V(t) = e^{-rt}v\big(t, S(t), Y(t)\big)$, where $V(t)$ is given by (7.4.5), is a martingale under $\widetilde{\mathbb{P}}$. We compute its differential and set the dt term equal to zero to obtain (7.4.6). However, when we do this, the term $dY(t)$ appears. This is different from the term $dS(t)$, because $S(t)$ has nonzero quadratic variation, whereas $Y(t)$ has zero quadratic variation. This is because $Y(t)$ is continuous and nondecreasing in t. Let $0 = t_0 < t_1 < \cdots < t_m = T$ be a partition of $[0,T]$. Then

$$\begin{aligned}
&\sum_{j=1}^{m} \big(Y(t_j) - Y(t_{j-1})\big)^2 \\
&\le \max_{j=1,\dots,m} \big(Y(t_j) - Y(t_{j-1})\big) \sum_{j=1}^{m} \big(Y(t_j) - Y(t_{j-1})\big) \\
&= \max_{j=1,\dots,m} \big(Y(t_j) - Y(t_{j-1})\big) \cdot \big(Y(T) - Y(0)\big),
\end{aligned} \tag{7.4.10}$$

and $\max_{j=1,\dots,m}\big(Y(t_j) - Y(t_{j-1})\big)$ has limit zero as $\max_{j=1,\dots,m}(t_j - t_{j-1})$ goes to zero because $Y(t)$ is continuous. We conclude that $Y(t)$ accumulates zero quadratic variation on $[0,T]$, a fact we record by writing

$$dY(t)\,dY(t) = 0. \tag{7.4.11}$$

This argument works because $Y(t_j) - Y(t_{j-1})$ is nonnegative, and hence we do not need to take the absolute value of these terms in (7.4.10). This argument shows that on any interval in which a function is continuous and nondecreasing, it will accumulate zero quadratic variation.

On the other hand, $dY(t)$ is not a dt term: there is no process $\Theta(t)$ such that $dY(t) = \Theta(t)\,dt$. In other words, we cannot write $Y(t)$ as

$$Y(t) = Y(0) + \int_0^t \Theta(u)\,du. \tag{7.4.12}$$

If we could, then $\Theta(u)$ would be zero whenever u is in a "flat spot" of $Y(t)$, which occurs whenever $S(t)$ drops below its maximum to date (see Figure 7.4.1). Figure 7.4.1 suggests that there are time intervals in which $Y(t)$ is strictly increasing, but in fact no such interval exists. Such an interval can occur only if $S(t)$ is strictly increasing on the interval, and if there were such an interval, then $S(t)$ would accumulate zero quadratic variation on the interval (see the argument in the previous paragraph). This is not the case because $dS(t)\,dS(t) = \sigma S^2(t)\,dt$ is positive for all t. Thus, despite the suggestion of Figure 7.4.1, the lengths of the "flat spots" of $Y(t)$ on any time interval $[0,T]$ sum to T. Therefore, if (7.4.12) were to hold, we would need to have $\Theta(u) = 0$ for Lebesgue almost every u in $[0,T]$. This would result in $Y(t) = Y(0)$ for $0 \leq t \leq T$. But in fact $Y(t) > Y(0)$ for all $t > 0$. We conclude that $Y(t)$ cannot be represented in the form (7.4.12); $dY(t)$ is not a dt term.

The paths of $Y(t)$ increase over time, but they do so on a set of times having zero Lebesgue measure. Each time interval $[0,T]$ contains a sequence of subintervals whose lengths sum to T, and on each of these subintervals, $Y(t)$ is constant. The particular subintervals depend on the path, but regardless of the path, the lengths of these subintervals sum to T. A similar situation is described in Appendix A, Section A.3. In the case discussed there, $T = 1$ and the subintervals are explicitly exhibited. Their union is the Cantor set. It is verified that although the lengths of these subintervals sum to 1, there are uncountably many points not contained in these intervals. The function $F(x)$ described in Section A.3 increases, but only on the complement of the Cantor set. Furthermore, $F(x)$ is continuous. Functions of this kind are said to be *singularly continuous*.

Fortunately, we can work with the differential of $Y(t)$. We have already argued that $dY(t)\,dY(t) = 0$. Similarly, we have

$$dY(t)\,dS(t) = 0 \tag{7.4.13}$$

(see Exercise 7.4). We now provide the proof of Theorem 7.4.1.

PROOF OF THEOREM 7.4.1: We use the Itô-Doeblin formula and (7.4.11) and (7.4.13) to differentiate the martingale $e^{-rt}v\big(t, S(t), Y(t)\big)$ to obtain

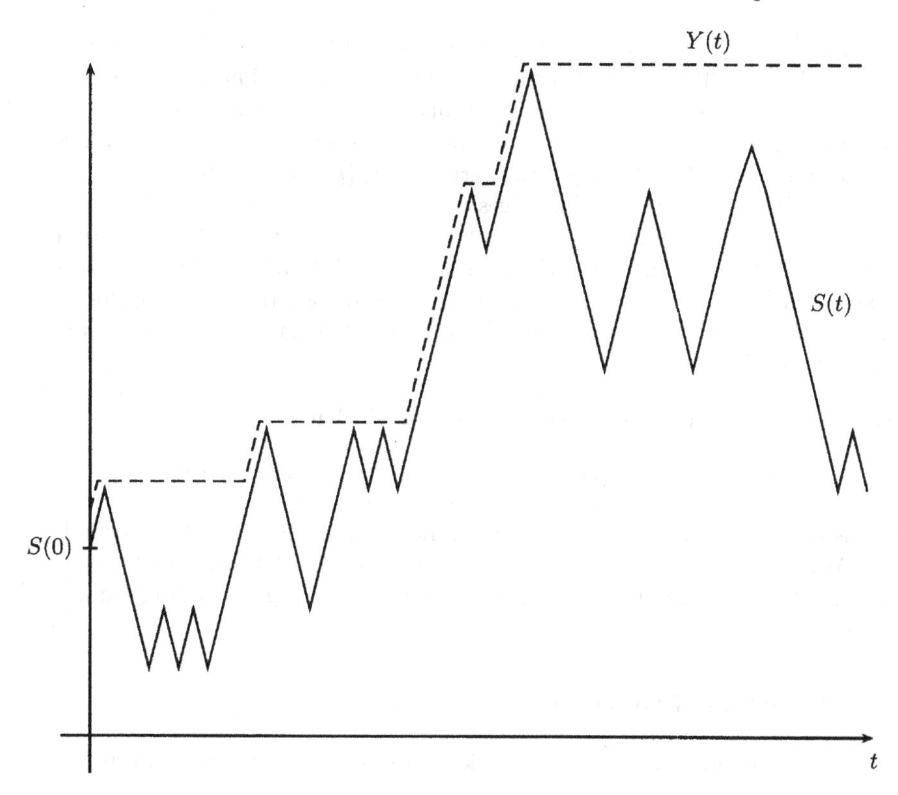

Fig. 7.4.1. Geometric Brownian motion and its maximum to date.

$$
\begin{aligned}
&d\left(e^{-rt}v\big(t,S(t),Y(t)\big)\right)\\
&= e^{-rt}\Big[-rv\big(t,S(t),Y(t)\big)\,dt + v_t\big(t,S(t),Y(t)\big)\,dt\\
&\qquad +v_x\big(t,S(t),Y(t)\big)\,dS(t) + \frac{1}{2}v_{xx}\big(t,S(t),Y(t)\big)\,dS(t)\,dS(t)\\
&\qquad +v_y\big(t,S(t),Y(t)\big)\,dY(t)\Big]\\
&= e^{-rt}\Big[-rv\big(t,S(t),Y(t)\big) + v_t\big(t,S(t),Y(t)\big) + rS(t)v_x\big(t,S(t),Y(t)\big)\\
&\qquad +\frac{1}{2}\sigma^2S^2(t)v_{xx}\big(t,S(t),Y(t)\big)\Big]\,dt\\
&\quad +e^{-rt}\sigma S(t)v_x\big(t,S(t),Y(t)\big)\,d\widetilde{W}(t)\\
&\quad +e^{-rt}v_y\big(t,S(t),Y(t)\big)\,dY(t). \qquad (7.4.14)
\end{aligned}
$$

In order to have a martingale, the dt term must be zero, and this gives us the Black-Scholes-Merton equation (7.4.6). The new feature is that the term

$e^{-rt}v_y(t, S(t), Y(t))\,dY(t)$ must also be zero. It cannot be canceled by the dt term nor by the $d\widetilde{W}(t)$ term because it is fundamentally different from both of these terms. The $dY(t)$ term is naturally zero on the "flat spots" of $Y(t)$ (i.e., when $S(t) < Y(t)$). However, at the times when $Y(t)$ increases, which are the times when $S(t) = Y(t)$, the term $e^{-rt}v_y\big(t, S(t), Y(t)\big)$ must be zero because $dY(t)$ is "positive." This gives us the boundary condition (7.4.8).

The boundary condition (7.4.9) is the payoff of the option. If at any time t we have $S(t) = 0$, then we will have $S(T) = 0$. Furthermore, Y will be constant on $[t, T]$; if $Y(t) = y$, then $Y(T) = y$ and the price of the option at time t is this value discounted from T back to t. This gives us the boundary condition (7.4.7). □

Remark 7.4.2. The proof of Theorem 7.4.1 shows that

$$d\left(e^{-rt}v(t, S(t), Y(t))\right) = e^{-rt}\sigma S(t) v_x\big(t, S(t), Y(t)\big)\,d\widetilde{W}(t).$$

Just as in Remark 7.3.3, this equation implies that the delta-hedging formula (7.3.15) works. In contrast to the situation in Remark 7.3.3, here the function $v(t, x, y)$ is continuous and we have no problems with large delta and gamma values. □

7.4.3 Reduction of Dimension

The price of the floating strike lookback option has a linear scaling property:

$$v(t, \lambda x, \lambda y) = \lambda v(t, x, y) \text{ for all } \lambda > 0. \tag{7.4.15}$$

This is because scaling both $S(t)$ and $Y(t)$ by the same positive constant at a time t prior to expiration results in the payoff $Y(T) - S(T)$ being scaled by the same constant. In particular, if we know the function of two variables

$$u(t, z) = v(t, z, 1), \quad 0 \le t \le T,\ 0 \le z \le 1, \tag{7.4.16}$$

then we can easily determine the function of three variables $v(t, x, y)$ by the formula

$$v(t, x, y) = yv\left(t, \frac{x}{y}, 1\right) = yu\left(t, \frac{x}{y}\right), \quad 0 \le t \le T,\ 0 \le x \le y,\ y > 0. \tag{7.4.17}$$

From (7.4.17), we can compute the partial derivatives:

$$v_t(t, x, y) = yu_t\left(t, \frac{x}{y}\right),$$
$$v_x(t, x, y) = yu_z\left(t, \frac{x}{y}\right) \cdot \frac{\partial}{\partial x}\left(\frac{x}{y}\right) = u_z\left(t, \frac{x}{y}\right),$$

$$v_{xx}(t,x,y) = u_{zz}\left(t,\frac{x}{y}\right)\cdot\frac{\partial}{\partial y}\left(\frac{x}{y}\right) = \frac{1}{y}u_{zz}\left(t,\frac{x}{y}\right),$$

$$\begin{aligned} v_y(t,x,y) &= u\left(t,\frac{x}{y}\right) + yu_z\left(t,\frac{x}{y}\right)\frac{\partial}{\partial y}\left(\frac{x}{y}\right) \\ &= u\left(t,\frac{x}{y}\right) - \frac{x}{y}u_z\left(t,\frac{x}{y}\right). \end{aligned}$$

Substitution into the Black-Scholes-Merton equation (7.4.6) yields

$$\begin{aligned} 0 &= -rv(t,x,y) + v_t(t,x,y) + rxv_x(t,x,y) + \frac{1}{2}\sigma^2x^2v_{xx}(t,x,y) \\ &= y\left[-ru\left(t,\frac{x}{y}\right) + u_t\left(t,\frac{x}{y}\right) + r\left(\frac{x}{y}\right)u_z\left(t,\frac{x}{y}\right)\right. \\ &\qquad \left. + \frac{1}{2}\sigma^2\left(\frac{x}{y}\right)^2 u_{zz}\left(t,\frac{x}{y}\right)\right]. \end{aligned}$$

Canceling y and making the change of variable $z = \frac{x}{y}$, we see that $u(t,z)$ satisfies the Black-Scholes-Merton equation

$$u_t(t,z) + rzu_z(t,z) + \frac{1}{2}\sigma^2z^2u_{zz}(t,z) = ru(t,z), \quad 0 \le t < T,\ 0 < z < 1. \tag{7.4.18}$$

Boundary conditions for $u(t,z)$ can be obtained from the boundary conditions (7.4.7)–(7.4.9) for $v(t,x,y)$. In particular,

$$e^{-r(T-t)}y = v(t,0,y) = yu\,(t,0)$$

implies

$$u(t,0) = e^{-r(T-t)}, \quad 0 \le t \le T. \tag{7.4.19}$$

Furthermore,

$$0 = v_y(t,y,y) = u(t,1) - u_z(t,1)$$

implies

$$u(t,1) = u_z(t,1), \quad 0 \le t < T. \tag{7.4.20}$$

Finally,

$$y - x = v(T,x,y) = yu\left(T,\frac{x}{y}\right)$$

implies

$$u(T,z) = 1 - z, \quad 0 \le z \le 1. \tag{7.4.21}$$

Equation (7.4.18) and the boundary conditions (7.4.19)–(7.4.21) uniquely determine the function $u(t,z)$. As a consequence, we see that the Black-Scholes-Merton equation and boundary conditions in Theorem 7.4.1 uniquely determine the function $v(t,x,y)$.

7.4.4 Computation of the Price of the Lookback Option

In this subsection, we compute the function $v(t,x,y)$ of Theorem 7.4.1. We do this for $0 \le t < T$ and $0 < x \le y$. Because $Y(t) \ge S(t)$ for all t, we do not need to compute values of $v(t,x,y)$ for $x > y$. The reader is invited in Exercise 7.5 to compute the partial derivatives of $v(t,x,y)$ and verify that the Black-Scholes-Merton equation and boundary conditions in Theorem 7.4.1 are satisfied.

For $0 \le t < T$ and $\tau = T - t$, we observe that

$$Y(T) = S(0)e^{\sigma\widehat{M}(t)}e^{\sigma(\widehat{M}(T)-\widehat{M}(t))} = Y(t)e^{\sigma(\widehat{M}(T)-\widehat{M}(t))}.$$

If $\max_{t\le u\le T}\widehat{W}(u) > \widehat{M}(t)$ (i.e., if $\widehat{W}$ attains a new maximum in $[t,T]$), then

$$\widehat{M}(T) - \widehat{M}(t) = \max_{t\le u\le T}\widehat{W}(u) - \widehat{M}(t).$$

On the other hand, if $\max_{t\le u\le T}\widehat{W}(u) \le \widehat{M}(t)$, then $\widehat{M}(T) = \widehat{M}(t)$ and

$$\widehat{M}(T) - \widehat{M}(t) = 0.$$

In either case, we have

$$\begin{aligned}\widehat{M}(T) - \widehat{M}(t) &= \left[\max_{t\le u\le T}\widehat{W}(u) - \widehat{M}(t)\right]^{+}\\ &= \left[\max_{t\le u\le T}\big(\widehat{W}(u) - \widehat{W}(t)\big) - \big(\widehat{M}(t) - \widehat{W}(t)\big)\right]^{+}.\end{aligned}$$

Multiplying this equation by σ and using (7.4.1) and (7.4.3), we obtain

$$\sigma\big(\widehat{M}(T) - \widehat{M}(t)\big) = \left[\max_{t\le u\le T}\sigma\big(\widehat{W}(u) - \widehat{W}(t)\big) - \log\frac{Y(t)}{S(t)}\right]^{+}. \tag{7.4.22}$$

Therefore, $V(t)$ in (7.4.5) is

$$\begin{aligned}V(t) = e^{-r\tau}\widetilde{\mathbb{E}}&\left[Y(t)\exp\left\{\left[\max_{t\le u\le T}\sigma\big(\widehat{W}(u) - \widehat{W}(t)\big) - \log\frac{Y(t)}{S(t)}\right]^{+}\right\}\Bigg|\,\mathcal{F}(t)\right]\\ &-e^{rt}\widetilde{\mathbb{E}}\left[e^{-rT}S(T)\big|\,\mathcal{F}(t)\right].\end{aligned} \tag{7.4.23}$$

Because the discounted asset price is a martingale under $\widetilde{\mathbb{P}}$, the second term in (7.4.23) is $-e^{rt}e^{-rt}S(t) = S(t)$. For the first term, we can "take out what is known" (see Theorem 2.3.2(ii)) to obtain

$$e^{-r\tau}Y(t)\widetilde{\mathbb{E}}\left[\exp\left\{\left[\max_{t\le u\le T}\sigma\big(\widehat{W}(u) - \widehat{W}(t)\big) - \log\frac{Y(t)}{S(t)}\right]^{+}\right\}\Bigg|\,\mathcal{F}(t)\right]. \tag{7.4.24}$$

Because $Y(t)$ and $S(t)$ are $\mathcal{F}(t)$-measurable and $\max_{t\le u\le T}\sigma\big(\widehat{W}(u)-\widehat{W}(t)\big)$ is independent of $\mathcal{F}(t)$, we can use the Independence Lemma, Lemma 2.3.4, to write the conditional expectation in (7.4.24) as $g\big(S(t),Y(t)\big)$, where

$$g(x,y)=\widetilde{\mathbb{E}}\exp\left\{\left[\max_{t\le u\le T}\sigma\big(\widehat{W}(u)-\widehat{W}(t)\big)-\log\frac{y}{x}\right]^+\right\}. \tag{7.4.25}$$

Note that the expectation in (7.4.25) is no longer conditioned on $\mathcal{F}(t)$. Putting this all together, we have

$$V(t)=e^{-r\tau}Y(t)g\big(S(t),Y(t)\big)-S(t)$$

or, equivalently,

$$v(t,x,y)=e^{-r\tau}yg(x,y)-x. \tag{7.4.26}$$

It remains to compute the function $g(x,y)$. Because

$$\max_{t\le u\le T}\sigma\big(\widehat{W}(u)-\widehat{W}(t)\big)=\sigma\max_{t\le u\le T}\big(\widehat{W}(u)-\widehat{W}(t)\big),$$

and $\max_{t\le u\le T}\big(\widehat{W}(u)-\widehat{W}(t)\big)$ has the same unconditional distribution under $\widetilde{\mathbb{P}}$ as $\max_{0\le u\le\tau}\big(\widehat{W}(u)-\widehat{W}(0)\big)=\widehat{M}(\tau)$, the function $g(x,y)$ of (7.4.25) can also be written as

$$\begin{aligned}g(x,y)&=\widetilde{\mathbb{E}}\exp\left\{\left[\sigma\widehat{M}(\tau)-\log\frac{y}{x}\right]^+\right\}\\&=\widetilde{\mathbb{P}}\left\{\widehat{M}(\tau)\le\frac{1}{\sigma}\log\frac{y}{x}\right\}+\frac{x}{y}\widetilde{\mathbb{E}}\left[e^{\sigma\widehat{M}(\tau)}\mathbb{I}_{\{\widehat{M}(\tau)\ge\frac{1}{\sigma}\log\frac{y}{x}\}}\right].\end{aligned} \tag{7.4.27}$$

We compute both terms on the right-hand side of (7.4.27).

In order to compute the first term on the right-hand side of (7.4.27), we use (7.2.6) with T replaced by τ and m replaced by $\frac{1}{\sigma}\log\frac{y}{x}$. With these replacements, the arguments of N appearing on the right-hand side of (7.2.6) are

$$\begin{aligned}\frac{1}{\sqrt{\tau}}\left[\frac{1}{\sigma}\log\frac{y}{x}-\alpha\tau\right]&=\frac{1}{\sigma\sqrt{\tau}}\left[\log\frac{y}{x}-\left(r-\frac{1}{2}\sigma^2\right)\tau\right]\\&=-\frac{1}{\sigma\sqrt{\tau}}\left[\log\frac{x}{y}+\left(r-\frac{1}{2}\sigma^2\right)\tau\right]\\&=-\delta_-\left(\tau,\frac{x}{y}\right),\\\frac{1}{\sqrt{\tau}}\left[-\frac{1}{\sigma}\log\frac{y}{x}-\alpha\tau\right]&=\frac{1}{\sigma\sqrt{\tau}}\left[-\log\frac{y}{x}-\left(r-\frac{1}{2}\sigma^2\right)\tau\right]\\&=-\delta_-\left(\tau,\frac{y}{x}\right),\end{aligned}$$

where $\delta_\pm(\tau, s)$ is defined by (7.3.18). The term $e^{2\alpha m}$ appearing on the right-hand side of (7.2.6) becomes

$$\exp\left\{\frac{2\alpha}{\sigma}\log\frac{y}{x}\right\} = \exp\left\{\left(\frac{2r}{\sigma^2}-1\right)\log\frac{y}{x}\right\} = \left(\frac{y}{x}\right)^{\frac{2r}{\sigma^2}-1}.$$

It follows from (7.2.6) that

$$\widetilde{\mathbb{P}}\left\{\widehat{M}(\tau) \le \frac{1}{\sigma}\log\frac{y}{x}\right\} = N\left(-\delta_-\left(\tau, \frac{x}{y}\right)\right) - \left(\frac{y}{x}\right)^{\frac{2r}{\sigma^2}-1} N\left(-\delta_-\left(\tau, \frac{y}{x}\right)\right). \tag{7.4.28}$$

The second term on the right-hand side of (7.4.27) is computed using the density for $\widehat{M}(\tau)$ under $\widetilde{\mathbb{P}}$ given by (7.2.7) with τ replacing T. Indeed,

$$\begin{aligned}
&\frac{x}{y}\widetilde{\mathbb{E}}\left[e^{\sigma\widehat{M}(\tau)}\mathbb{I}_{\{\widehat{M}(\tau)\ge\frac{1}{\sigma}\log\frac{y}{x}\}}\right]\\
&= \frac{x}{y}\int_{\frac{1}{\sigma}\log\frac{y}{x}}^{\infty} e^{\sigma m}\widetilde{f}_{\widehat{M}(\tau)}(m)\,dm\\
&= \frac{x}{y}\int_{\frac{1}{\sigma}\log\frac{y}{x}}^{\infty} \frac{2}{\sqrt{2\pi\tau}}e^{\sigma m-\frac{1}{2\tau}(m-\alpha\tau)^2}\,dm\\
&\qquad -\frac{x}{y}\int_{\frac{1}{\sigma}\log\frac{y}{x}}^{\infty} 2\alpha e^{(\sigma+2\alpha)m}N\left(\frac{-m-\alpha\tau}{\sqrt{\tau}}\right)dm.
\end{aligned} \tag{7.4.29}$$

We compute the first integral on the right-hand side of (7.4.29). Because

$$\begin{aligned}
r\tau &- \frac{1}{2\tau}(m-\alpha\tau-\sigma\tau)^2\\
&= r\tau - \frac{1}{2\tau}(m-\alpha\tau)^2 + \sigma(m-\alpha\tau) - \frac{1}{2}\sigma^2\tau\\
&= r\tau - \frac{1}{2\tau}(m-\alpha\tau)^2 + \sigma m - \left(r-\frac{1}{2}\sigma^2\right)\tau - \frac{1}{2}\sigma^2\tau\\
&= \sigma m - \frac{1}{2\tau}(m-\alpha\tau)^2,
\end{aligned}$$

we may write the first term on the right-hand side of (7.4.29) as

$$\begin{aligned}
&\frac{x}{y}\int_{\frac{1}{\sigma}\log\frac{y}{x}}^{\infty} \frac{2}{\sqrt{2\pi\tau}}e^{\sigma m-\frac{1}{2\tau}(m-\alpha\tau)^2}\,dm\\
&\qquad = \frac{2xe^{r\tau}}{y\sqrt{2\pi\tau}}\int_{\frac{1}{\sigma}\log\frac{y}{x}}^{\infty} e^{-\frac{1}{2\tau}(m-\alpha\tau-\sigma\tau)^2}\,dm.
\end{aligned} \tag{7.4.30}$$

We make the change of variable $\xi = \frac{\alpha\tau+\sigma\tau-m}{\sqrt{\tau}}$, so the lower limit of integration $\frac{1}{\sigma}\log\frac{y}{x}$ becomes

$$\frac{1}{\sqrt{\tau}}\left(\alpha\tau+\sigma\tau-\frac{1}{\sigma}\log\frac{y}{x}\right) = \frac{1}{\sigma\sqrt{\tau}}\left(\log\frac{x}{y}+r\tau+\frac{1}{2}\sigma^2\tau\right) = \delta_+\left(\tau, \frac{x}{y}\right).$$

With this change of variable in the integral on the right-hand side of (7.4.30), we obtain the following formula for the first term on the right-hand side of (7.4.29):

$$\frac{x}{y}\int_{\frac{1}{\sigma}\log\frac{y}{x}}^{\infty}\frac{2}{\sqrt{2\pi\tau}}e^{\sigma m-\frac{1}{2\tau}(m-\alpha\tau)^2}\,dm = \frac{2xe^{r\tau}}{y\sqrt{2\pi}}\int_{-\infty}^{\delta_+(\tau,\frac{x}{y})}e^{-\frac{1}{2}\xi^2}\,d\xi$$
$$= \frac{2xe^{r\tau}}{y}N\left(\delta_+\left(\tau,\frac{x}{y}\right)\right). \quad (7.4.31)$$

The second term on the right-hand side of (7.4.29) requires a reversal of the order of integration over the region shown in Figure 7.4.2. Because $\sigma+2\alpha=\frac{2r}{\sigma}$, this term is

$$-\frac{x}{y}\int_{\frac{1}{\sigma}\log\frac{x}{y}}^{\infty}2\alpha e^{(\sigma+2\alpha)m}N\left(\frac{-m-\alpha\tau}{\sqrt{\tau}}\right)dm$$
$$= -\frac{2\alpha x}{y\sqrt{2\pi}}\int_{\frac{1}{\sigma}\log\frac{y}{x}}^{\infty}\int_{-\infty}^{\frac{1}{\sqrt{\tau}}(-m-\alpha\tau)}e^{\frac{2}{\sigma}rm-\frac{1}{2}\xi^2}\,d\xi\,dm$$
$$= -\frac{2\alpha x}{y\sqrt{2\pi}}\int_{-\infty}^{-\delta_-(\tau,\frac{y}{x})}\int_{\frac{1}{\sigma}\log\frac{y}{x}}^{-\xi\sqrt{\tau}-\alpha\tau}e^{\frac{2}{\sigma}rm-\frac{1}{2}\xi^2}\,dm\,d\xi. \quad (7.4.32)$$

The inner integral in (7.4.32) can be evaluated. Indeed,

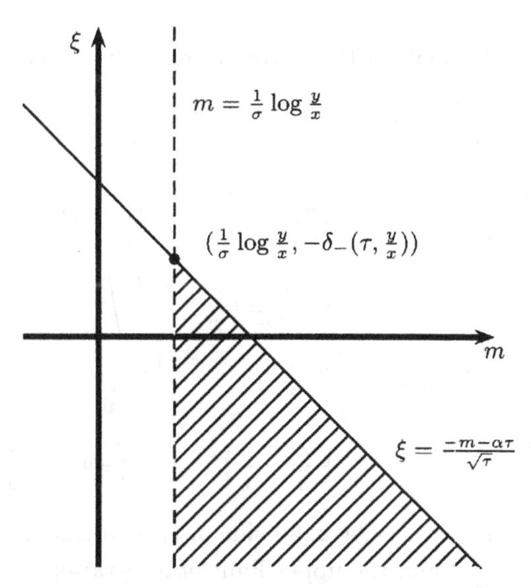

Fig. 7.4.2. Reversal of integration in (7.4.32).

$$\int_{\frac{1}{\sigma}\log\frac{y}{x}}^{-\xi\sqrt{\tau}-\alpha\tau} e^{\frac{2rm}{\sigma}-\frac{\xi^2}{2}}\,dm = \frac{\sigma}{2r}\,e^{\frac{2rm}{\sigma}-\frac{\xi^2}{2}}\Bigg|_{m=\frac{1}{\sigma}\log\frac{y}{x}}^{m=-\xi\sqrt{\tau}-\alpha\tau}$$
$$= \frac{\sigma}{2r}e^{\frac{2r}{\sigma}(-\xi\sqrt{\tau}-\alpha\tau)-\frac{1}{2}\xi^2} - \frac{\sigma}{2r}e^{\frac{2r}{\sigma^2}\log\frac{y}{x}-\frac{1}{2}\xi^2}.$$

But

$$\begin{aligned}\frac{2r}{\sigma}(-\xi\sqrt{\tau}-\alpha\tau)-\frac{\xi^2}{2} &= -\frac{\xi^2}{2}-\frac{2r\xi\sqrt{\tau}}{\sigma}-\frac{2r\alpha\tau}{\sigma}\\ &= -\frac{1}{2}\left(\xi+\frac{2r\sqrt{\tau}}{\sigma}\right)^2+\frac{2r^2\tau}{\sigma^2}-\frac{2r\alpha\tau}{\sigma}\\ &= -\frac{1}{2}\left(\xi+\frac{2r\sqrt{\tau}}{\sigma}\right)^2+\frac{2r\tau}{\sigma^2}(r-\sigma\alpha)\\ &= -\frac{1}{2}\left(\xi+\frac{2r\sqrt{\tau}}{\sigma}\right)^2+r\tau\end{aligned}$$

and

$$e^{\frac{2r}{\sigma^2}\log\frac{y}{x}-\frac{\xi^2}{2}} = \left(\frac{y}{x}\right)^{\frac{2r}{\sigma^2}} e^{-\frac{\xi^2}{2}}.$$

Therefore, the inner integral in (7.4.32) is

$$\int_{\frac{1}{\sigma}\log\frac{y}{x}}^{-\xi\sqrt{\tau}-\alpha\tau} e^{\frac{2rm}{\sigma}-\frac{\xi^2}{2}}\,dm = \frac{\sigma}{2r}e^{r\tau-\frac{1}{2}(\xi+\frac{2r\sqrt{\tau}}{\sigma})^2} - \frac{\sigma}{2r}\left(\frac{y}{x}\right)^{\frac{2r}{\sigma^2}} e^{-\frac{\xi^2}{2}}.$$

We continue (7.4.32), making this substitution for the inner integral:

$$\begin{aligned}-\frac{x}{y}&\int_{\frac{1}{\sigma}\log\frac{y}{x}}^{\infty} 2\alpha e^{(\sigma+2\alpha)m} N\left(\frac{-m-\alpha\tau}{\sqrt{\tau}}\right)dm\\ &= -\frac{\alpha\sigma x}{ry\sqrt{2\pi}}\int_{-\infty}^{-\delta_-(\tau,\frac{y}{x})} e^{r\tau-\frac{1}{2}(\xi+\frac{2r\sqrt{\tau}}{\sigma})^2}\,d\xi\\ &\quad+\frac{\alpha\sigma}{r\sqrt{2\pi}}\left(\frac{y}{x}\right)^{\frac{2r}{\sigma^2}-1}\int_{-\infty}^{-\delta_-(\tau,\frac{y}{x})} e^{-\frac{\xi^2}{2}}\,d\xi\\ &= -\frac{\alpha\sigma x e^{r\tau}}{ry\sqrt{2\pi}}\int_{-\infty}^{-\delta_-(\tau,\frac{y}{x})} e^{-\frac{1}{2}(\xi+\frac{2r\sqrt{\tau}}{\sigma})^2}\,d\xi\\ &\quad+\frac{\alpha\sigma}{r}\left(\frac{y}{x}\right)^{\frac{2r}{\sigma^2}-1} N\left(-\delta_-\left(\tau,\frac{y}{x}\right)\right).\end{aligned} \tag{7.4.33}$$

In the first integral on the right-hand side of (7.4.33), we make the change of variable $\eta = \xi + \frac{2r\sqrt{\tau}}{\sigma}$, and the upper limit of integration becomes

$$-\delta_-\left(\tau, \frac{y}{x}\right) + \frac{2r\sqrt{\tau}}{\sigma} = \frac{1}{\sigma\sqrt{\tau}}\left[-\log\frac{y}{x} - \left(r - \frac{1}{2}\sigma^2\right)\tau + 2r\tau\right]$$
$$= \frac{1}{\sigma\sqrt{\tau}}\left[\log\frac{x}{y} + \left(r + \frac{1}{2}\sigma^2\right)\tau\right]$$
$$= \delta_+\left(\tau, \frac{x}{y}\right).$$

We conclude that

$$-\frac{x}{y}\int_{\frac{1}{\sigma}\log\frac{y}{x}}^{\infty} 2\alpha e^{(\sigma+2\alpha)m} N\left(\frac{-m-\alpha\tau}{\sqrt{\tau}}\right) dm$$
$$= -\frac{\alpha\sigma x}{ry} e^{r\tau} N\left(\delta_+\left(\tau, \frac{x}{y}\right)\right) + \frac{\alpha\sigma}{r}\left(\frac{y}{x}\right)^{\frac{2r}{\sigma^2}-1} N\left(-\delta_-\left(\tau, \frac{y}{x}\right)\right). \tag{7.4.34}$$

We put all the pieces together. The function $v(t, x, y)$ for $0 \le t < T$ and $0 < x \le y$ is given by (7.4.26), where $g(x, y)$ is given by (7.4.27). We have computed both terms on the right-hand side of (7.4.27). The first term is given by (7.4.28), and the second term is itself the sum of the two terms in (7.4.29). These two terms are given by (7.4.31) and (7.4.34). Furthermore, the term $\frac{\alpha\sigma}{r}$ appearing in these formulas is equal to $1 - \frac{\sigma^2}{2r}$. We conclude that

$$\begin{aligned} v(t, x, y) = e^{-r\tau} y \Bigg[& N\left(-\delta_-\left(\tau, \frac{x}{y}\right)\right) - \left(\frac{y}{x}\right)^{\frac{2r}{\sigma^2}-1} N\left(-\delta_-\left(\tau, \frac{y}{x}\right)\right) \\ & + 2\left(\frac{x}{y}\right) e^{r\tau} N\left(\delta_+\left(\tau, \frac{x}{y}\right)\right) \\ & - \left(1 - \frac{\sigma^2}{2r}\right)\left(\frac{x}{y}\right) e^{r\tau} N\left(\delta_+\left(\tau, \frac{x}{y}\right)\right) \\ & + \left(1 - \frac{\sigma^2}{2r}\right)\left(\frac{y}{x}\right)^{\frac{2r}{\sigma^2}-1} N\left(-\delta_-\left(\tau, \frac{y}{x}\right)\right)\Bigg] - x. \end{aligned}$$

Simplification results in the formula

$$\begin{aligned} v(t, x, y) = & \left(1 + \frac{\sigma^2}{2r}\right) x N\left(\delta_+\left(\tau, \frac{x}{y}\right)\right) + e^{-r\tau} y N\left(-\delta_-\left(\tau, \frac{x}{y}\right)\right) \\ & - \frac{\sigma^2}{2r} e^{-r\tau}\left(\frac{y}{x}\right)^{\frac{2r}{\sigma^2}} x N\left(-\delta_-\left(\tau, \frac{y}{x}\right)\right) - x, \ 0 \le t < T, \ 0 < x \le y. \end{aligned} \tag{7.4.35}$$

The function u related to v by (7.4.16) satisfies

$$\begin{aligned} u\left(t, \frac{x}{y}\right) = & \left(1 + \frac{\sigma^2}{2r}\right)\left(\frac{x}{y}\right) N\left(\delta_+\left(\tau, \frac{x}{y}\right)\right) + e^{-r\tau} N\left(-\delta_-\left(\tau, \frac{x}{y}\right)\right) \\ & - \frac{\sigma^2}{2r} e^{-r\tau}\left(\frac{x}{y}\right)^{1-\frac{2r}{\sigma^2}} N\left(-\delta_-\left(\tau, \frac{y}{x}\right)\right) - \frac{x}{y}. \end{aligned}$$

Making the change of variable $z = \frac{x}{y}$, we obtain

$$u(t,z) = \left(1 + \frac{\sigma^2}{2r}\right) zN\big(\delta_+(\tau, z)\big) + e^{-r\tau} N\big(- \delta_-(\tau, z)\big)$$
$$- \frac{\sigma^2}{2r} e^{-r\tau} z^{1 - \frac{2r}{\sigma^2}} N\big(- \delta_-(\tau, z^{-1})\big) - z, \; 0 \le t < T, \; 0 < z \le 1. \tag{7.4.36}$$

7.5 Asian Options

An *Asian option* is one whose payoff includes a time average of the underlying asset price. The average may be over the entire time period between initiation and expiration or may be over some period of time that begins later than the initiation of the option and ends with the option's expiration. The average may be from continuous sampling,

$$\frac{1}{T} \int_0^T S(t)\, dt,$$

or may be from discrete sampling,

$$\frac{1}{m} \sum_{j=1}^{m} S(t_j),$$

where $0 < t_1 < t_2 \cdots < t_m = T$. The primary reason to base an option payoff on an average asset price is to make it more difficult for anyone to significantly affect the payoff by manipulation of the underlying asset price.

The price of Asian options is not known in closed form. Therefore, in this section we discuss two ways to derive partial differential equations for Asian option prices. The first of these was briefly presented in Example 6.6.1. The other method for computing Asian option prices is Monte Carlo simulation.

7.5.1 Fixed-Strike Asian Call

Once again, we begin with a geometric Brownian motion $S(t)$ given by

$$dS(t) = rS(t)\, dt + \sigma S(t)\, d\widetilde{W}(t), \tag{7.5.1}$$

where $\widetilde{W}(t)$, $0 \le t \le T$, is a Brownian motion under the risk-neutral measure $\widetilde{\mathbb{P}}$. Consider a *fixed-strike Asian call* whose payoff at time T is

$$V(T) = \left(\frac{1}{T} \int_0^T S(t)\, dt - K \right)^+, \tag{7.5.2}$$

where the strike price K is a nonnegative constant. The price at times t prior to the expiration time T of this call is given by the risk-neutral pricing formula

$$V(t) = \widetilde{\mathbb{E}}\left[e^{-r(T-t)}V(T)\Big|\,\mathcal{F}(t)\right], \quad 0 \le t \le T. \tag{7.5.3}$$

The usual iterated conditioning argument shows that

$$e^{-rt}V(t) = \widetilde{\mathbb{E}}\left[e^{-rT}V(T)\big|\,\mathcal{F}(t)\right], \quad 0 \le t \le T,$$

is a martingale under $\widetilde{\mathbb{P}}$. This is the quantity we wish to compute. In the next two subsections, we describe two different ways to undertake this.

7.5.2 Augmentation of the State

The Asian option payoff $V(T)$ in (7.5.2) is *path-dependent*. The price of the option at time t depends not only on t and $S(t)$, but also on the path that the asset price has followed up to time t. In particular, we cannot invoke the Markov property to claim that $V(t)$ is a function of t and $S(t)$ because $V(T)$ is not a function of T and $S(T)$; $V(T)$ depends on the whole path of S.

To overcome this difficulty, we *augment* the state $S(t)$ by defining a second process

$$Y(t) = \int_0^t S(u)\,du. \tag{7.5.4}$$

The stochastic differential equation for $Y(t)$ is thus

$$dY(t) = S(t)\,dt. \tag{7.5.5}$$

Because the pair of processes $(S(t), Y(t))$ is governed by the pair of stochastic differential equations (7.5.1) and (7.5.5), they constitute a two-dimensional Markov process (Corollary 6.3.2). Furthermore, the call payoff $V(T)$ is a function of T and the final value $\big(S(T), Y(T)\big)$ of this process. Indeed, $V(T)$ depends only on T and $Y(T)$, by the formula

$$V(T) = \left(\frac{1}{T}Y(T) - K\right)^+. \tag{7.5.6}$$

This implies that there must exist some function $v(t,x,y)$ such that the Asian call price (7.5.3) is given as

$$\begin{aligned} v\big(t, S(t), Y(t)\big) &= \widetilde{\mathbb{E}}\left[e^{-r(T-t)}\left(\frac{1}{T}Y(T) - K\right)^+\Bigg|\,\mathcal{F}(t)\right] \\ &= \widetilde{\mathbb{E}}\left[e^{-r(T-t)}V(T)\Big|\,\mathcal{F}(t)\right]. \end{aligned} \tag{7.5.7}$$

The function $v(t,x,y)$ satisfies a partial differential equation. This equation and three boundary conditions are provided in the next theorem. However, in order to numerically solve this equation, it would normally be necessary to also specify the behavior of $v(t,x,y)$ as x approaches ∞ and y approaches either ∞ or $-\infty$. This can be avoided by the method discussed in Subsection 7.5.3; see Remark 7.5.4 below.

Theorem 7.5.1. *The Asian call price function* $v(t,x,y)$ *of (7.5.7) satisfies the partial differential equation*

$$v_t(t,x,y) + rxv_x(t,x,y) + xv_y(t,x,y) + \frac{1}{2}\sigma^2 x^2 v_{xx}(t,x,y) = rv(t,x,y),$$
$$0 \le t < T,\ x \ge 0,\ y \in \mathbb{R}, \qquad (7.5.8)$$

and the boundary conditions

$$v(t,0,y) = e^{-r(T-t)}\left(\frac{y}{T} - K\right)^+,\ 0 \le t < T,\ y \in \mathbb{R}, \qquad (7.5.9)$$
$$\lim_{y \downarrow -\infty} v(t,x,y) = 0,\ 0 \le t < T,\ x \ge 0, \qquad (7.5.10)$$
$$v(T,x,y) = \left(\frac{y}{T} - K\right)^+,\ x \ge 0,\ y \in \mathbb{R}. \qquad (7.5.11)$$

PROOF: Using the stochastic differential equations (7.5.1) and (7.5.5) and noting that $dS(t)\,dY(t) = dY(t)\,dY(t) = 0$, we take the differential of the $\widetilde{\mathbb{P}}$-martingale $e^{-rt}V(t) = e^{-rt}v(t,S(t),Y(t))$. This differential is

$$\begin{aligned} d&\left(e^{-rt}v(t,S(t),Y(t))\right) \\ &= e^{-rt}\left[-rv\,dt + v_t\,dt + v_x\,dS + v_y\,dY + \frac{1}{2}v_{xx}\,dS\,dS\right] \\ &= e^{-rt}\left[-rv + v_t + rSv_x + Sv_y + \frac{1}{2}\sigma^2 S^2 v_{xx}\right]dt \\ &\quad + e^{-rt}\sigma S v_x\,d\widetilde{W}(t). \end{aligned} \qquad (7.5.12)$$

In order for this to be a martingale, the dt term must be zero, which implies

$$\begin{aligned} v_t(t,S(t),Y(t)) + rS(t)v_x(t,S(t),Y(t)) + S(t)v_y(t,S(t),Y(t)) \\ + \frac{1}{2}\sigma^2 S^2(t) v_{xx}(t,S(t),Y(t)) = rv(t,S(t),Y(t)). \end{aligned}$$

Replacing $S(t)$ by the dummy variable x and $Y(t)$ by the dummy variable y, we obtain (7.5.8).

We note that $S(t)$ must always be nonnegative, and so (7.5.8) holds for $x \ge 0$. If $S(t) = 0$ and $Y(t) = y$ for some value of t, then $S(u) = 0$ for all $u \in [t,T]$, and so $Y(u)$ is constant on $[t,T]$. Therefore, $Y(T) = y$, and the value of the Asian call at time t is $\left(\frac{y}{T} - K\right)^+$, discounted from T back to t. This gives us the boundary condition (7.5.9).

In contrast, it is not the case that if $Y(t) = 0$ for some time t, then $Y(u) = 0$ for all $u \ge 0$. Therefore, we cannot easily determine the value of

$v(t,x,0)$, and we do not provide a condition on the boundary $y=0$. Indeed, at least mathematically there is no problem with allowing y to be negative. If at time t we set $Y(t)=y$, then $Y(T)$ is defined by (7.5.5). In integrated form, this formula is

$$Y(T) = y + \int_t^T S(u)\,du. \tag{7.5.13}$$

Even if y is negative, this makes sense, and in this case we could still have $Y(T)>0$ or even $\frac{1}{T}Y(T)-K>0$, so that the call expires in the money. When using the differential equations (7.5.1) and (7.5.5) to describe the "state" processes $S(t)$ and $Y(t)$, there is no reason to require that $Y(t)$ be nonnegative. (We still require that $S(t)$ be nonnegative because $x=0$ is a natural boundary for $S(t)$.) For this reason, we do not restrict the values of y in the partial differential equation (7.5.8). The natural boundary for y is $y=-\infty$. If $Y(t)=y$, $S(t)=x$, and holding x fixed we let $y\to-\infty$, then $Y(T)$ approaches $-\infty$ (see (7.5.13)), the probability that the call expires in the money approaches zero, and the option price approaches zero. The natural boundary for y is $y=-\infty$, and the boundary condition there is (7.5.10).

The boundary condition (7.5.11) is just the payoff of the call. □

Remark 7.5.2. After we set the dt term in (7.5.12) equal to zero, we see that

$$d\left(e^{-rt}v\big(t,S(t),Y(t)\big)\right) = e^{-rt}\sigma S(t) v_x\big(t,S(t),Y(t)\big)\,d\widetilde{W}(t). \tag{7.5.14}$$

The discounted value of a portfolio that at each time t holds $\Delta(t)$ shares of the underlying asset is given by (see (5.2.27))

$$d\big(e^{-rt}X(t)\big) = e^{-rt}\sigma S(t)\Delta(t)\,d\widetilde{W}(t). \tag{7.5.15}$$

To hedge a short position in the Asian call, an agent should equate these two differentials, which leads to the delta-hedging formula

$$\Delta(t) = v_x\big(t,S(t),Y(t)\big).$$

7.5.3 Change of Numéraire

In this subsection we present a partial differential equation whose solution leads to Asian option prices. We work this out for both continuous and discrete averaging. The derivation of this equation involves a *change of numéraire*, a concept discussed systematically in Chapter 9. In this section, we derive formulas under the assumption that the interest rate r is not zero. The case $r=0$ is treated in Exercise 7.8.

We first consider the case of an Asian call with payoff

$$V(T) = \left(\frac{1}{c}\int_{T-c}^T S(t)\,dt - K\right)^+, \tag{7.5.16}$$

where c is a constant satisfying $0 < c \leq T$ and K is a nonnegative constant. If $c = T$, this is the Asian call (7.5.2) considered in Subsection 7.5.2. Here we also admit the possibility that the averaging is over less than the full time between initiation and expiration of the call.

To price this call, we create a portfolio process whose value at time T is

$$X(T) = \frac{1}{c}\int_{T-c}^{T} S(u)\,du - K.$$

We begin with a nonrandom function of time $\gamma(t)$, $0 \leq t \leq T$, which will be the number of shares of the risky asset held by our portfolio. There will be no Brownian motion term in $\gamma(t)$, and because of this it will satisfy $d\gamma(t)\,d\gamma(t) = d\gamma(t)\,dS(t) = 0$. This implies that

$$d\big(\gamma(t)S(t)\big) = \gamma(t)\,dS(t) + S(t)\,d\gamma(t), \tag{7.5.17}$$

which further implies

$$\begin{aligned} d\big(e^{r(T-t)}\gamma(t)S(t)\big) &= e^{r(T-t)}d\big(\gamma(t)S(t)\big) - re^{r(T-t)}\gamma(t)S(t)\,dt \\ &= e^{r(T-t)}\gamma(t)\,dS(t) + e^{r(T-t)}S(t)\,d\gamma(t) \\ &\qquad -re^{r(T-t)}\gamma(t)S(t)\,dt. \end{aligned} \tag{7.5.18}$$

Rearranging terms in (7.5.18), we obtain

$$e^{r(T-t)}\gamma(t)\big(dS(t) - rS(t)\,dt\big) = d\big(e^{r(T-t)}\gamma(t)S(t)\big) - e^{r(T-t)}S(t)\,d\gamma(t). \tag{7.5.19}$$

An agent who holds $\gamma(t)$ shares of the risky asset at each time t and finances this by investing or borrowing at the interest rate r will have a portfolio whose value evolves according to the equation

$$\begin{aligned} dX(t) &= \gamma(t)\,dS(t) + r\big(X(t) - \gamma(t)S(t)\,\big)dt \\ &= rX(t)\,dt + \gamma(t)\big(dS(t) - rS(t)\,dt\big). \end{aligned} \tag{7.5.20}$$

Using this equation and (7.5.19), we obtain

$$\begin{aligned} d\big(e^{r(T-t)}X(t)\big) &= -re^{r(T-t)}X(t)\,dt + e^{r(T-t)}\,dX(t) \\ &= e^{r(T-t)}\gamma(t)\big(S(t) - rS(t)\,dt\big) \\ &= d\big(e^{r(T-t)}\gamma(t)S(t)\big) - e^{r(T-t)}S(t)\,d\gamma(t). \end{aligned} \tag{7.5.21}$$

To study the Asian call with payoff (7.5.16), we take $\gamma(t)$ to be

$$\gamma(t) = \begin{cases} \dfrac{1}{rc}\left(1 - e^{-rc}\right), & 0 \leq t \leq T-c, \\ \dfrac{1}{rc}\left(1 - e^{-r(T-t)}\right), & T-c \leq t \leq T, \end{cases} \tag{7.5.22}$$

and we take the initial capital to be

$$X(0) = \frac{1}{rc}(1 - e^{-rc})S(0) - e^{-rT}K. \tag{7.5.23}$$

In the time interval $[0, T - c]$, the process $\gamma(t)$ mandates a buy-and-hold strategy. At time zero, we buy $\frac{1}{rc}(1 - e^{-rc})$ shares of the risky asset, which costs $\frac{1}{rc}(1 - e^{-rc})S(0)$. Our initial capital is insufficient to do this, and we must borrow $e^{-rT}K$ from the money market account. For $0 \le t \le T - c$, the value of our holdings in the risky asset is $\frac{1}{rc}(1 - e^{-rc})S(t)$ and we owe $e^{-r(T-t)}K$ to the money market account. Therefore,

$$X(t) = \frac{1}{rc}(1 - e^{-rc})S(t) - e^{-r(T-t)}K, \quad 0 \le t \le T - c. \tag{7.5.24}$$

In particular,

$$X(T - c) = \frac{1}{rc}(1 - e^{-rc})S(T - c) - e^{-rc}K. \tag{7.5.25}$$

For $T - c \le t \le T$, we have $d\gamma(t) = -\frac{1}{c}e^{-r(T-t)}$ and we compute $X(t)$ by first integrating (7.5.21) from $T - c$ to t and using (7.5.25) and (7.5.22) to obtain

$$\begin{aligned}
&e^{r(T-t)}X(t) \\
&= e^{rc}X(T-c) + \int_{T-c}^{t} d\big(e^{r(T-u)}\gamma(u)S(u)\big) - \int_{T-c}^{t} e^{r(T-u)}S(u)\,d\gamma(u) \\
&= \frac{1}{rc}e^{rc}(1 - e^{-rc})S(T-c) - K + e^{r(T-t)}\gamma(t)S(t) \\
&\qquad - \frac{1}{rc}e^{rc}(1 - e^{-rc})S(T-c) + \frac{1}{c}\int_{T-c}^{t} S(u)\,du \\
&= -K + e^{r(T-t)}\gamma(t)S(t) + \frac{1}{c}\int_{T-c}^{t} S(u)\,du.
\end{aligned}$$

Therefore,

$$X(t) = \frac{1}{rc}\big(1 - e^{-r(T-t)}\big)S(t) + e^{-r(T-t)}\frac{1}{c}\int_{T-c}^{t} S(u)\,du - e^{-r(T-t)}K, \quad T - c \le t \le T. \tag{7.5.26}$$

In particular,

$$X(T) = \frac{1}{c}\int_{T-c}^{T} S(u)\,du - K, \tag{7.5.27}$$

as desired, and

$$V(T) = X^{+}(T) = \max\{X(T), 0\}. \tag{7.5.28}$$

The price of the Asian call at time t prior to expiration is

$$V(t) = \widetilde{\mathbb{E}}\big[e^{-r(T-t)}V(T)\big|\mathcal{F}(t)\big] = \widetilde{\mathbb{E}}\big[e^{-r(T-t)}X^{+}(T)\big|\mathcal{F}(t)\big]. \tag{7.5.29}$$

The calculation of the right-hand side of (7.5.29) uses a change-of-numéraire argument, which we now exlain. Let us define

$$Y(t) = \frac{X(t)}{S(t)} = \frac{e^{-rt}X(t)}{e^{-rt}S(t)}.$$

This is the value of the portfolio denominated in units of the risky asset rather than in dollars. We have changed the numéraire, the unit of account, from dollars to the risky asset.

We work out the differential of $Y(t)$. Note first that

$$d\big(e^{-rt}S(t)\big) = -re^{-rt}S(t)\,dt + e^{-rt}\,dS(t) = \sigma e^{-rt}S(t)\,d\widetilde{W}(t). \tag{7.5.30}$$

Therefore,

$$\begin{aligned}
&d\left[\big(e^{-rt}S(t)\big)^{-1}\right]\\
&= -\big(e^{-rt}S(t)\big)^{-2}d\big(e^{-rt}S(t)\big) + \big(e^{-rt}S(t)\big)^{-3}d\big(e^{-rt}S(t)\big)\,d\big(e^{-rt}S(t)\big)\\
&= -\big(e^{-rt}S(t)\big)^{-2}\sigma\big(e^{-rt}S(t)\big)\,d\widetilde{W}(t) + \big(e^{-rt}S(t)\big)^{-3}\big(e^{-rt}S(t)\big)^2\sigma^2\,dt\\
&= -\sigma\big(e^{-rt}S(t)\big)^{-1}\,d\widetilde{W}(t) + \sigma^2\big(e^{-rt}S(t)\big)^{-1}\,dt.
\end{aligned}$$

On the other hand, (7.5.20) and (7.5.30) imply

$$\begin{aligned}
d\big(e^{-rt}X(t)\big) &= -re^{-rt}X(t)\,dt + e^{-rt}\,dX(t)\\
&= \gamma(t)e^{-rt}\big(dS(t) - rS(t)\big)\,dt\\
&= \gamma(t)\sigma e^{-rt}S(t)\,d\widetilde{W}(t).
\end{aligned}$$

Itô's product rule implies

$$\begin{aligned}
dY(t) &= d\left[\big(e^{-rt}X(t)\big)\big(e^{-rt}S(t)\big)^{-1}\right]\\
&= e^{-rt}X(t)\,d\left[\big(e^{-rt}S(t)\big)^{-1}\right] + \big(e^{-rt}S(t)\big)^{-1}\,d\big(e^{-rt}X(t)\big)\\
&\qquad + d\big(e^{-rt}X(t)\big)\,d\left[\big(e^{-rt}S(t)\big)^{-1}\right]\\
&= -\sigma Y(t)\,d\widetilde{W}(t) + \sigma^2 Y(t)\,dt + \sigma\gamma(t)\,d\widetilde{W}(t) - \sigma^2\gamma(t)\,dt\\
&= \sigma\big[\gamma(t) - Y(t)\big]\big[d\widetilde{W}(t) - \sigma\,dt\big].
\end{aligned} \tag{7.5.31}$$

The process $Y(t)$ is not a martingale under $\widetilde{\mathbb{P}}$ because its differential (7.5.31) has a dt term. However, we can change measure so that $Y(t)$ is a martingale, and this will simplify (7.5.31). We set

$$\widetilde{W}^S(t) = \widetilde{W}(t) - \sigma t \tag{7.5.32}$$

and then have

$$dY(t) = \sigma\big[\gamma(t) - Y(t)\big]\, d\widetilde{W}^S(t). \tag{7.5.33}$$

According to Girsanov's Theorem, Theorem 5.2.3, we can change the measure so that $\widetilde{W}^S(t)$, $0 \le t \le T$, is a Brownian motion. In this situation, $-\sigma$ plays the role of Θ in Theorem 5.2.3, and $\widetilde{W}$ and $\widetilde{\mathbb{P}}$ play the roles of W and $\mathbb{P}$. The Radon-Nikodým derivative process of (5.2.11) is

$$Z(t) = \exp\left\{\sigma\widetilde{W}(t) - \frac{1}{2}\sigma^2 t\right\}.$$

In other words,

$$Z(t) = \frac{e^{-rt}S(t)}{S(0)}. \tag{7.5.34}$$

Under the probability measure $\widetilde{\mathbb{P}}^S$ defined by

$$\widetilde{\mathbb{P}}^S(A) = \int_A Z(T)\, d\widetilde{\mathbb{P}} \text{ for all } A \in \mathcal{F},$$

$\widetilde{W}^S(t)$ is a Brownian motion and $Y(t)$ is a martingale.

Under the probability measure $\widetilde{\mathbb{P}}^S$, the process $Y(t)$ is Markov. It is given by the stochastic differential equation (7.5.33), and because $\gamma(t)$ is nonrandom, the term multiplying $d\widetilde{W}^S(t)$ in (7.5.33) is a function of t and $Y(t)$ and has no source of randomness other than $Y(t)$. Equation (7.5.33) is a stochastic differential equation of the type (6.2.1), and solutions to such equations are Markov (see Corollary 6.3.2).

We return to the option price $V(t)$ of (7.5.29) and use Lemma 5.2.2 to write (7.5.29) as

$$\begin{aligned}
V(t) &= e^{rt}\widetilde{\mathbb{E}}\big[e^{-rT}X^+(T)\big|\mathcal{F}(t)\big] \\
&= \frac{S(t)}{e^{-rt}S(t)}\widetilde{\mathbb{E}}\left[e^{-rT}S(T)\left(\frac{e^{-rT}X(T)}{e^{-rT}S(T)}\right)^+\middle|\,\mathcal{F}(t)\right] \\
&= \frac{S(t)}{Z(t)}\widetilde{\mathbb{E}}\left[Z(T)Y^+(T)\middle|\,\mathcal{F}(t)\right] \\
&= S(t)\widetilde{\mathbb{E}}^S\left[Y^+(T)\,\middle|\,\mathcal{F}(t)\right],
\end{aligned} \tag{7.5.35}$$

where $\widetilde{\mathbb{E}}^S[\cdots|\mathcal{F}(t)]$ denotes conditional expectation under the probability measure $\widetilde{\mathbb{P}}^S$. Because Y is Markov under $\widetilde{\mathbb{P}}^S$, there must be some function $g(t,y)$ such that

$$g\big(t, Y(t)\big) = \widetilde{\mathbb{E}}^S\left[Y^+(T)\middle|\,\mathcal{F}(t)\right]. \tag{7.5.36}$$

From (7.5.36), we see that

$$g\big(T, Y(T)\big) = \widetilde{\mathbb{E}}^S\left[Y^+(T)\middle|\,\mathcal{F}(T)\right] = Y^+(T). \tag{7.5.37}$$

We note that $Y(T) = \frac{X(T)}{S(T)}$ can take any value since the numerator $X(T)$, given by (7.5.27), can be either positive or negative, and the denominator $S(T)$ can be any positive number. Therefore, (7.5.37) leads to the boundary condition

$$g(T,y) = y^+, \quad y \in \mathbb{R}. \tag{7.5.38}$$

The usual iterated conditioning argument shows that the right-hand side of (7.5.36) is a martingale under $\widetilde{\mathbb{P}}^S$, and so the differential of $g\big(t, Y(t)\big)$ should have only a $d\widetilde{W}^S(t)$ term. This differential is

$$\begin{aligned} dg\big(t,Y(t)\big) &= g_t\big(t,Y(t)\big)\,dt + g_y\big(t,Y(t)\big)\,dY(t) \\ &\quad + \frac{1}{2} g_{yy}\big(t,Y(t)\big)\,dY(t)\,dY(t) \\ &= \left[g_t\big(t,Y(t)\big) + \frac{1}{2}\sigma^2\big(\gamma(t) - Y(t)\big)^2 g_{yy}\big(t,Y(t)\big) \right] dt \\ &\quad + \sigma\big(\gamma(t) - Y(t)\big) g_y\big(t,Y(t)\big)\,d\widetilde{W}^S(t). \end{aligned}$$

We conclude that $g(t,y)$ satisfies the partial differential equation

$$g_t(t,y) + \frac{1}{2}\sigma^2\big(\gamma(t) - y\big)^2 g_{yy}(t,y) = 0, \quad 0 \le t < T,\ y \in \mathbb{R}. \tag{7.5.39}$$

We summarize this discussion with the following theorem.

Theorem 7.5.3 (Večeř). *For $0 \le t \le T$, the price $V(t)$ at time t of the continuously averaged Asian call with payoff (7.5.16) at time T is*

$$V(t) = S(t)\, g\left(t, \frac{X(t)}{S(t)}\right), \tag{7.5.40}$$

where $g(t,y)$ satisfies (7.5.39) and $X(t)$ is given by (7.5.24) and (7.5.26). The boundary conditions for $g(t,y)$ are (7.5.38) and

$$\lim_{y \to -\infty} g(t,y) = 0, \quad \lim_{y \to \infty} \big[g(t,y) - y\big] = 0, \quad 0 \le t \le T. \tag{7.5.41}$$

Remark 7.5.4 (Boundary conditions). Let $0 \le t \le T$ be given. The first boundary condition in (7.5.41) can be derived from the fact that when $Y(t)$ is very negative, the probability that $Y(T)$ also is negative is near one and therefore the probability that $Y^+(T) = 0$ is near one. This causes $g(t, Y(t))$ in (7.5.36) to be near zero. The second boundary condition in (7.5.41) is a consequence of that fact that when $Y(t)$ is large, then the probability that $Y(T) > 0$ is near one. Therefore, $g\big(t, Y(t)\big)$ given by (7.5.36) is approximately equal to $\widetilde{\mathbb{E}}^S\big[Y(T)\big|\mathcal{F}(t)\big]$, and because $Y(T)$ is a martingale under $\widetilde{\mathbb{P}}^S$, this conditional expectation is $Y(t)$.

It is easier to derive these boundary conditions at $y = \pm\infty$ for $g(t,y)$ than it is to derive the boundary conditions for $v(t,x,y)$ in Theorem 7.5.1 because

$v(t, x, y)$ has two variables, x and y, that can become large. For example, it is not at all clear how $v(t, x, y)$ behaves as $x \to \infty$ and $y \to -\infty$. The reduction of the Asian option pricing problem provided by Theorem 7.5.3 reduces the dimensionality of the problem and simplifies the boundary conditions. It also removes a so-called "degeneracy" in equation (7.5.8) created by the absence of the $v_{yy}(t, x, y)$ term. This degeneracy complicates the numerical solution of (7.5.8). □

In the remainder of this subsection, we adapt the arguments just given to treat a *discretely sampled Asian call.* Assume we are given times $0 = t_0 < t_1 < t_2 \cdots < t_m = T$ and the Asian call payoff is

$$V(T) = \left(\frac{1}{m} \sum_{j=1}^{m} S(t_j) - K \right)^+. \tag{7.5.42}$$

We wish to create a portfolio process so that

$$X(T) = \frac{1}{m} \sum_{j=1}^{m} S(t_j) - K.$$

In place of (7.5.22), we define

$$\gamma(t_j) = \frac{1}{m} \sum_{i=j}^{m} e^{-r(T-t_i)}, \quad j = 0, 1, \ldots, m. \tag{7.5.43}$$

Then

$$\gamma(t_j) = \gamma(t_{j-1}) - \frac{1}{m} e^{-r(T-t_{j-1})}, \quad j = 1, \ldots, m, \tag{7.5.44}$$

and $\gamma(T) = \gamma(t_m) = \frac{1}{m}$. We complete the definition of $\gamma(t)$ by setting

$$\gamma(t) = \gamma(t_j), \quad t_{j-1} < t \leq t_j. \tag{7.5.45}$$

This defines $\gamma(t)$ for all $t \in [0, T]$. In this situation, (7.5.21) still holds, but now $d\gamma(t) = 0$ in each subinterval (t_{j-1}, t_j). Integrating (7.5.21) from t_{j-1} to t_j and using (7.5.44) and the fact that $\gamma(t) = \gamma(t_j)$ for $t \in (t_{j-1}, t_j]$, we obtain

$$\begin{aligned}
&e^{r(T-t_j)} X(t_j) - e^{r(T-t_{j-1})} X(t_{j-1}) \\
&= \gamma(t_j) \left[e^{r(T-t_j)} S(t_j) - e^{r(T-t_{j-1})} S(t_{j-1}) \right] \\
&= \gamma(t_j) e^{r(T-t_j)} S(t_j) - \left(\gamma(t_{j-1}) - \frac{1}{m} e^{-r(T-t_{j-1})} \right) e^{r(T-t_{j-1})} S(t_{j-1}) \\
&= \gamma(t_j) e^{r(T-t_j)} S(t_j) - \gamma(t_{j-1}) e^{r(T-t_{j-1})} S(t_{j-1}) + \frac{1}{m} S(t_{j-1}).
\end{aligned}$$

Summing this equation from $j = 1$ to $j = k$, we see that

$$
\begin{aligned}
&e^{r(T-t_k)}X(t_k) - e^{rT}X(0) \\
&= \gamma(t_k)e^{r(T-t_k)}S(t_k) - \gamma(0)e^{rT}S(0) + \frac{1}{m}\sum_{j=1}^{k} S(t_{j-1}) \\
&= \gamma(t_k)e^{r(T-t_k)}S(t_k) + \frac{1}{m}\sum_{i=1}^{k-1} S(t_i) + \left(-\gamma(0)e^{rT} + \frac{1}{m}\right) S(0).
\end{aligned}
$$

We set

$$
X(0) = e^{-rT}\left[\gamma(0)e^{rT} - \frac{1}{m}\right] S(0) - e^{-rT}K,
$$

so this equation becomes

$$
e^{r(T-t_k)}X(t_k) = \gamma(t_k)e^{r(T-t_k)}S(t_k) + \frac{1}{m}\sum_{i=1}^{k-1} S(t_i) - K
$$

or, equivalently,

$$
X(t_k) = \gamma(t_k)S(t_k) + e^{-r(T-t_k)}\frac{1}{m}\sum_{i=1}^{k-1} S(t_i) - e^{-r(T-t_k)}K. \tag{7.5.46}
$$

In particular,

$$
X(T) = X(t_m) = \frac{1}{m}\sum_{i=1}^{m} S(t_i) - K \tag{7.5.47}
$$

as desired.

To determine $X(t)$ for $t_k \leq t \leq t_{k+1}$, we integrate (7.5.21) from t_k to t to obtain

$$
\begin{aligned}
e^{r(T-t)}X(t) &= e^{r(T-t_k)}X(t_k) + \gamma(t_{k+1})\left[e^{r(T-t)}S(t) - e^{r(T-t_k)}S(t_k)\right] \\
&= \gamma(t_k)e^{r(T-t_k)}S(t_k) + \frac{1}{m}\sum_{i=1}^{k-1} S(t_i) - K + \gamma(t_{k+1})e^{r(T-t)}S(t) \\
&\quad - \left(\gamma(t_k) - \frac{1}{m}e^{-r(T-t_k)}\right) e^{r(T-t_k)}S(t_k) \\
&= \gamma(t_{k+1})e^{r(T-t)}S(t) + \frac{1}{m}\sum_{i=1}^{k} S(t_i) - K.
\end{aligned}
$$

Therefore,

$$
X(t) = \gamma(t_{k+1})S(t) + e^{-r(T-t)}\frac{1}{m}\sum_{i=1}^{k} S(t_i) - e^{-r(T-t)}K, \quad t_k \leq t \leq t_{k+1}. \tag{7.5.48}
$$

We now proceed with the change of numéraire as before. This leads again to Theorem 7.5.3 for the discretely sampled Asian call with payoff (7.5.42).

The price at time t is given by (7.5.40), where $g(t,x)$ satisfies (7.5.39) with boundary conditions (7.5.38) and (7.5.41). The only difference is that now the nonrandom function $\gamma(t)$ appearing in (7.5.39) is given by (7.5.43) and (7.5.45) and the process $X(t)$ in (7.5.40) is given by (7.5.48).

7.6 Summary

Three specific exotic options on a geometric Brownian motion have been considered: an up-and-out barrier call, a lookback call, and an Asian call. In each case, the discounted option price is a martingale under the risk-neutral measure, and this leads to a partial differential equation of the Black-Scholes-Merton type. However, the lookback call and the Asian call equations have an additional state variable in this equation.

For the barrier call and the lookback call, the option price was computed explicitly. The Asian option pricing problem was transformed by a change of numéraire to an equation with a single state variable. This transformation was done both for the continuously sampled and the discretely sampled Asian options.

7.7 Notes

There are scores of different exotic options, and the search for explicit pricing formulas can lead to complex computations. Analysis of many exotic options is provided by Zhang [167] and Haug [80]. Papers by a variety of authors who treat exotic options, including some of those cited below, have been collected by Lipton [110]. Exotic options are prevalent in foreign exchange markets. Analysis of several instruments appearing in these markets is provided by Hakala and Wystup [76]. Many exotic pricing formulas can be derived from the formulas for distributions related to Brownian motion collected by Borodin and Salminen [18].

The analysis of barrier options presented here follows Rubinstein and Reiner [142]. Monte Carlo simulation of barrier options normally obtains the price for the case when barrier crossing is checked only at discrete times. Broadie, Glasserman and Kou [22] provide a correction term to adjust this result to obtain the price for an option in which the barrier is monitored continuously. The problem of large delta and gamma values for barrier options near expiration near the barrier can be ameliorated by placing an a priori constraint on the hedging strategy and pricing this constraint into the option; see Schmock, Shreve, and Wystup [148].

The change-of-numéraire approach to Asian options, explained in Subsection 7.5.3, is due to Večeř [155], [156]. This methodology was extended to jump processes by Večeř and Xu [157]. Other partial differential equations for pricing Asian options are provided by Andreasen [4], Lipton [109], and Rogers and Shi [139].

Geman and Yor [71] obtain a closed-form formula for a Laplace transform of the Asian option price. Fu, Madan, and Wang [67] compare Monte Carlo and Laplace transform methods for Asian option pricing.

7.8 Exercises

Exercise 7.1 (Black-Scholes-Merton equation for the up-and-out call). This exercise shows by direct calculation that the function $v(t, x)$ of (7.3.20) satisfies the Black-Scholes-Merton equation (7.3.4).

(i) Recall that $\tau = T - t$, so $\frac{d\tau}{dt} = -1$. Show that $\delta_\pm(\tau, s)$ given by (7.3.18) satisfies

$$\frac{\partial}{\partial t}\delta_\pm(\tau, s) = -\frac{1}{2\tau}\delta_\pm\left(\tau, \frac{1}{s}\right). \tag{7.8.1}$$

(ii) Show that for any positive constant c,

$$\frac{\partial}{\partial x}\delta_\pm\left(\tau, \frac{x}{c}\right) = \frac{1}{x\sigma\sqrt{\tau}}, \quad \frac{\partial}{\partial x}\delta_\pm\left(\tau, \frac{c}{x}\right) = -\frac{1}{x\sigma\sqrt{\tau}}. \tag{7.8.2}$$

(iii) Show that

$$\frac{N'\big(\delta_+(\tau, s)\big)}{N'\big(\delta_-(\tau, s)\big)} = \frac{e^{-r\tau}}{s}$$

and hence

$$e^{-r\tau}N'\big(\delta_-(\tau, s)\big) = sN'\big(\delta_+(\tau, s)\big). \tag{7.8.3}$$

(iv) Show that

$$\frac{N'\big(\delta_\pm(\tau, s)\big)}{N'\big(\delta_\pm(\tau, s^{-1})\big)} = s^{-\left(\frac{2r}{\sigma^2}\pm 1\right)}$$

and hence

$$N'\big(\delta_\pm(\tau, s^{-1})\big) = s^{\frac{2r}{\sigma^2}\pm 1}N'\big(\delta_\pm(\tau, s)\big). \tag{7.8.4}$$

(v) Show that

$$\delta_+(\tau, s) - \delta_-(\tau, s) = \sigma\sqrt{\tau}. \tag{7.8.5}$$

(vi) Show that

$$\delta_\pm(\tau, s) - \delta_\pm(\tau, s^{-1}) = \frac{2}{\sigma\sqrt{\tau}}\log s. \tag{7.8.6}$$

(vii) Show that

$$N''(y) = -yN'(y). \tag{7.8.7}$$

(viii) Use (i) to compute $v_t(t,x)$ and (7.8.3)–(7.8.5) to simplify it, obtaining

$$\begin{aligned}
&v_t(t,x)\\
&= -\frac{x\sigma}{2\sqrt{\tau}}N'\left(\delta_+\left(\tau,\frac{x}{K}\right)\right) - \frac{x(B-K)}{B\sigma\tau\sqrt{\tau}}\log\frac{x}{B}N'\left(\delta_+\left(\tau,\frac{x}{B}\right)\right)\\
&\quad+\frac{B\sigma}{2\sqrt{\tau}}\left(\frac{x}{B}\right)^{-\frac{2r}{\sigma^2}}N'\left(\delta_+\left(\tau,\frac{B^2}{Kx}\right)\right)\\
&\quad-re^{-r\tau}K\left[N\left(\delta_-\left(\tau,\frac{x}{K}\right)\right)-N\left(\delta_-\left(\tau,\frac{x}{B}\right)\right)\right]\\
&\quad+re^{-r\tau}K\left(\frac{x}{B}\right)^{-\frac{2r}{\sigma^2}+1}\left[N\left(\delta_-\left(\tau,\frac{B^2}{Kx}\right)\right)-N\left(\delta_-\left(\tau,\frac{B}{x}\right)\right)\right].
\end{aligned}\tag{7.8.8}$$

(ix) Use (ii) to compute $v_x(t,x)$ and (7.8.3) and (7.8.4) to simplify it, obtaining

$$\begin{aligned}
&v_x(t,x)\\
&= \left[N\left(\delta_+\left(\tau,\frac{x}{K}\right)\right)-N\left(\delta_+\left(\tau,\frac{x}{B}\right)\right)\right] - \frac{2(B-K)}{B\sigma\sqrt{\tau}}N'\left(\delta_+\left(\tau,\frac{x}{B}\right)\right)\\
&\quad+\frac{2r}{\sigma^2}\left(\frac{x}{B}\right)^{-\frac{2r}{\sigma^2}-1}\left[N\left(\delta_+\left(\tau,\frac{B^2}{Kx}\right)\right)-N\left(\delta_+\left(\tau,\frac{B}{x}\right)\right)\right]\\
&\quad+\frac{e^{-r\tau}K}{B}\left(-\frac{2r}{\sigma^2}+1\right)\left(\frac{x}{B}\right)^{-\frac{2r}{\sigma^2}}\\
&\qquad\times\left[N\left(\delta_-\left(\tau,\frac{B^2}{Kx}\right)\right)-N\left(\delta_-\left(\tau,\frac{B}{x}\right)\right)\right].
\end{aligned}\tag{7.8.9}$$

(x) Use (ii) and (7.8.9) to compute $v_{xx}(t,x)$ and (7.8.3) and (7.8.4) to simplify it, obtaining

$$\begin{aligned}
&v_{xx}(t,x)\\
&= \frac{1}{x\sigma\sqrt{\tau}}N'\left(\delta_+\left(\tau,\frac{x}{K}\right)\right) - \frac{1}{B\sigma\sqrt{\tau}}\left(\frac{x}{B}\right)^{-\frac{2r}{\sigma^2}-2}N'\left(\delta_+\left(\tau,\frac{B^2}{Kx}\right)\right)\\
&\quad+\frac{2(B-K)}{xB\sigma\sqrt{\tau}}\left(\frac{2r}{\sigma^2}+\frac{1}{\sigma^2\tau}\log\frac{x}{B}\right)N'\left(\delta_+\left(\tau,\frac{x}{B}\right)\right)\\
&\quad-\frac{2r}{B\sigma^2}\left(\frac{2r}{\sigma^2}+1\right)\left(\frac{x}{B}\right)^{-\frac{2r}{\sigma^2}-2}\\
&\qquad\times\left[N\left(\delta_+\left(\tau,\frac{B^2}{Kx}\right)\right)-N\left(\delta_+\left(\tau,\frac{B}{x}\right)\right)\right]\\
&\quad-\frac{e^{-r\tau}K}{B^2}\left(\frac{2r}{\sigma^2}\right)\left(-\frac{2r}{\sigma^2}+1\right)\left(\frac{x}{B}\right)^{-\frac{2r}{\sigma^2}-1}\\
&\qquad\times\left[N\left(d_-\left(\tau,\frac{B^2}{Kx}\right)\right)-N\left(\delta_-\left(\tau,\frac{B}{x}\right)\right)\right].
\end{aligned}\tag{7.8.10}$$

(xi) Now verify that $v(t,x)$ satisfies the Black-Scholes-Merton equation (7.3.4).

Exercise 7.2 (Boundary conditions for the up-and-out call). In this exercise, it is verified that the up-and-out call price $v(t,x)$ given by (7.3.20) satisfies the boundary condition (7.3.6). Furthermore, the limit as $x \downarrow 0$ satisfies (7.3.5) and the limit as $t \uparrow T$ satisfies (7.3.7).

(i) Verify by direct substitution into (7.3.20) that (7.3.6) is satisfied.
(ii) Show that, for any positive constant c,

$$\lim_{x\downarrow 0} \delta_\pm\left(\tau, \frac{x}{c}\right) = -\infty, \quad \lim_{x\downarrow 0} \delta_\pm\left(\tau, \frac{c}{x}\right) = \infty. \tag{7.8.11}$$

Use this to show that for any $p \in \mathbb{R}$ and positive constants c_1 and c_2, we have

$$\lim_{x\downarrow 0} x^p \left[N\left(\delta_\pm\left(\tau, \frac{x}{c_1}\right)\right) - N\left(\delta_\pm\left(\tau, \frac{x}{c_2}\right)\right)\right] = 0, \tag{7.8.12}$$

$$\lim_{x\downarrow 0} x^p \left[N\left(\delta_\pm\left(\tau, \frac{c_1}{x}\right)\right) - N\left(\delta_\pm\left(\tau, \frac{c_2}{x}\right)\right)\right] = 0. \tag{7.8.13}$$

If $p \geq 0$, (7.8.12) and (7.8.13) are immediate consequences of (7.8.11). However, if $p < 0$, one should first use L'Hôpital's rule and then show that

$$\lim_{x\downarrow 0} x^p \exp\left\{-\frac{1}{2}\delta_\pm^2\left(\tau, \frac{x}{c_i}\right)\right\} = 0, \quad \lim_{x\downarrow 0} x^p \exp\left\{-\frac{1}{2}\delta_\pm^2\left(\tau, \frac{c_i}{x}\right)\right\} = 0. \tag{7.8.14}$$

To establish (7.8.14), you may wish to prove and use the inequality

$$\frac{1}{2}a^2 - b^2 \leq (a+b)^2 \text{ for all } a, b \in \mathbb{R}. \tag{7.8.15}$$

Conclude that $\lim_{x\downarrow 0} v(t,x) = 0$ for $0 \leq t < T$.
(iii) Show that, for any positive c,

$$\lim_{\tau\downarrow 0} \delta_\pm(\tau, c) = \begin{cases} -\infty & \text{if } 0 < c < 1, \\ 0 & \text{if } c = 1, \\ \infty & \text{if } c > 1. \end{cases} \tag{7.8.16}$$

Use this to show that $\lim_{\tau\downarrow 0} v(t,x) = (x-K)^+$ for $0 < x < B$.

Exercise 7.3 (Markov property for geometric Brownian motion and its maximum to date). Recall the geometric Brownian motion $S(t)$ of (7.4.1) and its maximum-to-date process $Y(t)$ of (7.4.3). According to Definition 2.3.6, in order to show that the pair of processes $\big(S(t), Y(t)\big)$ is Markov, we must show that whenever $0 \leq t \leq T$ and $f(x,y)$ is a function, there exists another function $g(x,y)$ such that

$$\mathbb{E}\big[f\big(S(T), Y(T)\big)\big|\mathcal{F}(t)\big] = g\big(S(t), Y(t)\big). \tag{7.8.17}$$

Use the Independence Lemma, Lemma 2.3.4, to show that such a function $g(x,y)$ exists.

Exercise 7.4 (Cross variation of geometric Brownian motion and its maximum to date). Let $S(t)$ be the geometric Brownian motion (7.4.1) and let $Y(t)$ be the maximum-to-date process (7.4.3). Let T be fixed and let $0 = t_0 < t_1 < \dots t_m = T$ be a partition of $[0, T]$. Show that as the number of partition points m approaches infinity and the length of the longest subinterval $\max_{j=1,\dots,m} t_j - t_{j-1}$ approaches zero, the sum

$$\sum_{j=1}^{m} \big(Y(t_j) - Y(t_{j-1})\big)\big(S(t_j) - S(t_{j-1})\big)$$

has limit zero.

Exercise 7.5 (Black-Scholes-Merton equation for lookback option). We wish to verify by direct computation that the function $v(t, x, y)$ of (7.4.35) satisfies the Black-Scholes-Merton equation (7.4.6). As we saw in Subsection 7.4.3, this is equivalent to showing that the function u defined by (7.4.36) satisfies the Black-Scholes-Merton equation (7.4.18). We verify that $u(t, z)$ satisfies (7.4.18) in the following steps. Let $0 \le t < T$ be given, and define $\tau = T - t$.

(i) Use (7.8.1) to compute $u_t(t, z)$, and use (7.8.3) and (7.8.4) to simplify the result, thereby showing that

$$u_t(t, z) = re^{-r\tau} N\big(-\delta_-(\tau, z)\big) - \frac{1}{2}\sigma^2 e^{-r\tau} z^{1-\frac{2r}{\sigma^2}} N\big(-\delta_-(\tau, z^{-1})\big) - \frac{\sigma z}{\sqrt{\tau}} N'\big(\delta_+(\tau, z)\big). \tag{7.8.18}$$

(ii) Use (7.8.2) to compute $u_z(t, z)$, and use (7.8.3) and (7.8.4) to simplify the result, thereby showing that

$$u_z(t, z) = \left(1 + \frac{\sigma^2}{2r}\right) N\big(\delta_+(\tau, z)\big) + \left(1 - \frac{\sigma^2}{2r}\right) e^{-r\tau} z^{-\frac{2r}{\sigma^2}} N\big(-\delta_-(\tau, z^{-1})\big) - 1. \tag{7.8.19}$$

(iii) Use (7.8.19) and (7.8.2) to compute $u_{zz}(t, z)$, and use (7.8.3) and (7.8.4) to simplify the result, thereby showing that

$$u_{zz}(t, z) = \left(1 - \frac{2r}{\sigma^2}\right) e^{-r\tau} z^{-\frac{2r}{\sigma^2}-1} N\big(-\delta_-(\tau, z^{-1})\big) + \frac{2}{z\sigma\sqrt{\tau}} N'\big(\delta_+(\tau, z)\big). \tag{7.8.20}$$

(iv) Verify that $u(t, z)$ satisfies the Black-Scholes-Merton equation (7.4.18).

(v) Verify that $u(t, z)$ satisfies the boundary condition (7.4.20).

Exercise 7.6 (Boundary conditions for lookback option). The lookback option price $v(t, x, y)$ of (7.4.35) must satisfy the boundary conditions

(7.4.7)–(7.4.9). As we saw in Subsection 7.4.3, this is equivalent to the function $u(t,z)$ of (7.4.16) given by (7.4.36),

$$u(t,z) = \left(1 + \frac{\sigma^2}{2r}\right) zN\big(\delta_+(\tau,z)\big) + e^{-r\tau}N\big(-\delta_-(\tau,z)\big)$$
$$-\frac{\sigma^2}{2r}e^{-r\tau}z^{1-\frac{2r}{\sigma^2}}N\big(-\delta_-(\tau,z^{-1})\big) - z,\ 0 \le t < T,\ 0 < z \le 1,$$

satisfying the boundary conditions (7.4.19)–(7.4.21). This function was shown to satisfy boundary condition (7.4.20) in Exercise 7.5(v). Here we verify by direct computation that the limit of $u(t,z)$ as $z \downarrow 0$ satisfies (7.4.19) and the limit of $u(t,z)$ as $t \uparrow T$ ($\tau \downarrow 0$) satisfies (7.4.21).

(i) If you have not worked Exercise 7.2, then verify (7.8.11), the second equality in (7.8.14) and (7.8.16).
(ii) Use (7.8.11) and the second part of (7.8.14) to show that $\lim_{z\downarrow 0} u(t,z) = e^{-r\tau}$ for $0 \le t < T$.
(iii) Use (7.8.16) to show that $\lim_{\tau\downarrow 0} u(t,z) = 1 - z$ for $0 < z \le 1$.

Exercise 7.7 (Zero-strike Asian call). Consider a zero-strike Asian call whose payoff at time T is

$$V(T) = \frac{1}{T}\int_0^T S(u)\,du.$$

(i) Suppose at time t we have $S(t) = x \ge 0$ and $\int_0^t S(u)\,du = y \ge 0$. Use the fact that $e^{-ru}S(u)$ is a martingale under $\widetilde{\mathbb{P}}$ to compute

$$e^{-r(T-t)}\widetilde{\mathbb{E}}\left[\frac{1}{T}\int_0^T S(u)\,du \,\middle|\, \mathcal{F}(t)\right].$$

Call your answer $v(t,x,y)$.
(ii) Verify that the function $v(t,x,y)$ you obtained in (i) satisfies the Black-Scholes-Merton equation (7.5.8) and the boundary conditions (7.5.9) and (7.5.11) of Theorem 7.5.1. (We do not try to verify (7.5.10) because the computation of $v(t,x,y)$ outlined here works only for $y \ge 0$.)
(iii) Determine explicitly the process $\Delta(t) = v_x\big(t,S(t),Y(t)\big)$, and observe that it is not random.
(iv) Use the Itô-Doeblin formula to show that if you begin with initial capital $X(0) = v\big(0,S(0),0\big)$ and at each time you hold $\Delta(t)$ shares of the underlying asset, investing or borrowing at the interest rate r in order to do this, then at time T the value of your portfolio will be

$$X(T) = \frac{1}{T}\int_0^T S(u)\,du.$$

Exercise 7.8. Consider the continuously sampled Asian option of Subsection 7.5.3, but assume now that the interest rate is $r = 0$. Find an initial capital $X(0)$ and a nonrandom function $\gamma(t)$ to replace (7.5.22) so that

$$X(T) = \frac{1}{c} \int_{T-c}^{T} S(u)\,du - K \tag{7.5.27}$$

still holds. Give the formula for the resulting process $X(t)$, $0 \leq t \leq T$, to replace (7.5.24) and (7.5.26). With this function $\gamma(t)$ and process $X(t)$, Theorem 7.5.3 still holds.

Exercise 7.9. Let $g(t, y)$ be the function in Theorem 7.5.3. Then the value of the Asian option at time t is $V(t) = v(t, S(t), X(t))$, where $v(t, s, x) = sg(t, y)$ and $y = \frac{x}{s}$. The process $S(t)$ is given by (7.5.1). For the sake of specificity, we consider the case of continuous sampling with $r \neq 0$, so $\gamma(t)$ is given by (7.5.22) and $X(t)$ is given by (7.5.24) and (7.5.26).

(i) Verify the derivative formulas

$$\begin{aligned}
v_t(t, s, x) &= sg_t(t, y), \\
v_s(t, s, x) &= g(t, y) - yg_y(t, y), \\
v_x(t, s, x) &= g_y(t, y), \\
v_{ss}(t, s, x) &= \frac{y^2}{s} g_{yy}(t, y), \\
v_{sx}(t, s, x) &= -\frac{y}{s} g_{yy}(t, y), \\
v_{xx}(t, s, x) &= \frac{1}{s} g_{yy}(t, y).
\end{aligned}$$

(ii) Show that $e^{-rt} v\big(t, S(t), X(t)\big)$ is a martingale under $\widetilde{\mathbb{P}}$ by computing its differential, writing the differential in terms of dt and $d\widetilde{W}$, and verifying that the dt term is zero. (Hint: Use the fact that $g(t, y)$ satisfies (7.5.39).)

(iii) Suppose we begin with initial capital $v(0, S(0), X(0))$ and at each time t take a position $\Delta(t)$ in the risky asset, investing or borrowing at the interest rate r in order to finance this. We want to do this so that the portfolio value at the final time is $\left(\frac{1}{c} \int_{T-c}^{T} S(u)\,du - K\right)^+$. Give a formula for $\Delta(t)$ in terms of the function v and the processes $S(t)$ and $X(t)$. (Warning: The process $X(t)$ appearing in Theorem 7.5.3 and in this problem is not the value of the hedging portfolio. For example, $X(0)$ is given by (7.5.23), and this is different from $v(0, S(0), X(0))$, the initial value of the hedging portfolio.)

8

American Derivative Securities

8.1 Introduction

European option contracts specify an expiration date, and if the option is to be exercised at all, the exercise must occur on the expiration date. An option whose owner can choose to exercise at any time up to and including the expiration date is called *American*. Because of this early exercise feature, such an option is at least as valuable as its European counterpart. Sometimes the difference in value is negligible or even zero, and then American and European options are close or exact substitutes. We shall see in this chapter that the early exercise feature for a call on a stock paying no dividends is worthless; American and European calls on such a stock have the same price. In other cases, most notably put options, the value of this early exercise feature, the so-called *early exercise premium*, can be substantial. An intermediate option between American and European is *Bermudan*, an option that permits early exercise but only on a contractually specified finite set of dates.

Because an American option can be exercised at any time prior to its expiration, it can never be worth less than the payoff associated with immediate exercise. This is called the *intrinsic value* of the option.

In contrast to the case for a European option, whose discounted price process is a martingale under the risk-neutral measure, the discounted price process of an American option is a supermartingale under this measure. The holder of this option may fail to exercise at the optimal exercise date, and in this case the discounted option price has a tendency to fall; hence, the supermartingale property. During any period of time in which it is not optimal to exercise, however, the discounted price process behaves as a martingale.

To price an American option, just as with a European option, we could imagine selling the option in exchange for some initial capital and then consider how to use this capital to hedge the short position in the option. In this case, we would need to be ready to pay off the option at all times prior to the expiration date because we do not know when it will be exercised. We could determine when, from our point of view, is the worst time for the owner to exercise the option. From the owner's point of view, this would be the *optimal exercise time*, and we shall call it that. We could then compute the initial capital we need in order to be hedged against exercise at the optimal exercise time. Finally, we could show how to invest this capital so that we are hedged even if the owner exercises at a nonoptimal time. In the subsequent sections, we do all these things but begin the analysis at a different point than for European options. We define the price of American options using a risk-neutral pricing formula and then show that this price is the smallest initial capital that permits construction of the hedge just described.

For the binomial model, the program described above was carried out in Chapter 4 of Volume I. Here we revisit these matters in a continuous-time setting. We treat first the perpetual American put (Section 8.3), which is not actually traded. The analysis of this option provides lessons that we apply in the subsequent sections. In Section 8.4, we discuss the finite-expiration American put, an option that is traded. Section 8.5 treats the American call. In the case of a non-dividend-paying stock, we show that the American and European calls have the same price. However, if the stock pays dividends, these prices can differ. We show how to compute the American call price in this latter case.

8.2 Stopping Times

Throughout this chapter, we need the concept of stopping times. These were defined and discussed in the binomial model in Section 4.3 of Volume I. A stopping time is a random variable τ that takes values in $[0, \infty]$. The stopping times we shall encounter are the times at which an American option is exercised. The decision of an agent to exercise this option may depend on all the information available at that time but may not depend on future information. We provide a mathematical formulation of this property in Definition 8.2.1 below. Before stating this definition, we seek to motivate it.

In the N-period model of Volume I, where the filtration is generated by coin tossing and there are only finitely many dates, we defined a stopping time to be a random variable τ taking values $0, 1, \ldots, N$ or ∞ and having the property that if $\tau(\omega_1 \ldots \omega_n \omega_{n+1} \ldots \omega_N) = n$, then $\tau(\omega_1 \ldots \omega_n \omega'_{n+1} \ldots \omega'_N) = n$ for all $\omega'_{n+1} \ldots \omega'_N$. This condition guarantees that the decision to stop at time n does not depend on the coin tosses that come after time n.

One way to try to capture this same idea in continuous time is to require that for each nonrandom $t \geq 0$, the set $\{\tau = t\} = \{\omega \in \Omega; \tau(\omega) = t\}$ should be in $\mathcal{F}(t)$ (i.e., the agent stops (exercises the option) at time t based on the information available at time t). However, we shall be interested in sets of ωs of the form $\{\omega \in \Omega; T_1 \leq \tau(\omega) \leq T_2\}$, and these cannot be gotten by taking countable unions of sets of the form $\{\omega \in \Omega; \tau(\omega) = t\}$. Therefore, we impose the slightly stronger condition of Definition 8.2.1 below.

Definition 8.2.1. *A* stopping time τ *is a random variable taking values in* $[0, \infty]$ *and satisfying*

$$\{\tau \leq t\} \in \mathcal{F}(t) \text{ for all } t \geq 0. \tag{8.2.1}$$

Remark 8.2.2. Let $t \geq 0$ be given. Note that (8.2.1) and the properties of σ-algebras imply that $\left\{\tau > t - \frac{1}{n}\right\} = \{\tau \leq t - \frac{1}{n}\}^c \in \mathcal{F}\left(t - \frac{1}{n}\right)$ for all positive integers n. Since every set in $\mathcal{F}\left(t - \frac{1}{n}\right)$ is also in $\mathcal{F}(t)$, we conclude that $\left\{\tau > t - \frac{1}{n}\right\}$ is in $\mathcal{F}(t)$ for every n, and hence

$$\{\tau = t\} = \{\tau \leq t\} \cap \left(\bigcap_{n=1}^{\infty} \left\{ \tau > t - \frac{1}{n} \right\} \right)$$

is also in $\mathcal{F}(t)$. In other words, by Definition 8.2.1, a stopping time τ has the property that the decision to stop at time t must be based on information available at time t.

Example 8.2.3 (First passage time for a continuous process). Let $X(t)$ be an adapted process with continuous paths, let m be a number, and set

$$\tau_m = \min\{t \geq 0; X(t) = m\}. \tag{8.2.2}$$

This is the first time the process $X(t)$ reaches the level m. If $X(t)$ never reaches the level m, then we interpret τ_m to be ∞. Intuitively, τ_m must be a stopping time because the value of τ_m is determined by the path of $X(t)$ up to time τ_m. An agent can exercise an option the first time the underlying asset price reaches a level; this exercise strategy does not require information about the underlying price movements after the exercise time.

We use Definition 8.2.1 and the properties of σ-algebras to show mathematically that τ_m is a stopping time. Let $t \geq 0$ be given. We need to show that $\{\tau \leq t\}$ is in $\mathcal{F}(t)$.

If $t = 0$, then $\{\tau \leq t\} = \{\tau = 0\}$ is either Ω or $\emptyset$, depending on whether $X(0) = m$ or $X(0) \neq m$. In either case, $\{\tau \leq 0\} \in \mathcal{F}(0)$.

We consider the case $t > 0$. Suppose $\omega \in \Omega$ satisfies $\tau(\omega) \leq t$. Then there is some number $s \leq t$ such that $X(s, \omega) = m$, where we indicate explicitly the dependence of X on ω. For each positive integer n, there is an open interval of time containing s for which the process X is in $\left(m - \frac{1}{n}, m + \frac{1}{n}\right)$. In this interval, there is a rational number $q \leq s \leq t$. Therefore, ω is in the set

$$A = \bigcap_{n=1}^{\infty} \bigcup_{0 \leq q \leq t, q \text{ rational}} \left\{ m - \frac{1}{n} < X(q) < m + \frac{1}{n} \right\}.$$

We have shown that $\{\tau \leq t\} \subset A$.

On the other hand, if $\omega \in A$, then for every positive integer n there is a rational number $q_n \leq t$ such that

$$m - \frac{1}{n} < X(q_n, \omega) < m + \frac{1}{n}.$$

The infinite sequence $\{q_n\}_{n=1}^{\infty}$ must have an accumulation point s in the closed, bounded interval $[0, t]$. In other words, there must exist a number $s \in [0, t]$ and a subsequence $\{q_{n_k}\}_{k=1}^{n}$ such that $\lim_{k \to \infty} q_{n_k} = s$. But

$$m - \frac{1}{n_k} < X(q_{n_k}, \omega) < m + \frac{1}{n_k} \text{ for all } k = 1, 2, \ldots.$$

Letting $k \to \infty$ in these inequalities and using the fact that X has continuous paths, we see that $X(s, \omega) = m$. It follows that $\tau(\omega) \leq t$. We have shown that $A \subset \{\tau \leq t\}$. Therefore $A = \{\tau \leq t\}$.

Because X is adapted to the filtration, for each positive integer n and rational $q \in [0, t]$, the set

$$\left\{ m - \frac{1}{n} < X(q) < m + \frac{1}{n} \right\}$$

is in $\mathcal{F}(q)$ and hence in the larger σ-algebra $\mathcal{F}(t)$. Because there are only countably many rational numbers q in $[0, t]$, they can be arranged in a sequence, and the union

$$B_n = \bigcup_{0 \leq q \leq t, q \text{ rational}} \left\{ m - \frac{1}{n} < X(q) < m + \frac{1}{n} \right\}$$

is really a union of a sequence of sets in $\mathcal{F}(t)$. The set B_n must therefore also be in $\mathcal{F}(t)$. Because B_n is in $\mathcal{F}(t)$ for every positive integer n, the intersection $\cap_{n=1}^{\infty} B_n = A$ is also in $\mathcal{F}(t)$. We have already shown that $A = \{\tau \leq t\}$. We conclude that $\{\tau \leq t\} \in \mathcal{F}(t)$. □

Suppose now that we have an adapted process $X(t)$ and a stopping time τ. We define the *stopped process* $X(t \wedge \tau)$, where $\wedge$ denotes the minimum of two quantities (i.e., $t \wedge \tau = \min\{t, \tau\}$). The stopped process $X(t \wedge \tau)$ agrees with $X(t)$ up to time τ, and thereafter it is frozen at the value of $X(\tau)$. See Figure 8.2.1.

Theorem 8.2.4 (Optional sampling). *A martingale stopped at a stopping time is a martingale. A supermartingale (or submartingale) stopped at a stopping time is a supermartingale (or submartingale, respectively).*

While the proof of Theorem 8.2.4 is technical and will not be given here, the intuition is clear. If $M(t)$ is a martingale, then the stopped process $M(t \wedge \tau)$ agrees with $M(t)$ before time τ and thus is also a martingale. After time τ, the stopped process is frozen (i.e., it no longer changes with time), and this is a trivial martingale. A martingale goes neither up nor down "on average." After being frozen, a process goes neither up nor down, path-by-path. The only way the martingale property could be violated is if the stopping decision

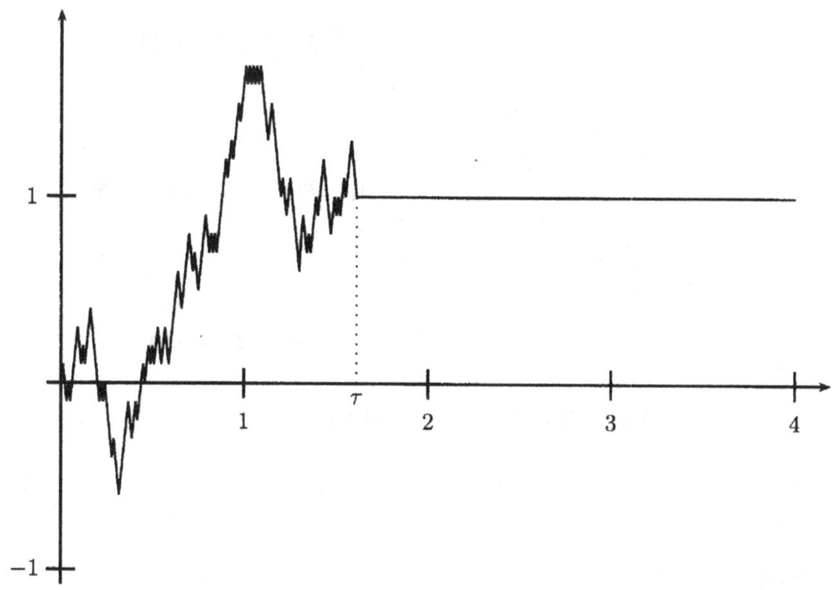

Fig. 8.2.1. A stopped process.

looked ahead. Suppose that a martingale is stopped (frozen) if it will go up in the near future but is allowed to continue if it will go down. Then stopping introduces a downward bias by removing the upward possibility. Figure 8.2.2 shows a martingale in a discrete-time model under the assumption that the probability of H (an up move) is $\tilde{p} = \frac{1}{2}$ and the probability of T (a down move) is $\tilde{q} = \frac{1}{2}$. Figures 8.2.2–8.2.4 are taken from Section 4.3 of Volume I, where the martingale in Figure 8.2.2 is a discounted stock price under a risk-neutral measure. Figure 8.2.3 shows a random time ρ that is not a stopping time; this random time ρ causes stopping at time 0 if there is an H on the first toss (an up move) but lets the process continue if there is a T on the first toss. Similarly, if there is a T on the first toss and an H on the second toss, ρ stops the martingale at time 1 but lets it continue to time 2 if there is a T on the first toss and an H on the second toss. The stopped martingale is shown in Figure 8.2.4, and it is not a martingale. For example,

$$\widetilde{\mathbb{E}}M_{2\wedge\rho} = \frac{1}{4}(4 + 4 + 1.60 + 0.64) = 2.56 < M_0 = 4,$$

whereas the expectation of a martingale does not change over time. Our definition of stopping time rules out this kind of stopping.

Similar intuition applies to supermartingales. A stopped supermartingale is a supermartingale before being frozen, and after being frozen it is a martingale, which is a special case of a supermartingale. The situation with submartingales is analogous. Again, the stopping must be done at a stopping time.

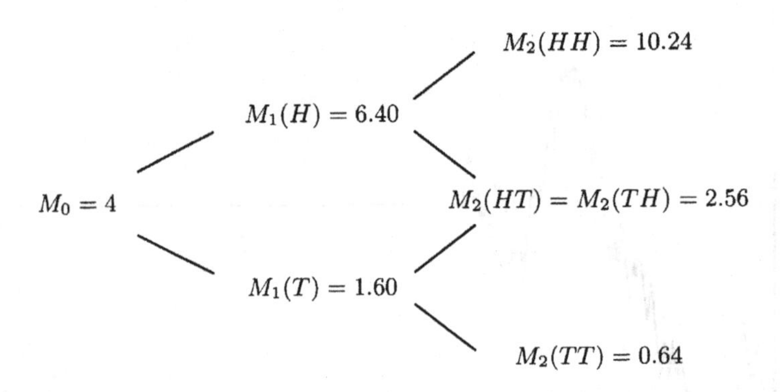

Fig. 8.2.2. Martingale under $\tilde{p} = \tilde{q} = \frac{1}{2}$.

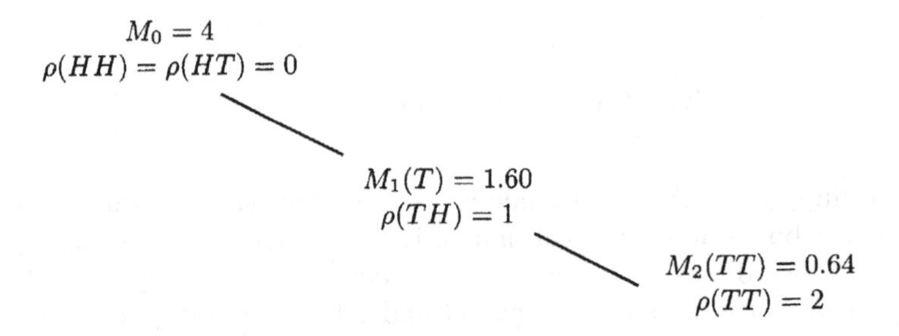

Fig. 8.2.3. Non-stopping time ρ.

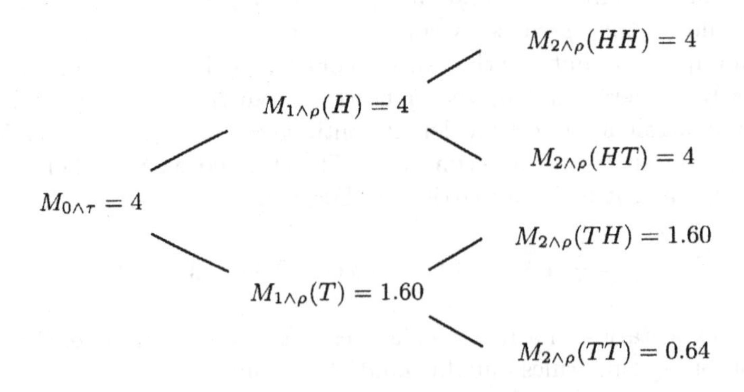

Fig. 8.2.4. Martingale stopped at the non-stopping time ρ.

Looking ahead to make the stopping decision can ruin the supermartingale (respectively, submartingale) property.

8.3 Perpetual American Put

The simplest interesting American option is the *perpetual American put.* It is interesting because the optimal exercise policy is not obvious, and it is simple because this policy can be determined explicitly. Although this is not a traded option, we begin our discussion with it in order to present in a simple context the ideas behind the subsequent analysis of more realistic options.

The underlying asset in most of this chapter (except in Subsection 8.5.2, where the asset pays dividends) has the price process $S(t)$ given by

$$dS(t) = rS(t)\,dt + \sigma S(t)\,d\widetilde{W}(t), \tag{8.3.1}$$

where the interest rate r and the volatility σ are strictly positive constants and $\widetilde{W}(t)$ is a Brownian motion under the risk-neutral probability measure $\widetilde{\mathbb{P}}$. The perpetual American put pays $K - S(t)$ if it is exercised at time t. This is its intrinsic value.

Definition 8.3.1. *Let $\mathcal{T}$ be the set of all stopping times. The* price of the perpetual American put *is defined to be*

$$v_*(x) = \max_{\tau\in\mathcal{T}} \widetilde{\mathbb{E}}\left[e^{-r\tau}(K - S(\tau))\right], \tag{8.3.2}$$

where $x = S(0)$ in (8.3.2) is the initial stock price. In the event that $\tau = \infty$, we interpret $e^{-r\tau}(K - S(\tau))$ to be zero.

The idea behind Definition 8.3.1 is that the owner of the perpetual American put can choose an exercise time τ, subject only to the condition that she may not look ahead to determine when to exercise. The mathematical formulation of this "not look ahead" restriction is that τ must be a stopping time. The price of the option at time zero is the risk-neutral expected payoff of the option, discounted from the exercise time back to time zero. If the option is never exercised, its payoff is zero. This explains the term under the expectation on the right-hand side of (8.3.2). The owner of the option should choose the exercise strategy that maximizes this expected payoff, discounted back to time zero, and thus we define the price of the option to be the maximum over $\tau \in \mathcal{T}$ of the discounted expected payoffs.

This risk-neutral pricing definition of the perpetual American put price appears to differ from the construction of the price of a European call in Section 4.5. There we took the price to be the initial capital required by an agent holding a short position in the option in order for this agent to hedge the short position (i.e., invest in the stock and money market account in such a way that at expiration of the option the resulting portfolio value is the payoff of the option). It turns out that $v_*(x)$ defined above is the initial capital required for an agent to hedge a short position in the American put *regardless of the exercise strategy* τ *used by the owner of the put*; see Corollaries 8.3.6 and 8.3.7.

The owner of the perpetual American put can exercise at any time. In particular, there is no expiration date after which the put can no longer be exercised. This makes every date like every other date; the time remaining to expiration is always the same (i.e., infinity). Because every date is like every other date, it is reasonable to expect that the optimal exercise policy depends only on the value of $S(t)$ and not on the time variable t. The owner of the put should exercise as soon as $S(t)$ falls "far enough" below K. In other words, it is reasonable to expect that the optimal exercise policy is of the form

"Exercise the put as soon as $S(t)$ falls to the level L_*."

We have two questions to answer:

(i) What is the value of L_* and how do we know it corresponds to optimal exercise?
(ii) What is the value of the put?

For the perpetual American put, we can base the answers to these questions on explicit computations.

8.3.1 Price Under Arbitrary Exercise

Theorem 8.3.2 (Laplace transform for first passage time of drifted Brownian motion). *Let $\widetilde{W}(t)$ be a Brownian motion under a probability measure $\widetilde{\mathbb{P}}$, let μ be a real number, and let m be a positive number. Define $X(t) = \mu t + \widetilde{W}(t)$, and set*

$$\tau_m = \min\{t \geq 0; X(t) = m\},$$

so that τ_m is the stopping time of Example 8.2.3. If $X(t)$ never reaches the level m, then we interpret τ_m to be ∞. Then

$$\widetilde{\mathbb{E}} e^{-\lambda \tau_m} = e^{-m(-\mu + \sqrt{\mu^2 + 2\lambda})} \text{ for all } \lambda > 0, \tag{8.3.3}$$

where we interpret $e^{-\lambda \tau_m}$ to be zero if $\tau_m = \infty$.

PROOF: Define $\sigma = -\mu + \sqrt{\mu^2 + 2\lambda}$ so that $\sigma > 0$ and

$$\begin{aligned}
\sigma\mu + \frac{1}{2}\sigma^2 &= -\mu^2 + \mu\sqrt{\mu^2 + 2\lambda} + \frac{1}{2}\left(-\mu + \sqrt{\mu^2 + 2\lambda}\right)^2 \\
&= -\mu^2 + \mu\sqrt{\mu^2 + 2\lambda} + \frac{1}{2}\mu^2 - \mu\sqrt{\mu^2 + 2\lambda} + \frac{1}{2}\mu^2 + \lambda \\
&= \lambda.
\end{aligned}$$

Then

$$e^{\sigma X(t)-\lambda t} = e^{\sigma\mu t+\sigma\widetilde{W}(t)-\sigma\mu t-\frac{1}{2}\sigma^2 t} = e^{\sigma\widetilde{W}(t)-\frac{1}{2}\sigma^2 t},$$

which is a martingale under $\widetilde{\mathbb{P}}$ (its differential has a $d\widetilde{W}(t)$ term and no dt term). According to Theorem 8.2.4 (optional sampling), the stopped martingale

$$M(t) = e^{\sigma\widetilde{W}(t\wedge\tau_m)-\frac{1}{2}\sigma^2(t\wedge\tau_m)}$$

is also a martingale. Therefore, for each positive integer n,

$$\begin{aligned} 1 = M(0) &= \widetilde{\mathbb{E}}M(n) \\ &= \widetilde{\mathbb{E}}\left[e^{\sigma X(n\wedge\tau_m)-\lambda(n\wedge\tau_m)}\right] \\ &= \widetilde{\mathbb{E}}\left[e^{\sigma m-\lambda\tau_m}\mathbb{I}_{\{\tau_m\leq n\}}\right] + \widetilde{\mathbb{E}}\left[e^{\sigma X(n)-\lambda n}\mathbb{I}_{\{\tau_m>n\}}\right]. \end{aligned} \tag{8.3.4}$$

The nonnegative random variables $e^{\sigma m-\lambda\tau_m}\mathbb{I}_{\{\tau_m\leq n\}}$ increase with n, and their limit is $e^{\sigma m-\lambda\tau_m}\mathbb{I}_{\{\tau_m<\infty\}}$. In other words,

$$0 \leq e^{\sigma m-\lambda\tau_m}\mathbb{I}_{\{\tau_m\leq 1\}} \leq e^{\sigma m-\lambda\tau_m}\mathbb{I}_{\{\tau_m\leq 2\}} \leq \ldots \text{ almost surely,}$$

and

$$\lim_{n\to\infty} e^{\sigma m-\lambda\tau_m}\mathbb{I}_{\{\tau_m\leq n\}} = e^{\sigma m-\lambda\tau_m}\mathbb{I}_{\{\tau_m<\infty\}} \text{ almost surely.}$$

The Monotone Convergence Theorem, Theorem 1.4.5, implies

$$\lim_{n\to\infty}\widetilde{\mathbb{E}}\left[e^{\sigma m-\lambda\tau_m}\mathbb{I}_{\{\tau_m\leq n\}}\right] = \widetilde{\mathbb{E}}\left[e^{\sigma m-\lambda\tau_m}\mathbb{I}_{\{\tau_m<\infty\}}\right]. \tag{8.3.5}$$

On the other hand, the random variable $e^{\sigma X(n)-\lambda n}\mathbb{I}_{\{\tau_m>n\}}$ satisfies

$$0 \leq e^{\sigma X(n)-\lambda n}\mathbb{I}_{\{\tau_m>n\}} \leq e^{\sigma m-\lambda n} \leq e^{\sigma m} \text{ almost surely}$$

because $X(n) \leq m$ for $n < \tau_m$ and σ is positive. Because λ is positive, we have

$$\lim_{n\to\infty} e^{\sigma X(n)-\lambda n}\mathbb{I}_{\{\tau_m>n\}} \leq \lim_{n\to\infty} e^{\sigma m-\lambda n} = 0.$$

According to the Dominated Convergence Theorem, Theorem 1.4.9,

$$\lim_{n\to\infty}\mathbb{E}\left[e^{\sigma X(n)-\lambda n}\mathbb{I}_{\{\tau_m>n\}}\right] = 0. \tag{8.3.6}$$

Taking the limit in (8.3.4) and using (8.3.5) and (8.3.6), we obtain

$$1 = \widetilde{\mathbb{E}}\left[e^{\sigma m-\lambda\tau_m}\mathbb{I}_{\{\tau_m<\infty\}}\right]$$

or, equivalently,

$$\widetilde{\mathbb{E}}\left[e^{-\lambda\tau_m}\mathbb{I}_{\{\tau_m<\infty\}}\right] = e^{-\sigma m} = e^{-m(-\mu+\sqrt{\mu^2+2\lambda})} \text{ for all } \lambda > 0. \tag{8.3.7}$$

This is (8.3.3) when we interpret $e^{-\lambda\tau_m}$ to be zero if $\tau_m = \infty$. □

Remark 8.3.3. We used the strict positivity of λ to derive (8.3.7), but now that we have it, we can take the limit as $\lambda \downarrow 0$. The random variables $e^{-\lambda\tau_m}\mathbb{I}_{\{\tau_m<\infty\}}$ are nonnegative and increase to $\mathbb{I}_{\{\tau_m<\infty\}}$ as $\lambda \downarrow 0$, and the Monotone Convergence Theorem allows us to conclude that

$$\widetilde{\mathbb{P}}\{\tau_m < \infty\} = \widetilde{\mathbb{E}}\mathbb{I}_{\{\tau_m<\infty\}} = \lim_{\lambda\downarrow 0} e^{-m(-\mu+\sqrt{\mu^2+2\lambda})} = e^{m\mu - m|\mu|}.$$

If $\mu \geq 0$, the drift in $X(t)$ is zero or upward, toward level m, and $\widetilde{\mathbb{P}}\{\tau_m < \infty\} = 1$; the level $X(t)$ is reached with probability one. On the other hand, if $\mu < 0$, the drift in $X(t)$ is downward, away from level m, and $\widetilde{\mathbb{P}}\{\tau_m < \infty\} = e^{-2m|\mu|} < 1$; there is a positive probability of never reaching m. □

The solution to (8.3.1) is

$$S(t) = S(0)\exp\left\{\sigma\widetilde{W}(t) + \left(r - \frac{1}{2}\sigma^2\right)t\right\}. \tag{8.3.8}$$

Suppose the owner of the perpetual American put sets a positive level $L < K$ and resolves to exercise the put the first time the stock price falls to L. If the initial stock price is at or below L, she exercises immediately (at time zero). The value of the put in this case is $v_L(S(0)) = K - S(0)$. If the initial stock price is above L, she exercises at the stopping time

$$\tau_L = \min\{t \geq 0; S(t) = L\}, \tag{8.3.9}$$

where τ_L is set equal to ∞ if the stock price never reaches the level L. At the time of exercise, the put pays $K - S(\tau_L) = K - L$. Discounting this back to time zero and taking the risk-neutral expected value, we compute the value of the put under this exercise strategy to be

$$v_L(S(0)) = (K - L)\widetilde{\mathbb{E}}e^{-r\tau_L} \text{ for all } S(0) \geq L. \tag{8.3.10}$$

On those paths where $\tau_L = \infty$, we interpret $e^{-r\tau_L}$ to be zero. (Recall our assumption at the beginning of this section that r is strictly positive.) Although not explicitly indicated by the notation, the distribution of τ_L depends on the initial stock price $S(0)$, so the right-hand side (8.3.10) is a function of $S(0)$.

Lemma 8.3.4. *The function $v_L(x)$ is given by the formula*

$$v_L(x) = \begin{cases} K - x, & 0 \leq x \leq L, \\ (K - L)\left(\frac{x}{L}\right)^{-\frac{2r}{\sigma^2}}, & x \geq L. \end{cases} \tag{8.3.11}$$

PROOF: We only need to establish the second line of (8.3.11). If $x = L$, then $\tau_L = 0$ and (8.3.10) implies $v_L(x) = K - L$.

We consider the case $S(0) = x > L$. The stopping time τ_L is the first time

$$S(t) = x\exp\left\{\sigma\widetilde{W}(t) + \left(r - \frac{1}{2}\sigma^2\right)\right\}$$

reaches the level L. But $S(t) = L$ if and only if

$$-\widetilde{W}(t) - \frac{1}{\sigma}\left(r - \frac{1}{2}\sigma^2\right)t = \frac{1}{\sigma}\log\frac{x}{L}.$$

We now apply Theorem 8.3.2 with $X(t)$ in that theorem replaced by $-\widetilde{W}(t) - \frac{1}{\sigma}\left(r - \frac{1}{2}\sigma^2\right)t$ (the processes $\widetilde{W}(t)$ and $-\widetilde{W}(t)$ are both Brownian motions under $\widetilde{\mathbb{P}}$), with λ replaced by r, with μ replaced by $-\frac{1}{\sigma}\left(r - \frac{1}{2}\sigma^2\right)$, and with m replaced by $\frac{1}{\sigma}\log\frac{x}{L}$, which is positive. With these replacements, τ_m in Theorem 8.3.2 is τ_L and

$$\begin{aligned}\mu^2 + 2\lambda &= \frac{1}{\sigma^2}\left(r^2 - r\sigma^2 + \frac{1}{4}\sigma^4\right) + 2r\\ &= \frac{1}{\sigma^2}\left(r^2 + r\sigma^2 + \frac{1}{4}\sigma^4\right)\\ &= \frac{1}{\sigma^2}\left(r + \frac{1}{2}\sigma^2\right)^2.\end{aligned}$$

Therefore,

$$-\mu + \sqrt{\mu^2 + 2\lambda} = \frac{1}{\sigma}\left(r - \frac{1}{2}\sigma^2\right) + \frac{1}{\sigma}\left(r + \frac{1}{2}\sigma^2\right) = \frac{2r}{\sigma}.$$

Equation (8.3.3) implies

$$\widetilde{\mathbb{E}}^{-r\tau_L} = \exp\left\{-\frac{1}{\sigma}\log\frac{x}{L}\cdot\frac{2r}{\sigma}\right\} = \left(\frac{x}{L}\right)^{-\frac{2r}{\sigma^2}}.$$

The second line in (8.3.11) follows. □

8.3.2 Price Under Optimal Exercise

Figure 8.3.1 shows the function $v_L(x)$ for three different values of L. The function $v_{L_1}(x)$ in that figure actually lies below the intrinsic value $K - x$ for x between L_1 and L_2. If the initial stock price is between L_1 and L_2, then the strategy of exercising the first time the stock price falls to L_1 is obviously a poor one; it would be better to exercise at time zero and receive the intrinsic value. The function $v_{L_2}(x)$ agrees with the intrinsic value for $0 \le x \le L_2$ and follows the indicated curve for $x \ge L_2$. The function $v_{L_*}(x)$ agrees with the intrinsic value for $0 \le x \le L_*$ and follows the indicated curve for $x \ge L_*$. For $x \ge L_*$, the function $v_{L_*}(x)$ is strictly larger than the function $v_{L_2}(x)$, and hence the strategy of exercising the first time the stock price falls to L_* is better than exercising the first time the stock price falls to L_2.

As Figure 8.3.1 suggests, for any value of L smaller than L_*, the function $v_L(x)$ agrees with the intrinsic value for $0 \le x \le L$, lies below the intrinsic

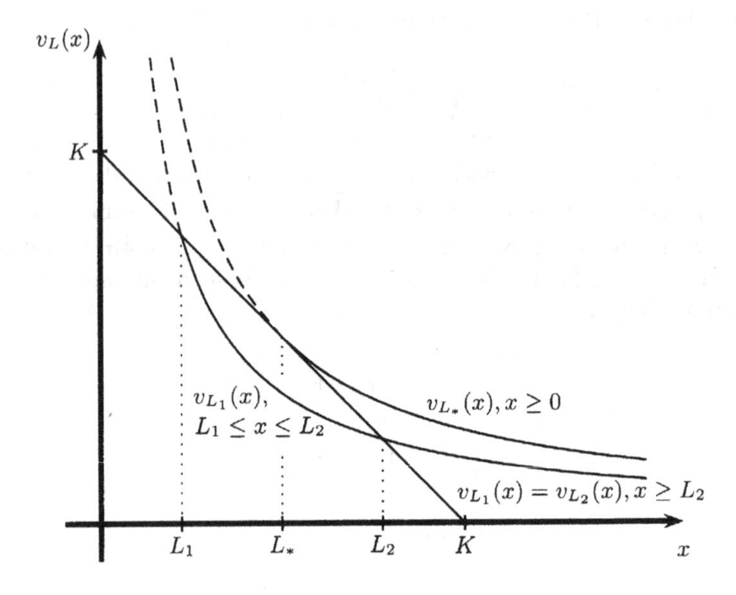

Fig. 8.3.1. $(K - L)\left(\frac{x}{L}\right)^{-\frac{2r}{\sigma^2}}$ for three values of L.

value immediately to the right of L, and lies below $v_{L_*}(x)$ everywhere to the right of L. For any value of L larger than L_*, the function $v_L(x)$ agrees with the intrinsic value for $0 \le x \le L$ and lies below $v_{L_*}(x)$ for all $x \ge L_*$. Thus, among those exercise policies of the form

"Exercise the put as soon as $S(t)$ falls to the level L,"

the best one is obtained by choosing $L = L_*$. We expect therefore that $v_{L_*}(x)$ is the price of the put $v_*(x)$ of Definition 8.3.1. We prove this below.

We must first determine the value of L_*. We note that

$$v_L(x) = (K - L)L^{\frac{2r}{\sigma^2}} x^{-\frac{2r}{\sigma^2}} \text{ for all } x \ge L.$$

From Figure 8.3.1, we know that L_* is the value of L that maximizes this quantity when we hold x fixed. We thus define

$$g(L) = (K - L)L^{\frac{2r}{\sigma^2}}$$

and seek the value of L that maximizes this function over $L \ge 0$. Because $\frac{2r}{\sigma^2}$ is strictly positive, we have $g(0) = 0$ and $\lim_{L\to\infty} g(L) = -\infty$. Moreover,

$$g'(L) = -L^{\frac{2r}{\sigma^2}} + \frac{2r}{\sigma^2}(K - L)L^{\frac{2r}{\sigma^2}-1} = -\frac{2r + \sigma^2}{\sigma^2} L^{\frac{2r}{\sigma^2}} + \frac{2r}{\sigma^2} K L^{\frac{2r}{\sigma^2}-1}.$$

Setting this equal to zero, we solve for

$$L_* = \frac{2r}{2r+\sigma^2}K. \tag{8.3.12}$$

This is a number between 0 and K. Furthermore,

$$g(L_*) = \frac{\sigma^2}{2r+\sigma^2}\left(\frac{2r}{2r+\sigma^2}\right)^{\frac{2r}{\sigma^2}} K^{\frac{2r+\sigma^2}{\sigma^2}}$$

is strictly positive. Therefore, the graph of $y = g(L)$ must be as shown in Figure 8.3.2, and L_* given by (8.3.12) is the point where $g(L)$ attains its maximum.

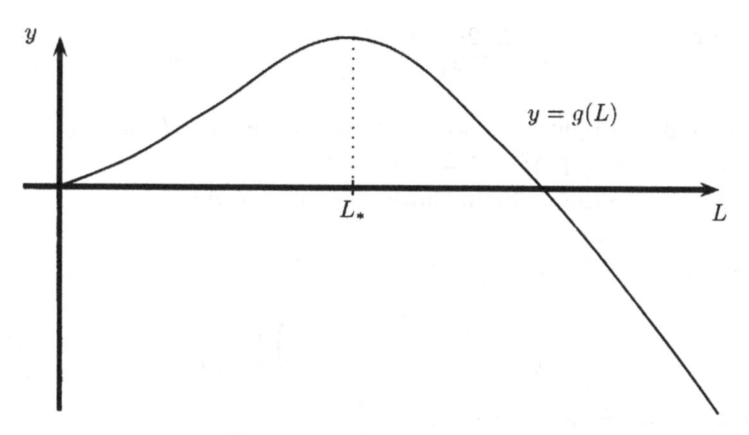

Fig. 8.3.2. Graph of $g(L)$.

8.3.3 Analytical Characterization of the Put Price

We have

$$v_{L_*}(x) = \begin{cases} K - x, & 0 \le x \le L_*, \\ (K - L_*)\left(\frac{x}{L_*}\right)^{-\frac{2r}{\sigma^2}}, & x \ge L_*, \end{cases} \tag{8.3.13}$$

so that

$$v'_{L_*}(x) = \begin{cases} -1, & 0 \le x \le L_*, \\ -(K - L_*)\frac{2r}{\sigma^2 x}\left(\frac{x}{L_*}\right)^{-\frac{2r}{\sigma^2}}, & x \ge L_*. \end{cases} \tag{8.3.14}$$

If we evaluate the second line in (8.3.14) at $x = L_*$, we get the right-hand derivative

$$v'_{L_*}(L_*+) = -\frac{2r}{\sigma^2 L_*}(K - L_*) = -\frac{2rK}{\sigma^2 L_*} + \frac{2r}{\sigma^2} = -\frac{2r}{\sigma^2}\cdot\frac{2r+\sigma^2}{2r} + \frac{2r}{\sigma^2} = -1,$$

which agrees with the left-hand derivative $v'_{L_*}(L_*-) = -1$ provided by the first line in (8.3.14). The derivative of $v_{L_*}(x)$ is continuous at $x = L_*$. This is known as *smooth pasting.* The two parts of the definition of $v_{L_*}(x)$ fit together at $x = L_*$ so that both $v_{L_*}(x)$ and $v'_{L_*}(x)$ are continuous. This is because the graph of the function $y = (K - L_*)(\frac{x}{L_*})^{-\frac{2r}{\sigma^2}}$ is tangent to the line $y = K - x$ at $x = L_*$, as one can see from Figure 8.3.1. In fact, we could have used the smooth pasting condition to solve for L_* (see Exercise 8.1).

The second derivative of $v(x)$ has a jump at $x = L_*$, and hence is undefined at this point. Indeed,

$$v''_{L_*}(x) = \begin{cases} 0, & 0 \le x < L_*, \\ (K - L_*)\dfrac{2r(2r+\sigma^2)}{\sigma^4 x^2}\left(\dfrac{x}{L_*}\right)^{-\frac{2r}{\sigma^2}}, & x > L_*. \end{cases} \tag{8.3.15}$$

The left-hand and right-hand second derivatives at $x = L_*$ are $v(L_*-) = 0$ and $v''(L_*+) = (K - L_*)\frac{2r(2r+\sigma^2)}{\sigma^4 L_*^2} > 0$.

For $x > L_*$, we can verify by direct computation that

$$\begin{aligned} & rv_{L_*}(x) - rxv'_{L_*}(x) - \frac{1}{2}\sigma^2 x^2 v''_{L_*}(x) \\ &= (K - L_*)\left(r + \frac{2r^2}{\sigma^2} - \frac{r(2r+\sigma^2)}{\sigma^2}\right)\left(\frac{x}{L_*}\right)^{-\frac{2r}{\sigma^2}} = 0. \end{aligned} \tag{8.3.16}$$

On the other hand, for $0 \le x < L_*$,

$$rv_{L_*}(x) - rxv'_{L_*}(x) - \frac{1}{2}\sigma^2 x^2 v''_{L_*}(x) = r(K - x) + rx = rK. \tag{8.3.17}$$

In particular, we see that $v_{L_*}(x)$ satisfies the so-called *linear complementarity conditions*

$$v(x) \ge (K - x)^+ \text{ for all } x \ge 0, \tag{8.3.18}$$

$$rv(x) - rxv'(x) - \frac{1}{2}\sigma^2 x^2 v''(x) \ge 0 \text{ for all } x \ge 0, \text{ and} \tag{8.3.19}$$

$$\text{for each } x \ge 0, \text{ equality holds in either (8.3.18) or (8.3.19).} \tag{8.3.20}$$

The point L_* is slightly problematical in (8.3.19) since $v''_{L_*}(L_*)$ is undefined. However, if we replace $v''_{L_*}(L_*)$ in (8.3.19) by either $v''_{L_*}(L_*-)$ or $v''_{L_*}(L_*+)$, the inequality holds.

The linear complementarity conditions (8.3.18)–(8.3.20) determine the function $v_{L_*}(x)$. More precisely, the function $v_{L_*}(x)$ given by (8.3.13) is the only bounded continuous function having a continuous derivative that satisfies these conditions; see Exercise 8.3.

8.3.4 Probabilistic Characterization of the Put Price

Theorem 8.3.5. *Let $S(t)$ be the stock price given by (8.3.1) and let τ_{L_*} be given by (8.3.9) with $L = L_*$. Then $e^{-rt}v_{L_*}(S(t))$ is a supermartingale under $\widetilde{\mathbb{P}}$, and the stopped process $e^{-r(t\wedge\tau_{L_*})}v_{L_*}(S(t\wedge\tau_{L_*}))$ is a martingale.*

PROOF: Fortunately, the Itô-Doeblin formula applies to functions whose second derivatives have jumps, provided the first derivative is continuous (see Exercise 4.20 for a discussion related to this). We may thus compute

$$\begin{aligned} &d\big[e^{-rt}v_{L_*}(S(t))\big]\\ &\quad= e^{-rt}\left[-rv_{L_*}(S(t))\,dt + v'_{L_*}(S(t))\,dS(t) + \frac{1}{2}v''_{L_*}(S(t))\,dS(t)\,dS(t)\right]\\ &\quad= e^{-rt}\left[-rv_{L_*}(S(t)) + rS(t)v'_{L_*}(S(t)) + \frac{1}{2}\sigma^2S^2(t)v''_{L_*}(S(t))\right]dt\\ &\qquad\quad+e^{-rt}\sigma S(t)v'_{L_*}(S(t))\,d\widetilde{W}(t). \end{aligned}$$

Because of (8.3.16) and (8.3.17), the dt term in this expression is either 0 or $-rK$, depending on whether $S(t) > L_*$ or $S(t) < L_*$. If $S(t) = L_*$, $v''_{L_*}(S(t))$ is undefined, but the probability $S(t) = L_*$ is zero so this does not matter. We thus have

$$d\big[e^{-rt}v_{L_*}(S(t))\big] = -e^{-rt}rK\mathbb{I}_{\{S(t)<L^*\}}\,dt + e^{-rt}\sigma S(t)v'_{L_*}(S(t))\,d\widetilde{W}(t). \tag{8.3.21}$$

Because the dt term in (8.3.21) is less than or equal to zero, $e^{-rt}v_{L_*}(S(t))$ is a supermartingale; when $S(t) < L_*$ it has a downward tendency. If the initial stock price is above L_*, then prior to the time τ_{L_*} when the stock price first reaches L_*, the dt term in (8.3.21) is zero and hence $e^{-r(t\wedge\tau_{L_*})}v(S(t\wedge\tau_{L_*}))$ is a martingale. Indeed, integration of (8.3.21) yields

$$e^{-r(t\wedge\tau_{L_*})}v_{L_*}(S(t\wedge\tau_{L_*})) = v_{L_*}(0) + \int_0^{t\wedge\tau_{L_*}} e^{-ru}\sigma S(u)v'_{L_*}(S(u))\,d\widetilde{W}(u).$$

Itô integrals are martingales, and hence the Itô integral above stopped at the stopping time τ_{L_*} is a martingale. □

Corollary 8.3.6. *Recall that $\mathcal{T}$ is the set of all stopping times, not just those of the form (8.3.9). We have*

$$v_{L_*}(x) = \max_{\tau\in\mathcal{T}}\widetilde{\mathbb{E}}\left[e^{-r\tau}(K - S(\tau))\right],$$

where $x = S(0)$ is the initial stock price. In other words, $v_{L_*}(x)$ is the perpetual American put price of Definition 8.3.1.

PROOF: Because $e^{-rt}v_{L_*}(S(t))$ is a supermartingale under $\widetilde{\mathbb{P}}$, we have from Theorem 8.2.4 (optional sampling) that, for every stopping time $\tau\in\mathcal{T}$,

$$v_{L_*}(x) = v_{L_*}(S(0)) \geq \widetilde{\mathbb{E}}\left[e^{-r(t\wedge\tau)}v_{L_*}(S(t\wedge\tau))\right]. \tag{8.3.22}$$

Because $v_{L_*}(S(t\wedge\tau))$ is bounded, we may let $t\to\infty$ in (8.3.22), using the Dominated Convergence Theorem, Theorem 1.4.9, to conclude that

$$v_{L_*}(x) \geq \widetilde{\mathbb{E}}\left[e^{-r\tau}v_{L_*}(S(\tau))\right] \geq \widetilde{\mathbb{E}}\left[e^{-r\tau}(K-S(\tau))\right],$$

where we have gotten the last inequality from (8.3.18). Because this inequality holds for every $\tau\in\mathcal{T}$, we have

$$v_{L_*}(x) \geq \max_{\tau\in\mathcal{T}}\widetilde{\mathbb{E}}\left[e^{-r\tau}(K-S(\tau))\right].$$

On the other hand, if we replace τ by τ_{L_*}, we obtain equality in (8.3.22) because $e^{-r(t\wedge\tau_{L_*})}v(S(t\wedge\tau_{L_*}))$ is a martingale under $\widetilde{\mathbb{P}}$. Letting $t\to\infty$ and using the Dominated Convergence Theorem, we obtain

$$v_{L_*}(x) = \widetilde{\mathbb{E}}\left[e^{-r\tau_{L_*}}v_{L_*}(S(\tau_{L_*}))\right].$$

Since

$$e^{-r\tau_{L_*}}v_{L_*}(S(\tau_{L_*})) = e^{-r\tau_{L_*}}v_{L_*}(L_*) = e^{-r\tau_{L_*}}(K-L_*) = e^{-r\tau_{L_*}}(K-S(\tau_{L_*}))$$

if $\tau_{L_*}<\infty$ (and is interpreted to be zero if $\tau_{L_*}=\infty$), we see that

$$v_{L_*}(x) = \widetilde{\mathbb{E}}\left[e^{-r\tau_{L_*}}(K-S(\tau_{L_*}))\right]. \tag{8.3.23}$$

It follows that

$$v_{L_*}(x) \leq \max_{\tau\in\mathcal{T}}\widetilde{\mathbb{E}}\left[e^{-r\tau}(K-S(\tau))\right]. \qquad \square$$

Discounted European option prices are martingales under the risk-neutral probability measure. Discounted American option prices are martingales up to the time they should be exercised. If they are not exercised when they should be, they tend downward. Since a martingale is a special case of a supermartingale, and processes that tend downward are supermartingales, discounted American option prices are supermartingales. An agent who is short an American option can hedge that short position in the usual way during the time the discounted option price is a martingale. If the option is not exercised when it should be, then the agent can continue the hedge and take money off the table. The following corollary illustrates this for the perpetual American put of this section.

Corollary 8.3.7. *Consider an agent with initial capital $X(0)=v_{L_*}(S(0))$, the initial perpetual American put price. Suppose this agent uses the portfolio process $\Delta(t)=v'_{L_*}(S(t))$ and consumes cash at rate $C(t)=rK\mathbb{I}_{\{S(t)<L^*\}}$ (i.e., consumes cash at rate rK whenever $S(t)<L^*$). Then the value $X(t)$ of the agent's portfolio agrees with the option price $v_{L_*}(S(t))$ for all times t until the option is exercised. In particular, $X(t)\geq(K-S(t))^+$ for all t until the option is exercised, so the agent can pay off a short option position regardless of when the option is exercised.*

PROOF: The differential of the agent's portfolio value process is

$$dX(t) = \Delta(t)\, dS(t) + r\big(X(t) - \Delta(t)S(t)\big)\, dt - C(t)\, dt,$$

so the differential of the discounted portfolio value process is

$$\begin{aligned} d\big(e^{-rt}X(t)\big) &= e^{-rt}\big(-rX(t)\, dt + dX(t)\big) \\ &= e^{-rt}\big(\Delta(t)\, dS(t) - r\Delta(t)S(t)\, dt - C(t)\, dt\big) \\ &= e^{-rt}\big(\Delta(t)\sigma S(t)\, d\widetilde{W}(t) - C(t)\, dt\big). \end{aligned} \tag{8.3.24}$$

Substituting $\Delta(t) = v'_{L_*}(S(t))$ and $C(t) = rK\mathbb{I}_{\{S(t)<L^*\}}$ into (8.3.24) and comparing it to (8.3.21), we see that $d(e^{-rt}X(t)) = d\big[e^{-rt}v_{L_*}(S(t))\big]$. Integrating both sides of this equation and using the initial equality $X(0) = v_{L_*}(S(0))$, we obtain $X(t) = v_{L_*}(S(t))$ for all t prior to exercise. □

Remark 8.3.8. During any period in which $S(t) < L^*$, the agent in Corollary 8.3.7 has stock position $\Delta(t) = v'_{L_*}(S(t)) = -1$ (i.e., is short one share of stock) and has a total portfolio value $X(t) = v_{L_*}(S(t)) = K - S(t)$. Therefore, the agent has K invested in the money market. If the owner of the put exercises, the agent in Corollary 8.3.7 receives a share of stock, which covers his short position, and pays out K from his money market account. If the owner of the put does not exercise, the agent holds his position and consumes the interest from the money market investment (i.e., consumes cash at rate rK per unit time). □

The argument in Corollary 8.3.7 applies generally. In a complete market, whenever some discounted price process is a supermartingale, it is possible to construct a hedging portfolio whose value tracks the price process. This portfolio may sometimes consume. In the case of the perpetual American put, the supermartingale property for the discounted put price follows from (8.3.19). If, in addition, the price process dominates some intrinsic value (see (8.3.18) for the perpetual American put), then a short position in the American option with that intrinsic value can be hedged. There are always two conditions on the price of any American option, corresponding to (8.3.18) and (8.3.19). These conditions guarantee that the price is sufficient to satisfy the seller of the put.

However, conditions (8.3.18) and (8.3.19) alone are not enough to determine the price of the perpetual American put. There can be functions that satisfy these conditions but are strictly greater than the price $v_{L_*}(x)$ we constructed in (8.3.13) (see Exercise 8.2). There must be some additional condition that guarantees that the price is satisfactory for the purchaser of the put. One version of this condition for the perpetual American put is (8.3.20). Condition (8.3.20) guarantees that there exists an exercise strategy that permits the owner of the put to capture the full value of the put. It says that if we divide the half-line $[0, \infty)$ into two sets, the *stopping set*

$$\mathcal{S} = \{x \geq 0; v_{L_*}(x) = (K - x)^+\} \tag{8.3.25}$$

and the *continuation set*

$$\mathcal{C} = \{x \geq 0; v_{L_*}(x) > (K - x)^+\}, \tag{8.3.26}$$

then equality holds in (8.3.19) for $x \in \mathcal{C}$. If the initial stock price is in $\mathcal{S}$, then the owner of the put can get full value by exercising it immediately. On the other hand, if the initial stock price is in $\mathcal{C}$, then the put is more valuable than its intrinsic value, and the owner of the put can capture this extra value by waiting until the stock price enters $\mathcal{S}$ to exercise, if it ever does enter $\mathcal{S}$. The time of entry into the set $\mathcal{S}$ is in fact τ_{L_*} in Theorem 8.3.5. We saw in (8.3.23) that

$$v(S(0)) = \widetilde{\mathbb{E}}\left[e^{-r\tau^*} v(S(\tau^*))\right] = \widetilde{\mathbb{E}}\left[e^{-r\tau^*}(K - S(\tau^*))\right].$$

In conclusion, the three linear complementarity conditions have counterparts that can be stated probabilistically rather than analytically (i.e., without writing conditions on the derivatives of $v(x)$). Let $V(t) = e^{-rt}v(S(t))$ be the value of the perpetual American put. The stochastic process $V(t)$ satisfies the following three conditions:

(i) $V(t) \geq \big(K - S(t)\big)^+$ for all $t \geq 0$,
(ii) $e^{-rt}V(t)$ is a supermartingale under $\widetilde{\mathbb{P}}$, and
(iii) there exists a stopping time τ_* such that

$$V(0) = \widetilde{\mathbb{E}}\left[e^{-r\tau_*}\big(K - S(\tau_*)\big)^+\right].$$

These three conditions determine the value of $V(0)$.

8.4 Finite-Expiration American Put

In this section, we consider an American put on a stock whose price is the geometric Brownian motion (8.3.1), but now the put has a finite expiration time T.

Definition 8.4.1. *Let $0 \leq t \leq T$ and $x \geq 0$ be given. Assume $S(t) = x$. Let $\mathcal{F}_u^{(t)}$, $t \leq u \leq T$, denote the σ-algebra generated by the process $S(v)$ as v ranges over $[t, u]$, and let $\mathcal{T}_{t,T}$ denote the set of stopping times for the filtration $\mathcal{F}_u^{(t)}$, $t \leq u \leq T$, taking values in $[t, T]$ or taking the value ∞. In other words, $\{\tau \leq u\} \in \mathcal{F}_u^{(t)}$ for every $u \in [t, T]$; a stopping time in $\mathcal{T}_{t,T}$ makes the decision to stop at a time $u \in [t, T]$ based only on the path of the stock price between* *times t and u. The* price at time t of the American put expiring at time T *is defined to be*[1]

$$v(t,x) = \max_{\tau \in \mathcal{T}_{t,T}} \widetilde{\mathbb{E}} \left[e^{-r(\tau - t)} (K - S(\tau)) \Big| S(t) = x \right]. \quad (8.4.1)$$

In the event that $\tau = \infty$, we interpret $e^{-r\tau}(K - S(\tau))$ to be zero. This is the case when the put expires unexercised.

In Subsection 8.4.1 we present without proof the primary analytical properties of the finite-expiration American put price $v(t,x)$. These are time-dependent versions of the properties developed in Section 8.3 for the perpetual American put. In Subsection 8.4.2, we show that the only function possessing the analytical properties presented in Subsection 8.4.1 is $v(t,x)$ defined by (8.4.1).

8.4.1 Analytical Characterization of the Put Price

The finite-expiration American put price function $v(t,x)$ satisfies the *linear complementarity conditions* (cf. (8.3.18)–(8.3.20))

$$v(t,x) \geq (K - x)^+ \text{ for all } t \in [0,T],\ x \geq 0, \quad (8.4.2)$$

$$rv(t,x) - v_t(t,x) - rxv_x(t,x) - \frac{1}{2}\sigma^2 x^2 v_{xx}(t,x) \geq 0$$
$$\text{for all } t \in [0,T),\ x \geq 0, \text{ and} \quad (8.4.3)$$

for each $t \in [0,T)$ and $x \geq 0$, equality holds in either (8.4.2) or (8.4.3). (8.4.4)

As with the perpetual American put, the owner of the finite-expiration American put should wait until the stock price falls to a certain level at or below K before exercising, but now this level $L(T-t)$ depends on the time to expiration $T-t$. The level L_* of (8.3.12) for the perpetual American put is $\lim_{T\to\infty} L(T)$. At the other extreme, $L(0) = K$; at expiration, one should exercise the put if the stock price is below K, one should not exercise if the stock price is above K, and one is indifferent between exercising and not exercising if the stock price is equal to K. No formula is known for the function $L(T-t)$, but this function can be determined numerically from the analytic characterization of the put price provided in the next subsection. It is known that $L(T)$ decreases wtih increasing T, as shown in Figure 8.4.1. The set $\{(t,x); 0 \leq t \leq T, x \geq 0\}$ can be divided into two regions, the *stopping set*

$$\mathcal{S} = \{(t,x); v(t,x) = (K-x)^+\} \quad (8.4.5)$$

and the *continuation set*

[1] Here we use $v(t,x)$ rather than $v_*(x)$ as in Section 8.3 to denote the put price because in this section we do not consider functions of t and x other than the put price itself.

$$\mathcal{C} = \{(t,x); v(t,x) > (K-x)^+\}. \tag{8.4.6}$$

The graph of the function $x = L(T-t)$ forms the boundary between $\mathcal{C}$ and $\mathcal{S}$ and belongs to $\mathcal{S}$. Because of (8.4.4), equality holds in (8.4.3) for (t,x) in $\mathcal{C}$, $t \neq T$. For (t,x) in $\mathcal{S}$, strict inequality holds in (8.4.3) except on the curve $x = L(T-t)$, where equality holds in (8.4.3). Because $v(t,x) = (K-x)^+ = K - x$ for $0 \le x \le L(T-t)$, we have (see Figure 8.4.1)

$$rv(t,x) - v_t(t,x) - rxv_x(t,x) - \frac{1}{2}\sigma^2 x^2 v_{xx}(t,x) = rK \text{ for } x \in \mathcal{C}.$$

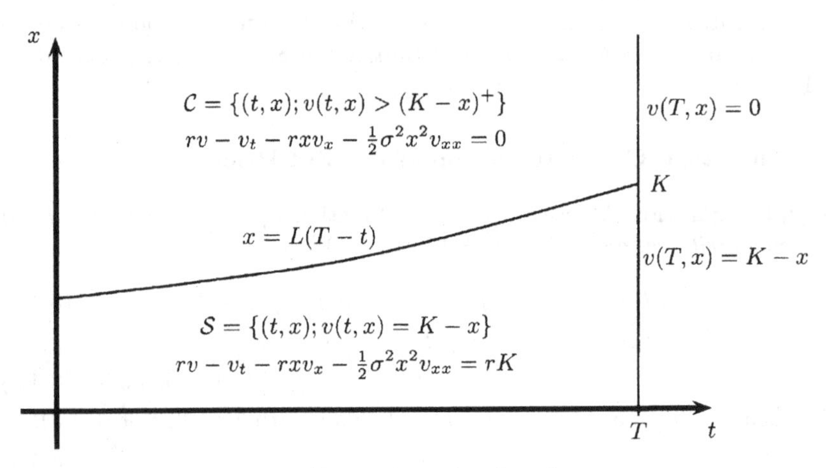

Fig. 8.4.1. Finite-expiration American put.

Because $v(t,x) = K - x$ for $0 \le x \le L(T-t)$, we also have the left-hand derivative $v_x(t,x-) = -1$ on the curve $x = L(T-t)$. The put price $v(t,x)$ satisfies the *smooth-pasting* condition that $v_x(t,x)$ is continuous, even at $x = L(T-t)$. In other words,

$$v_x(t,x+) = v_x(t,x-) = -1 \text{ for } x = L(T-t),\ 0 \le t < T. \tag{8.4.7}$$

The smooth-pasting condition does not hold at $t = T$. Indeed,

$$L(0) = K \text{ and } v(T,x) = (K-x)^+, \tag{8.4.8}$$

so $v_x(T,x-) = -1$, whereas $v_x(T,x+) = 0$ for $x = L(0)$. Also, $v_t(t,x)$ and $v_{xx}(t,x)$ are not continuous along the curve $x = L(T-t)$.

The equations

$$rv(t,x) - v_t(t,x) - rxv_x(t,x) - \frac{1}{2}\sigma^2 x^2 v_{xx}(t,x) = 0,\ x \ge L(T-t),$$
$$v(t,x) = K - x,\ 0 \le x \le L(T-t),$$

together with the smooth-pasting condition (8.4.7), the terminal condition (8.4.8), and the asymptotic condition

$$\lim_{x \to \infty} v(t,x) = 0, \tag{8.4.9}$$

determine the function $v(t,x)$. Using these equations, one can set up a finite-difference scheme to simultaneously compute $v(t,x)$ and $L(T-t)$.

8.4.2 Probabilistic Characterization of the Put Price

Theorem 8.4.2. *Let $S(u)$, $t \le u \le T$, be the stock price of (8.3.1) starting at $S(t) = x$ and with the stopping set $\mathcal{S}$ defined by (8.4.5). Let*

$$\tau_* = \min\{u \in [t,T]; \big(u, S(u)\big) \in \mathcal{S}\}, \tag{8.4.10}$$

where we interpret τ_* to be ∞ if $(u, S(u))$ doesn't enter $\mathcal{S}$ for any $u \in [t,T]$. Then $e^{-ru}v(u,S(u))$, $t \le u \le T$, is a supermartingale under $\widetilde{\mathbb{P}}$, and the stopped process $e^{-r(u \wedge \tau_*)}v(u, S(u \wedge \tau_*))$, $t \le u \le T$, is a martingale.

PROOF: The Itô-Doeblin formula applies to $e^{-ru}v(u,S(u))$, even though $v_u(u,x)$ and $v_{xx}(u,x)$ are not continuous along the curve $x = L(T-u)$ because the process $S(u)$ spends zero time on this curve. All that is needed for the Itô-Doeblin formula to apply is that $v_x(u,x)$ be continuous (see Exercise 4.20 for a discussion related to this), and this follows from the smooth-pasting condition (8.4.7). We may thus compute

$$\begin{aligned}
&d\big[e^{-ru}v\big(u,S(u)\big)\big]\\
&= e^{-ru}\Big[-rv\big(u,S(u)\big)\,du + v_u\big(u,S(u)\big)\,du + v_x\big(u,S(u)\big)\,dS(u)\\
&\qquad + \frac{1}{2}v_{xx}\big(u,S(u)\big)\,dS(u)\,dS(u)\Big]\\
&= e^{-ru}\Big[-rv\big(u,S(u)\big) + v_u\big(u,S(u)\big) + rS(u)v_x\big(u,S(u)\big)\\
&\qquad + \frac{1}{2}\sigma^2 S^2(u)v_{xx}\big(u,S(u)\big)\Big]\,du + e^{-ru}\sigma S(u)v_x\big(u,S(u)\big)\,d\widetilde{W}(u).
\end{aligned} \tag{8.4.11}$$

According to Figure 8.4.1, the du term in (8.4.11) is $-e^{-ru}rK\mathbb{I}_{\{S(u)<L(T-u)\}}$. This is nonpositive, and so $e^{-ru}v(u,S(u))$ is a supermartingale under $\widetilde{\mathbb{P}}$. In fact, starting from $u = t$ and up until time τ_*, we have $S(u) > L(T-u)$, so the du term is zero. Therefore, the stopped process $e^{-r(u\wedge\tau_*)}v(u\wedge\tau_*, S(u\wedge\tau_*))$, $t \le u \le T$, is a martingale. □

Corollary 8.4.3. *Consider an agent with initial capital $X(0) = v(0,S(0))$, the initial finite-expiration put price. Suppose this agent uses the portfolio process $\Delta(u) = v_x(u,S(u))$ and consumes cash at rate $C(u) = rK\mathbb{I}_{\{S(u)<L(T-u)\}}$*

per unit time. Then $X(u) = v(u, S(u))$ for all times u between $u = 0$ and the time the option is exercised or expires. In particular, $S(u) \geq (K - S(u))^+$ for all times u until the option is exercised or expires, so the agent can pay off a short option position regardless of when the option is exercised.

PROOF: The differential of the agent's discounted portfolio value is given by (8.3.24). Substituting for $\Delta(u)$ and $C(u)$ in this equation and comparing it to (8.4.11), we see that $d(e^{-ru}X(u)) = d[e^{-ru}v(u, S(u))]$. Integrating this equation and using $X(0) = v(0, S(0))$, we obtain $X(t) = v(t, S(t))$ for all times t prior to exercise or expiration. □

Remark 8.4.4. The proofs of Theorem 8.4.2 and Corollary 8.4.3 use the analytic characterization of the American put price captured in Figure 8.4.1 plus the smooth-pasting condition that guarantees that $v_x(t,x)$ is continuous even on the curve $x = L(T-t)$ so that the Itô-Doeblin formula can be applied. Here we show that the only function $v(t,x)$ satisfying these conditions is the function $v(t,x)$ defined by (8.4.1). To do this, we first fix t with $0 \leq t \leq T$. The supermartingale property for $e^{-rt}v(t, S(t))$ of Theorem 8.4.2 and Theorem 8.2.4 (optional sampling) implies that

$$e^{-r(t\wedge\tau)}v(t\wedge\tau, S(t\wedge\tau)) \geq \widetilde{\mathbb{E}}\left[e^{-r(T\wedge\tau)}v(T\wedge\tau, S(T\wedge\tau))\Big|\mathcal{F}(t)\right].$$

For $\tau \in \mathcal{T}_{t,T}$, we have $t \wedge \tau = t$, whereas $T \wedge \tau = \tau$ if $\tau < \infty$ and $T \wedge \tau = T$ if $\tau = \infty$. Therefore, for $\tau \in \mathcal{T}_{t,T}$,

$$\begin{aligned} e^{-rt}v(t, S(t)) &\geq \widetilde{\mathbb{E}}\left[e^{-r\tau}v(\tau, S(\tau))\mathbb{I}_{\{\tau<\infty\}} + e^{-rT}v(T, S(T))\mathbb{I}_{\{\tau=\infty\}}\big|\mathcal{F}(t)\right] \\ &\geq \widetilde{\mathbb{E}}\left[e^{-r\tau}v(\tau, S(\tau))\big|\mathcal{F}(t)\right], \end{aligned} \tag{8.4.12}$$

where, as usual, we interpret $e^{-r\tau}v(\tau, S(\tau)) = 0$ if $\tau = \infty$. Inequality (8.4.2) and the fact that $(K - S(t))^+ \geq K - S(t)$ imply that

$$\widetilde{\mathbb{E}}\left[e^{-r\tau}v(\tau, S(\tau))\big|\mathcal{F}(t)\right] \geq \widetilde{\mathbb{E}}\left[e^{-r\tau}(K - S(\tau))\big|\mathcal{F}(t)\right]. \tag{8.4.13}$$

Putting (8.4.12) and (8.4.13) together, we conclude that

$$e^{-rt}v(t, S(t)) \geq \widetilde{\mathbb{E}}\left[e^{-r\tau}(K - S(\tau))\big|\mathcal{F}(t)\right]. \tag{8.4.14}$$

Because $S(t)$ is a Markov process, the right-hand side of (8.4.14) is a function of t and $S(t)$. In particular, if we denote the value of $S(t)$ by x, we may rewrite (8.4.14) as

$$e^{-rt}v(t,x) = \widetilde{\mathbb{E}}\left[e^{-r\tau}(K - S(\tau))\big|S(t) = x\right]. \tag{8.4.15}$$

Since (8.4.15) holds for any $\tau \in \mathcal{T}_{t,T}$, we conclude that

$$v(t,x) \geq \max_{\tau\in\mathcal{T}_{t,T}} \widetilde{\mathbb{E}}\left[e^{-r(\tau-t)}(K - S(\tau))\Big|S(t) = x\right]. \tag{8.4.16}$$

For the reverse inequality, we recall from Theorem 8.4.2 that the stopped process $e^{-r(t\wedge\tau^*)}v(t\wedge\tau_*, S(t\wedge\tau_*))$ is a martingale, where τ_* defined by (8.4.10) is such that $v(\tau_*, S(\tau_*)) = K - S(\tau_*)$ if $\tau_* < \infty$. Replacing τ by τ_* in (8.4.12), we make the first inequality into an equality. If $\tau^* = \infty$, we have $(T, S(T)) \in \mathcal{C}$ (i.e., $S(T) > K$), so $v(T, S(T))\mathbb{I}_{\{\tau_*=\infty\}} = 0$. This makes the second inequality in (8.4.12) into an equality. Finally, because $v(\tau, S(\tau)) = K - S(\tau)$ on $\mathbb{I}_{\{\tau<\infty\}}$, the inequality in (8.4.13) is an equality, and hence (8.4.15) becomes

$$v(t,x) = \widetilde{\mathbb{E}}\left[e^{-r(\tau_*-t)}(K - S(\tau_*))\Big| S(t) = x\right]. \tag{8.4.17}$$

Equation (8.4.17) shows that equality must hold in (8.4.16), and this is (8.4.1). □

8.5 American Call

In this section, we treat the American call, first on the usual geometric Brownian motion asset of (8.3.1) and then on a variation of this asset that pays dividends at discrete dates. In the first case, presented in Subsection 8.5.1, we see that the American call price is the same as the European call price. In the second case, presented in Subsection 8.5.2, we provide a recursion formula for computing the American call price.

8.5.1 Underlying Asset Pays No Dividends

We begin with a case slightly more general than a call option. Consider a stock whose price process $S(t)$ is given by

$$dS(t) = rS(t)\,dt + \sigma S(t)\,d\widetilde{W}(t), \tag{8.5.1}$$

where the interest rate r and the volatility σ are strictly positive and $\widetilde{W}(t)$ is a Brownian motion under the risk-neutral probability measure $\widetilde{\mathbb{P}}$.

Lemma 8.5.1. *Let $h(x)$ be a nonnegative, convex function of $x \geq 0$ satisfying $h(0) = 0$. Then the discounted intrinsic value $e^{-rt}h(S(t))$ of the American derivative security that pays $h(S(t))$ upon exercise is a submartingale.*

Proof: Because $h(x)$ is convex, for $0 \leq \lambda \leq 1$ and $0 \leq x_1 \leq x_2$, we have

$$h\big((1-\lambda)x_1 + \lambda x_2\big) \leq (1-\lambda)h(x_1) + \lambda h(x_2). \tag{8.5.2}$$

See Figure 8.5.1 for the case of a call payoff, $h(x) = (x-K)^+$.

Taking $x_1 = 0$, $x_2 = x$, and using the fact that $h(0) = 0$, we obtain from (8.5.2) that

$$h(\lambda x) \leq \lambda h(x) \text{ for all } x \geq 0,\ 0 \leq \lambda \leq 1. \tag{8.5.3}$$

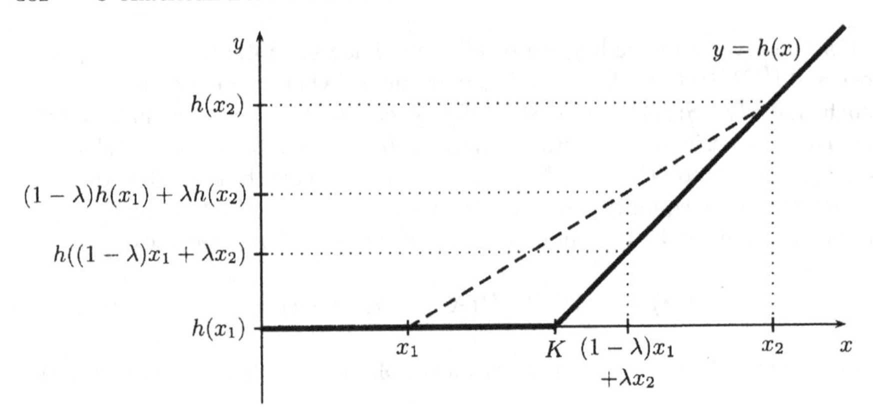

Fig. 8.5.1. The convex function $h(x) = (x - K)^+$.

For $0 \le u \le t \le T$, we have $0 \le e^{-r(t-u)} \le 1$, and (8.5.3) implies

$$\widetilde{\mathbb{E}}\left[e^{-r(t-u)}h(S(t))\Big|\mathcal{F}(u)\right] \ge \widetilde{\mathbb{E}}\left[h\left(e^{-r(t-u)}S(t)\right)\Big|\mathcal{F}(u)\right]. \tag{8.5.4}$$

The conditional Jensen's inequality (Theorem 2.3.2(v)) implies

$$\begin{aligned}\widetilde{\mathbb{E}}\left[h\left(e^{-r(t-u)}S(t)\right)\Big|\mathcal{F}(u)\right] &\ge h\left(\widetilde{\mathbb{E}}\left[e^{-r(t-u)}S(t)\Big|\mathcal{F}(u)\right]\right)\\ &= h\left(e^{ru}\widetilde{\mathbb{E}}\left[e^{-rt}S(t)\big|\mathcal{F}(u)\right]\right).\end{aligned} \tag{8.5.5}$$

Because $e^{-rt}S(t)$ is a martingale under $\widetilde{\mathbb{P}}$, we have

$$h\left(e^{ru}\widetilde{\mathbb{E}}\left[e^{-rt}S(t)\big|\mathcal{F}(u)\right]\right) = h(e^{ru}e^{-ru}S(u)) = h(S(u)). \tag{8.5.6}$$

Putting (8.5.4)–(8.5.6) together, we conclude that

$$\widetilde{\mathbb{E}}\left[e^{-r(t-u)}h(S(t))\Big|\mathcal{F}(u)\right] \ge h(S(u)) \tag{8.5.7}$$

or, equivalently,

$$\widetilde{\mathbb{E}}\left[e^{-rt}h(S(t))\big|\mathcal{F}(u)\right] \ge e^{-ru}h(S(u)). \tag{8.5.8}$$

This is the submartingale property for $e^{-rt}h(S(t))$. □

Theorem 8.5.2. *Let $h(x)$ be a nonnegative, convex function of $x \ge 0$ satisfying $h(0) = 0$. Then the price of the American derivative security expiring at time T and having intrinsic value $h(S(t))$, $0 \le t \le T$, is the same as the price of the European derivative security paying $h(S(T))$ at expiration T.*

PROOF: Replacing t by T in (8.5.7), we obtain

$$\widetilde{\mathbb{E}}\left[e^{-r(T-u)}h(S(T))\Big|\mathcal{F}(u)\right] \geq h(S(u)),\ 0 \leq u \leq T.$$

In other words, the European derivative security price always dominates the intrinsic value of the American derivative security. This shows that the option to exercise early is worthless, and the price of the American derivative security agrees with the price of the European security. □

Corollary 8.5.3. *The price of an American call on an asset not paying a dividend is the same as the price of the European call on the same asset with the same expiration.*

PROOF: Take $h(x) = (x - K)^+$ in Theorem 8.5.2. □

The idea behind Corollary 8.5.3 is that the discounted process $e^{-rt}(S(t) - K)^+$ is a submartingale under $\widetilde{\mathbb{P}}$ and hence tends to rise. Therefore, it is optimal to wait until expiration before deciding whether to exercise. There are two factors that contribute to the submartingale property for $e^{-rt}(S(t)-K)^+$. One is the discounting of the strike. In fact, $e^{-rt}(S(t)-K)$ (without the $^+$) is a submartingale because $e^{-rt}S(t)$ is a martingale under the risk-neutral measure $\widetilde{\mathbb{P}}$ and $-e^{-rt}K$ increases as t increases (throughout this chapter, we assume a strictly positive interest rate r). When we reinstate the $^+$, we are taking a convex function of a submartingale and, because of Jensen's inequality, this reinforces the upward trend.

The previous argument does not apply to the American put, whose discounted intrinsic value $e^{-rt}(K - S(t))$ (without the $^+$) is a supermartingale ($e^{-rt}K$ falls and $-e^{-rt}S(t)$ is a martingale). Jensen's inequality creates an upward trend that competes with this supermartingale property, and the analysis becomes complicated.

If the underlying asset pays a dividend, the case considered in the next subsection, the argument above no longer applies to the American call. In this case, $e^{-rt}S(t)$ is a supermartingale and tends to fall because of the dividend outflow.

8.5.2 Underlying Asset Pays Dividends

In this subsection, we consider an American call on an asset whose price process is a geometric Brownian motion governed by (8.5.1) between dividend payment dates. We assume there are times $0 < t_1 < t_2 < \cdots < t_n < T$, and at each time t_j the dividend paid is $a_j S(t_j-)$, where $S(t_j-)$ denotes the asset price just prior to the dividend payment. The asset price $S(t_j)$ after the dividend payment is the asset price before the dividend payment less the dividend payment:

$$S(t_j) = S(t_j-) - a_j S(t_j-) = (1 - a_j)S(t_j-). \tag{8.5.9}$$

We assume that each a_j, $j = 1, \dots, n$, is a number between 0 and 1. We set $t_0 = 0$, but this is not a dividend payment date. We also assume that T is not a dividend payment date, although it is not difficult to modify the analysis given below to handle the case when T is a dividend payment date.

We shall see that it is not optimal to exercise an American call on this asset except possibly immediately before a dividend payment. The price of the call will be seen to satisfy the Black-Scholes-Merton partial differential equation between dividend payment dates. At dividend payment dates, the price of the call is the maximum of the call's intrinsic value and the price of the call after the dividend is paid and the stock price is reduced by the amount of the payment. These observations lead to a recursive algorithm for determining the price, and that is developed in this subsection.

The asset price process in this section was considered in Subsection 5.5.4. For $t_j \le t < t_{j+1}$, we have

$$S(t) = S(t_j) \exp\left\{\sigma\big(\widetilde{W}(t) - \widetilde{W}(t_j)\big) + \left(r - \frac{1}{2}\sigma^2\right)(t - t_j)\right\},$$

which implies

$$S(t_{j+1}-) = S(t_j) \exp\left\{\sigma\big(\widetilde{W}(t_{j+1}) - \widetilde{W}(t_j)\big) + \left(r - \frac{1}{2}\sigma^2\right)(t_{j+1} - t_j)\right\} \tag{8.5.10}$$

and

$$\begin{aligned} &S(t_{j+1}) \\ &= (1 - a_{j+1}) S(t_j) \exp\left\{\sigma\big(\widetilde{W}(t_{j+1}) - \widetilde{W}(t_j)\big) + \left(r - \frac{1}{2}\sigma^2\right)(t_{j+1} - t_j)\right\}. \end{aligned} \tag{8.5.11}$$

We also have

$$S(T) = S(t_n) \exp\left\{\sigma\big(\widetilde{W}(T) - \widetilde{W}(t_n)\big) + \left(r - \frac{1}{2}\sigma^2\right)(T - t_n)\right\}. \tag{8.5.12}$$

We consider an American call expiring at time T with strike price K. For $t_n \le t \le T$, the discounted asset price $e^{-rt}S(t)$ is a martingale under $\widetilde{\mathbb{P}}$, and Lemma 8.5.1 can be invoked to show that $e^{-rt}(S(t) - K)^+$ is a submartingale. Therefore,

$$\widetilde{\mathbb{E}}\left[e^{-rT}\big(S(T) - K\big)^+ \Big| \mathcal{F}(t)\right] \ge e^{-rt}(S(t) - K)^+, \; t_n \le t \le T. \tag{8.5.13}$$

This shows that, for all $t \in [t_n, T]$, the price of the European call at time t,

$$c_n\big(t, S(t)\big) = \widetilde{\mathbb{E}}\left[e^{-r(T-t)}\big(S(T) - K\big)^+ \Big| \mathcal{F}(t)\right],$$

is greater than the intrinsic value of the American call, $(S(t) - K)^+$. Consequently, the early exercise feature of the American call is worthless, and the prices at time t of the European and American calls agree for $t_n \leq t \leq T$. This price is given by the Black-Scholes-Merton formula

$$c_n(t, x) = xN\big(d_+(T - t, x)\big) - Ke^{-r(T-t)}N\big(d_-(T - t, x)\big), \tag{8.5.14}$$

where

$$d_\pm(\tau, x) = \frac{1}{\sigma\sqrt{\tau}}\left[\log\frac{x}{K} + \left(r \pm \frac{1}{2}\sigma^2\right)\tau\right].$$

Although one cannot simply substitute $x = 0$ into (8.5.14), we have $c(t, 0) = 0$; see equation (4.5.17) and Exercise 4.9. Formula (8.5.14) can be determined by computing the conditional expectation in (8.5.13) under the condition $S(t) = x$. In the case $t = t_n$, using (8.5.12), this leads to

$$c_n(t_n, x) = \widetilde{\mathbb{E}}\left[e^{-r(T-t_n)}\left(x\exp\left\{\sigma\big(\widetilde{W}(T) - \widetilde{W}(t_n)\big) + \left(r - \frac{1}{2}\sigma^2\right)(T - t_n)\right\} - K\right)^+\right]. \tag{8.5.15}$$

The function $c_n(t, x)$ also satisfies the Black-Scholes-Merton differential equation

$$\frac{\partial}{\partial t}c_n(t, x) + rx\frac{\partial}{\partial x}c_n(t, x) + \frac{1}{2}\sigma^2x^2\frac{\partial^2}{\partial x^2}c_n(t, x) = rc_n(t, x), \; t_n \leq t < T, \; x \geq 0, \tag{8.5.16}$$

and the terminal condition

$$c_n(t, x) = (x - K)^+, \; x \geq 0. \tag{8.5.17}$$

The function $c_n(t_n, x)$ is convex in x. This is well-known, but we establish it here anyway to demonstrate a method we need later. To show convexity in x, we show that, whenever $0 \leq x_1 \leq x_2$ and $0 \leq \lambda \leq 1$, we have

$$c_n\big(t_n, (1 - \lambda)x_1 + \lambda x_2\big) \leq (1 - \lambda)c_n(t_n, x_1) + \lambda c_n(t_n, x_2). \tag{8.5.18}$$

We begin with the observation that, for any number α, the function $(\alpha x - K)^+$ is convex in x, and therefore

$$\left(x\exp\left\{\sigma\big(\widetilde{W}(T) - \widetilde{W}(t_n)\big) + \left(r - \frac{1}{2}\sigma^2\right)(T - t_n)\right\} - K\right)^+$$

is convex in x. It follows that

$$
\begin{aligned}
&c_n\big(t_n,(1-\lambda)x_1+\lambda x_2\big)\\
&=\widetilde{\mathbb{E}}\left[e^{-r(T-t_n)}\left(\left((1-\lambda)x_1+\lambda x_2\right)\exp\left\{\sigma\big(\widetilde{W}(T)-\widetilde{W}(t_n)\big)\right.\right.\right.\\
&\qquad\qquad\left.\left.\left.+\left(r-\frac{1}{2}\sigma^2\right)\right\}-K\right)^+\right]\\
&\le(1-\lambda)\widetilde{\mathbb{E}}\left[e^{-r(T-t_n)}\left(x_1\exp\left\{\sigma\big(\widetilde{W}(T)-\widetilde{W}(t_n)\big)+\left(r-\frac{1}{2}\sigma^2\right)\right\}-K\right)^+\right]\\
&\qquad+\lambda\widetilde{\mathbb{E}}\left[e^{-r(T-t_n)}\left(x_2\exp\left\{\sigma\big(\widetilde{W}(T)-\widetilde{W}(t_n)\big)+\left(r-\frac{1}{2}\sigma^2\right)\right\}-K\right)^+\right]\\
&=(1-\lambda)c_n(t_n,x_1)+\lambda c_n(t_n,x_2).
\end{aligned}\tag{8.5.19}
$$

This proves (8.5.18).

At time t_n, immediately before the dividend payment, the owner of the American call has two choices. She can exercise the option and receive $S(t_n-)-K$, or she can decline to exercise, permit the dividend to be paid (not to her) and the asset price to fall to $S(t_n)=(1-a_n)S(t_n-)$, and have an option valued at $c_n(t_n,(1-a_n)S(t_n-))$. The optimal decision is to exercise if $S(t_n-)-K>c_n(t_n,(1-a_n)S(t_n-))$ and to decline to exercise if $S(t_n-)-K<c_n(t_n,(1-a_n)S(t_n-))$. If $S(t_n-)-K=c_n(t_n,(1-a_n)S(t_n-))$, it does not matter whether she exercises or declines to exercise. Therefore, the call value at time t_n immediately before the dividend is paid is $h_n(S(t_n)-)$, where

$$h_n(x)=\max\{x-K,c_n(t_n,(1-a_n)x)\},\ x\ge 0. \tag{8.5.20}$$

We show that $h_n(x)$ satisfies the assumptions of Lemma 8.5.1. It is clear that $h_n(x)\ge 0$ for all $x\ge 0$ because $c_n(t_n,(1-a_n)x)\ge 0$ for all $x\ge 0$. It is also clear that $h_n(0)=0$ because $c_n(t_n,(1-a_n)0)=0$. To establish the convexity of $h_n(x)$, we recall from (8.5.18) that $c_n(t_n,x)$ is convex in x. For $0\le x_1\le x_2$ and $0\le\lambda\le 1$, we replace x_1 in (8.5.18) by $(1-a_n)x_1$ and replace x_2 by $(1-a_n)x_2$ to obtain

$$c_n\big(t_n,(1-a_n)((1-\lambda)x_1+\lambda x_2)\big)\le(1-\lambda)c_n\big(t_n,(1-a_n)x_1\big)+\lambda c_n\big(t_n,(1-a_n)x_2\big).$$

This shows that $c_n(t,(1-a_n)x)$ is a convex function of x. The maximum of two convex functions is convex (see Exercise 8.7), and therefore $h_n(x)$ defined by (8.5.20) is convex.

Starting from time t, where $t_{n-1}\le t<t_n$, the owner of the American call can exercise at any time $u\in[t,t_n)$, and if she does, she receives $S(u)-K$. If she does not exercise prior to t_n, then at time t_n, immediately before the dividend payment, she owns a call whose value we have just determined to be $h_n(S(t_n-))$. Therefore, for $t_{n-1}\le t<t_n$, the American call expiring at time T has the same price as the American call expiring immediately before the dividend payment at date t_n and paying $h_n(S(t_n-))$ upon expiration.

Because the underlying asset evolves as a geometric Brownian motion after the dividend is paid at time t_{n-1} until the dividend is paid at time t_n, Lemma 8.5.1 implies that $e^{-rt}h_n(S(t))$ is a submartingale for $t_{n-1} \le t < t_n$. In particular,

$$\widetilde{\mathbb{E}}\left[e^{-r(u-t)}h_n(S(u))\Big|\mathcal{F}(t)\right] \ge h_n(S(t)),\ t_{n-1} \le t \le u < t_n,$$

and letting $u \uparrow t_n$, we obtain

$$\widetilde{\mathbb{E}}\left[e^{-r(t_n-t)}h_n(S(t_n-))\Big|\mathcal{F}(t)\right] \ge h_n(S(t)). \tag{8.5.21}$$

By the definition of $h_n(x)$,

$$h_n(S(t)) \ge S(t) - K. \tag{8.5.22}$$

This shows that the value of the European call expiring at time t_n immediately before the dividend is paid and paying $h_n(S(t_n-))$ upon expiration, which is the left-hand side of (8.5.21), is greater than or equal to the intrinsic value of the American call, which is the right-hand side of (8.5.22). Therefore, the option to exercise the American call before time t_n is worthless, and the American call value is the same as the value of the European call just described.

Because $S(t)$ is a Markov process, there is some function $c_{n-1}(t,x)$ such that the left-hand side of (8.5.21), the European call value, is

$$c_{n-1}(t, S(t)) = \widetilde{\mathbb{E}}\left[e^{-r(t_n-t)}h_n(S(t_n-))\Big|\mathcal{F}(t)\right]. \tag{8.5.23}$$

The function $c_{n-1}(t,x)$ can be determined by computing the conditional expectation in (8.5.23) under the condition $S(t) = x$. In the case $t = t_{n-1}$, using (8.5.10), this leads to

$$\begin{aligned} &c_{n-1}(t_{n-1}, x)\\ &= \widetilde{\mathbb{E}}\Bigg[e^{-r(t_n-t_{n-1})} &(8.5.24)\\ &\quad \times h_n\left(x\exp\left\{\sigma(\widetilde{W}(t_n) - \widetilde{W}(t_{n-1})) + \left(r - \frac{1}{2}\sigma^2\right)(t_n - t_{n-1})\right\}\right)\Bigg]. &(8.5.25) \end{aligned}$$

The function $c_{n-1}(t,x)$ also satisfies the Black-Scholes-Merton differential equation

$$\frac{\partial}{\partial t}c_{n-1}(t,x) + rx\frac{\partial}{\partial x}c_{n-1}(t,x) + \frac{1}{2}\sigma^2x^2\frac{\partial^2}{\partial x^2}c_{n-1}(t,x) = rc_{n-1}(t,x),$$
$$t_{n-1} \le t < t_n,\ x \ge 0, \tag{8.5.26}$$

and the terminal condition

$$c_{n-1}(t_n, x) = h_n(t_n, x), \ x \geq 0. \tag{8.5.27}$$

We repeat this process, defining

$$h_{n-1}(x) = \max\left\{x - K, c_{n-1}\big(t_{n-1}, (1 - a_{n-1})x\big)\right\}, \ x \geq 0.$$

We can show as above that $h_{n-1}(x)$ satisfies the hypotheses of Lemma 8.5.1, and we continue.

In conclusion, we obtain an algorithm for the American call price on an asset paying dividends at the dates $t_1, t_2, \ldots, t_n$. Solve recursively for $j = n, n-1, \ldots, 0$, the partial differential equation

$$\frac{\partial}{\partial t} c_{j-1}(t, x) + rx \frac{\partial}{\partial x} c_{j-1}(t, x) + \frac{1}{2}\sigma^2 x^2 \frac{\partial^2}{\partial x^2} c_{j-1}(t, x) = rc_{j-1}(t, x),$$
$$t_{j-1} \leq t < t_j, \ x \geq 0, \tag{8.5.28}$$

with the terminal condition

$$c_{j-1}(t_j, x) = h_j(x), \ x \geq 0. \tag{8.5.29}$$

The functions $c_n(t, x)$ and $h_n(x)$ needed to get started are given by (8.5.14) and (8.5.20), and the function $h_{j-1}(x)$ needed for the next step is given by

$$h_{j-1}(x) = \max\left\{x - K, c_{j-1}\big(t_{j-1}, (1 - a_{j-1})x\big)\right\}, \ x \geq 0. \tag{8.5.30}$$

For $t_{j-1} \leq t < t_j$, if $S(t) = x$, then $c_{j-1}(t, x)$ is the American call price. Within each interval $[t_{j-1}, t_j)$, the American call price is actually the price of a European call expiring at t_j. The optimal exercise time is immediately prior to the dividend payment at the smallest time t_j for which $S(t_j-) - K$ exceeds $c_j(t_j, (1 - a_j)S(t_j-))$. If there is no t_j for which this condition is satisfied, then optimal exercise takes place at time T if $S(T) > K$, and otherwise the option should be allowed to expire unexercised.

8.6 Summary

This chaper discusses American puts and calls. To do this, we introduce the notions of stopping times and optional sampling in Section 8.2. The value of an American option can then be defined as the maximum over all stopping times of the discounted, risk-neutral payoff of the option evaluated at the stopping time. We do this for the perpetual American put in Section 8.3 and for the finite-horizon American put in Section 8.4. This definition of option value gives the no-arbitrage price. Starting with initial capital given by this definition, a person holding a short position in the option can hedge in such a way that, regardless of when the option is exercised, he will be able to pay off the short position. Furthermore, this definition of American option price is the smallest initial capital that permits such hedging. In particular, there is an optimal stopping time, and if the option owner exercises at this time, she captures the full value of the option.

The American put has an analytical characterization, which we present as linear complementarity conditions in Subsections 8.3.3 and 8.4.1. According to this characterization, there are two regions in the space of time and stock prices (t, x), one in which it is optimal to exercise the put (the stopping set) and another in which it is optimal not to exercise (the continuation set). The put price $v(t, x)$ and its first derivative $v_x(t, x)$ are continuous across the boundary between these two regions (smooth pasting), and this fact tells us that $v_x(t, x) = -1$ on this boundary. Using this smooth-pasting condition, one can solve numerically for the American put price.

The American call on a stock that pays no dividends has the same price as the corresponding European call; see Section 8.5.1. If the stock pays dividends, the American call can be more valuable than the European call. In Section 8.5.2, we work out an algorithm for the American call price when dividends are paid at discrete dates.

8.7 Notes

The use of stopping times with martingales was pioneered by Doob [53], who provided Theorem 8.2.4. A modern treatment can be found in many texts, including Chung [35] and Williams [161] in discrete time and Karatzas and Shreve [101] in continuous time.

The perpetual American put problem was first solved by McKean [119], who also wrote down the analytic characterization of the finite-horizon American put price. The fact that this analytic characterization determines the finite-horizon American put price follows from the optimal-stopping theory developed by van Moerbeke [153]. For the particular case of the American put, a simpler derivation of this fact is provided by Jacka [93], and this is presented in Section 2.7 of Karatzas and Shreve [102]. Although the price of the American put cannot be computed explicitly, it is possible to give a variety of characterizations of the early exercise premium, the difference between the American put price and the corresponding European put price; see Carr, Jarrow, and Myneni [27], Jacka [93], and Kim [103].

The probabilistic characterization of the American put price is due to Bensoussan [9] and Karatzas [100]. This is also reported in Section 2.5 of Karatzas and Shreve [102]. A survey of all these things, and a wealth of other references, are provided by Myneni [127]. Merton [122] observed that an American call on a stock paying no dividends has the same value as a European call.

There are two principal ways to compute option prices numerically: finite-difference schemes and Monte Carlo simulation. A finite-difference scheme for the American put is described in Wilmott, Howison, and Dewynne [165].

Monte Carlo methods are more difficult to develop because one must simultaneously determine the price of the put and determine the boundary between the stopping and continuation sets. A novel method to deal with this was recently provided by Longstaff and Schwartz [112] and Tsitsiklis and Van Roy [152]. Results on convergence of a modification of the Longstaff-Schwartz algorithm can be found in Clément, Lamberton, and Protter [37] and Glasserman and Yu [75]. Papers that use binomial trees and analytic approximations are listed in Section 2.8 of Karatzas and Shreve [102].

8.8 Exercises

Exercise 8.1 (Determination of L_* by smooth pasting). Consider the function $v_L(x)$ in (8.3.11). The first line in formula (8.3.11) implies that the left-hand derivative of $v_L(x)$ at $x = L$ is $v_L(L-) = -1$. Use the second line in formula (8.3.11) to compute the right-hand derivative $v_L'(L+)$. Show that the smooth-pasting condition

$$v_{L_*}'(L_*-) = v_{L_*}'(L_*+)$$

is satisfied only by L_* given by (8.3.12).

Exercise 8.2. Consider two perpetual American puts on the geometric Brownian motion (8.3.1). Suppose the puts have different strike prices, K_1 and K_2, where $0 < K_1 < K_2$. Let $v_1(x)$ and $v_2(x)$ denote their respective prices, as determined in Section 8.3.2. Show that $v_2(x)$ satisfies the first two linear complementarity conditions,

$$v_2(x) \geq (K_1 - x)^+ \text{ for all } x \geq 0, \tag{8.8.1}$$

$$rv_2(x) - rxv_2'(x) - \frac{1}{2}\sigma^2 x^2 v_2''(x) \geq 0 \text{ for all } x \geq 0, \tag{8.8.2}$$

for the perpetual American put price with strike K_1 but that $v_2(x)$ does not satisfy the third linear complementarity condition:

$$\text{for each } x \geq 0, \text{ equality holds in either (8.8.1) or (8.8.2) or both.} \tag{8.8.3}$$

Exercise 8.3 (Solving the linear complementarity conditions). Suppose $v(x)$ is a bounded continuous function having a continuous derivative and satisfying the linear complementarity conditions (8.3.18)–(8.3.20). This exercise shows that $v(x)$ must be the function $v_{L_*}(x)$ given by (8.3.13) with L_* given by (8.3.12). We assume that K is strictly positive.

(i) First consider an interval of x-values in which $v(x)$ satisfies (8.3.19) with equality, i.e., where

$$rv(x) - rxv'(x) - \frac{1}{2}\sigma^2 x^2 v''(x) = 0. \tag{8.8.4}$$

Equation (8.8.4) is a linear, second-order ordinary differential equation, and it has two solutions of the form x^p, the solutions differing because of different values of p. Substitute x^p into (8.8.4) and show that the only values of p that cause x^p to satisfy (8.8.4) are $p = -\frac{2r}{\sigma^2}$ and $p = 1$.

(ii) The functions $x^{-\frac{2r}{\sigma^2}}$ and x are said to be linearly independent solutions of (8.8.4), and every function that satisfies (8.8.4) on an interval must be of the form

$$f(x) = Ax^{-\frac{2r}{\sigma^2}} + Bx$$

for some constants A and B. Use this fact and the fact that both $v(x)$ and $v'(x)$ are continuous to show that there cannot be an interval $[x_1, x_2]$, where $0 < x_1 < x_2 < \infty$, such that $v(x)$ satisfies (8.3.19) with equality on $[x_1, x_2]$ and satisfies (8.3.18) with equality for x at and immediately to the left of x_1 and for x at and immediately to the right of x_2 unless $v(x)$ is identically zero on $[x_1, x_2]$.

(iii) Use the fact that $v(0)$ must equal K to show that there cannot be a number $x_2 > 0$ such that $v(x)$ satisfies (8.3.19) with equality on $[0, x_2]$.

(iv) Explain why $v(x)$ cannot satisfy (8.3.19) with equality for all $x \geq 0$.

(v) Explain why $v(x)$ cannot satisfy (8.3.18) with equality for all $x \geq 0$.

(vi) From (iv) and (v) and (8.3.20), we see that $v(x)$ sometimes satisfies (8.3.18) with equality and sometimes does not satisfy (8.3.18) with equality, in which case it must satisfy (8.3.19) with equality. From (ii) and (iii) we see that the region in which $v(x)$ does not satisfy (8.3.18) with equality and satisfies (8.3.19) with equality is not an interval $[x_1, x_2]$, where $0 \leq x_1 < x_2 < \infty$, nor can this region be a union of disjoint intervals of this form. Therefore, it must be a half-line $[x_1, \infty)$, where $x_1 > 0$. In the region $[0, x_1]$, $v(x)$ satisfies (8.3.18) with equality. Show that x_1 must equal L_* given by (8.3.12) and $v(x)$ must be $v_{L_*}(x)$ given by (8.3.13).

Exercise 8.4. It was asserted at the end of Subsection 8.3.3 and established in Exercise 8.3 that $v_{L_*}(x)$ given by (8.3.13) is the only *bounded* continuous function having a continuous derivative and satisfying the linear complementarity conditions (8.3.18)–(8.3.20). There are, however, unbounded functions that satisfy these conditions. Let $0 < L < K$ be given, and assume that

$$\frac{2r}{2r + \sigma^2} K > L. \tag{8.8.5}$$

(i) Show that, for any constants A and B, the function

$$f(x) = Ax^{-\frac{2r}{\sigma^2}} + Bx \tag{8.8.6}$$

satisfies the differential equation

$$rf(x) - rxf'(x) - \frac{1}{2}\sigma^2 x^2 f''(x) = 0 \text{ for all } x \geq 0. \tag{8.8.7}$$

(ii) Show that the constants A and B can be chosen so that

$$f(L) = K - L, \quad f'(L) = -1. \tag{8.8.8}$$

(iii) With the constants A and B you chose in (ii), show that $f(x) \geq (K-x)^+$ for all $x \geq L$.

(iv) Define

$$v(x) = \begin{cases} K - x, 0 \leq x \leq L, \\ f(x), \quad x \geq L. \end{cases}$$

Show that $v(x)$ satisfies the linear complementarity conditions (8.3.18)–(8.3.20), but $v(x)$ is not the function $v_{L_*}(x)$ given by (8.3.13).

(v) Every solution of the differential equation (8.8.7) is of the form (8.8.6). In order to have a bounded solution, we must have $B = 0$. Show that in order to have $B = 0$, we must have $L = \frac{2r}{2r+\sigma^2}K$, and in this case $v(x)$ agrees with the function $v_{L_*}(x)$ of (8.3.13).

Exercise 8.5 (Perpetual American put paying dividends). Consider a perpetual American put on a geometric Brownian motion asset price paying dividends at a constant rate $a > 0$. The differential of this asset is

$$dS(t) = (r - a)S(t)\,dt + \sigma S(t)\,d\widetilde{W}(t), \tag{8.8.9}$$

where $\widetilde{W}(t)$ is a Brownian motion under a risk-neutral measure $\widetilde{\mathbb{P}}$. (Equation (8.8.9) can be obtained by computing the differential in (5.5.8).)

(i) Suppose we adopt the strategy of exercising the put the first time the asset price is at or below L. What is the risk-neutral expected discounted payoff of this strategy? Write this as a function $v_L(x)$ of the initial asset price x. (Hint: Define the positive constant

$$\gamma = \frac{1}{\sigma^2}\left(r - a - \frac{1}{2}\sigma^2\right) + \frac{1}{\sigma}\sqrt{\frac{1}{\sigma^2}\left(r - a - \frac{1}{2}\sigma^2\right)^2 + 2r}$$

and write $v_L(x)$ using γ.)

(ii) Determine L_*, the value of L that maximizes the risk-neutral expected discounted payoff computed in (i).

(iii) Show that, for any initial asset price $S(0) = x$, the process $e^{-rt}v_{L_*}(S(t))$ is a supermartingale under $\widetilde{\mathbb{P}}$. Show that if $S(0) = x > L_*$ and $e^{-rt}v_{L_*}(S(t))$ is stopped the first time the asset price reaches L_*, then the stopped supermartingale is a martingale. (Hint: Show that

$$r + (r - a)\gamma - \frac{1}{2}\sigma^2\gamma(\gamma + 1) = 0.) \tag{8.8.10}$$

(iv) Show that, for any initial asset price $S(0) = x$,

$$v_{L_*}(x) = \max_{\tau \in \mathcal{T}} \widetilde{\mathbb{E}}\left[e^{-r\tau}(K - S(\tau))\right]. \tag{8.8.11}$$

Exercise 8.6. There is a second part to Theorem 8.2.4 (optional sampling), which says the following.

Theorem 8.8.1 (Optional sampling – Part II). *Let $X(t)$, $t \geq 0$, be a submartingale, and let τ be a stopping time. Then $\mathbb{E}X(t \wedge \tau) \leq \mathbb{E}X(t)$. If $X(t)$ is a supermartingale, then $\mathbb{E}X(t \wedge \tau) \geq \mathbb{E}X(t)$. If $X(t)$ is a martingale, then $\mathbb{E}X(t \wedge \tau) = \mathbb{E}X(t)$.*

The proof is technical and is omitted. The idea behind the statement about submartingales is the following. Submartingales tend to go up. Since $t \wedge \tau \leq t$, we would expect this upward trend to result in the inequality $\mathbb{E}X(t \wedge \tau) \leq \mathbb{E}X(t)$. When τ is a stopping time, this intuition is correct. Once we have Theorem 8.8.1 for submartingales, we easily obtain it for supermartingales by using the fact that the negative of a supermartingale is a submartingale. Since a martingale is both a submartingale and a supermartingale, we obtain the equality $\mathbb{E}X(t \wedge \tau) = \mathbb{E}X(t)$ for martingales.

Use Theorem 8.8.1 and Lemma 8.5.1 to show in the context of Subsection 8.5.1 that

$$\widetilde{\mathbb{E}}\left[e^{-rT}\left(S(T) - K\right)^+\right] = \max_{\tau \in \mathcal{T}_{0,T}} \widetilde{\mathbb{E}}\left[e^{-r\tau}\left(S(\tau) - K\right)^+\right], \tag{8.8.12}$$

where as usual we interpret $e^{-r\tau}(S(\tau) - K)^+$ to be zero if $\tau = \infty$. The right-hand side is the American call price analogous to Definition 8.4.1 for the American put price. The left-hand side is the European call price.

Exercise 8.7. A function $f(x)$ defined for $x \geq 0$ is said to be *convex* if, for every $0 \leq x_1 \leq x_2$ and every $0 \leq \lambda \leq 1$, the inequality

$$f\big((1 - \lambda)x_1 + \lambda x_2\big) \leq (1 - \lambda)f(x_1) + \lambda f(x_2)$$

holds. Suppose $f(x)$ and $g(x)$ are convex functions defined for $x \geq 0$. Show that

$$h(x) = \max\{f(x), g(x)\}$$

is also convex.

9

Change of Numéraire

9.1 Introduction

A *numéraire* is the unit of account in which other assets are denominated. One usually takes the numéraire to be the currency of a country. One might change the numéraire by changing to the currency of another country. As this example suggests, in some applications one must change the numéraire in which one works because of finance considerations. We shall see that sometimes it is convenient to change the numéraire because of modeling considerations as well. A model can be complicated or simple, depending on the choice of the numéraire for the model.

In this chapter, we will work within the multidimensional market model of Section 5.4. In particular, our model will be driven by a d-dimensional Brownian motion $W(t) = (W_1(t), \dots, W_d(t))$, $0 \le t \le T$, defined on a probability space $(\Omega, \mathcal{F}, \mathbb{P})$. In particular, $W_1, \dots, W_d$ are independent Brownian motions. The filtration $\mathcal{F}(t)$, $0 \le t \le T$, is the one generated by this vector of Brownian motions. There is an adapted interest rate process $R(t)$, $0 \le t \le T$. This can be used to create a money market account whose price per share at time t is

$$M(t) = e^{\int_0^t R(u)du}.$$

This is the capital an agent would have if the agent invested one unit of currency in the money market account at time zero and continuously rolled over the capital at the short-term interest rate. We also define the discount process

$$D(t) = e^{-\int_0^t R(u)du} = \frac{1}{M(t)}.$$

There are m primary assets in the model of this chapter, and their prices satisfy equation (5.4.6), which we repeat here:

$$dS_i(t) = \alpha_i(t)S_i(t)\,dt + S_i(t)\sum_{j=1}^{d}\sigma_{ij}(t)\,dW_j(t), \quad i = 1, \dots, m. \qquad (9.1.1)$$

We assume there is a unique risk-neutral measure $\widetilde{\mathbb{P}}$ (i.e., there is a unique d-dimensional process $\Theta(t) = (\Theta_1(t), \dots, \Theta_d(t))$ satisfying the market price of risk equations (5.4.18)). The risk-neutral measure is constructed using the multidimensional Girsanov Theorem 5.4.1. Under $\widetilde{\mathbb{P}}$, the Brownian motions

$$\widetilde{W}_j(t) = W_j(t) + \int_0^t \Theta_j(u)\,du, \quad j = 1, \dots, d,$$

are independent of one another. According to the Second Fundamental Theorem of Asset Pricing, Theorem 5.4.9, the market is complete; every derivative security can be hedged by trading in the primary assets and the money market account.

Under $\widetilde{\mathbb{P}}$, the discounted asset prices $D(t)S_i(t)$ are martingales, and so the discounted value of every portfolio process is also a martingale. The risk-neutral measure $\widetilde{\mathbb{P}}$ is thus associated with the money market account price $M(t)$ in the following way. If we were to denominate the ith asset in terms of the money market account, its price would be $S_i(t)/M(t) = D(t)S_i(t)$. In other words, at time t, the ith asset is worth $D(t)S_i(t)$ shares of the money market account. This process, the value of the ith asset denominated in shares of the money market account, is a martingale under $\widetilde{\mathbb{P}}$. We say the measure $\widetilde{\mathbb{P}}$ is *risk-neutral for the money market account numéraire.*

When we change the numéraire, denominating the ith asset in some other unit of account, it is no longer a martingale under $\widetilde{\mathbb{P}}$. When we change the numéraire, we need to also change the risk-neutral measure in order to maintain risk neutrality. The details and some applications of this idea are developed in this chapter.

9.2 Numéraire

In principle, we can take any positively priced asset as a *numéraire* and denominate all other assets in terms of the chosen numéraire. Associated with each numéraire, we shall have a risk-neutral measure. When making this association, we shall take only non-dividend-paying assets as numéraires. In particular, we regard $\widetilde{\mathbb{P}}$ as the risk-neutral measure associated with the domestic money market account, not the domestic currency. Currency pays a dividend because it can be invested in the money market. In contrast, in our model, a share of the money market account increases in value without paying a dividend.

The numéraires we consider in this chapter are:

- Domestic money market account. We denote the associated risk-neutral measure by $\widetilde{\mathbb{P}}$. It is the one discussed in Section 9.1.
- Foreign money market account. We denote the associated risk-neutral measure by $\widetilde{\mathbb{P}}^f$. It is constructed in Section 9.3 below.

- A zero-coupon bond maturing at time T. We denote the associated risk-neutral measure by $\widetilde{\mathbb{P}}^T$. It is called the *T-forward measure* and is used in Section 9.4.

The asset we take as numéraire could be one of the primary assets given by (9.1.1) or it could be a derivative asset. Regardless of which asset we take, it has the stochastic representation provided by the following theorem.

Theorem 9.2.1 (Stochastic representation of assets). *Let N be a strictly positive price process for a non-dividend-paying asset, either primary or derivative, in the multidimensional market model of Section 9.1. Then there exists a* vector volatility process

$$\nu(t) = (\nu_1(t), \ldots, \nu_d(t))$$

such that

$$dN(t) = R(t)N(t)\,dt + N(t)\nu(t) \cdot d\widetilde{W}(t). \tag{9.2.1}$$

This equation is equivalent to each of the equations

$$d\big(D(t)N(t)\big) = D(t)N(t)\nu(t) \cdot d\widetilde{W}(t), \tag{9.2.2}$$

$$D(t)N(t) = N(0)\exp\left\{\int_0^t \nu(u) \cdot d\widetilde{W}(u) - \frac{1}{2}\int_0^t \|\nu(u)\|^2\,du\right\}, \tag{9.2.3}$$

$$N(t) = N(0)\exp\left\{\int_0^t \nu(u) \cdot d\widetilde{W}(u) + \int_0^t \Big(R(u) - \frac{1}{2}\|\nu(u)\|^2\Big)du\right\}. \tag{9.2.4}$$

In other words, under the risk-neutral measure, every asset has a mean return equal to the interest rate. *The realized risk-neutral return for assets is characterized solely by their volatility vector processes (because initial conditions have no effect on return).*

PROOF: Under the risk-neutral measure $\widetilde{\mathbb{P}}$, the discounted price process $D(t)N(t)$ must be a martingale. The risk-neutral measure is constructed to enforce this condition for primary assets, and it is a consequence of the risk-neutral pricing formula for derivative assets. According to the Martingale Representation Theorem, Theorem 5.4.2,

$$d\left(D(t)N(t)\right) = \sum_{j=1}^{d} \widetilde{\Gamma}_j(t)\,d\widetilde{W}_j(t) = \widetilde{\Gamma}(t) \cdot d\widetilde{W}(t)$$

for some adapted d-dimensional process $\widetilde{\Gamma}(t) = (\widetilde{\Gamma}_1(t), \ldots, \widetilde{\Gamma}_d(t))$. Because $N(t)$ is strictly positive, we can define the vector $\nu(t) = (\nu_1(t), \ldots, \nu_d(t))$ by

$$\nu_j(t) = \frac{\widetilde{\Gamma}_j(t)}{D(t)N(t)}.$$

Then

$$d\left(D(t)N(t)\right) = D(t)N(t)\nu(t) \cdot d\widetilde{W}(t),$$

which is (9.2.2).

The solution to (9.2.2) is (9.2.3), as we now show. Define

$$\begin{aligned} X(t) &= \int_0^t \nu(u) \cdot d\widetilde{W}(u) - \frac{1}{2}\int_0^t \|\nu(u)\|^2\,du \\ &= \sum_{j=1}^d \int_0^t \nu_j(u)\,d\widetilde{W}_j(u) - \frac{1}{2}\sum_{j=1}^d \int_0^t \nu_j^2(u)\,du, \end{aligned}$$

so that

$$\begin{aligned} dX(t) &= \nu(t) \cdot d\widetilde{W}(t) - \frac{1}{2}\|\nu(t)\|^2\,dt \\ &= \sum_{j=1}^d \nu_j(t)\,d\widetilde{W}_j(t) - \frac{1}{2}\sum_{j=1}^d \nu_j^2(t)\,dt. \end{aligned}$$

Then

$$dX(t)\,dX(t) = \sum_{j=1}^d \nu_j^2(t)\,dt = \|\nu(t)\|^2\,dt.$$

Let $f(x) = N(0)e^x$, and compute

$$\begin{aligned} df(X(t)) &= f'(X(t))\,dX(t) + \frac{1}{2}f''(X(t))\,dX(t)\,dX(t) \\ &= f(X(t))\nu(t) \cdot d\widetilde{W}(t). \end{aligned}$$

We see that $f(X(t))$ solves (9.2.2), $f(X(t))$ has the desired initial condition $f(X(0)) = N(0)$, and $f(X(t))$ is the right-hand side of (9.2.3).

From (9.2.3), we have immediately that (9.2.4) holds. Applying the Itô-Doeblin formula to (9.2.4), we obtain (9.2.1). □

According to the multidimensional Girsanov Theorem, Theorem 5.4.1, we can use the volatility vector of $N(t)$ to change the measure. Define

$$\widetilde{W}_j^{(N)}(t) = -\int_0^t \nu_j(u)\,du + \widetilde{W}_j(t),\ j = 1, \ldots, d, \tag{9.2.5}$$

and a new probability measure

$$\widetilde{\mathbb{P}}^{(N)}(A) = \frac{1}{N(0)}\int_A D(T)N(T)\,d\widetilde{\mathbb{P}} \text{ for all } A \in \mathcal{F}. \tag{9.2.6}$$

We see from (9.2.3) that $\frac{D(T)N(T)}{N(0)}$ is the random variable $Z(T)$ appearing in (5.4.1) of the multidimensional Girsanov Theorem if we replace $\Theta_j(t)$ by

$-\nu_j(t)$ for $j = 1, \ldots, m$. Here we are using the probability measure $\widetilde{\mathbb{P}}$ in place of $\mathbb{P}$ in Theorem 5.4.1 and using the d-dimensional Brownian motion $(\widetilde{W}_1(t), \ldots, \widetilde{W}_d(t))$ under $\widetilde{\mathbb{P}}$ in place of the d-dimensional Brownian motion $(W_1(t), \ldots, W_d(t))$ under $\mathbb{P}$.

With these replacements, Theorem 5.4.1 implies that, under $\widetilde{\mathbb{P}}^{(N)}$, the process $\widetilde{W}^{(N)}(t) = (\widetilde{W}_1^{(N)}(t), \ldots, \widetilde{W}_d^{(N)}(t))$ is a d-dimensional Brownian motion. In particular, under $\widetilde{\mathbb{P}}^{(N)}$, the Brownian motions $\widetilde{W}_1^{(N)}, \ldots \widetilde{W}_d^{(N)}$ are independent of one another. The expected value of an arbitrary random variable X under $\widetilde{\mathbb{P}}^{(N)}$ can be computed by the formula

$$\widetilde{\mathbb{E}}^{(N)} X = \frac{1}{N(0)} \widetilde{\mathbb{E}}\big[X D(T) N(T)\big]. \tag{9.2.7}$$

More generally,

$$\frac{D(t)N(t)}{N(0)} = \widetilde{\mathbb{E}}\left[\frac{D(T)N(T)}{N(0)} \,\middle|\, \mathcal{F}(t) \right], \quad 0 \leq t \leq T,$$

is the Radon-Nikodým derivative process $Z(t)$ in the Theorem 5.4.1, and Lemma 5.2.2 implies that for $0 \leq s \leq t \leq T$ and Y an $\mathcal{F}(t)$-measurable random variable,

$$\widetilde{\mathbb{E}}^{(N)}[Y|\mathcal{F}(s)] = \frac{1}{D(s)N(s)} \widetilde{\mathbb{E}}\big[Y D(t) N(t) \big| \mathcal{F}(s)\big]. \tag{9.2.8}$$

Theorem 9.2.2 (Change of risk-neutral measure). *Let $S(t)$ and $N(t)$ be the prices of two assets denominated in a common currency, and let $\sigma(t) = (\sigma_1(t), \ldots, \sigma_d(t))$ and $\nu(t) = (\nu_1(t), \ldots, \nu_d(t))$ denote their respective volatility vector processes:*

$$d\big(D(t)S(t)\big) = D(t)S(t)\sigma(t) \cdot d\widetilde{W}(t), \quad d\big(D(t)N(t)\big) = D(t)N(t)\nu(t) \cdot d\widetilde{W}(t).$$

Take $N(t)$ as the numéraire, so the price of $S(t)$ becomes $S^{(N)}(t) = \frac{S(t)}{N(t)}$. Under the measure $\widetilde{\mathbb{P}}^{(N)}$, the process $S^{(N)}(t)$ is a martingale. Moreover,

$$dS^{(N)}(t) = S^{(N)}(t) \left[\sigma(t) - \nu(t)\right] \cdot d\widetilde{W}^{(N)}(t). \tag{9.2.9}$$

Remark 9.2.3. Equation (9.2.9) says that the volatility vector of $S^{(N)}(t)$ is the difference of the volatility vectors of $S(t)$ and $N(t)$. In particular, after the change of numéraire, the price of the numéraire becomes identically 1,

$$N^{(N)}(t) = \frac{N(t)}{N(t)} = 1,$$

and this has zero volatility vector:

$$dN^{(N)}(t) = N^{(N)}(t)[\nu(t) - \nu(t)] \cdot d\widetilde{W}^{(N)}(t) = 0.$$

We are not saying that volatilities subtract when we change the numéraire. We are saying that volatility *vectors* subtract. The process $N(t)$ in Theorem 9.2.2 has the stochastic differential representation (9.2.1), which we may rewrite as

$$dN(t) = R(t)N(t)\,dt + \|\nu(t)\|N(t)dB^N(t), \tag{9.2.10}$$

where

$$B^N(t) = \int_0^t \sum_{j=1}^d \frac{\nu_j(u)}{\|\nu(u)\|}\,d\widetilde{W}_u(t).$$

According to Lévy's Theorem, Theorem 4.6.4, $B^N(t)$ is a one-dimensional Brownian motion. From (9.2.10), we see that the volatility (not the volatility vector) of $N(t)$ is $\|\nu(t)\|$. Similarly, the volatility of $S(t)$ in Theorem 9.2.2 is $\|\sigma(t)\|$. Application of the same argument to equation (9.2.9) shows that the volatility of $S^{(N)}(t)$ is $\|\sigma(t)-\nu(t)\|$. This is not the difference of the volatilities $\|\sigma(t)\| - \|\nu(t)\|$ unless the volatility vector $\sigma(t)$ is a positive multiple of the volatility vector $\nu(t)$.

Remark 9.2.4. If we take the money market account as the numéraire in Theorem 9.2.2 (i.e., $N(t) = M(t) = \frac{1}{D(t)}$), then we have $d\big(D(t)N(t)\big) = 0$. The volatility vector for the money market account is $\nu(t) = 0$, and the volatility vector for an asset $S^{(N)}(t)$ denominated in units of money market account is the same as the volatility vector of the asset denominated in units of currency. Discounting an asset using the money market account does not affect its volatility vector.

Remark 9.2.5. Theorem 9.2.2 is a special case of a more general result. Whenever $M_1(t)$ and $M_2(t)$ are martingales under a measure $\mathbb{P}$, $M_2(0) = 1$, and $M_2(t)$ takes only positive values, then $M_1(t)/M_2(t)$ is a martingale under the measure $\mathbb{P}^{(M_2)}$ defined by

$$\mathbb{P}^{(M_2)}(A) = \int_A M_2(T)\,d\mathbb{P}.$$

See Exercise 9.1.

PROOF OF THEOREM 9.2.2: We have

$$D(t)S(t) = S(0)\exp\left\{\int_0^t \sigma(u)\cdot d\widetilde{W}(u) - \frac{1}{2}\int_0^t \|\sigma(u)\|^2\,du\right\},$$

$$D(t)N(t) = N(0)\exp\left\{\int_0^t \nu(u)\cdot d\widetilde{W}(u) - \frac{1}{2}\int_0^t \|\nu(u)\|^2\,du\right\},$$

and hence

$$S^{(N)}(t) = \frac{S(0)}{N(0)}\exp\left\{\int_0^t (\sigma(u)-\nu(u))\cdot d\widetilde{W}(u) - \frac{1}{2}\int_0^t \left(\|\sigma(u)\|^2 - \|\nu(u)\|^2\right)du\right\}.$$

To apply the Itô-Doeblin formula to this, we first define

$$X(t) = \int_0^t \big(\sigma(u) - \nu(u)\big) \cdot d\widetilde{W}(u) - \frac{1}{2} \int_0^t \big(\|\sigma(u)\|^2 - \|\nu(u)\|^2\big)\, du,$$

so that

$$\begin{aligned}
dX(t) &= \big(\sigma(t) - \nu(t)\big) \cdot d\widetilde{W}(t) - \frac{1}{2}\big(\|\sigma(t)\|^2 - \|\nu(t)\|^2\big)\, dt \\
&= \sum_{j=1}^d \big(\sigma_j(t) - \nu_j(t)\big)\, d\widetilde{W}_j(t) - \frac{1}{2}\sum_{j=1}^d \big(\sigma_j^2(t) - \nu_j^2(t)\big)\, dt, \\
dX(t)\, dX(t) &= \sum_{j=1}^d \big(\sigma_j(t) - \nu_j(t)\big)^2 dt \\
&= \sum_{j=1}^d \big(\sigma_j^2(t) - 2\sigma_j(t)\nu_j(t) + \nu_j^2(t)\big)\, dt \\
&= \|\sigma(t)\|^2\, dt - 2\sigma(t)\cdot\nu(t)\, dt + \|\nu(t)\|^2\, dt.
\end{aligned}$$

With $f(x) = \frac{S(0)}{N(0)} e^x$, we have $S^{(N)}(t) = f(X(t))$ and

$$\begin{aligned}
dS^{(N)}(t) &= df(X(t)) \\
&= f'(X)\, dX + \frac{1}{2} f''(X)\, dX\, dX \\
&= S^{(N)} \Big[(\sigma - \nu)\cdot d\widetilde{W} - \frac{1}{2}\|\sigma\|^2\, dt + \frac{1}{2}\|\nu\|^2\, dt \\
&\qquad + \frac{1}{2}\|\sigma\|^2\, dt - \sigma\cdot\nu\, dt + \frac{1}{2}\|\nu\|^2\, dt\Big] \\
&= S^{(N)} \Big[(\sigma - \nu)\cdot d\widetilde{W} - \nu\cdot(\sigma - \nu)\, dt\Big] \\
&= S^{(N)} (\sigma - \nu)\cdot(-\nu\, dt + d\widetilde{W}) \\
&= S^{(N)} (\sigma - \nu)\cdot d\widetilde{W}^{(N)}.
\end{aligned}$$

Since $\widetilde{W}^{(N)}(t)$ is a d-dimensional Brownian motion under $\widetilde{\mathbb{P}}^{(N)}$, the process $S^{(N)}(t)$ is a martingale under this measure. □

9.3 Foreign and Domestic Risk-Neutral Measures

9.3.1 The Basic Processes

We now apply the ideas of the previous section to a market with two currencies, which we call foreign and domestic. This model is driven by

$$W(t) = (W_1(t), W_2(t)),$$

a two-dimensional Brownian motion on some $(\Omega, \mathcal{F}, \mathbb{P})$. In particular, we are assuming that W_1 and W_2 are independent under $\mathbb{P}$. We begin with a stock whose price in domestic currency, $S(t)$, satisfies

$$dS(t) = \alpha(t)S(t)\,dt + \sigma_1(t)S(t)\,dW_1(t). \tag{9.3.1}$$

There is a domestic interest rate $R(t)$, which leads to a domestic money market account price and domestic discount process

$$M(t) = e^{\int_0^t R(u)du}, \quad D(t) = e^{-\int_0^t R(u)du}.$$

There is also a foreign interest rate $R^f(t)$, which leads to a foreign money market account price and foreign discount process

$$M^f(t) = e^{\int_0^t R^f(u)du}, \quad D^f(t) = e^{-\int_0^t R^f(u)du}.$$

Finally, there is an exchange rate $Q(t)$, which gives units of domestic currency per unit of foreign currency. We assume this satisfies

$$dQ(t) = \gamma(t)Q(t)\,dt + \sigma_2(t)Q(t)\left[\rho(t)\,dW_1(t) + \sqrt{1-\rho^2(t)}\,dW_2(t)\right]. \tag{9.3.2}$$

We define

$$W_3(t) = \int_0^t \rho(u)\,dW_1(u) + \int_0^t \sqrt{1-\rho^2(u)}\,dW_2(u). \tag{9.3.3}$$

By Lévy's Theorem, Theorem 4.6.4, $W_3(t)$ is a Brownian motion under $\widetilde{\mathbb{P}}$. We may rewrite (9.3.2) as

$$dQ(t) = \gamma(t)Q(t)\,dt + \sigma_2(t)Q(t)\,dW_3(t), \tag{9.3.4}$$

from which we see that $Q(t)$ has volatility $\sigma_2(t)$.

We assume $R(t)$, $R^f(t)$, $\sigma_1(t)$, $\sigma_2(t)$, and $\rho(t)$ are processes adapted to the filtration $\mathcal{F}(t)$ generated by the two-dimensional Brownian motion $W(t) = (W_1(t), W_2(t))$, and

$$\sigma_1(t) > 0, \quad \sigma_2(t) > 0, \quad -1 < \rho(t) < 1$$

for all t almost surely. Because

$$\frac{dS(t)}{S(t)} \cdot \frac{dQ(t)}{Q(t)} = \rho(t)\sigma_1(t)\sigma_2(t)\,dt,$$

the process $\rho(t)$ is the instantaneous correlation between relative changes in $S(t)$ and $Q(t)$.

9.3.2 Domestic Risk-Neutral Measure

There are three assets that can be traded: the domestic money market account, the stock, and the foreign money market account. We shall price each of these in domestic currency and discount at the domestic interest rate. The result is the price of each of them in units of the domestic money market account. Under the domestic risk-neutral measure, all three assets priced in units of the domestic money market account must be martingales. We use this observation to find the domestic risk-neutral measure.

We note that the first asset, the domestic money market account, when priced in units of the domestic money market, has constant price 1. This is always a martingale, regardless of the measure being used.

The second asset, the stock, in units of the domestic money market account has price $D(t)S(t)$, and this satisfies the stochastic differential equation

$$d\big(D(t)S(t)\big) = D(t)S(t)\big[\big(\alpha(t) - R(t)\big)\,dt + \sigma_1(t)\,dW_1(t)\big]. \tag{9.3.5}$$

We would like to construct a process

$$\widetilde{W}_1(t) = \int_0^t \Theta_1(u)\,du + W_1(t)$$

that permits us to rewrite (9.3.5) as

$$d\big(D(t)S(t)\big) = \sigma_1(t)D(t)S(t)\,d\widetilde{W}_1(t). \tag{9.3.6}$$

Equating the right-hand sides of (9.3.5) and (9.3.6), we see that $\Theta_1(t)$ must be chosen to satisfy the *first market price of risk equation*

$$\sigma_1(t)\Theta_1(t) = \alpha(t) - R(t). \tag{9.3.7}$$

The third asset available in the domestic market is the following. One can invest in the foreign money market account and convert that investment to domestic currency. The value of the foreign money market account in domestic currency is $M^f(t)Q(t)$, and its discounted value is $D(t)M^f(t)Q(t)$. The differential of this price is

$$\begin{aligned} d\big(D(t)M^f(t)Q(t)\big) &= D(t)M^f(t)Q(t)\big[\big(R^f(t) - R(t) + \gamma(t)\big)\,dt \\ &\quad + \sigma_2(t)\rho(t)\,dW_1(t) + \sigma_2(t)\sqrt{1-\rho^2(t)}\,dW_2(t)\big]. \end{aligned} \tag{9.3.8}$$

One can derive this using the fact that

$$d\left(M^f(t)\right) = R^f(t)M^f(t)\,dt,$$

using Itô's product rule to compute

$$\begin{aligned} d\big(M^f(t)Q(t)\big) &= M^f(t)Q(t)\big[\big(R^f(t) + \gamma(t)\big)\,dt \\ &\quad + \sigma_2(t)\rho(t)\,dW_1(t) + \sigma_2(t)\sqrt{1-\rho^2(t)}\,dW_2(t)\big], \end{aligned}$$

and then using Itô's product rule again on $D(t) \cdot M^f(t)Q(t)$ to obtain (9.3.8). The mean rate of change of $Q(t)$ is $\gamma(t)$. When we inflate this at the foreign interest rate and discount it at the domestic interest rate, (9.3.8) shows that the mean rate of return changes to $R^f(t) - R(t) + \gamma(t)$. The volatility terms are unchanged.

In addition to the process $\widetilde{W}_1(t)$, we would like to construct a process

$$\widetilde{W}_2(t) = \int_0^t \Theta_2(u)\,du + W_2(t)$$

so that (9.3.8) can be written as

$$\begin{aligned} &d\big(D(t)M^f(t)Q(t)\big) \\ &\quad = D(t)M^f(t)Q(t)\big[\sigma_2(t)\rho(t)\,d\widetilde{W}_1(t) + \sigma_2(t)\sqrt{1-\rho^2(t)}\,d\widetilde{W}_2(t)\big]. \end{aligned} \tag{9.3.9}$$

Equating the right-hand sides of (9.3.9) and (9.3.8), we obtain the *second market price of risk equation*

$$\sigma_2(t)\rho(t)\Theta_1(t) + \sigma_2(t)\sqrt{1-\rho^2(t)}\,\Theta_2(t) = R^f(t) - R(t) + \gamma(t). \tag{9.3.10}$$

The market price of risk equations (9.3.7) and (9.3.10) determine processes $\Theta_1(t)$ and $\Theta_2(t)$. We can solve explicitly for these processes by first solving (9.3.7) for $\Theta_1(t)$, substituting this into (9.3.10), and then solving (9.3.10) for $\Theta_2(t)$. The conditions $\sigma_1(t) > 0$, $\sigma_2(t) > 0$, and $-1 < \rho(t) < 1$ are needed to do this.

The particular formulas for $\Theta_1(t)$ and $\Theta_2(t)$ are irrelevant. What matters is that the market price of risk equations have one and only one solution, and so there is a unique risk-neutral measure $\widetilde{\mathbb{P}}$ given by the multidimensional Girsanov Theorem. Under this measure, $\widetilde{W}(t) = (\widetilde{W}_1(t), \widetilde{W}_2(t))$ is a two-dimensional Brownian motion and the processes 1, $D(t)S(t)$, and $D(t)M^f(t)Q(t)$ are martingales. In the spirit of (9.3.3), we may also define

$$\widetilde{W}_3(t) = \int_0^t \rho(u)\,d\widetilde{W}_1(u) + \int_0^t \sqrt{1-\rho^2(u)}\,d\widetilde{W}_2(t). \tag{9.3.11}$$

Then $\widetilde{W}_3(t)$ is a Brownian motion under $\widetilde{\mathbb{P}}$, and

$$d\widetilde{W}_1(t)\,d\widetilde{W}_3(t) = \rho(t)\,dt, \quad d\widetilde{W}_2(t)\,d\widetilde{W}_3(t) = \sqrt{1-\rho^2(t)}\,dt. \tag{9.3.12}$$

We can write the price processes 1, $D(t)S(t)$ and $D(t)M^f(t)Q(t)$ in undiscounted form by multiplying them by $M(t) = \frac{1}{D(t)}$ and using the formula $dM(t) = R(t)M(t)\,dt$ and Itô's product rule. This leads to the formulas

$$dM(t) = R(t)M(t)\,dt, \tag{9.3.13}$$

$$dS(t) = S(t)\big[R(t)\,dt + \sigma_1(t)\,d\widetilde{W}_1(t)\big], \tag{9.3.14}$$

$$\begin{aligned} d\big(M^f(t)Q(t)\big) &= M^f(t)Q(t)\big[R(t)\,dt + \sigma_2(t)\rho(t)\,d\widetilde{W}_1(t) \\ &\qquad + \sigma_2(t)\sqrt{1-\rho^2(t)}\,d\widetilde{W}_2(t)\big] \\ &= M^f(t)Q(t)\big[R(t)\,dt + \sigma_2(t)\,d\widetilde{W}_3(t)\big]. \end{aligned} \tag{9.3.15}$$

All these price processes have mean rate of return $R(t)$ under the domestic risk-neutral measure $\widetilde{\mathbb{P}}$. We constructed the domestic risk-neutral measure so this is the case.

We may multiply $M^f(t)Q(t)$ by $D^f(t)$ and use Itô's product rule again to obtain

$$\begin{aligned} dQ(t) &= Q(t)\big[\big(R(t)-R^f(t)\big)\,dt + \sigma_2(t)\rho(t)\,d\widetilde{W}_1(t) + \sigma_2(t)\sqrt{1-\rho^2(t)}\,d\widetilde{W}_2(t)\big] \\ &= Q(t)\big[\big(R(t)-R^f(t)\big)\,dt + \sigma_2(t)\,d\widetilde{W}_3(t)\big]. \qquad (9.3.16) \end{aligned}$$

Under the domestic risk-neutral measure, the mean rate of change of the exchange rate is the difference between the domestic and foreign interest rates $R(t)-R^f(t)$. In particular, it is not $R(t)$, as would be the case for an asset. If one regards the exchange rate as an asset (i.e., hold a unit of foreign currency whose value is always $Q(t)$), then it is a *dividend-paying* asset. The unit of foreign currency can and should be invested in the foreign money market, and this pays out a continuous dividend at rate $R^f(t)$. If this dividend is reinvested in the foreign money market, then we get the asset in (9.3.15), which has mean rate of return $R(t)$; if the dividend is not reinvested, then the rate of return is reduced by $R^f(t)$ and we have (9.3.16) (cf. (5.5.6)).

It is important to note that (9.3.16) tells us about the mean rate of change of the exchange rate under the domestic risk-neutral measure. Under the actual probability measure $\mathbb{P}$, the mean rate of change of the exchange rate can be anything. There are no restrictions on the process $\gamma(t)$ in (9.3.2).

9.3.3 Foreign Risk-Neutral Measure

In this model, we have three assets: the domestic money market account, the stock, and the foreign money market account. We list these assets across the top of Figure 9.3.1, and down the side of the figure we list the four ways of denominating them.

In the previous subsection, we constructed the domestic risk-neutral measure $\widetilde{\mathbb{P}}$ under which the three entries in the second line of Figure 9.3.1 are martingales. In this subsection, we construct the foreign risk-neutral measure under which the entries in the fourth line are martingales. (We cannot make all the entries in the first line be martingales because every path of the process $M(t)$ is increasing, and thus this process is not a martingale under any measure. The same applies to the entries in the third line, which contains the increasing process $M^f(t)$.)

We observe that the fourth line in Figure 9.3.1 is obtained by dividing each entry of the second line by $D(t)M^f(t)Q(t)$. In other words, to find the foreign risk-neutral measure, we take the foreign money market account as the numéraire. Its value at time t, denominated in units of the domestic money market account, is $D(t)M^f(t)Q(t)$, and denominated in units of domestic currency, it is $M^f(t)Q(t)$. The differential of $M^f(t)Q(t)$ is given in (9.3.15), and from that formula we see that its volatility vector is

$$(\nu_1(t), \nu_2(t)) = \left(\sigma_2(t)\rho(t), \sigma_2(t)\sqrt{1-\rho^2(t)}\right),$$

the same as the volatility vector of $Q(t)$.

	Domestic money market	Stock	Foreign money market
Domestic currency	$M(t)$	$S(t)$	$M^f(t)Q(t)$
Domestic money market	1	$D(t)S(t)$	$D(t)M^f(t)Q(t)$
Foreign currency	$M(t)/Q(t)$	$S(t)/Q(t)$	$M^f(t)$
Foreign money market	$M(t)D^f(t)/Q(t)$	$D^f(t)S(t)/Q(t)$	1

Fig. 9.3.1. Prices under different numéraires.

According to Theorem 9.2.2, the risk-neutral measure associated with the numéraire $M^f(t)Q(t)$ is given by

$$\widetilde{\mathbb{P}}^f(A) = \frac{1}{Q(0)} \int_A D(T)M^f(T)Q(T)\, d\widetilde{\mathbb{P}} \text{ for all } A \in \mathcal{F}, \tag{9.3.17}$$

where we have used the fact that $D(0) = M^f(0) = 1$. Furthermore, the process $\widetilde{W}^f(t) = (\widetilde{W}_1^f(t), \widetilde{W}_2^f(t))$ given by

$$\widetilde{W}_1^f(t) = -\int_0^t \sigma_2(u)\rho(u)\, du + \widetilde{W}_1(t), \tag{9.3.18}$$

$$\widetilde{W}_2^f(t) = -\int_0^t \sigma_2(u)\sqrt{1-\rho^2(u)}\, du + \widetilde{W}_2(t), \tag{9.3.19}$$

is a two-dimensional Brownian motion under $\widetilde{\mathbb{P}}^f$. We call $\widetilde{\mathbb{P}}^f$ the *foreign risk-neutral measure.* Following (9.3.11), we may also define

$$\begin{aligned}
\widetilde{W}_3^f(t) &= \int_0^t \rho(u)\, d\widetilde{W}_1^f(u) + \int_0^t \sqrt{1-\rho^2(u)}\, d\widetilde{W}_2^f(t) \\
&= \int_0^t \rho(u)\big(-\sigma_2(u)\rho(u)\, du + d\widetilde{W}_1(u)\big) \\
&\qquad + \int_0^t \sqrt{1-\rho^2(u)}\big(-\sigma_2(u)\sqrt{1-\rho^2(u)} + d\widetilde{W}_2(u)\big) \\
&= -\int_0^t \sigma_2(u)\, du + \int_0^t \big(\rho(u)\, d\widetilde{W}_1(u) + \sqrt{1-\rho^2(u)}\, d\widetilde{W}_2(u)\big) \\
&= -\int_0^t \sigma_2(u)\, du + \widetilde{W}_3(t).
\end{aligned} \tag{9.3.20}$$

Then $\widetilde{W}_3^f(t)$ is a Brownian motion under $\widetilde{\mathbb{P}}^f$, and

$$d\widetilde{W}_1^f(t)\,d\widetilde{W}_3^f(t) = \rho(t)\,dt, \quad d\widetilde{W}_2^f(t)\,d\widetilde{W}_3^f(t) = \sqrt{1-\rho^2(t)}\,dt. \tag{9.3.21}$$

Instead of relying on Theorem 9.2.2, one can verify directly by Itô calculus that the first two entries in the last row of Figure 9.3.1 are martingales under $\widetilde{\mathbb{P}}^f$ (the third entry, 1, is obviously a martingale). One can verify by direct computation that

$$\begin{aligned} d\left(\frac{M(t)D^f(t)}{Q(t)}\right) &= \frac{M(t)D^f(t)}{Q(t)}\big[-\sigma_2(t)\rho(t)\,d\widetilde{W}_1^f(t) \\ &\qquad -\sigma_2(t)\sqrt{1-\rho^2(t)}\,d\widetilde{W}_2^f(t)\big] \\ &= -\frac{M(t)D^f(t)}{Q(t)}\sigma_2(t)\,d\widetilde{W}_3^f(t), \end{aligned} \tag{9.3.22}$$

$$\begin{aligned} d\left(\frac{D^f(t)S(t)}{Q(t)}\right) &= \frac{D^f(t)S(t)}{Q(t)}\big[(\sigma_1(t)-\sigma_2(t)\rho(t))\,d\widetilde{W}_1^f(t) \\ &\qquad -\sigma_2(t)\sqrt{1-\rho^2(t)}\,d\widetilde{W}_2^f(t)\big] \\ &= \frac{D^f(t)S(t)}{Q(t)}\big[\sigma_1(t)\,d\widetilde{W}_1^f(t) - \sigma_2(t)\,d\widetilde{W}_3^f(t)\big], \end{aligned} \tag{9.3.23}$$

Because $\widetilde{W}_1^f(t)$, $\widetilde{W}_2^f(t)$, and $\widetilde{W}_3^f(t)$ are Brownian motions under $\widetilde{\mathbb{P}}^f$, the processes above are martingales under this measure. The Brownian motions $\widetilde{W}_1^f(t)$ and $\widetilde{W}_2^f(t)$ are independent under $\widetilde{\mathbb{P}}^f$, whereas $\widetilde{W}_3^f(t)$ has instantaneous correlations with $\widetilde{W}_1^f(t)$ and $\widetilde{W}_2^f(t)$ given by (9.3.21).

9.3.4 Siegel's Exchange Rate Paradox

In (9.3.16), we saw that under the domestic risk-neutral measure $\widetilde{\mathbb{P}}$, the mean rate of change for the exchange rate $Q(t)$ is $R(t) - R^f(t)$. From the foreign perspective, the exchange rate is $\frac{1}{Q(t)}$, and one should expect the mean rate of change of $\frac{1}{Q(t)}$ to be $R^f(t) - R(t)$. In other words, one might expect that if the average rate of change of the dollar against the euro is 5%, then the average rate of change of the euro against the dollar should be -5%. This turns out not to be as straight forward as one might expect because of the convexity of the function $f(x) = \frac{1}{x}$.

For example, an exchange rate of 0.90 euros to the dollar would be 1.1111 dollars to the euro. If the dollar price of euro falls by 5%, then price of the euro would be only $0.95 \times 1.1111 = 1.0556$ dollars. This is an exchange rate of 0.9474 euros to the dollar. The change from 0.90 euros to the dollar to 0.9474 euros to the dollar is a 5.26% increase in the euro price of the dollar, not a 5% increase.

The convexity effect seen in the previous paragraph makes itself felt when we compute the differential of $\frac{1}{Q(t)}$. We take $f(x) = \frac{1}{x}$ so that $f'(x) = -\frac{1}{x^2}$ and $f''(x) = \frac{2}{x^3}$. Using (9.3.16), we obtain

$$
\begin{aligned}
d\left(\frac{1}{Q}\right) &= df(Q)\\
&= f'(Q)\,dQ + \frac{1}{2}f''(Q)\,dQ\,dQ\\
&= \frac{1}{Q}\left[(R^f - R)\,dt - \sigma_2 d\widetilde{W}_3\right] + \frac{1}{Q}\sigma_2^2\,d\widetilde{W}_3\,d\widetilde{W}_3\\
&= \frac{1}{Q(t)}\left[(R^f - R + \sigma_2^2)\,dt - \sigma_2 d\widetilde{W}_3\right]. \qquad (9.3.24)
\end{aligned}
$$

The mean rate of change under the domestic risk-neutral measure is $R^f(t) - R(t) + \sigma_2^2$, not $R^f(t) - R(t)$.

However, the asymmetry introduced by the convexity of $f(x) = \frac{1}{x}$ is resolved if we switch to the foreign risk-neutral measure, which is the appropriate one for derivative security pricing in the foreign currency. First recall the relationship (9.3.20)

$$
d\widetilde{W}_3^f(t) = -\sigma_2(t)\,dt + d\widetilde{W}_3(t).
$$

In terms of $\widetilde{W}_3^f(t)$, we may rewrite (9.3.24) as

$$
d\left(\frac{1}{Q}\right) = \left(\frac{1}{Q}\right)\left[(R^f - R)\,dt - \sigma_2 d\widetilde{W}_3^f\right]. \qquad (9.3.25)
$$

Under the foreign risk-neutral measure, the mean rate of change for $\frac{1}{Q}$ is $R^f - R$, as expected.

Under the actual probability measure $\mathbb{P}$, however, the asymmetry remains. When we begin with (9.3.4), which shows the mean rate of change of the exchange rate to be $\gamma(t)$ under $\mathbb{P}$ and is repeated below as (9.3.26), and then use the Itô-Doeblin formula as we did in (9.3.24), we obtain the formula (9.3.27) below:

$$
dQ(t) = \gamma(t)Q(t)\,dt + \sigma_2(t)Q(t)\,dW_3(t), \qquad (9.3.26)
$$

$$
d\left(\frac{1}{Q(t)}\right) = \frac{1}{Q(t)}\big(-\gamma(t) + \sigma_2^2(t)\big)dt - \frac{1}{Q(t)}\sigma_2(t)\,dW_3(t). \qquad (9.3.27)
$$

Both Q and $\frac{1}{Q}$ have the same volatility. (A change of sign in the volatility does not affect volatility because Brownian motion is symmetric.) However, the mean rates of change of Q and $\frac{1}{Q}$ are not negatives of one another.

9.3.5 Forward Exchange Rates

We assume in this subsection that the domestic and foreign interest rates are constant and denote these constants by r and r^f, respectively. Recall that Q is units of domestic currency per unit of foreign currency. The exchange rate from the *domestic* viewpoint is governed by the stochastic differential equation (9.3.16)

$$dQ(t) = Q(t)\left[(r - r^f)\,dt + \sigma_2(t)\rho(t)\,d\widetilde{W}_1(t) + \sigma_2(t)\sqrt{1-\rho^2(t)}\,d\widetilde{W}_2(t)\right].$$

Therefore

$$e^{-(r-r^f)t}Q(t)$$

is a martingale under $\widetilde{\mathbb{P}}$, the *domestic* risk-neutral measure.

At time zero, the (domestic currency) forward price F for a unit of foreign currency, to be delivered at time T, is determined by the equation

$$\widetilde{\mathbb{E}}\left[e^{-rT}(Q(T) - F)\right] = 0.$$

The left-hand side is the risk-neutral pricing formula applied to the derivative security that pays $Q(T)$ in exchange for F at time T. Setting this equal to zero determines the forward price. We may solve this equation for F by observing that it implies

$$e^{-rT}F = \widetilde{\mathbb{E}}\left[e^{-rT}Q(T)\right] = e^{-r^fT}\widetilde{\mathbb{E}}\left[e^{-(r-r^f)T}Q(T)\right] = e^{-r^fT}Q(0),$$

which gives the T-forward (domestic per unit of foreign) exchange rate

$$F = e^{(r-r^f)T}Q(0).$$

The exchange rate from the foreign viewpoint is given by the stochastic differential equation (9.3.25)

$$d\left(\frac{1}{Q(t)}\right)$$
$$= \left(\frac{1}{Q(t)}\right)\left[(r^f - r)\,dt - \sigma_2(t)\rho(t)\,d\widetilde{W}_1^f(t) - \sigma_2(t)\sqrt{1-\rho^2(t)}\,d\widetilde{W}_2^f(t)\right].$$

Therefore,

$$e^{-(r^f-r)t}\frac{1}{Q(t)}$$

is a martingale under $\widetilde{\mathbb{P}}^f$, the *foreign* risk-neutral measure.

At time zero, the (foreign currency) forward price F^f for a unit of domestic currency to be delivered at time T is determined by the equation

$$\widetilde{\mathbb{E}}^f\left[e^{-r^fT}\left(\frac{1}{Q(T)} - F^f\right)\right] = 0.$$

The left-hand side is the risk-neutral pricing formula applied to the derivative security that pays $\frac{1}{Q(T)}$ in exchange for F^f (both denominated in foreign currency) at time T. Setting this equal to zero determines the forward price. We may solve this equation for F^f by observing that it implies

$$e^{-r^fT}F^f = \widetilde{\mathbb{E}}^f\left[e^{-r^fT}\frac{1}{Q(T)}\right] = e^{-rT}\widetilde{\mathbb{E}}^f\left[e^{-(r^f-r)T}\frac{1}{Q(T)}\right] = e^{-rT}\frac{1}{Q(0)},$$

which gives the T-forward (foreign per unit of domestic) exchange rate

$$F^f = e^{(r^f-r)T}\frac{1}{Q(0)} = \frac{1}{F}.$$

9.3.6 Garman-Kohlhagen Formula

In this section, we assume the domestic and foreign interest rates r and r^f and the volatility σ_2 are constant. Consider a call on a unit of foreign currency whose payoff in domestic currency is $(Q(T)-K)^+$. At time zero, the value of this is

$$\widetilde{\mathbb{E}}e^{-rT}(Q(T)-K)^+.$$

In this case, (9.3.16) becomes

$$dQ(t) = Q(t)\left[\left(r-r^f\right)dt + \sigma_2\, d\widetilde{W}_3(t)\right],$$

from which we conclude that

$$Q(T) = Q(0)\exp\left\{\sigma_2 \widetilde{W}_3(T) + \left(r - r^f - \frac{1}{2}\sigma_2^2\right)T\right\}.$$

Define

$$Y = -\frac{\widetilde{W}_3(T)}{\sqrt{T}},$$

so Y is a standard normal random variable under $\widetilde{\mathbb{P}}$. Then the price of the call is

$$\begin{aligned}
&\widetilde{\mathbb{E}}e^{-rT}\left(Q(T)-K\right)^+\\
&= \widetilde{\mathbb{E}}\left[e^{-rT}\left(Q(0)\exp\left\{-\sigma_2\sqrt{T}Y + \left(r - r^f - \frac{1}{2}\sigma_2^2\right)T\right\} - K\right)^+\right].
\end{aligned}$$

This expression is just like (5.5.10) with $\tau = T$, with $Q(0)$ in place of x, and with r^f in place of the dividend rate a. According to (5.5.12), the call price is

$$\widetilde{\mathbb{E}}e^{-rT}\left(Q(T)-K\right)^+ = e^{-r^f T}Q(0)N(d_+) - e^{-rT}KN(d_-), \qquad (9.3.28)$$

where

$$d_\pm = \frac{1}{\sigma_2\sqrt{T}}\left[\log\frac{Q(0)}{K} + \left(r - r^f \pm \frac{1}{2}\sigma_2^2\right)T\right]$$

and N is the cumulative standard normal distribution function. Equation (9.3.28) is called the *Garman-Kohlhagen formula*.

9.3.7 Exchange Rate Put–Call Duality

In this subsection, we develop a relationship between a call on domestic currency, denominated in foreign currency, and a put on a foreign currency, denominated in the domestic currency.

Recall the numéraire $M^f(t)Q(t)$, which is the domestic price of the foreign money market account. The Radon-Nikodým derivative of the foreign risk-neutral measure with respect to the domestic risk-neutral measure is (see (9.3.17))

$$\frac{d\widetilde{\mathbb{P}}^f}{d\widetilde{\mathbb{P}}} = \frac{D(T)M^f(T)Q(T)}{Q(0)}.$$

Thus, for any random variable X,

$$\widetilde{\mathbb{E}}^f X = \widetilde{\mathbb{E}}\left[\frac{D(T)M^f(T)Q(T)}{Q(0)} X\right].$$

A call struck at K on a unit of domestic currency denominated in the foreign currency pays off $\left(\frac{1}{Q(T)} - K\right)^+$ units of foreign currency at expiration time T. The foreign currency value of this at time zero, which is the foreign risk-neutral expected value of the discounted payoff, is

$$\begin{aligned}
&\widetilde{\mathbb{E}}^f\left[D^f(T)\left(\frac{1}{Q(T)} - K\right)^+\right] \\
&= \widetilde{\mathbb{E}}\left[\frac{D(T)M^f(T)Q(T)}{Q(0)} \cdot D^f(T)\left(\frac{1}{Q(T)} - K\right)^+\right] \\
&= \frac{1}{Q(0)}\widetilde{\mathbb{E}}\left[D(T)\big(1 - KQ(T)\big)^+\right] \\
&= \frac{K}{Q(0)}\widetilde{\mathbb{E}}\left[D(T)\left(\frac{1}{K} - Q(T)\right)^+\right].
\end{aligned}$$

This is the time-zero value in domestic currency of $\frac{K}{Q(0)}$ puts on the foreign exchange rate. More specifically, a put struck at $\frac{1}{K}$ on a unit of foreign currency denominated in the domestic currency pays off $\left(\frac{1}{K} - Q(T)\right)^+$ units of domestic currency at expiration time T. The domestic currency value of this put at time zero, which is the domestic risk-neutral expected value of the discounted payoff, is

$$\widetilde{\mathbb{E}}\left[D(T)\left(\frac{1}{K} - Q(T)\right)^+\right].$$

The call we began with is worth $\frac{K}{Q(0)}$ of these puts.

The foreign currency price of the put struck at $\frac{1}{K}$ on a unit of foreign currency is

$$\frac{1}{Q(0)}\widetilde{\mathbb{E}}\left[D(T)\left(\frac{1}{K} - Q(T)\right)^+\right].$$

The call we began with has a value K times this amount. When we denominate both the call and the put this way in foreign currency, we can then understand the final result. Indeed, we have seen that the option to exchange K units of foreign currency for one unit of domestic currency (the call) is the same as K options to exchange $\frac{1}{K}$ units of domestic currency for one unit of foreign currency (the put). Stated in this way, the result is almost obvious.

9.4 Forward Measures

Although there may be multiple Brownian motions driving the model of this section, in order to simplify the notation, we assume in this section that there is only one. It is not difficult to rederive the results presented here under the assumption that there are d Brownian motions.

9.4.1 Forward Price

We recall the discussion of Section 5.6.1. Consider a zero-coupon bond that pays 1 unit of currency (all currency is domestic in this section) at maturity T. According to the risk-neutral pricing formula, the value of this bond at time $t \in [0,T]$ is

$$B(t,T) = \frac{1}{D(t)}\widetilde{\mathbb{E}}\big[D(T)\big|\mathcal{F}(t)\big]. \tag{9.4.1}$$

In particular, $B(T,T) = 1$.

Consider now an asset whose price denominated in currency is $S(t)$. A *forward contract* that delivers one share of this asset at time T in exchange for K has a time-T payoff of $S(T) - K$. According to the risk-neutral pricing formula, the value of this contract at earlier times t is

$$V(t) = \frac{1}{D(t)}\widetilde{\mathbb{E}}\big[D(T)(S(T) - K)\big|\mathcal{F}(t)\big].$$

Because $D(t)S(t)$ is a martingale under $\widetilde{\mathbb{P}}$, this reduces to

$$V(t) = S(t) - \frac{K}{D(t)}\widetilde{\mathbb{E}}\big[D(T)\big|\mathcal{F}(t)\big] = S(t) - KB(t,T). \tag{9.4.2}$$

The *T-forward price* $\mathrm{For}_S(t,T)$ at time $t \in [0,T]$ of an asset is the value of K that causes the value of the forward contract in (9.4.2) to be zero:

$$\mathrm{For}_S(t,T) = \frac{S(t)}{B(t,T)}. \tag{9.4.3}$$

9.4.2 Zero-Coupon Bond as Numéraire

A zero-coupon bond is an asset, and therefore the discounted bond price $D(t)B(t,T)$ must be a martingale under the risk-neutral measure $\widetilde{\mathbb{P}}$. According to Theorem 9.2.1, there is a volatility process $\sigma^*(t,T)$ for the bond (a process in t; T is fixed) such that

$$d\big(D(t)B(t,T)\big) = -\sigma^*(t,T)D(t)B(t,T)\,d\widetilde{W}(t). \tag{9.4.4}$$

In (9.4.4), we write $-\sigma^*(t,T)$ rather than $\sigma^*(t,T)$ in order to be consistent with the notation used in our discussion of the Heath-Jarrow-Morton model in Chapter 10. This has no effect on the distribution of the bond price process since we could just as well write (9.4.4) as

$$d\big(D(t)B(t,T)\big) = \sigma^*(t,T)D(t)B(t,T)\,d\big(-\widetilde{W}(t)\big),$$

and, just like $\widetilde{W}(t)$, the process $-\widetilde{W}(t)$ is a Brownian motion under $\widetilde{\mathbb{P}}$.

Definition 9.4.1. *Let T be a fixed maturity date. We define the T-forward measure $\widetilde{\mathbb{P}}^T$ by*

$$\widetilde{\mathbb{P}}^T(A) = \frac{1}{B(0,T)}\int_A D(T)\,d\widetilde{\mathbb{P}} \text{ for all } A \in \mathcal{F}. \tag{9.4.5}$$

The T-forward measure corresponds to taking $N(t) = B(t,T)$ in (9.2.7) and (9.2.8). According to Theorem 9.2.2, the process

$$\widetilde{W}^T(t) = \int_0^t \sigma^*(u,T)\,du + \widetilde{W}(t)$$

is a Brownian motion under $\widetilde{\mathbb{P}}^T$. Furthermore, under the T-forward measure, all assets denominated in units of the zero-coupon bond maturing at time T are martingale. In other words,

T-forward prices are martingales under the T-forward measure $\widetilde{\mathbb{P}}^T$.

Furthermore, the volatility vector of the T-forward price of an asset is the difference between the volatility vector of the asset and the volatility vector of the T-maturity zero-coupon bond (see Remark 9.2.3).

The reason to introduce the T-forward measure is that it often simplifies the risk-neutral pricing formula. According to that formula, the value at time t of a contract that pays $V(T)$ at a later time T is

$$V(t) = \frac{1}{D(t)}\widetilde{\mathbb{E}}\big[D(T)V(T)\big|\mathcal{F}(t)\big]. \tag{9.4.6}$$

The computation of the right-hand side of this formula requires that we know something about the dependence between the discount factor $D(T)$ and the payoff $V(T)$ of the derivative security. Especially when the derivative security depends on the interest rate, this can be difficult to model. However, according to (9.2.8) (with t replacing s and T replacing t in that formula), we have

$$\widetilde{\mathbb{E}}^T[V(T)|\mathcal{F}(t)] = \frac{1}{D(t)B(t,T)}\widetilde{\mathbb{E}}\big[D(T)V(T)\big|\mathcal{F}(t)\big] = \frac{1}{B(t,T)}V(t).$$

This gives us the simple formula

$$V(t) = B(t,T)\widetilde{\mathbb{E}}^T\big[V(T)\big|\mathcal{F}(t)\big]. \tag{9.4.7}$$

If we can find a simple model for the evolution of assets under the T-forward measure, we can use (9.4.7), in which we only need to estimate $V(T)$, instead of using (9.4.6), which requires us to estimate $D(T)V(T)$. We give an example of the power of this approach in the next subsection.

9.4.3 Option Pricing with a Random Interest Rate

The classical Black-Scholes-Merton option-pricing formula assumes a constant interest rate. For options on bonds and other interest-rate-dependent instruments, movements in the interest rate are critical. For these "fixed income" derivatives, the assumption of a constant interest rate is inappropriate.

In this section, we present a generalized Black-Scholes-Merton option-pricing formula that permits the interest rate to be random. The classical Black-Scholes-Merton assumption that the volatility of the underlying asset is constant is here replaced by the assumption that the volatility of the forward price of the underlying asset is constant. Because the forward price is a martingale under the forward measure, and $\widetilde{W}^T(t)$ is the Brownian motion used to drive asset prices under the forward measure, the assumption of constant volatility for the forward price is equivalent to the assumption

$$d\text{For}_S(t,T) = \sigma \text{For}_S(t,T)\, d\widetilde{W}^T(t), \tag{9.4.8}$$

where σ is a constant. The bond maturity T is chosen to coincide with the expiration time T of the option.

Theorem 9.4.2 (Black-Scholes-Merton option pricing with random interest rate). *Let $S(t)$ be the price of an asset denominated in (domestic) currency, and assume the forward price of this asset satisfies (9.4.8) with a positive constant σ. The value at time $t \in [0,T]$ of a European call on this asset, expiring at time T with strike price K, is*

$$V(t) = S(t)N(d_+(t)) - KB(t,T)N(d_-(t)), \tag{9.4.9}$$

where the adapted processes $d_\pm(t)$ are given by

$$d_\pm(t) = \frac{1}{\sigma\sqrt{T-t}}\left[\log\frac{\text{For}_S(t,T)}{K} \pm \frac{1}{2}\sigma^2(T-t)\right]. \tag{9.4.10}$$

Furthermore, a short position in the option can be hedged by holding $N(d_+(t))$ shares of the asset and shorting $KN(d_-(t))$ T-maturity zero-coupon bonds at each time t.

Remark 9.4.3. If the interest rate is a constant r, then $B(t,T) = e^{-r(T-t)}$, $\text{For}_S(t,T) = e^{r(T-t)}S(t)$, and this theorem reduces to the usual Black-Scholes-Merton formula and hedging strategy.

PROOF OF THEOREM 9.4.2: We prove formula (9.4.9) for $t = 0$. It is not difficult to modify the proof to account for general t.

We observe that $\text{For}_S(0,T) = \frac{S(0)}{B(0,T)}$, and so the solution to (9.4.8) is

$$\text{For}_S(t,T) = \frac{S(0)}{B(0,T)} \exp\left\{\sigma\widetilde{W}^T(t) - \frac{1}{2}\sigma^2 t\right\}. \tag{9.4.11}$$

For each t, this has a log-normal distribution under $\widetilde{\mathbb{P}}^T$, the measure under which $\widetilde{W}^T(t)$ is a Brownian motion.

We need one more change of measure. Suppose we take the asset price $S(t)$ to be the numéraire. In terms of this numéraire, the asset price is identically 1. The risk-neutral measure for this numéraire is given by

$$\widetilde{\mathbb{P}}^S(A) = \frac{1}{S(0)} \int_A D(T)S(T)\, d\widetilde{\mathbb{P}} \text{ for all } A \in \mathcal{F}.$$

Denominated in units of $S(t)$, the zero-coupon bond is

$$\frac{B(t,T)}{S(t)} = \frac{1}{\mathrm{For}_S(t,T)},\quad 0 \le t \le T,$$

and, by Theorem 9.2.2, this is a martingale under $\widetilde{\mathbb{P}}^S$.

Indeed, we can compute the differential of $\frac{1}{\mathrm{For}_S(t,T)}$ using the Itô-Doeblin formula, the function $f(x) = \frac{1}{x}$, and (9.4.8). Since $f'(x) = -\frac{1}{x^2}$ and $f''(x) = \frac{2}{x^3}$, we have

$$\begin{aligned}
&d\left(\frac{1}{\mathrm{For}_S(t,T)}\right)\\
&= df(\mathrm{For}_S(t,T))\\
&= f'(\mathrm{For}_S(t,T))\, d\mathrm{For}_S(t,T) + \frac{1}{2} f''(\mathrm{For}_S(t,T))\, d\mathrm{For}_S(t,T)\, d\mathrm{For}_S(t,T)\\
&= -\frac{\sigma}{\mathrm{For}_S(t,T)}\, d\widetilde{W}^T(t) + \frac{\sigma^2}{\mathrm{For}_S(t,T)}\, dt\\
&= -\frac{\sigma}{\mathrm{For}_S(t,T)}\big(-\sigma\, dt + d\widetilde{W}^T\big).
\end{aligned} \tag{9.4.12}$$

Because we are guaranteed by Theorem 9.2.2 that $\frac{1}{\mathrm{For}_S(t,T)}$ is a martingale under $\widetilde{\mathbb{P}}^S$, we conclude that

$$\widetilde{W}^S(t) = -\sigma t + \widetilde{W}^T(t)$$

is a Brownian motion under $\widetilde{\mathbb{P}}^S$. We see also that $\frac{1}{\mathrm{For}_S(t,T)}$ has volatility σ. The solution to (9.4.12) is

$$\frac{1}{\mathrm{For}_S(t,T)} = \frac{B(0,T)}{S(0)} \exp\left\{-\sigma \widetilde{W}^S(t) - \frac{1}{2}\sigma^2 t\right\}. \tag{9.4.13}$$

For each t, this has a log-normal distribution under $\widetilde{\mathbb{P}}^S$, the measure under which $\widetilde{W}^S(t)$ is a Brownian motion.

At time zero, the value of a European call expiring at time T, according to the risk-neutral pricing formula, is

$$\begin{aligned}
V(0) &= \widetilde{\mathbb{E}}\left[D(T)(S(T)-K)^+\right] \\
&= \widetilde{\mathbb{E}}\left[D(T)S(T)\mathbb{I}_{\{S(T)>K\}}\right] - K\widetilde{\mathbb{E}}\left[D(T)\mathbb{I}_{\{S(T)>K\}}\right] \\
&= S(0)\widetilde{\mathbb{E}}\left[\frac{D(T)S(T)}{S(0)}\mathbb{I}_{\{S(T)>K\}}\right] - KB(0,T)\widetilde{\mathbb{E}}\left[\frac{D(T)}{B(0,T)}\mathbb{I}_{\{S(T)>K\}}\right] \\
&= S(0)\widetilde{\mathbb{P}}^S\{S(T)>K\} - KB(0,T)\widetilde{\mathbb{P}}^T\{S(T)>K\} \\
&= S(0)\widetilde{\mathbb{P}}^S\{\mathrm{For}_S(T,T)>K\} - KB(0,T)\widetilde{\mathbb{P}}^T\{\mathrm{For}_S(T,T)>K\} \\
&= S(0)\widetilde{\mathbb{P}}^S\left\{\frac{1}{\mathrm{For}_S(T,T)} < \frac{1}{K}\right\} - KB(0,T)\widetilde{\mathbb{P}}^T\{\mathrm{For}_S(T,T)>K\},
\end{aligned}$$

where in the next-to-last step we have used the fact that $\mathrm{For}_S(T,T) = S(T)$. Using the fact that $\widetilde{W}^S(T)$ is normal with mean zero and variance T under $\widetilde{\mathbb{P}}^S$, we compute

$$\begin{aligned}
&\widetilde{\mathbb{P}}^S\left\{\frac{1}{\mathrm{For}_S(T,T)} < \frac{1}{K}\right\} \\
&\quad = \widetilde{\mathbb{P}}^S\left\{-\sigma\widetilde{W}^S(T) - \frac{1}{2}\sigma^2 T < \log\frac{S(0)}{KB(0,T)}\right\} \\
&\quad = \widetilde{\mathbb{P}}^S\left\{\frac{-\widetilde{W}^S(T)}{\sqrt{T}} < \frac{1}{\sigma\sqrt{T}}\left[\log\frac{S(0)}{KB(0,T)} + \frac{1}{2}\sigma^2 T\right]\right\} \\
&\quad = N(d_+(0)).
\end{aligned}$$

Using the fact that $\widetilde{W}^T(T)$ is normal with mean zero and variance T under $\widetilde{\mathbb{P}}^T$, we obtain

$$\begin{aligned}
&\widetilde{\mathbb{P}}\{\mathrm{For}_S(T,T)>K\} \\
&\quad = \widetilde{\mathbb{P}}^T\left\{\sigma\widetilde{W}^T(T) - \frac{1}{2}\sigma^2 T > \log\frac{KB(0,T)}{S(0)}\right\} \\
&\quad = \widetilde{\mathbb{P}}^T\left\{\frac{\widetilde{W}^T(T)}{\sqrt{T}} > \frac{1}{\sigma\sqrt{T}}\log\left[\frac{KB(0,T)}{S(0)} + \frac{1}{2}\sigma^2 T\right]\right\} \\
&\quad = \widetilde{\mathbb{P}}^T\left\{-\frac{\widetilde{W}^T(T)}{\sqrt{T}} < \frac{1}{\sigma\sqrt{T}}\left[\log\frac{S(0)}{KB(0,T)} - \frac{1}{2}\sigma^2 T\right]\right\} \\
&\quad = N(d_-(0)).
\end{aligned}$$

This completes the proof of (9.4.9), at least for the case $t = 0$.

We now consider the hedge suggested by formula (9.4.9). It is easier to do this when we take the zero-coupon bond as the numéraire rather than when we use currency. Dividing (9.4.9) by $B(t,T)$, we obtain

$$\frac{V(t)}{B(t,T)} = \mathrm{For}_S(t,T)N(d_+(t)) - KN(d_-(t)). \tag{9.4.14}$$

This gives us the option price denominated in zero-coupon bonds. Suppose we hedge a short position in the option by holding $N(d_+(t))$ shares of the asset and shorting $KN(d_-(t))$ zero-coupon bonds at each time t. The value of this portfolio, denominated in units of zero-coupon bond, agrees with (9.4.14). To be sure this short option hedge works, however, we must verify that the portfolio just described is *self-financing.* In other words, we must be sure we do not need to infuse cash in order to maintain the positions just described. (A discussion related to this, passing from discrete to continuous time, is provided in Exercise 4.10 of Chapter 4.) The capital gains differential associated with this portfolio, again denominated in units of zero-coupon bond, is

$$N(d_+(t))\, d\mathrm{For}_S(t,T).$$

(When measuring wealth in units of zero-coupon bond, there is no capital gain from movements in the bond price.) The differential of the portfolio, according to Itô's formula, is

$$\begin{aligned} d\left(\frac{V(t)}{B(t,T)}\right) &= N(d_+(t))\, d\mathrm{For}_S(t,T) + \mathrm{For}_S(t,T)\, dN(d_+(t)) \\ &\quad + d\mathrm{For}_S(t,T)\, dN(d_+(t)) - K\, dN(d_-(t)). \end{aligned} \tag{9.4.15}$$

In order for the portfolio to be self-financing, we must have

$$\mathrm{For}_S(t,T)\, dN(d_+(t)) + d\mathrm{For}_S(t,T)\, dN(d_+(t)) - K\, dN(d_-(t)) = 0, \tag{9.4.16}$$

so that the change of value in the portfolio is entirely due to capital gains. The verification of (9.4.16) is Exercise 9.6. □

9.5 Summary

This chapter discusses the fact that when we change the units of account, the so-called *numéraire*, we must change the risk-neutral measure. Fortunately, the Radon-Nikodým derivative process needed to effect this change of measure is simple; it is the numéraire itself, discounted in order to be a martingale and normalized by its initial condition in order to have expected value 1. This is the content of Theorem 9.2.2.

In this chapter, we apply the change-of-numéraire idea in two cases: foreign exchange models and option pricing in the presence of a random interest rate. It was also used in the discussion of Asian options in Section 7.5.

In the context of foreign exchange models, we show that the mean rate of change of the exchange rate is the difference between the interest rates in the two economies *under the risk-neutral measure for the economy in which the exchange rate is being considered.* We show that one can derive other expected symmetries (e.g., the forward exchange rate in one currency is the reciprocal of the foreign exchange rate in the other currency), provided one is careful to use the appropriate risk-neutral measures.

When the interest rate is random, the classical Black-Scholes-Merton option-pricing formula does not apply. However, if one is willing to assume that the T-forward price of the underlying asset has constant volatility, then the price of a call expiring at time T has a simple formula and a simple hedging strategy (Theorem 9.4.2). This fact is exploited to build *LIBOR models* in Section 10.4.

9.6 Notes

The model of foreign and domestic markets presented in this chapter is a simplification of one in Musiela and Rutkowski [126]. The model in [126], drawn from Amin and Jarrow [2], permits foreign and domestic interest rates to be random. The Garman-Kohlhagen formula of Subsection 9.3.6 is taken from Garman and Kohlhagen [68]. The option to exchange one risky asset for another, of which Subsection 9.3.7 is a special case, was studied by Margrabe [117].

Theorem 9.4.2, option pricing with a random interest rate, is taken from Geman, El Karoui, and Rochet [70]. It traces back at least to Geman [69] and Jamshidian [94], who observed that the forward price of an asset is its price when denominated in the numéraire of the zero-coupon bond maturing at the delivery date. Even earlier, Merton [122] proposed hedging European options by using a bond maturing on the option expiration date.

9.7 Exercises

Exercise 9.1. This exercise provides an alternate proof of the main assertion of Theorem 9.2.2.

(i) Use Lemma 5.2.2 to prove Remark 9.2.5.
(ii) Let $S(t)$ and $N(t)$ be prices of two assets, denominated in a common currency, and assume $N(t)$ is always strictly positive. Let $\widetilde{\mathbb{P}}$ be the risk-neutral measure under which the discounted asset prices $D(t)S(t)$ and $D(t)N(t)$ are martingales. Apply Remark 9.2.5 to show that $S^{(N)}(t) = \frac{S(t)}{N(t)}$ is a martingale under $\widetilde{\mathbb{P}}^{(N)}$ defined by (9.2.6).

Exercise 9.2 (Portfolios under change of numéraire). Consider two assets with prices $S(t)$ and $N(t)$ given by

$$S(t) = S(0)\exp\left\{\sigma\widetilde{W}(t) + \left(r - \frac{1}{2}\sigma^2\right)t\right\},$$

$$N(t) = N(0)\exp\left\{\nu\widetilde{W}(t) + \left(r - \frac{1}{2}\nu^2\right)t\right\},$$

where $\widetilde{W}(t)$ is a one-dimensional Brownian motion under the risk-neutral measure $\widetilde{\mathbb{P}}$ and the volatilities $\sigma > 0$ and $\nu > 0$ are constant, as is the interest rate r. We define a third asset, the money market account, whose price per share at time t is $M(t) = e^{rt}$.

Let us now denominate prices in terms of the numéraire N, so that the redenominated first asset price is

$$\widehat{S}(t) = \frac{S(t)}{N(t)}$$

and the redenominated money market account price is

$$\widehat{M}(t) = \frac{M(t)}{N(t)}.$$

According to Theorem 9.2.2, $d\widehat{S}(t) = (\sigma-\nu)\widehat{S}(t)\,d\widehat{W}(t)$, where $\widehat{W}(t) = \widetilde{W}(t) - \nu t$.

(i) Compute the differential of $\frac{1}{N(t)}$.
(ii) Compute the differential of $\widehat{M}(t)$, expressing it in terms of $d\widehat{W}(t)$.

Consider a portfolio that at each time t holds $\Delta(t)$ shares of the first asset and finances this by investing in or borrowing from the money market. According to the usual formula, the differential of the value $X(t)$ of this portfolio is

$$dX(t) = \Delta(t)\,dS(t) + r\big(X(t) - \Delta(t)S(t)\big)\,dt.$$

We define

$$\Gamma(t) = \frac{X(t) - \Delta(t)S(t)}{M(t)}$$

to be the number of shares of money market account held by this portfolio at time t and can then rewrite the differential of $X(t)$ as

$$dX(t) = \Delta(t)\,dS(t) + \Gamma(t)\,dM(t). \tag{9.7.1}$$

Note also that by the definition of $\Gamma(t)$, we have

$$X(t) = \Delta(t)S(t) + \Gamma(t)M(t). \tag{9.7.2}$$

We redenominate the portfolio value, defining

$$\widehat{X}(t) = \frac{X(t)}{N(t)}, \tag{9.7.3}$$

so that (dividing (9.7.2) by $N(t)$) we have

$$\widehat{X}(t) = \Delta(t)\widehat{S}(t) + \Gamma(t)\widehat{M}(t). \tag{9.7.4}$$

(iii) Use stochastic calculus to show that

$$d\widehat{X}(t) = \Delta(t)\, d\widehat{S}(t) + \Gamma(t)\, d\widehat{M}(t).$$

This equation is the counterpart in the new numéraire of equation (9.7.1) and says that the change in $\widehat{X}(t)$ is solely due to changes in the prices of the assets held by the portfolio. (Hint: Start from equation (9.7.3) and use (9.7.1) and (9.7.4) along the way.)

Exercise 9.3 (Change in volatility caused by change of numéraire). Let $S(t)$ and $N(t)$ be the prices of two assets, denominated in a common currency, and let σ and ν denote their volatilities, which we assume are constant. We assume also that the interest rate r is constant. Then

$$dS(t) = rS(t)\, dt + \sigma S(t)\, d\widetilde{W}_1(t),$$
$$dN(t) = rN(t)\, dt + \nu N(t)\, d\widetilde{W}_3(t),$$

where $\widetilde{W}_1(t)$ and $\widetilde{W}_3(t)$ are Brownian motions under the risk-neutral measure $\widetilde{\mathbb{P}}$. We assume these Brownian motions are correlated, with $d\widetilde{W}_1(t)\, d\widetilde{W}_3(t) = \rho\, dt$ for some constant ρ.

(i) Show that $S^{(N)}(t) = \frac{S(t)}{N(t)}$ has volatility $\gamma = \sqrt{\sigma^2 - 2\rho\sigma\nu + \nu^2}$. In other words, show that there exists a Brownian motion $\widetilde{W}_4$ under $\widetilde{\mathbb{P}}$ such that

$$\frac{dS^{(N)}(t)}{S^{(N)}(t)} = (\text{Something})\, dt + \gamma\, d\widetilde{W}_4(t).$$

(ii) Show how to construct a Brownian motion $\widetilde{W}_2(t)$ under $\widetilde{\mathbb{P}}$ that is independent of $\widetilde{W}_1(t)$ such that $dN(t)$ may be written as

$$dN(t) = rN(t)\, dt + \nu N(t)\big[\rho\, d\widetilde{W}_1(t) + \sqrt{1-\rho^2}\, d\widetilde{W}_2(t)\big].$$

(iii) Using Theorem 9.2.2, determine the volatility vector of $S^{(N)}(t)$. In other words, find a vector (v_1, v_2) such that

$$dS^{(N)}(t) = S^{(N)}(t)\big[v_1\, d\widetilde{W}_1^{(N)}(t) + v_2\, d\widetilde{W}_2^{(N)}(t)\big],$$

where $\widetilde{W}_1(t)$ and $\widetilde{W}_2(t)$ are independent Brownian motions under $\widetilde{\mathbb{P}}^{(N)}$. Show that

$$\sqrt{v_1^2 + v_2^2} = \sqrt{\sigma^2 - 2\rho\sigma\nu + \nu^2}.$$

Exercise 9.4. From the differential formulas (9.3.14) and (9.3.15) for the stock and discounted exchange rate in terms of the Brownian motions under the domestic risk-neutral measure, derive the differential formulas (9.3.22) and (9.3.23) for the redenominated money market account and stock discounted at the foreign interest rate and written in terms of the Brownian motions under the *foreign* risk-neutral measure.

Exercise 9.5 (Quanto option). A *quanto option* pays off in one currency the price in another currency of an underlying asset without taking the currency conversion into account. For example, a quanto call on a British asset struck at \$25 would pay \$5 if the price of the asset upon expiration of the option is £30. To compute the payoff of the option, the price 30 is treated as if it were dollars, even though it is pounds sterling.

In this problem we consider a quanto option in the foreign exchange model of Section 9.3. We take the domestic and foreign interest rates to be constants r and r^f, respectively, and we assume that $\sigma_1 > 0$, $\sigma_2 > 0$, and $\rho \in (-1,1)$ are likewise constant.

(i) From (9.3.14), show that

$$S(t) = S(0)\exp\left\{\sigma_1 \widetilde{W}_1(t) + \left(r - \frac{1}{2}\sigma_1^2\right)t\right\}.$$

(ii) From (9.3.16), show that

$$Q(t) = Q(0)\exp\left\{\sigma_2\rho\widetilde{W}_1(t) + \sigma_2\sqrt{1-\rho^2}\,\widetilde{W}_2(t) + \left(r - r^f - \frac{1}{2}\sigma_2^2\right)t\right\}.$$

(iii) Show that

$$\frac{S(t)}{Q(t)} = \frac{S(0)}{Q(0)}\exp\left\{\sigma_4\widetilde{W}_4(t) + \left(r - a - \frac{1}{2}\sigma_4^2\right)t\right\},$$

where

$$\sigma_4 = \sqrt{\sigma_1^2 - 2\rho\sigma_1\sigma_2 + \sigma_2^2},$$
$$a = r - r^f + \rho\sigma_1\sigma_2 - \sigma_2^2,$$

and

$$\widetilde{W}_4(t) = \frac{\sigma_1 - \sigma_2\rho}{\sigma_4}\widetilde{W}_1(t) - \frac{\sigma_2\sqrt{1-\rho^2}}{\sigma_4}\widetilde{W}_2(t)$$

is a Brownian motion.

(iv) Consider a quanto call that pays off

$$\left(\frac{S(T)}{Q(T)} - K\right)^+$$

units of domestic currency at time T. (Note that $\frac{S(T)}{Q(T)}$ is denominated in units of foreign currency, but in this payoff it is treated as if it is a number of units of domestic currency.) Show that if at time $t \in [0,T]$ we have $\frac{S(t)}{Q(t)} = x$, then the price of the quanto call at this time is

$$q(t,x) = xe^{-a\tau}N(d_+(\tau,x)) - e^{-r\tau}KN(d_-(\tau,x)),$$

where $\tau = T - t$ and

$$d_{\pm}(\tau, x) = \frac{1}{\sigma_4\sqrt{\tau}}\left[\log\frac{x}{K} + \left(r - a \pm \frac{1}{2}\sigma_4^2\right)\tau\right].$$

(Hint: Argue that this is a case of formula (5.5.12).)

Exercise 9.6. Verify equation (9.4.16),

$$\mathrm{For}_S(t,T)\,dN(d_+(t)) + d\mathrm{For}_S(t,T)\,dN(d_+(t)) - K\,dN(d_-(t)) = 0,$$

in the following steps.

(i) Use (9.4.10) to show that

$$d_-(t) = d_+(t) - \sigma\sqrt{T-t}.$$

(ii) Use (9.4.10) to show that

$$d_+^2(t) - d_-^2(t) = 2\log\frac{\mathrm{For}_S(t,T)}{K}.$$

(iii) Use (ii) to show that

$$\mathrm{For}_S(t,T)e^{-d_+^2(t)/2} - Ke^{-d_-^2(t)/2} = 0.$$

(iv) Use (9.4.8) and the Itô-Doeblin formula to show that

$$dd_+(t) = \frac{1}{2\sigma(T-t)^{3/2}}\log\frac{\mathrm{For}_S(t,T)}{K}\,dt - \frac{3\sigma}{4\sqrt{T-t}}\,dt + \frac{1}{\sqrt{T-t}}\,d\widetilde{W}^T(t).$$

(v) Use (i) to show that

$$dd_-(t) = dd_+(t) + \frac{\sigma}{2\sqrt{T-t}}\,dt.$$

(vi) Use (iv) and (v) to show that

$$dd_+(t)\,dd_+(t) = dd_-(t)\,dd_-(t) = \frac{dt}{T-t}.$$

(vii) Use the Itô-Doeblin formula to show that

$$dN(d_+(t)) = \frac{1}{\sqrt{2\pi}}e^{-d_+^2(t)/2}\,dd_+(t) - \frac{d_+(t)}{2(T-t)\sqrt{2\pi}}e^{-d_+^2(t)/2}\,dt.$$

(viii) Use the Itô-Doeblin formula, (v), (i), and (vi) to show that

$$dN(d_-(t)) = \frac{1}{\sqrt{2\pi}}e^{-d_-^2(t)/2}\,dd_+(t) + \frac{\sigma}{\sqrt{2\pi(T-t)}}e^{-d_-^2(t)/2}\,dt$$
$$- \frac{d_+(t)}{2(T-t)\sqrt{2\pi}}e^{-d_-^2(t)/2}\,dt.$$

(ix) Use (9.4.8), (vii), and (iv) to show that

$$d\mathrm{For}_S(t,T)\,dN(d_+(t)) = \frac{\sigma\mathrm{For}_S(t,T)}{\sqrt{2\pi(T-t)}}e^{-d_+^2(t)/2}\,dt.$$

(x) Now prove (9.4.16).

10

Term-Structure Models

10.1 Introduction

Real markets do not have a single interest rate. Instead, they have bonds of different maturities, some paying coupons and others not paying coupons. From these bonds, *yields* to different maturities can be implied. More specifically, let $0 = T_0 < T_1 < T_2 < \cdots < T_n$ be a given set of dates, and let $B(0, T_j)$ denote the price at time zero of a zero-coupon bond paying 1 at maturity T_j. Consider a coupon-paying bond that makes fixed payments $C_1, C_2, \ldots, C_j$ at dates $T_1, T_2, \ldots, T_j$, respectively. Each of the numbers $C_1, C_2, \ldots, C_{j-1}$ represents a coupon (interest payment), and C_j represents the interest plus principal paid at the maturity T_j of the bond. The price of this bond at time zero can be decomposed as

$$\sum_{j=i}^{j} C_i B(0, T_i). \tag{10.1.1}$$

On the other hand, if one is given the price of a coupon-paying bond of each maturity $T_1, T_2, \ldots, T_n$, then using (10.1.1) one can solve recursively for $B(0, T_1), \ldots, B(0, T_n)$ by first observing that $B(0, T_1)$ is the price of the T_1-maturity bond divided by the payment it will make at T_1, then using this value of $B(0, T_1)$ and the price of the T_2-maturity bond to solve for $B(0, T_2)$, and continuing in this manner. This method of determining zero-coupon bond prices from coupon-paying bond prices is called *bootstrapping.*

In any event, from market data one can ultimately determine prices of zero-coupon bonds for a number of different maturity dates. Each of these bonds has a *yield* specific to its maturity, where yield is defined to be the constant continuously compounding interest rate over the lifetime of the bond that is consistent with its price:

$$\text{price of zero-coupon bond} = \text{face value} \times e^{-\text{yield} \times \text{time to maturity}}.$$

The *face value* of a zero-coupon bond is the amount it promises to pay upon maturity. The formula above implies that capital equal to the price of the bond, invested at a continuously compounded interest rate equal to the yield, would, over the lifetime of the bond, result in a final payment of the face value. In this chapter, we shall normalize zero-coupon bonds by taking the face value to be 1.

In summary, instead of having a single interest rate, real markets have a *yield curve*, which one can regard either as a function of finitely many yields plotted versus their corresponding maturities or more often as a function of a nonnegative real variable (time) obtained by interpolation from the finitely many maturity–yield pairs provided by the market. The interest rate (sometimes called the *short rate*) is an idealization corresponding to the shortest-maturity yield or perhaps the overnight rate offered by the government, depending on the particular application.

We assume throughout this chapter that the bonds have no risk of default. One generally regards U.S. government bonds to be nondefaultable.

Models for interest rates have already appeared in this text, most notably in Section 6.5, where the partial differential equation satisfied by zero-coupon bonds in a one-factor short-rate model was developed and the Hull-White and Cox-Ingersoll-Ross models were given as examples. In Section 10.2 of this chapter, we extend these models to permit finitely many factors. These are Markov models in which the state of the model at each time is a multidimensional vector.

Unlike the models for equities considered heretofore and the Heath-Jarrow-Morton model considered later, the multifactor models in Section 10.2 do not immediately provide a mechanism for evolution of the prices of tradeable assets. In the earlier models, we assume an evolution of the price of a primary asset or the prices of multiple primary assets under the actual measure and then determine the market prices of risk that enable us to switch to a risk-neutral measure. In the multifactor models of Section 10.2, we begin with the evolution of abstract "factors," and from these the interest rate is obtained. But the interest rate is not the price of an asset, and we cannot infer a market price of risk from the interest rate alone. If we also had prices of some primary assets, say zero-coupon bonds, we could determine market prices of risk. However, in the models of Section 10.2, the only way to get prices of zero-coupon bonds is to use the risk-neutral pricing formula, and this cannot be done until we have a risk-neutral measure. Therefore, we build these models under the risk-neutral measure from the outset. Zero-coupon bond prices are given by the risk-neutral pricing formula, which implies that discounted zero-coupon bond prices are martingales under the risk-neutral measure. This implies in turn that no arbitrage can be achieved by trading in the zero-coupon bonds and the money market. After these models are built, they are calibrated to market prices for zero-coupon bonds and probably also some fixed income derivatives. The actual probability measure and the market prices of risk never enter the picture.

In contrast to the models of Section 10.2, the *Heath-Jarrow-Morton (HJM) model* takes its state at each time to be the forward curve at that time. The forward rate $f(t,T)$, which is the state of the HJM model, is the instantaneous rate that can be locked in at time t for borrowing at time $T \geq t$. For fixed t, one calls the function $T \mapsto f(t,T)$, defined for $T \geq t$, the *forward rate curve.* The HJM model provides a mechanism for evolving this curve (a "curve" in the variable T) forward in time (the variable t). The forward rate curve can be deduced from the zero-coupon bond prices, and the zero-coupon bond prices can be deduced from the forward rate curve. Because zero-coupon bond prices are given directly by the HJM model rather than indirectly by the risk-neutral pricing formula, one needs to be careful that the model does not generate prices that admit arbitrage. Hence, HJM is more than a model because it provides a necessary and sufficient condition for a model driven by Brownian motion to be free of arbitrage. Every Brownian-motion-driven model must satisfy the HJM no-arbitrage condition, and to illustrate that point we provide Exercise 10.10 to verify that the Hull-White and Cox-Ingersoll-Ross models satisfy this condition.

For practical applications, it would be convenient to build a model where the forward rate had a log-normal distribution. Unfortunately, this is not possible. However, if one instead models the *simple interest rate* $L(t,T)$ that one can lock in at time t for borrowing over the interval T to $T+\delta$, where δ is a positive constant, this problem can be overcome. We call $L(t,T)$ *forward LIBOR (London interbank offered rate).* The constant δ is typically 0.25 (*three-month LIBOR*) or 0.50 (*six-month LIBOR*). The model that takes forward LIBOR as its state is often called the *forward LIBOR model*, the *market model*, or the *Brace-Gatarek-Musiela (BGM) model.* It is presented in Section 10.4.

10.2 Affine-Yield Models

The one-factor Cox-Ingersoll-Ross (CIR) and Hull-White models appearing in Section 6.5 are called *affine-yield models* because in these models the yield for zero-coupon bond prices is an affine (linear plus constant) function of the interest rate. In this section, we develop the two-factor, constant-coefficient versions of these models. (The constant-coefficient version of the Hull-White model is the Vasicek model.) Models with three or more factors can be developed along the lines of the two-factor models of this section.

It turns out that there are essentially three different two-factor affine-yield models, one in which both factors have constant diffusion terms (and hence are Gaussian processes, taking negative values with positive probability), one in which both factors appear under the square root in diffusion terms (and hence must be nonnegative at all times), and one in which only one factor appears under the square root in the diffusion terms (and only this factor is nonnegative at all times, whereas the other factor can become negative). We shall call these the *two-factor Vasicek*, the *two-factor CIR*, and the *two-factor mixed* term-structure models, respectively. For each of these types of models, there is a *canonical model* (i.e., a simplest way of writing the model).

Two-factor affine yield-models appearing in the literature, which often seem to be more complicated than the canonical models of this section, can always be obtained from one of the three canonical models by changing variables. It is desirable when calibrating a model to first change the variables to put the model into a form having the minimum number of parameters; otherwise, the calibration can be confounded by the fact that multiple sets of parameters yield the same result. The canonical models presented here have the minimmu number of parameters.

10.2.1 Two-Factor Vasicek Model

For the two-factor Vasicek model, we let the factors $X_1(t)$ and $X_2(t)$ be given by the system of stochastic differential equations

$$dX_1(t) = \big(a_1 - b_{11}X_1(t) - b_{12}X_2(t)\big)\,dt + \sigma_1\,d\widetilde{B}_1(t), \tag{10.2.1}$$

$$dX_1(t) = \big(a_2 - b_{21}X_1(t) - b_{22}X_2(t)\big)\,dt + \sigma_2\,d\widetilde{B}_2(t), \tag{10.2.2}$$

where the processes $\widetilde{B}_1(t)$ and $\widetilde{B}_2(t)$ are Brownian motions under a risk-neutral measure $\widetilde{\mathbb{P}}$ with constant correlation $\nu \in (-1,1)$ (i.e., $d\widetilde{B}_1(t)\,d\widetilde{B}_2(t) = \nu\,dt$). The constants σ_1 and σ_2 are assumed to be strictly positive. We further assume that the matrix

$$B = \begin{bmatrix} b_{11} & b_{12} \\ b_{21} & b_{22} \end{bmatrix}$$

has strictly positive eigenvalues λ_1 and λ_2. The positivity of these eigenvalues causes the factors $X_1(t)$ and $X_2(t)$, as well as the canonical factors $Y_1(t)$ and $Y_2(t)$ defined below, to be mean-reverting. Finally, we assume the interest rate is an affine function of the factors,

$$R(t) = \epsilon_0 + \epsilon_1 X_1(t) + \epsilon_2 X_2(t), \tag{10.2.3}$$

where ϵ_0, ϵ_1, and ϵ_2 are constants. This is the most general two-factor Vasicek model.

Canonical Form

As presented above, the two-factor Vasicek model is "overparametrized" (i.e., different choices of the parameters a_i, b_{ij}, σ_i, and ϵ_i can lead to the same distribution for the process $R(t)$). To eliminate this overparametrization, we reduce the model (10.2.1)–(10.2.3) to the *canonical two-factor Vasicek model*

$$dY_1(t) = -\lambda_1 Y_1(t)\,dt + d\widetilde{W}_1(t), \tag{10.2.4}$$

$$dY_2(t) = -\lambda_{21} Y_1(t)\,dt - \lambda_2 Y_2(t)\,dt + d\widetilde{W}_2(t), \tag{10.2.5}$$

$$R(t) = \delta_0 + \delta_1 Y_1(t) + \delta_2 Y_2(t), \tag{10.2.6}$$

where $\widetilde{W}_1(t)$ and $\widetilde{W}_2(t)$ are independent Brownian motions.

The canonical two-factor Vasicek model has six parameters:

$$\lambda_1 > 0,\ \lambda_2 > 0,\ \lambda_{21}\ \delta_0,\ \delta_1,\ \delta_2.$$

The parameters are used to calibrate the model. In practice, one often permits some of these parameters to be time-varying but nonrandom in order to make the model fit the initial yield curve; see Exercise 10.3.

To achieve this reduction, we first transform B to its Jordan canonical form by choosing a nonsingular matrix

$$P = \begin{bmatrix} p_{11} & p_{12} \\ p_{21} & p_{22} \end{bmatrix}$$

such that

$$K = PBP^{-1} = \begin{bmatrix} \lambda_1 & 0 \\ \kappa & \lambda_2 \end{bmatrix}.$$

If $\lambda_1 \neq \lambda_2$, then the columns of P^{-1} are eigenvectors of B and $\kappa = 0$ (i.e., K is diagonal). If $\lambda_1 = \lambda_2$, then κ might be zero, but it can also happen that $\kappa \neq 0$, in which case we may choose P so that $\kappa = 1$. Using the notation

$$X(t) = \begin{bmatrix} X_1(t) \\ X_2(t) \end{bmatrix},\ A = \begin{bmatrix} a_1 \\ a_2 \end{bmatrix},\ \Sigma = \begin{bmatrix} \sigma_1 & 0 \\ 0 & \sigma_2 \end{bmatrix},\ \widetilde{B}(t) = \begin{bmatrix} \widetilde{B}_1(t) \\ \widetilde{B}_2(t) \end{bmatrix},$$

we may rewrite (10.2.1) and (10.2.2) in vector notation:

$$dX(t) = A\,dt - BX(t)\,dt + \Sigma\,d\widetilde{B}(t).$$

Multiplying both sides by P and defining $\overline{X}(t) = PX(t)$, we obtain

$$d\overline{X}(t) = PA\,dt - K\overline{X}(t)\,dt + P\Sigma\,d\widetilde{B}(t),$$

which can be written componentwise as

$$\begin{aligned} d\overline{X}_1(t) &= (p_{11}a_1 + p_{12}a_2)\,dt - \lambda_1\overline{X}_1(t)\,dt \\ &\quad + p_{11}\sigma_1\,d\widetilde{B}_1(t) + p_{12}\sigma_2\,d\widetilde{B}_2(t), \qquad (10.2.7) \\ d\overline{X}_2(t) &= (p_{21}a_1 + p_{22}a_2)\,dt - \kappa\overline{X}_1(t)\,dt - \lambda_2\overline{X}_2(t)\,dt \\ &\quad + p_{21}\sigma_1\,d\widetilde{B}_1(t) + p_{22}\sigma_2\,d\widetilde{B}_2(t). \qquad (10.2.8) \end{aligned}$$

Lemma 10.2.1. *Under our assumptions that $\sigma_1 > 0$, $\sigma_2 > 0$, $-1 < \nu < 1$, and P is nonsingular, we have*

$$\gamma_i = p_{i1}^2\sigma_1^2 + 2\nu p_{i1}p_{i2}\sigma_1\sigma_2 + p_{i2}^2\sigma_2^2, \quad i = 1, 2, \qquad (10.2.9)$$

are strictly positive, and

$$\rho = \frac{1}{\sqrt{\gamma_1\gamma_2}}(p_{11}p_{21}\sigma_1^2 + \nu(p_{11}p_{22} + p_{12}p_{21})\sigma_1\sigma_2 + p_{12}p_{22}\sigma_2^2) \qquad (10.2.10)$$

is in $(-1, 1)$.

PROOF: Because $\nu \in (-1, 1)$, the matrix

$$N = \begin{bmatrix} 1 & \nu \\ \nu & 1 \end{bmatrix}$$

has a matrix square root. Indeed, one such square root is

$$\sqrt{N} = \begin{bmatrix} a & \sqrt{1-a^2} \\ \sqrt{1-a^2} & a \end{bmatrix},$$

where $a = \text{sign}(\nu)\sqrt{\frac{1}{2} + \frac{1}{2}\sqrt{1-\nu^2}}$. Verification of this uses the equation

$$\begin{aligned} 2a\sqrt{1-a^2} &= 2\text{sign}(\nu)\sqrt{\frac{1}{2} + \frac{1}{2}\sqrt{1-\nu^2}} \cdot \sqrt{\frac{1}{2} - \frac{1}{2}\sqrt{1-\nu^2}} \\ &= 2\text{sign}(\nu)\sqrt{\frac{1}{4} - \frac{1}{4}(1-\nu^2)} \\ &= 2\text{sign}(\nu) \cdot \frac{1}{2}|\nu| \;=\; \nu. \end{aligned}$$

The matrices $\sqrt{N}$, Σ, and P^{tr} are nonsingular, which implies nonsingularity of the matrix

$$\sqrt{N}\Sigma P^{\text{tr}} = \begin{bmatrix} p_{11}\sigma_1 a + p_{12}\sigma_2\sqrt{1-a^2} & p_{21}\sigma_1 a + p_{22}\sigma_2\sqrt{1-a^2} \\ p_{11}\sigma_1\sqrt{1-a^2} + p_{12}\sigma_2 a & p_{21}\sigma_1\sqrt{1-a^2} + p_{22}\sigma_2 a \end{bmatrix}.$$

Let c_1 be the first column of this matrix and c_2 the second column. Because of the nonsingularity of $\sqrt{N}\Sigma P^{\text{tr}}$, these vectors are linearly independent, and hence neither of them is the zero vector,

Therefore,

$$\gamma_i = \|c_i\|^2 > 0, \quad i = 1, 2.$$

For linearly independent vectors, the Cauchy-Schwarz inequality implies

$$-\|c_1\|\,\|c_2\| < c_1 \cdot c_2 < \|c_1\|\,\|c_2\|.$$

This is equivalent to $-1 < \rho < 1$. □

We define

$$\overline{B}_i(t) = \frac{1}{\sqrt{\gamma_i}}\big(p_{i1}\sigma_1\widetilde{B}_1(t) + p_{i2}\sigma_2\widetilde{B}_2(t)\big), \quad i = 1, 2.$$

The processes $\overline{B}_1(t)$ and $\overline{B}_2(t)$ are continuous martingales starting at zero. Furthermore,

$$d\overline{B}_1(t)\,d\overline{B}_1(t) = d\overline{B}_2(t)\,d\overline{B}_2(t) = dt.$$

According to Lévy's Theorem, Theorem 4.6.4, $\overline{B}_1(t)$ and $\overline{B}_2(t)$ are Brownian motions. Furthermore,

$$d\overline{B}_1(t)\, d\overline{B}_2(t) = \rho\, dt,$$

where ρ is defined by (10.2.10). We may rewrite (10.2.7) and (10.2.8) as

$$d\overline{X}_1(t) = (p_{11}a_1 + p_{12}a_2)\, dt - \lambda_1 \overline{X}_1(t)\, dt + \sqrt{\gamma_1}\, d\overline{B}_1(t), \quad (10.2.11)$$
$$d\overline{X}_2(t) = (p_{21}a_1 + p_{22}a_2)\, dt - \kappa \overline{X}_1(t)\, dt - \lambda_2 \overline{X}_2(t)\, dt + \sqrt{\gamma_2}\, d\overline{B}_2(t). \quad (10.2.12)$$

Setting

$$\widehat{X}_1(t) = \frac{1}{\sqrt{\gamma_1}}\left(\overline{X}_1(t) - \frac{p_{11}a_1 + p_{12}a_2}{\lambda_1}\right),$$
$$\widehat{X}_2(t) = \frac{1}{\sqrt{\gamma_2}}\left(\overline{X}_2(t) + \frac{\kappa(p_{11}a_1 + p_{12}a_2)}{\lambda_1\lambda_2} - \frac{p_{21}a_1 + p_{22}a_2}{\lambda_2}\right),$$

we may further rewrite (10.2.11) and (10.2.12) as

$$d\widehat{X}_1(t) = -\lambda_1 \widehat{X}_1(t)\, dt + d\overline{B}_1(t), \quad (10.2.13)$$
$$d\widehat{X}_2(t) = -\kappa\sqrt{\frac{\gamma_1}{\gamma_2}}\, \widehat{X}_1(t)\, dt - \lambda_2 \widehat{X}_2(t)\, dt + d\overline{B}_2(t). \quad (10.2.14)$$

As the last step, we define

$$\widetilde{W}_1(t) = \overline{B}_1(t), \quad \widetilde{W}_2(t) = \frac{1}{\sqrt{1-\rho^2}}\big[-\rho\overline{B}_1(t) + \overline{B}_2(t)\big].$$

Both $\widetilde{W}_1(t)$ and $\widetilde{W}_2(t)$ are continuous martingales, and it is easily verified that

$$d\widetilde{W}_1(t)\, d\widetilde{W}_1(t) = dt, \quad d\widetilde{W}_1(t)\, d\widetilde{W}_2(t) = 0, \quad d\widetilde{W}_2(t)\, d\widetilde{W}_2(t) = dt.$$

According to Lévy's Theorem, Theorem 4.6.4, $\widetilde{W}_1(t)$ and $\widetilde{W}_2(t)$ are independent Brownian motions. Setting

$$Y_1(t) = \widehat{X}_1(t), \quad Y_2(t) = \frac{-\rho\widehat{X}_1(t) + \widehat{X}_2(t)}{\sqrt{1-\rho^2}},$$

we have

$$\begin{aligned}
dY_2(t) &= \frac{1}{\sqrt{1-\rho^2}}\big[-\rho\, d\widehat{X}_1(t) + d\widehat{X}_2(t)\big]\\
&= \frac{1}{\sqrt{1-\rho^2}}\left[\left(\rho\lambda_1 - \kappa\sqrt{\frac{\gamma_1}{\gamma_2}}\right)\widehat{X}_1(t) - \lambda_2\widehat{X}_2(t)\right] dt\\
&\qquad + \frac{1}{\sqrt{1-\rho^2}}\big[-\rho\, d\overline{B}_1(t) + d\overline{B}_2(t)\big]\\
&= \frac{1}{\sqrt{1-\rho^2}}\left(\rho\lambda_1 - \rho\lambda_2 - \kappa\sqrt{\frac{\gamma_1}{\gamma_2}}\right) Y_1(t)\, dt - \lambda_2 Y_2(t)\, dt + d\widetilde{W}_2(t).
\end{aligned}$$

We may thus rewrite (10.2.13) and (10.2.14) as

$$dY_1(t) = -\lambda_1 Y_1(t)\,dt + d\widetilde{W}_1(t),$$
$$dY_2(t) = -\lambda_{21} Y_1(t)\,dt - \lambda_2 Y_2(t)\,dt + d\widetilde{W}_2(t),$$

where

$$\lambda_{21} = \frac{1}{\sqrt{1-\rho^2}}\left(-\rho\lambda_1 + \rho\lambda_2 + \kappa\sqrt{\frac{\gamma_1}{\gamma_2}}\right).$$

These are the canonical equations (10.2.4) and (10.2.5).

To obtain (10.2.6), we trace back through the changes of variables:

$$\begin{aligned}
Y_1(t) &= \widehat{X}_1(t)\\
&= \frac{1}{\sqrt{\gamma_1}}\left(\overline{X}_1(t) - \frac{p_{11}a_1 + p_{12}a_2}{\lambda_1}\right)\\
&= \frac{1}{\sqrt{\gamma_1}}\left(p_{11}X_1(t) + p_{12}X_2(t) - \frac{p_{11}a_1 + p_{12}a_2}{\lambda_1}\right),\\
Y_2(t) &= \frac{1}{\sqrt{1-\rho^2}}\big(-\rho\widehat{X}_1(t) + \widehat{X}_2(t)\big)\\
&= -\frac{\rho}{\sqrt{\gamma_1(1-\rho^2)}}\left(\overline{X}_1(t) - \frac{p_{11}a_1 + p_{12}a_2}{\lambda_1}\right)\\
&\quad + \frac{1}{\sqrt{\gamma_2(1-\rho^2)}}\left(\overline{X}_2(t) + \frac{\kappa(p_{11}a_1 + p_{12}a_2)}{\lambda_1\lambda_2} - \frac{p_{21}a_1 + p_{22}a_2}{\lambda_2}\right)\\
&= -\frac{\rho}{\sqrt{\gamma_1(1-\rho^2)}}\left(p_{11}X_1(t) + p_{12}X_2(t) - \frac{p_{11}a_1 + p_{12}a_2}{\lambda_1}\right)\\
&\quad + \frac{1}{\sqrt{\gamma_2(1-\rho^2)}}\left(p_{21}X_1(t) + p_{22}X_2(t) + \frac{\kappa(p_{11}a_1 + p_{12}a_2)}{\lambda_1\lambda_2}\right.\\
&\qquad\left. - \frac{p_{21}a_1 + p_{22}a_2}{\lambda_2}\right).
\end{aligned}$$

In vector notation,

$$Y(t) = \Gamma(PX(t) + V), \tag{10.2.15}$$

where

$$Y(t) = \begin{bmatrix} Y_1(t)\\ Y_2(t)\end{bmatrix},\quad \Gamma = \begin{bmatrix} \frac{1}{\sqrt{\gamma_1}} & 0\\ -\frac{\rho}{\sqrt{\gamma_1(1-\rho^2)}} & \frac{1}{\sqrt{\gamma_2(1-\rho^2}} \end{bmatrix},$$

$$V = \begin{bmatrix} -\frac{p_{11}a_1 + p_{12}a_2}{\lambda_1}\\ \frac{\kappa(p_{11}a_1 + p_{12}a_2)}{\lambda_1\lambda_2} - \frac{p_{21}a_1 + p_{22}a_2}{\lambda_2} \end{bmatrix}.$$

We solve (10.2.15) for $X(t)$:

$$X(t) = P^{-1}\big(\Gamma^{-1}Y(t) - V\big).$$

Therefore,

$$\begin{aligned}R(t) &= \epsilon_0 + [\epsilon_1\ \epsilon_2]X(t)\\ &= \epsilon_0 + [\epsilon_1\ \epsilon_2]P^{-1}\Gamma^{-1}Y(t) - [\epsilon_1\ \epsilon_2]P^{-1}V\\ &= \delta_0 + [\delta_1\ \delta_2]Y(t),\end{aligned}$$

where

$$\delta_0 = \epsilon_0 - [\epsilon_1\ \epsilon_2]P^{-1}V, \quad [\delta_1\ \delta_2] = [\epsilon_1\ \epsilon_2]P^{-1}\Gamma^{-1}.$$

We have obtained (10.2.6).

Bond Prices

We derive the formula for zero-coupon bond prices in the canonical two-factor Vasicek model. According to the risk-neutral pricing formula, the price at time t of a zero-coupon bond paying 1 at a later time T is

$$B(t,T) = \widetilde{\mathbb{E}}\left[e^{-\int_t^T R(u)du}\Big|\,\mathcal{F}(t)\right], \quad 0 \le t \le T.$$

Because $R(t)$ given by (10.2.6) is a function of the factors $Y_1(t)$ and $Y_2(t)$, and the solution of the system of stochastic differential equations (10.2.4) and (10.2.5) is Markov, there must be some function $f(t, y_1, y_2)$ such that

$$B(t,T) = f\big(t, Y_1(t), Y_2(t)\big). \tag{10.2.16}$$

The discount factor $D(t) = e^{-\int_0^t R(u)du}$ satisfies $dD(t) = -R(t)D(t)\,dt$ (see (5.2.18)). Iterated conditioning implies that the discounted bond price $D(t)B(t,T)$ is a martingale under $\widetilde{\mathbb{P}}$. Therefore, the differential of $D(t)B(t,T)$ has dt term zero. We compute this differential:

$$\begin{aligned}&d\big(D(t)B(t,T)\big)\\ &= d\Big(D(t)f\big(t,Y_1(t),Y_2(t)\big)\Big)\\ &= -R(t)D(t)f\big(t,Y_1(t),Y_2(t)\big)\,dt + D(t)\,df\big(t,Y_1(t),Y_2(t)\big)\\ &= D\bigg[-Rf\,dt + f_t\,dt + f_{y_1}\,dY_1 + f_{y_2}\,dY_2\\ &\qquad +\frac{1}{2}f_{y_1y_1}\,dY_1\,dY_1 + f_{y_1y_2}\,dY_1\,dY_2 + \frac{1}{2}f_{y_2y_2}\,dY_2\,dY_2\bigg].\end{aligned} \tag{10.2.17}$$

We use equations (10.2.4)–(10.2.6) to take the next step:

$$
\begin{aligned}
&d(D(t)B(t,T)) \\
&= D\Big[-(\delta_0+\delta_1 Y_1+\delta_2 Y_2)f + f_t - \lambda_1 Y_1 f_{y_1} - \lambda_{21} Y_1 f_{y_2} \\
&\qquad -\lambda_2 Y_2 f_{y_2} + \frac{1}{2} f_{y_1 y_1} + \frac{1}{2} f_{y_2 y_2}\Big]\,dt + D\left[f_{y_1}\,d\widetilde{W}_1 + f_{y_2}\,d\widetilde{W}_2\right].
\end{aligned}
$$

Setting the dt term equal to zero, we obtain the partial differential equation

$$
\begin{aligned}
&-(\delta_0+\delta_1 y_1+\delta_2 y_2)f(t,y_1,y_2) + f_t(t,y_1,y_2) \\
&\quad -\lambda_1 y_1 f_{y_1}(t,y_1,y_2) - \lambda_{21} y_1 f_{y_2}(t,y_1,y_2) - \lambda_2 y_2 f_{y_2}(t,y_1,y_2) \\
&\qquad +\frac{1}{2} f_{y_1 y_1}(t,y_1,y_2) + \frac{1}{2} f_{y_2 y_2}(t,y_1,y_2) = 0 \qquad (10.2.18)
\end{aligned}
$$

for all $t \in [0,T)$ and all $y_1 \in \mathbb{R}$, $y_2 \in \mathbb{R}$. We have also the terminal condition

$$
f(T,y_1,y_2) = 1 \text{ for all } y_1 \in \mathbb{R},\ y_2 \in \mathbb{R}. \qquad (10.2.19)
$$

To solve this equation, we seek a solution of the affine-yield form

$$
f(t,y_1,y_2) = e^{-y_1 C_1(T-t) - y_2 C_2(T-t) - A(T-t)} \qquad (10.2.20)
$$

for some functions $C_1(\tau)$, $C_2(\tau)$, and $A(\tau)$. Here we define $\tau = T - t$ to be the *relative maturity* (i.e., the time until maturity). So long as the model parameters do not depend on t, zero-coupon bond prices will depend on t and T only through τ. The terminal condition (10.2.19) implies that

$$
C_1(0) = C_2(0) = A(0) = 0. \qquad (10.2.21)
$$

We compute derivatives, where $'$ denotes differentiation with respect to τ. We use the fact $\frac{d}{dt}C_i(\tau) = C_i'(\tau)\cdot\frac{d}{dt}\tau = -C_i'(\tau)$, $i = 1, 2$, and the similar equation $\frac{d}{dt}A(\tau) = -A'(\tau)$ to obtain

$$
\begin{aligned}
f_t &= [y_1 C_1' + y_2 C_2' + A']f, \quad & f_{y_1} &= -C_1 f, \quad & f_{y_2} &= -C_2 f, \\
f_{y_1 y_1} &= C_1^2 f, & f_{y_1 y_2} &= C_1 C_2 f, & f_{y_2 y_2} &= C_2^2 f.
\end{aligned}
$$

Equation (10.2.18) becomes

$$
\begin{aligned}
&\Big[(C_1' + \lambda_1 C_1 + \lambda_{21} C_2 - \delta_1)y_1 + (C_2' + \lambda_2 C_2 - \delta_2)y_2 \\
&\qquad + \Big(A' + \frac{1}{2}C_1^2 + \frac{1}{2}C_2^2 - \delta_0\Big)\Big]f = 0. \qquad (10.2.22)
\end{aligned}
$$

Because (10.2.22) must hold for all y_1 and y_2, the term $C_1' + \lambda C_1 + \lambda_{21} C_2 - \delta_1$ multiplying y_1 must be zero. If it were not, and (10.2.22) held for one value of y_1, then a change in the value of y_1 would cause the equation to be violated. Similarly, the term $C_2' + \lambda_2 C_2 - \delta_2$ multiplying y_2 must be zero, and consequently the remaining term $A' + \frac{1}{2}C_1^2 + \frac{1}{2}C_2^2 - \delta_0$ must also be zero. This gives us a system of three ordinary differential equations:

$$C_1'(\tau) = -\lambda_1 C_1(\tau) - \lambda_{21} C_2(\tau) + \delta_1, \tag{10.2.23}$$

$$C_2'(\tau) = -\lambda_2 C_2(\tau) + \delta_2, \tag{10.2.24}$$

$$A'(\tau) = -\frac{1}{2}C_1^2(\tau) - \frac{1}{2}C_2^2(\tau) + \delta_0. \tag{10.2.25}$$

The solution of (10.2.24) satisfying the initial condition $C_2(0) = 0$ (see (10.2.21)) is

$$C_2(\tau) = \frac{\delta_2}{\lambda_2}\left(1 - e^{-\lambda_2 \tau}\right). \tag{10.2.26}$$

We substitute this into (10.2.23) and solve using the initial condition $C_1(0) = 0$. In particular, (10.2.23) implies

$$\begin{aligned}
\frac{d}{d\tau}\left(e^{\lambda_1 \tau} C_1(\tau)\right) &= e^{\lambda_1 \tau}\left(\lambda_1 C_1(\tau) + C_1'(\tau)\right) \\
&= e^{\lambda_1 \tau}\left(- \lambda_{21} C_2(\tau) + \delta_1\right) \\
&= e^{\lambda_1 \tau}\left(-\frac{\lambda_{21}\delta_2}{\lambda_2}\left(1 - e^{-\lambda_2 \tau}\right) + \delta_1\right).
\end{aligned}$$

If $\lambda_1 \neq \lambda_2$, integration from 0 to τ yields

$$C_1(\tau) = \frac{1}{\lambda_1}\left(\delta_1 - \frac{\lambda_{21}\delta_2}{\lambda_2}\right)\left(1 - e^{-\lambda_1 \tau}\right) + \frac{\lambda_{21}\delta_2}{\lambda_2(\lambda_1 - \lambda_2)}\left(e^{-\lambda_2 \tau} - e^{-\lambda_1 \tau}\right). \tag{10.2.27}$$

If $\lambda_1 = \lambda_2$, we obtain instead

$$C_1(\tau) = \frac{1}{\lambda_1}\left(\delta_1 - \frac{\lambda_{21}\delta_2}{\lambda_1}\right)\left(1 - e^{-\lambda_1 \tau}\right) + \frac{\lambda_{21}\delta_2}{\lambda_1}\tau e^{-\lambda_1 \tau}. \tag{10.2.28}$$

Finally, (10.2.25) and the initial condition $A(0) = 0$ imply

$$A(\tau) = \int_0^\tau \left[-\frac{1}{2}C_1^2(u) - \frac{1}{2}C_2^2(u) + \delta_0\right] du, \tag{10.2.29}$$

and this can be obtained in closed form by a lengthy but straightforward computation.

Short Rate and Long Rate

We fix a positive relative maturity $\overline{\tau}$ (say, thirty years) and call the yield at time t on the zero-coupon bond with relative maturity $\overline{\tau}$ (i.e., the bond maturing at date $t+\overline{\tau}$) the *long rate* $L(t)$. Once we have a model for evolution of the short rate $R(t)$ under the risk-neutral measure, then for each $t \geq 0$ the price of the $(t + \overline{\tau})$-maturity zero-coupon bond is determined by the risk-neutral pricing formula, and hence the short-rate model alone determines the long rate. We cannot therefore write down an arbitrary stochastic differential equation for the long rate. Nonetheless, in any affine-yield model, the long rate satisfies some stochastic differential equation, and we can work out this equation.

Consider the canonical two-factor Vasicek model. As we have seen in the previous discussion, zero-coupon bond prices in this model are of the form

$$B(t,T) = e^{-Y_1(t)C_1(T-t)-Y_2(t)C_2(T-t)-A(T-t)},$$

where $C_1(\tau)$, $C_2(\tau)$, and $A(\tau)$ are given by (10.2.26)–(10.2.29). Thus, the long rate at time t is

$$L(t) = -\frac{1}{\overline{\tau}} \log B(t, t+\overline{\tau}) = \frac{1}{\overline{\tau}}\big[C_1(\overline{\tau})Y_1(t) + C_2(\overline{\tau})Y_2(t) + A(\overline{\tau})\big], \quad (10.2.30)$$

which is an affine function of the canonical factors $Y_1(t)$ and $Y_2(t)$ at time t. Because the canonical factors do not have an economic interpretation, we may wish to use $R(t)$ and $L(t)$ as the model factors. We now show how to do this, obtaining a two-factor Vasicek model of the form (10.2.1), (10.2.2), and (10.2.3), where $X_1(t)$ is replaced by $R(t)$ and $X_2(t)$ is replaced by $L(t)$.

We begin by writing the formulas (10.2.6) and (10.2.30) in vector notation:

$$\begin{bmatrix} R(t) \\ L(t) \end{bmatrix} = \begin{bmatrix} \delta_1 & \delta_2 \\ \frac{1}{\overline{\tau}}C_1(\overline{\tau}) & \frac{1}{\overline{\tau}}C_2(\overline{\tau}) \end{bmatrix} \begin{bmatrix} Y_1(t) \\ Y_2(t) \end{bmatrix} + \begin{bmatrix} \delta_0 \\ \frac{1}{\overline{\tau}}A(\overline{\tau}) \end{bmatrix}. \quad (10.2.31)$$

We wish to solve this system for $(Y_1(t), Y_2(t))$.

Lemma 10.2.2. *The matrix*

$$D = \begin{bmatrix} \delta_1 & \delta_2 \\ \frac{1}{\overline{\tau}}C_1(\overline{\tau}) & \frac{1}{\overline{\tau}}C_2(\overline{\tau}) \end{bmatrix}$$

is nonsingular if and only if $\delta_2 \neq 0$ *and*

$$(\lambda_1 - \lambda_2)\delta_1 + \lambda_{21}\delta_2 \neq 0. \quad (10.2.32)$$

PROOF: Consider the function $f(x) = 1 - e^{-x} - xe^{-x}$, for which $f(0) = 0$ and $f'(x) = xe^{-x} > 0$ for all $x > 0$. We have $f(x) > 0$ for all $x > 0$. Define $h(x) = \frac{1}{x}(1 - e^{-x})$. Since $h'(x) = -x^{-2}f(x)$, which is strictly negative for all $x > 0$, $h(x)$ is strictly decreasing on $(0, \infty)$.

To examine the nonsingularity of D, we consider first the case $\lambda_1 \neq \lambda_2$. In this case, (10.2.26) and (10.2.27) imply

$$\begin{aligned} \det(D) &= \frac{1}{\overline{\tau}}\big[\delta_1 C_2(\overline{\tau}) - \delta_2 C_1(\overline{\tau})\big] \\ &= \frac{\delta_1\delta_2}{\lambda_2\overline{\tau}}\left(1 - e^{-\lambda_2\overline{\tau}}\right) - \frac{\delta_1\delta_2}{\lambda_1\overline{\tau}}\left(1 - e^{-\lambda_1\overline{\tau}}\right) \\ &\quad + \frac{\lambda_{21}\delta_2^2}{(\lambda_1 - \lambda_2)\lambda_1\lambda_2\overline{\tau}}\left[(\lambda_1 - \lambda_2)\left(1 - e^{-\lambda_1\overline{\tau}}\right) - \lambda_1 e^{\lambda_2\overline{\tau}} + \lambda_1 e^{\lambda_1\overline{\tau}}\right] \end{aligned}$$

$$= \delta_1\delta_2\left[\frac{1}{\lambda_2\overline{\tau}}\left(1-e^{-\lambda_2\overline{\tau}}\right) - \frac{1}{\lambda_1\overline{\tau}}\left(1-e^{-\lambda_1\overline{\tau}}\right)\right]$$
$$+\frac{\lambda_{21}\delta_2^2}{(\lambda_1-\lambda_2)\lambda_1\lambda_2\overline{\tau}}\left[\lambda_1\left(1-e^{-\lambda_2\overline{\tau}}\right) - \lambda_2\left(1-e^{-\lambda_1\overline{\tau}}\right)\right]$$
$$= \delta_2\left(\delta_1 + \frac{\lambda_{21}\delta_2}{(\lambda_1-\lambda_2)}\right)\left[\frac{1}{\lambda_2\overline{\tau}}\left(1-e^{-\lambda_2\overline{\tau}}\right) - \frac{1}{\lambda_1\overline{\tau}}\left(1-e^{-\lambda_1\overline{\tau}}\right)\right]$$
$$= \delta_2\left(\delta_1 + \frac{\lambda_{21}\delta_2}{(\lambda_1-\lambda_2)}\right)\left[h(\lambda_2\overline{\tau}) - h(\lambda_1\overline{\tau})\right].$$

Because $\lambda_1 \neq \lambda_2$ and h is strictly decreasing, $h(\lambda_2\overline{\tau}) \neq h(\lambda_2\overline{\tau})$. The determinant of D is nonzero if and only if $\delta_2 \neq 0$ and (10.2.32) holds.

Next consider the case $\lambda_1 = \lambda_2$. In this case, (10.2.26) and (10.2.28) imply

$$\det(D) = \frac{1}{\overline{\tau}}\left[\delta_1 C_2(\overline{\tau}) - \delta_2 C_1(\overline{\tau})\right]$$
$$= \frac{\delta_1\delta_2}{\lambda_1\overline{\tau}}\left(1-e^{-\lambda_1\overline{\tau}}\right) - \frac{\delta_2}{\lambda_1\overline{\tau}}\left(\delta_1 - \frac{\lambda_{21}\delta_2}{\lambda_1}\right)\left(1-e^{-\lambda_1\overline{\tau}}\right) - \frac{\lambda_{21}\delta_2^2}{\lambda_1}e^{-\lambda_1\overline{\tau}}$$
$$= \frac{\lambda_{21}\delta_2^2}{\lambda_1^2\overline{\tau}}f(\lambda_1\overline{\tau}).$$

Because $\lambda_1\overline{\tau}$ is positive, $f(\lambda_1\overline{\tau})$ is not zero. In this case, (10.2.32) is equivalent to $\delta_2 \neq 0$ and $\lambda_{21} \neq 0$. The determinant of D is nonzero if and only if (10.2.32) holds (in which case $\delta_2 \neq 0$ and $\lambda_{21} \neq 0$). □

Under the assumptions of Lemma 10.2.2, we can invert (10.2.31) to obtain

$$\begin{bmatrix} Y_1(t) \\ Y_2(t) \end{bmatrix} = \begin{bmatrix} \delta_1 & \delta_2 \\ \frac{1}{\overline{\tau}}C_1(\overline{\tau}) & \frac{1}{\overline{\tau}}C_2(\overline{\tau}) \end{bmatrix}^{-1}\left(\begin{bmatrix} R(t) \\ L(t) \end{bmatrix} - \begin{bmatrix} \delta_0 \\ \frac{1}{\overline{\tau}}A(\overline{\tau}) \end{bmatrix}\right). \quad (10.2.33)$$

We can compute the differential in (10.2.31) using (10.2.4) and (10.2.5). This leads to a formula in which $Y_1(t)$ and $Y_2(t)$ appear on the right-hand side, but we can then use (10.2.33) to rewrite the right-hand side in terms of $R(t)$, $L(t)$. These steps result in the equation

$$\begin{bmatrix} dR(t) \\ dL(t) \end{bmatrix}$$
$$= \begin{bmatrix} \delta_1 & \delta_2 \\ \frac{1}{\overline{\tau}}C_1(\overline{\tau}) & \frac{1}{\overline{\tau}}C_2(\overline{\tau}) \end{bmatrix}\begin{bmatrix} dY_1(t) \\ dY_2(t) \end{bmatrix}$$
$$= \begin{bmatrix} \delta_1 & \delta_2 \\ \frac{1}{\overline{\tau}}C_1(\overline{\tau}) & \frac{1}{\overline{\tau}}C_2(\overline{\tau}) \end{bmatrix}\left(-\begin{bmatrix} \lambda_1 & 0 \\ \lambda_{21} & \lambda_2 \end{bmatrix}\begin{bmatrix} Y_1(t) \\ Y_2(t) \end{bmatrix}dt + \begin{bmatrix} d\widetilde{W}_1(t) \\ d\widetilde{W}_2(t) \end{bmatrix}\right)$$

$$= \begin{bmatrix} \delta_1 & \delta_2 \\ \frac{1}{\overline{\tau}}C_1(\overline{\tau}) & \frac{1}{\overline{\tau}}C_2(\overline{\tau}) \end{bmatrix} \begin{bmatrix} \lambda_1 & 0 \\ \lambda_{21} & \lambda_2 \end{bmatrix} \begin{bmatrix} \delta_1 & \delta_2 \\ \frac{1}{\overline{\tau}}C_1(\overline{\tau}) & \frac{1}{\overline{\tau}}C_2(\overline{\tau}) \end{bmatrix}^{-1} \begin{bmatrix} \delta_0 \\ \frac{1}{\overline{\tau}}A(\overline{\tau}) \end{bmatrix} dt$$
$$- \begin{bmatrix} \delta_1 & \delta_2 \\ \frac{1}{\overline{\tau}}C_1(\overline{\tau}) & \frac{1}{\overline{\tau}}C_2(\overline{\tau}) \end{bmatrix} \begin{bmatrix} \lambda_1 & 0 \\ \lambda_{21} & \lambda_2 \end{bmatrix} \begin{bmatrix} \delta_1 & \delta_2 \\ \frac{1}{\overline{\tau}}C_1(\overline{\tau}) & \frac{1}{\overline{\tau}}C_2(\overline{\tau}) \end{bmatrix}^{-1} \begin{bmatrix} R(t) \\ L(t) \end{bmatrix} dt$$
$$+ \begin{bmatrix} \delta_1 & \delta_2 \\ \frac{1}{\overline{\tau}}C_1(\overline{\tau}) & \frac{1}{\overline{\tau}}C_2(\overline{\tau}) \end{bmatrix} \begin{bmatrix} d\widetilde{W}_1(t) \\ d\widetilde{W}_2(t) \end{bmatrix}.$$

This is the vector notation for a pair of equations of the form (10.2.1) and (10.2.2) for a two-factor Vasicek model for the short rate $R(t)$ and the long rate $L(t)$. The parameters a_1 and a_2 appearing in (10.2.1) and (10.2.2) are given by

$$\begin{bmatrix} a_1 \\ a_2 \end{bmatrix} = \begin{bmatrix} \delta_1 & \delta_2 \\ \frac{1}{\overline{\tau}}C_1(\overline{\tau}) & \frac{1}{\overline{\tau}}C_2(\overline{\tau}) \end{bmatrix} \begin{bmatrix} \lambda_1 & 0 \\ \lambda_{21} & \lambda_2 \end{bmatrix} \begin{bmatrix} \delta_1 & \delta_2 \\ \frac{1}{\overline{\tau}}C_1(\overline{\tau}) & \frac{1}{\overline{\tau}}C_2(\overline{\tau}) \end{bmatrix}^{-1} \begin{bmatrix} \delta_0 \\ \frac{1}{\overline{\tau}}A(\overline{\tau}) \end{bmatrix}.$$

The matrix B is

$$\begin{bmatrix} b_{11} & b_{12} \\ b_{21} & b_{22} \end{bmatrix} = \begin{bmatrix} \delta_1 & \delta_2 \\ \frac{1}{\overline{\tau}}C_1(\overline{\tau}) & \frac{1}{\overline{\tau}}C_2(\overline{\tau}) \end{bmatrix} \begin{bmatrix} \lambda_1 & 0 \\ \lambda_{21} & \lambda_2 \end{bmatrix} \begin{bmatrix} \delta_1 & \delta_2 \\ \frac{1}{\overline{\tau}}C_1(\overline{\tau}) & \frac{1}{\overline{\tau}}C_2(\overline{\tau}) \end{bmatrix}^{-1},$$

and the eigenvalues of B are $\lambda_1 > 0$, $\lambda_2 > 0$. With

$$\sigma_1 = \sqrt{\delta_1^2 + \delta_2^2}, \quad \sigma_2 = \frac{1}{\overline{\tau}}\sqrt{C_1^2(\overline{\tau}) + C_2^2(\overline{\tau})},$$

the processes

$$\widetilde{B}_1(t) = \frac{1}{\sigma_1}\big(\delta_1 \widetilde{W}_1(t) + \delta_2 \widetilde{W}_2(t)\big),$$
$$\widetilde{B}_2(t) = \frac{1}{\sigma_2 \overline{\tau}}\big(C_1(\overline{\tau})\widetilde{W}_1(t) + C_2(\overline{\tau})\widetilde{W}_2(t)\big),$$

are the Brownian motions appearing in (10.2.1) and (10.2.2). Finally, equation (10.2.3) takes the form

$$R(t) = 0 + 1 \cdot R(t) + 0 \cdot L(t)$$

(i.e., $\epsilon_0 = \epsilon_2 = 0$, $\epsilon_1 = 1$).

Gaussian Factor Processes

The canonical two-factor Vasicek model in vector notation is

$$dY(t) = -\Lambda Y(t) + d\widetilde{W}(t), \tag{10.2.34}$$

where

$$Y(t) = \begin{bmatrix} Y_1(t) \\ Y_2(t) \end{bmatrix}, \quad \Lambda = \begin{bmatrix} \lambda_1 & 0 \\ \lambda_{21} & \lambda_2 \end{bmatrix}, \quad \widetilde{W}(t) = \begin{bmatrix} \widetilde{W}_1(t) \\ \widetilde{W}_2(t) \end{bmatrix}.$$

Recall that $\lambda_1 > 0$, $\lambda_2 > 0$. There is a closed-form solution to this matrix differential equation. To derive this solution, we first form the matrix exponential $e^{\Lambda t}$ defined by

$$e^{\Lambda t} = \sum_{n=0}^{\infty} \frac{1}{n!} (\Lambda t)^n,$$

where $(\Lambda t)^0 = I$, the 2×2 identity matrix.

Lemma 10.2.3. *If $\lambda_1 \neq \lambda_2$, then*

$$e^{\Lambda t} = \begin{bmatrix} e^{\lambda_1 t} & 0 \\ \frac{\lambda_{21}}{\lambda_1 - \lambda_2}\left(e^{\lambda_1 t} - e^{\lambda_2 t}\right) & e^{\lambda_2 t} \end{bmatrix}. \tag{10.2.35}$$

If $\lambda_1 = \lambda_2$, then

$$e^{\Lambda t} = \begin{bmatrix} e^{\lambda_1 t} & 0 \\ \lambda_{21} t e^{\lambda_1 t} & e^{\lambda_1 t} \end{bmatrix}. \tag{10.2.36}$$

In either case,

$$\frac{d}{dt} e^{\Lambda t} = \Lambda e^{\Lambda t} = e^{\Lambda t} \Lambda, \tag{10.2.37}$$

where the derivative is defined componentwise, and

$$e^{-\Lambda t} = \left(e^{\Lambda t}\right)^{-1}, \tag{10.2.38}$$

where $e^{-\Lambda t}$ is obtained by replacing λ_1, λ_2, and λ_{21} in the formula for $e^{\Lambda t}$ by $-\lambda_1$, $-\lambda_2$, and $-\lambda_{21}$, respectively.

PROOF: We consider first the case $\lambda_1 \neq \lambda_2$. We claim that in this case

$$(\Lambda t)^n = \begin{bmatrix} (\lambda_1 t)^n & 0 \\ \lambda_{21} t^n \frac{\lambda_1^n - \lambda_2^n}{\lambda_1 - \lambda_2} & (\lambda_2 t)^n, \end{bmatrix}, \quad n = 0, 1, \ldots. \tag{10.2.39}$$

This equation holds for the base case $n = 0$: $(\Lambda t)^0 = \begin{bmatrix} 1 & 0 \\ 0 & 1 \end{bmatrix}$. We show by mathematical induction that the equation holds in general. Assume (10.2.39) is true for some value of n. Then

$$\begin{aligned} (\Lambda t)^{n+1} &= (\Lambda t)(\Lambda t)^n \\ &= \begin{bmatrix} \lambda_1 t & 0 \\ \lambda_{21} t & \lambda_2 t \end{bmatrix} \begin{bmatrix} (\lambda_1 t)^n & 0 \\ \lambda_{21} t^n \frac{\lambda_1^n - \lambda_2^n}{\lambda_1 - \lambda_2} & (\lambda_2 t)^n \end{bmatrix} \\ &= \begin{bmatrix} (\lambda_1 t)^{n+1} & 0 \\ \lambda_{21} t^{n+1} \left(\lambda_1^n + \lambda_2 \frac{\lambda_1^n - \lambda_2^n}{\lambda_1 - \lambda_2}\right) & (\lambda_2 t)^{n+1} \end{bmatrix} \\ &= \begin{bmatrix} (\lambda_1 t)^{n+1} & 0 \\ \lambda_{21} t^{n+1} \frac{\lambda_1^{n+1} - \lambda_2^{n+1}}{\lambda_1 - \lambda_2} & (\lambda_2 t)^{n+1} \end{bmatrix}, \end{aligned}$$

which is (10.2.39) with n replaced by $n+1$. Having thus established (10.2.39) for all values of n, we have

$$\begin{aligned} e^{At} &= \sum_{n=0}^{\infty} \frac{1}{n!}(At)^n \\ &= \begin{bmatrix} \sum_{n=0}^{\infty} \frac{1}{n!}(\lambda_1 t)^n & 0 \\ \frac{\lambda_{21}}{\lambda_1-\lambda_2}\left(\sum_{n=0}^{\infty} \frac{1}{n!}(\lambda_1 t)^n - \sum_{n=0}^{\infty} \frac{1}{n!}(\lambda_2 t)^n\right) & \sum_{n=0}^{\infty} \frac{1}{n!}(\lambda_2 t)^n \end{bmatrix} \\ &= \begin{bmatrix} e^{\lambda_1 t} & 0 \\ \frac{\lambda_{21}}{\lambda_1-\lambda_2}\left(e^{\lambda_1 t} - e^{\lambda_2 t}\right) & e^{\lambda_2 t} \end{bmatrix}. \end{aligned}$$

This is (10.2.35).

We next consider the case $\lambda_1 = \lambda_2$. We claim in this case that

$$(At)^n = \begin{bmatrix} (\lambda_1 t)^n & 0 \\ n\lambda_{21}\lambda_1^{n-1}t^n & (\lambda_1 t)^n \end{bmatrix}, \ n = 1, 2, \dots. \tag{10.2.40}$$

This equation holds for the base case $n = 0$. We again use mathematical induction to establish the equation for all n. Assume (10.2.40) holds for some value of n. Then

$$\begin{aligned} (At)^{n+1} &= (At)(At)^n \\ &= \begin{bmatrix} \lambda_1 t & 0 \\ \lambda_{21} t & \lambda_1 t \end{bmatrix} \begin{bmatrix} (\lambda_1 t)^n & 0 \\ n\lambda_{21}\lambda_1^{n-1}t^n & (\lambda_1 t)^n \end{bmatrix} \\ &= \begin{bmatrix} (\lambda_1 t)^{n+1} & 0 \\ (\lambda_{21}\lambda_1^n + n\lambda_{21}\lambda_1^n)t^{n+1} & (\lambda_1 t)^{n+1} \end{bmatrix} \\ &= \begin{bmatrix} (\lambda_1 t)^{n+1} & 0 \\ (n+1)\lambda_{21}\lambda_1^n t^{n+1} & (\lambda_1 t)^{n+1} \end{bmatrix}, \end{aligned}$$

which is (10.2.40) with n replaced by $n+1$. Having thus established (10.2.40) for all values of n, we have

$$e^{At} = \sum_{n=0}^{\infty} \frac{1}{n!}(At)^n = \begin{bmatrix} \sum_{n=0}^{\infty} \frac{1}{n!}(\lambda_1 t)^n & 0 \\ \lambda_{21} \sum_{n=0}^{\infty} \frac{n}{n!}\lambda_1^{n-1}t^n & \sum_{n=0}^{\infty} \frac{1}{n!}(\lambda_1 t)^n \end{bmatrix}. \tag{10.2.41}$$

But

$$\lambda_{21} \sum_{n=0}^{\infty} \frac{n}{n!}\lambda_1^{n-1}t^n = \lambda_{21} \frac{d}{d\lambda_1} \sum_{n=0}^{\infty} \frac{1}{n!}(\lambda_1 t)^n = \lambda_{21} \frac{d}{d\lambda_1} e^{\lambda_1 t} = \lambda_{21} t e^{\lambda_1 t}.$$

Substituting this into (10.2.41), we obtain (10.2.36).

When $\lambda_1 \neq \lambda_2$, we have

$$\frac{d}{dt} e^{At} = \begin{bmatrix} \lambda_1 e^{\lambda_1 t} & 0 \\ \frac{\lambda_{21}}{\lambda_1-\lambda_2}\left(\lambda_1 e^{\lambda_1 t} - \lambda_2 e^{\lambda_2 t}\right) & \lambda_2 e^{\lambda_2 t} \end{bmatrix}$$

and

$$e^{-\Lambda t} = \begin{bmatrix} e^{-\lambda_1 t} & 0 \\ \frac{\lambda_{21}}{\lambda_1 - \lambda_2}\left(e^{-\lambda_1 t} - e^{-\lambda_2 t}\right) & e^{-\lambda_2 t} \end{bmatrix}.$$

When $\lambda_1 = \lambda_2$,

$$\frac{d}{dt}e^{\Lambda t} = \begin{bmatrix} \lambda_1 e^{\lambda_1 t} & 0 \\ \lambda_{21}(1 + \lambda_1 t)e^{\lambda_1 t} & \lambda_1 e^{\lambda_1 t} \end{bmatrix}$$

and

$$e^{-\Lambda t} = \begin{bmatrix} e^{-\lambda_1 t} & 0 \\ -\lambda_{21} t e^{-\lambda_1 t} & e^{-\lambda_1 t} \end{bmatrix}.$$

The verification of (10.2.37) and (10.2.38) can be done by straightforward matrix multiplications. □

We use (10.2.34) to compute

$$d\left(e^{\Lambda t} Y(t)\right) = e^{\Lambda t}\left(\Lambda Y(t)\,dt + dY(t)\right) = e^{\Lambda t}\,d\widetilde{W}(t).$$

Integration from 0 to t yields

$$e^{\Lambda t} Y(t) = Y(0) + \int_0^t e^{\Lambda u}\,d\widetilde{W}(u).$$

We solve for

$$\begin{aligned} Y(t) &= e^{-\Lambda t} Y(0) + e^{-\Lambda t}\int_0^t e^{\Lambda u}\,d\widetilde{W}(u) \\ &= e^{-\Lambda t} Y(0) + \int_0^t e^{-\Lambda(t-u)}\,d\widetilde{W}(u). \end{aligned} \tag{10.2.42}$$

If $\lambda_1 \neq \lambda_2$, equation (10.2.42) may be written componentwise as

$$Y_1(t) = e^{-\lambda_1 t} Y_1(0) + \int_0^t e^{-\lambda_1(t-u)}\,d\widetilde{W}_1(u), \tag{10.2.43}$$

$$\begin{aligned} Y_2(t) &= \frac{\lambda_{21}}{\lambda_1 - \lambda_2}\left(e^{-\lambda_1 t} - e^{-\lambda_2 t}\right) Y_1(0) + e^{-\lambda_2 t} Y_2(0) \\ &\quad + \frac{\lambda_{21}}{\lambda_1 - \lambda_2}\int_0^t \left(e^{-\lambda_1(t-u)} - e^{-\lambda_2(t-u)}\right) d\widetilde{W}_1(u) \\ &\quad + \int_0^t e^{-\lambda_2(t-u)}\,d\widetilde{W}_2(u). \end{aligned} \tag{10.2.44}$$

If $\lambda_1 = \lambda_2$, then the componentwise form of (10.2.42) is

$$Y_1(t) = e^{-\lambda_1 t} Y_1(0) + \int_0^t e^{-\lambda_1(t-u)}\,d\widetilde{W}_1(u), \tag{10.2.45}$$

$$\begin{aligned} Y_2(t) &= -\lambda_{21} t e^{-\lambda_1 t} Y_1(0) + e^{-\lambda_1 t} Y_2(0) \\ &\quad -\lambda_{21}\int_0^t (t-u)e^{-\lambda_1(t-u)}\,d\widetilde{W}_1(u) + \int_0^t e^{-\lambda_1(t-u)}\,d\widetilde{W}_2(u). \end{aligned} \tag{10.2.46}$$

Being nonrandom quantities plus Itô integrals of nonrandom integrands, the processes $Y_1(t)$ and $Y_2(t)$ are Gaussian, and so $R(t) = \delta_0 + \delta_1 Y_1(t) + \delta_2 Y_2(t)$ is normally distributed. The statistics of $Y_1(t)$ and $Y_2(t)$ are provided in Exercise 10.1.

10.2.2 Two-Factor CIR Model

In the two-factor Vasicek model, the canonical factors $Y_1(t)$ and $Y_2(t)$ are jointly normally distributed. Because these factors are driven by independent Brownian motions, they are not perfectly correlated and hence, for all $t > 0$,

$$R(t) = \delta_0 + \delta_1 Y_1(t) + \delta_2 Y_2(t) \tag{10.2.47}$$

is a normal random variable with positive variance except in the degenerate case $\delta_1 = \delta_2 = 0$. In particular, for each $t > 0$, there is a positive probability that $R(t)$ is strictly negative.

In the *two-factor Cox-Ingersoll-Ross model (CIR)* of this subsection, both factors are guaranteed to be nonnegative at all times almost surely. We again define the interest rate by (10.2.47) but now assume that

$$\delta_0 \geq 0, \quad \delta_1 > 0, \quad \delta_2 > 0. \tag{10.2.48}$$

We take the initial interest rate $R(0)$ to be nonnegative, and then we have $R(t) \geq 0$ for all $t \geq 0$ almost surely.

The evolution of the factor processes in the canonical two-factor CIR model is given by

$$dY_1(t) = \big(\mu_1 - \lambda_{11} Y_1(t) - \lambda_{12} Y_2(t)\big)\,dt + \sqrt{Y_1(t)}\,d\widetilde{W}_1(t), \tag{10.2.49}$$

$$dY_2(t) = \big(\mu_2 - \lambda_{21} Y_1(t) - \lambda_{22} Y_2(t)\big)\,dt + \sqrt{Y_2(t)}\,d\widetilde{W}_2(t). \tag{10.2.50}$$

In addition to (10.2.48), we assume

$$\mu_1 \geq 0,\ \mu_2 \geq 0,\ \lambda_{11} > 0,\ \lambda_{22} > 0,\ \lambda_{12} \leq 0,\ \lambda_{21} \leq 0. \tag{10.2.51}$$

These conditions guarantee that although the drift term $\mu_1 - \lambda_{11} Y_1(t) - \lambda_{12} Y_2(t)$ can be negative, it is nonnegative whenever $Y_1(t) = 0$ and $Y_2(t) \geq 0$. Similarly, the drift term $\mu_2 - \lambda_{21} Y_1(t) - \lambda_{22} Y_2(t)$ is nonnegative whenever $Y_2(t) = 0$ and $Y_1(t) \geq 0$. Starting with $Y_1(0) \geq 0$ and $Y_2(0) \geq 0$, we have $Y_1(t) \geq 0$ and $Y_2(t) \geq 0$ for all $t \geq 0$ almost surely.

The Brownian motions $\widetilde{W}_1(t)$ and $\widetilde{W}_2(t)$ in (10.2.49) and (10.2.50) are assumed to be independent. We do not need this assumption to guarantee nonnegativity of $Y_1(t)$ and $Y_2(t)$ but rather to obtain the affine-yield result below; see Remark 10.2.4.

Bond Prices

We derive the formula for zero-coupon bond prices in the canonical two-factor CIR model. As in the two-factor Vasicek model, the price at time t of a zero-coupon bond maturing at a later time T must be of the form

$$B(t,T) = f\big(t, Y_1(t), Y_2(t)\big)$$

for some function $f(t, y_1, y_2)$. The discounted bond price has differential

$$\begin{aligned}
&d\big(D(t)B(t,T)\big)\\
&= d\Big(D(t)f\big(t, Y_1(t), Y_2(t)\big)\Big)\\
&= -R(t)D(t)f\big(t, Y_1(t), Y_2(t)\big)\,dt + D(t)\,df\big(t, Y_1(t), Y_2(t)\big)\\
&= D\Big[-Rf\,dt + f_t\,dt + f_{y_1}\,dY_1 + f_{y_2}\,dY_2\\
&\qquad + \frac{1}{2}f_{y_1y_1}\,dY_1\,dY_1 + f_{y_1y_2}\,dY_1\,dY_2 + \frac{1}{2}f_{y_2y_2}\,dY_2\,dY_2\Big]\\
&= D\bigg[-(\delta_0 + \delta_1 Y_1 + \delta_2 Y_2)f + f_t + (\mu_1 - \lambda_{11}Y_1 - \lambda_{12}Y_2)f_{y_1}\\
&\qquad + (\mu_2 - \lambda_{21}Y_1 - \lambda_{22}Y_2)f_{y_2} + \frac{1}{2}Y_1 f_{y_1y_1} + \frac{1}{2}Y_2 f_{y_2y_2}\bigg]\,dt\\
&\quad + D\left[\sqrt{Y_1}\,f_{y_1}\,d\widetilde{W}_1 + \sqrt{Y_2}\,f_{y_2}\,d\widetilde{W}_2\right].
\end{aligned}$$

Setting the dt term equal to zero, we obtain the partial differential equation

$$\begin{aligned}
&-(\delta_0 + \delta_1 y_1 + \delta_2 y_2)f(t,y_1,y_2) + f_t(t,y_1,y_2)\\
&\quad + (\mu_1 - \lambda_{11}y_1 - \lambda_{12}y_2)f_{y_1}(t,y_1,y_2) + (\mu_2 - \lambda_{21}y_1 - \lambda_{22}y_2)f_{y_2}(t,y_1,y_2)\\
&\qquad + \frac{1}{2}y_1 f_{y_1y_1}(t,y_1,y_2) + \frac{1}{2}y_2 f_{y_2y_2}(t,y_1,y_2) = 0 \quad (10.2.52)
\end{aligned}$$

for all $t \in [0,T)$ and all $y_1 \geq 0$, $y_2 \geq 0$. To solve this equation, we seek a solution of the affine-yield form

$$f(t, y_1, y_2) = e^{-y_1 C_1(T-t) - y_2 C_2(T-t) - A(T-t)} \quad (10.2.53)$$

for some functions $C_1(\tau)$, $C_2(\tau)$, and $A(\tau)$, where $\tau = T - t$. The terminal condition

$$f\big(T, Y_1(T), Y_2(T)\big) = B(T,T) = 1$$

implies

$$C_1(0) = C_2(0) = A(0) = 0. \quad (10.2.54)$$

With $'$ denoting differentiation with respect to τ, we have $\frac{d}{dt}C_i(\tau) = -C_i'(\tau)$, $i = 1, 2$, $\frac{d}{dt}A(\tau) = -A'(\tau)$, and (10.2.52) becomes

$$\left[\left(C_1' + \lambda_{11}C_1 + \lambda_{21}C_2 + \frac{1}{2}C_1^2 - \delta_1\right)y_1 + \left(C_2' + \lambda_{12}C_1 + \lambda_{22}C_2 + \frac{1}{2}C_2^2 - \delta_2\right)y_2 + (A' - \mu_1 C_1 - \mu_2 C_2 - \delta_0)\right] f = 0. \quad (10.2.55)$$

Because (10.2.55) must hold for all $y_1 \geq 0$ and $y_2 \geq 0$, the term $C_1' + \lambda_{11}C_1 + \lambda_{21}C_2 + \frac{1}{2}C_1^2 - \delta_1$ multiplying y_1 must be zero. Similarly, the term $C_2' + \lambda_{12}C_1 + \lambda_{22}C_2 + \frac{1}{2}C_2^2 - \delta_2$ multiplying y_2 must be zero, and consequently the remaining term $A' - \mu_1 C_2 - \mu_2 C_2 - \delta_0$ must also be zero. This gives us a system of three ordinary differential equations:

$$C_1'(\tau) = -\lambda_{11}C_1(\tau) - \lambda_{21}C_2(\tau) - \frac{1}{2}C_1^2(\tau) + \delta_1, \quad (10.2.56)$$

$$C_2'(\tau) = -\lambda_{12}C_1(\tau) - \lambda_{22}C_2(\tau) - \frac{1}{2}C_2^2(\tau) + \delta_2, \quad (10.2.57)$$

$$A'(\tau) = \mu_1 C_1(\tau) + \mu_2 C_2(\tau) + \delta_0. \quad (10.2.58)$$

The solution to these equations satisfying the initial condition (10.2.54) can be found numerically. Solving this system of ordinary differential equations numerically is simpler than solving the partial differential equation (10.2.52).

Remark 10.2.4. We note that if the Brownian motions $\widetilde{W}_1(t)$ and $\widetilde{W}_2(t)$ in (10.2.49) and (10.2.50) were correlated with some correlation coefficient $\rho \neq 0$, then the partial differential equation (10.2.52) would have the additional term $\rho\sqrt{y_1 y_2}\, f_{y_1 y_2}$ on the left-hand side. This term would ruin the argument that led to the system of ordinary differential equations (10.2.56)–(10.2.58). For this reason, we assume at the outset that these Brownian motions are independent.

10.2.3 Mixed Model

Both factors in the two-factor CIR model are always nonnegative. In the two-factor Vasicek model, both factors can become negative. In the two-factor mixed model, one of the factors is always nonnegative and the other can become negative.

The *canonical two-factor mixed model* is

$$dY_1(t) = \big(\mu - \lambda_1 Y_1(t)\big)\,dt + \sqrt{Y_1(t)}\,d\widetilde{W}_1(t), \quad (10.2.59)$$

$$dY_2(t) = -\lambda_2 Y_2(t)\,dt + \sigma_{21}\sqrt{Y_1(t)}\,d\widetilde{W}_1(t) + \sqrt{\alpha + \beta Y_1(t)}\,d\widetilde{W}_2(t). \quad (10.2.60)$$

We assume $\mu \geq 0$, $\lambda_1 > 0$, $\lambda_2 > 0$, $\alpha \geq 0$, $\beta \geq 0$, and $\sigma_{21} \in \mathbb{R}$. The Brownian motions $\widetilde{W}_1(t)$ and $\widetilde{W}_2(t)$ are independent. We assume $Y_1(0) \geq 0$, and we have $Y_1(t) \geq 0$ for all $t \geq 0$ almost surely. On the other hand, even if $Y_2(t)$ is positive, $Y_2(t)$ can take negative values for $t > 0$. The interest rate is defined by

$$R(t) = \delta_0 + \delta_1 Y_1(t) + \delta_2 Y_2(t). \tag{10.2.61}$$

In this model, zero-coupon bond prices have the affine-yield form

$$B(t,T) = e^{-Y_1(t)C_1(T-t) - Y_2(t)C_2(T-t) - A(T-t)}. \tag{10.2.62}$$

Just as in the two-factor Vasicek model and the two-factor CIR model, the functions $C_1(\tau)$, $C_2(\tau)$, and $A(\tau)$ must satisfy the terminal condition

$$C_1(0) = C_2(0) = A(0). \tag{10.2.63}$$

Exercise 10.2 derives the system of ordinary differential equations that determine the functions $C_1(\tau)$, $C_2(\tau)$, and $A(\tau)$.

10.3 Heath-Jarrow-Morton Model

The Heath-Jarrow-Morton (HJM) model of this section evolves the whole yield curve forward in time. There are several possible ways to represent the yield curve, and the one chosen by the HJM model is in terms of the *forward rates* that can be locked in at one time for borrowing at a later time. In this section, we first discuss forward rates, then write down the HJM model for their evolution, discuss how to guarantee that the resulting model does not admit arbitrage, and conclude with a procedure for calibrating the HJM model.

10.3.1 Forward Rates

Let us fix a time horizon $\overline{T}$ (say 50 years). All bonds in the following discussion will mature at or before time $\overline{T}$. For $0 \leq t \leq T \leq \overline{T}$, as before, we denote by $B(t,T)$ the price at time t of a zero-coupon bond maturing at time T and having face value 1. We assume this bond bears no risk of default. We assume further that, for every t and T satisfying $0 \leq t \leq T \leq \overline{T}$, the bond price $B(t,T)$ is defined. If the interest rate is strictly positive between times t and T, then $B(t,T)$ must be strictly less than one whenever $t < T$. This is the situation to keep in mind, although some implementations of the HJM model violate it.

At time t, we can engage in *forward investing* at the later time T by setting up the following portfolio. Here δ is a small positive number.

- Take a short position of size 1 in T-maturity bonds. This generates income $B(t,T)$.
- Take a long position of size $\frac{B(t,T)}{B(t,T+\delta)}$ in $(T+\delta)$-maturity bonds. This costs $B(t,T)$.

The net cost of setting up this portfolio at time t is zero. At the later time T, holding this portfolio requires that we pay 1 to cover the short position in the T-maturity bond. At the still later time $T + \delta$, we receive $\frac{B(t,T)}{B(t,T+\delta)}$ from the long position in the $T+\delta$-maturity bond. In other words, we have invested 1 at time T and received more than 1 at time $T + \delta$. The yield that explains the surplus received at time $T + \delta$ is

$$\frac{1}{\delta} \log \frac{B(t,T)}{B(t,T+\delta)} = -\frac{\log B(t,T+\delta) - \log B(t,T)}{\delta}. \tag{10.3.1}$$

This is the continuously compounding rate of interest that, applied to the 1 invested at time T, would return $\frac{B(t,T)}{B(t,T+\delta)}$ at time $T+\delta$. If the bond $B(t,T+\delta)$ with the longer time to maturity has the smaller price, as it would if the interest rate is strictly positive, then the quotient $\frac{B(t,T)}{B(t,T+\delta)}$ is strictly greater than 1 and the yield is strictly positive.

Note that the yield in (10.3.1) is $\mathcal{F}(t)$-measurable. Although it is an interest rate for investing at time T, it can be "locked in" at the earlier time t. In fact, if someone were to propose any other interest rate for investing (or borrowing) at time T that is set at the earlier time t, then by accepting this interest rate and setting up the portfolio described above or its opposite, one could create an arbitrage.

We define the *forward rate at time t for investing at time T* to be

$$\begin{aligned} f(t,T) &= -\lim_{\delta \downarrow 0} \frac{\log B(t,T+\delta) - \log B(t,T)}{\delta} \\ &= -\frac{\partial}{\partial T} \log B(t,T). \end{aligned} \tag{10.3.2}$$

This is the limit of the yield in (10.3.1) as $\delta \downarrow 0$ and can thus be regarded as the *instantaneous interest rate* at time T that can be locked in at the earlier time t.

If we know $f(t,T)$ for all values of $0 \le t \le T \le \overline{T}$, we can recover $B(t,T)$ for all values of $0 \le t \le T \le \overline{T}$ by the formula

$$\int_t^T f(t,v)\, dv = -\big[\log B(t,T) - \log B(t,t)\big] = -\log B(t,T),$$

where we have used the fact that $B(t,t) = 1$. Therefore,

$$B(t,T) = \exp\left\{ -\int_t^T f(t,v)\, dv \right\}, \quad 0 \le t \le T \le \overline{T}. \tag{10.3.3}$$

From bond prices, we can determine forward rates from (10.3.2). From forward rates, we can determine bond prices from (10.3.3). Therefore, at least theoretically, it does not appear to matter whether we build a model for forward rates or for bond prices. In fact, the no-arbitrage condition works out to have a simple form when we model forward rates. From a practical point of view, forward rates are a more difficult object to determine from market data because the differentiation in (10.3.2) is sensitive to small changes in the bond prices. On the other hand, once we have forward rates, bond prices are easy to determine because the integration in (10.3.3) is not sensitive to small changes in the forward rates.

The interest rate at time t is

$$R(t) = f(t,t). \tag{10.3.4}$$

This is the instantaneous rate we can lock in at time t for borrowing at time t.

10.3.2 Dynamics of Forward Rates and Bond Prices

Assume that $f(0,T)$, $0 \leq T \leq \overline{T}$, is known at time 0. We call this the *initial forward rate curve.* In the HJM model, the forward rate at later times t for investing at still later times T is given by

$$f(t,T) = f(0,T) + \int_0^t \alpha(u,T)\,du + \int_0^t \sigma(u,T)\,dW(u). \tag{10.3.5}$$

We may write this in differential form as

$$df(t,T) = \alpha(t,T)\,dt + \sigma(t,T)\,dW(t), \quad 0 \leq t \leq T. \tag{10.3.6}$$

Here and elsewhere in this section, d indicates the differential with respect to the variable t; the variable T is being held constant in (10.3.6).

Here the process $W(u)$ is a Brownian motion under the actual measure $\mathbb{P}$. In particular, $\alpha(t,T)$ is the drift of $f(t,T)$ under the actual measure. The processes $\alpha(t,T)$ and $\sigma(t,T)$ may be random. For each fixed T, they are adapted processes in the t variable. To simplify the notation, we assume that the forward rate is driven by a single Brownian motion. The case when the forward rate is driven by multiple Brownian motions is addressed in Exercise 10.9.

From (10.3.6), we can work out the dynamics of the bond prices given by (10.3.3). Note first that because $-\int_t^T f(t,v)\,dv$ has a t-variable in two places, its differential has two terms. Indeed,

$$d\left(-\int_t^T f(t,v)\,dv\right) = f(t,t)\,dt - \int_t^T df(t,v)\,dv.$$

The first term on the right-hand side is the result of taking the differential with respect to the lower limit of integration t. The fact that this is the *lower* limit produces a minus sign, which cancels the minus sign on the left-hand side. The other term is the result of taking the differential with respect to the t under the integral sign. Using (10.3.4) and (10.3.6), we see that

$$d\left(-\int_t^T f(t,v)\,dv\right) = R(t)\,dt - \int_t^T \left[\alpha(t,v)\,dt + \sigma(t,v)\,dW(t)\right] dv.$$

We next reverse the order of the integration (see Exercise 10.8), writing

$$\int_t^T \alpha(t,v)\,dt\,dv = \int_t^T \alpha(t,v)\,dv\,dt = \alpha^*(t,T)\,dt, \tag{10.3.7}$$

$$\int_t^T \sigma(t,v)\,dW(t)\,dv = \int_t^T \sigma(t,v)\,dv\,dW(t) = \sigma^*(t,T)\,dW(t), \tag{10.3.8}$$

where

$$\alpha^*(t,T) = \int_t^T \alpha(t,v)\,dv, \quad \sigma^*(t,T) = \int_t^T \sigma(t,v)\,dv. \tag{10.3.9}$$

In conclusion, we have

$$d\left(-\int_t^T f(t,v)\,dv\right) = R(t)\,dt - \alpha^*(t,T)\,dt - \sigma^*(t,T)\,dW(t). \tag{10.3.10}$$

Let $g(x) = e^x$, so that $g'(x) = e^x$ and $g''(x) = e^x$. According to (10.3.3),

$$B(t,T) = g\left(-\int_t^T f(t,v)\,dv\right).$$

The Itô-Doeblin formula implies

$$\begin{aligned} dB(t,T) &= g'\left(-\int_t^T f(t,v)\,dv\right) d\left(-\int_t^T f(t,v)\,dv\right) \\ &\quad + \frac{1}{2} g''\left(-\int_t^T f(t,v)\,dv\right) \left[d\left(-\int_t^T f(t,v)\,dv\right)\right]^2 \\ &= B(t,T)\left[R(t)\,dt - \alpha^*(t,T)\,dt - \sigma^*(t,T)\,dW(t)\right] \\ &\quad + \frac{1}{2} B(t,T)\left(\sigma^*(t,T)\right)^2 dt \\ &= B(t,T)\left[R(t) - \alpha^*(t,T) + \frac{1}{2}\left(\sigma^*(t,T)\right)^2\right] dt \\ &\quad - \sigma^*(t,T) B(t,T)\,dW(t). \end{aligned} \tag{10.3.11}$$

10.3.3 No-Arbitrage Condition

The HJM model has a zero-coupon bond with maturity T for every $T \in [0, \overline{T}]$. We need to make sure there is no opportunity for arbitrage by trading in these bonds. The First Fundamental Theorem of Asset Pricing, Theorem 5.4.7, says that, in order to guarantee this, we should seek a probability measure $\widetilde{\mathbb{P}}$ under which each discounted bond price

$$D(t)B(t,T) = \exp\left\{-\int_0^t R(u)\,du\right\} B(t,T), \quad 0 \le t \le T,$$

is a martingale. Because $dD(t) = -R(t)D(t)\,dt$, we have the differential

$$\begin{aligned} &d\big(D(t)B(t,T)\big)\\ &= -R(t)D(t)B(t,T)\,dt + D(t)\,dB(t,T)\\ &= D(t)B(t,T)\left[\left(-\alpha^*(t,T) + \frac{1}{2}(\sigma^*(t,T))^2\right)dt - \sigma^*(t,T)\,dW(t)\right]. \end{aligned} \quad (10.3.12)$$

We want to write the term in square brackets as

$$-\sigma^*(t,T)[\Theta(t)\,dt + dW(t)],$$

and we can then use Girsanov's Theorem, Theorem 5.2.3, to change to a probability measure $\widetilde{\mathbb{P}}$ under which

$$\widetilde{W}(t) = \int_0^t \Theta(u)\,du + W(t) \quad (10.3.13)$$

is a Brownian motion. Using this Brownian motion, we may rewrite (10.3.12) as

$$d\big(D(t)B(t,T)\big) = -D(t)B(t,T)\sigma^*(t,T)\,d\widetilde{W}(t). \quad (10.3.14)$$

It would then follow that $D(t)B(t,T)$ is a martingale under $\widetilde{\mathbb{P}}$ (i.e., $\widetilde{\mathbb{P}}$ would be risk-neutral).

For the program above to work, we must solve the equation

$$\begin{aligned} &\left[\left(-\alpha^*(t,T) + \frac{1}{2}(\sigma^*(t,T))^2\right)dt - \sigma^*(t,T)\,dW(t)\right]\\ &\qquad\qquad = -\sigma^*(t,T)\big[\Theta(t)\,dt + dW(t)\big] \end{aligned}$$

for $\Theta(t)$. In other words, we must find a process $\Theta(t)$ satisfying

$$-\alpha^*(t,T) + \frac{1}{2}\big(\sigma^*(t,T)\big)^2 = -\sigma^*(t,T)\Theta(t). \quad (10.3.15)$$

Actually, (10.3.15) represents infinitely many equations, one for each maturity $T \in (0,\overline{T}]$. These are the *market price of risk equations*, and we have one such equation for each bond (maturity). However, there is only one process $\Theta(t)$. This process is the *market price of risk*, and we have as many such processes as there are sources of uncertainty. In this case, there is only one Brownian motion driving the model.

To solve (10.3.15), we recall from (10.3.9) that

$$\frac{\partial}{\partial T}\alpha^*(t,T) = \alpha(t,T), \quad \frac{\partial}{\partial T}\sigma^*(t,T) = \sigma(t,T).$$

Differentiating (10.3.15) with respect to T, we obtain

$$-\alpha(t,T) + \sigma^*(t,T)\sigma(t,T) = -\sigma(t,T)\Theta(t)$$

or, equivalently,

$$\alpha(t,T) = \sigma(t,T)\big[\sigma^*(t,T) + \Theta(t)\big]. \tag{10.3.16}$$

Theorem 10.3.1 (Heath-Jarrow-Morton no-arbitrage condition). *A term-structure model for zero-coupon bond prices of all maturities in $(0,\overline{T}]$ and driven by a single Brownian motion does not admit arbitrage if there exists a process $\Theta(t)$ such that (10.3.16) holds for all $0 \le t \le T \le \overline{T}$. Here $\alpha(t,T)$ and $\sigma(t,T)$ are the drift and diffusion, respectively, of the forward rate (i.e., the processes satisfying (10.3.6)), $\sigma^*(t,T) = \int_t^T \sigma(t,v)\,dv$, and $\Theta(t)$ is the market price of risk.*

PROOF: It remains only to check that if $\Theta(t)$ solves (10.3.16), then it also satisfies (10.3.15), for then we can use Girsanov's Theorem as described above to construct a risk-neutral measure. The existence of a risk-neutral measure guarantees the absence of arbitrage.

Suppose $\Theta(t)$ solves (10.3.16). We rewrite this equation, replacing T by v:

$$\alpha(t,v) = \sigma(t,v)\big[\sigma^*(t,v) + \Theta(t)\big].$$

Integrating with respect to v from $v = t$ to $v = T$, we obtain

$$\alpha^*(t,v)\Big|_{v=t}^{v=T} = \frac{1}{2}\big(\sigma^*(t,v)\big)^2\Big|_{v=t}^{v=T} + \sigma^*(t,v)\Theta(t)\Big|_{v=t}^{v=T}.$$

But because $\alpha^*(t,t) = \sigma^*(t,t) = 0$, this reduces to

$$\alpha^*(t,T) = \frac{1}{2}\big(\sigma^*(t,T)\big)^2 + \sigma^*(t,T)\Theta(t),$$

which is (10.3.15). □

So long as $\sigma(t,T)$ is nonzero, we can solve (10.3.16) for $\Theta(t)$:

$$\Theta(t) = \frac{\alpha(t,T)}{\sigma(t,T)} - \sigma^*(t,T), \quad 0 \le t \le T. \tag{10.3.17}$$

This shows that $\Theta(t)$ is unique, and hence the risk-neutral measure is unique. In this case, the Second Fundamental Theorem of Asset Pricing, Theorem 5.4.9, guarantees that the model is complete (i.e., all interest rate derivatives can be hedged by trading in zero-coupon bonds).

10.3.4 HJM Under Risk-Neutral Measure

We began with the formula (10.3.5) for the evolution of the forward rate, and the driving process $W(u)$ appearing in (10.3.5) is a Brownian motion under the actual measure $\mathbb{P}$. Assuming the model satisfies the HJM no-arbitrage condition (10.3.16), we may rewrite (10.3.5) as

$$\begin{aligned} df(t,T) &= \alpha(t,T)\,dt + \sigma(t,T)\,dW(t) \\ &= \sigma(t,T)\sigma^*(t,T)\,dt + \sigma(t,T)\Big[\Theta(t) + dW(t)\Big] \\ &= \sigma(t,T)\sigma^*(t,T)\,dt + \sigma(t,T)\,d\widetilde{W}(t), \end{aligned}$$

where $\widetilde{W}(t)$ is given by (10.3.13). To conclude that there is no arbitrage, we need the drift of the forward rate under the risk-neutral measure to be $\sigma(t,T)\sigma^*(t,T)$. We saw in the proof of Theorem 10.3.1 that the no-arbitrage condition (10.3.16) implies (10.3.15), and using (10.3.15) we may rewrite the differential of the discounted bond price (10.3.12) as

$$\begin{aligned} d\big(D(t)B(t,T)\big) &= -\sigma^*(t,T)D(t)B(t,T)\big[\Theta(t)\,dt + dW(t)\big] \\ &= -\sigma^*(t,T)D(t)B(t,T)\,d\widetilde{W}(t). \end{aligned}$$

Because $d\frac{1}{D(t)} = \frac{R(t)}{D(t)}\,dt$, the differential of the undiscounted bond price is

$$\begin{aligned} dB(t,T) &= d\left(\frac{1}{D(t)}\cdot D(t)B(t,T)\right) \\ &= \frac{R(t)}{D(t)}D(t)B(t,T)\,dt - \sigma^*(t,T)\frac{1}{D(t)}D(t)B(t,T)\,d\widetilde{W}(t) \\ &= R(t)B(t,T)\,dt - \sigma^*(t,T)B(t,T)\,d\widetilde{W}(t). \end{aligned}$$

The following theorem summarizes this discussion.

Theorem 10.3.2 (Term-structure evolution under risk-neutral measure). *In a term-structure model satisfying the HJM no-arbitrage condition of Theorem 10.3.1, the forward rates evolve according to the equation*

$$df(t,T) = \sigma(t,T)\sigma^*(t,T)\,dt + \sigma(t,T)\,d\widetilde{W}(t), \tag{10.3.18}$$

and the zero-coupon bond prices evolve according to the equation

$$dB(t,T) = R(t)B(t,T)\,dt - \sigma^*(t,T)B(t,T)\,d\widetilde{W}(t), \tag{10.3.19}$$

where $\widetilde{W}(t)$ is a Brownian motion under a risk-neutral measure $\widetilde{\mathbb{P}}$. Here $\sigma^*(t) = \int_t^T \sigma(t,v)\,dv$ and $R(t) = f(t,t)$ is the interest rate. The discounted bond prices satisfy

$$d\big(D(t)B(t,T)\big) = -\sigma^*(t,T)D(t)B(t,T)\,d\widetilde{W}(t), \tag{10.3.20}$$

where $D(t) = e^{-\int_0^t R(u)du}$ is the discount process. The solution to the stochastic differential equation (10.3.19) is

$$\begin{aligned}
&B(t,T)\\
&= B(0,T)\exp\left\{\int_0^t R(u)du - \int_0^t \sigma^*(u,T)\,d\widetilde{W}(u) - \frac{1}{2}\int_0^T (\sigma^*(u,T))^2\,du\right\}\\
&= \frac{B(0,T)}{D(t)}\exp\left\{-\int_0^t \sigma^*(u,T)\,d\widetilde{W}(u) - \frac{1}{2}\int_0^T \left(\sigma^*(u,T)\right)^2 du\right\}. \qquad (10.3.21)
\end{aligned}$$

10.3.5 Relation to Affine-Yield Models

Every term-structure model driven by Brownian motion is an HJM model. In any such model, there are forward rates. The drift and diffusion of the forward rates must satisfy the conditions of Theorem 10.3.1 in order for a risk-neutral measure to exist, which rules out arbitrage. Under these conditions, the formulas of Theorem 10.3.2 describe the evolution of the forward rates and bonds under the risk-neutral measure.

We illustrate this with the one-factor Hull-White and Cox-Ingersoll-Ross (CIR) models of Examples 6.5.1 and 6.5.2. For both these models, the interest rate dynamics are of the form

$$dR(t) = \beta\big(t, R(t)\big)\,dt + \gamma\big(t, R(t)\big)\,d\widetilde{W}(t),$$

where $\widetilde{W}(t)$ is a Brownian motion under a risk-neutral probability measure $\widetilde{\mathbb{P}}$. In the case of the Hull-White model,

$$\beta(t,r) = a(t) - b(t)r, \quad \gamma(t,r) = \sigma(t),$$

for some nonrandom positive functions $a(t)$, $b(t)$, and $\sigma(t)$. For the CIR model,

$$\beta(t,r) = a - br, \quad \gamma(t,r) = \sigma\sqrt{r}, \qquad (10.3.22)$$

for some positive constants a, b, and σ. The zero-coupon bond prices are of the form

$$B(t,T) = e^{-R(t)C(t,T)-A(t,T)}, \qquad (10.3.23)$$

where $C(t,T)$ and $A(t,T)$ are nonrandom functions. In the case of the Hull-White model, $C(t,T)$ and $A(t,T)$ are given by (6.5.10) and (6.5.11), which we repeat here:

$$C(t,T) = \int_t^T e^{-\int_t^s b(v)dv}\,ds, \qquad (10.3.24)$$

$$A(t,T) = \int_t^T \left(a(s)C(s,T) - \frac{1}{2}\sigma^2(s)C^2(s,T)\right) ds. \qquad (10.3.25)$$

In the case of the CIR model, $C(t,T)$ and $A(t,T)$ are given by (6.5.16) and (6.5.17). According to (10.3.2), the forward rates are

$$f(t,T) = -\frac{\partial}{\partial T}\log B(t,T) = R(t)\frac{\partial}{\partial T}C(t,T) + \frac{\partial}{\partial T}A(t,T).$$

With $C'(t,T)$ and $A'(t,T)$ denoting derivatives with respect to t, we have the forward rate differential

$$\begin{aligned} df(t,T) &= \frac{\partial}{\partial T}C(t,T)\,dR(t) + R(t)\frac{\partial}{\partial T}C'(t,T)\,dt + \frac{\partial}{\partial T}A'(t,T)\,dt \\ &= \left[\frac{\partial}{\partial T}C(t,T)\beta(t,R(t)) + R(t)\frac{\partial}{\partial T}C'(t,T) + \frac{\partial}{\partial T}A'(t,T)\right]dt \\ &\quad + \frac{\partial}{\partial T}C(t,T)\gamma(t,R(t))\,d\widetilde{W}(t). \end{aligned}$$

This is an HJM model with

$$\sigma(t,T) = \frac{\partial}{\partial T}C(t,T)\gamma\big(t,R(t)\big). \tag{10.3.26}$$

Since we are working under the risk-neutral measure, Theorem 10.3.2 implies that the drift term should be $\sigma(t,T)\sigma^*(t,T) = \sigma(t,T)\int_t^T \sigma(t,v)\,dv$. In other words, for these affine-yield models, the HJM no-arbitrage condition becomes

$$\begin{aligned} &\frac{\partial}{\partial T}C(t,T)\beta\big(t,R(t)\big) + R(t)\frac{\partial}{\partial T}C'(t,T) + \frac{\partial}{\partial T}A'(t,T) \\ &= \left(\frac{\partial}{\partial T}C(t,T)\right)\gamma\big(t,R(t)\big)\int_t^T \frac{\partial}{\partial v}C(t,v)\gamma\big(t,R(t)\big)\,dv \\ &= \left(\frac{\partial}{\partial T}C(t,T)\right)\gamma\big(t,R(t)\big)\big[C(t,T)-C(t,t)\big]\gamma\big(t,R(t)\big) \\ &= \left(\frac{\partial}{\partial T}C(t,T)\right)C(t,T)\gamma^2\big(t,R(t)\big). \end{aligned} \tag{10.3.27}$$

We verify (10.3.27) for the *Vasicek model*, which is the Hull-White model with constant a, b, and σ, and we leave the verification for the Hull-White and CIR models as Exercise 10.10. For the Vasicek model, (10.3.24) and (10.3.25) reduce to

$$\begin{aligned} C(t,T) &= \frac{1}{b}\left(1-e^{-b(T-t)}\right), \\ A'(t,T) &= -aC(t,T) + \frac{1}{2}\sigma^2 C^2(t,T), \end{aligned}$$

and hence

$$\begin{aligned} \frac{\partial}{\partial T}C(t,T) &= e^{-b(T-t)}, \\ \frac{\partial}{\partial T}A'(t,T) &= \left(\frac{\sigma^2}{b}-a\right)e^{-b(T-t)} - \frac{\sigma^2}{a}e^{-2b(T-t)}. \end{aligned}$$

Therefore,

$$\sigma(t,T) = \sigma e^{-b(T-t)},$$

and

$$\sigma^*(t,T) = \int_t^T \sigma(t,u)\,du = \sigma \int_t^T e^{-b(T-u)}\,du = \frac{\sigma}{b}\left(1 - e^{-b(T-t)}\right).$$

It follows that

$$\begin{aligned}
&\frac{\partial}{\partial T}C(t,T)\beta(t,R(t)) + R(t)\frac{\partial}{\partial T}C'(t,T) + \frac{\partial}{\partial T}A'(t,T)\\
&= e^{-b(T-t)}(a - bR(t)) + R(t)be^{-b(T-t)} + \left(\frac{\sigma^2}{b} - a\right)e^{-b(T-t)}\\
&\quad -\frac{\sigma^2}{b}e^{-2b(t-t)}\\
&= \frac{\sigma^2}{b}\left(e^{-b(T-t)} - e^{-2b(T-t)}\right)\\
&= \sigma(t,T)\sigma^*(t,T),
\end{aligned}$$

as expected.

10.3.6 Implementation of HJM

To implement an HJM model, we need to know $\sigma(t,T)$ for $0 \le t \le T \le \overline{T}$. We can use historical data to estimate this because the same diffusion process $\sigma(t,T)$ appears in both the stochastic differential equation (10.3.6) driven by the Brownian motion $W(t)$ under the actual probability measure $\mathbb{P}$ and in the stochastic differential equation (10.3.18) driven by the Brownian motion $\widetilde{W}(t)$ under the risk-neutral measure $\widetilde{\mathbb{P}}$. Once we have $\sigma(t,T)$, we can compute $\sigma^*(t,T) = \int_t^T \sigma(t,v)\,dv$. This plus the initial forward curve $f(0,T)$, $0 \le T \le \overline{T}$, permits us to determine all the terms appearing in the formulas in Theorem 10.3.2. In particular, we use the initial forward curve to compute

$$R(t) = f(t,t) = f(0,t) + \int_0^t \sigma(u,t)\sigma^*(u,t)\,du + \int_0^t \sigma(u,t)\,d\widetilde{W}(u). \quad (10.3.28)$$

Since all expectations required for pricing interest rate derivatives are computed under $\widetilde{\mathbb{P}}$, we need only the formulas in Theorem 10.3.2; the market price of risk $\Theta(t)$ and the drift of the forward rate $\alpha(t,T)$ in (10.3.6) are irrelevant to derivative pricing. They are relevant, however, if we want to estimate nondiffusion terms from historical data (e.g., the probability of credit class migration for defaultable bonds) or we want to compute a quantity such as Value-at-Risk that requires use of the actual measure.

Assume for the moment that $\sigma(t,T)$ is of the form

$$\sigma(t,T) = \widetilde{\sigma}(T-t)\min\{M, f(t,T)\} \quad (10.3.29)$$

for some nonrandom function $\widetilde{\sigma}(\tau)$, $\tau \geq 0$, and some positive constant M. In (10.3.29), we need to have the capped forward rate $\min\{M, f(t,T)\}$ on the right-hand side rather than the forward rate $f(t,T)$ itself to prevent explosion of the forward rate. This is discussed in more detail in Subsection 10.4.1. One consequence of this fact is that forward rates (recall we are working here with *continuously compounding* forward rates; see (10.3.2)) cannot be log-normal. This is a statement about forward rates, not about the HJM model. Section 10.4 discusses how to overcome this feature of continuously compounding forward rates by building a model for simple forward rates.

We choose $\widetilde{\sigma}(T-t)$ to match historical data. The forward rate evolves according to the continuous-time model

$$df(t,T) = \alpha(t,T)\,dt + \widetilde{\sigma}(T-t)\min\{M, f(t,T)\}\,dW(t).$$

Suppose we have observed this forward rate at times $t_1 < t_2 < \cdots < t_J < 0$ in the past, and the forward rate we observed at those times was for the relative maturities $\tau_1 < \tau_2 < \cdots < \tau_K$ (i.e., we have observed $f(t_j, t_j+\tau_k)$ for $j = 1,\dots,J$ and $k = 1,\dots,K$). Suppose further that for some small positive δ we have also observed $f(t_j+\delta, t_j+\tau_k)$. We assume that δ is sufficiently small that $t_j + \delta < t_{j+1}$ for $j = 1,\dots,J-1$ and $t_J + \delta \leq 0$. According to our model,

$$\begin{aligned}&f(t_j+\delta, t_j+\tau_k) - f(t_j, t_j+\tau_k)\\&\approx \delta\alpha(t_j, t_j+\tau_k) + \widetilde{\sigma}(\tau_k)\min\{M, f(t_j, t_j+\tau_k)\}\big(W(t_j+\delta) - W(t_j)\big).\end{aligned}$$

We identify $\widetilde{\sigma}$ by defining

$$D_{j,k} = \frac{f(t_j+\delta, t_j+\tau_k) - f(t_j, t_j+\tau_k)}{\sqrt{\delta}\,\min\{M, f(t_j, t_j+\tau_k)\}} \tag{10.3.30}$$

and observing that

$$D_{j,k} \approx \frac{\sqrt{\delta}\,\alpha(t_j, t_j+\tau_k)}{\min\{M, f(t_j, t_j+\tau_k)\}} + \widetilde{\sigma}(\tau_k)\frac{W(t_j+\delta) - W(t_j)}{\sqrt{\delta}}.$$

The first term on the right-hand side is small relative to the second term because the first term contains the factor $\sqrt{\delta}$. We define

$$X_j = \frac{W(t_j+\delta) - W(t_j)}{\sqrt{\delta}},\quad j = 1,\dots,J, \tag{10.3.31}$$

the expression appearing in the second term, which is a standard normal random variable. We conclude that

$$D_{j,k} \approx \widetilde{\sigma}(\tau_k)X_j. \tag{10.3.32}$$

Observe that not only are $X_1,\dots,X_J$ standard normal random variables but are also independent of one another. The approximation (10.3.32) permits us to regard $D_{1k}, D_{2k},\dots,D_{Jk}$ as independent observations taken at times $t_1, t_2, \ldots t_J$ on forward rates, all with the same relative maturity τ_k. We compute the empirical covariance

$$C_{k_1,k_2} = \frac{1}{J}\sum_{j=1}^{J} D_{j,k_1} D_{j,k_2}.$$

The theoretical covariance, computed from the right-hand side of (10.3.32), is

$$\mathbb{E}\big[\widetilde{\sigma}(\tau_{k_1})\widetilde{\sigma}(\tau_{k_2})X_j^2\big] = \widetilde{\sigma}(\tau_{k_1})\widetilde{\sigma}(\tau_{k_2}).$$

Ideally, we would find $\widetilde{\sigma}(\tau_1), \widetilde{\sigma}(\tau_2), \ldots \widetilde{\sigma}(\tau_K)$ so that

$$C_{k_1 k_2} = \widetilde{\sigma}(\tau_{k_1})\widetilde{\sigma}(\tau_{k_2}), \quad k_1, k_2 = 1, 2, \ldots, K. \tag{10.3.33}$$

However, we have K^2 equations and only K unknowns. (Actually, for different values of k_1 and k_2, the equations $C_{k_1,k_2} = \widetilde{\sigma}(\tau_{k_1})\widetilde{\sigma}(\tau_{k_2})$ and $C_{k_2,k_1} = \widetilde{\sigma}(\tau_{k_2})\widetilde{\sigma}(\tau_{k_1})$ are the same. By eliminating these duplicates, one can reduce the system to $\frac{1}{2}K(K+1)$ equations, but this is still more than the number of unknowns if $K \geq 2$.)

To determine a best choice of $\widetilde{\sigma}(\tau_1), \widetilde{\sigma}(\tau_2), \ldots \widetilde{\sigma}(\tau_K)$, we use *principal components analysis.* Set

$$D = \begin{bmatrix} D_{1,1} & D_{1,2} & \cdots & D_{1,K} \\ D_{2,1} & D_{2,2} & \cdots & D_{2,K} \\ \vdots & \vdots & & \vdots \\ D_{J,1} & D_{J,2} & \cdots & D_{J,K} \end{bmatrix}.$$

The J rows of D correspond to observation times, and the K columns correspond to relative maturities. Then

$$C = \begin{bmatrix} C_{1,1} & C_{1,2} & \cdots & C_{1,K} \\ C_{2,1} & C_{2,2} & \cdots & C_{2,K} \\ \vdots & \vdots & & \vdots \\ C_{K,1} & C_{K,2} & \cdots & C_{K,K} \end{bmatrix} = \frac{1}{J} D^{\mathrm{tr}} D$$

is symmetric and positive semidefinite. Every symmetric, positive semidefinite matrix has a principal component decomposition

$$C = \lambda_1 e_1 e_1^{\mathrm{tr}} + \lambda_2 e_2 e_2^{\mathrm{tr}} + \cdots + \lambda_K e_K e_K^{\mathrm{tr}},$$

where $\lambda_1 \geq \lambda_2 \geq \cdots \geq \lambda_K \geq 0$ are the eigenvalues of C and the column vectors $e_1, e_2, \ldots e_K$ are the orthogonal eigenvectors, all normalized to have length one. We want to write

$$C = \begin{bmatrix} \widetilde{\sigma}(\tau_1) \\ \widetilde{\sigma}(\tau_2) \\ \vdots \\ \widetilde{\sigma}(\tau_K) \end{bmatrix} \big[\widetilde{\sigma}(\tau_1), \widetilde{\sigma}(\tau_2), \ldots, \widetilde{\sigma}(\tau_K)\big].$$

However, this cannot be done exactly. The best approximation is

$$\begin{bmatrix} \widetilde{\sigma}(\tau_1) \\ \widetilde{\sigma}(\tau_2) \\ \vdots \\ \widetilde{\sigma}(\tau_K) \end{bmatrix} = \sqrt{\lambda_1}\, e_1.$$

To get a better approximation to C, we can introduce more Brownian motions into the equation driving the forward rates (see Exercise 10.9). Each of these has its own $\widetilde{\sigma}$ vector, and these can be chosen to be $\sqrt{\lambda_2}\, e_2$, $\sqrt{\lambda_3}\, e_3$, etc.

So far we have used only historical data. In the final step of the calibration, we introduce a nonrandom function $s(t)$ into the forward rate evolution under the risk-neutral measure, writing

$$df(t,T) = \sigma(t,T)\sigma^*(t,T)\,dt + s(t)\widetilde{\sigma}(T-t)\min\{M, f(t,T)\}\,d\widetilde{W}(t). \quad (10.3.34)$$

This is our final model. We use the values of $\widetilde{\sigma}(T-t)$ estimated from historical data under the assumption $s(t) \equiv 1$. We then allow the possibility that $s(t)$ is different from 1. We have $\sigma(t,T) = s(t)\widetilde{\sigma}(T-t)\min\{M, f(t,T)\}$. Therefore,

$$\sigma^*(t,T) = \int_t^T \sigma(t,v)\,dv = s(t)\int_t^T \widetilde{\sigma}(v-t)\min\{M, f(t,v)\}\,dv. \quad (10.3.35)$$

We substitute this function into (10.3.34) and evolve the forward rate. Even with this last-minute introduction of $s(t)$ into the model, the model is free of arbitrage when $\sigma^*(t,T)$ in (10.3.34) is defined by (10.3.35). Typically, one assumes that $s(t)$ is piecewise constant, and the values of these constants are free parameters that can be used to get the model to agree with market prices. Recalibrations of the model affect $s(t)$ only.

10.4 Forward LIBOR Model

In this section, we present the *forward LIBOR model*, which leads to the *Black caplet formula*. This requires us to build a model for *LIBOR* (London interbank offered rate) and use the *forward measures* introduced in Section 9.4. We begin by explaining why the continuously compounding forward rates of Section 10.3 are inadequate for the purposes of this section.

10.4.1 The Problem with Forward Rates

We have seen in Theorem 10.3.2 that in an arbitrage-free term-structure model, forward rates must evolve according to (10.3.18),

$$df(t,T) = \sigma(t,T)\sigma^*(t,T)\,dt + \sigma(t,T)\,d\widetilde{W}(t), \quad (10.3.18)$$

where $\widetilde{W}$ is a Brownian motion under a risk-neutral measure $\widetilde{\mathbb{P}}$. In order to adapt the Black-Scholes formula for equity options to fixed income markets, and thereby obtain the Black caplet formula (see Theorem 10.4.2 below), it would be desirable to build a model in which forward rates are log-normal under a risk-neutral measure. To do that, we should set $\sigma(t,T) = \sigma f(t,T)$ in (10.3.18), where σ is a positive constant. However, we would then have

$$\sigma^*(T,t) = \int_t^T \sigma(t,v)\,dv = \sigma \int_t^T f(t,v)\,dv,$$

and the dt term in (10.3.18) would be

$$\sigma^2 f(t,T) \int_t^T f(t,v)\,dv. \tag{10.4.1}$$

Heath, Jarrow, and Morton [83] show that this drift term causes forward rates to explode. For T near t, the dt term (10.4.1) is approximately equal to $\sigma^2(T-t)f^2(t,T)$, and the square of the forward rate creates the problem. With the drift term (10.4.1), equation (10.3.18) is similar to the deterministic ordinary differential equation

$$f'(t) = \sigma^2 f^2(t) \tag{10.4.2}$$

with a positive initial condition $f(0)$. The solution to (10.4.2) is

$$f(t) = \frac{f(0)}{1-\sigma^2 f(0)t},$$

as can easily be verified by computing

$$f'(t) = \frac{\sigma^2 f^2(0)}{\left(1-\sigma^2 f(0)t\right)^2}.$$

The function $f(t)$ explodes at time $t = \frac{1}{\sigma^2 f(0)}$. In fact, when the drift function (10.4.1) is used in (10.3.18), then (10.3.18) is worse than (10.4.2) because the randomness in (10.3.18) causes some paths to explode immediately no matter what initial condition is given. This difficulty with continuously compounding forward rates causes us to introduce *forward LIBOR*.

10.4.2 LIBOR and Forward LIBOR

Let $0 \le t \le T$ and $\delta > 0$ be given. We recall the discussion in Subsection 10.3.1 of how at time t one can lock in an interest rate for investing over the interval $[T, T+\delta]$ by taking a short position of size 1 in a T-maturity zero-coupon bond and a long position of size $\frac{B(t,T)}{B(t,T+\delta)}$ in $(T+\delta)$-maturity zero-coupon bonds. This position can be created at zero cost at time t, it calls for "investment" of 1 at time T to cover the short position, and it "repays" $\frac{B(t,T)}{B(t,T+\delta)}$ at time $T+\delta$. The continuously compounding interest rate that would explain this repayment on the investment of 1 over the time interval $[T, T+\delta]$ is given by (10.3.1). In this section, we study the *simple* interest rate that would explain this repayment, and this interest rate $L(t,T)$ is determined by the equation

$$\text{investment} \times \Big(1 + \text{duration of investment} \times \text{interest rate}\Big) = \text{repayment},$$

or in symbols:

$$1 + \delta L(t,T) = \frac{B(t,T)}{B(t,T+\delta)}. \tag{10.4.3}$$

We solve this equation for $L(t,T)$:

$$L(t,T) = \frac{B(t,T) - B(t,T+\delta)}{\delta B(t,T+\delta)}. \tag{10.4.4}$$

When $0 \leq t < T$, we call $L(t,T)$ *forward LIBOR*. When $t = T$, we call it *spot LIBOR*, or simply *LIBOR*, set at time T. The positive number δ is called the *tenor* of the LIBOR, and it is usually either 0.25 years or 0.50 years.

10.4.3 Pricing a Backset LIBOR Contract

An *interest rate swap* is an agreement between two parties A and B that A will make fixed interest rate payments on some "notional amount" to B at regularly spaced dates and B will make variable interest rate payments on the same notional amount on these same dates. The variable rate is often *backset LIBOR*, defined on one payment date to be the LIBOR set on the previous payment date. The no-arbitrage price of a payment of backset LIBOR on a notional amount of 1 is given by the following theorem.

Theorem 10.4.1 (Price of backset LIBOR). *Let $0 \leq t \leq T$ and $\delta > 0$ be given. The no-arbitrage price at time t of a contract that pays $L(T,T)$ at time $T+\delta$ is*

$$S(t) = \begin{cases} B(t,T+\delta)L(t,T), & 0 \leq t \leq T, \\ B(t,T+\delta)L(T,T), & T \leq t \leq T+\delta. \end{cases} \tag{10.4.5}$$

PROOF: There are two cases to consider. In the first case, $T \leq t \leq T+\delta$, LIBOR has been set at $L(T,T)$ and is known at time t. The value at time t of a contract that pays 1 at time $T+\delta$ is $B(t,T+\delta)$, so the value at time t of a contract that pays $L(T,T)$ at time $T+\delta$ is $B(t,T+\delta)L(T,T)$.

In the second case, $0 \leq t \leq T$, we note from (10.4.4) that

$$B(t,T+\delta)L(t,T) = \frac{1}{\delta}\big[B(t,T) - B(t,T+\delta)\big].$$

We must show that the right-hand side is the value at time t of the backset LIBOR contract. To do this, suppose at time t we have $\frac{1}{\delta}\big[B(t,T) - B(t,T+\delta)\big]$, and we use this capital to set up a portfolio that is:

- long $\frac{1}{\delta}$ bonds maturing at T;
- short $\frac{1}{\delta}$ bonds maturing at $T+\delta$.

At time T, we receive $\frac{1}{\delta}$ from the long position and use it to buy $\frac{1}{\delta B(T,T+\delta)}$ bonds maturing at time $T+\delta$, so that we now have a position of $\frac{1}{\delta B(T,T+\delta)} - \frac{1}{\delta}$ in $(T+\delta)$-maturity bonds. At time $T+\delta$, this portfolio pays

$$\frac{1}{\delta B(T,T+\delta)} - \frac{1}{\delta} = \frac{B(T,T) - B(T,T+\delta)}{\delta B(T,T+\delta)} = L(T,T).$$

We conclude that the capital $\frac{1}{\delta}\big[B(t,T) - B(t,T+\delta)\big]$ we used at time t to set up the portfolio must be the value at time t of the payment $L(T,T)$ at time $T+\delta$. □

We have proved Theorem 10.4.1 by a no-arbitrage argument. One can also obtain (10.4.5) from the risk-neutral pricing formula. For the case $t=0$, this is Exercise 10.12.

10.4.4 Black Caplet Formula

A common fixed income derivative security is an *interest rate cap*, a contract that pays the difference between a variable interest rate applied to a principal and a fixed interest rate (a *cap*) applied to the same principal whenever the variable interest rate exceeds the fixed rate. More specifically, let the *tenor* δ, the *principal* (also called the *notional amount*) P, and the *cap* K be fixed positive numbers. An interest rate cap pays $\big(\delta PL(\delta j, \delta j) - K\big)^+$ at time $\delta(j+1)$ for $j = 0, \dots, n$. To determine the price at time zero of the cap, it suffices to price one of the payments, a so-called *interest rate caplet*, and then sum these prices over the payments. We show here how to do this and obtain the *Black caplet formula*. We also note that each of these payments is of the form $\delta P(L(\delta j, \delta j) - K')^+$, where $K' = \frac{K}{\delta P}$. Thus, it suffices to determine the time-zero price of the payment $(L(T,T) - K)^+$ at time $T+\delta$ for an arbitrary T and $K > 0$.

Consider the contract that pays $L(T,T)$ at time $T+\delta$ whose price $S(t)$ at earlier times is given by Theorem 10.4.1. Suppose we use the zero-coupon bond $B(t,T+\delta)$ as the numéraire. In terms of this numéraire, the price of the contract paying backset LIBOR is

$$\frac{S(t)}{B(t,T+\delta)} = \begin{cases} L(t,T), & 0 \le t \le T, \\ L(T,T), & T \le T \le T+\delta. \end{cases} \tag{10.4.6}$$

Recalling Definition 5.6.1 and Theorem 5.6.2, at least for $0 \le t \le T$, we see that forward LIBOR $L(t,T)$ is the $(T+\delta)$-forward price of the contract paying backset LIBOR $L(T,T)$ at time $T+\delta$.

If we build a term-structure model driven by a single Brownian motion under the actual probability measure $\mathbb{P}$ and satisfying the Heath-Jarrow-Morton no-arbitrage condition of Theorem 10.3.1, then there is a Brownian motion $\widetilde{W}(t)$ under a risk-neutral probability measure $\widetilde{\mathbb{P}}$ such that forward rates are given by (10.3.18) and bond prices by (10.3.19). Theorem 9.2.2 implies that the risk-neutral measure corresponding to numéraire $B(t, T+\delta)$ is given by

$$\widetilde{\mathbb{P}}^{T+\delta}(A) = \frac{1}{B(0, T+\delta)} \int_A D(T+\delta)\, d\widetilde{\mathbb{P}} \text{ for all } A \in \mathcal{F} \tag{10.4.7}$$

and

$$\widetilde{W}^{T+\delta}(t) = \int_0^t \sigma^*(u, T+\delta)\, du + \widetilde{W}(t) \tag{10.4.8}$$

is a Brownian motion under $\widetilde{\mathbb{P}}^{T+\delta}$. We call $\widetilde{\mathbb{P}}^{T+\delta}$ the $(T+\delta)$-forward measure.

Theorem 9.2.2 implies that $\frac{S(t)}{B(t,T+\delta)}$ is a martingale under $\widetilde{\mathbb{P}}^{T+\delta}$. (See the discussion in Subsections 9.4.1 and 9.4.2.) According to the Martingale Representation Theorem (see Corollary 5.3.2), there must exist some process $\gamma(t, T)$, a process in $t \in [0, T]$ for each fixed T, such that

$$dL(t, T) = \gamma(t, T) L(t, T)\, d\widetilde{W}^{T+\delta}(t),\ 0 \le t \le T. \tag{10.4.9}$$

We relate this process to the zero-coupon bond volatilities in Subsection 10.4.5. The point of (10.4.9) is that there is no dt term, which was the term causing the problem with forward rates in Subsection 10.4.1. The dt term has been removed by changing to the $(T+\delta)$-forward measure, under which $L(t, T)$ is a martingale.

The forward LIBOR model is constructed so that $\gamma(t, T)$, defined for $0 \le t \le T \le \overline{T}$, is nonrandom. When $\gamma(t, T)$ is nonrandom, forward LIBOR $L(t, T)$ will be log-normal under the forward measure $\widetilde{\mathbb{P}}^{T+\delta}$. This leads to the following pricing result.

Theorem 10.4.2 (Black caplet formula). *Consider a* caplet *that pays* $\big(L(T, T) - K\big)^+$ *at time* $T+\delta$, *where* K *is some nonnegative constant. Assume forward LIBOR is given by (10.4.9) and* $\gamma(t, T)$ *is nonrandom. Then the price of the caplet at time zero is*

$$B(0, T+\delta)\big[L(0, T) N(d_+) - K N(d_-)\big], \tag{10.4.10}$$

where

$$d_\pm = \frac{1}{\sqrt{\int_0^T \gamma^2(t, T)\, dt}} \left[\log \frac{L(0, T)}{K} \pm \frac{1}{2} \int_0^T \gamma^2(t, T)\, dt\right]. \tag{10.4.11}$$

PROOF: According to the risk-neutral pricing formula, the price of the caplet at time zero is the discounted risk-neutral (under $\widetilde{\mathbb{P}}$) expected value of the payoff, which is

$$
\begin{aligned}
&\widetilde{\mathbb{E}}\left[D(T+\delta)\left(L(T,T)-K\right)^{+}\right]\\
&=B(0,T+\delta)\widetilde{\mathbb{E}}\left[\frac{D(T+\delta)}{B(0,T+\delta)}\left(L(T,T)-K\right)^{+}\right]\\
&=B(0,T+\delta)\widetilde{\mathbb{E}}^{T+\delta}\left(L(T,T)-K\right)^{+}. \qquad (10.4.12)
\end{aligned}
$$

The solution to the stochastic differential equation (10.4.9) is

$$
L(T,T)=L(0,T)\exp\left\{\int_0^T \gamma(t,T)\,d\widetilde{W}^{T+\delta}(t)-\frac{1}{2}\int_0^t \gamma^2(t,T)\,dt\right\}.
$$

Let us define $\overline{\gamma}(T)=\sqrt{\frac{1}{T}\int_0^T \gamma^2(t,T)\,dt}$. According to Example 4.7.3, the Itô integral $\int_0^T \gamma(t,T)\,d\widetilde{W}^{T+\delta}(t)$ is a normal random variable under $\widetilde{\mathbb{P}}^{T+\delta}$ with mean zero and variance $\overline{\gamma}^2(T)T$; we may thus write it as $-\overline{\gamma}(T)\sqrt{T}\,X$, where $X=-\frac{1}{\overline{\gamma}(T)\sqrt{T}}\int_0^T \gamma(t,T)\,d\widetilde{W}^{T+\delta}(t)$ is a standard normal random variable under $\widetilde{\mathbb{P}}^{T+\delta}$. In this notation,

$$
L(T,T)=L(0,T)e^{-\overline{\gamma}(T)\sqrt{T}\,X-\frac{1}{2}\overline{\gamma}^2(T)T},
$$

and

$$
\widetilde{\mathbb{E}}^{T+\delta}\left(L(T,T)-K\right)^{+}=\widetilde{\mathbb{E}}^{T+\delta}\left[\left(L(0,T)e^{-\overline{\gamma}(T)\sqrt{T}\,X-\frac{1}{2}\overline{\gamma}^2(T)T}-K\right)^{+}\right].
$$

This is the same computation as in (5.2.35), which led to (5.2.36). Therefore,

$$
\begin{aligned}
\widetilde{\mathbb{E}}^{T+\delta}(L(T,T)-K)^{+}&=BS\big(T,L(0,T);K,0,\overline{\gamma}(T)\big)\\
&=L(0,T)N(d_{+})-KN(d_{-}),
\end{aligned}
$$

and the risk-neutral price of the caplet (10.4.12) is (10.4.10). □

10.4.5 Forward LIBOR and Zero-Coupon Bond Volatilities

Recall that forward LIBOR is determined by the equation (10.4.3), which we can rewrite as

$$
L(t,T)+\frac{1}{\delta}=\frac{B(t,T)}{\delta B(t,T+\delta)}.
$$

We work out the evolution of $L(t,T)$ under the forward measure $\mathbb{P}^{T+\delta}$. According to Theorem 10.3.2,

$$
\begin{aligned}
&D(t)B(t,T)\\
&=B(0,T)\exp\left\{-\int_0^t \sigma^*(u,T)\,d\widetilde{W}(u)-\frac{1}{2}\int_0^t (\sigma^*(u,T))^2\,du\right\},\\
&D(t)B(t,T+\delta)\\
&=B(0,T+\delta)\exp\left\{-\int_0^t \sigma^*(u,T+\delta)\,d\widetilde{W}(u)-\frac{1}{2}\int_0^t (\sigma^*(u,T+\delta))^2\,du\right\}.
\end{aligned}
$$

This implies

$$\begin{aligned}
&L(t,T)+\frac{1}{\delta}\\
&=\frac{B(t,T)}{\delta B(t,T+\delta)}\\
&=\frac{B(0,T)}{\delta B(0,T+\delta)}\exp\left\{\int_0^t\left[\sigma^*(u,T+\delta)-\sigma^*(u,T)\right]d\widetilde{W}(u)\right.\\
&\qquad\left.+\frac{1}{2}\int_0^t\left[(\sigma^*(u,T+\delta))^2-(\sigma^*(u,T))^2\right]du\right\}.
\end{aligned}$$

The Itô-Doeblin formula implies

$$\begin{aligned}
&dL(t,T)\\
&=\left(L(t,T)+\frac{1}{\delta}\right)\Big\{\left[\sigma^*(t,T+\delta)-\sigma^*(t,T)\right]d\widetilde{W}(t)\\
&\qquad+\frac{1}{2}\left[(\sigma^*(t,T+\delta))^2-(\sigma^*(t,T))^2\right]dt\\
&\qquad+\frac{1}{2}\left[\sigma^*(t,T+\delta)-\sigma^*(t,T)\right]^2d\widetilde{W}(t)\,d\widetilde{W}(t)\Big\}\\
&=\left(L(t,T)+\frac{1}{\delta}\right)\Big\{\left[\sigma^*(t,T+\delta)-\sigma^*(t,T)\right]d\widetilde{W}(t)\\
&\qquad+\frac{1}{2}\Big[(\sigma^*(t,T+\delta))^2-(\sigma^*(t,T))^2+(\sigma^*(t,T+\delta))^2\\
&\qquad\qquad-2\sigma^*(t,T+\delta)\sigma^*(t,T)+(\sigma^*(t,T))^2\Big]dt\Big\}\\
&=\left(L(t,T)+\frac{1}{\delta}\right)\Big\{\left[\sigma^*(t,T+\delta)-\sigma^*(t,T)\right]d\widetilde{W}(t)\\
&\qquad+\left[(\sigma^*(t,T+\delta))^2-\sigma^*(t,T+\delta)\sigma^*(t,T)\right]dt\Big\}\\
&=\left(L(t,T)+\frac{1}{\delta}\right)\left[\sigma^*(t,T+\delta)-\sigma^*(t,T)\right]\left[\sigma^*(t,T+\delta)dt+d\widetilde{W}(t)\right].
\end{aligned}$$

From (10.4.8), we have

$$d\widetilde{W}^{T+\delta}(t)=\sigma^*(t,T+\delta)\,dt+d\widetilde{W}(t). \tag{10.4.13}$$

Therefore,

$$dL(t,T)=\frac{1}{\delta}\big(1+\delta L(t,T)\big)\left[\sigma^*(t,T+\delta)-\sigma^*(t,T)\right]d\widetilde{W}^{T+\delta}(t). \tag{10.4.14}$$

Comparing this with (10.4.9), we conclude that the forward LIBOR volatility $\gamma(t,T)$ of (10.4.9) and the $(T+\delta)$- and T-maturity zero-coupon bond volatilities $\sigma^*(t,T+\delta)$ and $\sigma^*(t,T)$ are related by the formula

$$\gamma(t,T) = \frac{1+\delta L(t,T)}{\delta L(t,T)}\big[\sigma^*(t,T+\delta) - \sigma^*(t,T)\big]. \tag{10.4.15}$$

10.4.6 A Forward LIBOR Term-Structure Model

The Black caplet formula of Theorem 10.4.2 is used to calibrate the forward LIBOR model. However, this calibration does not determine all the parameters needed to have a full term-structure model. In this section, we discuss the calibration and display some of the choices left open by it. We begin by collecting the equations appearing earlier in this section that we need for this subsection:

$$1+\delta L(t,T) = \frac{B(t,T)}{B(t,T+\delta)}, \quad 0 \le t \le T \le \overline{T} - \delta, \tag{10.4.3}$$

$$\widetilde{\mathbb{P}}^{T+\delta}(A) = \frac{1}{B(0,T+\delta)} \int_A D(T+\delta)\,d\widetilde{\mathbb{P}} \text{ for all } A \in \mathcal{F},\, 0 \le T \le \overline{T} - \delta, \tag{10.4.7}$$

$$d\widetilde{W}^{T+\delta}(t) = \sigma^*(t,T+\delta)\,dt + d\widetilde{W}(t), \quad 0 \le t \le T \le \overline{T} - \delta, \tag{10.4.8}$$

$$dL(t,T) = \gamma(t,T)L(t,T)\,d\widetilde{W}^{T+\delta}(t), \quad 0 \le t \le T \le \overline{T} - \delta, \tag{10.4.9}$$

$$\gamma(t,T) = \frac{1+\delta L(t,T)}{\delta L(t,T)}\big[\sigma^*(t,T+\delta) - \sigma^*(t,T)\big], \quad 0 \le t \le T \le \overline{T} - \delta. \tag{10.4.15}$$

Suppose now, at time zero, that market data allow us to determine caplet prices for maturity dates $T_j = j\delta$ for $j = 1, \ldots, n$. We can then imply the volatilities $\overline{\gamma}(T_j)$, $j = 1, \ldots, n$, appearing in the proof of Theorem 10.4.2. We wish to build a term structure model consistent with these data. We begin by setting $\overline{T}$ in the equations above equal to $(n+1)\delta$.

- We choose nonrandom nonnegative functions

$$\gamma(t,T_j), \quad 0 \le t \le T_j, \ j = 1, \ldots, n,$$

  so that $\sqrt{\frac{1}{T_j}\int_0^{T_j} \gamma^2(t,T_j)\,dt} = \overline{\gamma}(T_j)$.

For example, we could take $\gamma(t,T_j) = \overline{\gamma}(T_j)$ for $0 \le t \le T_j$.

With these volatility functions $\gamma(t,T_j)$, we can evolve forward LIBORs by equation (10.4.9), at least for $T = T_j$, $j = 1, \ldots, n$, and the forward LIBORs we obtain will agree with the market cap prices. However, (10.4.9) with $T = T_j$ gives us a formula for forward LIBOR $L(t,T_j)$ in terms of the forward Brownian motion $\widetilde{W}^{T_{j+1}}(t)$, and these are different for different values of j. Before we use (10.4.9) to evolve forward LIBORs, we must determine the relationship among these different equations.

Construction of Forward LIBOR Processes

Observe from (10.4.8) that

$$d\widetilde{W}^{T_j}(t) = \sigma^*(t,T_j)\,dt + d\widetilde{W}(t), \quad 0 \le t \le T_j.$$

Similarly,

$$d\widetilde{W}^{T_{j+1}}(t) = \sigma^*(t,T_{j+1})\,dt + d\widetilde{W}(t), \quad 0 \le t \le T_{j+1}.$$

Subtracting these equations, we obtain

$$\begin{aligned} d\widetilde{W}^{T_j}(t) &= \left[\sigma^*(t,T_j) - \sigma^*(t,T_{j+1})\right] dt + d\widetilde{W}^{T_{j+1}}(t) \\ &= -\frac{\delta\gamma(t,T_j)L(t,T_j)}{1+\delta L(t,T_j)}\,dt + d\widetilde{W}^{T_{j+1}}(t), \quad 0 \le t \le T_j, \end{aligned} \tag{10.4.16}$$

where we have used (10.4.15) for the second equality. Setting $j = n$ in (10.4.16), we have

$$d\widetilde{W}^{T_n}(t) = -\frac{\delta\gamma(t,T_n)L(t,T_n)}{1+\delta L(t,T_n)}\,dt + d\widetilde{W}^{T_{n+1}}(t), \quad 0 \le t \le T_n. \tag{10.4.17}$$

Setting $j = n-1$ in (10.4.16) and using (10.4.17), we obtain

$$\begin{aligned} d\widetilde{W}^{T_{n-1}}(t) &= -\frac{\delta\gamma(t,T_{n-1})L(t,T_{n-1})}{1+\delta L(t,T_{n-1})}\,dt + d\widetilde{W}^{T_n}(t) \\ &= -\frac{\delta\gamma(t,T_{n-1})L(t,T_{n-1})}{1+\delta L(t,T_{n-1})}\,dt - \frac{\delta\gamma(t,T_n)L(t,T_n)}{1+\delta L(t,T_n)}\,dt + d\widetilde{W}^{T_{n+1}}(t), \\ &\qquad 0 \le t \le T_{n-1}. \end{aligned}$$

Repeating this process, we conclude that

$$d\widetilde{W}^{T_{j+1}}(t) = -\sum_{i=j+1}^{n} \frac{\delta\gamma(t,T_i)L(t,T_i)}{1+\delta L(t,T_i)}\,dt + d\widetilde{W}^{T_{n+1}}(t), \quad 0 \le t \le T_{j+1}. \tag{10.4.18}$$

Equation (10.4.18) holds for $j = 0,\ldots,n$, provided we interpret $\sum_{i=n+1}^{n}$ to be zero.

We return to (10.4.9), using (10.4.18) to write

$$dL(t,T_j) = \gamma(t,T_j)L(t,T_j)\left[-\sum_{i=j+1}^{n} \frac{\delta\gamma(t,T_i)L(t,T_i)}{1+\delta L(t,T_i)}\,dt + d\widetilde{W}^{T_{n+1}}(t)\right],$$
$$0 \le t \le T_j,\ j = 1,\ldots,n. \tag{10.4.19}$$

Now we have a single Brownian motion driving all n equations. Thus, to construct the forward LIBOR model, we choose a Brownian motion, which we call $\widetilde{W}^{T_{n+1}}(t)$, $0 \le t \le T_{n+1}$, under a probability measure we call $\widetilde{\mathbb{P}}^{T_{n+1}}$. That is, we start with a probability space $(\Omega, \mathcal{F}, \widetilde{\mathbb{P}}^{T_{n+1}})$ on which is defined a Brownian motion $\widetilde{W}^{T_{n+1}}(t)$, $0 \le t \le T_{n+1}$. We assume the initial forward LIBORs $L(0, T_j)$, $j = 1, \dots, n+1$, are known from market data. With these initial conditions, (10.4.19) generates the forward LIBOR processes $L(t, T_j)$, $0 \le t \le T_j$, generating first $L(t, T_n)$, which has no drift in (10.4.19), then using $L(t, T_n)$ in the differential equation for $L(t, T_{n-1})$ to generate that process, then using $L(t, T_n)$ and $L(t, T_{n-1})$ in the differential equation for $L(t, T_{n-2})$ to generate that process, and so on. Implicit in this computation is a dependence among these different forward LIBOR processes.

Construction of T_j-Maturity Discounted Bond Prices

We construct the volatility $\sigma^*(t, T_j)$ for the zero-coupon bond maturing at T_j, $j = 1, \dots, n+1$. The forward LIBOR model has a tenor $\delta > 0$, and while it puts constraints on the cumulative effect of processes between set points T_j, it does not provide fine detail about what happens between set points. In particular, we are free to choose the bond volatilities $\sigma^*(t, T_j)$ for $T_{j-1} \le t < T_j$. The only constraint is that

$$\lim_{t \uparrow T_j} \sigma^*(t, T_j) = \sigma^*(T_j, T_j) = 0. \tag{10.4.20}$$

This constraint is present because the bond price $B(t, T_j)$ converges to 1 as $t \uparrow T_j$, and so the volatility must vanish. This is also apparent in the second formula in (10.3.9).

- For each $j = 1, \dots, n+1$, choose $\sigma^*(t, T_j)$ for $T_{j-1} \le t < T_j$ so that (10.4.20) is satisfied.

We show that this determines $\sigma(t, T_j)$ for all values of $t \in [0, T_j)$. (Again, we know from the outset that $\sigma(T_j, T_j) = 0$; that does not need to be chosen or determined.)

First of all, the initial choice of $\sigma^*(t, T_1)$ determines this function for all relevant values of t, namely, for $0 \le t < T_1$. From (10.4.15), we have

$$\sigma^*(t, T_2) = \sigma^*(t, T_1) + \frac{\delta \gamma(t, T_1) L(t, T_1)}{1 + \delta L(t, T_1)},$$

and since $\sigma^*(t, T_1)$ has been chosen for $0 \le t < T_1$, the function $\sigma(t, T_2)$ is determined by this equation for $0 \le t < T_1$. For $T_1 \le t < T_2$, $\sigma(t, T_2)$ has already been chosen. Therefore, $\sigma^*(t, T_2)$ is determined for $0 \le t < T_2$. From (10.4.15), we also have

$$\sigma^*(t, T_3) = \sigma^*(t, T_2) + \frac{\delta \gamma(t, T_2) L(t, T_2)}{1 + \delta L(t, T_2)},$$

and since $\sigma^*(t, T_2)$ has been determined for $0 \le t < T_2$, the function $\sigma(t, T_3)$ is determined by this equation for $0 \le t < T_2$. For $T_2 \le t < T_3$, $\sigma(t, T_3)$

has already been chosen. Therefore, $\sigma^*(t, T_3)$ is determined for $0 \le t < T_3$. Continuing in this way, we determine $\sigma(t, T_j)$ for all $j = 1, \ldots, n+1$ and $0 \le t < T_j$.

Using the bond volatilities $\sigma^*(t,T)$ and (10.4.8), we may write the zero-coupon bond price formula (10.3.19) of Theorem 10.3.2 as

$$\begin{aligned} dB(t,T_j) &= R(t)B(t,T_j)\,dt - \sigma^*(t,T_j)B(t,T_j)\,d\widetilde{W}(t) \\ &= R(t)B(t,T_j)\,dt + \sigma^*(t,T_j)\sigma^*(t,T_{n+1})B(t,T_j)\,dt \\ &\quad -\sigma^*(t,T_j)B(t,T_j)\,d\widetilde{W}^{T_{n+1}}(t). \end{aligned}$$

However, we have not yet determined an interest rate process $R(t)$, and so we prefer to write this equation in discounted form. For $j = 1, \ldots, n+1$,

$$\begin{aligned} d\big(D(t)B(t,T_j)\big) &= \sigma^*(t,T_j)\sigma^*(t,T_{n+1})D(t)B(t,T_j)\,dt \\ &\quad -\sigma^*(t,T_j)D(t)B(t,T_j)\,d\widetilde{W}^{T_{n+1}}(t),\ 0 \le t \le T_j. \end{aligned} \tag{10.4.21}$$

The initial condition can be obtained from (10.4.3):

$$D(0)B(0,T_j) = B(0,T_j) = \prod_{i=0}^{j-1} \frac{B(0,T_{i+1})}{B(0,T_i)} = \prod_{i=0}^{j-1} \big(1 + \delta L(0,T_i)\big)^{-1}. \tag{10.4.22}$$

This permits us to generate the discounted bond prices $D(t)B(t,T_j)$, $j = 1, \ldots, n+1$. Indeed, the solution to (10.4.21) is

$$\begin{aligned} D(t)B(t,T_j) = B(0,T_j)\exp\bigg\{ &-\int_0^t \sigma^*(u,T_j)\,d\widetilde{W}^{T_{n+1}}(u) \\ &- \int_0^t \left[\frac{1}{2}\big(\sigma^*(u,T_j)\big)^2 - \sigma^*(u,T_j)\sigma^*(u,T_{n+1})\right] du \bigg\}. \end{aligned} \tag{10.4.23}$$

Remark 10.4.3. Equation (10.4.23) does not determine the discount process $D(t)$ and the bond price $B(t,T_j)$ separately, except when $t = T_j$ for some j. In the case when $t = T_j$, we have $B(T_j,T_j) = 1$, so

$$\begin{aligned} D(T_j) &= D(T_j)B(T_j,T_j) \\ &= B(0,T_j)\exp\bigg\{ -\int_0^{T_j} \sigma^*(u,T_j)\,d\widetilde{W}^{T_{n+1}}(u) \\ &\qquad - \int_0^{T_j} \left[\frac{1}{2}\big(\sigma^*(u,T_j)\big)^2 - \sigma^*(u,T_j)\sigma^*(u,T_{n+1})\right] du \bigg\}. \end{aligned} \tag{10.4.24}$$

In the special case when $j = n+1$, we obtain

$$D(T_{n+1}) = B(0,T_{n+1}) \exp\left\{ - \int_0^{T_{n+1}} \sigma^*(u,T_{n+1})\, d\widetilde{W}^{T_{n+1}}(u) + \frac{1}{2}\int_0^{T_{n+1}} \big(\sigma^*(u,T_{n+1})\big)^2\, du \right\}. \quad (10.4.25)$$

Risk-Neutral Measure

The risk-neutral measure $\widetilde{\mathbb{P}}$ is related to the forward measure $\widetilde{\mathbb{P}}^{T_{n+1}}$ by (10.4.7),

$$\widetilde{\mathbb{P}}^{T_{n+1}}(A) = \int_A \frac{D(T_{n+1})}{B(0,T_{n+1})}\, d\widetilde{\mathbb{P}} \text{ for all } A \in \mathcal{F},$$

or, equivalently,

$$\widetilde{\mathbb{P}}(A) = \int_A \frac{B(0,T_{n+1})}{D(T_{n+1})}\, d\widetilde{\mathbb{P}}^{T_{n+1}} \text{ for all } A \in \mathcal{F}. \quad (10.4.26)$$

Because we have begun with the measure $\widetilde{\mathbb{P}}^{T_{n+1}}$ rather than $\widetilde{\mathbb{P}}$, we use (10.4.26) to define $\widetilde{\mathbb{P}}$. According to (10.4.25),

$$\frac{B(0,T_{n+1})}{D(T_{n+1})} = \exp\left\{ \int_0^{T_{n+1}} \sigma^*(u,T_{n+1})\, d\widetilde{W}^{T_{n+1}}(u) - \frac{1}{2}\int_0^{T_{n+1}} \sigma^*(u,T_{n+1})\, du \right\}, \quad (10.4.27)$$

and so the terms appearing on the right-hand side of (10.4.26) are defined. The following theorem justifies calling $\widetilde{\mathbb{P}}$ the risk-neutral measure.

Theorem 10.4.4. *Under $\widetilde{\mathbb{P}}$ given by (10.4.26), the discounted zero-coupon bond prices given by (10.4.21) and (10.4.22), or equivalently by (10.4.23), are martingales.*

PROOF: With

$$\widetilde{W}(t) = \widetilde{W}^{T_{n+1}}(t) - \int_0^t \sigma^*(u,T_{n+1})\, du, \quad 0 \le t \le T_{n+1},$$

(10.4.21) may be written as

$$d\big(D(t)B(t,T_j)\big) = -\sigma^*(t,T_j)D(t)B(t,T_j)\, d\widetilde{W}(t). \quad (10.4.28)$$

It suffices to show that $\widetilde{W}(t)$ is a Brownian motion under $\widetilde{\mathbb{P}}$ defined by (10.4.26). According to Girsanov's Theorem, Theorem 5.2.3, with $\Theta(u) = -\sigma^*(u,T_{n+1})$, since $\widetilde{W}^{T_{n+1}}(t)$ is a Brownian motion under $\widetilde{\mathbb{P}}^{T_{n+1}}$, then $\widetilde{W}(t)$ is a Brownian motion under a measure $\widehat{\mathbb{P}}$ defined by

$$\widehat{\mathbb{P}}(A) = \int_A Z(T_{n+1})\, d\widetilde{\mathbb{P}}^{T_{n+1}} \text{ for all } A \in \mathcal{F},$$

where

$$Z(T_{n+1}) = \exp\left\{ -\int_0^{T_{n+1}} \Theta(u)\, d\widetilde{W}^{T_{n+1}}(u) - \frac{1}{2}\int_0^{T_{n+1}} \Theta^2(u)\, du \right\}.$$

From (10.4.27), we see that $Z(T_{n+1}) = \frac{B(0,T_{n+1})}{D(T_{n+1})}$, so $\widehat{\mathbb{P}} = \widetilde{\mathbb{P}}$. □

Remark 10.4.5. In order to complete the determination of a full term-structure model with bond prices for all maturities T, a discount process, and forward rates, it is necessary to choose $\gamma(t,T)$ for $0 \le t \le T$ and $T \in (0, T_{n+1}) \setminus \{T_1, \dots, T_n\}$ and to also make some choices in order to determine bond volatility $\sigma^*(t,T)$ for $0 \le t \le T$ and $T \in (0, T_{n+1}) \setminus \{T_1, \dots, T_n\}$. This can be done, and thus the forward LIBOR model is consistent with a full term-structure model. However, the model obtained by exercising these choices arbitrarily is not a reliable vehicle for pricing instruments that depend on these choices.

10.5 Summary

We have presented three types of term-structure models: finite-factor Markov models for the short rate, the Heath-Jarrow-Morton model, and the forward LIBOR model.

There are many finite-factor short-rate models. For all of them, one writes down a stochastic differential equation or system of stochastic differential equations for the "factors", and then provides a formula for the interest rate as a function of these factors. One then uses the risk-neutral pricing formula to obtain prices of bonds and fixed income derivatives. In particular, these models begin under the risk-neutral measure, for otherwise there is no way to infer prices of assets from the factor processes and the interest rate.

Affine-yield models belong to the class of finite-factor short-rate models, and we have presented the two-factor affine-yield models. In these models, the interest rate is given by an equation of the form

$$R(t) = \delta_0 + \delta_1 Y_1(t) + \delta_2 Y_2(t), \tag{10.2.6}$$

where δ_0, δ_1, and δ_2 are either constants (as in the text) or nonrandom functions of time (as in Exercise 10.3), and $Y_1(t)$ and $Y_2(t)$ are the factor processes. When regarded as a two-dimensional process, $\big(Y_1(t), Y_2(t)\big)$ is Markov, and hence bond prices and the prices of interest rate derivatives are functions of these processes. These functions can be determined by solving partial differential equations with boundary conditions depending on the particular instrument being priced. For the boundary condition associated with zero-coupon bonds, the partial differential equations reduce to a system of ordinary differential equations, which permits rapid calibration of the models.

For the two-factor affine-yield models, the price at time t of a zero-coupon bond maturing at a later time T and paying 1 upon maturity is of the form

$$B(t,T) = e^{-Y_1(t)C_1(t,T)-Y_2(t)C_2(t,T)-A(t,T)}. \tag{10.5.1}$$

The nonrandom functions $C_1(t,T)$, $C_2(t,T)$, and $A(t,T)$ are given by a system of ordinary differential equations in the t variable and the boundary condition

$$C_1(T,T) = C_2(T,T) = A(T,T) = 0.$$

When the model coefficients, both δ_0, δ_1, and δ_2 in (10.2.6) and the coefficients in the differential equations satisfied by the factor processes, are constant, the functions $C_1(t,T)$, $C_2(t,T)$, and $A(t,T)$ depend on t and T only through their difference $\tau = T - t$.

The affine-yield models are calibrated by choosing the coefficients in (10.2.6) and/or in the stochastic differential equations for the factor processes. To introduce more variables for the calibration, it is customary to take the coefficients to be nonrandom, often piecewise constant, functions of time. It is helpful before beginning the calibration to make sure that the models are written in their most parsimonious form so that one cannot obtain the same model statistics from two different sets of parameter choices. The canonical forms presented here are "most parsimonious" in this sense.

There are three canonical two-factor affine-yield models, which we call the *two-factor Vasicek model*, the *two-factor Cox-Ingersoll-Ross model*, and the *two-factor mixed model*. In the first of these, both factors can become negative. In the second, both factors are guaranteed to be nonnegative. In the third, one factor is guaranteed to be nonnegative and the other can become negative. All three of these models are driven by independent Brownian motions $\widetilde{W}_1(t)$, $\widetilde{W}_2(t)$ under a risk-neutral measure $\widetilde{\mathbb{P}}$.

The canonical two-factor Vasicek model is

$$dY_1(t) = -\lambda_1 Y_1(t)\,dt + d\widetilde{W}_1(t), \tag{10.2.4}$$

$$dY_2(t) = -\lambda_{21} Y_1(t)\,dt - \lambda_2 Y_2(t)\,dt + d\widetilde{W}_1(t), \tag{10.2.5}$$

where $\lambda_1 > 0$ and $\lambda_2 > 0$. These factors are Gaussian processes, and their statistics and the statistics of the resulting interest rate $R(t)$ can be determined (Exercise 10.2). The functions $C_1(T-t)$, $C_2(T-t)$, and $A(T-t)$ in (10.5.1) are determined by the system of ordinary differential equations (10.2.23)–(10.2.25), and the solution to this system is given by (10.2.26)–(10.2.29). The canonical two-factor Cox-Ingersoll-Ross model is

$$dY_1(t) = \big(\mu_1 - \lambda_{11} Y_1(t) - \lambda_{12} Y_2(t)\big)\,dt + \sqrt{Y_1(t)}\,d\widetilde{W}_1(t), \tag{10.2.49}$$

$$dY_2(t) = \big(\mu_2 - \lambda_{21} Y_1(t) - \lambda_{22} Y_2(t)\big)\,dt + \sqrt{Y_1(t)}\,d\widetilde{W}_2(t), \tag{10.2.50}$$

where $\mu_1 \geq 0$, $\mu_2 \geq 0$, $\lambda_{11} > 0$, $\lambda_{22} > 0$, $\lambda_{12} \leq 0$, and $\lambda_{21} \leq 0$. The system of ordinary differential equations (10.2.56)–(10.2.58) determines the functions

$C_1(T-t)$, $C_2(T-t)$, and $A(T-t)$ in (10.5.1). The canonical two-factor mixed model is

$$dY_1(t) = (\mu - \lambda_1 Y_1(t))\,dt + \sqrt{Y_1(t)}\,d\widetilde{W}_1(t), \tag{10.2.59}$$

$$dY_2(t) = -\lambda_2 Y_2(t)\,dt + \sigma_{21}\sqrt{Y_1(t)}\,d\widetilde{W}_1(t) + \sqrt{\alpha + \beta Y_1(t)}\,d\widetilde{W}_2(t), \tag{10.2.60}$$

where $\mu \geq 0$, $\lambda_1 > 0$, $\lambda_2 > 0$, $\alpha \geq 0$, and $\beta \geq 0$. The functions $C_1(T-t)$, $C_2(T-t)$, and $A(T-t)$ in (10.5.1) are determined by the system of differential equations (10.7.4)–(10.7.6). When the model coefficients depend on time, the differential equations in all three cases are modified by replacing the constant coefficients by time-varying coefficients and replacing C_i' in these equations (which is the derivative of C_i with respect to $\tau = T - t$) by $-\frac{\partial}{\partial t}C_i(t,T)$ and making the similar replacement for A'.

The Heath-Jarrow-Morton (HJM) model evolves the whole yield curve forward in time rather than a finite set of factors. The yield curve is an infinite-dimensional object. Note, however, that the HJM model is driven by finitely many Brownian motions (in fact, by one Brownian motion in Section 10.3 but by multiple Brownian motions in Exercise 10.9). As a result, the HJM model is "finite-dimensional" in the sense that not every possible yield curve can be obtained from the model.

The yield curve in the HJM model is characterized by *forward rates*. The forward rate $f(t,T)$ is the instantaneous interest rate that can be locked in at time t for borrowing at a later time T. The HJM model begins under the actual probability measure $\mathbb{P}$ and derives a condition on the drift $\alpha(t,T)$ and diffusion $\sigma(t,T)$ of $f(t,T)$ that guarantees the existence of a risk-neutral measure $\widetilde{\mathbb{P}}$ and hence guarantees the absence of arbitrage. This condition is that there must exist a *market price of risk process* $\Theta(t)$ that does not depend on T and that satisfies

$$\alpha(t,T) = \sigma(t,T)\big[\sigma^*(t,T) + \Theta(t)\big], \quad 0 \leq t \leq T; \tag{10.3.16}$$

see Theorem 10.3.1. Although this condition was developed within the HJM model, one would not encounter in practice an arbitrage-free term-structure model driven by a single Brownian motion and not satisfying this condition. For term-structure models driven by multiple Brownian motions, the analogous condition appears in Exercise 10.9(i).

In terms of the Brownian motion $\widetilde{W}(t)$ under the risk-neutral measure, bond prices in the HJM model satisfy

$$dB(t,T) = R(t)B(t,T)\,dt - \sigma^*(t,T)B(t,T)\,d\widetilde{W}(t),$$

where $\sigma^*(t,T) = \int_t^T \sigma(t,v)\,dv$; see Theorem 10.3.2. A calibration procedure for the HJM model is provided in Subsection 10.3.6.

In contrast to the continuously compounding forward rate $f(t,T)$, which is the basis of the HJM model, forward LIBOR $L(t,T)$ is the simple interest rate that can be locked in at time t for borrowing at a later time T over the interval $[T, T+\delta]$. Here δ is a positive constant, and although not indicated by the notation, $L(t,T)$ depends on the choice of this constant.

Section 10.4 introduces a model for forward LIBOR. One can build this model so that forward LIBOR $L(t,T)$ is log-normal under the forward measure $\widetilde{\mathbb{P}}^{T+\delta}$, and this permits a mathematically rigorous derivation of the *Black caplet formula.* This formula is similar to the Black-Scholes-Merton formula for equities but used in fixed income markets in which the essence of the market is that the interest rate is random, in contrast to the Black-Scholes-Merton assumption.

10.6 Notes

The Vasicek model appears in [154] and the Cox-Ingersoll-Ross model in [41]. Hull and White generalized the Vasicek model in [88]. The general concept of multifactor affine-yield models is developed in Duffie and Kan [57], [58]. The reduction of affine-yield models to canonical versions is due to Dai and Singleton [44]. A sampling of other articles related to affine-yield models includes Ait-Sahalia [1], Balduzzi, Das, Foresi, and Sundaram [7], Chen [29], Chen and Scott [30], [31], [32], Collin-Dufresne and Goldstein [38], [39], Duffee [55], and Piazzesi [132]. Maghsoodi [116] provides a detailed study of the one-dimensional CIR equation when the parameters are time-varying.

Although affine-yield models have simple bond price formulas, the prices for fixed income derivatives are more complicated. However, numerical solution of partial differential equations can be avoided by Fourier transform analysis; see, Duffie, Pan, and Singleton [59].

Some other common short rate models are those of Black, Derman, and Toy [15], Black and Karasinski [16], and Longstaff and Schwartz [111]. An empirical comparison of various short rate models is provided by Chan et al. [28].

Ho and Lee [85] developed a discrete-time model for the evolution of the yield curve. The continuous-time limit of the Ho-Lee model is a constant-diffusion forward rate. In particular, the interest rate behaves like that in a Vasicek model and can become negative.

An arbitrage-free framework for the evolution of the yield curve in continuous time was developed by Heath, Jarrow, and Morton [83]. Related papers are [81] and [82]. The HJM framework presented in this chapter is general, but it can be specialized to obtain a Markov implementation; see Brace and Musiela [20], Cheyette [34], and Hunt, Kennedy, and Pelsser [90]. Filipović [66] examines the issue of making the yield curves generated by the HJM model consistent with the scheme used to generate the initial yield curve. Jara [96] considers an HJM-type model but for interest rate futures rather than forward rates. The advantage is that the drift term causing the explosion discussed in Subsection 10.4.1 does not appear in such a model.

The switch from continuously compounding forward rates to simple forward rates in order to remove the explosion problem described in Section 10.4.1 was proposed by Sandmann and Sondermann [146], [147]. The use of a log-normal simple interest rate to price caps and floors was worked out by Miltersen, Sandmann, and Sondermann [125]. This idea was embedded in a full forward LIBOR term-structure model by Brace, Gątarek, and Musiela [19]. This was the first full term-structure model consistent with the heuristic formula provided by Black [13] in 1976 and in common use since then.

Recently, a variation on forward LIBOR models has been developed for swaps markets; see Jamshidian [95] and the three books cited below. Term-structure models with jumps have been studied by Björk, Kabanov, and Runggaldier [12], Das [46], Das and Foresi [47], Glasserman and Kou [73], Glasserman and Merener [74], and Shirakawa [149].

Three recent books by authors with practical experience in term-structure modeling are Pelsser [131], Brigo and Mercurio [21], and Rebonato [137]. Pelsser's text [131] is succinct but comprehensive, Brigo and Mercurio's text [21] contains considerably more detail, and Rebonato's book [137] is devoted to forward LIBOR models.

10.7 Exercises

Exercise 10.1 (Statistics in the two-factor Vasicek model). According to Example 4.7.3, $Y_1(t)$ and $Y_2(t)$ in (10.2.43)–(10.2.46) are Gaussian processes.

(i) Show that

$$\widetilde{\mathbb{E}}Y_1(t) = e^{-\lambda_1 t}Y_1(0), \tag{10.7.1}$$

that when $\lambda_1 \neq \lambda_2$, then

$$\widetilde{\mathbb{E}}Y_2(t) = \frac{\lambda_{21}}{\lambda_1 - \lambda_2}\left(e^{-\lambda_1 t} - e^{-\lambda_2 t}\right)Y_1(0) + e^{-\lambda_2 t}Y_2(0), \tag{10.7.2}$$

and when $\lambda_1 = \lambda_2$, then

$$\widetilde{\mathbb{E}}Y_2(t) = -\lambda_{21}te^{-\lambda_1 t}Y_1(0) + e^{-\lambda_1 t}Y_2(0). \tag{10.7.3}$$

We can write

$$Y_1(t) - \widetilde{\mathbb{E}}Y_1(t) = e^{-\lambda_1 t}I_1(t),$$

when $\lambda_1 \neq \lambda_2$,

$$Y_2(t) - \mathbb{E}Y_2(t) = \frac{\lambda_{21}}{\lambda_1 - \lambda_2}\left(e^{-\lambda_1 t}I_1(t) - e^{-\lambda_2 t}I_2(t)\right) - e^{-\lambda_2 t}I_3(t),$$

and when $\lambda_1 = \lambda_2$,

$$Y_2(t) - \widetilde{\mathbb{E}}Y_2(t) = -\lambda_{21}te^{-\lambda_1 t}I_1(t) + \lambda_{21}e^{-\lambda_1 t}I_4(t) + e^{-\lambda_1 t}I_3(t),$$

where the Itô integrals

$$I_1(t) = \int_0^t e^{\lambda_1 u}\,d\widetilde{W}_1(u),\ I_2(t) = \int_0^t e^{\lambda_2 u}\,d\widetilde{W}_1(u),$$
$$I_3(t) = \int_0^t e^{\lambda_2 u}\,d\widetilde{W}_2(u),\ I_4(t) = \int_0^t ue^{\lambda_1 u}\,d\widetilde{W}_1(u),$$

all have expectation zero under the risk-neutral measure $\widetilde{\mathbb{P}}$. Consequently, we can determine the variances of $Y_1(t)$ and $Y_2(t)$ and the covariance of $Y_1(t)$ and $Y_2(t)$ under the risk-neutral measure from the variances and covariances of $I_j(t)$ and $I_k(t)$. For example, if $\lambda_1 = \lambda_2$, then

$$\begin{aligned}
&\mathrm{Var}\big(Y_1(t)\big)\\
&= e^{-2\lambda_1 t}\widetilde{\mathbb{E}}I_1^2(t),\\
&\mathrm{Var}\big(Y_2(t)\big)\\
&= \lambda_{21}^2 t^2 e^{-2\lambda_1 t}\widetilde{\mathbb{E}}I_1^2(t) + \lambda_{21}^2 e^{-2\lambda_1 t}\widetilde{\mathbb{E}}I_4^2(t) + e^{-2\lambda_1 t}\widetilde{\mathbb{E}}I_3^2(t)\\
&\quad -2\lambda_{21}^2 te^{-2\lambda_1 t}\widetilde{\mathbb{E}}\big[I_1(t)I_4(t)\big] - 2\lambda_{21}te^{-2\lambda_1 t}\widetilde{\mathbb{E}}\big[I_1(t)I_3(t)\big]\\
&\quad +2\lambda_{21}e^{-2\lambda_1 t}\widetilde{\mathbb{E}}\big[I_4(t)I_3(t)\big],\\
&\mathrm{Cov}\big(Y_1(t), Y_2(t)\big)\\
&= -\lambda_{21}te^{-2\lambda_1 t}\widetilde{\mathbb{E}}I_1^2(t) + \lambda_{21}e^{-2\lambda_1 t}\widetilde{\mathbb{E}}\big[I_1(t)I_4(t)\big] + e^{-2\lambda_1 t}\widetilde{\mathbb{E}}\big[I_1(t)I_3(t)\big],
\end{aligned}$$

where the variances and covariance above are under the risk-neutral measure $\widetilde{\mathbb{P}}$.

(ii) Compute the five terms

$$\widetilde{\mathbb{E}}I_1^2(t),\ \widetilde{\mathbb{E}}\big[I_1(t)I_2(t)\big],\ \widetilde{\mathbb{E}}\big[I_1(t)I_3(t)\big],\ \widetilde{\mathbb{E}}\big[I_1(t)I_4(t)\big],\ \widetilde{\mathbb{E}}\big[I_4^2(t)\big].$$

The five other terms, which you are not being asked to compute, are

$$\begin{aligned}
\mathbb{E}I_2^2(t) &= \frac{1}{2\lambda_2}\big(e^{2\lambda_2 t} - 1\big),\\
\mathbb{E}\big[I_2(t)I_3(t)\big] &= 0,\\
\mathbb{E}\big[I_2(t)I_4(t)\big] &= \frac{t}{\lambda_1 + \lambda_2}e^{(\lambda_1+\lambda_2)t} + \frac{1}{(\lambda_1+\lambda_2)^2}\big(1 - e^{(\lambda_1+\lambda_2)t}\big),\\
\mathbb{E}I_3^2(t) &= \frac{1}{\lambda_2}\big(e^{2\lambda_2 t} - 1\big),\\
\mathbb{E}\big[I_3(t)I_4(t)\big] &= 0.
\end{aligned}$$

(iii) Some derivative securities involve *time spread* (i.e., they depend on the interest rate at two different times). In such cases, we are interested in the joint statistics of the factor processes at different times. These are still jointly normal and depend on the statistics of the Itô integrals I_j at different times. Compute $\widetilde{\mathbb{E}}\big[I_1(s)I_2(t)\big]$, where $0 \le s < t$. (Hint: Fix $s \ge 0$ and define

$$J_1(t) = \int_0^t e^{\lambda_1 u} \mathbb{I}_{\{u \le s\}}\, d\widetilde{W}_1(u),$$

where $\mathbb{I}_{\{u \le s\}}$ is the function of u that is 1 if $u \le s$ and 0 if $u > s$. Note that $J_1(t) = I_1(s)$ when $t \ge s$.)

Exercise 10.2 (Ordinary differential equations for the mixed affine-yield model). In the mixed model of Subsection 10.2.3, as in the two-factor Vasicek model and the two-factor Cox-Ingersoll-Ross model, zero-coupon bond prices have the affine-yield form

$$f(t, y_1, y_2) = e^{-y_1 C_1(T-t) - y_2 C_2(T-t) - A(T-t)},$$

where $C_1(0) = C_2(0) = A(0) = 0$.

(i) Find the partial differential equation satisfied by $f(t, y_1, y_2)$.
(ii) Show that C_1, C_2, and A satisfy the system of ordinary differential equations

$$C_1' = -\lambda_1 C_1 - \frac{1}{2} C_1^2 - \sigma_{21} C_1 C_2 - (1+\beta) C_2^2 + \delta_1, \tag{10.7.4}$$

$$C_2' = -\lambda_2 C_2 + \delta_2, \tag{10.7.5}$$

$$A' = \mu C_1 - \frac{1}{2} \alpha C_2^2 + \delta_0. \tag{10.7.6}$$

Exercise 10.3 (Calibration of the two-factor Vasicek model). Consider the canonical two-factor Vasicek model (10.2.4), (10.2.5), but replace the interest rate equation (10.2.6) by

$$R(t) = \delta_0(t) + \delta_1 Y_1(t) + \delta_2 Y_2(t), \tag{10.7.7}$$

where δ_1 and δ_2 are constant but $\delta_0(t)$ is a nonrandom function of time. Assume that for each T there is a zero-coupon bond maturing at time T. The price of this bond at time $t \in [0, T]$ is

$$B(t, T) = \widetilde{\mathbb{E}}\left[e^{-\int_t^T R(u)du} \,\Big|\, \mathcal{F}(t) \right].$$

Because the pair of processes $\big(Y_1(t), Y_2(t)\big)$ is Markov, there must exist some function $f(t, T, y_1, y_2)$ such that $B(t, T) = f\big(t, T, Y_1(t), Y_2(t)\big)$. (We indicate the dependence of f on the maturity T because, unlike in Subsection 10.2.1, here we shall consider more than one value of T.)

(i) The function $f(t, T, y_1, y_2)$ is of the affine-yield form

$$f(t, T, y_1, y_2) = e^{-y_1 C_1(t,T) - y_2 C_2(t,T) - A(t,T)}. \tag{10.7.8}$$

Holding T fixed, derive a system of ordinary differential equations for $\frac{d}{dt} C_1(t, T)$, $\frac{d}{dt} C_2(t, T)$, and $\frac{d}{dt} A(t, T)$.

(ii) Using the terminal conditions $C_1(T,T) = C_2(T,T) = 0$, solve the equations in (i) for $C_1(t,T)$ and $C_2(t,T)$. (As in Subsection 10.2.1, the functions C_1 and C_2 depend on t and T only through the difference $\tau = T - t$; however, the function A discussed in part (iii) below depends on t and T separately.)

(iii) Using the terminal condition $A(T,T) = 0$, write a formula for $A(t,T)$ as an integral involving $C_1(u,T)$, $C_2(u,T)$, and $\delta_0(u)$. You do not need to evaluate this integral.

(iv) Assume that the model parameters $\lambda_1 > 0$ $\lambda_2 > 0$, λ_{21}, δ_1, and δ_2 and the initial conditions $Y_1(0)$ and $Y_2(0)$ are given. We wish to choose a *function* δ_0 so that the zero-coupon bond prices given by the model match the bond prices given by the market at the initial time zero. In other words, we want to choose a function $\delta(T)$, $T \geq 0$, so that

$$f\big(0, T, Y_1(0), Y_2(0)\big) = B(0,T), \quad T \geq 0.$$

In this part of the exercise, we regard both t and T as variables and use the notation $\frac{\partial}{\partial t}$ to indicate the derivative with respect to t when T is held fixed and the notation $\frac{\partial}{\partial T}$ to indicate the derivative with respect to T when t is held fixed. Give a formula for $\delta_0(T)$ in terms of $\frac{\partial}{\partial T} \log B(0,T)$ and the model parameters. (Hint: Compute $\frac{\partial}{\partial T} A(0,T)$ in two ways, using (10.7.8) and also using the formula obtained in (iii). Because $C_i(t,T)$ depends only on t and T through $\tau = T - t$, there are functions $\overline{C}_i(\tau)$ such that $\overline{C}_i(\tau) = \overline{C}_i(T-t) = C_i(t,T)$, $i = 1, 2$. Then

$$\frac{\partial}{\partial t} C_i(t,T) = -\overline{C}_i'(\tau), \quad \frac{\partial}{\partial T} C_i(t,T) = \overline{C}_i'(\tau),$$

where $'$ denotes differentiation with respect to τ. This shows that

$$\frac{\partial}{\partial T} C_i(t,T) = -\frac{\partial}{\partial t} C_i(t,T), \; i = 1, 2, \tag{10.7.9}$$

a fact that you will need.)

Exercise 10.4. Hull and White [89] propose the two-factor model

$$dU(t) = -\lambda_1 U(t)\,dt + \sigma_1\,d\widetilde{B}_2(t), \tag{10.7.10}$$

$$dR(t) = \big[\theta(t) + U(t) - \lambda_2 R(t)\big]\,dt + \sigma_2\,d\widetilde{B}_1(t), \tag{10.7.11}$$

where λ_1, λ_2, σ_1, and σ_2 are positive constants, $\theta(t)$ is a nonrandom function, and $\widetilde{B}_1(t)$ and $\widetilde{B}_2(t)$ are correlated Brownian motions with $d\widetilde{B}_1(t)\,d\widetilde{B}_2(t) = \rho\,dt$ for some $\rho \in (-1,1)$. In this exercise, we discuss how to reduce this to the two-factor Vasicek model of Subsection 10.2.1, except that, instead of (10.2.6), the interest rate is given by (10.7.7), in which $\delta_0(t)$ is a nonrandom function of time.

(i) Define

$$X(t) = \begin{bmatrix} U(t) \\ R(t) \end{bmatrix}, \quad K = \begin{bmatrix} \lambda_1 & 0 \\ -1 & \lambda_2 \end{bmatrix}, \quad \Sigma = \begin{bmatrix} \sigma_1 & 0 \\ 0 & \sigma_2 \end{bmatrix}$$
$$\Theta(t) = \begin{bmatrix} 0 \\ \theta(t) \end{bmatrix}, \quad \widetilde{B}(t) = \begin{bmatrix} \widetilde{B}_1(t) \\ \widetilde{B}_2(t) \end{bmatrix},$$

so that (10.7.10) and (10.7.11) can be written in vector notation as

$$dX(t) = \Theta(t)\,dt - KX(t)\,dt + \Sigma\,d\widetilde{B}(t). \tag{10.7.12}$$

Now set

$$\widehat{X}(t) = X(t) - e^{-Kt}\int_0^t e^{Ku}\Theta(u)\,du.$$

Show that

$$d\widehat{X}(t) = -K\widehat{X}(t)\,dt + \Sigma d\widetilde{B}(t). \tag{10.7.13}$$

(ii) With

$$C = \begin{bmatrix} \frac{1}{\sigma_1} & 0 \\ -\frac{\rho}{\sigma_1\sqrt{1-\rho^2}} & \frac{1}{\sigma_2\sqrt{1-\rho^2}} \end{bmatrix},$$

define $Y(t) = C\widehat{X}(t)$, $\widetilde{W}(t) = C\Sigma\widetilde{B}(t)$. Show that the components of $\widetilde{W}_1(t)$ and $\widetilde{W}_2(t)$ are independent Brownian motions and

$$dY(t) = -\Lambda Y(t) + d\widetilde{W}(t), \tag{10.7.14}$$

where

$$\Lambda = CKC^{-1} = \begin{bmatrix} \lambda_1 & 0 \\ \frac{\rho\sigma_2(\lambda_2 - \lambda_1) - \sigma_1}{\sigma_2\sqrt{1-\rho^2}} & \lambda_2. \end{bmatrix}.$$

Equation (10.7.14) is the vector form of the canonical two-factor Vasicek equations (10.2.4) and (10.2.5).

(iii) Obtain a formula for $R(t)$ of the form (10.7.7). What are $\delta_0(t)$, δ_1, and δ_2?

Exercise 10.5 (Correlation between long rate and short rate in the one-factor Vasicek model). The one-factor Vasicek model is the one-factor Hull-White model of Example 6.5.1 with constant parameters,

$$dR(t) = \big(a - bR(t)\big)\,dt + \sigma\,d\widetilde{W}(t), \tag{10.7.15}$$

where a, b, and σ are positive constants and $\widetilde{W}(t)$ is a one-dimensional Brownian motion. In this model, the price at time $t \in [0,T]$ of the zero-coupon bond maturing at time T is

$$B(t,T) = e^{-C(t,T)R(t)-A(t,T)},$$

where $C(t,T)$ and $A(t,T)$ are given by (6.5.10) and (6.5.11):

$$C(t,T) = \int_t^T e^{-\int_t^s b\,dv}\,ds = \frac{1}{b}\left(1-e^{-b(T-t)}\right),$$

$$A(t,T) = \int_t^T \left(aC(s,T) - \frac{1}{2}\sigma^2 C^2(s,T)\right) ds$$

$$= \frac{2ab-\sigma^2}{2b^2}(T-t) + \frac{\sigma^2-ab}{b^3}\left(1-e^{-b(T-t)}\right) - \frac{\sigma^2}{4b^3}\left(1-e^{-2b(T-t)}\right).$$

In the spirit of the discussion of the short rate and the long rate in Subsection 10.2.1, we fix a positive relative maturity $\overline{\tau}$ and define the long rate $L(t)$ at time t by (10.2.30):

$$L(t) = -\frac{1}{\overline{\tau}}\log B(t, t+\overline{\tau}).$$

Show that changes in $L(t)$ and $R(t)$ are perfectly correlated (i.e., for any $0 \leq t_1 < t_2$, the correlation coefficient between $L(t_2)-L(t_1)$ and $R(t_2)-R(t_1)$ is one). This characteristic of one-factor models caused the development of models with more than one factor.

Exercise 10.6 (Degenerate two-factor Vasicek model). In the discussion of short rates and long rates in the two-factor Vasicek model of Subsection 10.2.1, we made the assumptions that $\delta_2 \neq 0$ and $(\lambda_1 - \lambda_2)\delta_1 + \lambda_{21}\delta_2 \neq 0$ (see Lemma 10.2.2). In this exercise, we show that if either of these conditions is violated, the two-factor Vasicek model reduces to a one-factor model, for which long rates and short rates are perfectly correlated (see Exercise 10.5).

(i) Show that if $\delta_2 = 0$ (and $\delta_0 > 0$, $\delta_1 > 0$), then the short rate $R(t)$ given by the system of equations (10.2.4)–(10.2.6) satisfies the one-dimensional stochastic differential equation

$$dR(t) = \big(a - bR(t)\big)\,dt + d\widetilde{W}_1(t). \tag{10.7.16}$$

Define a and b in terms of the parameters in (10.2.4)–(10.2.6).

(ii) Show that if $(\lambda_1 - \lambda_2)\delta_1 + \lambda_{21}\delta_2 = 0$ (and $\delta_0 > 0$, $\delta_1^2 + \delta_2^2 \neq 0$), then the short rate $R(t)$ given by the system of equations (10.2.4)–(10.2.6) satisfies the one-dimensional stochastic differential equation

$$dR(t) = \big(a - bR(t)\big)\,dt + \sigma\,d\widetilde{B}(t). \tag{10.7.17}$$

Define a and b in terms of the parameters in (10.2.4)–(10.2.6) and define the Brownian motion $\widetilde{B}(t)$ in terms of the independent Brownian motions $\widetilde{W}_1(t)$ and $\widetilde{W}_2(t)$ in (10.2.4) and (10.2.5).

Exercise 10.7 (Forward measure in the two-factor Vasicek model). Fix a maturity $T > 0$. In the two-factor Vasicek model of Subsection 10.2.1, consider the T-forward measure $\widetilde{\mathbb{P}}^T$ of Definition 9.4.1:

$$\widetilde{\mathbb{P}}^T(A) = \frac{1}{B(0,T)} \int_A D(T)\, d\widetilde{\mathbb{P}} \text{ for all } A \in \mathcal{F}.$$

(i) Show that the two-dimensional $\widetilde{\mathbb{P}}^T$-Brownian motions $\widetilde{W}_1^T(t)$, $\widetilde{W}_2^T(t)$ of (9.2.5) are

$$\widetilde{W}_j^T(t) = \int_0^t C_1(T-u)\, du + \widetilde{W}_j(t), \quad j = 1, 2, \tag{10.7.18}$$

where $C_1(\tau)$ and $C_2(\tau)$ are given by (10.2.26)–(10.2.28).

(ii) Consider a call option on a bond maturing at time $\overline{T} > T$. The call expires at time T and has strike price K. Show that at time zero the risk-neutral price of this option is

$$B(0,T)\widetilde{\mathbb{E}}^T\left[\left(e^{-C_1(\overline{T}-T)Y_1(T)-C_2(\overline{T}-T)Y_2(T)-A(\overline{T}-T)} - K\right)^+\right]. \tag{10.7.19}$$

(iii) Show that, under the T-forward measure $\widetilde{\mathbb{P}}^T$, the term

$$X = -C_1(\overline{T}-T)Y_1(T) - C_2(\overline{T}-T)Y_1(T) - A(\overline{T}-T)$$

appearing in the exponent in (10.7.19) is normally distributed.

(iv) It is a straightforward but lengthy computation, like the computations in Exercise 10.1, to determine the mean and variance of the term X. Let us call its variance σ^2 and its mean $\mu - \frac{1}{2}\sigma^2$, so that we can write X as

$$X = \mu - \frac{1}{2}\sigma^2 - \sigma Z,$$

where Z is a standard normal random variable under $\widetilde{\mathbb{P}}^T$. Show that the call option price in (10.7.19) is

$$B(0,T)\big(e^{\mu}N(d_+) - KN(d_-)\big),$$

where

$$d_\pm = \frac{1}{\sigma}\left(\mu - \log K \pm \frac{1}{2}\sigma^2\right).$$

Exercise 10.8 (Reversal of order of integration in forward rates). The forward rate formula (10.3.5) with v replacing T states that

$$f(t,v) = f(0,v) + \int_0^t \alpha(u,v)\, du + \int_0^t \sigma(u,v)\, dW(u).$$

Therefore,

$$-\int_t^T f(t,v)\, dv = -\int_t^T \left[f(0,v) + \int_0^t \alpha(u,v)\, du + \int_0^t \sigma(u,v)\, dW(u)\right] dv. \tag{10.7.20}$$

(i) Define

$$\widehat{\alpha}(u,t,T) = \int_t^T \alpha(u,v)\,dv, \quad \widehat{\sigma}(u,t,T) = \int_t^T \sigma(u,v)\,dv.$$

Show that if we reverse the order of integration in (10.7.20), we obtain the equation

$$\begin{aligned} &-\int_t^T f(t,v)\,dv \\ &= -\int_t^T f(0,v)\,dv - \int_0^t \widehat{\alpha}(u,t,T)\,du - \int_0^t \widehat{\sigma}(u,t,T)\,dW(u). \end{aligned} \tag{10.7.21}$$

(In one case, this is a reversal of the order of two Riemann integrals, a step that uses only the theory of ordinary calculus. In the other case, the order of a Riemann and an Itô integral are being reversed. This step is justified in the appendix of [83]. You may assume without proof that this step is legitimate.)

(ii) Take the differential with respect to t in (10.7.21), remembering to get two terms from each of the integrals $\int_0^t \widehat{\alpha}(u,t,T)\,du$ and $\int_0^t \widehat{\sigma}(u,t,T)\,dW(u)$ because one must differentiate with respect to each of the two ts appearing in these integrals.

(iii) Check that your formula in (ii) agrees with (10.3.10).

Exercise 10.9 (Multifactor HJM model). Suppose the Heath-Jarrow-Morton model is driven by a d-dimensional Brownian motion, so that $\sigma(t,T)$ is also a d-dimensional vector and the forward rate dynamics are given by

$$df(t,T) = \alpha(t,T)\,dt + \sum_{j=1}^{d} \sigma_j(t,T)\,dW_j(t).$$

(i) Show that (10.3.16) becomes

$$\alpha(t,T) = \sum_{j=1}^{d} \sigma_j(t,T)\left[\sigma_j^*(t,T) + \Theta_j(t)\right].$$

(ii) Suppose there is an adapted, d-dimensional process

$$\Theta(t) = (\Theta_1(t), \ldots, \Theta_d(t))$$

satisfying this equation for all $0 \le t \le T \le \overline{T}$. Show that if there are maturities $T_1, \ldots, T_d$ such that the $d \times d$ matrix $\left(\sigma_j(t,T_i)\right)_{i,j}$ is nonsingular, then $\Theta(t)$ is unique.

Exercise 10.10. (i) Use the ordinary differential equations (6.5.8) and (6.5.9) satisfied by the functions $A(t,T)$ and $C(t,T)$ in the one-factor Hull-White model to show that this model satisfies the HJM no-arbitrage condition (10.3.27).

(ii) Use the ordinary differential equations (6.5.14) and (6.5.15) satisfied by the functions $A(t,T)$ and $C(t,T)$ in the one-factor Cox-Ingersoll-Ross model to show that this model satisfies the HJM no-arbitrage condition (10.3.27).

Exercise 10.11. Let $\delta > 0$ be given. Consider an interest rate swap paying a fixed interest rate K and receiving backset LIBOR $L(T_{j-1}, T_{j-1})$ on a principal of 1 at each of the payment dates $T_j = \delta j$, $j = 1, 2, \ldots, n+1$. Show that the value of the swap is

$$\delta K \sum_{j=1}^{n+1} B(0,T_j) - \delta \sum_{j=1}^{n+1} B(0,T_j) L(0,T_{j-1}). \tag{10.7.22}$$

Remark 10.7.1. The *swap rate* is defined to be the value of K that makes the initial value of the swap equal to zero. Thus, the swap rate is

$$K = \frac{\sum_{j=1}^{n+1} B(0,T_j) L(0,T_{j-1})}{\sum_{j=1}^{n+1} B(0,T_j)}. \tag{10.7.23}$$

Exercise 10.12. In the proof of Theorem 10.4.1, we showed by an arbitrage argument that the value at time 0 of a payment of backset LIBOR $L(T,T)$ at time $T+\delta$ is $B(0,T+\delta)L(0,T)$. The risk-neutral price of this payment, computed at time zero, is

$$\widetilde{\mathbb{E}}\left[D(T+\delta)L(T,T)\right].$$

Use the definitions

$$L(T,T) = \frac{1 - B(T,T+\delta)}{\delta B(T,T+\delta)},$$
$$B(0,T+\delta) = \widetilde{\mathbb{E}}\left[D(T+\delta)\right],$$

and the properties of conditional expectations to show that

$$\widetilde{\mathbb{E}}\left[D(T+\delta)L(T,T)\right] = B(0,T+\delta)L(0,T).$$

11

Introduction to Jump Processes

11.1 Introduction

This chapter studies *jump-diffusion* processes. The "diffusion" part of the nomenclature refers to the fact that these processes can have a Brownian motion component or, more generally, an integral with respect to Brownian motion. In addition, the paths of these processes may have jumps. We consider in this chapter the special case when there are only finitely many jumps in each finite time interval.

One can also construct processes in which there are infinitely many jumps in a finite time interval, although for such processes it is necessarily the case that, for each positive threshold, only finitely many jumps can have a size exceeding the threshold in any finite time interval. The number exceeding the threshold can depend on the threshold and become arbitrarily large as the threshold approaches zero. Such processes are not considered here, although the theory provided here gives some idea of how such processes can be analyzed.

The fundamental pure jump process is the *Poisson process*, and this is presented in Section 11.2. All jumps of a Poisson process are of size one. A *compound Poisson process* is like a Poisson process, except that the jumps are of random size. Compound Poisson processes are the subject of Section 11.3.

In Section 11.4, we define a *jump process* to be the sum of a nonrandom initial condition, an Itô integral with respect to a Brownian motion $dW(t)$, a Riemann integral with respect to dt, and a *pure jump* process. A pure jump process begins at zero, has finitely many jumps in each finite time interval, and is constant between jumps. Section 11.4 defines stochastic integrals with respect to jump processes. These stochastic integrals are themselves jump processes. Section 11.4 also examines the quadratic variation of jump processes and their stochastic integrals.

In Section 11.5, we present the stochastic calculus for jump processes. The key result is the extension of the Itô-Doeblin formula to cover these processes.

In Section 11.6, we take up the matter of changing measures for Poisson processes and for compound Poisson processes. We conclude with a discussion of how to simultaneously change the measure for a Brownian motion and a compound Poisson process. The effect of this change is to adjust the drift of the Brownian motion and to adjust the *intensity* (average rate of jump arrival) and the distribution of the jump sizes for the compound Poisson process.

In Section 11.7, we apply this theory to the problem of pricing and partially hedging a European call in a jump-diffusion model.

11.2 Poisson Process

In the way that Brownian motion is the basic building block for continuous-path processes, the Poisson process serves as the starting point for jump processes. In this section, we construct the Poisson process and develop its basic properties.

11.2.1 Exponential Random Variables

Let τ be a random variable with density

$$f(t) = \begin{cases} \lambda e^{-\lambda t}, & t \geq 0, \\ 0, & t < 0, \end{cases} \tag{11.2.1}$$

where λ is a positive constant. We say that τ has the *exponential distribution* or simply that τ is an *exponential random variable*.

The expected value of τ can be computed by an integration by parts:

$$\begin{aligned} \mathbb{E}\tau = \int_0^\infty t f(t)\,dt = \lambda \int_0^\infty t e^{-\lambda t}\,dt &= -te^{-\lambda t}\Big|_{t=0}^{t=\infty} + \int_0^\infty e^{-\lambda t}\,dt \\ &= 0 - \frac{1}{\lambda} e^{-\lambda t}\Big|_{t=0}^{t=\infty} = \frac{1}{\lambda}. \end{aligned}$$

For the cumulative distribution function, we have

$$F(t) = \mathbb{P}\{\tau \leq t\} = \int_0^t \lambda e^{-\lambda u}\,du = -e^{-\lambda u}\Big|_{u=0}^{u=t} = 1 - e^{-\lambda t}, \quad t \geq 0,$$

and hence

$$\mathbb{P}\{\tau > t\} = e^{-\lambda t}, \quad t \geq 0. \tag{11.2.2}$$

Suppose we are waiting for an event, such as default of a bond, and we know that the distribution of the time of this event is exponential with mean $\frac{1}{\lambda}$ (i.e., it has the density (11.2.1)). Suppose we have already waited s time units, and we are interested in the probability that we will have to wait an additional t time units (conditioned on knowing that the event has not occurred during the time interval $[0, s]$). This probability is

$$\mathbb{P}\{\tau > t+s | \tau > s\} = \frac{\mathbb{P}\{\tau > t+s \text{ and } \tau > s\}}{\mathbb{P}\{\tau > s\}}$$
$$= \frac{\mathbb{P}\{\tau > t+s\}}{\mathbb{P}\{\tau > s\}} = \frac{e^{-\lambda(t+s)}}{e^{-\lambda s}} = e^{-\lambda t}. \tag{11.2.3}$$

In other words, after waiting s time units, the probability that we will have to wait an additional t time units is the same as the probability of having to wait t time units when we were starting at time 0. The fact that we have already waited s time units does not change the distribution of the remaining time. This property for the exponential distribution is called *memorylessness*.

11.2.2 Construction of a Poisson Process

To construct a Poisson process, we begin with a sequence $\tau_1, \tau_2, \ldots$ of independent exponential random variables, all with the same mean $\frac{1}{\lambda}$. We will build a model in which an event, which we call a "jump," occurs from time to time. The first jump occurs at time τ_1, the second occurs τ_2 time units after the first, the third occurs τ_3 time units after the second, etc. The τ_k random variables are called the *interarrival times*. The *arrival times* are

$$S_n = \sum_{k=1}^{n} \tau_k \tag{11.2.4}$$

(i.e., S_n is the time of the nth jump). The *Poisson process* $N(t)$ counts the number of jumps that occur at or before time t. More precisely,

$$N(t) = \begin{cases} 0 \text{ if } 0 \le t < S_1, \\ 1 \text{ if } S_1 \le t < S_2, \\ \vdots \\ n \text{ if } S_n \le t < S_{n+1}, \\ \vdots \end{cases}$$

Note that at the jump times $N(t)$ is defined so that it is *right-continuous* (i.e., $N(t) = \lim_{s \downarrow t} N(s)$). We denote by $\mathcal{F}(t)$ the σ-algebra of information acquired by observing $N(s)$ for $0 \le s \le t$.

Because the expected time between jumps is $\frac{1}{\lambda}$, the jumps are arriving at an average rate of λ per unit time. We say the Poisson process $N(t)$ has *intensity* λ. Figure 11.2.1 shows one path of a Poisson process.

11.2.3 Distribution of Poisson Process Increments

In order to determine the distribution of the increments of a Poisson process, we must first determine the distribution of the jump times $S_1, S_2, \ldots$.

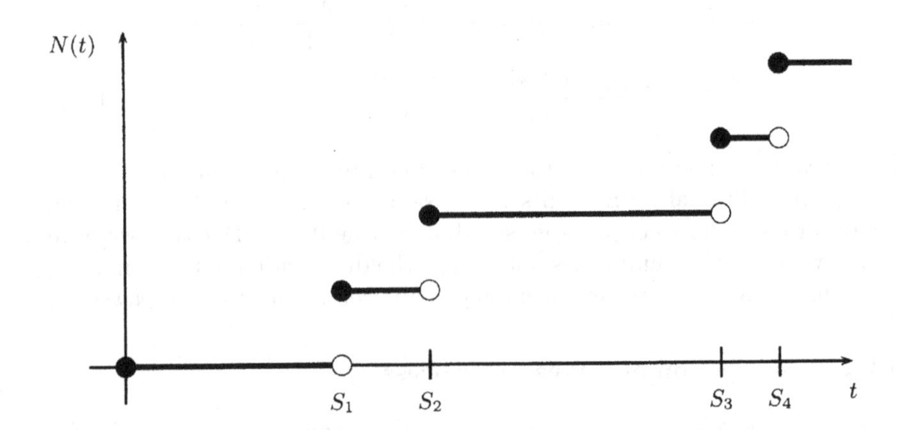

Fig. 11.2.1. One path of a Poisson process.

Lemma 11.2.1. *For $n \geq 1$, the random variable S_n defined by (11.2.4) has the* gamma density

$$g_n(s) = \frac{(\lambda s)^{n-1}}{(n-1)!} \lambda e^{-\lambda s}, \quad s \geq 0. \tag{11.2.5}$$

PROOF: We prove (11.2.5) by induction on n. For $n = 1$, we have that $S_1 = \tau_1$ is exponential, and (11.2.5) becomes the exponential density

$$g_1(s) = \lambda e^{-\lambda s}, \quad s \geq 0.$$

(Recall that 0! is defined to be 1.) Having thus established the base case, let us assume that (11.2.5) holds for some value of n and prove it for $n+1$. In other words, we assume S_n has density $g_n(s)$ given in (11.2.5) and we want to compute the density of $S_{n+1} = S_n + \tau_{n+1}$. Since S_n and τ_{n+1} are independent, the density of S_{n+1} can be computed by the convolution

$$\begin{aligned}
\int_0^s g_n(v) f(s-v)\, dv &= \int_0^s \frac{(\lambda v)^{n-1}}{(n-1)!} \lambda e^{-\lambda v} \cdot \lambda e^{-\lambda(s-v)}\, dv \\
&= \frac{\lambda^{n+1} e^{-\lambda s}}{(n-1)!} \int_0^s v^{n-1}\, ds = \frac{\lambda^{n+1} e^{-\lambda s}}{n!} v^n \Big|_{v=0}^{v=s} \\
&= \frac{(\lambda s)^n}{n!} \lambda e^{-\lambda s} = g_{n+1}(s).
\end{aligned}$$

This completes the induction step and proves the lemma. □

Lemma 11.2.2. *The Poisson process $N(t)$ with intensity λ has the distribution*

$$\mathbb{P}\{N(t) = k\} = \frac{(\lambda t)^k}{k!} e^{-\lambda t}, \quad k = 0, 1, \ldots. \tag{11.2.6}$$

PROOF: For $k \geq 1$, we have $N(t) \geq k$ if and only if there are at least k jumps by time t (i.e., if and only if S_k, the time of the kth jump, is less than or equal to t). Therefore,

$$\mathbb{P}\{N(t) \geq k\} = \mathbb{P}\{S_k \leq t\} = \int_0^t \frac{(\lambda s)^{k-1}}{(k-1)!}\lambda e^{-\lambda s}\,ds.$$

Similarly,

$$\mathbb{P}\{N(t) \geq k+1\} = \mathbb{P}\{S_{k+1} \leq t\} = \int_0^t \frac{(\lambda s)^k}{k!}\lambda e^{-\lambda s}\,ds.$$

We integrate this last expression by parts to obtain

$$\begin{aligned}\mathbb{P}\{N(t) \geq k+1\} &= -\frac{(\lambda s)^k}{k!}e^{-\lambda s}\Big|_{s=0}^{s=t} + \int_0^t \frac{(\lambda s)^{k-1}}{(k-1)!}\lambda e^{-\lambda s}\,ds \\ &= -\frac{(\lambda t)^k}{k!}e^{-\lambda t} + \mathbb{P}\{N(t) \geq k\}.\end{aligned}$$

This implies that for $k \geq 1$,

$$\mathbb{P}\{N(t) = k\} = \mathbb{P}\{N(t) \geq k\} - \mathbb{P}\{N(t) \geq k+1\} = \frac{(\lambda t)^k}{k!}e^{-\lambda t}.$$

For $k = 0$, we have from (11.2.2)

$$\mathbb{P}\{N(t) = 0\} = \mathbb{P}\{S_1 > t\} = \mathbb{P}\{\tau_1 > t\} = e^{-\lambda t},$$

which is (11.2.6) with $k = 0$. □

Suppose we observe the Poisson process up to time s and then want to know the distribution of $N(t+s) - N(s)$, conditioned on knowing what has happened up to and including time s. It turns out that the information about what has happened up to and including time s is irrelevant. This is a consequence of the memorylessness of exponential random variables (see (11.2.3)). Because $N(t+s)-N(s)$ is the number of jumps in the time interval $(s, t+s]$, in order to compute the distribution of $N(t+s) - N(s)$, we are interested in the time of the next jump after s. At time s, we know the time since the last jump, but the time between s and the next jump does not depend on this. Indeed, the time between s and the first jump after s has an exponential distribution with mean $\frac{1}{\lambda}$, independent of everything that has happened up to time s. The time between that jump and the one after it is also exponentially distributed with mean $\frac{1}{\lambda}$, independent of everything that has happened up to time s. The same applies for all subsequent jumps. Consequently, $N(t+s) - N(s)$ is independent of $\mathcal{F}(s)$. Furthermore, the distribution of $N(t+s) - N(s)$ is the same as the distribution of $N(t)$. In both cases, one is simply counting the number of jumps that occur in a time interval of length t, and the jumps are independent and exponentially distributed with mean $\frac{1}{\lambda}$. When a process has the property that the distribution of the increment depends only on the difference between the two time points, the increments are said to be *stationary*. Both the Poisson process and Brownian motion have stationary independent increments.

Theorem 11.2.3. *Let $N(t)$ be a Poisson process with intensity $\lambda > 0$, and let $0 = t_0 < t_1 < \cdots < t_n$ be given. Then the increments*

$$N(t_1) - N(t_0),\ N(t_2) - N(t_1), \ldots, N(t_n) - N(t_{n-1})$$

are stationary and independent, and

$$\mathbb{P}\{N(t_{j+1}) - N(t_j) = k\} = \frac{\lambda^k (t_{j+1} - t_j)^k}{k!} e^{-\lambda(t_{j+1} - t_j)}, \quad k = 0, 1, \ldots. \tag{11.2.7}$$

OUTLINE OF PROOF: Let $\mathcal{F}(t)$ be the σ-algebra of information acquired by observing $N(s)$ for $0 \le s \le t$. As we just discussed, $N(t_n) - N(t_{n-1})$ is independent of $\mathcal{F}(t_{n-1})$ and has the same distribution as $N(t_n - t_{n-1})$, which by Lemma 11.2.2 is the distribution given by (11.2.7) with $j = n - 1$. Since the other increments $N(t_1) - N(t_0), \ldots, N(t_{n-1}) - N(t_{n-2})$ are $\mathcal{F}(t_{n-1})$-measurable, these increments are independent of $N(t_n) - N(t_{n-1})$. We now repeat the argument for the next-to-last increment $N(t_{n-1}) - N(t_{n-2})$, then the increment before that, etc. □

11.2.4 Mean and Variance of Poisson Increments

Let $0 \le s < t$ be given. According to Theorem 11.2.3, the Poisson increment $N(t) - N(s)$ has distribution

$$\mathbb{P}\{N(t) - N(s) = k\} = \frac{\lambda^k (t-s)^k}{k!} e^{-\lambda(t-s)}, \quad k = 0, 1, \ldots. \tag{11.2.8}$$

Recall the exponential power series, which we shall use in the three different forms given below:

$$e^x = \sum_{k=0}^{\infty} \frac{x^k}{k!} = \sum_{k=1}^{\infty} \frac{x^{k-1}}{(k-1)!} = \sum_{k=2}^{\infty} \frac{x^{k-2}}{(k-2)!}.$$

We note first of all from this that

$$\sum_{k=0}^{\infty} \mathbb{P}\{N(t) - N(s) = k\} = e^{-\lambda(t-s)} \sum_{k=0}^{\infty} \frac{\lambda^k (t-s)^k}{k!} = e^{-\lambda(t-s)} \cdot e^{\lambda(t-s)} = 1,$$

as we would expect. We next compute the expected increment

$$\begin{aligned}
\mathbb{E}\big[N(t) - N(s)\big] &= \sum_{k=0}^{\infty} k \frac{\lambda^k (t-s)^k}{k!} e^{-\lambda(t-s)} \\
&= \lambda(t-s) e^{-\lambda(t-s)} \sum_{k=1}^{\infty} \frac{\lambda^{k-1}(t-s)^{k-1}}{(k-1)!} \\
&= \lambda(t-s) \cdot e^{-\lambda(t-s)} \cdot e^{\lambda(t-s)} \\
&= \lambda(t-s). \qquad (11.2.9)
\end{aligned}$$

This is consistent with our observation at the end of Subsection 11.2.2 that jumps are arriving at an average rate of λ per unit time. Therefore, the average number of jumps between times s and t is $\mathbb{E}\big[N(t) - N(s)\big] = \lambda(t-s)$.

Finally, we compute the second moment of the increment

$$\begin{aligned}
\mathbb{E}\big[\big(N(t) - N(s)\big)^2\big] &= \sum_{k=0}^{\infty} k^2 \frac{\lambda^k (t-s)^k}{k!} e^{-\lambda(t-s)} \\
&= e^{-\lambda(t-s)} \sum_{k=1}^{\infty} (k-1+1) \frac{\lambda^k (t-s)^k}{(k-1)!} \\
&= e^{-\lambda(t-s)} \sum_{k=2}^{\infty} \frac{\lambda^k (t-s)^k}{(k-2)!} + e^{-\lambda(t-s)} \sum_{k=1}^{\infty} \frac{\lambda^k (t-s)^k}{(k-1)!} \\
&= \lambda^2 (t-s)^2 e^{-\lambda(t-s)} \sum_{k=2}^{\infty} \frac{\lambda^{k-2}(t-s)^{k-2}}{(k-2)!} \\
&\qquad + \lambda(t-s) e^{-\lambda(t-s)} \sum_{k=1}^{\infty} \frac{\lambda^{k-1}(t-s)^{k-1}}{(k-1)!} \\
&= \lambda^2 (t-s)^2 + \lambda(t-s).
\end{aligned}$$

This implies

$$\begin{aligned}
\mathrm{Var}\big[N(t) - N(s)\big] &= \mathbb{E}\big[\big(N(t) - N(s)\big)^2\big] - \big(\mathbb{E}\big[N(t) - N(s)\big]\big)^2 \\
&= \lambda^2 (t-s)^2 + \lambda(t-s) - \lambda^2 (t-s)^2 \\
&= \lambda(t-s); \qquad (11.2.10)
\end{aligned}$$

the variance is the same as the mean.

11.2.5 Martingale Property

Theorem 11.2.4. *Let $N(t)$ be a Poisson process with intensity λ. We define the* compensated Poisson process *(see Figure 11.2.2)*

$$M(t) = N(t) - \lambda t.$$

Then $M(t)$ is a martingale.

PROOF: Let $0 \leq s < t$ be given. Because $N(t) - N(s)$ is independent of $\mathcal{F}(s)$ and has expected value $\lambda(t-s)$, we have

$$\begin{aligned}\mathbb{E}\big[M(t)\big|\mathcal{F}(s)\big] &= \mathbb{E}\big[M(t) - M(s)\big|\mathcal{F}(s)\big] + \mathbb{E}\big[M(s)\big|\mathcal{F}(s)\big] \\ &= \mathbb{E}\big[N(t) - N(s) - \lambda(t-s)\big|\mathcal{F}(s)\big] + M(s) \\ &= \mathbb{E}\big[N(t) - N(s)\big] - \lambda(t-s) + M(s) \\ &= M(s).\end{aligned}$$

□

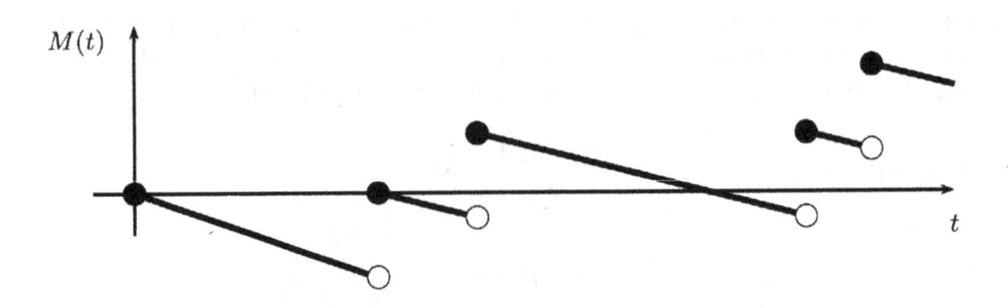

Fig. 11.2.2. One path of a compensated Poisson process.

11.3 Compound Poisson Process

When a Poisson process or a compensated Poisson process jumps, it jumps up one unit. For models of financial markets, we need to allow the jump size to be random. We introduce random jump sizes in this section.

11.3.1 Construction of a Compound Poisson Process

Let $N(t)$ be a Poisson process with intensity λ, and let $Y_1, Y_2, \ldots$ be a sequence of identically distributed random variables with mean $\beta = \mathbb{E}Y_i$. We assume the random variables $Y_1, Y_2, \ldots$ are independent of one another and also independent of the Poisson process $N(t)$. We define the *compound Poisson process*

$$Q(t) = \sum_{i=1}^{N(t)} Y_i, \quad t \geq 0. \tag{11.3.1}$$

The jumps in $Q(t)$ occur at the same times as the jumps in $N(t)$, but whereas the jumps in $N(t)$ are always of size 1, the jumps in $Q(t)$ are of random size. The first jump is of size Y_1, the second of size Y_2, etc. Figure 11.3.1 shows one path of a compound Poisson process.

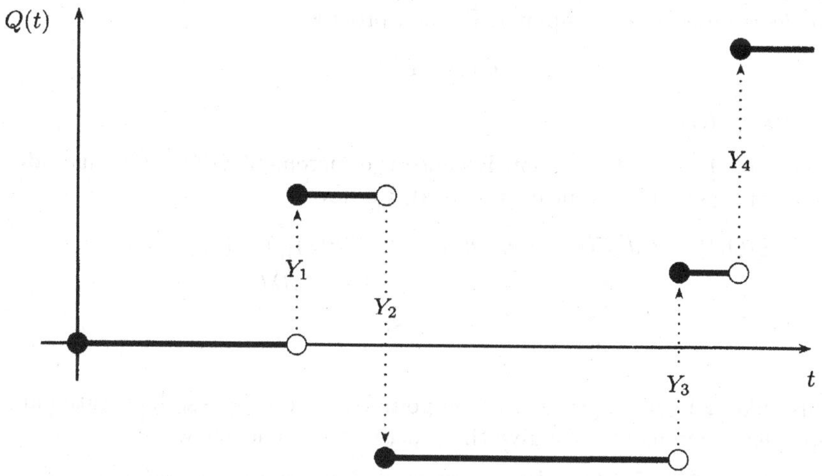

Fig. 11.3.1. One path of a compound Poisson process.

Like the simple Poisson process $N(t)$, the increments of the compound Poisson process $Q(t)$ are independent. In particular, for $0 \leq s < t$,

$$Q(s) = \sum_{i=1}^{N(s)} Y_i,$$

which sums up the first $N(s)$ jumps, and

$$Q(t) - Q(s) = \sum_{i=N(s)+1}^{N(t)} Y_i,$$

which sums up jumps $N(s)+1$ to $N(t)$, are independent. Moreover, $Q(t)-Q(s)$ has the same distribution as $Q(t-s)$ because $N(t) - N(s)$ has the same distribution as $N(t-s)$.

The mean of the compound Poisson process is

$$\begin{aligned}
\mathbb{E}Q(t) &= \sum_{k=0}^{\infty} \mathbb{E}\Big[\sum_{i=1}^{k} Y_i \Big| N(t) = k\Big] \mathbb{P}\{N(t) = k\} \\
&= \sum_{k=0}^{\infty} \beta k \frac{(\lambda t)^k}{k!} e^{-\lambda t} = \beta \lambda t e^{-\lambda t} \sum_{k=1}^{\infty} \frac{(\lambda t)^{k-1}}{(k-1)!} = \beta \lambda t.
\end{aligned}$$

On average, there are λt jumps in the time interval $[0, t]$, the average jump size is β, and the number of jumps is independent of the size of the jumps. Hence, $\mathbb{E}Q(t)$ is the product $\beta \lambda t$.

Theorem 11.3.1. *Let $Q(t)$ be the compound Poisson process defined above. Then the* compensated compound Poisson process

$$Q(t) - \beta\lambda t$$

is a martingale.

PROOF: Let $0 \leq s < t$ be given. Because the increment $Q(t) - Q(s)$ is independent of $\mathcal{F}(s)$ and has mean $\beta\lambda(t-s)$, we have

$$\begin{aligned}\mathbb{E}\big[Q(t) - \beta\lambda t \big| \mathcal{F}(s)\big] &= \mathbb{E}\big[Q(t) - Q(s) \big| \mathcal{F}(s)\big] + Q(s) - \beta\lambda t \\ &= \beta\lambda(t-s) + Q(s) - \beta\lambda t \\ &= Q(s) - \beta\lambda s. \end{aligned}$$

□

Just like a Poisson process, a compound Poisson process has stationary independent increments. We give the precise statement below.

Theorem 11.3.2. *Let $Q(t)$ be a compound Poisson process and let $0 = t_0 < t_1 < \cdots < t_n$ be given. The increments*

$$Q(t_1) - Q(t_0), Q(t_2) - Q(t_1), \dots, Q(t_n) - Q(t_{n-1}),$$

are independent and stationary. In particular, the distribution of $Q(t_j) - Q(t_{j-1})$ is the same as the distribution of $Q(t_j - t_{j-1})$.

11.3.2 Moment-Generating Function

In Theorem 11.3.2, we did not write an explicit formula for the distribution of $Q(t_j - t_{j-1})$ because the formula for the density or probability mass function of this random variable is quite complicated. However, the formula for its moment-generating function is simple. For this reason, we use moment generating functions rather than densities or probability mass functions in much of what follows.

Let $Q(t)$ be the compound Poisson process defined by (11.3.1). Denote the moment-generating function of the random variable Y_i by

$$\varphi_Y(u) = \mathbb{E}e^{uY_i}.$$

This does not depend on the index i because $Y_1, Y_2, \dots$ all have the same distribution. The moment generating function for the compound Poisson process $Q(t)$ is

$$\begin{aligned}\varphi_{Q(t)}(u) &= \mathbb{E}e^{uQ(t)} \\ &= \mathbb{E}\exp\left\{u\sum_{i=1}^{N(t)} Y_i\right\} \\ &= \mathbb{P}\{N(t)=0\} + \sum_{k=1}^{\infty}\mathbb{E}\exp\left\{u\sum_{i=1}^{k} Y_i \Big| N(t)=k\right\}\mathbb{P}\{N(t)=k\}\end{aligned}$$

$$
\begin{aligned}
&= \mathbb{P}\{N(t)=0\} + \sum_{k=1}^{\infty} \mathbb{E}\exp\Big\{u\sum_{i=1}^{k} Y_i\Big\}\mathbb{P}\{N(t)=k\} \\
&= e^{-\lambda t} + \sum_{k=1}^{\infty} \mathbb{E}e^{uY_1}\mathbb{E}e^{uY_2}\cdots\mathbb{E}e^{uY_k}\frac{(\lambda t)^k}{k!}e^{-\lambda t} \\
&= e^{-\lambda t} + e^{-\lambda t}\sum_{k=1}^{\infty}\frac{\big(\varphi_Y(u)\lambda t\big)^k}{k!} \\
&= e^{-\lambda t}\sum_{k=0}^{\infty}\frac{\big(\varphi_Y(u)\lambda t\big)^k}{k!} \\
&= \exp\big\{\lambda t\big(\varphi_Y(u)-1\big)\big\}. \qquad (11.3.2)
\end{aligned}
$$

If the random variables Y_i are not really random but rather always take the constant value y, then the compound Poisson process $Q(t)$ is actually $yN(t)$ and $\varphi_Y(u) = e^{uy}$. It follows that y times a Poisson process has the moment-generating function

$$\varphi_{yN(t)}(u) = \mathbb{E}e^{uyN(t)} = \exp\{\lambda t(e^{uy}-1)\}. \qquad (11.3.3)$$

When $y = 1$, we have the Poisson process, whose moment-generating function is thus

$$\varphi_{N(t)}(u) = \mathbb{E}e^{uN(t)} = \exp\{\lambda t(e^{u}-1)\}. \qquad (11.3.4)$$

Finally, consider the case when Y_i takes one of finitely many possible nonzero values $y_1, y_2, \ldots, y_M$, with $p(y_m) = \mathbb{P}\{Y_i = y_m\}$ so that $p(y_m) > 0$ for every m and $\sum_{m=1}^{M} p(y_m) = 1$. Then $\varphi_Y(u) = \sum_{m=1}^{M} p(y_m)e^{uy_m}$. It follows from (11.3.2) that

$$
\begin{aligned}
\varphi_{Q(t)}(u) &= \exp\Big\{\lambda t\Big(\sum_{m=1}^{M} p(y_m)e^{uy_m} - 1\Big)\Big\} \\
&= \exp\Big\{\lambda t\sum_{m=1}^{M} p(y_m)\big(e^{uy_m}-1\big)\Big\} \\
&= \exp\{\lambda p(y_1)t(e^{uy_1}-1)\}\exp\{\lambda p(y_2)t(e^{uy_2}-1)\}\cdots \\
&\qquad \cdots\exp\{\lambda p(y_M)t(e^{uy_M}-1)\}. \qquad (11.3.5)
\end{aligned}
$$

This last expression is the product of the moment generating-functions for M scaled Poisson processes, the mth process having intensity $\lambda p(y_m)$ and jump size y_m (see (11.3.3)). This observation leads to the following theorem.

Theorem 11.3.3 (Decomposition of a compound Poisson process). *Let $y_1, y_2, \ldots, y_M$ be a finite set of nonzero numbers, and let $p(y_1), p(y_2), \ldots, p(y_M)$ be positive numbers that sum to 1. Let $\lambda > 0$ be given, and let $\overline{N}_1(t), \overline{N}_2(t), \ldots, \overline{N}_M(t)$ be independent Poisson processes, each $\overline{N}_m(t)$ having intensity $\lambda p(y_m)$. Define*

$$\overline{Q}(t) = \sum_{m=1}^{M} y_m \overline{N}_m(t), \quad t \geq 0. \tag{11.3.6}$$

Then $\overline{Q}(t)$ is a compound Poisson process. In particular, if $\overline{Y}_1$ is the size of the first jump of $\overline{Q}(t)$, $\overline{Y}_2$ is the size of the second jump, etc., and

$$\overline{N}(t) = \sum_{m=1}^{M} \overline{N}_m(t), \quad t \geq 0,$$

is the total number of jumps on the time interval $(0,t]$, then $\overline{N}(t)$ is a Poisson process with intensity λ, the random variables $\overline{Y}_1, \overline{Y}_2, \ldots$ are independent with $\mathbb{P}\{\overline{Y}_i = y_m\} = p(y_m)$ for $m = 1, \ldots, M$, the random variables $\overline{Y}_1, \ldots, \overline{Y}_M$ are independent of $\overline{N}(t)$, and

$$\overline{Q}(t) = \sum_{i=0}^{\overline{N}(t)} \overline{Y}_i, \quad t \geq 0.$$

OUTLINE OF PROOF: According to (11.3.3), for each m, the characteristic function of $y_m \overline{N}_m(t)$ is

$$\varphi_{y_m \overline{N}_m(t)}(u) = \exp\{\lambda p(y_m) t (e^{u y_m} - 1)\}.$$

With $\overline{Q}(t)$ defined by (11.3.6), we use the fact that $\overline{N}_1(t), \overline{N}_2(t), \ldots, \overline{N}_M(t)$ are independent of one another to write

$$\begin{aligned}
\varphi_{\overline{Q}(t)}(u) &= \mathbb{E} \exp\left\{ u \sum_{m=1}^{M} y_m \overline{N}_m(t) \right\} \\
&= \mathbb{E} e^{u y_1 \overline{N}_1(t)} \mathbb{E} e^{u y_2 \overline{N}_2(t)} \cdots \mathbb{E} e^{u y_M \overline{N}_M(t)} \\
&= \varphi_{y_1 \overline{N}_1(t)}(u) \varphi_{y_2 \overline{N}_2(t)}(u) \cdots \varphi_{y_M \overline{N}_M(t)}(u) \\
&= \exp\{\lambda p(y_1) t (e^{u y_1} - 1)\} \exp\{\lambda p(y_2) t (e^{u y_2} - 1)\} \cdots \\
&\qquad \cdots \exp\{\lambda p(y_M) t (e^{u y_M} - 1)\},
\end{aligned}$$

which is the right-hand side of (11.3.5). It follows that the random variable $\overline{Q}(t)$ of (11.3.6) has the same distribution as the random variable $Q(t)$ appearing on the left-hand side of (11.3.5). With a bit more work, one can show that the distribution of the whole path of $\overline{Q}$ defined by (11.3.6) agrees with the distribution of the whole path of the process Q appearing on the left-hand side of (11.3.5).

Recall that the process Q appearing on the left-hand side of (11.3.5) is the compound Poisson process defined by (11.3.1). For this process $N(t)$, the total number of jumps by time t is Poisson with intensity λ, and the sizes of the jumps, $Y_1, Y_2, \ldots$, are identically distributed random variables, independent of one another and independent of $N(t)$, and with $\mathbb{P}\{Y_i = y_m\} = p(y_m)$ for

$m = 1, \ldots, M$. Because the processes Q and $\overline{Q}$ have the same distribution, these statements must also be true for the total number of jumps and the sizes of the jumps of the process $\overline{Q}$ of (11.3.6), which is what the theorem asserts. □

The substance of Theorem 11.3.3 is that there are two equivalent ways of regarding a compound Poisson process that has only finitely many possible jump sizes. It can be thought of as a single Poisson process in which the size-one jumps are replaced by jumps of random size. Alternatively, it can be regarded as a sum of independent Poisson processes in each of which the size-one jumps are replaced by jumps of a fixed size. We restate Theorem 11.3.3 in a way designed to make this more clear.

Corollary 11.3.4. *Let $y_1, \ldots, y_M$ be a finite set of nonzero numbers, and let $p(y_1), \ldots, p(y_M)$ be positive numbers that sum to 1. Let $Y_1, Y_2, \ldots$ be a sequence of independent, identically distributed random variables with $\mathbb{P}\{Y_i = y_m\} = p(y_m)$, $m = 1, \ldots, M$. Let $N(t)$ be a Poisson process and define the compound Poisson process*

$$Q(t) = \sum_{i=1}^{N(t)} Y_i.$$

For $m = 1, \ldots, M$, let $N_m(t)$ denote the number of jumps in Q of size y_m up to and including time t. Then

$$N(t) = \sum_{m=1}^{M} N_m(t) \text{ and } Q(t) = \sum_{m=1}^{M} y_m N_m(t).$$

The processes $N_1, \ldots, N_M$ defined this way are independent Poisson processes, and each N_m has intensity $\lambda p(y_m)$.

11.4 Jump Processes and Their Integrals

In this section, we introduce the stochastic integral when the integrator is a process with jumps, and we develop properties of this integral. We shall have a Brownian motion and Poisson and compound Poisson processes. There will always be a single filtratoin associated with all of them, in the sense of the following definition.

Definition 11.4.1. *Let $(\Omega, \mathcal{F}, \mathbb{P})$ be a probability space, and let $\mathcal{F}(t)$, $t \geq 0$, be a filtration on this space. We say that a Brownian motion W is a* Brownian motion relative to this filtration *if $W(t)$ is $\mathcal{F}(t)$-measurable for every t and for every $u > t$ the increment $W(u) - W(t)$ is independent of $\mathcal{F}(t)$. Similarly, we say that a Poisson process N is a* Poisson process relative to this filtration *if $N(t)$ is $\mathcal{F}(t)$-measurable for every t and for every $u > t$ the increment*

$N(u)-N(t)$ is independent of $\mathcal{F}(t)$. Finally, we say that a compound Poisson process Q is a compound Poisson process relaative to this filtration *if $Q(t)$ is $\mathcal{F}(t)$-measurable for every t and for every $u>t$ the increment $Q(u)-Q(t)$ is independent of $\mathcal{F}(t)$.*

11.4.1 Jump Processes

We wish to define the stochastic integral

$$\int_0^t \Phi(s)\,dX(s),$$

where the integrator X can have jumps. Let $(\Omega,\mathcal{F},\mathbb{P})$ be a probability space on which is given a filtration $\mathcal{F}(t)$, $t\geq 0$. All processes will be adapted to this filtration. Furthermore, the integrators we consider in this section will be right-continuous and of the form

$$X(t)=X(0)+I(t)+R(t)+J(t). \tag{11.4.1}$$

In (11.4.1), $X(0)$ is a nonrandom initial condition. The process

$$I(t)=\int_0^t \Gamma(s)\,dW(s) \tag{11.4.2}$$

is an *Itô integral* of an adapted process $\Gamma(s)$ with respect to a Brownian motion relative to the filtration. We shall call $I(t)$ the *Itô integral part of X*. The process $R(t)$ in (11.4.1) is a *Riemann integral*[1]

$$R(t)=\int_0^t \Theta(s)\,ds \tag{11.4.3}$$

for some adapted process $\Theta(t)$. We shall call $R(t)$ the *Riemann integral part of X*. The *continuous part* of $X(t)$ is defined to be

$$X^c(t)=X(0)+I(t)+R(t)=X(0)+\int_0^t \Gamma(s)\,dW(s)+\int_0^t \Theta(s)\,ds.$$

The quadratic variation of this process is

$$[X^c,X^c](t)=\int_0^t \Gamma^2(s)\,ds,$$

an equation that we write in differential form as

[1] One usually takes this to be a Lebesgue integral with respect to dt, but for all the cases we consider, the Riemann integral is defined and agrees with the Lebesgue integral.

$$dX^c(t)\,dX^c(t) = \Gamma^2(t)\,dt.$$

Finally, in (11.4.1), $J(t)$ is an adapted, right-continuous *pure jump process* with $J(0) = 0$. By *right-continuous*, we mean that $J(t) = \lim_{s \downarrow t} J(s)$ for all $t \geq 0$. The *left-continuous* version of such a process will be denoted $J(t-)$. In other words, if J has a jump at time t, then $J(t)$ is the value of J immediately after the jump, and $J(t-)$ is its value immediately before the jump. We assume that J does not jump at time zero, has only finitely many jumps on each finite time interval $(0, T]$, and is constant between jumps. The constancy between jumps is what justifies calling $J(t)$ a *pure* jump process. A Poisson process and a compound Poisson process have this property. A compensated Poisson process does not because it decreases between jumps. We shall call $J(t)$ the *pure jump part of* X.

Definition 11.4.2. *A process $X(t)$ of the form (11.4.1), with Itô integral part $I(t)$, Riemann integral part $R(t)$, and pure jump part $J(t)$ as described above, will be called a* jump process. *The* continuous part *of this process is $X^c(t) = X(0) + I(t) + R(t)$.*

A jump process in this book is not the most general possible because we permit only finitely many jumps in finite time. For many applications, these processes are sufficient. Furthermore, the stochastic calculus for these processes gives a good indication of how the stochastic calculus works for the more general case.

A jump process $X(t)$ is right-continuous and adapted. Because both $I(t)$ and $R(t)$ are continuous, the left-continuous version of $X(t)$ is

$$X(t-) = X(0) + I(t) + R(t) + J(t-).$$

The jump size of X at time t is denoted

$$\Delta X(t) = X(t) - X(t-).$$

If X is continuous at t, then $\Delta X(t) = 0$. If X has a jump at time t, then $\Delta X(t)$ is the size of this jump, which is also $\Delta J(t) = J(t) - J(t-)$, the size of the jump in J. Whenever $X(0-)$ appears in the formulas below, we mean it to be $X(0)$. In particular, $\Delta X(0) = 0$; there is no jump at time zero.

Definition 11.4.3. *Let $X(t)$ be a jump process of the form (11.4.1)–(11.4.3) and let $\Phi(s)$ be an adapted process. The* stochastic integral *of Φ with respect to X is defined to be*

$$\int_0^t \Phi(s)\,dX(s) = \int_0^t \Phi(s)\Gamma(s)\,dW(s) + \int_0^t \Phi(s)\Theta(s)\,ds + \sum_{0<s\leq t} \Phi(s)\Delta J(s). \tag{11.4.4}$$

In differential notation,

$$\begin{aligned}\Phi(t)dX(t) &= \Phi(t)\,dI(t) + \Phi(t)\,dR(t) + \Phi(t)\,dJ(t)\\ &= \Phi(t)\,dX^c(t) + \Phi(t)\,dJ(t),\end{aligned}$$

where

$$\Phi(t)\,dI(t) = \Phi(t)\Gamma(t)\,dW(t),\ \ \Phi(t)\,dR(t) = \Phi(t)\Theta(t)\,dt,$$
$$\Phi(t)\,dX^c(t) = \Phi(t)\Gamma(t)\,dW(t) + \Phi(t)\Theta(t)\,dt.$$

Example 11.4.4. Let $X(t) = M(t) = N(t) - \lambda t$, where $N(t)$ is a Poisson process with intensity λ so that $M(t)$ is the compensated Poisson process of Theorem 11.2.4. In the terminology of Definition 11.4.2, $I(t) = 0$, $X^c(t) = R(t) = -\lambda t$, and $J(t) = N(t)$. Let $\Phi(s) = \Delta N(s)$ (i.e., $\Phi(s)$ is 1 if N has a jump at time s, and $\Phi(s)$ is zero otherwise). For $s \in [0,t]$, $\Phi(s)$ is zero except for finitely many values of s, and thus

$$\int_0^t \Phi(s)\,dX^c(s) = \int_0^t \Phi(s)\,dR(s) = -\lambda\int_0^t \Phi(s)\,ds = 0.$$

However,

$$\int_0^t \Phi(s)\,dN(s) = \sum_{0<s\leq t} \left(\Delta N(s)\right)^2 = N(t).$$

Therefore,

$$\int_0^t \Phi(s)\,dM(s) = -\lambda\int_0^t \Phi(s)\,ds + \int_0^t \Phi(s)\,dN(s) = N(t). \tag{11.4.5}$$

□

For Brownian motion $W(t)$, we defined the stochastic integral

$$I(t) = \int_0^t \Gamma(s)\,dW(s)$$

in a way that caused $I(t)$ to be a martingale. To define the stochastic integral, we approximated the integrand $\Gamma(s)$ by simple integrands $\Gamma_n(s)$, wrote down a formula for

$$I_n(t) = \int_0^t \Gamma_n(s)\,dW(s),$$

and verified that, for each n, $I_n(t)$ is a martingale. We defined $I(t)$ as the limit of $I_n(t)$ as $n \to \infty$ and, because it is the limit of martingales, $I(t)$ also is a martingale. The only conditions we needed on $\Gamma(s)$ for this construction were that it be adapted and that it satisfy the technical condition $\mathbb{E}\int_0^t \Gamma^2(s)\,ds < \infty$ for every $t > 0$.

This construction makes sense for finance because we ultimately replace $\Gamma(s)$ by a position in an asset and replace $W(s)$ by the price of that asset. If the asset price is a martingale (i.e., it is pure volatility with no underlying trend), then the gain we make from investing in the asset should also be a martingale. The stochastic integral is this gain.

In the context of processes that can jump, we still want the stochastic integral with respect to a martingale to be a martingale. However, we see in Example 11.4.4 that this is not always the case. The integrator $M(t)$ in that example is a martingale (see Theorem 11.2.4), but the integral $N(t)$ in (11.4.5) is not because it goes up but cannot go down.

An agent who invests in the compensated Poisson process $M(t)$ by choosing his position according to the formula $\Phi(s) = \Delta N(s)$ has created an arbitrage. To do this, he is holding a zero position at all times except the jump times of $N(s)$, which are also the jump times of $M(s)$, at which times he holds a position one. Because the jumps in $M(s)$ are always up and our investor holds a long position at exactly the jump times, he will reap the upside gain from all these jumps and have no possibility of loss.

In reality, the portfolio process $\Phi(s) = \Delta N(s)$ cannot be implemented because investors must take positions before jumps occur. No one without insider information can arrange consistently to take a position exactly at the jump times. However, $\Phi(s)$ depends only on the path of the underlying process M up to and including at time s and does not depend on the future of the path. That is the definition of adapted we used when constructing stochastic integrals with respect to Brownian motion. Here we see that it is not enough to require the integrand to be adapted. A mathematically convenient way of formulating the extra condition is to insist that our integrands be *left-continuous*. That rules out $\Phi(s) = \Delta N(s)$. In the time interval between jumps, this process is zero, and a left-continuous process that is zero between jumps must also be zero at the jump times.

We give the following theorem without proof.

Theorem 11.4.5. *Assume that the jump process $X(s)$ of (11.4.1)–(11.4.3) is a martingale, the integrand $\Phi(s)$ is* left-continuous *and adapted, and*

$$\mathbb{E}\int_0^t \Gamma^2(s)\Phi^2(s)\,ds < \infty \text{ for all } t \geq 0.$$

Then the stochastic integral $\int_0^t \Phi(s)\,dX(s)$ is also a martingale.

The mathematical literature on integration with respect to jump processes gives a slightly more general version of Theorem 11.4.5 in which the integrand is required only to be *predictable*. Roughly speaking, such processes are those that can be gotten as the limit of left-continuous processes. We shall not need this more general concept.

Note that although we require the integrand $\Phi(s)$ to be left-continuous in Theorem 11.4.5, the integrator $X(t)$ is always taken to be right-continuous, and so the integral $\int_0^t \Phi(s)\,dX(s)$ will be right-continuous in the upper limit of integration t. The integral jumps whenever X jumps and Φ is simultaneously not zero. The value of the integral at time t includes the jump at time t if there is a jump; see (11.4.4).

Example 11.4.6. Let $N(t)$ be a Poisson process with intensity λ, let $M(t) = N(t) - \lambda t$ be the compensated Poisson process, and let

$$\Phi(s) = \mathbb{I}_{[0,S_1]}(s)$$

be 1 up to and including the time of the first jump and zero thereafter. Note that Φ is left-continuous. We have

$$\begin{aligned}\int_0^t \Phi(s)\,dM(s) &= \begin{cases} -\lambda t, & 0 \le t < S_1, \\ 1 - \lambda S_1, & t \ge S_1 \end{cases} \\ &= \mathbb{I}_{[S_1,\infty)}(t) - \lambda(t \wedge S_1). \end{aligned} \tag{11.4.6}$$

The notation $t \wedge S_1$ in (11.4.6) denotes the minimum of t and S_1. See Figure 11.4.1.

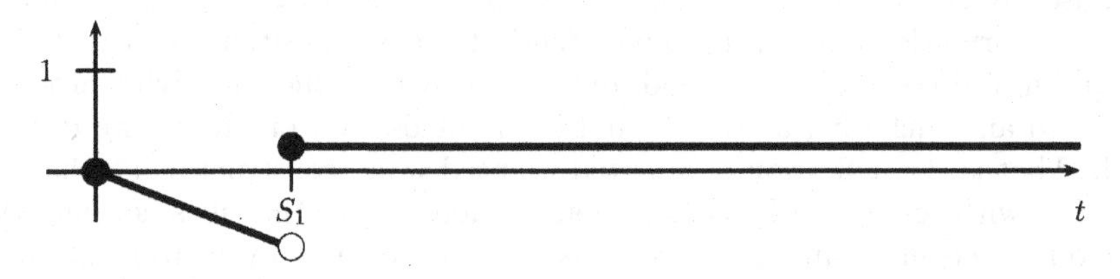

Fig. 11.4.1. $\mathbb{I}_{[S_1,\infty)}(t) - \lambda(t \wedge S_1)$.

We verify the martingale property for the process $\mathbb{I}_{[S_1,\infty)}(t) - \lambda(t \wedge S_1)$ by direct computation. For $0 \le s < t$, we have

$$\mathbb{E}\big[\mathbb{I}_{[S_1,\infty)}(t) - \lambda(t \wedge S_1)\big|\mathcal{F}(s)\big] = \mathbb{P}\{S_1 \le t|\mathcal{F}(s)\} - \lambda\mathbb{E}\big[t \wedge S_1\big|\mathcal{F}(s)\big]. \tag{11.4.7}$$

If $S_1 \le s$, then at time s we know the value of S_1 and the conditional expectations above give us the random variables being estimated. In particular, the right-hand side of (11.4.7) is $1 - \lambda S_1 = \mathbb{I}_{[S_1,\infty)}(s) - \lambda(s \wedge S_1)$, and the martingale property is satisfied. On the other hand, if $S_1 > s$, then

$$\mathbb{P}\{S_1 \le t|\mathcal{F}(s)\} = 1 - \mathbb{P}\{S_1 > t|S_1 > s\} = 1 - e^{-\lambda(t-s)}, \tag{11.4.8}$$

where we have used the fact that S_1 is exponentially distributed and used the memorylessness (11.2.3) of exponential random variables. In fact, the memorylessness says that, conditioned on $S_1 > s$, the density of S_1 is

$$-\frac{\partial}{\partial u}\mathbb{P}\{S_1 > u|S_1 > s\} = -\frac{\partial}{\partial u}e^{-\lambda(u-s)} = \lambda e^{-\lambda(u-s)}, \quad u > s.$$

It follows that, when $S_1 > s$,

$$\begin{aligned}\lambda\mathbb{E}\big[t\wedge S_1\big|\mathcal{F}(s)\big] &= \lambda\mathbb{E}\big[t\wedge S_1\big|S_1>s\big]\\&=\lambda^2\int_s^\infty (t\wedge u)e^{-\lambda(u-s)}\,du\\&=\lambda^2\int_s^t ue^{-\lambda(u-s)}du+\lambda^2\int_t^\infty te^{-\lambda(u-s)}du\\&=-\lambda ue^{-\lambda(u-s)}\Big|_{u=s}^{u=t}+\lambda\int_s^t e^{-\lambda(u-s)}du-\lambda te^{-\lambda(u-s)}\Big|_{u=t}^{u=\infty}\\&=\lambda s-\lambda te^{-\lambda(t-s)}-e^{-\lambda(u-s)}\Big|_{u=s}^{u=t}+\lambda te^{-\lambda(t-s)}\\&=\lambda s-e^{-\lambda(t-s)}+1.\end{aligned}\tag{11.4.9}$$

Subtracting (11.4.9) from (11.4.8), we obtain in the case $S_1 > s$ that

$$\mathbb{E}\big[\mathbb{I}_{[S_1,\infty)}(t)-\lambda(t\wedge S_1)\big|\mathcal{F}(s)\big]=-\lambda s=\mathbb{I}_{[S_1,\infty)}(s)-\lambda(s\wedge S_1).$$

This completes the verification of the martingale property for the stochastic integral in (11.4.6).

Note that if we had taken the integrand in (11.4.6) to be $\mathbb{I}_{[0,S_1)}(t)$, which is right-continuous rather than left-continuous at S_1, then we would have gotten

$$\int_0^t \mathbb{I}_{[0,S_1)}(u)dM(u)=-\lambda(t\wedge S_1).\tag{11.4.10}$$

According to (11.4.9) with $s=0$,

$$\mathbb{E}\big[-\lambda(t\wedge S_1)\big]=e^{-\lambda t}-1,$$

which is strictly decreasing in t. Consequently, the integral (11.4.10) obtained from the right-continuous integrand $\mathbb{I}_{[0,S_1)}(t)$ is not a martingale. □

11.4.2 Quadratic Variation

In order to write down the Itô-Doeblin formula for processes with jumps, we need to discuss *quadratic variation.* Let $X(t)$ be a jump process. To compute the quadratic variation of X on $[0,T]$, we choose $0=t_0<t_1<t_2<\cdots<t_n=T$, denote the set of these times by $\Pi=\{t_0,t_1,\ldots,t_n\}$, denote the length of the longest subinterval by $\|\Pi\|=\max_j(t_{j+1}-t_j)$, and define

$$Q_\Pi(X)=\sum_{j=0}^{n-1}\big(X(t_{j+1})-X(t_j)\big)^2.$$

The *quadratic variation* of X on $[0,T]$ is defined to be

$$[X,X](T)=\lim_{\|\Pi\|\to 0}Q_\Pi(X),$$

where of course as $\|\Pi\| \to 0$ the number of points in Π must approach infinity.

In general, $[X,X](T)$ can be random (i.e., can depend on the path of X). However, in the case of Brownian motion, we know that $[W,W](T) = T$ does not depend on the path. In the case of an Itô integral $I(T) = \int_0^T \Gamma(s)dW(s)$ with respect to Brownian motion, $[I,I](T) = \int_0^T \Gamma^2(s)ds$ can depend on the path because $\Gamma(s)$ can depend on the path.

We will also need the concept of *cross variation*. Let $X_1(t)$ and $X_2(t)$ be jump processes. We define

$$C_\Pi(X_1,X_2) = \sum_{j=0}^{n-1} \big(X_1(t_{j+1}) - X_1(t_j)\big)\big(X_2(t_{j+1}) - X_2(t_j)\big)$$

and

$$[X_1,X_2](T) = \lim_{\|\Pi\|\to 0} C_\Pi(X_1,X_2).$$

Theorem 11.4.7. *Let $X_1(t) = X_1(0) + I_1(t) + R_1(t) + J_1(t)$ be a jump process, where $I_1(t) = \int_0^t \Gamma_1(s)\,dW(s)$, $R_1(t) = \int_0^t \Theta_1(s)\,ds$, and $J_1(t)$ is a right-continuous pure jump process. Then $X_1^c(t) = X_1(0) + I_1(t) + R_1(t)$ and*

$$[X_1,X_1](T) = [X_1^c,X_1^c](T) + [J_1,J_1](T) = \int_0^T \Gamma_1^2(s)\,ds + \sum_{0<s\le T} \big(\Delta J_1(s)\big)^2. \tag{11.4.11}$$

Let $X_2(t) = X_2(0) + I_2(t) + R_2(t) + J_2(t)$ be another jump process, where $I_2(t) = \int_0^t \Gamma_2(s)\,dW(s)$, $R_2(t) = \int_0^t \Theta_2(s)\,ds$, and $J_2(t)$ is a right-continuous pure jump process. Then $X_2^c(t) = X_2(0) + I_2(t) + R_2(t)$, and

$$\begin{aligned}[X_1,X_2](T) &= [X_1^c,X_2^c](T) + [J_1,J_2](T)\\ &= \int_0^T \Gamma_1(s)\Gamma_2(s)\,ds + \sum_{0<s\le T} \Delta J_1(s)\Delta J_2(s). \end{aligned} \tag{11.4.12}$$

PROOF: We only need to prove (11.4.12) since (11.4.11) is the special case of (11.4.12) in which $X_2 = X_1$. We have

$$\begin{aligned} C_\Pi(X_1,X_2) &= \sum_{j=0}^{n-1} \big(X_1(t_{j+1}) - X_1(t_j)\big)\big(X_2(t_{j+1}) - X_2(t_j)\big)\\ &= \sum_{j=0}^{n-1} \big(X_1^c(t_{j+1}) - X^c(t_j) + J_1(t_{j+1}) - J_1(t_j)\big)\\ &\qquad \times \big(X_2^c(t_{j+1}) - X_2^c(t_j) + J_2(t_{j+1}) - J_2(t_j)\big) \end{aligned}$$

$$
\begin{aligned}
&= \sum_{j=0}^{n-1} \big(X_1^c(t_{j+1}) - X_1^c(t_j)\big)\big(X_2^c(t_{j+1}) - X_2^c(t_j)\big) \\
&\quad + \sum_{j=0}^{n-1} \big(X_1^c(t_{j+1}) - X_1^c(t_j)\big)\big(J_2(t_{j+1}) - J_2(t_j)\big) \\
&\quad + \sum_{j=0}^{n-1} \big(J_1(t_{j+1}) - J_1(t_j)\big)\big(X_2^c(t_{j+1}) - X_2^c(t_j)\big) \\
&\quad + \sum_{j=0}^{n-1} \big(J_1(t_{j+1}) - J_1(t_j)\big)\big(J_2(t_{j+1}) - J_2(t_j)\big). \quad (11.4.13)
\end{aligned}
$$

We know from the theory of continuous processes that

$$
\begin{aligned}
\lim_{\|\Pi\|\to 0} \sum_{j=0}^{n-1} \big(X_1^c(t_{j+1}) - X_1^c(t_j)\big)\big(X_2^c(t_{j+1}) - X_2^c(t_j)\big) &= \big[X_1^c, X_2^c\big](T) \\
&= \int_0^T \Gamma_1(s)\Gamma_2(s)\,ds.
\end{aligned}
$$

We shall show that the second and third terms appearing on the right-hand side of (11.4.13) have limit zero as $\|\Pi\| \to 0$, and the fourth term has limit

$$
[J_1, J_2](T) = \sum_{0<s\leq T} \Delta J_1(s)\Delta J_2(s).
$$

We consider the second term on the right-hand side of (11.4.13):

$$
\begin{aligned}
&\left| \sum_{j=0}^{n-1} \big(X_1^c(t_{j+1}) - X_1^c(t_j)\big)\big(J_2(t_{j+1}) - J_2(t_j)\big) \right| \\
&\leq \max_{0\leq j\leq n-1} \big|X_1^c(t_{j+1}) - X_1^c(t_j)\big| \cdot \sum_{j=0}^{n-1} \big|J_2(t_{j+1}) - J_2(t_j)\big| \\
&\leq \max_{0\leq j\leq n-1} \big|X_1^c(t_{j+1}) - X_1^c(t_j)\big| \cdot \sum_{0<s\leq T} |\Delta J_2(s)|.
\end{aligned}
$$

As $\|\Pi\| \to 0$, the factor $\max_{0\leq j\leq n-1} \big|X_1^c(t_{j+1}) - X_1^c(t_j)\big|$ has limit zero, whereas $\sum_{0<s\leq T} |\Delta J_2(s)|$ is a finite number not depending on Π. Hence, the second term on the right-hand side of (11.4.13) has limit zero as $\|\Pi\| \to 0$. Similarly, the third term on the right-hand side of (11.4.13) has limit zero.

Let us fix an arbitrary $\omega \in \Omega$, which fixes the paths of these processes, and choose the time points in Π so close together that there is at most one jump of J_1 in each interval $(t_j, t_{j+1}]$, at most one jump of J_2 in each interval $(t_j, t_{j+1}]$, and if J_1 and J_2 have a jump in the same interval, then these jumps are simultaneous. Let A_1 denote the set of indices j for which $(t_j, t_{j+1}]$ contains a jump of J_1, and let A_2 denote the set of indices j for which $(t_j, t_{j+1}]$ contains a jump of J_2. The fourth term on the right-hand side of (11.4.13) is

$$\begin{aligned}\sum_{j=0}^{n-1} &\big(J_1(t_{j+1}) - J_1(t_j)\big)\big(J_2(t_{j+1}) - J_2(t_j)\big)\\ &= \sum_{j\in A_1\cap A_2} \big(J_1(t_{j+1}) - J_1(t_j)\big)\big(J_2(t_{j+1}) - J_2(t_j)\big)\\ &= \sum_{0<s\le t} \Delta J_1(s)\Delta J_2(s).\end{aligned}$$

This completes the proof. □

Remark 11.4.8. In differential notation, equation (11.4.12) of Theorem 11.4.7 says that if

$$X_1(t) = X_1(0) + X_1^c(t) + J_1(t), \quad X_2(t) = X_2(0) + X_2^c(t) + J_2(t),$$

then

$$dX_1(t)\,dX_2(t) = dX_1^c(t)\,dX_2^c(t) + dJ_1(t)\,dJ_2(t).$$

In particular,

$$dX_1^c(t)\,dJ_2(t) = dX_2^c(t)\,dJ_1(t) = 0;$$

the cross variation between a continuous process and a pure jump process is zero. It follows that the cross variation between a Brownian motion and a Poisson process is zero.

More generally, the cross variation between two processes is zero if one of them is continuous and the other has no Itô integral part. In order to get a nonzero cross variation, both processes must have a dW term or the processes must have simultaneous jumps. This means that the cross variation between a Brownian motion and a compensated Poisson process is also zero. We state this last fact as a corollary. □

Corollary 11.4.9. *Let $W(t)$ be a Brownian motion and $M(t) = N(t) - \lambda t$ be a compensated Poisson process relative to the same filtration $\mathcal{F}(t)$ (Definition 11.4.1). Then*

$$[W, M](t) = 0, \quad t \ge 0.$$

PROOF: In Theorem 11.4.7, take $I_1(t) = W(t)$, $R_1(t) = J_1(t) = 0$ and take $I_2(t) = 0$, $R_2(t) = -\lambda t$, and $J_2(t) = N(t)$. □

We shall see in Corollary 11.5.3 that the equation $[W, M](t) = 0$ implies that W and M are independent, and hence W and N are independent. *A Brownian motion and a Poisson process relative to the same filtration must be independent.*

Corollary 11.4.10. *For $i = 1, 2$, let $X_i(t)$ be an adapted, right-continuous jump process. In other words, $X_i(t) = X_i(0) + I_i(t) + R_i(t) + J_i(t)$, where $I_i(t) = \int_0^t \Gamma_i(s)\,dW(s)$, $R_i(t) = \int_0^t \Theta_i(s)\,ds$, and $J_i(t)$ is a pure jump process. Let $\widetilde{X}_i(0)$ be a constant, let $\Phi_i(s)$ be an adapted process, and set*

$$\widetilde{X}_i(t) = \widetilde{X}_i(0) + \int_0^t \Phi_i(s)\,dX_i(s).$$

By definition,

$$\widetilde{X}_i(t) = \widetilde{X}_i(0) + \widetilde{I}_i(t) + \widetilde{R}_i(t) + \widetilde{J}_i(t),$$

where

$$\widetilde{I}_i(t) = \textstyle\int_0^t \Phi_i(s)\Gamma_i(s)\,dW(s), \quad \widetilde{R}_i(t) = \int_0^t \Phi_i(s)\Theta_i(s)\,ds,$$
$$\widetilde{J}_i(t) = \textstyle\sum_{0<s\le t} \Phi_i(s)\Delta J_i(s).$$

Note that $\widetilde{X}_i(t)$ is a jump process with continuous part $\widetilde{X}_i^c(t) = \widetilde{X}_i(0) + \widetilde{I}_i(t) + \widetilde{R}_i(t)$ and pure jump part $\widetilde{J}_i(t)$. We have

$$\begin{aligned}
&[\widetilde{X}_1, \widetilde{X}_2](t) \\
&= [\widetilde{X}_1^c, \widetilde{X}_2^c](t) + [\widetilde{J}_1, \widetilde{J}_2](t) \\
&= \int_0^t \Phi_1(s)\Phi_2(s)\Gamma_1(s)\Gamma_2(s)\,ds + \sum_{0<s\le t} \Phi_1(s)\Phi_2(s)\Delta J_1(s)\Delta J_2(s) \\
&= \int_0^t \Phi_1(s)\Phi_2(s)\,d[X_1, X_2](s).
\end{aligned}$$

Remark 11.4.11. Corollary 11.4.10 may be rewritten using differential notation. The corollary says that if

$$d\widetilde{X}_1(t) = \Phi_1(t)\,dX_1(t) \text{ and } d\widetilde{X}_2(t) = \Phi_2(t)\,dX_2(t),$$

then

$$d\widetilde{X}_1(t)\,d\widetilde{X}_2(t) = \Phi_1(t)\Phi_2(t)\,dX_1(t)\,dX_2(t).$$

11.5 Stochastic Calculus for Jump Processes

11.5.1 Itô-Doeblin Formula for One Jump Process

For a continuous-path process, the Itô-Doeblin formula is the following. Let

$$X^c(t) = X^c(0) + \int_0^t \Gamma(s)\,dW(s) + \int_0^t \Theta(s)\,ds, \tag{11.5.1}$$

where $\Gamma(s)$ and $\Theta(s)$ are adapted processes. In differential notation, we write

$$dX^c(s) = \Gamma(s)\,dW(s) + \Theta(s)\,ds, \quad dX^c(s)\,dX^c(s) = \Gamma^2(s)\,ds.$$

Let $f(x)$ be a function whose first and second derivatives are defined and continuous. Then

$$\begin{aligned} df\big(X^c(s)\big) &= f'\big(X^c(s)\big)\,dX^c(s) + \frac{1}{2}f''\big(X^c(s)\big)\,dX^c(s)\,dX^c(s) \\ &= f'\big(X^c(s)\big)\Gamma(s)\,dW(s) + f'\big(X^c(s)\big)\Theta(s)\,ds \\ &\quad + \frac{1}{2}f''\big(X^c(s)\big)\Gamma^2(s)\,ds. \end{aligned} \tag{11.5.2}$$

We write this in integral form as

$$\begin{aligned} f\big(X^c(t)\big) = f\big(X^c(0)\big) &+ \int_0^t f'\big(X^c(s)\big)\Gamma(s)\,dW(s) + \int_0^t f'\big(X^c(s)\big)\Theta(s)\,ds \\ &+ \frac{1}{2}\int_0^t f''\big(X^c(s)\big)\Gamma^2(s)\,ds. \end{aligned}$$

We now add a right-continuous pure jump term J into (11.5.1), setting

$$X(t) = X(0) + I(t) + R(t) + J(t),$$

where $I(t) = \int_0^t \Gamma(s)\,dW(s)$ and $R(t) = \int_0^t \Theta(s)\,ds$. As usual, we denote by $X^c(t) = X(0) + I(t) + R(t)$ the continuous part of $X(t)$. Between jumps of J, the analogue of (11.5.2) holds:

$$\begin{aligned} df\big(X(s)\big) &= f'\big(X(s)\big)\,dX(s) + \frac{1}{2}f''\big(X(s)\big)\,dX(s)\,dX(s) \\ &= f'\big(X(s)\big)\Gamma(s)\,dW(s) + f'\big(X(s)\big)\Theta(s)\,ds \\ &\quad + \frac{1}{2}f''\big(X(s)\big)\Gamma^2(s)\,ds \\ &= f'\big(X(s)\big)\,dX^c(s) + \frac{1}{2}f''\big(X(s)\big)\,dX^c(s)\,dX^c(s). \end{aligned} \tag{11.5.3}$$

When there is a jump in X from $X(s-)$ to $X(s)$, there is typically also a jump in $f(X)$ from $f\big(X(s-)\big)$ to $f\big(X(s)\big)$. When we integrate both sides of (11.5.3) from 0 to t, we must add in all the jumps that occur between these two times. This leads to the following theorem.

Theorem 11.5.1 (Itô-Doeblin formula for one jump process). *Let $X(t)$ be a jump process and $f(x)$ a function for which $f'(x)$ and $f''(x)$ are defined and continuous. Then*

$$\begin{aligned} f\big(X(t)\big) = f\big(X(0)\big) &+ \int_0^t f'\big(X(s)\big)\,dX^c(s) + \frac{1}{2}\int_0^t f''\big(X(s)\big)\,dX^c(s)\,dX^c(s) \\ &+ \sum_{0<s\le t}\big[f\big(X(s)\big) - f\big(X(s-)\big)\big]. \end{aligned} \tag{11.5.4}$$

PROOF: Fix $\omega \in \Omega$, which fixes the path of X, and let $0 < \tau_1 < \tau_2 < \cdots < \tau_{n-1} < t$ be the jump times in $[0, t)$ of this path of the process X. We set $\tau_0 = 0$, which is not a jump time, and $\tau_n = t$, which may or may not be a jump time. Whenever $u < v$ are both in the same interval (τ_j, τ_{j+1}), there is no jump between times u and v, and the Itô-Doeblin formula (11.5.3) for continuous processes applies. We thus have

$$f(X(v)) - f(X(u)) = \int_u^v f'(X(s))\, dX^c(s) + \frac{1}{2}\int_u^v f''(X(s))\, dX^c(s)\, dX^c(s).$$

Letting $u \downarrow \tau_j$ and $v \uparrow \tau_{j+1}$ and using the right-continuity of X, we conclude that

$$\begin{aligned} &f(X(\tau_{j+1}-)) - f(X(\tau_j)) \\ &= \int_{\tau_j}^{\tau_{j+1}} f'(X(s))\, dX^c(s) + \frac{1}{2}\int_{\tau_j}^{\tau_{j+1}} f''(X(s))\, dX^c(s)\, dX^c(s). \end{aligned} \tag{11.5.5}$$

(Note here that

$$\lim_{v \uparrow \tau_{j+1}} \int_u^v f'(X(s))\, dX^c(s) = \int_v^{\tau_{j+1}} f'(X(s))\, dX^c(s),$$

but this is not the case if we replace $dX^c(s)$ by $dX(s)$ in this equation. If we made this replacement, the jump in X at time τ_{j+1} would appear on the right-hand side of the equation but not on the left-hand side. It is for this reason that we integrate with respect to $dX^c(s)$ in (11.5.5).) We now add the jump in $f(X)$ at time τ_{j+1} into (11.5.5), obtaining thereby

$$\begin{aligned} &f(X(\tau_{j+1})) - f(X(\tau_j)) \\ &= \int_{\tau_j}^{\tau_{j+1}} f'(X(s))\, dX^c(s) + \frac{1}{2}\int_{\tau_j}^{t_{j+1}} f''(X(s))\, dX^c(s)\, dX^c(s) \\ &\qquad + f(X(\tau_{j+1})) - f(X(\tau_{j+1}-)). \end{aligned}$$

Summing over $j = 0, \ldots, n-1$, we obtain

$$\begin{aligned} &f(X(t)) - f(X(0)) \\ &= \sum_{j=0}^{n-1} \left[f(X(\tau_{j+1})) - f(X(\tau_j)) \right] \\ &= \int_0^t f'(X(s))\, dX^c(s) + \frac{1}{2}\int_0^t f''(X(s))\, dX^c(s)\, dX^c(s) \\ &\qquad + \sum_{j=0}^{n-1} \left[f(X(\tau_{j+1})) - f(X(\tau_{j+1}-)) \right], \end{aligned}$$

which is (11.5.4). Note in this connection that if there is no jump at $\tau_n = t$, then the last term in the sum on the right-hand side, $f\big(X(\tau_n)\big) - f\big(X(\tau_n-)\big)$, is zero. □

It is not always possible to rewrite (11.5.4) in differential form because it is not always possible to find a differential form for the sum of jumps. We provide one case in which this can be done in the next example.

Example 11.5.2 (Geometric Poisson process). Consider the geometric Poisson process

$$S(t) = S(0)\exp\big\{N(t)\log(\sigma+1) - \lambda\sigma t\big\} = S(0)e^{-\lambda\sigma t}(\sigma+1)^{N(t)}, \quad (11.5.6)$$

where $\sigma > -1$ is a constant. If $\sigma > 0$, this process jumps up and moves down between jumps; if $-1 < \sigma < 0$, it jumps down and moves up between jumps. We show that the process is a martingale.

We may write $S(t) = S(0) f\big(X(t)\big)$, where $f(x) = e^x$ and

$$X(t) = N(t)\log(\sigma+1) - \lambda\sigma t$$

has continuous part $X^c(t) = -\lambda\sigma t$ and pure jump part $J(t) = N(t)\log(\sigma+1)$. According to the Itô-Doeblin formula for jump processes,

$$\begin{aligned} S(t) &= f\big(X(t)\big) \\ &= f\big(X(0)\big) - \lambda\sigma \int_0^t f'\big(X(u)\big)\,du + \sum_{0<u\leq t}\big[f\big(X(u)\big) - f\big(X(u-)\big)\big] \\ &= S(0) - \lambda\sigma\int_0^t S(u)\,du + \sum_{0<u\leq t}\big[S(u) - S(u-)\big]. \end{aligned} \quad (11.5.7)$$

If there is a jump at time u, then $S(u) = (\sigma+1)S(u-)$. Therefore,

$$S(u) - S(u-) = \sigma S(u-) \quad (11.5.8)$$

whenever there is a jump at time u, and of course $S(u) - S(u-) = 0$ if there is no jump at time u. In either case, we have

$$S(u) - S(u-) = \sigma S(u-)\Delta N(u).$$

This observation permits us to rewrite the sum on the right-hand side of (11.5.7) as

$$\sum_{0<u\leq t}\big[S(u) - S(u-)\big] = \sum_{0<u\leq t}\sigma S(u-)\Delta N(u) = \sigma\int_0^t S(u-)\,dN(u).$$

It does not matter whether we write the Riemann integral on the right-hand side of (11.5.7) as $\int_0^t S(u)\,du$ or as $\int_0^t S(u-)\,du$. The integrands in these two integrals differ at only finitely many times, and when we integrate with respect to du, these differences do not matter. Therefore, we may rewrite (11.5.7) as

$$\begin{aligned} S(t) &= S(0) - \lambda\sigma \int_0^t S(u-)\,du + \sigma \int_0^t S(u-)\,dN(u) \\ &= S(0) + \sigma \int_0^t S(u-)\,dM(u), \end{aligned}$$

where M is the compensated Poisson process $M(u) = N(u) - \lambda u$, which is a martingale. Because the integrand $S(u-)$ is left-continuous, Theorem 11.4.5 guarantees that $S(t)$ is a martingale.

In this case, the Itô-Doeblin formula (11.5.7) has a differential form, namely,

$$dS(t) = \sigma S(t-)\,dM(t) = -\lambda\sigma S(t)\,dt + \sigma S(t-)\,dN(t). \tag{11.5.9}$$

We were able to obtain this differential form because in (11.5.8) we were able to write the jump in $f(X)$ (i.e., the jump in S) at time u in terms of $f\big(X(u-)\big)$ (i.e., in terms of $S(u-)$). □

Corollary 11.5.3. *Let $W(t)$ be a Brownian motion and let $N(t)$ be a Poisson process with intensity $\lambda > 0$, both defined on the same probability space $(\Omega, \mathcal{F}, \mathbb{P})$ and relative to the same filtration $\mathcal{F}(t)$, $t \geq 0$. Then the processes $W(t)$ and $N(t)$ are independent.*

KEY STEP IN PROOF: Let u_1 and u_2 be fixed real numbers and define

$$Y(t) = \exp\left\{u_1 W(t) + u_2 N(t) - \frac{1}{2}u_1^2 t - \lambda\big(e^{u_2} - 1\big)t\right\}.$$

We use the Itô-Doeblin formula to show that Y is a martingale.

To do this, we define

$$X(s) = u_1 W(s) + u_2 N(s) - \frac{1}{2}u_1^2 s - \lambda\big(e^{u_2} - 1\big)s$$

and $f(x) = e^x$, so that $Y(s) = f\big(X(s)\big)$. The process $X(s)$ has Itô integral part $I(s) = u_1 W(s)$, Riemann integral part $R(s) = -\frac{1}{2}u_1^2 s - \lambda\big(e^{u_2} - 1\big)s$, and pure jump part $J(s) = u_2 N(s)$. In particular,

$$dX^c(s) = u_1\,dW(s) - \frac{1}{2}u_1^2\,ds - \lambda\big(e^{u_2} - 1\big)\,ds, \quad dX^c(s)\,dX^c(s) = u_1^2\,ds.$$

We next observe that if Y has a jump at time s, then

$$Y(s) = \exp\left\{u_1 W(s) + u_2(N(s-) + 1) - \frac{1}{2}u_1^2 s - \lambda\big(e^{u_2} - 1\big)s\right\} = Y(s-)e^{u_2}.$$

Therefore,

$$Y(s) - Y(s-) = (e^{u_2} - 1)Y(s-)\Delta N(s).$$

According to the Itô-Doeblin formula for jump processes,

$$\begin{aligned}
Y(t) &= f(X(t)) \\
&= f(X(0)) + \int_0^t f'(X(s))\,dX^c(s) + \frac{1}{2}\int_0^t f''(X(s))\,dX^c(s)\,dX^c(s) \\
&\quad + \sum_{0<s\le t} [f(X(s)) - f(X(s-))] \\
&= 1 + u_1\int_0^t Y(s)\,dW(s) - \frac{1}{2}u_1^2\int_0^t Y(s)\,ds - \lambda(e^{u_2}-1)\int_0^t Y(s)\,ds \\
&\quad + \frac{1}{2}u_1^2\int_0^t Y(s)\,ds + \sum_{0<s\le t} [Y(s) - Y(s-)] \\
&= 1 + u_1\int_0^t Y(s)\,dW(s) - \lambda(e^{u_2}-1)\int_0^t Y(s-)\,ds \\
&\quad + (e^{u_2}-1)\int_0^t Y(s-)\,dN(s) \\
&= 1 + u_1\int_0^t Y(s)\,dW(s) + (e^{u_2}-1)\int_0^t Y(s-)\,dM(s), \qquad (11.5.10)
\end{aligned}$$

where $M(s) = N(s) - \lambda s$ is a compensated Poisson process. Here we have used the fact that because Y has only finitely many jumps, $\int_0^t Y(s)\,ds = \int_0^t Y(s-)\,ds$. The Itô integral $\int_0^t Y(s)\,dW(s)$ in the last line of (11.5.10) is a martingale, and the integral of the left-continuous process $Y(s-)$ with respect to the martingale $M(s)$ is also. Therefore, Y is a martingale.

Because $Y(0) = 1$ and Y is a martingale, we have $\mathbb{E}Y(t) = 1$ for all t. In other words,

$$\mathbb{E}\exp\left\{u_1W(t) + u_2N(t) - \frac{1}{2}u_1^2t - \lambda(e^{u_2}-1)t\right\} = 1 \text{ for all } t \ge 0.$$

We have obtained the joint moment-generating function formula

$$\mathbb{E}e^{u_1W(t)+u_2N(t)} = \exp\left\{\frac{1}{2}u_1^2t\right\}\cdot\exp\left\{\lambda t(e^{u_2}-1)\right\}.$$

This is the product of the moment-generating function $\mathbb{E}e^{u_1W(t)} = \exp\left\{\frac{1}{2}u_1^2t\right\}$ for $W(t)$ (see Exercise 1.6(i)) and the moment-generating function $\mathbb{E}e^{u_2N(t)} = \exp\left\{\lambda t(e^{u_2}-1)\right\}$ for $N(t)$ (see (11.3.4)). Since the joint moment-generating function factors into the product of moment-generating functions, the random variables $W(t)$ and $N(t)$ are independent.

The corollary asserts more than the independence between $N(t)$ and $W(t)$ for fixed t, saying that the *processes* N and W are independent (i.e., anything depending only on the path of W is independent of anything depending only on the path of N). For example, the corollary asserts that $\max_{0\leq s\leq t} W(s)$ is independent of $\int_0^t N(s)\,ds$. The first step in the proof of this statement is the one just given, which shows that the *random variables* $W(t)$ and $N(t)$ are independent of each fixed t. The next step, which we omit, is to show that for any finite set of times $0 \leq t_1 < t_2 < \cdots < t_n$, the *vector* of random variables $(W(t_1), W(t_2), \ldots, W(t_n))$ is independent of the *vector* of random variables $(N(t_1), N(t_2), \ldots, N(t_n))$. The assertion of the corollary follows from this. □

11.5.2 Itô-Doeblin Formula for Multiple Jump Processes

There is a multidimensional version of the Itô-Doeblin formula for jump processes. We give the two-dimensional version. The formula for higher dimensions follows the same pattern.

Theorem 11.5.4 (Two-dimensional Itô-Doeblin formula for processes with jumps). *Let $X_1(t)$ and $X_2(t)$ be jump processes, and let $f(t, x_1, x_2)$ be a function whose first and second partial derivatives appearing in the following formula are defined and are continuous. Then*

$$\begin{aligned}
&f(t, X_1(t), X_2(t))\\
&= f(0, X_1(0), X_2(0)) + \int_0^t f_t(s, X_1(s), X_2(s))\,ds\\
&\quad + \int_0^t f_{x_1}(s, X_1(s), X_2(s))\,dX_1^c(s) + \int_0^t f_{x_2}(s, X_1(s), X_2(s))\,dX_2^c(s)\\
&\quad + \frac{1}{2}\int_0^t f_{x_1,x_1}(s, X_1(s), X_2(s))\,dX_1^c(s)\,dX_1^c(s)\\
&\quad + \int_0^t f_{x_1,x_2}(s, X_1(s), X_2(s))\,dX_1^c(s)\,dX_2^c(s)\\
&\quad + \frac{1}{2}\int_0^t f_{x_2,x_2}(s, X_1(s), X_2(s))\,dX_2^c(s)\,dX_2^c(s)\\
&\quad + \sum_{0<s\leq t}\left[f(s, X_1(s), X_2(s)) - f(s, X_1(s-), X_2(s-))\right].
\end{aligned}$$

Corollary 11.5.5 (Itô's product rule for jump processes). *Let $X_1(t)$ and $X_2(t)$ be jump processes. Then*

$$\begin{aligned}
X_1(t)X_2(t) &= X_1(0)X_2(0) + \int_0^t X_2(s)\,dX_1^c(s) + \int_0^t X_1(s)\,dX_2^c(s)\\
&\quad + [X_1^c, X_2^c](t) + \sum_{0<s\leq t}\left[X_1(s)X_2(s) - X_1(s-)X_2(s-)\right]\\
&= X_1(0)X_2(0) + \int_0^t X_2(s-)\,dX_1(s) + \int_0^t X_1(s-)\,dX_2(s)\\
&\quad + [X_1, X_2](t). \qquad (11.5.11)
\end{aligned}$$

PROOF: Take $f(x_1, x_2) = x_1 x_2$ so that

$$f_{x_1} = x_2,\ f_{x_2} = x_1,\ f_{x_1x_1} = 0,\ f_{x_1x_2} = 1,\ f_{x_2x_2} = 0.$$

The two-dimensional Itô-Doeblin formula implies

$$\begin{aligned} X_1(t)X_2(t) &= X_1(0)X_2(0) + \int_0^t X_2(s)\,dX_1^c(s) + \int_0^t X_1(s)\,dX_2^c(s) \\ &\quad + \int_0^t 1\,dX_1^c(s)\,dX_2^c(s) + \sum_{0<s\le t}\Big[X_1(s)X_2(s) - X_1(s-)X_2(s-)\Big]. \end{aligned} \tag{11.5.12}$$

The notation $\int_0^t 1\,dX_1^c(s)\,dX_2^c(s)$ in (11.5.12) means $[X_1^c, X_2^c](t)$ (see Remark 11.4.8). This establishes the first equality in (11.5.11).

To obtain the second equality, we denote by $J_1(t) = X_1(t) - X_1^c(t)$ and $J_2(t) = X_2(t) - X_2^c(t)$ the pure jump parts of $X_1(t)$ and $X_2(t)$, respectively, and begin with the last line of (11.5.11), using (11.4.12) to compute

$$\begin{aligned} & X_1(0)X_2(0) + \int_0^t X_2(s-)\,dX_1(s) + \int_0^t X_1(s-)\,dX_2(s) + [X_1, X_2](t) \\ &= X_1(0)X_2(0) + \int_0^t X_2(s-)\,dX_1^c(s) + \int_0^t X_2(s-)\,dJ_1(s) \\ &\quad + \int_0^t X_1(s-)\,dX_2^c(s) + \int_0^t X_1(s-)\,dJ_2(s) \\ &\quad + [X_1^c, X_2^c](t) + \sum_{0<s\le t} \Delta J_1(s)\Delta J_2(s) \\ &= X_1(0)X_2(0) + \int_0^t X_2(s)\,dX_1^c(s) + \int_0^t X_1(s)\,dX_2^c(s) + [X_1^c, X_2^c](t) \\ &\quad + \sum_{0<s\le t} \big[X_2(s-)\Delta X_1(s) + X_1(s-)\Delta X_2(s) + \Delta X_1(s)\Delta X_2(s)\big]. \end{aligned} \tag{11.5.13}$$

We have also used the fact that the jumps in $X_i(t)$ are the same as the jumps in $J_i(t)$. It remains to show that this last sum is the same as the sum

$$\sum_{0<s\le t} \big[X_1(s)X_2(s) - X_1(s-)X_2(s-)\big]$$

in the second line of (11.5.11). We expand the typical term in the sum in the second line of (11.5.11):

$$
\begin{aligned}
&X_1(s)X_2(s) - X_1(s-)X_2(s-)\\
&= \big(X_1(s-) + \Delta X_1(s)\big)\big(X_2(s-) + \Delta X_2(s)\big) - X_1(s-)X_2(s-)\\
&= X_1(s-)X_2(s-) + X_1(s-)\Delta X_2(s) + \Delta X_1(s)X_2(s-) + \Delta X_1(s)\Delta X_2(s)\\
&\qquad -X_1(s-)X_2(s-)\\
&= X_1(s-)\Delta X_2(s) + \Delta X_1(s)X_2(s-) + \Delta X_1(s)\Delta X_2(s).
\end{aligned}
$$

This is the typical term in the sum appearing at the end of (11.5.13). □

For stochastic calculus without jumps, Girsanov's Theorem tells us how to change the measure using the Radon-Nikodým derivative process

$$Z(t) = \exp\Big\{ - \int_0^t \Gamma(s)\,dW(s) - \frac{1}{2}\int_0^t \Gamma^2(s)\,ds\Big\}.$$

This process satisfies the stochastic differential equation

$$dZ(t) = -\Gamma(t)Z(t)\,dW(t) = Z(t)\,dX^c(t),$$

where $X^c(t) = -\int_0^t \Gamma(s)\,dW(s)$ and $[X^c, X^c](t) = \int_0^t \Gamma^2(s)\,ds$. We may rewrite $Z(t)$ as

$$Z(t) = \exp\left\{X^c(t) - \frac{1}{2}[X^c, X^c](t)\right\}. \tag{11.5.14}$$

In stochastic calculus for processes with jumps, the analogous stochastic differential equation is

$$dZ^X(t) = Z^X(t-)\,dX(t), \tag{11.5.15}$$

where the integrator X is now allowed to have jumps. The solution to (11.5.15) is like (11.5.14), except now, whenever there is a jump in X, (11.5.15) says there is a jump in Z^X of size

$$\Delta Z^X(s) = Z^X(s-)\Delta X(s).$$

Therefore,

$$Z^X(s) = Z^X(s-) + \Delta Z^X(s) = Z^X(s-)\big(1 + \Delta X(s)\big).$$

The following corollary presents the result.

Corollary 11.5.6. *Let $X(t)$ be a jump process. The* Doleans-Dade exponential *of X is defined to be the process*

$$Z^X(t) = \exp\left\{X^c(t) - \frac{1}{2}[X^c, X^c](t)\right\} \prod_{0<s\le t} \big(1 + \Delta X(s)\big).$$

This process is the solution to the stochastic differential equation (11.5.15) with initial condition $Z^X(0) = 1$, which in integral form is

$$Z^X(t) = 1 + \int_0^t Z^X(s-)\,dX(s). \tag{11.5.16}$$

PROOF: We may write $X(t)$ as $X(t) = X^c(t) + J(t)$, where

$$X^c(t) = \int_0^t \Gamma(s)\,dW(s) + \int_0^t \Theta(s)\,ds \tag{11.5.17}$$

is the continuous part of X and $J(t)$ is the pure jump part. We define

$$\begin{aligned} Y(t) &= \exp\left\{\int_0^t \Gamma(s)\,dW(s) + \int_0^t \Theta(s)\,ds - \frac{1}{2}\int_0^t \Gamma^2(s)\,ds\right\} \\ &= \exp\left\{X^c(t) - \frac{1}{2}[X^c, X^c](t)\right\}. \end{aligned} \tag{11.5.18}$$

From the Itô-Doeblin formula for continuous processes, we know that

$$dY(t) = Y(t)\,dX^c(t) = Y(t-)\,dX^c(t). \tag{11.5.19}$$

We next define $K(t) = 1$ for t between 0 and the time of the first jump of X, and we set

$$K(t) = \prod_{0<s\leq t} (1 + \Delta X(s)) \tag{11.5.20}$$

for t greater than or equal to the first jump time of X. The process $K(t)$ is a pure jump process, and $Z^X(t) = Y(t)K(t)$. If X has a jump at time t, then $K(t) = K(t-)\big(1 + \Delta X(t)\big)$. Therefore,

$$\Delta K(t) = K(t) - K(t-) = K(t-)\Delta X(t). \tag{11.5.21}$$

Because $Y(t)$ is continuous and $K(t)$ is a pure jump process, $[Y, K](t) = 0$. We now use Itô's product rule for jump processes to obtain

$$\begin{aligned} Z^X(t) &= Y(t)K(t) \\ &= Y(0) + \int_0^t K(s-)\,dY(s) + \int_0^t Y(s-)\,dK(s) \\ &= 1 + \int_0^t Y(s-)K(s-)\,dX^c(s) + \sum_{0<s\leq t} Y(s-)K(s-)\Delta X(s) \\ &= 1 + \int_0^t Y(s-)K(s-)\,dX(s) \\ &= 1 + \int_0^t Z^X(s-)\,dX(s). \end{aligned} \tag{11.5.22}$$

This is (11.5.16). □

11.6 Change of Measure

Just as we can use Girsanov's Theorem to change the measure so that a Brownian motion with drift becomes a Brownian motion without drift, we can change the measure for Poisson processes and compound Poisson processes. For a Poisson process, the change of measure affects the intensity. For a compound Poisson process, the change of measure can affect both the intensity and the distribution of the jump sizes. We treat these two situations in the next two subsections, and in the third subsection we also include a Brownian motion component in the process under consideration.

11.6.1 Change of Measure for a Poisson Process

Let $N(t)$ be a Poisson process on a probability space $(\Omega, \mathcal{F}, \mathbb{P})$ relative to a filtration $\mathcal{F}(t)$, $t \geq 0$. We denote the intensity of $N(t)$ by λ, a positive constant (i.e., $\mathbb{E}N(t) = \lambda t$). The compensated Poisson process $M(t) = N(t) - \lambda t$ is a martingale under $\mathbb{P}$ (Theorem 11.2.4). Let $\tilde{\lambda}$ be a positive number. We define

$$Z(t) = e^{(\lambda - \tilde{\lambda})t} \left(\frac{\tilde{\lambda}}{\lambda} \right)^{N(t)}. \tag{11.6.1}$$

We fix a time $T > 0$ and will use $Z(T)$ to change to a new measure $\widetilde{\mathbb{P}}$ under which $N(t)$, $0 \leq t \leq T$, has intensity $\tilde{\lambda}$ rather than λ. It is clear that $Z(T) > 0$ almost surely. In order to use $Z(T)$ to change the measure, we also need to verify that $\mathbb{E}Z(T) = 1$.

Lemma 11.6.1. *The process $Z(t)$ of (11.6.1) satisfies*

$$dZ(t) = \frac{\tilde{\lambda} - \lambda}{\lambda} Z(t-)\, dM(t). \tag{11.6.2}$$

In particular, $Z(t)$ is a martingale under $\mathbb{P}$ and $\mathbb{E}Z(t) = 1$ for all t.

PROOF: Define $X(t) = \frac{\tilde{\lambda}-\lambda}{\lambda} M(t)$, which is a martingale with continuous part $X^c(t) = (\lambda - \tilde{\lambda})t$ and pure jump part $J(t) = \frac{\tilde{\lambda}-\lambda}{\lambda} N(t)$. Then $[X^c, X^c](t) = 0$, and if there is a jump at time t, then $\Delta X(t) = \frac{\tilde{\lambda}-\lambda}{\lambda}$, so

$$1 + \Delta X(t) = \frac{\tilde{\lambda}}{\lambda}.$$

Therefore, the process in (11.6.1) may be written as

$$Z(t) = \exp\left\{ X^c(t) - \frac{1}{2}[X^c, X^c](t) \right\} \prod_{0<s\leq t} (1 + \Delta X(s)).$$

We see from this formula that $Z(t)$ is the Doleans-Dade exponential $Z^X(t)$ of Corollary 11.5.6. In particular,

$$Z(t) = 1 + \int_0^t Z(s-)\, dX(s).$$

Since X is a martingale and $Z(s-)$ is left-continuous, $Z(t)$ is a martingale. Because $Z(t)$ is a martingale and $Z(0) = 1$, we know that $\mathbb{E}Z(t) = 1$ for all $t \geq 0$. □

We may now fix a positive time T and use $Z(T)$ to change the measure. We define

$$\widetilde{\mathbb{P}}(A) = \int_A Z(T)\, d\mathbb{P} \quad \text{for all } A \in \mathcal{F}. \tag{11.6.3}$$

Theorem 11.6.2 (Change of Poisson intensity). *Under the probability measure $\widetilde{\mathbb{P}}$, the process $N(t)$, $0 \leq t \leq T$, is Poisson with intensity $\tilde{\lambda}$.*

KEY STEP IN PROOF: We compute the moment-generating function of $N(t)$ under $\widetilde{\mathbb{P}}$. For $0 \leq t \leq T$, we can change the $\widetilde{\mathbb{E}}$ expectation of $e^{uN(t)}$ to the $\mathbb{E}$ expectation by using $Z(t)$ as the Radon-Nikodým derivative rather than $Z(T)$ (see Lemma 5.2.1). Using the formula for $Z(t)$ and the moment-generating function formula (11.3.4), we obtain

$$\begin{aligned}
\mathbb{E}\left[e^{uN(t)} Z(t)\right] &= e^{(\lambda-\tilde{\lambda})t}\, \mathbb{E}\left[e^{uN(t)} \left(\frac{\tilde{\lambda}}{\lambda}\right)^{N(t)}\right] \\
&= e^{(\lambda-\tilde{\lambda})t}\, \mathbb{E}\left[\exp\left\{\left(u + \log\frac{\tilde{\lambda}}{\lambda}\right) N(t)\right\}\right] \\
&= e^{(\lambda-\tilde{\lambda})t} \exp\left\{\lambda t \left(e^{u+\log(\tilde{\lambda}/\lambda)} - 1\right)\right\} \\
&= \exp\left\{\tilde{\lambda} t (e^u - 1)\right\},
\end{aligned}$$

which is the moment generating function for a Poisson process with intensity $\tilde{\lambda}$ (see again (11.3.4)). □

Example 11.6.3. Consider a stock modeled as a geometric Poisson process

$$S(t) = S(0) \exp\{\alpha t + N(t) \log(\sigma + 1) - \lambda\sigma t\} = S(0) e^{(\alpha-\lambda\sigma)t} (\sigma + 1)^{N(t)},$$

where $\sigma > -1$, $\sigma \neq 0$, and $N(t)$ is a Poisson process with intensity λ under the actual probability measure $\mathbb{P}$. We saw in Example 11.5.2 that $e^{-\alpha t} S(t)$ is a martingale under $\mathbb{P}$, and hence $S(t)$ has mean rate of return α. Indeed, in place of (11.5.9), we now have

$$dS(t) = \alpha S(t)\, dt + \sigma S(t-)\, dM(t), \tag{11.6.4}$$

where $M(t)$ is the compensated Poisson process $M(t) = N(t) - \lambda t$. We would like to change to a probability measure $\widetilde{\mathbb{P}}$ under which

$$dS(t) = rS(t)\, dt + \sigma S(t-)\, d\widetilde{M}(t), \tag{11.6.5}$$

where r is the interest rate, $N(t)$ is a Poisson process with intensity $\tilde{\lambda}$ under $\widetilde{\mathbb{P}}$, and $\widetilde{M}(t) = N(t) - \tilde{\lambda} t$ is a compensated Poisson process under $\widetilde{\mathbb{P}}$. Then, under $\widetilde{\mathbb{P}}$, the geometric Poisson process would have mean rate of return equal to the interest rate, and $\widetilde{\mathbb{P}}$ would be the risk-neutral measure.

To accomplish this, we note that the "dt" term in (11.6.4) is

$$(\alpha - \lambda\sigma)S(t)\,dt \tag{11.6.6}$$

(recall that $dM(t) = dN(t) - \lambda\,dt$) and the "$dt$" term in (11.6.5) is

$$(r - \tilde{\lambda}\sigma)S(t)\,dt. \tag{11.6.7}$$

(Here again we are using the fact that $S(t-)\,dt$ and $S(t)\,dt$ have the same integrals, and we can thus use them interchangeably.) We set (11.6.6) and (11.6.7) equal and solve for

$$\tilde{\lambda} = \lambda - \frac{\alpha - r}{\sigma}.$$

We then change to the risk-neutral measure by formula (11.6.3) with $Z(T)$ defined by (11.6.1).

To make the change of measure, we must have $\tilde{\lambda} > 0$, which is equivalent to

$$\lambda > \frac{\alpha - r}{\sigma}. \tag{11.6.8}$$

If condition (11.6.8) does not hold, then there is no risk-neutral measure and hence there must be an arbitrage. Indeed, if $\sigma > 0$ and (11.6.8) fails, then

$$S(t) \geq S(0)e^{rt}(\sigma + 1)^{N(t)} \geq S(0)e^{rt},$$

and borrowing at the interest rate r to invest in the stock is an arbitrage. If $-1 < \sigma < 0$, the inequalities are reversed and the arbitrage consists of shorting the stock to invest in the money market account. □

11.6.2 Change of Measure for a Compound Poisson Process

Let $N(t)$ be a Poisson process with intensity λ, and let $Y_1, Y_2, \ldots$ be a sequence of identically distributed random variables defined on a probability space $(\Omega, \mathcal{F}, \mathbb{P})$. We assume the random variables $Y_1, Y_2, \ldots$ are independent of one another and also independent of the Poisson process $N(t)$. We define the *compound Poisson process*

$$Q(t) = \sum_{i=1}^{N(t)} Y_i. \tag{11.6.9}$$

Note for future reference that if N jumps at time t, then Q jumps at time t and

$$\Delta Q(t) = Y_{N(t)}. \tag{11.6.10}$$

Our goal is to change the measure so that the intensity of $N(t)$ and the distribution of the jump sizes $Y_1, Y_2, \ldots$ both change. We first consider the case when the jump-size random variables have a discrete distribution (i.e., each Y_i takes one of finitely many possible nonzero values $y_1, y_2, \ldots, y_M$). Let $p(y_m)$ denote the probability that a jump is of size y_m:

$$p(y_m) = \mathbb{P}\{Y_i = y_m\},\ m = 1, \ldots, M.$$

This does not depend on i since $Y_1, Y_2, \ldots$ are identically distributed. We assume that $p(y_m) > 0$ for every m and, of course, that $\sum_{m=1}^{M} p(y_m) = 1$.

Let $N_m(t)$ denote the number of jumps in $Q(t)$ of size y_m up to and including time t, so that

$$N(t) = \sum_{m=1}^{M} N_m(t) \text{ and } Q(t) = \sum_{m=1}^{M} y_m N_m(t).$$

According to Corollary 11.3.4, $N_1, \ldots, N_M$ are independent Poisson processes and each N_m has intensity $\lambda_m = \lambda p(y_m)$.

Let $\tilde{\lambda}_1, \ldots, \tilde{\lambda}_M$ be given positive numbers, and set

$$Z_m(t) = e^{(\lambda_m - \tilde{\lambda}_m)t} \left(\frac{\tilde{\lambda}_m}{\lambda_m}\right)^{N_m(t)} \quad \text{and} \quad Z(t) = \prod_{m=1}^{M} Z_m(t). \tag{11.6.11}$$

Lemma 11.6.4. *The process $Z(t)$ of (11.6.11) is a martingale. In particular, $\mathbb{E}Z(t) = 1$ for all t.*

PROOF: From Lemma 11.6.1, we have

$$dZ_m(t) = \frac{\tilde{\lambda}_m - \lambda_m}{\lambda_m} Z_m(t-)\, dM_m(t), \tag{11.6.12}$$

where

$$M_m(t) = N_m(t) - \lambda_m\, dt.$$

Because the integrand in (11.6.12) is left-continuous and the compensated Poisson process is a martingale, the process Z_m is a martingale (Theorem 11.4.5).

For $m \neq n$, the Poisson processes N_m and N_n have no simultaneous jumps, and hence $[Z_m, Z_n] = 0$. Itô's product rule (Corollary 11.5.5) implies that

$$d\big(Z_1(t)Z_2(t)\big) = Z_2(t-)\, dZ_1(t) + Z_1(t-)\, dZ_2(t). \tag{11.6.13}$$

Because both Z_1 and Z_2 are martingales and the integrands in (11.6.13) are left-continuous, the process $Z_1 Z_2$ is a martingale. Because $Z_1 Z_2$ has no jumps simultaneous with the jumps of Z_3, Itô's product rule further implies

$$d\big(Z_1(t)Z_2(t)Z_3(t)\big) = Z_3(t-)\, d\big(Z_1(t)Z_2(t)\big) + \big(Z_1(t-)Z_2(t-)\big)\, dZ_3(t).$$

Once again, the integrators are martingales and the integrands are left-continuous. Therefore, $Z_1Z_2Z_3$ is a martingale. Continuing this process, we eventually conclude that $Z(t) = Z_1(t)Z_2(t)\cdots Z_m(t)$ is a martingale. □

Fix $T > 0$. Because $Z(T) > 0$ almost surely and $\mathbb{E}Z(T) = 1$, we can use $Z(T)$ to change the measure, defining

$$\widetilde{\mathbb{P}}(A) = \int_A Z(T)\,dP \text{ for all } Z \in \mathcal{F}.$$

Theorem 11.6.5 (Change of compound Poisson intensity and jump distribution for finitely many jump sizes). *Under $\widetilde{\mathbb{P}}$, $Q(t)$ is a compound Poisson process with intensity $\tilde{\lambda} = \sum_{m=1}^{M} \tilde{\lambda}_m$, and $Y_1, Y_2, \ldots$ are independent, identically distributed random variables with*

$$\widetilde{\mathbb{P}}\{Y_i = y_m\} = \tilde{p}(y_m) = \frac{\tilde{\lambda}_m}{\tilde{\lambda}}. \tag{11.6.14}$$

KEY STEP IN PROOF: We use the independence of $N_1, \ldots, N_M$ under $\mathbb{P}$ to compute the moment-generating function of $Q(t)$ under $\widetilde{\mathbb{P}}$. For $0 \le t \le T$, Lemma 5.2.1 and the moment-generating function formula (11.3.4) imply

$$\begin{aligned}
\widetilde{\mathbb{E}}\left[e^{uQ(t)}\right] &= \mathbb{E}\left[e^{uQ(t)}Z(t)\right] \\
&= \mathbb{E}\left[\exp\left\{u\sum_{m=1}^{M} y_m N_m(t)\right\}\cdot \prod_{m=1}^{M} e^{(\lambda_m - \tilde{\lambda}_m)t}\left(\frac{\tilde{\lambda}_m}{\lambda_m}\right)^{N_m(t)}\right] \\
&= \prod_{m=1}^{M} \exp\{(\lambda_m - \tilde{\lambda}_m)t\}\cdot \mathbb{E}\left[\exp\left\{\left(uy_m + \log\frac{\tilde{\lambda}_m}{\lambda_m}\right)N_m(t)\right\}\right] \\
&= \prod_{m=1}^{M} \exp\{(\lambda_m - \tilde{\lambda}_m)t\}\exp\left\{\lambda_m t\left(e^{uy_m + \log(\tilde{\lambda}_m/\lambda_m)} - 1\right)\right\} \\
&= \prod_{m=1}^{M} \exp\left\{(\lambda_m - \tilde{\lambda}_m)t + \tilde{\lambda}_m t e^{uy_m} - \lambda_m t\right\} \\
&= \prod_{m=1}^{M} \exp\left\{\tilde{\lambda}_m t(e^{uy_m} - 1)\right\} \\
&= \prod_{m=1}^{M} \exp\left\{\tilde{\lambda} t \tilde{p}(y_m) e^{uy_m} - \tilde{\lambda}_m t\right\} \\
&= \exp\left\{\tilde{\lambda} t\left(\sum_{m=1}^{M} \tilde{p}(y_m)e^{uy_m} - 1\right)\right\}.
\end{aligned}$$

According to (11.3.5), this is the moment-generating function for a compound Poisson process with intensity $\tilde{\lambda}$ and jump-size distribution (11.6.14). □

The Radon-Nikodým derivative process $Z(t)$ of (11.6.11) may be written as

$$Z(t) = \exp\left\{\sum_{m=1}^{M}(\lambda_m - \tilde{\lambda}_m)t\right\} \cdot \prod_{m=1}^{M}\left(\frac{\tilde{\lambda}\tilde{p}(y_m)}{\lambda p(y_m)}\right)^{N_m(t)} = e^{(\lambda-\tilde{\lambda})t}\prod_{i=1}^{N(t)}\frac{\tilde{\lambda}\tilde{p}(Y_i)}{\lambda p(Y_i)}.$$

This suggests that if $Y_1, Y_2, \ldots$ are not discrete but instead have a common density $f(y)$, then we could change the measure so that $Q(t)$ has intensity $\tilde{\lambda}$ and $Y_1, Y_2, \ldots$ have a different density $\widetilde{f}(y)$ by using the Radon-Nikodým derivative process

$$Z(t) = e^{(\lambda-\tilde{\lambda})t}\prod_{i=1}^{N(t)}\frac{\tilde{\lambda}\widetilde{f}(Y_i)}{\lambda f(Y_i)}. \tag{11.6.15}$$

This is in fact the case, although the proof, given below, is harder than the one just given for the case of a discrete jump-size distribution.

To avoid division by zero in (11.6.15), we assume that $\widetilde{f}(y) = 0$ whenever $f(y) = 0$. This means that if a certain set of jump sizes has probability zero under $\mathbb{P}$, then it will also have probability zero under $\widetilde{\mathbb{P}}$ considered in Theorem 11.6.7 below.

Lemma 11.6.6. *The process $Z(t)$ of (11.6.15) is a martingale. In particular, $\mathbb{E}Z(t) = 1$ for all $t \geq 0$.*

PROOF: We define the pure jump process

$$J(t) = \prod_{i=1}^{N(t)}\frac{\tilde{\lambda}\widetilde{f}(Y_i)}{\lambda f(Y_i)}. \tag{11.6.16}$$

At the jump times of Q, which are also the jump times of N and J, we have (recall (11.6.10))

$$J(t) = J(t-)\frac{\tilde{\lambda}\widetilde{f}(Y_{N(t)})}{\lambda f(Y_{N(t)})} = J(t-)\frac{\tilde{\lambda}\widetilde{f}(\Delta Q(t))}{\lambda f(\Delta Q(t))},$$

and hence

$$\Delta J(t) = J(t) - J(t-) = \left[\frac{\tilde{\lambda}\widetilde{f}(\Delta Q(t))}{\lambda f(\Delta Q(t))} - 1\right]J(t-) \tag{11.6.17}$$

at the jump times of Q.

We define the compound Poisson process

$$H(t) = \sum_{i=1}^{N(t)}\frac{\tilde{\lambda}\widetilde{f}(Y_i)}{\lambda f(Y_i)} \tag{11.6.18}$$

for which

$$\Delta H(t) = \frac{\tilde{\lambda}\tilde{f}(\Delta Q(t))}{\lambda f(\Delta Q(t))}. \tag{11.6.19}$$

Because

$$\mathbb{E}\left[\frac{\tilde{\lambda}\tilde{f}(Y_i)}{\lambda f(Y_i)}\right] = \frac{\tilde{\lambda}}{\lambda}\int_{-\infty}^{\infty}\frac{\tilde{f}(y)}{f(y)}f(y)\,dy = \frac{\tilde{\lambda}}{\lambda}\int_{-\infty}^{\infty}\tilde{f}(y)\,dy = \frac{\tilde{\lambda}}{\lambda},$$

the compensated compound Poisson process $H(t) - \tilde{\lambda}t$ is a martingale (Theorem 11.3.1 with $\beta = \frac{\tilde{\lambda}}{\lambda}$). We may rewrite (11.6.17) as

$$\Delta J(t) = J(t-)\Delta H(t) - J(t-)\Delta N(t), \tag{11.6.20}$$

and because all these terms are zero if there is no jump at t, this equation holds at all times t, not just at the jump times of Q. Because J, H, and N are all pure jump processes, we may also write (11.6.20) as

$$dJ(t) = J(t-)\,dH(t) - J(t-)\,dN(t).$$

Because $J(t)$ is a pure jump process and $e^{(\lambda-\tilde{\lambda})t}$ is continuous, the cross variation between these two processes is zero. Therefore, Itô's product rule for jump processes (Corollary 11.5.5) implies that $Z(t) = e^{(\lambda-\tilde{\lambda})t}J(t)$ may be written as

$$\begin{aligned}
Z(t) &= Z(0) + \int_0^t J(s-)(\lambda - \tilde{\lambda})e^{(\lambda-\tilde{\lambda})s}\,ds + \int_0^t e^{(\lambda-\tilde{\lambda})s}\,dJ(s)\\
&= 1 + \int_0^t e^{(\lambda-\tilde{\lambda})s}J(s-)(\lambda-\tilde{\lambda})\,ds + \int_0^t e^{(\lambda-\tilde{\lambda})s}J(s-)\,dH(s)\\
&\quad - \int_0^t e^{(\lambda-\tilde{\lambda})s}J(s-)\,dN(s)\\
&= 1 + \int_0^t e^{(\lambda-\tilde{\lambda})s}J(s-)\,d\big(H(s)-\tilde{\lambda}s\big) - \int_0^t e^{(\lambda-\tilde{\lambda})s}J(s-)\,d\big(N(s)-\lambda s\big)\\
&= 1 + \int_0^t Z(s-)\,d\big(H(s)-\tilde{\lambda}s\big) - \int_0^t Z(s-)\,d\big(N(s)-\lambda s\big).
\end{aligned} \tag{11.6.21}$$

Theorem 11.4.5 implies that $Z(t)$ is a martingale. Since $Z(t)$ is a martingale and $Z(0) = 1$, we have $\mathbb{E}Z(t) = 1$ for all t. □

For future reference, we rewrite (11.6.21) in the differential form

$$dZ(t) = Z(t-)\,d\big(H(t)-\tilde{\lambda}t\big) - Z(t-)\,d\big(N(t)-\lambda t\big).$$

This equation implies

$$\Delta Z(t) = Z(t-)\Delta H(t) - Z(t-)\Delta N(t). \tag{11.6.22}$$

Fix a positive T and define

$$\widetilde{\mathbb{P}}(A) = \int_A Z(T)\,d\mathbb{P} \quad \text{for all } A \in \mathcal{F}. \tag{11.6.23}$$

Theorem 11.6.7 (Change of compound Poisson intensity and jump distribution for a continuum of jump sizes). *Under the probability measure $\widetilde{\mathbb{P}}$, the process $Q(t)$, $0 \le t \le T$, of (11.6.9) is a compound Poisson process with intensity $\tilde{\lambda}$. Furthermore, the jumps in $Q(t)$ are independent and identically distributed with density $\widetilde{f}(y)$.*

KEY STEP IN PROOF: We need to show that, under $\widetilde{\mathbb{P}}$, the process $Q(t)$ has the moment-generating function corresponding to a compound Poisson process with intensity $\tilde{\lambda}$ and jump density $\widetilde{f}(y)$. In other words, we must show that (see (11.3.2))

$$\widetilde{\mathbb{E}}e^{uQ(t)} = \exp\left\{\tilde{\lambda}t\big(\widetilde{\varphi}_Y(u) - 1\big)\right\}, \tag{11.6.24}$$

where

$$\widetilde{\varphi}_Y(u) = \int_{-\infty}^{\infty} e^{uy}\widetilde{f}(y)\,dy. \tag{11.6.25}$$

We define

$$X(t) = \exp\left\{uQ(t) - \tilde{\lambda}t\big(\widetilde{\varphi}_Y(u) - 1\big)\right\}$$

and show that $X(t)Z(t)$ is a martingale under $\mathbb{P}$. At jump times of Q,

$$X(t) = X(t-)e^{u\Delta Q(t)},$$

and hence

$$\Delta X(t) = X(t) - X(t-) = X(t-)\left(e^{u\Delta Q(t)} - 1\right). \tag{11.6.26}$$

We introduce the compound Poisson process

$$V(t) = \sum_{i=1}^{N(t)} e^{uY_i}\frac{\tilde{\lambda}\widetilde{f}(Y_i)}{\lambda f(Y_i)}.$$

Because

$$\mathbb{E}\left[e^{uY_i}\frac{\tilde{\lambda}\widetilde{f}(Y_i)}{\lambda f(Y_i)}\right] = \frac{\tilde{\lambda}}{\lambda}\int_{-\infty}^{\infty} e^{uy}\frac{\widetilde{f}(y)}{f(y)}f(y)\,dy = \frac{\tilde{\lambda}}{\lambda}\widetilde{\varphi}_Y(u),$$

the compensated compound Poisson process $V(t) - \tilde{\lambda}t\widetilde{\varphi}_Y(u)$ is a martingale (see Theorem 11.3.1 with $\beta = \frac{\tilde{\lambda}}{\lambda}\widetilde{\varphi}_Y(u)$). At jump times of Q,

$$\Delta V(t) = e^{u\Delta Q(t)}\frac{\tilde{\lambda}\widetilde{f}(\Delta Q(t))}{\lambda f(\Delta Q(t))} = e^{u\Delta Q(t)}\Delta H(t), \tag{11.6.27}$$

where $H(t)$, defined by (11.6.18), satisfies (11.6.19) at jump times of Q.

Because $X(t)$ and $Z(t)$ have no Itô integral components, (11.6.26), (11.6.22), and (11.6.27) imply

$$\begin{aligned}
[X,Z](t) &= \sum_{0<s\le t} \Delta X(s)\Delta Z(s)\\
&= \sum_{0<s\le t} X(s-)Z(s-)\big(e^{u\Delta Q(s)}-1\big)\Delta H(s)\\
&\quad - \sum_{0<s\le t} X(s-)Z(s-)\big(e^{u\Delta Q(s)}-1\big)\Delta N(s)\\
&= \sum_{0<s\le t} X(s-)Z(s-)\Delta V(s) - \sum_{0<s\le t} X(s-)Z(s-)\Delta H(s)\\
&\quad - \sum_{0<s\le t} X(s-)Z(s-)\big(e^{u\Delta Q(s)}-1\big). \qquad (11.6.28)
\end{aligned}$$

We have omitted $\Delta N(s)$ in the last term because it is always either 1 or 0, and when it is zero, $e^{u\Delta Q(s)}-1$ is also zero. In other words,

$$\big(e^{u\Delta Q(s)}-1\big)\Delta N(s) = \big(e^{u\Delta Q(s)}-1\big).$$

We use Itô's product rule for jump processes to write

$$X(t)Z(t) = 1 + \int_0^t X(s-)\,dZ(s) + \int_0^t Z(s-)\,dX(s) + [X,Z](t).$$

We show that the right-hand side is a martingale under $\mathbb{P}$. The integral $\int_0^t X(s-)\,dZ(s)$ is a martingale because the integrand is left-continuous and Z is a martingale. We examine the two other terms, using (11.6.26) and (11.6.28):

$$\begin{aligned}
&\int_0^t Z(s-)\,dX(s) + [X,Z](t)\\
&= \int_0^t Z(s-)\,dX^c(s) + \sum_{0<s\le t} Z(s-)\Delta X(s) + [X,Z](t)\\
&= -\tilde{\lambda}\big(\widetilde{\varphi}_Y(u)-1\big)\int_0^t X(s-)Z(s-)ds + \sum_{0<s\le t} X(s-)Z(s-)\big(e^{u\Delta Q(s)}-1\big)\\
&\quad + \sum_{0<s\le t} X(s-)Z(s-)\Delta V(s) - \sum_{0<s\le t} X(s-)Z(s-)\Delta H(s)\\
&\quad - \sum_{0<s\le t} X(s-)Z(s-)\big(e^{u\Delta Q(s)}-1\big)\\
&= \int_0^t X(s-)Z(s-)\,d\big(V(s)-\tilde{\lambda}s\widetilde{\varphi}_Y(u)\,ds\big) - \int_0^t X(s-)Z(s-)\,d\big(H(s)-\tilde{\lambda}s\big).
\end{aligned}$$

This is a martingale because the processes $V(t)-\tilde{\lambda}t\widetilde{\varphi}_Y(u)$ and $H(t)-\tilde{\lambda}t$ are martingales and the integrands are left-continuous.

We can now prove (11.6.24). Using Lemma 5.2.1, we may write

$$\widetilde{\mathbb{E}}\big[e^{uQ(t)}\big] = \mathbb{E}\big[e^{uQ(t)}Z(t)\big]. \tag{11.6.29}$$

But the martingale $X(t)Z(t)$ has constant expectation 1, which implies

$$\begin{aligned} 1 &= \mathbb{E}\big[X(t)Z(t)\big] \\ &= \exp\big\{-\tilde{\lambda}t(\widetilde{\varphi}_Y(u)-1)\big\}\cdot\mathbb{E}\big[e^{uQ(t)}Z(t)\big]. \end{aligned} \tag{11.6.30}$$

Combining (11.6.29) and (11.6.30), we obtain (11.6.24). □

11.6.3 Change of Measure for a Compound Poisson Process and a Brownian Motion

Suppose now that we have a probability space $(\Omega, \mathcal{F}, \mathbb{P})$ on which is defined a Brownian motion $W(t)$. Suppose that on this same probability space there is defined a compound Poisson process

$$Q(t) = \sum_{i=1}^{N(t)} Y_i$$

as in (11.3.1) with intensity λ and jumps having density function $f(y)$. We assume there is a single filtration $\mathcal{F}(t)$, $t \geq 0$, for both the Brownian motion and the compound Poisson process. In this case, the Brownian motion and compound Poisson process must be independent. (See Corollary 11.4.9 for the case of a Brownian motion and a Poisson process. The case of a Brownian motion and a compound Poisson process is Exercise 11.6.)

Let $\tilde{\lambda}$ be a positive number, let $\widetilde{f}(y)$ be another density function with the property that $\widetilde{f}(y) = 0$ whenever $f(y) = 0$, and let $\Theta(t)$ be an adapted process. We define

$$Z_1(t) = \exp\left\{-\int_0^t \Theta(u)\,dW(u) - \frac{1}{2}\int_0^t \Theta^2(u)\,du\right\}, \tag{11.6.31}$$

$$Z_2(t) = e^{(\lambda-\tilde{\lambda})t}\prod_{i=1}^{N(t)}\frac{\tilde{\lambda}\widetilde{f}(Y_i)}{\lambda f(Y_i)}, \tag{11.6.32}$$

$$Z(t) = Z_1(t)Z_2(t). \tag{11.6.33}$$

Lemma 11.6.8. *The process $Z(t)$ of (11.6.33) is a martingale. In particular, $\mathbb{E}Z(t) = 1$ for all $t \geq 0$.*

PROOF: We know from stochastic calculus for continuous processes that $Z_1(t)$ is a martingale and from Lemma 11.6.6 that $Z_2(t)$ is a martingale. Since $Z_1(t)$ is continuous and $Z_2(t)$ has no Itô integral part, $[Z_1, Z_2](t) = 0$. Itô's product rule for jump processes thus implies

$$Z_1(t)Z_2(t) = Z_1(0)Z_2(0) + \int_0^t Z_1(s-)\,dZ_2(s) + \int_0^t Z_2(s-)\,dZ_1(s), \quad (11.6.34)$$

and both integrals are martingales because of Theorem 11.4.5. This implies that $Z(t)$ is a martingale, and because $Z(0) = 1$, we have $\mathbb{E}Z(t) = 1$ for all $t \geq 0$. □

Fix a positive T and define $\widetilde{\mathbb{P}}(A) = \int_A Z(T)\,d\mathbb{P}$ for all $A \in \mathcal{F}$. We have the following.

Theorem 11.6.9. *Under the probability measure* $\widetilde{\mathbb{P}}$, *the process*

$$\widetilde{W}(t) = W(t) + \int_0^t \Theta(s)\,ds$$

is a Brownian motion, $Q(t)$ *is a compound Poisson process with intensity* $\tilde{\lambda}$ *and independent, identically distributed jump sizes having density* $\tilde{f}(y)$, *and the processes* $\widetilde{W}(t)$ *and* $Q(t)$ *are independent.*

The key step in the proof of Theorem 11.6.9 is to show that $\widetilde{W}(t)$ and $Q(t)$ have the correct joint moment-generating function under $\widetilde{\mathbb{P}}$. In other words, we must show

$$\widetilde{\mathbb{E}}\big[e^{u_1\widetilde{W}(t)+u_2Q(t)}\big] = \exp\left\{\frac{1}{2}u_1^2t\right\}\cdot\exp\left\{\tilde{\lambda}t\big(\tilde{\varphi}_Y(u_2)-1\big)\right\}, \qquad (11.6.35)$$

where $\tilde{\varphi}_Y(u_2)$ is given by (11.6.25). Since $e^{\frac{1}{2}u_1^2t}$ is the moment-generating function for a normal random variable with mean zero and variance t, $\exp\left\{\tilde{\lambda}t\big(\tilde{\varphi}_Y(u_2)-1\big)\right\}$ is the moment-generating function for a compound Poisson process with intensity $\tilde{\lambda}$ and jump density $\tilde{f}(y)$, and since the joint moment-generating function factors into the product of these two moment-generating functions, we would then know that $\widetilde{W}(t)$ and $Q(t)$ have the right distributions under $\widetilde{\mathbb{P}}$ and are independent.

If the process $\Theta(t)$ is independent of the process $Q(t)$, then Z_1 is independent of Q and we can obtain (11.6.35) from the following independence-based computation:

$$\begin{aligned}\widetilde{\mathbb{E}}\big[e^{u_1\widetilde{W}(t)+u_2Q(t)}\big] &= \mathbb{E}\big[e^{u_1\widetilde{W}(t)}Z_1(t)\cdot e^{u_2Q(t)}Z_2(t)\big]\\ &= \mathbb{E}\big[e^{u_1\widetilde{W}(t)}Z_1(t)\big]\cdot\mathbb{E}\big[e^{u_2Q(t)}Z_2(t)\big].\end{aligned}$$

Girsanov's Theorem from stochastic calculus for continuous processes implies

$$\mathbb{E}\big[e^{u_1\widetilde{W}(t)}Z_1(t)\big] = \exp\left\{\frac{1}{2}u_1^2t\right\},$$

and (11.6.30) implies

$$\mathbb{E}\big[e^{u_2Q(t)}Z_2(t)\big] = \exp\left\{\tilde{\lambda}t\big(\tilde{\varphi}_Y(u_2)-1\big)\right\}.$$

Equation (11.6.35) follows.

The surprising fact is that (11.6.35) and hence the conclusion of Theorem 11.6.9 hold even if $\Theta(t)$ is allowed to depend on $Q(t)$. Indeed, we could have $\Theta(t)$ *equal to* $Q(t)$. We give the proof of this fact.

PROOF OF (11.6.35): We define

$$X_1(t) = \exp\left\{u_1\widetilde{W}(t) - \frac{1}{2}u_1^2 t\right\},$$
$$X_2(t) = \exp\left\{u_2 Q(t) - \tilde{\lambda}t\big(\widetilde{\varphi}_Y(u_2) - 1\big)\right\},$$

and show below that $X_1(t)Z_1(t)$, $X_2(t)Z_2(t)$, and $X_1(t)Z_1(t)X_2(t)Z_2(t)$ are martingales under $\mathbb{P}$.

The Itô-Doeblin formula for continuous processes implies

$$\begin{aligned} dX_1(t) &= X_1(t)\Big(u_1\, d\widetilde{W}(t) - \frac{1}{2}u_1^2\, dt\Big) + \frac{1}{2}u_1^2 X_1(t)\, dt \\ &= u_1 X_1(t)\, d\widetilde{W}(t) \\ &= u_1 X_1(t)\, dW(t) + u_1\Theta(t)X_1(t)\, dt. \end{aligned}$$

The Itô-Doeblin formula also implies

$$dZ_1(t) = -\Theta(t)Z_1(t)\, dW(t).$$

Itô's product rule yields

$$\begin{aligned} d\big(X_1(t)Z_1(t)\big) &= X_1(t)\, dZ_1(t) + Z_1(t)\, dX_1(t) + dX_1(t)\, dZ_1(t) \\ &= -\Theta(t)X_1(t)Z_1(t)\, dW(t) + u_1X_1(t)Z_1(t)\, dW(t) \\ &\qquad + u_1\Theta(t)X_1(t)Z_1(t)\, dt - u_1\Theta(t)X_1(t)Z_1(t)\, dt \\ &= \big(u_1 - \Theta(t)\big)X_1(t)Z_1(t)\, dW(t). \end{aligned}$$

Because its differential has no dt term, $X_1(t)Z_1(t)$ is a martingale.

We showed in the proof of Theorem 11.6.7 that $X_2(t)Z_2(t)$ is a martingale.

Finally, because $X_1(t)Z_1(t)$ is continuous and $X_2(t)Z_2(t)$ has no Itô integral part, $[X_1Z_1, X_2Z_2](t) = 0$. Therefore, Itô's product rule implies

$$\begin{aligned} X_1(t)Z_1(t)X_2(t)Z_2(t) = 1 &+ \int_0^t X_1(s-)Z_1(s-)\, d\big(X_2(s)Z_2(s)\big) \\ &+ \int_0^t X_2(s-)Z_2(s-)\, d\big(X_1(s)Z_1(s)\big), \end{aligned}$$

and Theorem 11.4.5 implies that $X_1(t)Z_1(t)X_2(t)Z_2(t)$ is a martingale. It follows that

$$\mathbb{E}\big[X_1(t)Z_1(t)X_2(t)Z_2(t)\big] = 1;$$

this gives us (11.6.35). □

Suppose a compound Poisson process $Q(t)$ has jumps $Y_1, Y_2, \ldots$ that take only finitely many nonzero values $y_1, y_2, \ldots, y_M$, with $p(y_m) = \mathbb{P}\{Y_i = y_m\}$ so that $p(y_m) > 0$ and $\sum_{m=1}^{M} p_m = 1$. Let $\tilde{\lambda}$ be a positive constant and let $\tilde{p}(y_1), \ldots, \tilde{p}(y_M)$ be positive numbers that sum to 1. In place of (11.6.32), we now define

$$Z_2(t) = e^{(\lambda-\tilde{\lambda})t} \prod_{i=1}^{N(t)} \frac{\tilde{\lambda}\tilde{p}(Y_i)}{\lambda p(Y_i)}$$

and then define $Z(t)$ by (11.6.33). Lemma 11.6.8 still applies and permits us to define the probability measure $\widetilde{\mathbb{P}}$ by the formula $\widetilde{\mathbb{P}}(A) = \int_A Z(T)\, d\mathbb{P}$ for all $Z \in \mathcal{F}$. A straightforward modification of the proof of Theorem 11.6.9 gives the following result.

Theorem 11.6.10. *Under the probability measure $\widetilde{\mathbb{P}}$, the process*

$$\widetilde{W}(t) = W(t) + \int_0^t \Theta(s)\, ds$$

is a Brownian motion, $Q(t)$ is a compound Poisson process with intensity $\tilde{\lambda}$ and independent, identically distributed jump sizes satisfying $\widetilde{\mathbb{P}}\{Y_i = y_m\} = \tilde{p}(y_m)$ for all i and $m = 1, \ldots, M$, and the processes $\widetilde{W}(t)$ and $Q(t)$ are independent.

11.7 Pricing a European Call in a Jump Model

In this section, we consider the problem of pricing a European call when the underlying asset is a jump process. We work out the details for two cases: (1) the underlying asset is driven by a single Poisson process, and (2) the underlying asset is driven by a Brownian motion and a compound Poisson process. The market is complete in the first case and incomplete in the second. We discuss the nature of the incompleteness in the second case.

11.7.1 Asset Driven by a Poisson Process

We return to Example 11.6.3, in which the underlying asset price is given by

$$\begin{aligned} S(t) &= S(0) \exp\left\{\alpha t + N(t) \log(\sigma + 1) - \lambda\sigma t\right\} \\ &= S(0) e^{(\alpha - \lambda\sigma)t} (\sigma + 1)^{N(t)}, \end{aligned} \qquad (11.7.1)$$

for which the differential is

$$dS(t) = \alpha S(t)\, dt + \sigma S(t-)\, dM(t).$$

In this model, $N(t)$ is a Poisson process with intensity $\lambda > 0$ on a probability space $(\Omega, \mathcal{F}, \mathbb{P})$, and $M(t) = N(t) - \lambda t$ is the compensated Poisson process.

We fix a positive time T and wish to price a European call whose payoff at time T is

$$V(T) = \big(S(T) - K\big)^{+}.$$

We saw in Example 11.6.3 that we must assume $\lambda > \frac{\alpha - r}{\sigma}$ in order to rule out arbitrage. Under this assumption,

$$\tilde{\lambda} = \lambda - \frac{\alpha - r}{\sigma}$$

is positive, and there is a risk-neutral measure given by

$$\widetilde{\mathbb{P}}(A) = \int_A Z(T)\,d\mathbb{P} \text{ for all } A \in \mathcal{F},$$

where $Z(t) = e^{(\lambda - \tilde{\lambda})t}\left(\frac{\tilde{\lambda}}{\lambda}\right)^{N(t)}$. This risk-neutral measure is in fact unique; see Remark 11.7.2 below.

Under the risk-neutral measure, the compensated Poisson process $\widetilde{M}(t) = N(t) - \tilde{\lambda}t$ is a martingale, and

$$dS(t) = rS(t)\,dt + \sigma S(t-)\,d\widetilde{M}(t) \tag{11.7.2}$$

or, equivalently,

$$d\big(e^{-rt}S(t)\big) = \sigma e^{-rt}S(t-)\,d\widetilde{M}(t).$$

The discounted asset price is a martingale under $\widetilde{\mathbb{P}}$. In terms of $\tilde{\lambda}$, we may rewrite the second line in (11.7.1) as

$$S(t) = S(0)e^{(r - \tilde{\lambda}\sigma)t}(\sigma + 1)^{N(t)}.$$

For $0 \le t \le T$, let $V(t)$ denote the risk-neutral price of a European call paying $V(T) = \big(S(T) - K\big)^{+}$ at time T. The discounted call price is a martingale under the risk-neutral measure. In other words, the call price $V(t)$ satisfies

$$e^{-rt}V(t) = \widetilde{\mathbb{E}}\big[e^{-rT}V(T)\big|\mathcal{F}(t)\big] = \widetilde{\mathbb{E}}\big[e^{-rT}\big(S(T) - K\big)^{+}\big|\mathcal{F}(t)\big].$$

We have

$$\begin{aligned}
S(T) &= S(0)e^{(r-\tilde{\lambda}\sigma)t}(\sigma+1)^{N(t)} \cdot e^{(r-\tilde{\lambda}\sigma)(T-t)}(\sigma+1)^{N(T)-N(t)} \\
&= S(t) \cdot e^{(r-\tilde{\lambda}\sigma)(T-t)}(\sigma+1)^{N(T)-N(t)}.
\end{aligned}$$

It follows that

$$\begin{aligned}
V(t) &= \widetilde{\mathbb{E}}\big[e^{-r(T-t)}\big(S(T) - K\big)^{+}\big|\mathcal{F}(t)\big] \\
&= \widetilde{\mathbb{E}}\left[e^{-r(T-t)}\Big(S(t)e^{(r-\tilde{\lambda}\sigma)(T-t)}(\sigma+1)^{N(T)-N(t)} - K\Big)^{+}\Big|\mathcal{F}(t)\right].
\end{aligned}$$

The random variable $S(t)$ is $\mathcal{F}(t)$-measurable, whereas

$$e^{(r-\tilde{\lambda}\sigma)(T-t)}(\sigma+1)^{N(T)-N(t)}$$

is independent of $\mathcal{F}(t)$. According to the Independence Lemma, Lemma 2.3.4,

$$V(t) = c\big(t, S(t)\big),$$

where

$$\begin{aligned} c(t,x) &= \widetilde{\mathbb{E}}\Big[e^{-r(T-t)}\Big(xe^{(r-\tilde{\lambda}\sigma)(T-t)}(\sigma+1)^{N(T)-N(t)} - K\Big)^+\Big] \\ &= \sum_{j=0}^{\infty} e^{-r(T-t)}\Big(xe^{(r-\tilde{\lambda}\sigma)(T-t)}(\sigma+1)^j - K\Big)^+ \frac{\tilde{\lambda}^j(T-t)^j}{j!}e^{-\tilde{\lambda}(T-t)} \\ &= \sum_{j=0}^{\infty} \Big(xe^{-\tilde{\lambda}\sigma(T-t)}(\sigma+1)^j - Ke^{-r(T-t)}\Big)^+ \frac{\tilde{\lambda}^j(T-t)^j}{j!}e^{-\tilde{\lambda}(T-t)}. \end{aligned} \tag{11.7.3}$$

From this formula, the risk-neutral price of the call $c(t,x)$ can be computed. The $j=0$ term in (11.7.3) is

$$\big(xe^{-\tilde{\lambda}\sigma(T-t)} - Ke^{-r(T-t)}\big)^+ e^{-\tilde{\lambda}(T-t)}.$$

When $t=T$, this term is $(x-K)^+$, and it is the only nonzero term in the sum in (11.7.3) when $t=T$. Therefore, the function c satisfies the terminal condition

$$c(T,x) = (x-K)^+ \text{ for all } x \geq 0. \tag{11.7.4}$$

We next derive the "partial differential equation" that $c(t,x)$ must satisfy. The usual iterated conditioning argument shows that

$$e^{-rt}c\big(t,S(t)\big) = e^{-rt}V(t) = \widetilde{\mathbb{E}}\big[e^{-rT}\big(S(T)-K\big)^+\big|\mathcal{F}(t)\big]$$

is a martingale under $\widetilde{\mathbb{P}}$. Therefore, we compute $d\big(e^{-rt}c(t,S(t))\big)$ and set the "dt" term equal to zero. The stochastic differential equation (11.7.2) may be rewritten as

$$dS(t) = (r-\tilde{\lambda}\sigma)S(t)\,dt + \sigma S(t-)\,dN(t), \tag{11.7.5}$$

which shows that the continuous part of the stock price satisfies

$$dS^c(t) = (r-\tilde{\lambda}\sigma)S(t)\,dt.$$

On the other hand, if the stock price jumps at time t, then

$$\Delta S(t) = S(t) - S(t-) = \sigma S(t-), \quad S(t) = (\sigma+1)S(t-).$$

The Itô-Doeblin formula implies

$$
\begin{aligned}
&e^{-rt}c(t,S(t)) \\
&\quad = c(0,S(0)) + \int_0^t e^{-ru}\big[-rc(u,S(u))\,du + c_t(u,S(u))\,du \\
&\qquad\qquad + c_x(u,S(u))\,dS^c(u)\big] \\
&\qquad + \sum_{0<u\le t} e^{-ru}\big[c(u,S(u)) - c(u,S(u-))\big] \\
&\quad = c(0,S(0)) + \int_0^t e^{-ru}\big[-rc(u,S(u)) + c_t(u,S(u)) \\
&\qquad\qquad + (r-\tilde{\lambda}\sigma)S(u)c_x(u,S(u))\big]\,du \\
&\qquad + \int_0^t e^{-ru}\big[c(u,(\sigma+1)S(u-)) - c(u,S(u-))\big]\,dN(u) \\
&\quad = c(0,S(0)) + \int_0^t e^{-ru}\big[-rc(u,S(u)) + c_t(u,S(u)) \\
&\qquad\qquad + (r-\tilde{\lambda}\sigma)S(u)c_x(u,S(u))\big]\,du \\
&\qquad + \int_0^t e^{-ru}\big[c(u,(\sigma+1)S(u-)) - c(u,S(u-))\big]\tilde{\lambda}\,du \\
&\qquad + \int_0^t e^{-ru}\big[c(u,(\sigma+1)S(u-)) - c(u,S(u-))\big]\,d\widetilde{M}(u).
\end{aligned}
$$

However, the integral

$$
\int_0^t e^{-ru}\big[c(u,(\sigma+1)S(u-)) - c(u,S(u-))\big]\tilde{\lambda}\,du
$$

is the same as the integral

$$
\int_0^t e^{-ru}\big[c(u,(\sigma+1)S(u)) - c(u,S(u))\big]\tilde{\lambda}\,du.
$$

We have shown that

$$
\begin{aligned}
&e^{-rt}c(t,S(t)) \\
&\quad = c(0,S(0)) \\
&\qquad + \int_0^t e^{-ru}\big[-rc(u,S(u)) + c_t(u,S(u)) + (r-\tilde{\lambda}\sigma)S(u)c_x(u,S(u)) \\
&\qquad\qquad + \tilde{\lambda}\big(c(u,(\sigma+1)S(u)) - c(u,S(u))\big)\big]\,du \\
&\qquad + \int_0^t e^{-ru}\big[c(u,(\sigma+1)S(u-)) - c(u,S(u-))\big]\,d\widetilde{M}(u).
\end{aligned} \tag{11.7.6}
$$

The last integral is a martingale because the integrator $\widetilde{M}(u)$ is a martingale and the integrand is left-continuous. Because left-hand side of (11.7.6), $e^{-rt}c(t,S(t))$, is also a martingale we can then solve for

$$c(0,S(0)) + \int_0^t e^{-ru}\big[-rc(u,S(u)) + c_t(u,S(u)) + (r-\tilde{\lambda}\sigma)S(u)c_x(u,S(u)) \\ +\tilde{\lambda}\big(c(u,(\sigma+1)S(u)) - c(u,S(u))\big)\big]\,du$$

and see that it is the difference of two martingales and hence is itself a martingale. This can only happen if the integrand is zero:

$$-rc(t,S(t)) + c_t(t,S(t)) + (r-\tilde{\lambda}\sigma)S(t)c_x(t,S(t)) \\ +\tilde{\lambda}\big(c(t,(\sigma+1)S(t)) - c(t,S(t))\big) = 0. \quad (11.7.7)$$

The way we have in the past argued for (11.7.7) using (11.7.6) (see the discussion preceding Theorem 6.4.3) is by first taking the differential in (11.7.6) to obtain

$$\begin{aligned} &d\big(e^{-rt}c(t,S(t))\big) \\ &= e^{-rt}\big[-rc(t,S(t)) + c_t(t,S(t)) + (r-\tilde{\lambda}\sigma)S(t)c_x(t,S(t)) \\ &\qquad +\tilde{\lambda}\big(c(t,(\sigma+1)S(t)) - c(t,S(t))\big)\big]\,dt \\ &\qquad +e^{-rt}\big[c(t,(\sigma+1)S(t-)) - c(t,S(t-))\big]\,d\widetilde{M}(t) \end{aligned}$$

and then setting the dt term equal to zero. This still works, provided we make sure the non-dt term has a martingale integrator, and if this integrator has jumps, then the integrand for this martingale is left-continuous. In particular, we also have

$$\begin{aligned} &d\big(e^{-rt}c(t,S(t))\big) \\ &= e^{-rt}\big[-rc(t,S(t)) + c_t(t,S(t)) + (r-\tilde{\lambda}\sigma)S(t)c_x(t,S(t))\big]\,dt \\ &\qquad +e^{-rt}\big[c(t,(\sigma+1)S(t-)) - c(t,S(t-))\big]\,dN(t), \end{aligned} \quad (11.7.8)$$

but setting the "dt" term

$$e^{-rt}\big[-rc(t,S(t)) + c_t(t,S(t)) + (r-\tilde{\lambda}\sigma)S(t)c_x(t,S(t))\big]\,dt$$

in this expression equal to zero gives an incorrect result because the non-dt term has integrator $dN(t)$ and $N(t)$ is not a martingale.

We conclude by replacing the stock price process $S(t)$ in (11.7.7) by a dummy variable x. This gives the equation

$$-rc(t,x)+c_t(t,x)+(r-\tilde{\lambda}\sigma)xc_x(t,x)+\tilde{\lambda}\big(c(t,(\sigma+1)x)-c(t,x)\big) = 0, \quad (11.7.9)$$

which must hold for $0 \le t < T$ and $x \ge 0$. This is sometimes called a *differential-difference* equation because it involves c at two different values of the stock price, namely x and $(\sigma+1)x$. The function $c(t,x)$ defined by (11.7.3) satisfies this equation because, by its construction, $e^{-rt}c(t,S(t))$ is a martingale under $\widetilde{\mathbb{P}}$.

Returning to (11.7.6) and using equation (11.7.9), we see that for $0 \leq t \leq T$,

$$\begin{aligned} & e^{-rt}c(t,S(t)) \\ &= c(0,S(0)) + \int_0^t e^{-ru}\big[c(u,(\sigma+1)S(u-)) - c(u,S(u-))\big]\,d\widetilde{M}(u). \quad (11.7.10) \end{aligned}$$

In particular,

$$\begin{aligned} & e^{-rT}\big(S(T)-K\big)^+ \\ &= e^{-rT}c(T,S(T)) \\ &= c(0,S(0)) + \int_0^T e^{-ru}\big[c(u,(\sigma+1)S(u-)) - c(u,S(u-))\big]\,d\widetilde{M}(u). \quad (11.7.11) \end{aligned}$$

We use this observation to construct the hedge for a short position in the call.

Suppose we sell the call at time zero in exchange for initial capital $X(0) = c(0,S(0))$. We want to invest in the stock and money market account so that $X(t) = c(t,S(t))$ for all t or, equivalently,

$$e^{-rt}X(t) = e^{-rt}c(t,S(t)) \text{ for all } t \in [0,T].$$

To accomplish this, we match differentials. From (11.7.10), we see that the differential of $e^{-rt}c(t,S(t))$ is

$$d\big(e^{-rt}c(t,S(t))\big) = e^{-rt}\big[c(t,(\sigma+1)S(t-)) - c(t,S(t-))\big]\,d\widetilde{M}(t). \quad (11.7.12)$$

The differential of the value $X(t)$ of a portfolio that at each time t holds $\Gamma(t)$ shares of stock (we use $\Gamma(t)$ rather than $\Delta(t)$ to denote the number of shares of stock held in the hedging portfolio to avoid confusion with the use of Δ as the size of the jump in a process) is

$$dX(t) = \Gamma(t-)\,dS(t) + r\big[X(t) - \Gamma(t)S(t)\big]\,dt.$$

Therefore,

$$\begin{aligned} d\big(e^{-rt}X(t)\big) &= e^{-rt}[-rX(t)\,dt + dX(t)] \\ &= e^{-rt}[\Gamma(t-)\,dS(t) - r\Gamma(t)S(t)\,dt] \\ &= e^{-rt}\sigma\Gamma(t-)S(t-)\,d\widetilde{M}(t), \quad (11.7.13) \end{aligned}$$

where we have used (11.7.2) in the last step. We are interested in determining the value of $\Gamma(t-)$, the position held just before any jump that may occur at time t. Comparing (11.7.12) and (11.7.13), we conclude that we should take

$$\Gamma(t-) = \frac{c(t,(\sigma+1)S(t-)) - c(t,S(t-))}{\sigma S(t-)}. \quad (11.7.14)$$

This is the hedging position we should hold at all times, whether they are jump times or not. More specifically, if we define

$$\Gamma(t) = \frac{c(t,(\sigma+1)S(t)) - c(t,S(t))}{\sigma S(t)} \text{ for all } t \in [0,T], \tag{11.7.15}$$

then (11.7.14) will also hold and integration of (11.7.13) yields

$$e^{-rt}X(t) \\ = X(0) + \int_0^t e^{-ru}\big[c(u,(\sigma+1)S(u-)) - c(u,S(u-))\big]\,d\widetilde{M}(u). \tag{11.7.16}$$

Comparison of (11.7.10) with (11.7.16) shows that $X(t) = c(t,S(t))$ for all t. In particular, (11.7.11) shows that $X(T) = \big(S(T)-K\big)^+$; the short position in the European call has been hedged.

Remark 11.7.1 (Sanity check). To convince ourselves that the hedge (11.7.15) really works, we consider separately the cases when the stock jumps at time t and when the stock does not jump at time t. In the event of a jump, the change in the option price is $c(t,(\sigma+1)S(t-)) - c(t,S(t-))$. The change in the hedging portfolio value is

$$\Gamma(t-)\big(S(t)-S(t-)\big) = \Gamma(t-)\sigma S(t-) = c(t,(\sigma+1)S(t-)) - c(t,S(t-)),$$

which agrees with the change in the option price.

On the other hand, if the stock price does not jump at time t, then the stock price follows equation (11.7.5) without the $dN(t)$ term at time t:

$$dS(t) = (r - \tilde{\lambda}\sigma)S(t)\,dt.$$

At this time, (11.7.8) shows that the discounted option price has the differential

$$\begin{aligned} &d\big(e^{-rt}c(t,S(t))\big) \\ &= e^{-rt}\big[-rc(t,S(t)) + c_t(t,S(t)) + (r-\tilde{\lambda}\sigma)S(t)c_x(t,S(t))\big]\,dt \\ &= -e^{-rt}\tilde{\lambda}\big[c(t,(\sigma+1)S(t)) - c(t,S(t))\big]\,dt, \end{aligned}$$

where we have used the differential-difference equation (11.7.9) to obtain the second equality. The differential of the discounted portfolio value at this time is (from (11.7.13) without the $dN(t)$ term implicit in $d\widetilde{M}(t)$)

$$\begin{aligned} d\big(e^{-rt}X(t)\big) &= e^{-rt}\sigma\Gamma(t)S(t)(-\tilde{\lambda}\,dt) \\ &= -e^{-rt}\tilde{\lambda}\big[c(t,(\sigma+1)S(t)) - c(t,S(t))\big]\,dt. \end{aligned}$$

Once again, the discounted portfolio value tracks the discounted option price.
□

Remark 11.7.2 (Completeness). In this subsection, we have constructed the price and hedge for a European call on a stock driven by a single Poisson process. It is clear from the analysis that this same argument would work for an arbitrary European derivative security with payoff $h(S(T))$ at time T written on a stock modeled this way. One could simply replace the call payoff by the function h in equation (11.7.3). The differential-difference equation (11.7.9) would still apply, although now with terminal condition $c(T, x) = h(x)$ replacing (11.7.4), and the hedging formula (11.7.15) would still be correct.

The model is complete and the risk-neutral measure is unique if and only if every derivative security can be hedged (Second Fundamental Theorem of Asset Pricing, Theorem 5.4.9). "Every" derivative security means also those derivative securities that are path-dependent. We have not considered path-dependent derivative securities in this subsection, but one can show that they also can be hedged, and thus the model is complete.

11.7.2 Asset Driven by a Brownian Motion and a Compound Poisson Process

Let $(\Omega, \mathcal{F}, \mathbb{P})$ be a probability space on which is defined a Brownian motion $W(t)$, $0 \leq t \leq T$, and M independent Poisson processes $N_1(t), \dots, N_M(t)$, $0 \leq t \leq T$. Let $\mathcal{F}(t)$, $0 \leq t \leq T$, be the filtration generated by the Brownian motion and the M Poisson processes.

Let $\lambda_m > 0$ be the intensity of the mth Poisson process and let $-1 < y_1 < \cdots < y_M$ be nonzero numbers. Set

$$N(t) = \sum_{m=1}^{M} N_m(t), \quad Q(t) = \sum_{m=1}^{M} y_m N_m(t).$$

Then N is a Poisson process with intensity $\lambda = \sum_{m=1}^{M} \lambda_m$ and Q is a compound Poisson process. Let Y_i denote the size of the ith jump of Q. Then the Y_i random variables take values in the set $\{y_1, \dots, y_M\}$, and $Q(t)$ can be written as

$$Q(t) = \sum_{i=1}^{N(t)} Y_i.$$

Define

$$p(y_m) = \frac{\lambda_m}{\lambda}.$$

The random variables $Y_1, Y_2, \dots$ are independent and identically distributed, with $\mathbb{P}\{Y_i = y_m\} = p(y_m)$. These assertions all follow from Theorem 11.3.3.

Set

$$\beta = \mathbb{E}Y_i = \sum_{m=1}^{M} y_m p(y_m) = \frac{1}{\lambda} \sum_{m=1}^{M} \lambda_m y_m. \tag{11.7.17}$$

According to Theorem 11.3.1,

$$Q(t) - \beta\lambda t = Q(t) - t\sum_{m=1}^{M} \lambda_m y_m$$

is a martingale.

In this subsection, the stock price will be modeled by the stochastic differential equation

$$\begin{aligned} dS(t) &= \alpha S(t)\,dt + \sigma S(t)\,dW(t) + S(t-)d\big(Q(t) - \beta\lambda t\big) \\ &= (\alpha - \beta\lambda)S(t)\,dt + \sigma S(t)\,dW(t) + S(t-)\,dQ(t). \end{aligned} \quad (11.7.18)$$

Under the original probability measure $\mathbb{P}$, the mean rate of return on the stock is α. The assumption that $y_i > -1$ for $i = 1, \ldots, M$ guarantees that although the stock price can jump down, it cannot jump from a positive to a negative value or to zero. We begin with a positive initial stock price $S(0)$, and the stock price is positive at all subsequent times; see (11.7.19) below. If $S(0) = 0$, then $S(t) = 0$ for all t.

Theorem 11.7.3. *The solution to (11.7.18) is*

$$S(t) = S(0)\exp\left\{\sigma W(t) + \left(\alpha - \beta\lambda - \frac{1}{2}\sigma^2\right)t\right\}\prod_{i=1}^{N(t)}(Y_i + 1). \quad (11.7.19)$$

PROOF: We show that $S(t)$ defined by the right-hand side of (11.7.19) satisfies the stochastic differential equation (11.7.18). Toward this end, define the continuous stochastic process

$$X(t) = S(0)\exp\left\{\sigma W(t) + \left(\alpha - \beta\lambda - \frac{1}{2}\sigma^2\right)t\right\}$$

and the pure jump process

$$J(t) = \prod_{i=1}^{N(t)}(Y_i + 1).$$

Then $S(t) = X(t)J(t)$. We show that $S(t) = X(t)J(t)$ is a solution to the stochastic differential equation (11.7.18).

The Itô-Doeblin formula for a continuous process says that

$$dX(t) = (\alpha - \beta\lambda)X(t)\,dt + \sigma X(t)\,dW(t). \quad (11.7.20)$$

At the time of the ith jump, $J(t) = J(t-)(Y_i + 1)$ and hence

$$\Delta J(t) = J(t) - J(t-) = J(t-)Y_i = J(t-)\Delta Q(t).$$

The equation $\Delta J(t) = J(t-)\Delta Q(t)$ also holds at nonjump times, with both sides equal to zero. Therefore,

$$dJ(t) = J(t-)\,dQ(t). \tag{11.7.21}$$

Itô's product rule for jump processes implies that

$$S(t) = X(t)J(t) = S(0) + \int_0^t X(s-)\,dJ(s) + \int_0^t J(s)\,dX(s) + [X, J](t). \tag{11.7.22}$$

Since J is a pure jump process and X is continuous, $[X, J](t) = 0$. Substituting (11.7.20) and (11.7.21) into (11.7.22), we obtain

$$\begin{aligned} S(t) &= X(t)J(t) \\ &= S(0) + \int_0^t X(s-)J(s-)\,dQ(s) + (\alpha - \beta\lambda)\int_0^t J(s)X(s)\,ds \\ &\quad + \sigma\int_0^t J(s)X(s)\,dW(s), \end{aligned}$$

which in differential form is

$$\begin{aligned} dS(t) &= d\big(X(t)J(t)\big) \\ &= X(t-)J(t-)\,dQ(t) + (\alpha - \beta\lambda)J(t)X(t)\,dt + \sigma J(t)X(t)\,dW(t) \\ &= S(t-)\,dQ(t) + (\alpha - \beta\lambda)S(t)\,dt + \sigma S(t)\,dW(t). \end{aligned}$$

This is (11.7.18). □

We now undertake to construct a risk-neutral measure. Let θ be a constant and let $\tilde{\lambda}_1, \dots, \tilde{\lambda}_M$ be positive constants.[2] Define

$$\begin{aligned} Z_0(t) &= \exp\Big\{ -\theta W(t) - \frac{1}{2}\theta^2 t\Big\}, \\ Z_m(t) &= e^{(\lambda_m - \tilde{\lambda}_m)t}\left(\frac{\tilde{\lambda}_m}{\lambda_m}\right)^{N_m(t)}, \quad m = 1, \dots, M, \\ Z(t) &= Z_0(t)\prod_{m=1}^M Z_m(t), \\ \widetilde{\mathbb{P}}(A) &= \int_A Z(T)\,d\mathbb{P} \text{ for all } A \in \mathcal{F}. \end{aligned}$$

The following assertions follow from Theorem 11.6.10 and Corollary 11.3.4. Independence under $\widetilde{\mathbb{P}}$ between $\widetilde{W}$ and each of the Poisson processes N_m, asserted in (iii) below, follows from Corollary 11.5.3. Under the probability measure $\widetilde{\mathbb{P}}$,

[2] One could create more risk-neutral measures than we consider here by letting θ and $\tilde{\lambda}_1, \dots, \tilde{\lambda}_M$ be adapted stochastic processes.

(i) the process

$$\widetilde{W}(t) = W(t) + \theta t \tag{11.7.23}$$

is a Brownian motion,

(ii) each N_m is a Poisson process with intensity $\tilde{\lambda}_m$, and

(iii) $\widetilde{W}$ and $N_1, \dots, N_m$ are independent of one another.

Define

$$\tilde{\lambda} = \sum_{m=1}^{M} \tilde{\lambda}_m, \quad \tilde{p}(y_m) = \frac{\tilde{\lambda}_m}{\tilde{\lambda}}.$$

Under $\widetilde{\mathbb{P}}$, the process $N(t) = \sum_{m=1}^{M} N_m(t)$ is Poisson with intensity $\tilde{\lambda}$, the jump-size random variables $Y_1, Y_2, \dots$ are independent and identically distributed with $\widetilde{\mathbb{P}}\{Y_i = y_m\} = \tilde{p}(y_m)$, and $Q(t) - \tilde{\beta}\tilde{\lambda}t$ is a martingale, where

$$\tilde{\beta} = \widetilde{\mathbb{E}}Y_i = \sum_{m=1}^{M} y_m \tilde{p}(y_m) = \frac{1}{\tilde{\lambda}} \sum_{m=1}^{M} \tilde{\lambda}_m y_m.$$

The probability measure $\widetilde{\mathbb{P}}$ is risk-neutral if and only if the mean rate of return of the stock under $\widetilde{\mathbb{P}}$ is the interest rate r. In other words, $\widetilde{\mathbb{P}}$ is risk-neutral if and only if

$$\begin{aligned} dS(t) &= (\alpha - \beta\lambda)S(t)\,dt + \sigma S(t)\,dW(t) + S(t-)\,dQ(t) \\ &= rS(t)\,dt + \sigma S(t)\,d\widetilde{W}(t) + S(t-)d(Q(t) - \tilde{\beta}\tilde{\lambda}t). \end{aligned} \tag{11.7.24}$$

This is equivalent to the equation

$$\alpha - \beta\lambda = r + \sigma\theta - \tilde{\beta}\tilde{\lambda}, \tag{11.7.25}$$

which is the *market price of risk equation* for this model. Recalling the definitions of β and $\tilde{\beta}$, we may rewrite the market price of risk equation (11.7.25) as

$$\begin{aligned} \alpha - r &= \sigma\theta + \beta\lambda - \tilde{\beta}\tilde{\lambda} \\ &= \sigma\theta + \sum_{m=1}^{M} (\lambda_m - \tilde{\lambda}_m) y_m. \end{aligned} \tag{11.7.26}$$

Because there is one equation and $M + 1$ unknowns, $\theta, \tilde{\lambda}_1, \dots, \tilde{\lambda}_M$, there are multiple risk-neutral measures.

Extra stocks would help determine a unique risk-neutral measure. We illustrate this point by taking $M = 2$ in the following example.

Example 11.7.4 (Three stocks and two Poisson processes). With one Brownian motion W and two independent Poisson processes N_1 and N_2, define three compound Poisson processes

$$Q_i(t) = y_{i,1}N_1(t) + y_{i,2}N_2(t), \; i = 1, 2, 3,$$

where $y_{i,m} > -1$ for $i = 1, 2, 3$ and $m = 1, 2$. Set

$$\beta_i = \frac{1}{\lambda}(\lambda_1 y_{i,1} + \lambda_2 y_{i,2}), \; i = 1, 2, 3,$$

where λ_1 and λ_2 are the intensities of N_1 and N_2, respectively, under the original measure $\mathbb{P}$. For $i = 1, 2, 3$, we have a stock process modeled by

$$dS_i(t) = (\alpha_i - \beta_i\lambda)S_i(t)\,dt + \sigma_i S_i(t)\,dW(t) + S_i(t-)\,dQ_i(t).$$

In this model, there is a market price of risk equation analogous to (11.7.26) for each stock. The market price of risk equations are

$$\begin{aligned}
\alpha_1 - r &= \sigma_1\theta + (\lambda_1 - \tilde{\lambda}_1)y_{1,1} + (\lambda_2 - \tilde{\lambda}_2)y_{1,2},\\
\alpha_2 - r &= \sigma_2\theta + (\lambda_1 - \tilde{\lambda}_1)y_{2,1} + (\lambda_2 - \tilde{\lambda}_2)y_{2,2},\\
\alpha_3 - r &= \sigma_3\theta + (\lambda_1 - \tilde{\lambda}_1)y_{3,1} + (\lambda_2 - \tilde{\lambda}_2)y_{3,2}.
\end{aligned}$$

These are three equations in the three unknowns θ, $\tilde{\lambda}_1$, and $\tilde{\lambda}_2$. If they have a unique solution, then there is a unique risk-neutral measure. In that case, the market would be complete and free of arbitrage. □

We return to the discussion of the model with a single stock given by (11.7.18) and (11.7.19). Let us choose some θ and $\tilde{\lambda}_1, \dots, \tilde{\lambda}_M$ satisfying the market price of risk equations (11.7.26). Then, in the notation of (11.7.24), we have

$$\begin{aligned}
dS(t) &= rS(t) + \sigma S(t)\,d\widetilde{W}(t) + S(t-)d(Q(t) - \tilde{\beta}\tilde{\lambda}t)\\
&= (r - \tilde{\beta}\tilde{\lambda})\,dt + \sigma S(t)\,d\widetilde{W}(t) + S(t-)dQ(t). \qquad (11.7.27)
\end{aligned}$$

This is like equation (11.7.18), and just as (11.7.19) is the solution to (11.7.18), the solution to (11.7.27) is

$$S(t) = S(0)\exp\left\{\sigma\widetilde{W}(t) + \left(r - \tilde{\beta}\tilde{\lambda} - \frac{1}{2}\sigma^2\right)t\right\}\prod_{i=1}^{N(t)}(Y_i + 1). \qquad (11.7.28)$$

Indeed, it is a straightforward matter to use (11.7.25) to verify that (11.7.19) and (11.7.28) are in fact the same equation. We have not changed the stock price process; we have changed only its distribution.

We compute the risk-neutral price of a call on the stock with price process given by (11.7.28). Because θ does not appear explicitly in (11.7.28), it will not appear in our pricing formula. However,

$$\tilde{\beta}\tilde{\lambda} = \sum_{m=1}^{M}\tilde{\lambda}_m y_m$$

will appear in this formula, and we can choose the risk-neutral intensities $\tilde{\lambda}_1, \dots, \tilde{\lambda}_M$ to be any positive constants and subsequently choose θ so that the market price of risk equation (11.7.25) is satisfied. We assume for the remainder of this section that some choice has been made. Our pricing formula will depend on the choice. It is common to use these free parameters to calibrate the model to market data.

For the next step, we need some notation. Define

$$\kappa(\tau, x) = xN(d_+(\tau, x)) - Ke^{-r\tau}N(d_-(\tau, x)), \tag{11.7.29}$$

where

$$d_\pm(\tau, x) = \frac{1}{\sigma\sqrt{\tau}}\left[\log\frac{x}{K} + \left(r \pm \frac{1}{2}\sigma^2\right)\tau\right]$$

and

$$N(y) = \frac{1}{\sqrt{2\pi}}\int_{\infty}^{y} e^{-\frac{1}{2}z^2}\,dz$$

is the cumulative standard normal distribution function. In other words, $\kappa(\tau, x)$ is the standard Black-Scholes-Merton call price on a geometric Brownian motion with volatility σ when the current stock price is x, the expiration date is τ time units in the future, the interest rate is r, and the strike price is K. We have

$$\kappa(\tau, x) = \widetilde{\mathbb{E}}\left[e^{-r\tau}\left(x\exp\left\{-\sigma\sqrt{\tau}Y + \left(r - \frac{1}{2}\sigma^2\right)\tau\right\} - K\right)^+\right],$$

where Y is a standard normal random variable under $\widetilde{\mathbb{P}}$; see Subsection 5.2.5.

Theorem 11.7.5. *For $0 \le t < T$, the risk-neutral price of a call,*

$$V(t) = \widetilde{\mathbb{E}}\left[e^{-r(T-t)}(S(T) - K)^+ \big| \mathcal{F}(t)\right],$$

is given by $V(t) = c(t, S(t))$, where

$$c(t, x) = \sum_{j=0}^{\infty} e^{-\tilde{\lambda}(T-t)}\frac{\tilde{\lambda}^j(T-t)^j}{j!}\,\widetilde{\mathbb{E}}\,\kappa\left(T-t, xe^{-\tilde{\beta}\tilde{\lambda}(T-t)}\prod_{i=1}^{j}(Y_i+1)\right). \tag{11.7.30}$$

PROOF: Let $t \in [0, T)$ be given and define $\tau = T - t$. From (11.7.28), we see that

$$S(T) = S(t)\exp\left\{\sigma(\widetilde{W}(T) - \widetilde{W}(t)) + \left(r - \tilde{\beta}\tilde{\lambda} - \frac{1}{2}\sigma^2\right)\tau\right\}\prod_{i=N(t)+1}^{N(T)}(Y_i + 1). \tag{11.7.31}$$

The term $S(t)$ is $\mathcal{F}(t)$-measurable, and the other term appearing on the right-hand side of (11.7.31) is independent of $\mathcal{F}(t)$. Therefore, the Independence Lemma, Lemma 2.3.4, implies that

$$V(t) = \widetilde{\mathbb{E}}\big[e^{-r\tau}(S(T)-K)^+\big|\mathcal{F}(t)\big] = c(t,S(t)),$$

where

$$\begin{aligned}
c(t,x)& \\
&= \widetilde{\mathbb{E}}\left[e^{-r\tau}\left(x\exp\left\{\sigma(\widetilde{W}(T)-\widetilde{W}(t))+\left(r-\tilde{\beta}\tilde{\lambda}-\frac{1}{2}\sigma^2\right)\tau\right\}\right.\right.\\
&\qquad\left.\left.\times\prod_{i=N(t)+1}^{N(T)}(Y_i+1)-K\right)^+\right]\\
&= \widetilde{\mathbb{E}}\left[\widetilde{\mathbb{E}}\left[e^{-r\tau}\left(x\exp\left\{\sigma(\widetilde{W}(T)-\widetilde{W}(t))+\left(r-\tilde{\beta}\tilde{\lambda}-\frac{1}{2}\sigma^2\right)\tau\right\}\right.\right.\right.\\
&\qquad\left.\left.\left.\times\prod_{i=N(t)+1}^{N(T)}(Y_i+1)-K\right)^+\middle|\,\sigma\left(\prod_{i=N(t)+1}^{N(T)}(Y_i+1)\right)\right]\right]\\
&= \widetilde{\mathbb{E}}\left[\widetilde{\mathbb{E}}\left[e^{-r\tau}\left(xe^{-\tilde{\beta}\tilde{\lambda}}\exp\left\{-\sigma\sqrt{\tau}Y+\left(r-\frac{1}{2}\sigma^2\right)\tau\right\}\right.\right.\right.\\
&\qquad\left.\left.\left.\times\prod_{i=N(t)+1}^{N(T)}(Y_i+1)-K\right)^+\middle|\,\sigma\left(\prod_{i=N(t)+1}^{N(T)}(Y_i+1)\right)\right]\right],
\end{aligned}$$

where

$$Y = -\frac{\widetilde{W}(T)-\widetilde{W}(t)}{\sqrt{\tau}}$$

is a standard normal random variable under $\widetilde{\mathbb{P}}$, and where the conditioning σ-algebra $\sigma\left(\prod_{i=N(t)+1}^{N(T)}(Y_i+1)\right)$ is the one generated by the random variable $\prod_{i=N(t)+1}^{N(T)}(Y_i+1)$. Because $\prod_{i=N(t)+1}^{N(T)}(Y_i+1)$ is $\sigma\left(\prod_{i=N(t)+1}^{N(T)}(Y_i+1)\right)$-measurable and Y is independent of $\sigma\left(\prod_{i=N(t)+1}^{N(T)}(Y_i+1)\right)$, we may use the Independence Lemma, Lemma 2.3.4, again to obtain

$$\begin{aligned}
&\widetilde{\mathbb{E}}\left[e^{-r\tau}\left(xe^{-\tilde{\beta}\tilde{\lambda}}\exp\left\{-\sigma\sqrt{\tau}Y+\left(r-\frac{1}{2}\sigma^2\right)\tau\right\}\right.\right.\\
&\qquad\left.\left.\times\prod_{i=N(t)+1}^{N(T)}(Y_i+1)-K\right)^+\middle|\,\sigma\left(\prod_{i=N(t)+1}^{N(T)}(Y_i+1)\right)\right]\\
&\qquad\qquad= \kappa\left(\tau, xe^{-\tilde{\beta}\tilde{\lambda}\tau}\prod_{i=N(t)+1}^{N(T)}(Y_i+1)\right).
\end{aligned}$$

It follows that

$$c(t,x) = \mathbb{E}\,\kappa\left(\tau, xe^{-\tilde{\beta}\tilde{\lambda}\tau} \prod_{i=N(t)+1}^{N(T)} (Y_i + 1)\right). \tag{11.7.32}$$

To see that (11.7.32) agrees with (11.7.30), we note that conditioned on $N(T) - N(t) = j$, the random variable $\prod_{i=N(t)+1}^{N(T)}(Y_i+1)$ has the same distribution as $\prod_{i=1}^{j}(Y_i+1)$. Furthermore,

$$\mathbb{P}\{N(T) - N(t) = j\} = e^{-\tilde{\lambda}\tau}\frac{\tilde{\lambda}^j \tau^j}{j!}.$$

□

Remark 11.7.6 (Continuous jump distribution). Suppose the jump sizes Y_i have a density $f(y)$ rather than a probability mass function $p(y_1), \dots, p(y_m)$, and this density is strictly positive on a set $B \subset (-1, \infty)$ and zero elsewhere. In this case, we replace (11.7.17) by the formula

$$\beta = \mathbb{E}Y_i = \int_{-1}^{\infty} y f(y)\, dy.$$

For the risk-neutral measure, we can choose θ, $\tilde{\lambda} > 0$ and any density $\tilde{f}(y)$ that is strictly positive on B and zero elsewhere so that the market price of risk equation (see (11.7.26))

$$\alpha - r = \sigma\theta + \beta\lambda - \tilde{\beta}\tilde{\lambda}$$

is satisfied, where now

$$\tilde{\beta} = \widetilde{\mathbb{E}}Y_i = \int_{-1}^{\infty} y \tilde{f}(y)\, dy.$$

Under these conditions, Theorem 11.7.5 still holds. □

We return to the model with discrete jump sizes. The following theorem provides the differential-difference equation satisfied by the call price.

Theorem 11.7.7. *The call price $c(t,x)$ of (11.7.30) satisfies the equation*

$$-rc(t,x) + c_t(t,x) + (r - \tilde{\beta}\tilde{\lambda})xc_x(t,x) + \frac{1}{2}\sigma^2 x^2 c_{xx}(t,x)$$
$$+\tilde{\lambda}\Big[\sum_{m=1}^{M} \tilde{p}(y_m)c(t,(y_m+1)x) - c(t,x)\Big] = 0,\ 0 \le t < T,\ x \ge 0, \tag{11.7.33}$$

and the terminal condition

$$c(T,x) = (x-K)^+,\ x \ge 0.$$

PROOF: From (11.7.27), we see that the continuous part of the stock price satisfies $dS^c(t) = (r - \tilde{\beta}\tilde{\lambda})S(t)\,dt + \sigma S(t)\,d\widetilde{W}(t)$. Therefore, the Itô-Doeblin formula implies

$$\begin{aligned}
&e^{-rt}c(t,S(t)) - c(0,S(0))\\
&= \int_0^t e^{-ru}\Big[-rc(u,S(u)) + c_t(u,S(u)) + (r-\tilde{\beta}\tilde{\lambda})S(u)c_x(u,S(u))\\
&\qquad + \frac{1}{2}\sigma^2 S^2(u)c_{xx}(u,S(u))\Big]\,du + \int_0^t e^{-ru}\sigma S(u)c_x(u,S(u))\,d\widetilde{W}(u)\\
&\quad + \sum_{0<u\le t} e^{-ru}\big[c(u,S(u)) - c(u,S(u-))\big]. \qquad (11.7.34)
\end{aligned}$$

We examine the last term in (11.7.34). If u is a jump time of the mth Poisson process N_m, the stock price satisfies $S(u) = (y_m+1)S(u-)$. Therefore,

$$\begin{aligned}
&\sum_{0<u\le t} e^{-ru}\big[c(u,S(u)) - c(u,S(u-))\big]\\
&= \sum_{m=1}^{M}\sum_{0<u\le t} e^{-ru}\big[c(u,(y_m+1)S(u-)) - c(u,S(u-))\big]\Delta N_m(u)\\
&= \sum_{m=1}^{M}\int_0^t e^{-ru}\big[c(u,(y_m+1)S(u-)) - c(u,S(u-))\big]\,d(N_m(u)-\tilde{\lambda}_m u)\\
&\qquad + \int_0^t e^{-ru}\Big[\sum_{m=1}^{M}\frac{\tilde{\lambda}_m}{\tilde{\lambda}}c(u,(y_m+1)S(u)) - c(u,S(u))\Big]\tilde{\lambda}\,du\\
&= \sum_{m=1}^{M}\int_0^t e^{-ru}\big[c(u,(y_m+1)S(u-)) - c(u,S(u-))\big]\,d(N_m(u)-\tilde{\lambda}_m u)\\
&\qquad + \int_0^t e^{-ru}\tilde{\lambda}\sum_{m=1}^{M}\big[\tilde{p}(y_m)c(u,(y_m+1)S(u)) - c(u,S(u))\big]\Big\}\,du.
\end{aligned}$$

Substituting this into (11.7.34) and taking differentials, we obtain

$$\begin{aligned}
&d\big(e^{-rt}c(t,S(t))\big)\\
&= e^{-rt}\Big\{-rc(t,S(t)) + c_t(t,S(t)) + (r-\tilde{\beta}\tilde{\lambda})S(t)c_x(t,S(t))\\
&\qquad + \frac{1}{2}\sigma^2 S^2(t)c_{xx}(t,S(t))\\
&\qquad + \tilde{\lambda}\sum_{m=1}^{M}\big[\tilde{p}(y_m)c(t,(y_m+1)S(t)) - c(t,S(t))\big]\Big\}\,dt\\
&\quad + e^{-rt}\sigma S(t)c_x(t,S(t))\,d\widetilde{W}(t)\\
&\quad + \sum_{m=1}^{M} e^{-rt}\big[c(t,(y_m+1)S(t-)) - c(t,S(t-))\big]\,d(N_m(t)-\tilde{\lambda}_m t). \qquad (11.7.35)
\end{aligned}$$

The integrators $N_m(t) - \tilde{\lambda}_m t$ in the last term are martingales under $\widetilde{\mathbb{P}}$, and the integrands $e^{-rt}[c(t,(y_m+1)S(t-)) - c(t,S(t-))]$ are left-continuous. Therefore, the integral of this term is a martingale. Likewise, the integral of the next-to-last term $e^{-rt}c_x(t,S(t))\,d\widetilde{W}(t)$ is a martingale. Since the discounted option price appearing on the left-hand side of (11.7.35) is also a martingale, the remaining term in (11.7.35) is a martingale as well. Because the remaining term is a dt term, it must be zero. Replacing the price process $S(t)$ by the dummy variable x in the integrand of this term, we obtain (11.7.33). □

Corollary 11.7.8. *The call price $c(t,x)$ of (11.7.30) satisfies*

$$\begin{aligned}
&d\big(e^{-rt}c(t,S(t))\big)\\
&= e^{-rt}\sigma S(t)c_x(t,S(t))\,d\widetilde{W}(t)\\
&\quad+\sum_{m=1}^{M} e^{-rt}\big[c(t,(y_m+1)S(t-)) - c(t,S(t-))\big]\,d(N_m(t) - \tilde{\lambda}_m t)\\
&= e^{-rt}\sigma S(t)c_x(t,S(t))\,d\widetilde{W}(t)\\
&\quad+e^{-rt}\big[c(t,S(t)) - c(t,S(t-))\big]\,dN(t)\\
&\quad-e^{-rt}\tilde{\lambda}\left[\sum_{m=1}^{M}\tilde{p}(y_m)c\big(t,(y+1)S(t-)\big) - c\big(t,S(t-)\big)\right]dt. \qquad (11.7.36)
\end{aligned}$$

PROOF: We use (11.7.33) to cancel the dt term in (11.7.35) and obtain the first equality in (11.7.36). For the second equality, recall that $N(t) = \sum_{m=1}^{M} N_m(t)$, $\tilde{\lambda} = \sum_{m=1}^{M}\tilde{\lambda}_m$, and $\tilde{\lambda}\tilde{p}(y_m) = \tilde{\lambda}_m$. □

Remark 11.7.9 (Continuous jump distribution). There are modifications of Theorem 11.7.7 and Corollary 11.7.8 for the case when the jump sizes Y_i have a density $\tilde{f}(y)$ under the risk-neutral measure $\widetilde{\mathbb{P}}$. In (11.7.33), the term $\sum_{m=1}^{M}\tilde{p}(y_m)c(t,(y_m+1)x)$ would be replaced by $\int_{-1}^{\infty} c(t,(y+1)x)\tilde{f}(y)\,dy$. In (11.7.36), we would use the second formula for $d\big(e^{-rt}c(t,S(t))\big)$, which is written in terms of the total number of jumps (i.e., in terms of the Poisson process $N(t) = \sum_{n=1}^{M} N_m(t)$) rather than in terms of the individual Poisson processes N_m, and replace $\sum_{m}^{M}\tilde{p}(y_m)c\big(t,(y_m+1)S(t-)\big)$ by $\int_{-1}^{\infty} c(t,y+1)S(t-))\tilde{f}(y)\,dy$ □

Finally, we think about hedging a short position in the European call whose discounted price satisfies (11.7.36). Suppose we begin with a short call position and a hedging portfolio whose initial capital is $X(0) = c(0,S(0))$. We compare the differential of the discounted call price with the differential of the discounted value of the hedging portfolio. If $\Gamma(t)$ shares of stock are held by the hedging portfolio at each time t, then

$$dX(t) = \Gamma(t-)\,dS(t) + r\big[X(t) - \Gamma(t)S(t)\big]\,dt$$

and

$$\begin{aligned} d\big(e^{-rt}X(t)\big) &= e^{-rt}\big[-rX(t)\,dt + dX(t)\big] \\ &= e^{-rt}\big[\Gamma(t-)\,dS(t) - r\Gamma(t)S(t)\,dt\big] \\ &= e^{-rt}\big[\Gamma(t)\sigma S(t)\,d\widetilde{W}(t) + \Gamma(t-)S(t-)\,d(Q(t) - \tilde{\beta}\lambda t)\big] \\ &= e^{-rt}\big[\Gamma(t)\sigma S(t)\,d\widetilde{W}(t) \\ &\qquad + \Gamma(t-)S(t-)\sum_{m=1}^{M} y_m\,(dN_m(t) - \tilde{\lambda}_m\,dt)\big], \end{aligned} \tag{11.7.37}$$

where we have used (11.7.27). It is natural to try the "delta-hedging" strategy

$$\Gamma(t) = c_x(t, S(t)).$$

This equates the $d\widetilde{W}(t)$ terms in (11.7.36) and (11.7.37) (i.e., it provides a perfect hedge against the risk introduced by the Brownian motion).

However, the delta hedge leaves us with

$$\begin{aligned} &d\big[e^{-rt}c(t, S(t)) - e^{-rt}X(t)\big] \\ &= \sum_{m=1}^{M} e^{-rt}\big[c(t, (y_m+1)S(t-)) - c(t, S(t-)) - y_m S(t-)c_x(t, S(t-))\big] \\ &\qquad \times (dN_m(t) - \tilde{\lambda}_m\,dt). \end{aligned} \tag{11.7.38}$$

The function $c(t,x)$ is strictly convex in x. This is a consequence of the strict convexity of the function $\kappa(\tau, x)$ of (11.7.29) and equation (11.7.30). From strict convexity, we have

$$c(t, x_2) - c(t, x_1) > (x_2 - x_1)c_x(t, x_1)$$

for all $x_1 \geq 0$, $x_2 \geq 0$ such that $x_1 \neq x_2$. Therefore,

$$c(t, (y_m+1)S(t-)) - c(t, S(t-)) > y_m S(t-)c_x(t, S(t-)), \tag{11.7.39}$$

the strict inequality being a consequence of the assumption that each y_m is greater than -1 and different from 0. It follows from (11.7.39) and (11.7.38) that between jumps

$$d\big[e^{-rt}c(t, S(t)) - e^{-rt}X(t)\big] < 0.$$

Between jumps, the hedging portfolio outperforms the option. However, at jump times, the option outperforms the hedging portfolio.

Because both $e^{-rt}c(t, S(t))$ and $e^{-rt}X(t)$ are martingales under $\widetilde{\mathbb{P}}$, so is their difference. Furthermore, at the initial time, the difference is $c(0, S(0)) - X(0) = 0$. Therefore, the expected value of the difference is always zero:

$$\widetilde{\mathbb{E}}\big[e^{-rt}c(t, S(t))\big] = \widetilde{\mathbb{E}}\big[e^{-rt}X(t)\big], \; 0 \leq t \leq T.$$

"On average," the delta-hedging formula hedges the option, where the average is computed under the risk-neutral measure we have chosen. This provides some justification for choosing $\tilde{\lambda}_m = \lambda_m$, so that, at least as far as the jumps are concerned, the average under the risk-neutral measure we are using is also the average under the actual probability measure.

Remark 11.7.10 (Continuous jump distribution). When the risk-neutral distribution of the jumps Y_i has density $\tilde{f}(y)$, (11.7.38) becomes

$$\begin{aligned}
&d\big[e^{-rt}c(t,S(t)) - e^{-rt}X(t)\big]\\
&= e^{-rt}\big[c(t,S(t)) - c(t,S(t-)) - (S(t)-S(t-))c_x(t,S(t-))\big]\,dN(t)\\
&\quad -e^{-rt}\tilde{\lambda}\int_{-1}^{\infty}\big[c(t,(y+1)S(t-)) - c(t,S(t-))\\
&\qquad\qquad -yS(t-)c_x(t,S(t-))\big]\tilde{f}(y)\,dy\,dt. \qquad (11.7.40)
\end{aligned}$$

Equation (11.7.40) can be interpreted just as (11.7.38) was. Because

$$c(t,(y+1)S(t-)) - c(t,S(t-)) - yS(t-)c_x(t,S(t-)) > 0$$

for all $y > -1$, $y \neq 0$, between jumps

$$d\big[e^{-rt}c(t,S(t)) - e^{-rt}X(t)\big] < 0,$$

the hedging portfolio outperforms the option. At jump times, the option outperforms the hedging portfolio because

$$c(t,S(t)) - c(t,S(t-)) - (S(t)-S(t-))c_x(t,S(t-)) > 0.$$

On "average," where the average is computed under the risk-neutral measure we have chosen, these two effects cancel one another.

11.8 Summary

The fundamental pure jump process is the *Poisson process*. Like Brownian motion, the Poisson process is Markov, but unlike Brownian motion, it is not a martingale. The Possion process only jumps up, and between jumps it is constant. To obtain a martingale, one must subtract away the mean of the Poisson process to obtain a *compensated Poisson process* (Theorem 11.2.4).

All jumps of a Poisson process are of size one. A *compound Poisson process* is like a Poisson process, except that the jumps are of random size. Like the Poisson process, a compound Poisson process is Markov (Exercise 11.7), and although it is generally not a martingale, one can obtain a martingale by subtracting away its mean (Theorem 11.3.1). A compound Poisson process that has only finitely many, say M, possible jump sizes can be decomposed into a sum of M independent scaled Poisson processes (Theorem 11.3.3 and Corollary 11.3.4).

A *jump process* has four components: an initial condition, an Itô integral, a Riemann integral, and a pure jump process. The sum of the first three constitute the *continuous part* of the jump process. Stochastic integrals and stochastic calculus for the continuous part of a jump process were treated in Chapter 4. In this chapter, the pure jump part is a right-continuous process that has finitely many jumps in each finite time interval and is constant between jumps. Stochastic integrals with respect to such processes are straightforward. The quadratic variation of such a process over a time interval is the sum of the squares of the jumps within that time interval, and the quadratic variation of a (nonpure) jump process is the quadratic variation of the continuous part plus the quadratic variation of the pure jump part. These observations lead to a version of the Itô-Doeblin formula for jump processes (Theorems 11.5.1 and 11.5.4). One of the consequences of these theorems is that a Brownian motion and a Poisson process relative to the same filtration must be independent (Corollary 11.5.3) and that two Poisson processes are independent if and only if they have no simultaneous jumps (Exercises 11.4 and 11.5).

If we integrate an adapted process with respect to a jump process that is a martingale, the resulting stochastic integral can fail to be a martingale. However, if the integrand is left-continuous, then the stochastic integral will be a martingale.

For compound Poisson processes, one can change the measure in order to obtain an arbitrary positive *intensity* (average rate of jump arrival) and an arbitrary distribution of jump sizes, subject to the condition that every jump size that was impossible before the change of measure is still impossible after the change of measure. This provides a great deal of freedom when constructing risk-neutral measures. In particular, if there are M possible jump sizes, there are $M-1$ degrees of freedom in the assignment of probabilities to these jump sizes (the probabilities must sum to one, and thus there are not M degrees of freedom). In order to have a complete market, there must be a money market account and as many nonredundant securities as there are sources of uncertainty. Each possible jump size counts as a source of uncertainty. If there is no Brownian motion and only one possible jump size, a single security in addition to the money market account will make the model complete (Section 11.7.1). If there are two possible jump sizes and an additional source of uncertainty due to a Brownian motion, three securities in addition to the money market account are required (Example 11.7.4). If there are infinitely many possible jump sizes, infinitely many securities would be required to make the model complete.

As the discussion above suggests, jump-diffusion models are generally incomplete and there are typically multiple risk-neutral measures in such models. The practice is to consider a parametrized class of such measures and then calibrate the model to market prices to determine values for the parameters. One can then apply the risk-neutral pricing formula to price derivative securities, but this formula can no longer be justified by a hedging argument. It is instead an elaborate interpolation procedure by which prices of nontraded securities are computed based on prices of traded ones. One can use this formula to examine the effectiveness of various hedging techniques. This is done for the delta-hedging rule in Subsection 11.7.2 following Remark 11.7.9.

11.9 Notes

A text on Poisson and compound Possion processes, but that does not include the ideas of change of measure, is Ross [141]. The easiest place to read about stochastic calculus for processes with jumps is Protter [133].

In Section 11.7, we consider a European call in two models, one in which the driving process for the underlying asset is a single Poisson process and the other in which the underlying asset is driven by a Brownian motion and multiple Poisson processes. In both these models, there are only finitely many jump sizes, but the analogous results for models with a continuous jump distribution are presented in Remarks 11.7.6, 11.7.9, and 11.7.10. Such a model was first treated by Merton [123], who considered the case in which one plus the jump size has a log-normal distribution. Some of the more recent works on option pricing in models with jumps are Brockhaus et al. [23], Elliott and Kopp [63], Madan, Carr, and Chang [113], Madan and Milne [114], Madan and Seneta [115], Mercurio and Runggaldier [120], and Overhaus et al [130]. Term-structure models with jumps are treated by Björk, Kabanov and Runggaldier [12], Das [46], Das and Foresi [47], Glasserman and Kou [73], and Glasserman and Merener [74].

11.10 Exercises

Exercise 11.1. Let $M(t)$ be the compensated Poisson process of Theorem 11.2.4.

(i) Show that $M^2(t)$ is a submartingale.
(ii) Show that $M^2(t) - \lambda t$ is a martingale.

Exercise 11.2. Suppose we have observed a Poisson process up to time s, have seen that $N(s) = k$, and are interested in the value of $N(s+t)$ for small positive t. Show that

$$\begin{aligned}
&\mathbb{P}\{N(s+t) = k \mid N(s) = k\} = 1 - \lambda t + O(t^2),\\
&\mathbb{P}\{N(s+t) = k+1 \mid N(s) = k\} = \lambda t + O(t^2),\\
&\mathbb{P}\{N(s+t) \geq k+2 \mid N(s) = k\} = O(t^2),
\end{aligned}$$

where $O(t^2)$ is used to denote terms involving t^2 and higher powers of t.

Exercise 11.3 (Geometric Poisson process). Let $N(t)$ be a Poisson process with intensity $\lambda > 0$, and let $S(0) > 0$ and $\sigma > -1$ be given. Using Theorem 11.2.3 rather than the Itô-Doeblin formula for jump processes, show that

$$S(t) = \exp\{N(t)\log(\sigma+1) - \lambda\sigma t\} = (\sigma+1)^{N(t)}e^{-\lambda\sigma t}$$

is a martingale.

Exercise 11.4. Suppose $N_1(t)$ and $N_2(t)$ are Poisson processes with intensities λ_1 and λ_2, respectively, both defined on the same probability space $(\Omega, \mathcal{F}, P)$ and relative to the same filtration $\mathcal{F}(t)$, $t \geq 0$. Show that almost surely $N_1(t)$ and $N_2(t)$ can have no simultaneous jump. (Hint: Define the compensated Poisson processes $M_1(t) = N_1(t) - \lambda_1 t$ and $M_2(t) = N_2(t) - \lambda_2 t$, which like N_1 and N_2 are independent. Use Itô's product rule for jump processes to compute $M_1(t)M_2(t)$ and take expectations.)

Exercise 11.5. Suppose $N_1(t)$ and $N_2(t)$ are Poisson processes defined on the same probability space $(\Omega, \mathcal{F}, \mathbb{P})$ relative to the same filtration $\mathcal{F}(t)$, $t \geq 0$. Assume that almost surely $N_1(t)$ and $N_2(t)$ have no simultaneous jump. Show that, for each fixed t, the random variables $N_1(t)$ and $N_2(t)$ are independent. (Hint: Adapt the proof of Corollary 11.5.3.) (In fact, the whole path of N_1 is independent of the whole path of N_2, although you are not being asked to prove this stronger statement.)

Exercise 11.6. Let $W(t)$ be a Brownian motion and let $Q(t)$ be a compound Poisson process, both defined on the same probability space $(\Omega, \mathcal{F}, \mathbb{P})$ and relative to the same filtration $\mathcal{F}(t)$, $t \geq 0$. Show that, for each t, the random variables $W(t)$ and $Q(t)$ are independent. (In fact, the whole path of W is independent of the whole path of Q, although you are not being asked to prove this stronger statement.)

Exercise 11.7. Use Theorem 11.3.2 to prove that a compound Poisson process is Markov. In other words, show that, whenever we are given two times $0 \leq t \leq T$ and a function $h(x)$, there is another function $g(t,x)$ such that

$$\mathbb{E}\big[h(Q(T))\big|\mathcal{F}(t)\big] = g(t, Q(t)).$$

A

Advanced Topics in Probability Theory

This appendix to Chapter 1 examines more deeply some of the topics touched upon in that chapter. It is intended for readers who desire a fuller explanation. The material in this appendix is not used in the text.

A.1 Countable Additivity

It is tempting to believe that the finite-additivity condition (1.1.5) can be used to obtain the countable-additivity condition (1.1.2). However, the right-hand side of (1.1.5) is a finite sum, whereas the right-hand side of (1.1.2) is an infinite sum. An infinite sum is not really a sum at all but rather a limit of finite sums:

$$\sum_{n=1}^{\infty} \mathbb{P}(A_n) = \lim_{N\to\infty} \sum_{n=1}^{N} \mathbb{P}(A_n). \tag{A.1.1}$$

Because of this fact, there is no way to get condition (1.1.2) from condition (1.1.5), and so we build the stronger condition (1.1.2) into the definition of probability space.

In fact, condition (1.1.2) is so strong that it is not possible to define $\mathbb{P}(A)$ for every subset A of an uncountably infinite sample space Ω so that (1.1.2) holds. Because of this, we content ourselves with defining $P(A)$ for every set A in a σ-algebra $\mathcal{F}$ that contains all the sets we will need for our analysis but omits some of the pathological sets that a determined mathematician can construct.

There are two other consequences of (1.1.2) that we often use implicitly, and these are provided by the next theorem.

Theorem A.1.1. *Let $(\Omega, \mathcal{F}, \mathbb{P})$ be a probability space and let $A_1, A_2, A_3, \ldots$ be a sequence of sets in $\mathcal{F}$.*

(i) If $A_1 \subset A_2 \subset A_3 \subset \ldots$, then

$$\mathbb{P}(\cup_{k=1}^{\infty} A_k) = \lim_{n\to\infty} \mathbb{P}(A_n).$$

(ii) If $A_1 \supset A_2 \supset A_3 \supset \dots$, then

$$\mathbb{P}(\cap_{k=1}^{\infty} A_k) = \lim_{n\to\infty} \mathbb{P}(A_n).$$

PROOF: In the first case, we define

$$B_1 = A_1,\ B_2 = A_2 \setminus A_1,\ B_3 = A_3 \setminus A_2, \dots,$$

where $A_{k+1} \setminus A_k = A_{k+1} \cap A_k^c$. Then $B_1, B_2, B_3, \dots$ are disjoint sets, and

$$A_n = \bigcup_{k=1}^{n} A_k = \bigcup_{k=1}^{n} B_k, \quad \bigcup_{k=1}^{\infty} A_k = \bigcup_{k=1}^{\infty} B_k.$$

Condition (1.1.2) used to justify the second equality below and (1.1.5) used to justify the fourth imply

$$\begin{aligned}\mathbb{P}(\cup_{k=1}^{\infty} A_k) = \mathbb{P}(\cup_{k=1}^{\infty} B_k) &= \sum_{k=1}^{\infty} \mathbb{P}(B_k) = \lim_{n\to\infty} \sum_{k=1}^{n} \mathbb{P}(B_k) \\ &= \lim_{n\to\infty} \mathbb{P}(\cup_{k=1}^{n} B_k) = \lim_{n\to\infty} \mathbb{P}(A_n).\end{aligned}$$

This concludes the proof of (i).

Let us now assume $A_1 \supset A_2 \supset A_3 \supset \dots$. We define $C_k = A_k^c$, so that $C_1 \subset C_2 \subset C_3 \subset \dots$ and $\cap_{k=1}^{\infty} A_k = (\cup_{k=1} C_k)^c$. Then (1.1.6) and (i) imply

$$\begin{aligned}\mathbb{P}(\cap_{k=1}^{\infty} A_k) = 1 - \mathbb{P}(\cup_{k=1}^{\infty} C_k) &= 1 - \lim_{n\to\infty} \mathbb{P}(C_n) \\ &= \lim_{n\to\infty} (1 - \mathbb{P}(C_n)) = \lim_{n\to\infty} \mathbb{P}(A_n).\end{aligned}$$

Thus we have (ii). □

Property (i) of Theorem A.1.1 was used in (1.2.6) at the step

$$\lim_{n\to\infty} \mathbb{P}\{-n \le X \le n\} = \mathbb{P}\{X \in \mathbb{R}\}.$$

Property (ii) of this theorem was used in (1.2.4). Property (ii) can also be used in the following example.

Example A.1.2. We continue Example 1.1.3, the uniform measure on $[0,1]$. Recall the σ-algebra $\mathcal{B}[0,1]$ of Borel subsets of $[0,1]$, obtained by beginning with the closed intervals and adding all other sets necessary in order to have a σ-algebra. A complicated but instructive example of a set in $\mathcal{B}[0,1]$ is the *Cantor set*, which we now construct. We also compute its probability, where the probability measure $\mathbb{P}$ we use is the uniform measure, assigning a probability to each interval $[a,b] \subset [0,1]$ equal to its length $b-a$.

From the interval $[0, 1]$, remove the middle third (i.e., the open interval $(\frac{1}{3}, \frac{2}{3})$). The remaining set is

$$C_1 = \left[0, \frac{1}{3}\right] \cup \left[\frac{2}{3}, 1\right],$$

which has two pieces, each with probability $\frac{1}{3}$, and the whole set C_1 has probability $\frac{2}{3}$. From each of the two pieces of C_1, remove the middle third (i.e., remove the open intervals $(\frac{1}{9}, \frac{2}{9})$ and $(\frac{7}{9}, \frac{8}{9})$). The remaining set is

$$C_2 = \left[0, \frac{1}{9}\right] \cup \left[\frac{2}{9}, \frac{1}{3}\right] \cup \left[\frac{2}{3}, \frac{7}{9}\right] \cup \left[\frac{8}{9}, 1\right],$$

which has four pieces, each with probability $\frac{1}{9}$, and the whole set C_2 has probability $\frac{4}{9}$. See Figure A.1.1.

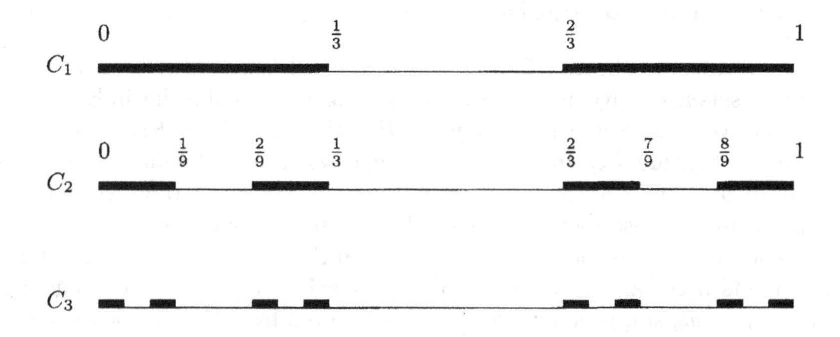

Fig. A.1.1. Constructing the Cantor set.

Continue this process so at stage k we have a set C_k that has 2^k pieces, each with probability $\frac{1}{3^k}$, and the whole set C_k has probability $\left(\frac{2}{3}\right)^k$. The Cantor set is defined to be $C = \cap_{k=1}^{\infty} C_k$. From Theorem A.1.1(ii), we see that

$$\mathbb{P}(C) = \lim_{k \to \infty} \mathbb{P}(C_k) = \lim_{k \to \infty} \left(\frac{2}{3}\right)^k = 0.$$

Despite the fact that it has zero probability, the Cantor set has infinitely many points. It certainly contains the points $0, \frac{1}{3}, \frac{2}{3}, 1, \frac{1}{9}, \frac{2}{9}, \frac{7}{9}, \frac{8}{9}, \frac{1}{27}, \frac{2}{27}, \dots$, which are the endpoints of the intervals appearing at the successive stages, because these are never removed. This is a countably infinite set of points. In fact, the Cantor set has uncountably many points. To see this, assume that all the points in the Cantor set can be listed in a sequence $x_1, x_2, x_3, \dots$. Let K_1 denote the piece of C_1, either $\left[0, \frac{1}{3}\right]$ or $\left[\frac{2}{3}, 1\right]$, that does not contain x_1. Let K_2 be a piece of $K_1 \cap C_2$ that does not contain x_2. For example, if $K_1 = \left[0, \frac{1}{3}\right]$

and $x_2 \in \left[\frac{2}{9}, \frac{1}{3}\right]$, we take $K_2 = \left[0, \frac{1}{9}\right]$. If $x_2 \neq K_1$, it does not matter whether we take $K_2 = \left[0, \frac{1}{9}\right]$ or $K_2 = \left[\frac{2}{9}, \frac{1}{3}\right]$. Next let K_3 be a piece of $K_2 \cap C_3$ that does not contain x_3. Continue this process. Then

$$K_1 \supset K_2 \supset K_3 \supset \ldots, \tag{A.1.2}$$

and $x_1 \notin K_1$, $x_2 \notin K_2$, $x_3 \notin K_3, \ldots$. In particular, $\cap_{n=1}^{\infty} K_n$ does not contain any point in the sequence $x_1, x_2, x_3, \ldots$. But the intersection of a sequence of nonempty closed intervals that are "nested" as described by (A.1.2) must contain something, and so there is a point y satisfying $y \in \cap_{n=1}^{\infty} K_n$. But $\cap_{n=1}^{\infty} K_n \subset C$, and so the point y is in the Cantor set but not on the list $x_1, x_2, x_3, \ldots$. This shows that the list cannot include every point in the Cantor set. The set of all points in the Cantor set cannot be listed in a sequence, which means that the Cantor set is uncountably infinite. □

A.2 Generating σ-algebras

We often have some collection $\mathcal{C}$ of subsets of a sample space Ω and want to put in all other sets necessary in order to have a σ-algebra. We did this in Example 1.1.3 when we constructed the σ-algebra $\mathcal{B}[0,1]$ and again in Example 1.1.4 when we constructed $\mathcal{F}_\infty$. In the former case, $\mathcal{C}$ was the collection of all closed intervals $[a, b] \subset [0, 1]$; in the latter case, $\mathcal{C}$ was the collection of all subsets of Ω_∞ that could be described in terms of finitely many coin tosses.

In general, when we begin with a collection $\mathcal{C}$ of subsets of Ω and put in all other sets necessary in order to have a σ-algebra, the resulting σ-algebra is called the *σ-algebra generated by* $\mathcal{C}$ and is denoted by $\sigma(\mathcal{C})$. The description just given of $\sigma(\mathcal{C})$ is not mathematically precise because it is difficult to determine how and whether the process of "putting in all other sets necessary in order to have a σ-algebra" terminates. We provide a precise mathematical definition at the end of this discussion.

The precise definition of $\sigma(\mathcal{C})$ works from the outside in rather than the inside out. In particular, we define $\sigma(\mathcal{C})$ to be the "smallest" σ-algebra containing all the sets in $\mathcal{C}$ in the following sense. Put in $\sigma(\mathcal{C})$ every set that is in every σ-algebra that is "bigger" than $\mathcal{C}$ (i.e., that contains all the sets in $\mathcal{C}$). There is at least one σ-algebra containing all the sets in $\mathcal{C}$, the σ-algebra of all subsets of Ω. If this is the only σ-algebra bigger than $\mathcal{C}$, then we put every subset of Ω into $\sigma(\mathcal{C})$ and we are done. If there are other σ-algebras bigger than $\mathcal{C}$, then we put into $\sigma(\mathcal{C})$ only those sets that are in *every* such σ-algebra. We note the following items.

(i) The empty set $\emptyset$ is in $\sigma(\mathcal{C})$ because it is in every σ-algebra bigger than $\mathcal{C}$.
(ii) If $A \in \sigma(\mathcal{C})$, then A is in every σ-algebra bigger than $\mathcal{C}$. Therefore, A^c is in every such σ-algebra, which implies that A^c is in $\sigma(\mathcal{C})$.
(iii) If $A_1, A_2, A_3, \ldots$ is a sequence of sets in $\sigma(\mathcal{C})$, then this sequence is in every σ-algebra bigger than $\mathcal{C}$, and so the union $\cup_{n=1}^{\infty} A_n$ is also in every such σ-algebra. This shows that the union is in $\sigma(\mathcal{C})$.

(iv) By definition, every set in $\mathcal{C}$ is in every σ-algebra bigger than $\mathcal{C}$ and so is in $\sigma(\mathcal{C})$.
(v) Suppose $\mathcal{G}$ is a σ-algebra bigger than $\mathcal{C}$. By definition, every set in $\sigma(\mathcal{C})$ is also in $\mathcal{G}$.

Properties (i)–(iii) show that $\sigma(\mathcal{C})$ is a σ-algebra. Property (iv) shows that $\sigma(\mathcal{C})$ contains all the sets in $\mathcal{C}$. Property (v) shows that $\sigma(\mathcal{C})$ is the "smallest" σ-algebra containing all the sets in $\mathcal{C}$.

Definition A.2.1 *Let $\mathcal{C}$ be a collection of subsets of a nonempty set Ω. The* σ-algebra generated by $\mathcal{C}$, *denoted $\sigma(\mathcal{C})$, is the collection of sets that belong to all σ-algebras bigger than $\mathcal{C}$ (i.e., all σ-algebras containing all the sets in $\mathcal{C}$).*

A.3 Random Variable with Neither Density nor Probability Mass Function

Using the notation of Example 1.2.5, let us define

$$Y = \sum_{n=1}^{\infty} \frac{2Y_n}{3^n}.$$

If $Y_1 = 0$, which happens with probability $\frac{1}{2}$, then $0 \le Y \le \frac{1}{3}$. If $Y_1 = 1$, which also happens with probability $\frac{1}{2}$, then $\frac{2}{3} \le Y \le 1$. If $Y_1 = 0$ and $Y_2 = 0$, which happens with probability $\frac{1}{4}$, then $0 \le Y \le \frac{1}{9}$. If $Y_1 = 0$ and $Y_2 = 1$, which also happens with probability $\frac{1}{4}$, then $\frac{2}{9} \le Y \le \frac{1}{3}$. This pattern continues. Indeed, when we consider the first n tosses we see that the random variable Y takes values in the set C_n defined in Example A.1.2, and hence Y can only take values in the Cantor set $C = \cap_{n=1}^{\infty} C_n$.

We first argue that Y cannot have a density. If it did, then the density f would have to be zero except on the set C. But C has zero Lebesgue measure, and so f is almost everywhere zero and $\int_0^1 f(x)\,dx = 0$ (i.e., the function f would not integrate to one, as is required of a density).

We next argue that Y cannot have a probability mass function. If it did, then for some number $x \in C$ we would have $\mathbb{P}(Y = x) > 0$. But x has a unique base-three expansion

$$x = \sum_{n=1}^{\infty} \frac{x_n}{3^n},$$

where each x_n is either 0, 1, or 2 unless x is of the form $\frac{k}{3^n}$ for some positive integers k and n. In the latter case, x has two base-three expansions. For example, $\frac{7}{9}$ can be written as both

$$\frac{7}{9} = \frac{2}{3} + \frac{1}{9} + \frac{0}{27} + \frac{0}{81} + \frac{0}{243} + \dots$$

and

$$\frac{7}{9} = \frac{2}{3} + \frac{0}{9} + \frac{1}{27} + \frac{1}{81} + \frac{1}{243} + \cdots.$$

In either case, there are at most two choices of $\omega \in \Omega_\infty$ for which $Y(\omega) = x$. In other words, the set $\{\omega \in \Omega; Y(\omega) = x\}$ has either one or two elements. The probability of a set with one element is zero, and the probability of a set with two elements is $0 + 0 = 0$. Hence $\mathbb{P}\{Y = x\} = 0$.

The cumulative distribution function $F(x) = \mathbb{P}\{Y \le x\}$ satisfies (see Figure A.3.1 for a partial rendition of $F(x)$)

$$F(0) = 0,\ F(1) = 1,\quad F(x) = \frac{1}{2} \text{ for } \frac{1}{3} \le x \le \frac{2}{3},$$
$$F(x) = \frac{1}{4} \text{ for } \frac{1}{9} \le x \le \frac{2}{9},\quad F(x) = \frac{3}{4} \text{ for } \frac{7}{9} \le x \le \frac{8}{9},$$
$$F(x) = \frac{1}{8} \text{ for } \frac{1}{27} \le x \le \frac{2}{27},\quad F(x) = \frac{3}{8} \text{ for } \frac{7}{27} \le x \le \frac{8}{27},$$
$$F(x) = \frac{5}{8} \text{ for } \frac{19}{27} \le x \le \frac{20}{27},\quad F(x) = \frac{7}{8} \text{ for } \frac{25}{27} \le x \le \frac{26}{27},$$
$$\vdots$$

and, because $\mathbb{P}\{Y = x\} = 0$ for every x, F is continuous. Furthermore, $F'(x) = 0$ for every $x \in [0,1] \setminus C$, which is almost every $x \in [0,1]$. A nonconstant continuous function whose derivative is almost everywhere zero is said to be *singularly continuous*.

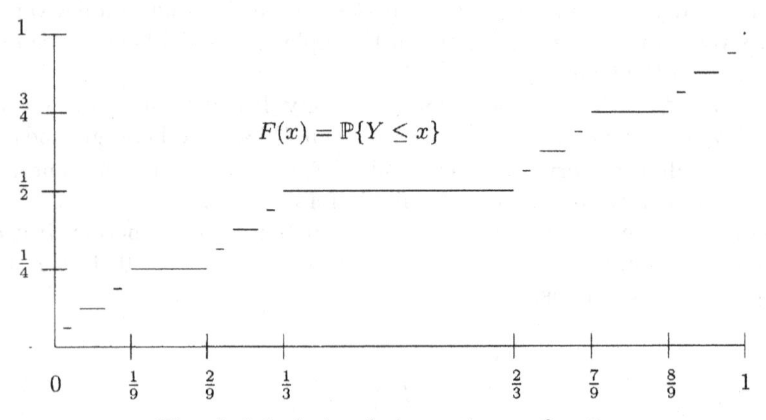

Fig. A.3.1. A singularly continuous function.

B

Existence of Conditional Expectations

This appendix uses the Radon-Nikodým Theorem, Theorem 1.6.7, to establish the existence of the conditional expectation of a random variable X with respect to a σ-algebra $\mathcal{G}$. Here we treat the case when X is nonnegative *and* integrable. If X is only integrable, one can decompose it in the usual way as $X = X^+ - X^-$, the difference of nonnegative integrable random variables, and then apply Theorem B.1 below to X^+ and X^- separately. If X is only nonnegative, one can write it as the limit of a nondecreasing sequence of nonnegative integrable random variables and use the Monotone Convergence Theorem, Theorem 1.4.5, to extend Theorem B.1 below to cover this case.

Theorem B.1. *Let $(\Omega, \mathcal{F}, \mathbb{P})$ be a probability space, let $\mathcal{G}$ be a sub-σ-algebra of $\mathcal{F}$, and let X be an integrable nonnegative random variable. Then there exists a $\mathcal{G}$-measurable random variable Y such that*

$$\int_A Y(\omega)\, d\mathbb{P}(\omega) = \int_A X(\omega)\, d\mathbb{P}(\omega) \text{ for every } A \in \mathcal{G}. \tag{B.1}$$

In light of Definition 2.3.1, the random variable Y in the theorem above is the conditional expectation $\mathbb{E}[X|\mathcal{G}]$.

PROOF OF THEOREM B.1: We define a probability measure by

$$\widetilde{\mathbb{P}}(A) = \int_A \frac{X(\omega)+1}{\mathbb{E}[X+1]}\, d\mathbb{P}(\omega) \text{ for every } A \in \mathcal{F}.$$

Because the integrand $\frac{X+1}{\mathbb{E}[X+1]}$ is strictly positive almost surely and has expectation 1, $\mathbb{P}$ and $\widetilde{\mathbb{P}}$ are equivalent probability measures (see Theorem 1.6.1 and the comment following Definition 1.6.3).

The probabilities $\mathbb{P}(A)$ and $\widetilde{\mathbb{P}}(A)$ are defined for every subset A of Ω that is in $\mathcal{F}$. We define two equivalent probability measures on the smaller σ-algebra $\mathcal{G}$. The first is simply $\mathbb{P}$ restricted to $\mathcal{G}$ (i.e., we define $\mathbb{Q}(A) = \mathbb{P}(A)$ for every

$A \in \mathcal{G}$, and we leave $\mathbb{Q}(A)$ undefined for $A \notin \mathcal{G}$). The second is $\widetilde{\mathbb{P}}$ restricted to $\mathcal{G}$ (i.e., we define $\widetilde{\mathbb{Q}}(A) = \widetilde{\mathbb{P}}(A)$ for every $A \in \mathcal{G}$, and we leave $\widetilde{\mathbb{Q}}(A)$ undefined for $A \notin \mathcal{G}$). We now have two probability spaces, $(\Omega, \mathcal{G}, \mathbb{Q})$ and $(\Omega, \mathcal{G}, \widetilde{\mathbb{Q}})$, which differ only by their probability measures $\mathbb{Q}$ and $\widetilde{\mathbb{Q}}$. Moreover, $\mathbb{Q}$ and $\widetilde{\mathbb{Q}}$ are equivalent. The Radon-Nikodým Theorem, Theorem 1.6.7, implies the existence of a random variable Z such that

$$\widetilde{\mathbb{Q}}(A) = \int_A Z(\omega)\, d\mathbb{Q}(\omega) \text{ for every } A \in \mathcal{G}.$$

However, since we are now working on probability spaces with σ-algebra $\mathcal{G}$, the random variable Z whose existence is guaranteed by the Radon-Nikodým Theorem will be $\mathcal{G}$-measurable rather than $\mathcal{F}$-measurable. (Recall from Definition 1.2.1 that every random variable is measurable with respect to the σ-algebra in the space on which it is defined.)

Since $\widetilde{\mathbb{Q}}$ and $\mathbb{Q}$ agree with $\widetilde{\mathbb{P}}$ and $\mathbb{P}$ on $\mathcal{G}$, we may rewrite the formula above as

$$\widetilde{\mathbb{P}}(A) = \int_A Z(\omega)\, d\mathbb{P}(\omega) \text{ for every } A \in \mathcal{G}$$

or, equivalently,

$$\int_A \frac{X(\omega)+1}{\mathbb{E}[X+1]}\, d\mathbb{P}(\omega) = \int_A Z(\omega)\, d\mathbb{P}(\omega) \text{ for every } A \in \mathcal{G}.$$

Multiplication by $\mathbb{E}[X+1]$ leads to the equation

$$\int_A X(\omega)\, d\mathbb{P}(\omega) + \int_A 1\, d\mathbb{P}(\omega) = \int_A \mathbb{E}[X+1] Z(\omega)\, d\mathbb{P}(\omega) \text{ for every } A \in \mathcal{G}.$$

We conclude that

$$\int_A X(\omega)\, d\mathbb{P}(\omega) = \int_A \big(\mathbb{E}[X+1] Z(\omega) - 1\big)\, d\mathbb{P}(\omega) \text{ for every } A \in \mathcal{G}.$$

Taking $Y(\omega) = \mathbb{E}[X+1] Z(\omega) - 1$, we have (B.1). Because Z is $\mathcal{G}$-measurable and $\mathbb{E}[X+1]$ is constant, Y is also $\mathcal{G}$-measurable. □

C

Completion of the Proof of the Second Fundamental Theorem of Asset Pricing

This appendix provides a lemma that is the last step in the proof of the Second Fundamental Theorem of Asset Pricing, Theorem 5.4.9 of Chapter 5.

Lemma C.1 *Let A be an $m \times d$-dimensional matrix, b an m-dimensional vector, and c a d-dimensional vector. If the equation*

$$Ax = b \tag{C.1}$$

has a unique solution x_0, a d-dimensional vector, then the equation

$$A^{\text{tr}}y = c \tag{C.2}$$

has at least one solution y_0, an m-dimensional vector. (Here, A^{tr} denotes the transpose of the matrix A.)

PROOF: We regard A as a mapping from $\mathbb{R}^d$ to $\mathbb{R}^m$ and define the *kernel of* A to be

$$K(A) = \{x \in \mathbb{R}^d : Ax = 0\}.$$

If x_0 solves (C.1) and $x \in K(A)$, then $x_0 + x$ also solves (C.1). Thus, the assumption of a unique solution to (C.1) implies that $K(A)$ contains only the d-dimensional zero vector.

The rank of A is defined to be the number of linearly independent columns of A. Because $K(A)$ contains only the d-dimensional zero vector, the rank must be d. Otherwise, we could find a linear combination of these columns that would be the m-dimensional zero vector, and the coefficients in this linear combination would give us a non-zero vector in $K(A)$. But any matrix and its transpose have the same rank, and so the rank of A^{tr} is d as well. The rank of a matrix is also the dimension of its range space. The range space of A^{tr} is

$$R(A^{\text{tr}}) = \{z \in \mathbb{R}^d : z = A^{\text{tr}}y \text{ for some } y \in \mathbb{R}^m\}.$$

Because the dimension of this space is d and it is a subspace of $\mathbb{R}^d$, it must in fact be equal to $\mathbb{R}^d$. In other words, for every $z \in \mathbb{R}^d$, there is some $y \in \mathbb{R}^m$ such that $z = A^{\text{tr}}y$. Hence, (C.2) has a solution $y_0 \in \mathbb{R}^m$. □

References

1. Ait-Sahalia, Y. (1996) Testing continuous-time models of the spot interest rate, *Rev. Fin. Stud.* **9**, 385–426.
2. Amin, K. and Jarrow, R. (1991) Pricing foreign currency options under stochastic interest rates, *J. Int. Money Fin.* **10**, 310–329.
3. Amin, K. and Khanna, A. (1994) Convergence of American option values from discrete- to continuous-time financial models, *Math. Fin.* **4**, 289–304.
4. Andreasen, J. (1998) The pricing of discretely sampled Asian and lookback options: a change of numéraire approach, *J. Comput. Fin.* **2**, 5–30.
5. Arrow, K. and Debreu, G. (1954) Existence of equilibrium for a competitive economy, *Econometrica* **22**, 265–290.
6. Bachelier, L. (1900) Théorie de la spéculation, *Ann. Sci. École Norm. Sup.* **17**, 21–86.
7. Balduzzi, P., Das, S., Foresi, S., and Sundaram, R. (1996) A simple approach to three-factor term structure models, *J. Fixed Income* **6**, 43–53.
8. Baxter, M. W. and Rennie, A. (1996) *Financial Calculus: An Introduction to Derivative Pricing*, Cambridge University Press, Cambridge, UK.
9. Bensoussan, A. (1984) On the theory of option pricing, *Acta Appl. Math.* **2**, 139–158.
10. Billingsley, P. (1986) *Probability and Measure*, 2nd ed., Wiley, New York.
11. Björk, T. (1998) *Arbitrage Theory in Continuous Time*, Oxford University Press, Oxford, UK.
12. Björk, T., Kabanov, Y., and Runggaldier, W. (1997) Bond market structure in the presence of marked point processes, *Math. Fin.* **7**, 211–239.
13. Black, F. (1976) The pricing of commodity contracts, *J. Fin. Econ.* **3**, 167–179.
14. Black, F. (1986) Noise, *J. Fin.* **41**, 529–543.
15. Black, F., Derman, E., and Toy, W. (1990) A one-factor model of interest rates and its applications to treasury bond options, *Fin. Anal. J.* **46**(1), 33–39.
16. Black, F. and Karasinski, P. (1991) Bond and option pricing when short rates are lognormal, *Fin. Anal. J.* **47**(4), 52–59.
17. Black, F. and Scholes, M. (1973) The pricing of options and corporate liabilities, *J. Polit. Econ.* **81**, 637–659.
18. Borodin, A. and Salminen, P. (1996) *Handbook of Brownian Motion – Facts and Formulae*, Birkhäuser, Boston.

19. BRACE, A., GĄTAREK, D., AND MUSIELA, M. (1997) The market model of interest rate dynamics, *Math. Fin.* **4**, 127–155.
20. BRACE, A. AND MUSIELA, M. (1994) A multifactor Gauss-Markov implementation of Heath, Jarrow and Morton, *Math. Fin.* **2**, 259–283.
21. BRIGO, D. AND MERCURIO, F. (2001) *Interest Rate Models: Theory and Practice*, Springer-Verlag, Berlin.
22. BROADIE, M., GLASSERMAN, P., AND KOU, S. (1999) Connecting discrete and continuous path-dependent options, *Fin. Stochastics* **3**, 55–82.
23. BROCKHAUS, O., FARKAS, M., FERRARIS, A., LONG, D., AND OVERHAUS, M. (2000) *Equity Derivatives and Market Risk Models*, Risk Books, London.
24. BRU, B. (2000) Un hiver en campagne, *C. R. Ser. I* **331**, 1037–1058.
25. BRU, B. AND YOR, M. (2002) Comments on the life and mathematical legacy of Wolfgang Doeblin, *Fin. Stochastics* **6**, 3–47.
26. CAMERON, R. H. AND MARTIN, W. T. (1944) Transformation of Wiener integrals under translations, *Ann. Math.* **45**, 386–396.
27. CARR, P., JARROW, R., AND MYNENI, R. (1992) Alternative characterizations of American put options, *Math. Fin.* **2**(2), 87–106.
28. CHAN, K., KAROLY, A., LONGSTAFF, F., AND SANDERS, A. (1992) An empirical comparison of alternative models of the short-term interest rate, *J. Fin.* **47**, 1209–1227.
29. CHEN, L. (1996) *Stochastic Mean and Stochastic Volatility – A Three-Factor Model of the Term Structure of Interest Rates and Its Application to the Pricing of Interest Rate Derivatives*, Blackwell, Oxford and Cambridge.
30. CHEN, R. AND SCOTT, L. (1992) Pricing interest rate options in a two-factor Cox-Ingersoll-Ross model of the term structure, *Rev. Fin. Stud.* **5**, 613–636.
31. CHEN, R. AND SCOTT, L. (1993) Maximum likelihood estimation for a multifactor equilibrium model of the term structure of interest rates, *J. Fixed Income* **3**, 14–31.
32. CHEN, R. AND SCOTT, L. (1995) Interest rate options in multifactor Cox-Ingersoll-Ross models of the term structure, *J. Derivatives* **3**, 53–72.
33. CHERIDITO, P. (2003) Arbitrage in fractional Brownian motion models, *Fin. Stochastics* **7**, 533–553.
34. CHEYETTE, O. (1996) Markov representation of the Heath-Jarrow-Morton model, BARRA, Berkeley, CA.
35. CHUNG, K. L. (1968) *A Course in Probability Theory*, Academic Press, Orlando.
36. CHUNG, K. L. AND WILLIAMS, R. J. (1983) *Introduction to Stochastic Integration*, Birkhäuser, Boston.
37. CLÉMENT, E., LAMBERTON, D., AND PROTTER, P. (2002) An analysis of a least squares regression algorithm for American option pricing, *Fin. Stochastics* **6**, 449–471.
38. COLLIN-DUFRESNE, P. AND GOLDSTEIN, R. (2002) Pricing swaptions in the affine framework, *J. Derivatives* **10**, 9–26.
39. COLLIN-DUFRESNE, P. AND GOLDSTEIN, R. (2001) Generalizing the affine framework to HJM and random fields, Graduate School of Industrial Administration, Carnegie Mellon University.
40. COX, J. C., INGERSOLL, J. E., AND ROSS, S. (1981) The relation between forward prices and futures prices, *J. Fin. Econ.* **9**, 321–346.
41. COX, J. C., INGERSOLL, J. E., AND ROSS, S. (1985) A theory of the term structure of interest rates, *Econometrica* **53**, 373–384.

42. Cox, J. C., Ross, S. A., and Rubinstein, M. (1979) Option pricing: a simplified approach, *J. Fin. Econ.* **7**, 229–263.
43. Cox, J. C. and Rubinstein, M. (1985) *Options Markets*, Prentice-Hall, Englewood Cliffs, NJ.
44. Dai, Q. and Singleton, K. (2000) Specification analysis of affine term structure models, *J. Fin.* **55**, 1943–1978.
45. Dalang, R. C., Morton, A., and Willinger, W. (1990) Equivalent martingale measures and no-arbitrage in stochastic security market models, *Stochastics* **29**, 185–201.
46. Das, S. (1999) A discrete-time approach to Poisson-Gaussian bond option pricing in the Heath-Jarrow-Morton model, *J. Econ. Dynam. Control* **23**, 333–369.
47. Das, S. and Foresi, S. (1996) Exact solutions for bond and option prices with systematic jump risk, *Rev. Derivatives Res.* **1**, 7–24.
48. DeGroot, M. (1986) *Probability and Statistics*, 2nd ed., Addison-Wesley, Reading, MA.
49. Delbaen, F. and Schachermayer, W. (1997) Non-arbitrage and the fundamental theorem of asset pricing: summary of main results, *Proceedings of Symposia in Applied Mathematics*, American Mathematical Society, Providence, RI.
50. Derman, E. and Kani, I. (1994) Riding on a smile, *Risk* **7** (2), 98–101.
51. Derman, E., Kani, I., and Chriss, N. (1996) Implied binomial trees of the volatility smile, *J. Derivatives* **3**, 7–22.
52. Doeblin, W. (1940) Sur l'équation de Kolmogoroff, *C. R. Ser. I* **331**, 1059–1102.
53. Doob, J. (1942) *Stochastic Processes*, Wiley, New York.
54. Dothan, M. U. (1990) *Prices in Financial Markets*, Oxford University Press, New York.
55. Duffee, G. (2002) Term premia and interest rate forecasts in affine models, *J. Fin.* **57**, 405–444.
56. Duffie, D. (1992) *Dynamic Asset Pricing Theory*, Princeton University Press, Princeton, NJ.
57. Duffie, D. and Kan, R. (1994) Multi-factor term structure models, *Philos. Trans. R. Soc. London, Ser. A* **347**, 577–586.
58. Duffie, D. and Kan, R. (1994) A yield-factor model of interest rates, *Math. Fin.* **6**, 379–406.
59. Duffie, D., Pan, J., and Singleton, K. (2000) Transform analysis and option pricing for affine jump-diffusions, *Econometrica* **68**, 1343–1376.
60. Duffie, D. and Protter, P. (1992) From discrete- to continuous-time finance; weak convergence of the financial gain process, *Math. Fin.* **2**, 1–15.
61. Dupire, B. (1994) Pricing with a smile, *Risk* **9** (3), 18–20.
62. Einstein, A. (1905) On the movement of small particles suspended in a stationary liquid demanded by the molecular-kinetic theory of heat, *Ann. Phys.* **17**.
63. Elliott, R. and Kopp, P. (1990) Option pricing and hedge portfolios for Poisson processes, *Stochastic Anal. Appl.* **9**, 429–444.
64. Fama, E. (1965) The behavior of stock-market prices, *J. Business* **38**, 34–104.
65. Feynman, R. (1948) Space-time approach to nonrelativistic quantum mechanics, *Rev. Mod. Phys.* **20**, 367–387.

66. FILIPOVIĆ, D. (2001) *Consistency Problems for Heath-Jarrow-Morton Interest Rate Models*, Springer, Berlin.
67. FU, M., MADAN, D., AND WANG, T. (1998/1999) Pricing continuous Asian options: a comparison of Monte Carlo and Laplace transform inversion methods, *J. Comput. Fin.*, **2**(2), 49–74.
68. GARMAN, M. AND KOHLHAGEN, S. (1983) Foreign currency option values, *J. Int. Money Fin.* **2**, 231–237.
69. GEMAN, H. (1989) The importance of the forward neutral probability in a stochastic approach of interest rates, Working paper, ESSEC.
70. GEMAN, H., EL KAROUI, N., AND ROCHET, J. (1995) Changes of numéraire, change of probability measure and option pricing, *J. Appl. Prob.* **32**, 443–458.
71. GEMAN, H. AND YOR, M. (1993) Bessel processes, Asian options, and perpetuities, *Math. Fin.* **3**, 349–375.
72. GIRSANOV, I. V. (1960) On transforming a certain class of stochastic processes by absolutely continuous substitution of measures, *Theory Prob. Appl.* **5**, 285–301.
73. GLASSERMAN, P. AND KOU, S. (2003) The term structure of simple forward rates with jump risk, *Math. Fin.* **13**, 383–410.
74. GLASSERMAN, P. AND MERENER, N. (2003) Numerical solution of jump diffusion LIBOR market models, *Fin. Stochastics* **7**, 1–27.
75. GLASSERMAN, P. AND YU, B. Number of paths versus number of basis functions in American option pricing, *Fin. Stochastics* to appear.
76. HAKALA, J. AND WYSTUP, U. (2002) *Foreign Exchange Risk*, Risk Books, London.
77. HARRISON, J. M. AND KREPS, D. M. (1979) Martingales and arbitrage in multiperiod security markets, *J. Econ. Theory* **20**, 381–408.
78. HARRISON, J. M. AND PLISKA, S. R. (1981) Martingales and stochastic integrals in the theory of continuous trading, *Stochastic Processes Appl.* **11**, 215–260.
79. HARRISON, J. M. AND PLISKA, S. R. (1983) A stochastic calculus model of continuous trading: complete markets, *Stochastic Processes Appl.* **15**, 313–316.
80. HAUG, E. (1998) *The Complete Guide to Option Pricing Formulas*, McGraw-Hill, New York.
81. HEATH, D., JARROW, R., AND MORTON, A. (1990) Bond pricing and the term structure of interest rates: A discrete time approximation, *J. Fin. Quant. Anal.* **25**, 419–440.
82. HEATH, D., JARROW, R., AND MORTON, A. (1990) Contingent claim valuation with a random evolution of interest rates, *Rev. Futures Markets* **9**, 54–76.
83. HEATH, D., JARROW, R., AND MORTON, A. (1992) Bond pricing and the term structure of interest rates: a new methodology, *Econometrica* **60**, 77–105.
84. HESTON, S. (1993) A closed-form solution for options with stochastic volatility and applications to bond and currency options, *Rev. Fin. Stud.* **6**, 327–343.
85. HO, T. AND LEE, S. (1986) Term structure movements and pricing interest rate contingent claims, *J. Fin.* **41**, 1011–1029.
86. HUANG, C.-F. AND LITZENBERGER, R. (1988) *Foundations for Financial Economics*, North Holland, Amsterdam.
87. HULL, J. (2002) *Options, Futures, and other Derivative Securities*, 5th ed., Prentice-Hall, Englewood Cliffs, NJ.
88. HULL, J. AND WHITE, A. (1990) Pricing interest rate derivative securities, *Rev. Fin. Stud.* **3**, 573–592.

89. HULL, J. AND WHITE, A. (1994) Numerical procedures for implementing term structure models II: two-factor models, *J. Derivatives* **2**, 37–47.
90. HUNT, P., KENNEDY, J., AND PELSSER, A. (2000) Markov-functional interest rate models, *Fin. Stochastics* **4**, 391–408.
91. INGERSOLL, J. E. (1987) *Theory of Financial Decision Making*, Rowman and Littlefield, Savage, MD.
92. ITÔ, K. (1944) Stochastic integral, *Proc. Imperial Acad. Tokyo* **20**, 519–524.
93. JACKA, S. (1991) Optimal stopping and the American put, *Math. Fin.* **1**(2), 1–14.
94. JAMSHIDIAN, F. (1989) An exact bond option formula, *J. Fin.* **44**, 205–209.
95. JAMSHIDIAN, F. (1990) LIBOR and swap market models and measures, *Fin. Stochastics* **1**, 293–330.
96. JARA, D. (2000) An extension of Lévy's theorem and applications to financial models based on futures prices, Ph.D. dissertation, Dept. of Mathematical Sciences, Carnegie Mellon University.
97. JARROW, R. (1988) *Finance Theory*, Prentice-Hall, Englewood Cliffs, NJ.
98. JARROW, R. A. AND OLDFIELD, G. S. (1981) Forward contracts and futures contracts, *J. Fin. Econ.* **9**, 373–382.
99. KAC, M. (1951) On some connections between probability theory and differential and integral equations, *Proceedings of the 2nd Berkeley Symposium on Mathematical Statistics and Probability*, 189–215, University of California Press, Berkeley
100. KARATZAS, I. (1988) On the pricing of American options, *Appl. Math. Optim.* **17**, 37–60.
101. KARATZAS, I. AND SHREVE, S. E. (1991) *Brownian Motion and Stochastic Calculus*, Springer-Verlag, New York.
102. KARATZAS, I. AND SHREVE, S. (1998) *Methods of Mathematical Finance*, Springer-Verlag, New York.
103. KIM, I. J. (1990) The analytic valuation of American options, *Rev. Fin. Stud.* **3**, 547–572.
104. KOLMOGOROV, A. N. (1933) Grundbegriffe der Wahrscheinlichkeitsrechnung, *Ergeb. Math.* **2**, No. 3. Reprinted by Chelsea Publishing Company, New York, 1946. English translation: *Foundations of Probability Theory*, Chelsea Publishing Co., New York, 1950.
105. LAMBERTON, D. AND LAPEYRE, B. (1996) *Introduction to Stochastic Calculus Applied to Finance*, Chapman and Hall, London.
106. LEROY, S. F. (1989) Efficient capital markets and martingales, *J. Econ. Lit.* **27**, 1583–1621.
107. LÉVY, P. (1939) Sur certains processus stochastiques homogénes, *Composition Math.* **7**, 283–339.
108. LÉVY, P. (1948) *Processus Stochastiques et Mouvement Brownian*, Gauthier-Villars, Paris.
109. LIPTON, A. (1999) Similarities via self-similarities, *Risk* **12**(9), 101–105.
110. LIPTON, A. (2003) *Exotic Options: Technical Papers Published in Risk 1999–2003*, Risk Books, London.
111. LONGSTAFF, F. AND SCHWARTZ, E. (1992) Interest rate volatility and the term structure: a two-factor general equilibrium model, *J. Fin.* **47**, 1259–1282.
112. LONGSTAFF, F. AND SCHWARTZ, E. (2001) Valuing American options by simulation: a simple least-squares approach, *Rev. Fin. Stud.* **14**, 113–147.

113. Madan, D., Carr, P., and Chang, E. (1998) The variance gamma process and option pricing, *Eur. Fin. Rev.* **2**, 79–105.
114. Madan, D. and Milne, F. (1991) Option pricing with V.G. martingale components, *Math. Fin.* **1** (4), 39–56.
115. Madan, D. and Seneta, E. (1990) The V.G. model for share returns, *J. Business* **63**, 511–524.
116. Maghsoodi, Y. (1996) Solution of the extended CIR term structure and bond option valuation, *Math. Fin.* **6**, 89–109.
117. Margrabe, W. (1978) The value of an option to exchange one asset for another, *J. Fin.* **33**, 177–186.
118. Margrabe, W. (1978) A theory of forward and futures prices, preprint, Wharton School, University of Pennsylvania.
119. McKean, H. (1965) A free-boundary problem for the heat equation arising from a problem in mathematical economics, *Ind. Manage. Rev.* **6**, 32–39. Appendix to [144].
120. Mercurio, F. and Runggaldier, W. (1993) Option pricing for jump diffusions: approximations and their interpretation, *Math. Fin.* **3**, 191–200.
121. Merton, R. C. (1969) Lifetime portfolio selection under uncertainty: the continuous-time case, *Rev. Econ. Stat.* **51**, 247–257.
122. Merton, R. C. (1973) Theory of rational option pricing, *Bell J. Econ. Manage. Sci.* **4**, 141–183.
123. Merton, R. C. (1976) Option pricing when underlying stock returns are discontinuous, *J. Fin. Econ.* **3**, 125–144.
124. Merton, R. C. (1990) *Continuous-Time Finance*, Basil Blackwell, Oxford and Cambridge.
125. Miltersen, K., Sandmann, S., and Sondermann, D. (1997) Closed form solutions for term structure derivatives with log-normal interest rates, *J. Fin.* **52**, 409–430.
126. Musiela, M. and Rutkowski, M. (1997) *Martingale Methods in Financial Modelling*, Springer-Verlag, New York.
127. Myneni, R. (1992) The pricing of the American option, *Ann. Appl. Prob.* **2**, 1–23.
128. Novikov, A. A. (1971) On moment inequalities for stochastic integrals, *Theory Prob. Appl.* **17**, 717–720.
129. Øksendal, B. (1995) *Stochastic Differential Equations*, 4th ed., Springer-Verlag, New York.
130. Overhaus, M., Ferraris, A., Knudsen, T., Milward, R., Nguyen-Ngoc, L., and Schindlmayr, G. (2002) *Equity Derivatives: Theory and Applications*, Wiley, New York.
131. Pelsser, A. (2000) *Efficient Methods for Valuing Interest Rate Derivatives*, Springer, Berlin.
132. Piazzesi, M. (2003) Affine term structure models. In *Handbook of Financial Econometrics* (Y. Ait-Sahalia and L. P. Hansen, eds.), North Holland, Amsterdam.
133. Protter, P. (2004) *Stochastic Integration and Differential Equations*, 2nd ed., Springer-Verlag, New York.
134. Protter, P. (1999) Featured review: recent books on mathematical finance, *SIAM Rev.* **41**, 167–173.
135. Protter, P. (2000) *Arbitrage Theory in Continuous Time* by T. Björk, *J. Fin.* **55**, 518.

136. PROTTER, P. (2001), *Stochastic Calculus and Financial Applications*, *SIAM Rev.* **43**, 731–733.
137. REBONATO, R. (2002) *Modern Pricing of Interest-Rate Derivatives: The LIBOR Market Model and Beyond*, Princeton University Press, Princeton, NJ.
138. ROGERS, L. C. G. (1997) Arbitrage with fractional Brownian motion, *Math. Fin.* **7** (1), 95–105.
139. ROGERS, L. C. G. AND SHI, Z. (1995) The value of an Asian option, *J. Appl. Prob.* **32**, 1077–1088.
140. ROSS, S. (1976) The arbitrage theory of capital asset pricing, *J. Econ. Theory* **13**, 341–360.
141. ROSS, S. M. (1983) *Stochastic Processes*, Wiley, New York.
142. RUBINSTEIN, M. AND REINER, E. (1991) Breaking down barriers, *Risk* **4**(9), 28–35.
143. SAMUELSON, P. A. (1965) Proof that properly anticipated prices fluctuate randomly, *Ind. Manage. Rev.* **6**, 41–50.
144. SAMUELSON, P. A. (1965) Rational theory of warrant pricing, *Ind. Manage. Rev.* **6**, 13–31.
145. SAMUELSON, P. A. (1973) Mathematics of speculative prices, *SIAM Rev.* **15**, 1–42.
146. SANDMANN, K. AND SONDERMANN, D. (1993) A term structure model and the pricing of interest rate derivatives, *Rev. Futures Markets* **12**, 391–423.
147. SANDMANN, K. AND SONDERMANN, D. (1997) A note on the stability of lognormal interest rate models and the pricing of Eurodollar futures, *Math. Fin.* **7**, 119–128.
148. SCHMOCK, U., SHREVE, S., AND WYSTUP, U. (2002) Valuation of exotic options under shortselling constraints, *Fin. Stochastics* **6**, 143–172.
149. SHIRAKAWA, H. (1991) Interest rate option pricing with Poisson-Gaussian forward rate curve processes, *Math. Fin.* **1**, 77–94.
150. STEELE, J. M. (2001) *Stochastic Calculus and Financial Applications*, Springer-Verlag, New York.
151. STROOCK, D. AND VARADHAN, S. R. S. (1969) Diffusion processes with continuous coefficients, I and II, *Communications Pure Appl. Math.* **22**, 245–400 and 479–530.
152. TSITSIKLIS, J. AND VAN ROY, B. (2001) Regression methods for pricing complex American-style options, *IEEE Trans. Neural Networks* **12**, 694–703.
153. VAN MOERBEKE, P. (1979) On optimal stopping and free boundary problems, *Arch. Rational Mech. Anal.* **60**, 101–148.
154. VASICEK, O. (1977) An equilibrium characterization of the term structure, *J. Fin. Econ.* **5**, 177–188.
155. VEČEŘ, J. (2001) A new PDE approach for pricing arithmetic average Asian options, *J. Comput. Fin.* **4**(4), 105–113.
156. VEČEŘ, J. (2002) Unified Asian pricing, *Risk* **15**(6), 113–116.
157. VEČEŘ, J. AND XU, M. (2004) Pricing Asian options in a semimartingale model, *Quant. Fin.* **4**, 170–175.
158. VILLE, J. (1939) *Étude Critique de la Notion du Collectif*, Gauthier-Villars, Paris.
159. WIENER, N. (1923) Differential spaces, *J. Math. Phys.* **2**, 131–174.
160. WIENER, N. (1924) Un probléme de probabilités dénombrables, *Bull. Soc. Math. France* **52**, 569–578.

161. Williams, D. (1991) *Probability with Martingales*, Cambridge University Press, Cambridge, UK.
162. Willinger, W. and Taqqu, M. (1991) Towards a convergence theory for continuous stochastic securities market models, *Math. Fin.* **1**, 55–99.
163. Willinger, W., Taqqu, M., and Teverovsky, V. (1999) Stock market prices and long-range dependence, *Fin. Stochastics* **3**, 1–14.
164. Wilmott, P. (1998) *Derivatives*, Wiley, New York.
165. Wilmott, P., Howison, S., and Dewynne, J. (1995) *The Mathematics of Financial Derivatives: A Student Introduction*, Cambridge University Press, Cambridge, UK.
166. Yor, M. (2000) Présentation du pli cacheté, *C. R. Ser. I* **331**, 1033–1036.
167. Zhang, P. (1997) *Exotic Options*, World Scientific, Singapore.

Index